sical Constants

Name	Symbol	Value/Units
Vacuum dielectric constant	ε_0	8.85 aF/μm
Silicon dielectric constant	ε_{si}	$11.7\varepsilon_0$
SiO$_2$ dielectric constant	ε_{ox}	$3.97\varepsilon_0$
SiN$_3$ dielectric constant	ε_{Ni}	$16\varepsilon_0$
Boltzmann's constant	k	1.38×10^{-23} J/K
Electronic charge	q	1.6×10^{-19} C
Temperature	T	K
Thermal voltage	V_T	kT/q = 26 mV @ 300 K

Equations

Parameter	NMOS equations in terms of BSIM1 parameters
	for PMOS use V_{SD}, V_{SG}, V_{BS}, and V_{THP}
V_{THN}	$VFB + PHI + K1 \cdot \sqrt{PHI + V_{SB}} - K2 \cdot (PHI + V_{SB})$
C'_{ox}	ε_{ox}/TOX
KP	$MUZ \cdot C'_{ox}$
β	$KP \cdot (W/L)$
I_D (triode) $V_{DS} \leq V_{GS} - V_{THN}$	$\beta\left((V_{GS} - V_{THN})V_{DS} - \dfrac{V_{DS}^2}{2}\right)$ (for Long L)
I_D (saturation) $V_{DS} \geq V_{GS} - V_{THN}$	$\dfrac{\beta}{2}(V_{GS} - V_{THN})^2[1 + \lambda(V_{DS} - V_{DS,sat})]$ (for Long L)
g_m	$\beta(V_{GS} - V_{THN}) = \sqrt{2\beta I_D}$ or $I_D/(V_T \cdot N0)$
η	$K1 \cdot \left(2\sqrt{PHI + V_{SB}}\right)^{-1} - K2$ and $g_{mb} = g_m \cdot \eta$
r_o	$1/(\lambda I_D)$

CMOS

CMOS
Circuit Design, Layout, and Simulation

R. Jacob Baker, Harry W. Li and David E. Boyce
Department of Electrical Engineering
Microelectronics Research Center
The University of Idaho

IEEE Press Series on Microelectronic Systems
Stuart K. Tewksbury, *Series Editor*

IEEE Circuits & Systems Society, *Sponsor*
IEEE Solid-State Circuits Society, *Sponsor*

IEEE PRESS

The Institute of Electrical and Electronics Engineers, Inc., New York

This book and other books may be purchased at a discount
from the publisher when ordered in bulk quantities. Contact:

IEEE Press Marketing
Attn: Special Sales
Piscataway, NJ 08855-1331
Fax: (732) 981-9334

For more information about IEEE PRESS products,
visit the IEEE Home Page: http://www.ieee.org/

Printed in the United States of America
10   9   8   7   6   5

ISBN 0-7803-3416-7
IEEE Order Number: PC5689

Library of Congress Cataloging-in-Publication Data

Baker, R. Jacob (date)
    CMOS circuit design, layout, and simulation / R. Jacob Baker,
    Harry W. Li, and David E. Boyce.
    p.   cm. -- (IEEE Press series on microelectronic systems)
    Includes bibliographical references and index.
    ISBN 0-7803-3416-7
    1. Metal oxide semiconductors, Complementary--Design and
construction.   2. Integrated circuits--Design and construction.
3. Metal oxide semiconductor field-effect transistors.   I. Li,
Harry W.   II. Boyce, David E.   III. Title.   IV. Series
TK7871.99.M44B35   1997
621.3815--DC21   97-21906
                                              CIP

To
Julie and Melanie

Contents

Preface

Over the last ten years the electronics industry has exploded. A recent report by the Semiconductor Industry Association (SIA) [1] proclaimed that in 1995 alone, world chip revenues increased by 41.7 percent and for the past five years the growth had been exponential. By the year 1999, the report estimates that world chip sales will surpass $234.5 billion, up from $154 billion in 1996. The largest portion of total worldwide sales is dominated by the MOS market. Composed primarily of memory, micro and logic sales, the total combined MOS revenue contributed approximately 75 percent of total world-wide sales ($114.2 billion), illustrating the strength of CMOS technology. The percentage of MOS sales relative to all chip revenues is expected to remain constant through 1999, when MOS sales will total $178 billion.

CMOS technology continues to mature, with minimum feature sizes now approaching 0.1 μm. Texas Instruments recently announced a 0.18 μm process [2] in which the equivalent of 20 high-performance microprocessors could exist on the same substrate, with a transistor density of 125 million transistors. This high density allows for true system-level integration on a chip, with digital signal processors, microprocessors or microcontroller cores, memory, analog or mixed-signal functions all residing on the same die.

As educators we are often asked by our students, "Isn't analog dead? I thought everything was going digital!" How untrue! The prediction of the future demise of analog electronics has been around since the mid-1970s. According to the SIA report [1], the revenues generated by analog products closely parallel the MOS logic market and achieved a 22.5 percent increase in 1995. The analog market expects to reach $18.2 billion in 1996 (a 9.5 percent increase) with double-digit growth projected for the next three years. In 1999, the total revenues generated by analog sales is forecasted to peak at $26.6 billion (11.3 percent of total chip sales!). However, while there is still demand for analog designers, their role is definitely changing. As was communicated by Paul Gray in [3], the days of pure analog design are over, meaning that very few systems remain purely analog. More and more systems are integrated, with increased functionality being performed in the digital domain. He goes on to state that the analog designer should become broad-based, with analog transistor-level design as the core skill. This means that the analog designer should also

- Have a good understanding of digital very large scale integration (VLSI) and be competent at using the latest computer-aided design (CAD) tools.

- know how to apply digital signal processing (DSP), analog signal processing (ASP), and filtering concepts to system-level design.

- possess insight into system implications of component-level performance.

For example, DSP and transistor-level analog design skills are needed for oversampling applications such as data converters, filtering, and a host of relatively new circuit topologies based on sigma-delta modulation. Being able to design both analog and digital circuits, as well as understand the interactions between the two domains, will provide an added dimension to a designer's portfolio that is difficult to match. Analog designers are in demand more than ever, simply because the end limitations of digital electronics need to be examined under the "analog" microscope to fully understand the mechanisms that are occurring. Therefore, this text attempts to combine digital and analog IC design in one complete reference.

Layout is the process of physically defining the layers that compose an integrated circuit. Typically, layouts are constructed using a computer-aided design program. CAD companies such as Mentor Graphics, Synopsis, and Cadence specialize in providing extremely powerful CAD software for the entire integrated circuit (IC) design process, including design, synthesis, simulation, and layout tools within an integrated framework. These workstation-based software tools can literally cost millions of dollars, but provide convenient and powerful features found nowhere else. CAD tools also exist for the PC. Tanner Tool's L-Edit provides a complete IC design CAD program for the personal computer. The program discussed in this book, LAyout System for Individuals (LASI) (pronounced "LAZY"), also provides the student with the ability to lay out ICs on a PC and includes design rule checking and design verification capability. It is distributed as shareware, free for educational purposes.

With decreases in feature size come added complexities in the design. Layouts must now be considered heavily in the design process as matching and parasitic effects become the limiting factors in many precision and high-speed applications. The more the designer knows about the process with respect to layout and modeling, the more performance the engineer can "squeeze" out the design. However, performance is not the only reason to consider the layout. The economic impact of IC layouts can be detrimental to the circuit's marketing potential. In some cases a 20 percent increase in chip area can reduce the profits of a chip by several hundreds of thousands of dollars. Chip area should be considered as premium real estate. Therefore, much of the first ten chapters of this book is devoted to fundamental layout issues, with other issues presented as the need arises.

Modeling is also a key issue. A simulation is only as accurate as its model. Although the Berkeley Short-channel IGFET Model (BSIM) model has become the industry standard its relatively nonintuitive structure makes hand analysis using BSIM model parameters an intimidating process. To many students (and engineers), the BSIM parameters are nothing more than sets of numbers at the end of their SPICE decks. However, some very useful information can be gleaned from the BSIM model which helps make the hand analysis more closely resemble the simulated result. Chapter 6 provides a great deal of information that relates the BSIM model to first-order hand-analysis equations.

A successful CMOS integrated circuit design engineer has knowledge in the areas of device operation, circuit design, layout, and simulation. Students learning CMOS IC design should be trained at a fundamental level in these areas. In the past, courses on CMOS integrated circuits dealt mainly with circuit design or analysis. Little to no time was spent on layout of the integrated circuits. This may have been justified. It is difficult to find a reason to lay out an entire chip and then not have the chip fabricated. However, through the use of the MOSIS[1] program, students can submit their chip designs for fabrication through one of the MOSIS contracted vendors. In approximately ten weeks the chips are returned to the university for evaluation. The MOSIS program is an outstanding way of introducing students to the design of ICs.

Although many texts [4-32] are available covering some aspects of CMOS analog or digital circuit design, none integrates the coverage of both topics with layout and includes layout software as is done in this text. Our focus, when writing this text, was on the fundamentals of custom CMOS integrated circuit design. It was our goal that a student who studies and masters the material in this text will possess the fundamental skills needed to design high-performance analog and digital CMOS circuits and have the basic understanding and problem-solving skills needed to enhance the performance of an IC or to determine why an IC doesn't function as simulated.

Use of This Text

This text can be used for two courses. Both courses can be offered at the senior/first-year graduate level. The first course concentrates on the physical design of CMOS digital integrated circuits with prerequisites of junior level Electronics I and a course on digital logic design. A possible semester course outline is as follows.

Week 1	Chs. 1 & 2, introduction, course requirements, layout and SPICE demonstrations, the n-well, sheet resistance.
Week 2	Chs. 2 & 3, the n-well, pn junction, capacitance, resistance, delay through the well, introduction to the metal layers.

---

[1] MOSIS - MOS Implementation System through the Information Sciences Institute at the University of Southern California, (310) 822-1511 or http://www.mosis.org

Week 3   Chs. 3 &4, the metal layers, parasitics, electomigration, layout of the padframe, active/poly layers, layout of the MOSFET and standard frame.

Week 4   Ch. 5, MOSFET operation

Week 5   Chs. 5 & 6, completion of MOSFET operation, discussion of modeling using the BSIM model.

Week 6   Chs. 6 & 7, completion of BSIM model, layout of a capacitor, MOS temp dependence.

Week 7   Chs. 10 & 11, digital models and the inverter.

Week 8   Ch. 11, the inverter, switching point voltage and switching times, layout, latch-up, and design.

Week 9   Ch. 12, static logic gates, switching point voltages, speed, and layout.

Week 10   Chs. 13 & 14, the transmission gate, flip-flops, and dynamic logic gates.

Week 11   Chs. 15 & 16, VLSI layout and BiCMOS logic.

Week 12   Ch. 17, memory circuits, basic memory cells, and organization.

Week 13   Ch. 18, special-purpose digital circuits.

Week 14   Ch. 19, introduction to digital phase locked loops, phase detectors, VCOs.

Week 15   Ch. 19, digital PLL design.

The second course concentrates on CMOS analog circuit design. A possible semester course outline is as follows.

Weeks 1 & 2   Review of Chs. 1-6.

Week 3   Chs. 7, CMOS passive elements, noise characteristics.

Week 4   Ch. 9, analog MOSFET models.

Week 5   Ch. 20, current sources and sinks.

Week 6   Ch. 21, references.

Week 7   Ch. 22, amplifiers.

Week 8   Ch. 23, selected topics in feedback amplifier design.

Week 9   Ch. 24, differential amplifiers.

Weeks 10-12   Ch. 25, operational amplifiers.

Week 13   Ch. 26, nonlinear analog circuits.

Week 14   Ch. 27, dynamic analog circuits.

Week 15   Chs. 28 & 29, selected topics in data converter design.

This text can also be used as an accompanying text in a VLSI systems course that focuses on the implementation of systems rather than circuits. Use of the text in this manner is benefited by inclusion of the LASI layout software.

REFERENCES

[1]   *Revised Forecast for World Chip Market Shows Growth of 6.7% in 1996, 19% by 1999*, Semiconductor Forcast Summary 1995-1998, Semiconductor Industry Association.

[2]   "New TI Technology Doubles Transistor Density," *Texas Instruments Integration Newsletter*, Vol. 13, No. 5, July 1995.

[3]   P. Gray, "Possible Analog IC Scenarios for the 90's," http://kabuki. eecs.berkeley.edu/slides.html

Digital Circuits and VLSI System Design

[4]   C. Mead and L. Conway, *Introduction to VLSI Systems*, Addison-Wesley, 1980.

[5]   Glasser and Dopperpuhl, *The Design and Analysis of VLSI Circuits*, Addison Wesley, 1985.

[6]   M. Annaratone, *Digital CMOS Circuit Design*, Kluwer, 1986.

[7]   A. Mukherjee, *Introduction to NMOS and CMOS VLSI Systems Design*, Prentice-Hall Publishers, 1986. ISBN 0-13-490947-X

[8]   D. A. Hodges and H. G. Jackson, *Analysis and Design of Digital Integrated Circuits*, McGraw-Hill , 2nd ed., 1988. ISBN 0 - 07 - 029158 - 6.

[9]   M. Shoji, *CMOS Digital Circuit Technology*, Prentice-Hall, 1988. ISBN 0-13-138850-9.

[10]   J. P. Uyemura, *Fundamentals of MOS Digital Integrated Circuits*, Addison-Wesley, 1988. ISBN 0-201-13318-0.

[11]   N. Wang, *Digital Mos Integrated Circuits : Design and Applications* Prentice-Hall, 1989. ISBN 0-132-13109-9.

[12]   R. L. Geiger, P. E. Allen, and N. R. Strader, *VLSI - Design Techniques for Analog and Digital Circuits*, McGraw-Hill, 1990. ISBN 0-07-023253-9.

[13]   J. Y. Chen, *CMOS Devices and Technology for VLSI*, Prentice-Hall, 1990. ISBN 0-13-138082-6.

[14]   E. Fabricius, *Introduction to VLSI Design*, McGraw-Hill, 1990. ISBN 0-07019-948-5

[15]   M. I. Elmasry, *Digital MOS Integrated Circuits II*, IEEE Press, 1992. ISBN 0-87942-275-0.

[16]   N.H.E. Weste and K. Eshraghian, *Principles of CMOS VLSI Design*, Addison Wesley, 2nd ed., 1993. ISBN 0-201-53376-6.

[17]   J. P. Uyemura, *Circuit Design for Digital CMOS VLSI*, Kluwer, 1992.

[18]   D. A. Pucknell and K. Eshraghian, *Basic VLSI Design*, 3rd ed., Prentice Hall Publishers, 1994. ISBN 0-13-079153-9

[19]   W. Wolf, *Modern VLSI Design: A Systems Approach*, Prentice Hall, 1994. ISBN: 0-13-588377-6

[20]   S. Kang and Y. Leblebici, *CMOS Digital Integrated Circuits - Analysis and Design*, McGraw-Hill, 1996. ISBN 0-07-038046-5.

[21]   K. Gopalan, *Introduction to Digital Microelectronic Circuits*, Irwin, 1996. ISBN 0-256-12089-7.

[22]   J. M. Rabaey, *Digital Integrated Circuits - A Design Perspective*, Prentice Hall, 1996, ISBN 0-13-178609-1.

Analog Circuits

[23]   A. B. Grebene, *Bipolar and MOS Analog Integrated Circuit Design*, John-Wiley, 1984. ISBN 0-471-08529-4

[24]   R. Gregorian and G. C. Temes, *Analog MOS Integrated Circuits for Signal Processing*, John Wiley, 1986. ISBN 0-471-09797-7.

[25]   P. E. Allen and D. R. Holberg, *CMOS Analog Circuit Design*, Holt, Rinehart and Winston, 1987. ISBN 0-03-006587-9.

[26]   P. R. Gray, B. A. Wooley and R. W. Broderson, *Analog MOS Integrated Circuits II*, IEEE Press. ISBN 0-87942-246-7.

[27]   R. L. Geiger, P. E. Allen, and N. R. Strader, *VLSI - Design Techniques for Analog and Digital Circuits*, McGraw-Hill, 1990. ISBN 0-07-023253-9.

[28]   P. R. Gray and R. G. Meyer, *Analysis and Design of Analog Integrated Circuits*, 3rd ed., John Wiley, Inc., 1993. ISBN 0-471-57495-3.

[29]   M. Ismail and T. Fiez, *Analog VLSI - Signal and Information Processing*, McGraw-Hill, Inc. 1994. ISBN 0-07-032386-0.

[30]   K. R. Laker and W. Sansen, *Design of Analog Integrated Circuits and Systems*, McGraw-Hill, 1994. ISBN 0-07-036060-X.

[31]   G.A.S. Machado, *Low-Power HF Microelectronics a unified approach*, IEE, 1996. ISBN 0-85296-874-4.

[32]   D. Johns and K. Martin, *Analog Integrated Circuit Design*, John Wiley, 1997. ISBN 0-471-14448-7.

Acknowledgments

We would like to thank the reviewers, contributors, and colleagues who helped make this book possible; Dr. Phil Allen, Savoula Amanatidis, Ben Ba, Jan Bissey, Dr. William Black, Jeff Bruce, Alan Buchholz, Dr. Joseph Cavallaro, Brian P. Lum Shue Chan, Irfan Chaudhry, Lisa Dayne, Dr. Ian Galton, Dr. Randall Geiger, John Griffin, Wes Hansford, Aaron Huntsinger, Dr. Bruce Johnson, David Kao, Dr. Joe Karniewicz, Brent Keeth, Dr. William Kuhn, Wen Li, Dr. H. Alan Mantooth, Dr. Richard Marks, Dean Moriarty, Dr. Ken Noren, Dr. Adrian Ong, Dr. James Rochelle, Dr. Terry Sculley, Joseph P. Skudlarek, Dr. Stuart Tewksbury, Dr. Don Thelen, Dr. Axel Thomsen, Dr. Vance Tyree, Scott Ward, Veronica Wilson, Dr. Jeff Wu, and Dr. Kwang S. Yoon.

The authors would also like to thank Orbit Semiconductor, MOSIS and HP for allowing us to include their process information and design rules with the text.

R. Jacob Baker

Harry W. Li

David E. Boyce

CMOS Fundamentals

Introduction

This chapter discusses the CMOS (complementary metal oxide semiconductor) integrated circuit (IC) design process, how to set up the LASI (LAyout System for Individuals) layout software, and fabrication of CMOS integrated circuits through MOSIS (MOS Implementation Service).

1.1 The CMOS IC Design Process

The CMOS circuit design process consists of defining circuit inputs and outputs, hand calculations, circuit simulations, layout of the circuit, simulations including parasitics, reevaluation of the circuit inputs and outputs, fabrication, and testing. A flowchart of this process is shown in Fig. 1.1. The circuit specifications are rarely set in concrete; that is, they can change as the project matures. This can be the result of tradeoffs made between cost and performance, changes in the marketability of the chip, or simply changes in the customer's needs. In almost all cases, major changes after the chip has gone into production are not possible.

This text concentrates on custom IC design. A custom-designed chip is often called an ASIC (application-specific integrated circuit). Other (noncustom) methods of designing chips, including field-programmable-gate-arrays (FPGAs) and standard cell libraries, are used when low volume and quick-design turnaround are important. Most chips that are mass produced, including microprocessors and memory, are examples of chips that are custom designed.

The task of laying out the IC is often given to a draftsman. However, it is extremely important that the engineer be able to lay out a chip (and direct the draftsman on how to lay the chip out) and understand the parasitics involved in the layout. Parasitics are the stray capacitances, inductances, pn junctions, and bipolar transistors, with the associated problems (breakdown, stored charge, latch-up, etc.). A fundamental understanding of these problems is important in precision/high-speed design.

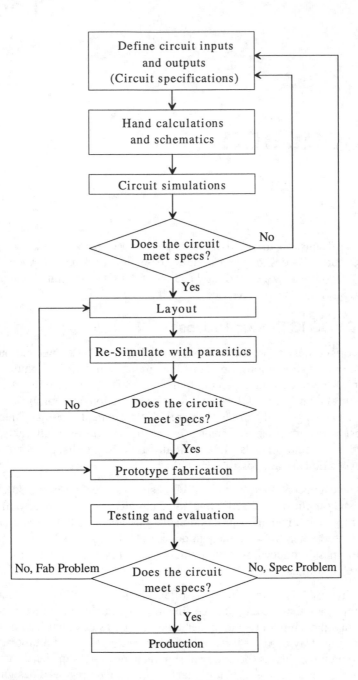

Figure 1.1 Flowchart for the CMOS IC design process.

1.1.1 Fabrication

CMOS integrated circuits are fabricated on thin circular slices of silicon called wafers. Each wafer contains several individual chips or "die" (Fig. 1.2). For production purposes each die on a wafer is usually identical. Added to the wafer are test structures and process monitor plugs (sections of the wafer used to monitor process parameters).

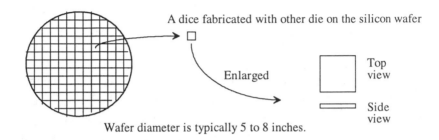

Figure 1.2 CMOS integrated circuits are fabricated on and in a silicon wafer.

The ICs we design and lay out using LASI can be fabricated through MOSIS [1] on what is called a multiproject wafer; that is, the wafer consists of chip designs of varying sizes from different sources (educational, private, government, etc.). MOSIS combines multiple chips on a wafer to split the fab cost among several designs to keep the cost low. MOSIS subcontracts the fabrication of the chip designs out to one of many commercial manufacturers, including Orbit and HP.

The view we see when laying out an IC is always a top view of the die. When laying out the chip, we draw boxes or polygons on differing layers indicating how to assemble the circuit. We may specify a box on layer 1 (n-well) from the coordinates (0,0) to the coordinates (10,10). The coordinates of this box and other shapes defining the circuitry are specified in a binary file using Calma Stream Format (abbreviated CSF, GDSII, or simply GDS). This file describes the completed chip design. When using the LASI program, TransportabLe LASI Cell files (TLC files) are used to store the design information. When the design is finished, the TLC files can be converted to a GDS file and sent via the Internet to MOSIS for fabrication.

1.2 Using the Windows LASI Program

The LASI program discussed in this text is a powerful CAD (Computer-aided design) package used in the design of integrated circuits. This section discusses the installation and operation of LASI in the Windows environment.

Installing LASI

To install LASI, follow the instructions on the Windows LASI download webpage linked to:

http://www.mrc.uidaho.edu/vlsi/cad_free.html

on the internet. After the installation is complete, assuming you installed LASI on C:, your hard disk should have a directory C:\Lasi6, (containing all executable files and directories in the LASI system) C:\Lasi6\Wcn20, C:\Lasi6\Wmosis, C:\Lasi6\W2uchip and C:\Lasi6\Wclib containing the setups for the CN20 process, MOSIS, MOSIS 2 μm processes, and a cell library using the MOSIS design rules respectively.

Drawing Directories

Under the C:\Lasi6 directory should also be a directory \Tutor containing the example of a bipolar op-amp supplied with LASI. *All chip designs should reside in a directory other than the* C:\Lasi6 *directory.* The C:\Lasi6 directory is used only for the executable programs in the layout system. The directory of a chip design can be a subdirectory of C:\Lasi6, similar to \Tutor above, or a directory somewhere else on the hard disk.

Throughout the book we will be using Orbit's CN20 CMOS process. Using Windows Explorer verify that the setups were successfully copied into the C:\Lasi6\Wcn20 directory (the directory shouldn't be empty or absent.) Setup an icon for this directory following the instructions given on the Windows LASI download webpage. By double clicking on this icon the screen in Fig. 1.3 will appear.

The System Menu

By pressing the **Sys** button on the top of the LASI drawing window in Fig. 1.3 we can enter the LASI system menu. This menu is useful in transfering LASI drawing cells to other platforms and will be discussed in greater detail below. You can return to the drawing window by pressing the **Return** button.

1.2.1 Cells in LASI

Complex IC designs are made from simpler objects called cells. A cell might be a logic gate or an op-amp. To show a listing of the cells in this drawing directory, select **List** (or Alt-l) from the top menu in the LASI drawing window. The cell collection subwindow will be empty. Click the **Cancel** button to return to the drawing window.

TLC Files

Cells in LASI are backed up using transportable LASI cell files (or TLC files for short). The setup files that were copied into the Wcn20 directory during installation contain several TLC files. By selecting the **TLCin** command from the system menu (running LASI from the C:\Lasi6\Wcn20 directory), we can convert these TLC files into a binary format (files with the *.BP6 and *.CL6 extensions) for use with the LASI program.

Pressing the **TLCin** command button at the system menu will begin this process (do this now). Leaving the source and destination paths blank indicates the current drawing directory. Specifying "*" for the cell names will convert all TLC files in the directory into *.BP6 and *.CL6 format. The cells in the LASI drawing directory are backed up using the **TLCout** command button. In Fig. 1.4, pressing **OK** in the **TLCin**

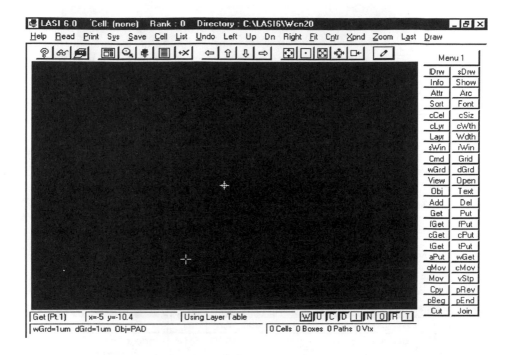

Figure 1.3 LASI startup screen.

window and then **Yes** to replace any existing or lesser cells in the drawing directory (at this point, none should exist) will cause several cells to be read into LASI. Using the **List** command button (after closing the LASI system menu window) on the LASI drawing screen and clicking on the cell "rule1" will display the contents of the rule1 cell. Using the **Fit** command button will center the contents of the cell in the window. The result is seen in Fig. 1.5.

Toggling Menus

Notice, in Fig. 1.5, that clicking the RIGHT mouse button in the drawing area causes some of the menu items on the right of the display to change or toggle between different commands. Again, a menu item can be selected using the LEFT mouse button.

Creating a Cell

To begin the layout of a new cell, the **Cell** command is used. The cell is assigned a name, for example, AND, and a rank. Since the AND gate is a basic building block, we will assign a rank of 1 (the lowest rank). If a cell is created that uses the AND gate cell, then the new cell will be assigned a rank of 2 or higher. That is, a cell with a rank of 2 can contain cells with a rank of 1. If the cell has a rank of 5, then it can contain any cell (or cells) with a rank of 4 or lower. Another analogy is to consider a chip, a printed

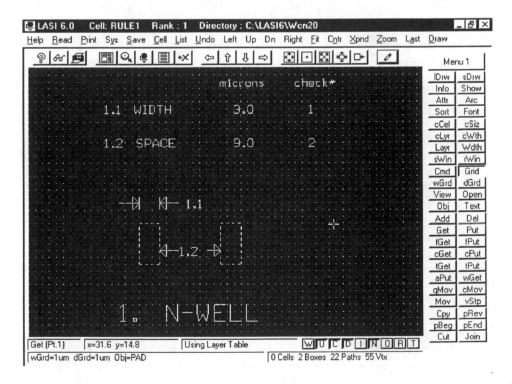

Figure 1.4 Using the **TLCin** command.

Figure 1.5 LASI display after opening the file containing rule 1.

circuit board, and a computer. A chip can be put on a circuit board, and the circuit board can be put into a computer. Using the ranking analogy, the chip has a rank of 1, the circuit board has a rank of 2, and the computer has a rank of 3. Therefore, the computer with rank 3 cannot be put into the chip with rank 1.

Create a cell called "test" for experimentation by selecting the command **Cell** and entering the name "test" with a rank 1. The bare test cell is shown in Fig. 1.6. Notice in the top of the display that the cell name and rank are displayed.

1.2.2 Navigating LASI

Notice in Fig. 1.6 that the cross-hairs show the location of the origin. Pressing **R** in the bottom right corner (or typing **r** on the keyboard) of the display turns this reference designator on or off. Now hit the **Draw** command (or Alt-d). This command redraws the screen with (or without) the reference cross-hairs showing. Press **r** so that the reference designator is showing (followed by the **Draw** command). Placing the mouse cursor over the origin causes the distance indicator at the bottom of the display to read (0,0). The distance indicator shows the distance away from the origin. If the indicator is showing distances of 10^6 or larger (10^6 μm is 1 meter), you are viewing the entire drawing universe. Selecting the **Fit** command with nothing drawn could easily result in

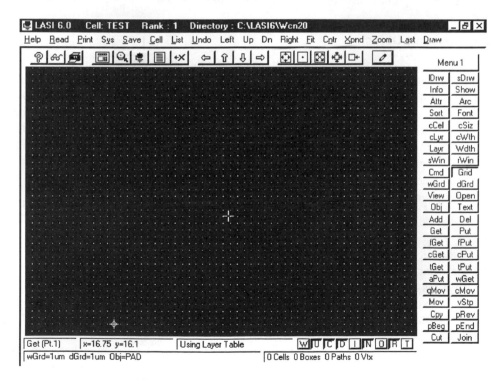

Figure 1.6 LASI display for the test cell (see text).

this situation. To get back to a display that shows tens of microns, simply select the **Zoom** command (or Alt-z) and then double click, with the LEFT mouse button, on the reference designator. If you cannot find the reference designator, select **Fit,** keeping in mind that **R** in the bottom right of the display should be depressed. One other useful command is the **Orig** command (in the command buttons on the right of the display). This command allows the user to set the origin at any location in the drawing display.

Grid

The grid can be turned on and off by clicking on the **Grid** command button on the right of the display. The grid will not display when "zoomed out."

Cursor

The cursor can be changed between a small-cross and full-screen cross-hairs by pressing the **Tab** button on the keyboard.

Measurements

By pressing **z** on the keyboard, a zero reference point is established independent of the origin. When the **spacebar** is pressed, the distance between the mouse cursor and the zero reference point is displayed. Pressing **w** causes the mouse to snap to the working grid.

1.2.3 Adding Objects

Select the **Layr**[1] command (on the right) to display the layer table used in this process. From the layer table with the LEFT mouse button, select layer 1 (NWEL 1), the n-well. Notice that the selected layer is displayed at the bottom of the screen. Now select **Obj** or the object command. Using the object command will allow the user to select drawing objects. Boxes, polygons/paths or other lowering ranking cells can be selected. Select a box by doubling clicking on the word box in the object window. Now the bottom of the drawing display should show that the working and dot grids are 1 µm, the object is a BOX, and the layer "NWEL." We are now ready to begin drawing the layout.

Click on the **Add** command with Obj = BOX and Layr = NWEL. Click the LEFT mouse button in the drawing area over the origin once. Move the mouse toward the top right corner of the display until you get a box similar to the one shown in Fig. 1.7 and click the LEFT mouse button again. Using the **arrow keys** on the keyboard or the command buttons, change the display view. Now select **Fit** (Alt-f) and notice that the n-well is centered in the drawing display. The **Xpnd** (Alt-x) command will expand the view. Notice how the viewing area increases. Use the **Zoom** (Alt-z) command to zoom in on the corner of the n-well box by drawing an imaginary box around one of the corners of the n-well box. When drawing, it is often useful to show a grid in the drawing area. The commands **wGrd** and **dGrd** are used to change between the

---

[1] Again, remember that if the command is not showing, press the RIGHT mouse button in the drawing area (or the **Menu** button) to toggle between menus.

working (the grid the cursor snaps to) and dot (the one you see) grids. For the setups included with this book, both grids are set to 1 μm. Using the **Set** command, the user can set the number of working and dot grids as well as the grid spacing. It is not recommended that any of the settings be changed until the user becomes very familiar with LASI.

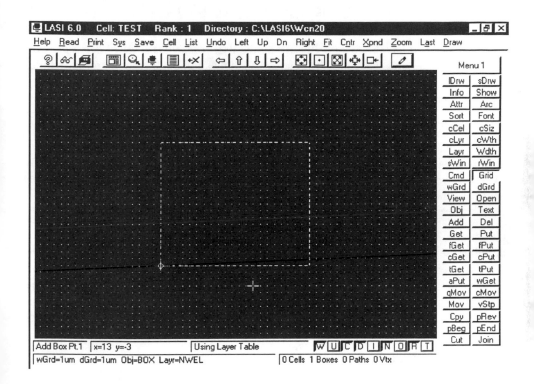

Figure 1.7 Drawing a box in LASI.

1.2.4 Editing Objects

Only one basic operation, moving, can be performed on a Box after it has been added to a cell. The entire box or any of the four sides can be moved. This operation consists of *getting* the object or side to be moved and performing the *move*, and then *putting* the object back (de-selecting the object).

Using the previously generated n-well layout as an example, use the **Get** command and select the right side of the n-well box using the mouse. This is done by clicking the LEFT mouse button on the left of the side and then again on the right, ensuring the box drawn by the mouse intersects the side. Objects that are selected using **Get** should be highlighted. Next, select the **Mov** command. Using the LEFT mouse button, click once somewhere in the display. Move the mouse a small distance to the

left and click the LEFT mouse button again. Notice that the highlighted side moved an
equal distance to the left of its original position. Use the **Put** command to unselect the
highlighted side. A simpler method of unselecting all of the active elements is to use
the **aPut** (all put) command.

Sometimes we may want to get the entire box. The **fGet** (full get) command
allows dragging the mouse through a portion of the object to select the entire object.
Try selecting the entire box using **fGet** and then unselecting it using **aPut**.

LASI also has the feature that allows the user to execute a set of commands by
pressing a single key. This is easily accomplished by adding a line to the form.dbd file.
This feature is useful when the users wants to perform one or a series of operations very
quickly. For more information, read the on-line help.

Viewing and Editing Specific Layers

Suppose that it is desired to view only a few layers of a complicated layout. This can be
accomplished by selecting the desired layers using the **View** command. The user needs
to be warned, however, that if the invisible layers are not made visible again, frustration
could abound.

The **Open** command determines which layers can be selected using the **Get**
command. This feature allows the user to make certain layers uneditable. In the current
drawing, select the **Open** command and unselect the n-well layer.   Return to the
drawing area by clicking OK. Attempt to **Get** any part of the n-well box and notice that
none of the sides can be highlighted or edited in any fashion. Clicking again on the
n-well layer in the **Open** command window will then allow any editing operations on
that layer. This command could also be frustrating if the user forgets that certain layers
are unselectable.

1.2.5 Placing Cells

Select **Cell** from the command button and make another cell. Call the new cell test2
and assign a rank of 2. Next, using the **Obj** command, use the previously laid out cell
(test) as the new object. Using the **Add** command, add the cell, test, to the cell test2 by
simply clicking the LEFT mouse button in the drawing area. Repeat this several times
until the layout is similar to Fig. 1.8.

Viewing Complex Layouts

Cells can be drawn as outlines using **Outl** (Fig. 1.9). The mouse is used to enclose the
cell you want drawn as an outline. Using the **Full** command draws the entire cell. This
becomes helpful when a large number of cells are present and the redraw time is long.
Also, cells will be drawn as outlines if the ESC key is pressed when LASI is redrawing
the screen (and this feature, pressing ESC, can be used to abort a long redraw time or a
command). The **Dpth** command is used to draw cells as outlines if they are nested
deeper than the cells depth setting.

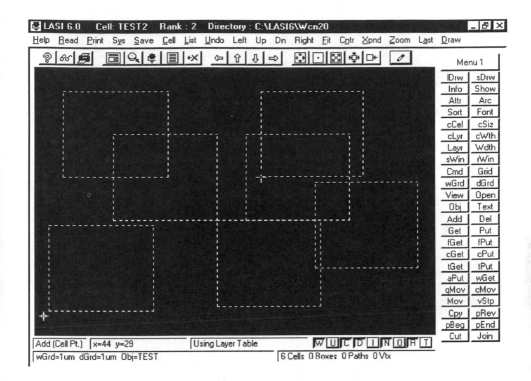

Figure 1.8 Inserting the cell, test, using the **Add** command.

Moving Cells

Moving a cell consists of getting the cell using the **cGet** (cell get) command, moving the cell using the **Mov** command, and putting the cell using the **aPut** (all put) or **cPut** (cell put) commands.

Displaying Cells

When cells are drawn as outlines, using the **Outl** command, the names, on the outline, can be displayed or hidden by pressing **n** on the keyboard (indicated by **N** on the bottom right corner of the drawing display). If the cells are not drawn as outlines, pressing **i** on the keyboard draws a dotted line around the cell. This is useful as a reminder of which parts of the layout are cells and which parts are boxes or polygons. A **Draw** command (Alt-d) must follow pressing **n** or **i** for the results to be seen.

A Note on Editing

Only objects that are drawn in the current cell can be edited. The objects resulting from adding a cell to the current (open) cell cannot be edited. Trying to change the size of any of the n-well boxes of Fig. 1.8 from the *test2* cell (the open cell) would result in frustration. To change the size of these boxes, we must first open the *test* cell.

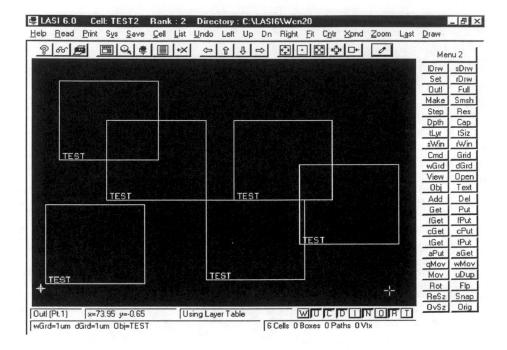

Figure 1.9 Using the **Outl** command.

Saving Your Work

LASI can automatically generate a TLC file in the drawing directory each time the cell mode is entered or exited. This is specified by setting the "TLCout when Saving Cell" check box using the **Set** command. A backup to floppy should also be made periodically using the **TLCout** command button on the system mode.

1.2.6 Common Problems

After adding an object, the object cannot be seen.

Check the **View** layers in the drawing display to ensure that the layer is not in hidden mode. The **Draw** command must be used after using **View**.

*Cannot **Get** an object.*

(1) Check, using the **Open** command, that the layer can be opened (or moved). (2) Verify that the object is not part of another cell. (3) When trying to get an object made using the path or polygon object, make sure the cursor encompasses a vertex.

*The menu isn't accessible when the **Layr** button is pressed.*

The layer table is not being used. Press CNTRL-ENTER so that the bottom middle of the display shows "Using Layer Table."

Cells are drawn as outlines, or the perimeter of a cell has a dashed line.

(1) Use the **Full** command to show the contents of a cell. (2) Use the **Dpth** command to limit the depth of the cells shown. Increasing the depth to the rank of the cell shows all cells. (3) Press **i** on the keyboard to force an outline to be drawn around a cell. This is indicated by the **I** in the bottom right corner of the drawing display.

*The **Fit** command causes the drawing window to expand much larger than the current cell.*

There is an unknown object someplace in the cell. Use the **fGet** command to get any objects outside the main cell area. Use the **Del** command to delete the unknown object.

The command buttons or layout are not displaying correctly.

Verify that windows is using small fonts. Reduce the amount of hardware acceleration used with the display (this is changed in the windows control panel).

The cursor movement is not smooth.

The cursor may be in the octagonal mode. Press "o" on the keyboard or the button "O" in the bottom right portion of the display to toggle this mode on and off.

1.3 MOSIS

The MOSIS[2] IC fabrication service is available to universities provided they have access to the Internet, software for layout and simulation, and the capability to test the completed designs. The instructor must contact MOSIS and submit a proposal for funding. If the proposal is funded, the university will have an account set up with the number of chips to be fabricated and the process used determined by the number of students in the CMOS course and whether the course is introductory or advanced. At the present time, funding from the NSF[3] for introductory classes is one 2.0 micron "tiny chip" per two students. A tiny chip measures 2.2 mm by 2.2 mm edge to edge. A minimum quantity of four chips is supplied with each order. Commercial companies and non-US univerisities may also fabricate ICs through MOSIS but receive no US government funding.

    After an account has been established by MOSIS and the university has an account number, completed chip designs can be submitted in UUEncoded GDS (UUGDS[4]) format or CIF (CalTech Intermediate Form). This text will only describe the fabrication process using GDS format. For information regarding CIF, contact MOSIS. To translate a TLC file into a GDS format, simply select the **Tlc2Gds** command button

---

[2] MOSIS - MOS Implementation Service through the Information Sciences Institute at the University of Southern California, (310) 822-1511 ext. 403 or http://www.mosis.org

[3] National Science Foundation

[4] UUEncoding is used to change a binary file such as the GDS file generated with the **Tlc2Gds** into an ASCII file for transmission over the Internet. Here UUGDS is a UUEncoded GDS file.

on the LASI system (**Sys**) menu. Selecting the **Setup** command causes the screen shown in Fig. 1.10 to appear. Here, the name of the highest ranking cell, that is, the cell to be fabricated, is QCELL.TLC. All cells used in QCELL.TLC will be converted into GDS format and placed in the binary file QCELL.GDS (this is transparent to the user). Of course, the **TLCout** command must be executed prior to calling the **Tlc2Gds** utility. The conversion process is started by selecting **Go** on the TLC2GDS command window. Additional information concerning a TLC to GDS conversion can be found by pressing the **Help** button on the **Tlc2Gds** command window.

Main Setup

Main TLC File to be Converted	QCELL.TLC	
Name of GDS File to Make	QCELL.GDS	
GDS Library Name	DEFAULT.DB	Scale Unit LSB Correction |0
Physical Unit	um	Lambda Size in um |1.
LASI Units per Phys Unit		GDS Units per Phys Unit |1000
Datatype to Use |0	Default Path Width |2.e-003
Layer 64 to Layer |0	Layer Filter |-

Conversion Options
- ☑ Sort Cells in Ascending Rank Order   ☑ Convert Text   ☐ Lowercase Cell Names
- ☑ Check for Proper GDS IC Protocol   ☑ Check Poly Intersects / Modulus |1

Report File Name | TLC2GDS.RPT   OK   Cancel

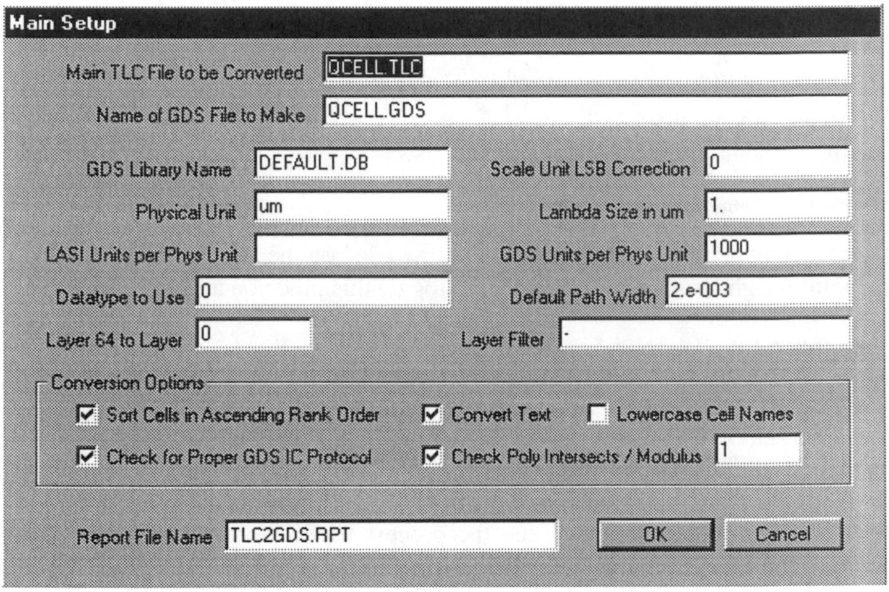

Figure 1.10 Converting TLC files to GDS format.

An Important Note

A note is in order here about polygons and the TLC file conversion into GDS (Graphical Design System, or GDS, is a derivative of the Calma Stream Format) format. It is possible to draw a polygon that is not closed, for example, a triangle with only two sides. These shapes are referred to as "open polygons". If the TLC file contains open polygons LASI will prompt the user to either cancel the conversion, so the user can go back to the layout program and fix the polygon, or to close the polygon. If the user selects "close" the **Tlc2Gds** converter will add the final segment in the polygon and thus change the layout. In almost all cases the users should fix the open polygon and run the design rule checking software again on the fixed cell. If the design file is translated into CIF, instead of GDS, the default action is to close the polygon. In other words, by default an open polygon will be closed when translating from TLC to CIF. MOSIS will reject GDS files and accept (and close the polygons) CIF files with open polygons.

To translate the binary GDS file (Qcell.gds) into an ASCII file (Qcell.uue or a UUGDS file) suitable for transmitting to MOSIS, the following command is used from the Windows Run line (or from a DOS prompt):

C:\Lasi6\uuen -j   C:\Lasi6\Mydesign\qcell.gds   C:\Lasi6\Mydesign\qcell.uue

The executable file Uuen.exe is located in the C:\Lasi6 directory. We have assumed that the cell to be fabricated is located in the design directory C:\Lasi6\Mydesign\. The switch "j" creates a Unix-compatible ASCII file for transmission to MOSIS.

The final step, before submission of the file to MOSIS, is to run the checksum program (Cksum.exe) located in C:\Lasi6 on Qcell.uue, or

C:\Lasi6\cksum C:\Lasi6\Mydesign\qcell.uue

The result is two numbers; the layout-checksum and the byte count. These numbers are used in the submission process (to MOSIS) discussed below.

Submission of Chips to MOSIS

The basic submission of chips to MOSIS consists of requesting an ID for a new project and submitting the project for check/fabrication. Checking the status of the chips and changing or canceling project parameters can also be performed before the chips are sent out, from MOSIS, for fabrication.

A project ID can be requested from MOSIS by sending an appropriate e-mail address to mosis@mosis.org for each of the chip designs that will be fabricated. The MOSIS command language syntax is used when communicating with MOSIS. When making this request, the user must specify several items, as illustrated in the following example. The project check is used to ensure that the ASCII file containing the UUGDS specifications of your chip does not get corrupted when transmitted over the Internet. If the GDS file is accepted, the user will be notified that the project is in the queue for fabrication.

Example 1.1
Submit a chip to MOSIS for fabrication using Orbit's 2.0 μm n-well process (CN20). Information on this process is given in Appendix A and is used throughout the text.

Step 1: Request New Project ID

Assuming that MOSIS specified an account number of 123-ABC, the password is WINFECT, and the instructor's name is SMITH, the first step in submitting a chip for fabrication is to send MOSIS an e-mail (mosis@mosis.org) requesting a new project ID (the e-mail message is shown below).

REQUEST:	NEW-PROJECT
ACCOUNT:	123-ABC
D-NAME:	SMITH

D-PASSWORD:	WINFECT
P-NAME:	CHIP1
P-PASSWORD:	UNIVER
PHONE:	(123) 456-7890
TECHNOLOGY:	FORESIGHT-CN20
SIZE:	2160 X 2160
PADS:	40
PACKAGE:	DIP40
DESCRIPTION:	HIGH SPEED OP-AMP
REQUEST:	END

The part name and password are defined by the user. It was assumed that the pads were designed by the students and measured 180 μm square; that is, the outline of the pads is 180 μm, while the actual pad is 100 μm on a side (see Ch. 3). The resulting padframe measures 2160 μm by 2160 μm. The "PADS:" line specifies the number of pads on the chip. The description line of the request file is also defined by the user. MOSIS replies to the above request with an ID number such as 876543. The MOSIS system is automated so that precise syntax is required.

Note that here we assumed the chip was laid out using the the Foresight design rules (from Orbit) given in Appendix A. If the chip were laid out using the MOSIS Scalable Design rules given in Appendix B we would use a Technology specification of "SCNA" (Scalable CMOS N-well Analog) and add a line, to the email message, specifying the scalable parameter Lamba, that is,

LAMBDA:	1.0

Note that in either case, whether using the Foresight or the MOSIS design rules, the process used (Orbit's 2 μm n-well process) is the same.

Step 2: GDS File Submission

To submit the chip for project check and fabrication, send the message shown below. The final cell that references all the other cells is the "TOP-STRUCTURE" cell. This cell is the chip that will be fabricated.

REQUEST:	FABRICATE
ID:	876543
P-PASSWORD:	UNIVER
LAYOUT-CHECKSUM:	123456 1234
LAYOUT-FORMAT:	UUGDS
TOP-STRUCTURE:	QCELL

LAYOUT:

Insert UUGDS file here (qcell.uue). Do not add characters.

REQUEST: END

If the file is received uncorrupted, MOSIS responds by sending an e-mail message to the user stating that the project is queued for fabrication.

For some chip designs, the UUGDS ASCII file will be very large. MOSIS has the capability to FTP large design files from the user. The LAYOUT statement can be changed, in the submission above, so that this is possible. The general form of the LAYOUT statement is

LAYOUT-FTP-PATH: !hostname!username!password!filename

where "!" is used as a delimiter and hostname is the name of the computer connected to the Internet where the design file resides. An example is

LAYOUT-FTP-PATH: !mycom.univ.edu!anonymous!guest!pub/chips/qcell.uue

where the entire statement must fit on one line. Also, it is possible to send the email message to MOSIS containing the commands and layout file, as shown in Step 2 above, as an attached text file (so that a word processor can be used to generate the message.)

When the chip is actually sent to the foundry, the user will be notified. Information about the status of the chip while being fabricated is available via anonymous FTP or the World Wide Web (http://www.mosis.org). Consult the MOSIS user's guide and the on-line information for additional information on submitting a chip for fabrication and the fabrication schedule. ■

To summarize the procedure for submitting a chip to MOSIS for fabrication, begin with a LASI-generated TLC file. Then

1.   Using the command button **Tlc2Gds** on the LASI system menu, generate a Calma Stream Format (GDS) binary file.

2.   Run uuen.exe on the GDS file. The result is a UUGDS (ASCII) file.

3.   Next, the checksum program is run on the UUGDS file. This results in the generation of two numbers: checksum and byte count.

4.   Send MOSIS a request for a new project ID (assuming you already have an account).

5.   Submit the UUGDS file to MOSIS for syntax check (not a design rule check) and fabrication.

6.   After MOSIS replies that the file has no fatal syntax errors the project will be queued for fabrication.

REFERENCES

[1]   W. Tanner, *MOSIS User Manual,* Release 4.0, August 1994. Also located at
       http://www.mosis.org/manual.html.

[2]   D. E. Boyce, *LASI Users Manual,* available as on-line help or as a printable
       manual by pressing Help while LASI is running.

PROBLEMS

For the following problems, use the LASI setups given in Appendix A and in the
C:\Lasi6\Wcn20 directory for the CN20 process.

1.1   Create a cell called test3 with a rank of 1 using LASI. In this cell, draw a 10 μm
           by 10 μm box using the pol1 layer. Place the lower left corner of the box at the
           origin. Use the "z" (used to set the zero point) and the spacebar to measure the
           distance between opposite corners.

1.2   Explain how the **qMov** command can be used to edit the box in Problem 1 so
           that it measures 5 x 8 μm². How would this be accomplished using **Get**, **Mov**,
           and **Put**?

1.3   What functions do the **sWin** and **rWin** commands perform?

1.4   What functions do the **cGet** and **cPut** commands perform?

1.5   The **Text** command allows text to be used for labeling in LASI. The **tLayr**
           selects which layer the text will be written in, while the **tSiz** sets the size of the
           text in 1.5 μm increments. Write the word "test" on the met1 layer with sizes of
           3, 9, and 24 μm in the test3 cell of Problem 1. Show the result without using the
           reference mark. (The reference mark is removed by pressing **t** on the keyboard or
           by selecting the **T** in the top right corner of the drawing display, remembering to
           execute a **Draw** command afterward.) Labeling is extremely important in
           layout.

1.6   Create a cell named test4 with a rank of 1. Generate the layout shown in Fig.
           P1.6a in this cell. The text and boxes are written using the met1 layer. Next
           create a cell named test5 with a rank of 2. Add the test4 cell into the test5 cell
           five times as shown in Fig. P1.6b. The **cGet** and **Mov** commands may come in
           handy at this point. Next draw the cells as outlines, shown in Fig. P1.6c, using
           the **Outl** command. Note that we could have used the **Cpy** command to copy the
           layout in test4 five times and avoid adding the test5 cell. The problem with this
           is that as the layout becomes complicated the memory required in a "flat" cell
           increases dramatically. The hierarchical layout using the nested cells keeps
           memory usage to a minimum. The **Cpy** command should be used as little as
           possible.

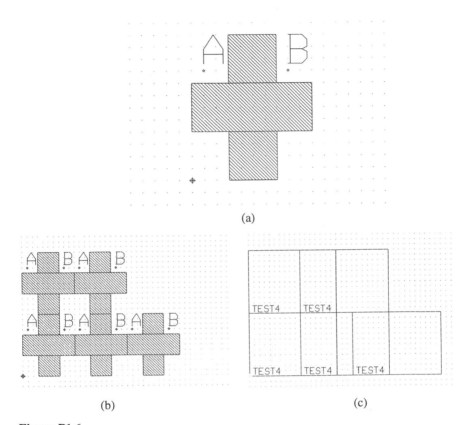

(a)

(b)   (c)

Figure P1.6

1.7   Polygons or paths can be drawn using LASI by setting the **Obj** to "p" (instead of "b" for box). Setting the **Wdth** size to 0 causes LASI to draw polygons, while setting the width greater than 0 causes LASI to draw paths. When finished drawing a path or polygon, use the **aPut** command. Using LASI, copy the layout shown in Fig. P1.6a using the polygon object.

1.8   What part of an object made using the polygon or path must be encompassed to **Get** the object?

1.9   Using the pol1 layer, draw a triangle that measures nominally 10 µm on each side. How many vertices does the object have?

1.10   Circles can be drawn using a polygon (path with zero width) and the **Arc** command. Consider the layout shown in Fig. P1.10. Copy this layout in LASI. Begin by adding a vertex at point A using the **Add** command. This is followed by selecting the **Arc** command, moving and clicking the mouse in the desired center of the circle, and returning and clicking the mouse at point A. The bottom of the display will then inquire how many segments should be used and which direction to draw, that is, counterclockwise or clockwise. Hit the Enter key to both of these questions, and LASI will draw the following circle.

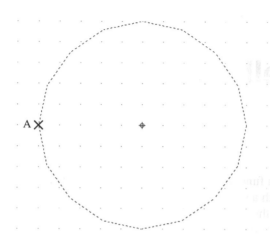

Figure P1.10

1.11 Pressing F1 on the keyboard within the LASI program calls the LASI help file. What would be added to the end of the form.dbd file in a drawing directory so that F2 performs the LASI command **Fit** and F3 performs an **aPut**? Frequently used commands can be executed using function keys to help speed up layout time.

1.12 Using the **Dpth** command, show how the cells in Fig. P1.6b can be drawn as outlines. What does the depth level mean? Show that, by pressing **i** on the keyboard (or top right corner of the drawing screen), the cells are also drawn as outlines.

1.13 What do the keyboard buttons **w, u, a, z,** and **space** do in LASI?

1.14 Describe how to add text in LASI and how to set the text size and layer.

1.15 Using LASI, show example layouts that show the difference between path objects and the poly objects. Use poly1 in your examples. How do you **Get** a poly or path object?

Chapter

2

The Well

In order to develop a fundamental understanding of CMOS integrated circuit layout and design, we begin with a study of the n-well. This approach will build a solid foundation for understanding the performance limitations and parasitics (the pn junctions, capacitances, and resistances inherent in a CMOS circuit) of the CMOS process.

2.1 The Substrate

CMOS circuits are fabricated on and in a silicon wafer as was discussed in Ch. 1. This wafer is doped with donor atoms, such as phosphorus for an n-type wafer, or acceptor atoms, such as boron for a p-type wafer. Our discussion will center around a p-type wafer. When designing CMOS integrated circuits with a p-type wafer, n-channel MOSFETs (NMOS for short) are fabricated directly in the p-type wafer, while the p-channel transistors, PMOS, are fabricated in an "n-well." The substrate or well are sometimes referred to as the bulk or body of a MOSFET. CMOS processes that fabricate MOSFETs in the bulk are known as "bulk CMOS processes." The well and the substrate are illustrated in Fig. 2.1, though not to scale.

Often an epitaxial layer is grown on the wafer. We will not make a distinction between this layer and the substrate. Some processes use a p-well or both n- and p-wells (sometimes called twin tub processes). A process that uses a p-type substrate with an n-well is called an "n-well process." The process described in Appendix A, CN20, is an n-well process. The n-well acts as the substrate or body of the p-channel transistors.

Another important consideration is that the n-well and the p-substrate form a diode (Fig. 2.2). In CMOS circuits, the substrate is usually tied to the lowest voltage in the circuit to keep this diode from forward biasing. Ideally, zero current flows through the substrate connection.

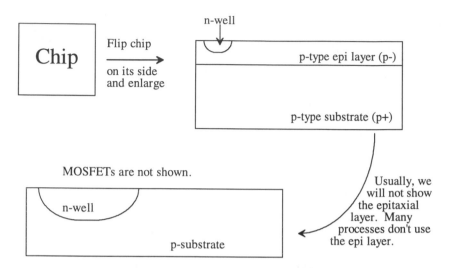

Figure 2.1 Illustration of the top and side view of a die.

Besides being used for fabricating p-channel transistors, the n-well can be used as a resistor. The voltage on either side of the resistor must be large enough to keep the substrate/well diode from forward biasing.

2.1.1 Patterning

CMOS integrated circuits are formed by patterning different layers on and in the CMOS wafer. Consider the following sequence of events that apply, in a fundamental way, to any layer we need to pattern. We start out with a clean, bare wafer as shown in Fig. 2.3a. The distance given by the line A to B will be used as a reference in Figs. 2.3b-j. Figures 2.3b-j are cross-sectional views of the dashed line shown in (a).

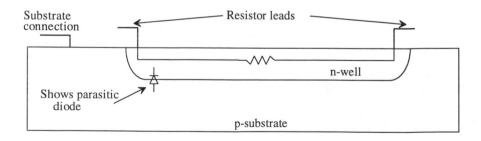

Figure 2.2 The n-well can be used as a resistor.

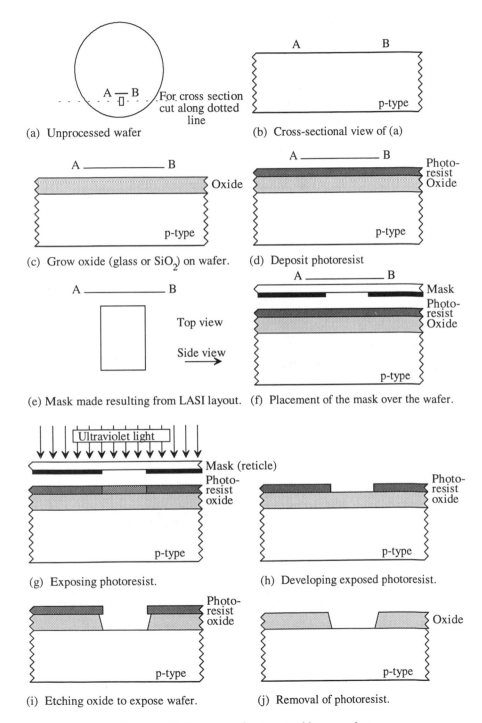

(a) Unprocessed wafer

(b) Cross-sectional view of (a)

(c) Grow oxide (glass or SiO$_2$) on wafer.

(d) Deposit photoresist

(e) Mask made resulting from LASI layout.

(f) Placement of the mask over the wafer.

(g) Exposing photoresist.

(h) Developing exposed photoresist.

(i) Etching oxide to expose wafer.

(j) Removal of photoresist.

Figure 2.3 Sequence of events used in patterning.

The first step in our generic patterning discussion is to grow an oxide, SiO_2 or glass, a very good insulator, on the wafer. Simply exposing the wafer to air yields the reaction $Si + O_2 \rightarrow SiO_2$. However, semiconductor processes must have tightly controlled conditions to precisely set the thickness and purity of the oxide. Therefore, we can grow the oxide using a reaction with steam, H_2O, or with O_2 alone. The oxide resulting from the reaction with steam is called a wet oxide, while the reaction with O_2 is a dry oxide. Both oxides are called thermal oxides due to the increased temperature used during oxide growth. The growth rate increases with temperature. The main benefit of the wet oxide is fast growing time. The main drawback of the wet oxide is the hydrogen byproduct. In general terms, the oxide grown using the wet techniques is not as pure as the dry oxide. The dry oxide, as we can conclude, generally takes a considerably longer time to grow. Both methods of growing oxide are found in CMOS processes.

An important observation we should make when looking at Fig. 2.3c is that the oxide growth actually consumes silicon. This is illustrated in Fig. 2.4. The overall thickness of the oxide is related to thickness of the consumed silicon by

$$t_{Si} = 0.45 \cdot t_{ox} \qquad (2.1)$$

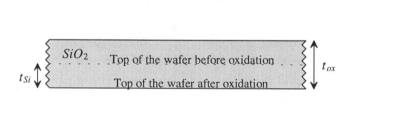

Figure 2.4 How growing oxide consumes silicon.

The next step of the CMOS patterning process is to deposit a photosensitive resist layer across the wafer (see Fig. 2.3d). Keep in mind that the dimensions of the layers, that is, oxide, resist, and the wafer, are not drawn to scale. The thickness of a wafer is typically 500 μm, while the thickness of a grown oxide or a deposited resist may be only a few μm or even less. After the resist is baked, the mask derived from the layout program, Figs. 2.3e and f, is used to selectively illuminate areas of the wafer, Fig. 2.3g. In practice, a single mask called a reticle, with openings several times larger than the final illuminated area on the wafer, is used to project the pattern and is stepped across the wafer with a machine called a stepper to generate the patterns needed to create multiple copies of a single chip. The light passing through the opening in the reticle is photographically reduced to illuminate the correct size area on the wafer.

The photoresist is developed (Fig. 2.3h), removing the areas that were illuminated. This process is called a positive resist process because the area that was illuminated was removed. A negative resist process removes the areas of resist that were not exposed to the light. Using both types of resist allows the process designer to cut down on the number of masks needed to define a CMOS process. Since creating the masks is expensive, lowering the number of masks is equated with lowering the cost of a process. This is also important in large manufacturing plants where fewer steps are equated with lower cost.

The next step in the patterning process is to remove the exposed oxide areas (Fig. 2.3i). Notice that the etchant etches under the resist, causing the opening in the oxide to be larger than what was specified by the mask. Some manufacturers intentionally bloat (make larger) or shrink (make smaller) the masks as specified by the layout program. Figure 2.3j shows the cross-sectional view of the opening after the resist has been removed.

2.1.2 Patterning the N-well

At this point we can make an n-well by diffusing donor atoms, those with 5 valence electrons, as compared to 4 for silicon, into the wafer. Referring to our generic patterning discussion given in Fig. 2.3, we begin by depositing a layer of resist directly on the wafer, Fig. 2.3d (without oxide). This is followed by exposing the resist to light through a mask (Figs. 2.3f and g) and developing or removing the resist (Fig. 2.3h). The mask used here can be generated with the LASI program. The next step in fabricating the n-well is to expose the wafer to donor atoms. The resist will block the diffusion of the atoms, while the openings will allow the donor atoms to penetrate into the wafer. This is shown in Fig. 2.5a. After a certain amount of time, dependent on the depth of the n-well desired, the diffusion source is removed (Fig. 2.5b). Notice that the n-well "outdiffuses" under the resist; that is, the final n-well size is not the same as the mask size. Again, the foundry where the chips are fabricated may bloat or shrink the mask to compensate for this *lateral diffusion*. The final step in making the n-well is the removal of the resist (Fig. 2.5c).

2.2 Laying out the N-well

When we lay out the n-well, we are viewing the chip from the top. The following example illustrates how to lay out an n-well of size 10 μm square.

Example 2.1
Using the LASI program, lay out an n-well that is 10 μm square. Sketch the cross-sectional view of the layout. Assume we are using the CN20 setups given in the last chapter and using the layer table select **Layr** from the drawing screen commands and the layer "nwel." If the origin marker is not showing, type "r." Now select **Add**. The bottom of the display should show that the object is a box and that the grids are 1 μm. Place the cursor in the display area on the origin

and press and release the LEFT mouse button. Move the mouse until the distance shows $ux = 10$ μm and $uy = 10$ μm. The resulting display is shown in the top of Fig. 2.6. Notice that the drawn size, both width and length, of the n-well differs from the actual size because of the lateral diffusion. ■

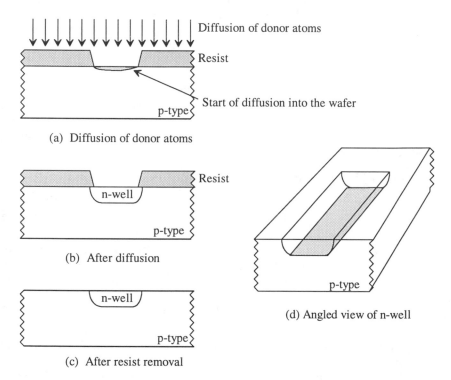

(a) Diffusion of donor atoms

(b) After diffusion

(c) After resist removal

(d) Angled view of n-well

Figure 2.5 Formation of the n-well.

2.2.1 Design Rules for the N-well

Now that we know how to lay out the n-well, we might ask the question, "Are there any limitations or constraints on the size and spacing of the n-wells?" That is to say, can we make the n-well 2 μm square? Can we make the distance between the n-wells 1 μm? As we might expect, there are minimum spacing and size requirements for all layers in a CMOS process. Process engineers, who design the integrated circuit process, specify the design rules. A complete listing of the CN20 design rules can be found in Appendix A.

Figure 2.7 shows the design rules for the n-well. The minimum width of any n-well is 3 μm, while the minimum spacing between different n-wells is 9 μm. As the layout becomes complicated, the need for a program that ensures that the design rules are not violated is needed.

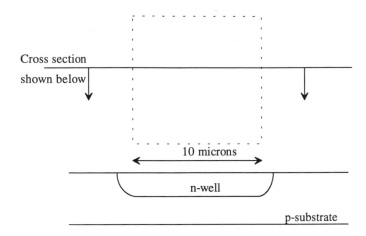

Figure 2.6 Layout and cross-sectional view of a 10 µm n-well.

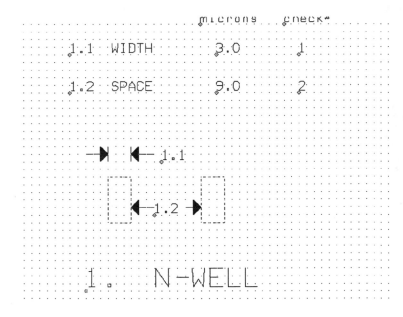

Figure 2.7 Design rules for layout of the n-well.

2.2.2 Using the LasiDrc Program

To perform a design rule check (DRC) of a layout, simply select **LasiDrc**, from the LASI system menu, followed by **Setup**. Enter the name of the cell (see below) to be checked and type cn20.drc as the name of the check file. Since we only know two design rules at this point, that is, from Fig. 2.7 checks 1 and 2, the starting check should be 1 and the finish check should be 2. To start the program, select **Go** on the top of the LASIDRC screen (after closing the setup screen). If there is an error a bit map of the layout will be generated (and can be viewed with the **Map** command) and the error will be reported in a report file (which can be read using the **Read** command). The DRC will be performed on the *section* of layout shown in the drawing display just prior to calling the LASIDRC program or after the **Save** command is used if the DRC program is already open. To DRC the entire cell, press the **Fit** command button on the DRC setup screen. This feature can be used to decrease the time it takes to perform a DRC by DRCing only the specific areas of interest.

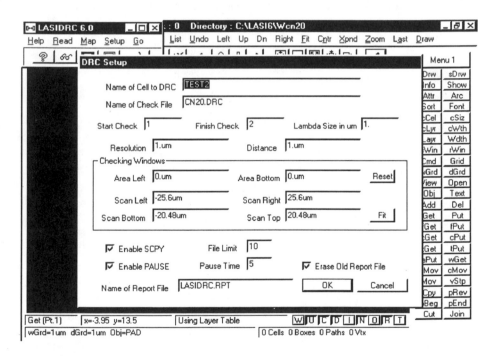

2.3 Resistance Calculation

In addition to serving as a region in which to build PMOS transistors, n-wells are sometimes used to create integrated resistors. The resistance of a material is a function of the materials resistivity, ρ, and the materials dimensions. For example, the slab of material in Fig. 2.8 between the two leads has a resistance given by

$$R = \frac{\rho}{t} \cdot \frac{L}{W} \tag{2.2}$$

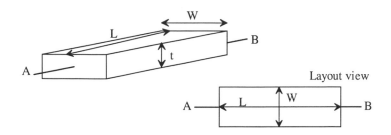

Figure 2.8 Calculation of the resistance of a rectangular block of material.

In semiconductor processing, all of the fabricated thicknesses, such as the n-well, are constant. We only have control over W and L. Also notice that the W and L are what we see from the top view, that is, the layout view. We can rewrite Eq. (2.2) as

$$R = R_{square} \cdot \frac{L}{W} \tag{2.3}$$

R_{square} is the sheet resistance of the material in Ω/square.

Example 2.2

Calculate the resistance of an n-well that is 10 μm wide and 100 μm long.

From the Orbit electrical parameters in Appendix A for CN20 we see that the sheet resistance of the n-well varies from a minimum of 2,000 Ω/square to a maximum of 3,000 Ω/square, with a typical number of 2,500 Ω/square. So the typical resistance between the ends of the n-well is

$$R = 2,500 \cdot \frac{100 \text{ μm}}{10 \text{ μm}} = 25 \text{ k}\Omega \quad \blacksquare$$

When laying out resistors using the "path" object, n-well, poly, or some other layer, LASI has a resistor calculator that will help in the calculation of nonrectangular resistances. Often, to minimize space, resistors are laid out in a serpentine pattern. The corners, that is, where the layer bends, are not rectangular. This is shown in Fig. 2.9a. All sections in Fig 2.9a are square, so the resistance of sections 1 and 3 is R_{square}. The equivalent resistance of section 2 between the adjacent sides, however, is approximately $0.6\ R_{square}$. The overall resistance between points A and B is therefore $2.6 \cdot R_{square}$.

The layout shown in Fig. 2.9b uses wires to connect separate sections of a resistor avoiding corners. Avoiding corners in a resistor is the generally preferred method of layout in analog circuit design where the ratio of two resistors is important.

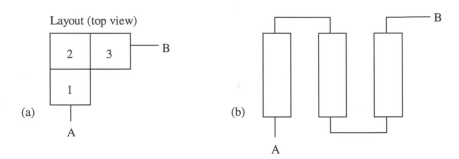

Figure 2.9 (a) Figure for the calculation of the resistance of a corner section and
(b) layout to avoid corners.

2.3.1 The N-well Resistor

At this point, it is appropriate to show the actual cross-sectional view of the n-well after all processing steps are completed (Fig. 2.10). The n+ and p+ implants are used to increase the threshold voltage of the field devices; more will be said on this later in Ch. 4. Notice in Appendix A, the Orbit electrical parameters, that the sheet resistance of the n-well is measured with the field implant in place, that is, with the n+ implant between the two metal connections in Fig. 2.10. Not shown in Fig. 2.10 is the connection to substrate. The field oxide (FOX; also known as ROX or recessed oxide) will be discussed in Ch. 4 when we discuss the active and poly layers.

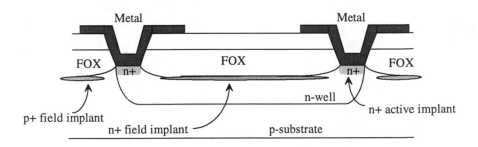

Figure 2.10 Cross-sectional view of n-well showing field implant. The field implantation is sometimes called the "channel stop implant".

2.4 The N-well/substrate Diode

Placing an n-well in the p-substrate forms a diode. Therefore, it is important to understand how to model a diode for hand calculations and in SPICE simulations. In particular, let's discuss diodes using the n-well/substrate pn junction as an example [2]. The DC characteristics of the diode are given by the Shockley diode equation, or

$$I_D = I_S\left(e^{\frac{V_d}{nV_T}} - 1\right) \tag{2.4}$$

The current I_D is the diode current; I_S is the scale (saturation) current; V_d is the voltage across the diode where the anode (p-type material) is assumed positive with respect to the cathode (n-type); and V_T is the thermal voltage which is given by $\frac{kT}{q}$ where k = Boltzmann's constant (1.3806×10^{-23} joules per degree kelvin), T is temperature in kelvin, n is the emission coefficient (a term that is related to the doping profile and affects the exponential behavior of the diode), and q is the electron charge of 1.6022×10^{-19} coulombs. The scale current and thus the overall diode current are related in SPICE by an area factor. The SPICE circuit simulation program assumes that the value of I_S supplied in the model statement was measured for a device with a reference area of 1. If an area factor of 2 is supplied for a diode, then I_S is doubled in Eq. (2.4).

2.4.1 Depletion Layer Capacitance

N-type silicon has a number of mobile electrons, while p-type silicon has a number of mobile holes (a vacancy of electrons in the valence band). Formation of a pn junction results in a depleted region at the p-n interface (Fig. 2.11). A depletion region is an area depleted of mobile holes or electrons. The mobile electrons move across the junction, leaving behind fixed donor atoms and thus a positive charge. The movement of holes across the junction, to the right in Fig. 2.11, occurs for the p-type semiconductor as well with a resulting negative charge. The fixed atoms on each side of the junction within the depleted region exert a force on the electrons or holes that have crossed the junction. This equalizes the charge distribution in the diode, preventing further charges from crossing the diode junction and also gives rise to a parasitic (depletion) capacitance.

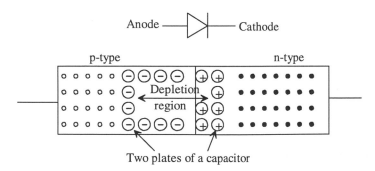

Figure 2.11 Simple illustration of depletion region formation in a pn junction.

The depletion capacitance, C_j, of a pn junction is given by

$$C_j = \frac{C_{j0}}{\left[1 - \left(\frac{V_d}{\varphi_0}\right)\right]^m} \tag{2.5}$$

C_{j0} is the zero-bias capacitance of the pn junction, that is, the capacitance when the voltage across the diode is zero. V_d is the voltage across the diode, m is the grading coefficient (showing how the silicon changes from n- to p-type), and ϕ_0 is the built-in potential given by

$$\phi_0 = V_T \cdot \ln\left(\frac{N_A N_D}{n_i^2}\right) \qquad (2.6)$$

where N_A and N_D are the dopings for the p- and n-type semiconductors, respectively, V_T is the thermal equivalent voltage $\frac{kT}{q}$ (26 mV @ room temperature), and n_i is the intrinsic carrier concentration of silicon ($n_i = 14.5 \times 10^9$ atoms/cm^3).

Example 2.3

Sketch schematically the depletion capacitance of an n-well/p-substrate diode 100×100 µm^2 square given that the substrate doping is 10^{16} atoms/cm^3 and the well doping is 10^{16} atoms/cm^3. The measured zero-bias depletion capacitance of the junction is 100 aF/µm^2, and the grading coefficient is 0.333. Assume the depth of the n-well is 3 µm.

We can begin this problem by calculating the built-in potential using Eq. (2.6):

$$\phi_0 = (.026) \cdot \ln\frac{10^{16} \cdot 10^{16}}{(14.5 \times 10^9)^2} = 0.7 \text{ V}$$

The depletion capacitance is made up of a *bottom* component and a *sidewall* component as shown in Fig. 2.12.

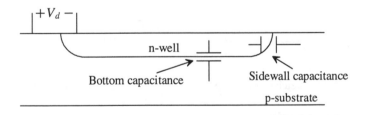

Figure 2.12 A pn junction on the bottom and sides of the junction.

The bottom zero-bias depletion capacitance, C_{j0b}, is given by

C_{j0b} = (Capacitance per Area) · (Bottom Area) which, for this example, is

$$C_{j0b} = (100 \text{ aF/µm}^2) \cdot (100 \text{ µm})^2 = 1 \text{ pF}$$

The sidewall zero-bias depletion capacitance, C_{j0s}, is given by

C_{j0s} = (Capacitance per Area) · (Depth of the Well) · (Perimeter of the Well)

or

$$C_{j0s} = (100 \text{ aF/}\mu\text{m}^2) \cdot (3 \ \mu\text{m}) \cdot (400 \ \mu\text{m}) = 120 \text{ fF}$$

The total diode depletion capacitance between the n-well and the p-substrate is the parallel combination of the bottom and sidewall capacitances, or

$$C_j = \frac{C_{j0b}}{\left[1 - \left(\frac{V_d}{\varphi_0}\right)\right]^m} + \frac{C_{j0s}}{\left[1 - \left(\frac{V_d}{\varphi_0}\right)\right]^m} = \frac{C_{j0b} + C_{j0s}}{\left[1 - \left(\frac{V_d}{\varphi_0}\right)\right]},$$

Substituting in the numbers, we get

$$C_j = \frac{1 \text{ pF} + 0.120 \text{ pF}}{\left(1 - \left(\frac{V_d}{0.7}\right)\right)^{0.33}}$$

A sketch of how this capacitance changes with reverse potential is given in Fig. 2.13. Notice that when we discuss the depletion capacitance of a diode, it is usually with regard to a reverse bias. When the diode becomes forward-biased minority carriers, electrons in the p material and holes in the n material, injected across the junction, form a stored charge in and around the junction and give rise to a storage capacitance. This capacitance is usually much larger than the depletion capacitance. Furthermore, the time it takes to remove this stored charge can be significant. ■

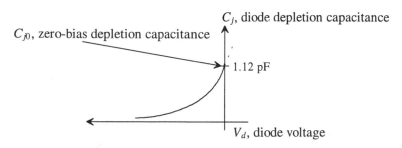

Figure 2.13 Sketch of diode depletion capacitance against diode reverse voltage.

2.4.2 Storage Capacitance

Consider the charge distribution of a forward-biased diode shown in Fig. 2.14. When the diode becomes forward biased, electrons from the n-type side of the junction are attracted to the p-type side (and vice versa for the holes). After an electron drifts across the junction, it starts to diffuse toward the metal contact. If the electron recombines, that is, falls into a hole, before it hits the metal contact, the diode is called a "long base diode." The time it takes an electron to diffuse from the junction to the point the electron recombines is called the carrier lifetime. For silicon this lifetime is on the order of 10 μs. If the distance between the junction and the metal contact is short, such that the electrons make it to the metal contact before recombining, the diode is said to be a "short base diode." In either case, the time between crossing the junction and

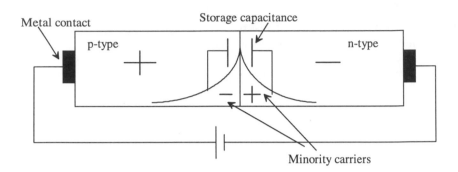

Figure 2.14 Charge distribution in a forward-biased diode.

recombining is called the transit time, τ_T. A capacitance is formed between the electrons diffusing into the p-side and the holes diffusing into the n-side, that is, formed between the minority carriers. (Electrons are the minority carriers in the p-type semiconductor.) This capacitance is often called a diffusion capacitance or storage capacitance (due to the presence of the stored minority carriers around the junction).

We can characterize the storage capacitance, C_S, in terms of the minority carrier lifetime. Under DC operating conditions, the storage capacitance is given by

$$C_S = \frac{I_D}{nV_T} \cdot \tau_T \tag{2.7}$$

I_D is the DC current flowing through the forward-biased junction given by Eq. (2.4). Looking at the diode capacitance in this way is very useful for analog AC small-signal analysis. However, for digital applications we are more interested in the large-signal switching behavior of the diode. It should be pointed out that in general, for a CMOS process, it is undesirable to have a forward-biased pn junction. If we do have a forward-biased junction, it usually means there is a problem, for example, electrostatic protection, capacitive feedthrough possibly causing latch-up, and so on. These topics are discussed in more detail later in this chapter.

Consider Fig. 2.15. In the following diode switching analysis, we will assume that $V_F \gg 0.7$, $V_R < 0$ and that the voltage source has been at V_F long enough to reach steady-state condition; that is, the minority carriers have diffused out to an equilibrium condition. At the time t_1 the input voltage source makes an abrupt transition from V_F to V_R, causing the current to change from $\frac{V_F}{R}$ to $\frac{V_R}{R}$. The diode voltage remains at 0.7 V since the diode contains a stored charge that must be removed . At time t_2 the stored charge is removed. At this point, the diode basically looks like a voltage-dependent capacitor that follows Eq. (2.5). In other words for $t > t_2$ the diode depletion capacitance is charged through R until the current in the circuit goes to zero and the voltage across the diode is V_R. This accounts for the exponential decay of the current and voltage shown in Fig. 2.15.

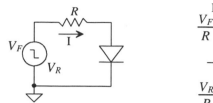 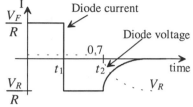

Figure 2.15 Diode test circuit.

The diode storage time, the time it takes to remove the stored charge, t_s, is simply the difference in t_2 and t_1, or

$$t_s = t_2 - t_1 \tag{2.8}$$

This time is also given by

$$t_S = \tau_T \cdot \ln \frac{i_F - i_R}{-i_R} \tag{2.9}$$

where $\frac{V_F}{R} = i_F$ and $\frac{V_R}{R} = i_R$ = a negative number in this discussion. Note that it is quite easy to determine the minority carrier lifetime using this test setup.

Defining a time t_3, where $t_3 > t_2$, when the current in the diode becomes 10 percent of $\frac{V_R}{R}$, we can define the diode reverse recovery time, or

$$t_{rr} = t_3 - t_1 \tag{2.10}$$

2.4.3 SPICE Modeling

The SPICE (simulation program with IC emphasis) diode model parameters are listed in Table 2.1. The series resistance, R_s, deserves some additional comment. This resistance results from the finite resistance of the semiconductor used in making the diode and the contact resistance, the resistance resulting from a metal contact to the semiconductor. At this point, we are only concerned with the resistance of the semiconductor. For a reverse-biased diode, the depletion layer width changes, increasing for larger reverse voltages (decreasing both the capacitance and series resistance, of the diode). However, when we model the series resistance, we use a constant value. In other words, SPICE will not show us the effects of a varying R_s.

Example 2.4
Using SPICE, explain what happens when a diode with a carrier lifetime of 30 ns is taken from the forward-biased region to the reverse-biased region. Use the circuit shown in Fig. Ex2.4 to illustrate your understanding.

We will assume a zero-bias depletion capacitance of 1 pF. The SPICE netlist for this circuit is shown below.

Figure Ex2.4

```
Diode storage time.
D1      1 0 TRR
R1      3 1 10k
Vin     3 0     DC 0 PULSE(10 -10 50n .1n .1n 50n 100n)
.Model TRR  D
+ IS=1.0E-15 TT=30E-9 CJO=1E-12 VJ=.7 M=0.33
.probe
.tran 1n 100n
.end
```

Figure 2.16 shows the current through the diode, the scaled input voltage step, and the scaled voltage across the diode. One of the interesting things to notice about this circuit is that current actually flows through the diode in the negative direction, even though the diode is forward biased. During this time, the stored minority carrier charge is removed from the junction. The storage time is given by

$$t_S = 30 \text{ ns} \cdot \ln \frac{0.93 \text{ mA} + 1.07 \text{ mA}}{1.07 \text{ mA}} = 18.8 \text{ ns}$$

which is close to the simulation results. Note that the input pulse doesn't change until 50 ns after the simulation starts. This ensures a steady-state condition when the input changes from 10 to −10 V. ∎

Name	SPICE	
I_S	IS	Saturation current
R_S	RS	Series resistance
n	N	Emission coefficient
V_{bd}	BV	Breakdown voltage
I_{bd}	IBV	Current which flows during V_{bd}
C_{j0}	CJ0	Zero-bias pn junction capacitance
φ_0	VJ	Built-in potential
m	M	Grading coefficient
τ_T	TT	Carrier transit time

Table 2.1 SPICE parameters related to diode.

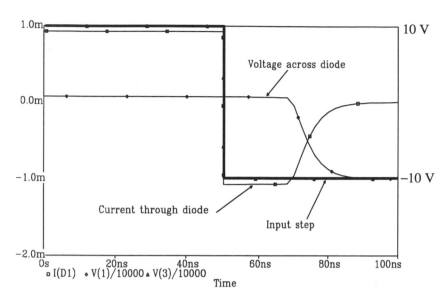

Figure 2.16 Results of Example 2.4 showing current and scaled voltages.

2.5 The RC Delay Through an N-well

At this point, we know that the n-well can be used as a resistor and as a diode when used with the substrate. Figure 2.17a shows the parasitic capacitance and resistance associated with the n-well. Since there is a depletion capacitance from the n-well to the substrate, we could sketch the equivalent symbol for the n-well resistor as shown in Fig. 2.17b. This is the basic form of an RC transmission line. If we put a step into one side of the n-well resistor a finite time later, called the delay time and measured at the 50 percent points of the pulses, the pulse will appear.

The delay can be calculated by knowing the resistance, r, per unit length, the capacitance, c, per unit length, and l, the number of unit lengths using the following [3]:

$$t_d = 0.35rcl^2 \qquad (2.11)$$

Example 2.5

Estimate the delay through a 250 kΩ resistor made using an n-well with a width of 3 μm and a length of 300 μm. Verify your answer with SPICE.

If we divide the resistor up into 100 squares each 3 μm wide and 3 μm long, we can define the number of unit lengths, l, as 100. The resistance of one of these squares is 2.5 kΩ ($= r$) per unit length. Now we are faced with determining the capacitance to substrate of one of these squares. Since the capacitance is a function of the voltage, we can get a worst-case estimate by selecting the zero-bias depletion capacitance (see Fig. 2.13). From Ex. 2.3 the zero-bias

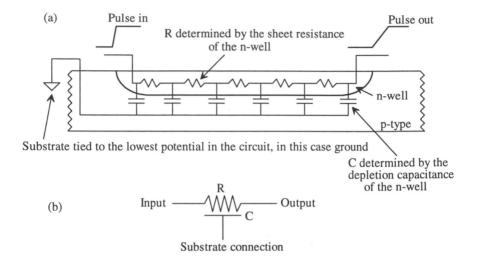

Figure 2.17 (a) Parasitic resistance and capacitance of the n-well and (b) schematic symbol.

depletion capacitance between the n-well and substrate is 100 aF/μm^2. The capacitance per unit length is the sum of the bottom and sidewall capacitances. However, since each square, except for the first and last squares, has only two sides contributing to the depletion capacitance and it is desirable to keep the number of calculations to a minimum in an estimate, we will neglect the sidewall capacitance. The capacitance per unit length is given by

$$c = C_{j0b} = 100\frac{\text{aF}}{\mu\text{m}^2}(3 \cdot 3)\mu\text{m}^2 = 900\ a\text{F}$$

The delay is now estimated by

$$t_d = 0.35 \cdot rcl^2 = 0.35 \cdot 2.5\text{k} \cdot 900\text{ aF} \cdot 100^2 = 7.88\text{ ns}$$

The SPICE netlist and the resulting output are shown in Fig. 2.18. Note that this is a SPICE3 netlist and not a PSPICE netlist. Also note that the nodes were labeled with names (i.e., Vin and Vout) rather than numbers. ■

We can simplify Eq. (2.11) by realizing that the products $r \cdot l$ and $c \cdot l$ are the total resistance and capacitance to substrate of the n-well resistor. Using this result on the previous example gives $R = r \cdot l = 2,500 \cdot 100 = 250$ kΩ and $C = c \cdot l = 900$ aF$\cdot 100 = 90$ fF and therefore

$$t_d = 0.35 \cdot RC = 0.35 \cdot 250\text{k} \cdot 90\text{ fF} = 7.88\text{ ns} \qquad (2.12)$$

which is the same result given the example. The important thing to notice here is that we can totally avoid the unit length parameter l. The resistance R is the resistance of

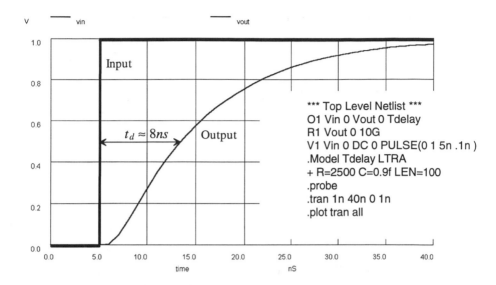

The netlist shown in the figure:

```
*** Top Level Netlist ***
O1 Vin 0 Vout 0 Tdelay
R1 Vout 0 10G
V1 Vin 0 DC 0 PULSE(0 1 5n .1n )
.Model Tdelay LTRA
+ R=2500 C=0.9f LEN=100
.probe
.tran 1n 40n 0 1n
.plot tran all
```

Figure 2.18 Simulation results from Example 2.5.

the resistor (= 250 kΩ above), and the capacitance C is simply the product of the bottom area of the resistor with the zero-bias depletion capacitance (= $3 \cdot 300 \cdot 100$ aF = 90 fF).

REFERENCES

[1] D. E. Boyce, *LASI Users Manual,* available as on-line help or as a printable manual by pressing Help while LASI is running.

[2] D. A. Hodges and H. G. Jackson, *Analysis and Design of Digital Integrated Circuits,* McGraw-Hill Publishing Company, 2nd ed., 1988. ISBN 0-07-029158-6.

[3] N. H. E. Weste and K. Eshraghian, *Principles of CMOS VLSI Design,* Addison-Wesley, 2nd ed., 1993. ISBN 0-201-53376-6.

PROBLEMS

2.1 Figure P2.1 is a section of n-well laid out using the path object with a width of 4 μm. Sketch the cross-sectional view (see Ex. 2.1, at the positions indicated in the figure). Copy this layout using the LASI program. Using the **Res** command (before using the **Res** command, use **fget** on the path) with a sheet resistance of 2,500 Ω/square (the typical sheet resistance of the n-well) and an end correction of 0.6, determine the resistance of the section. How does this compare with the $2.6R_{square}$ used for the layout in Fig. 2.9a?

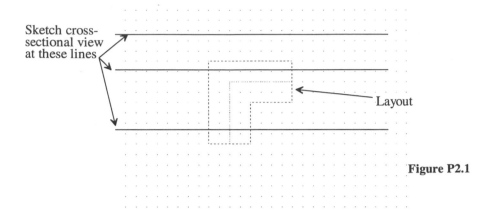

Sketch cross-
sectional view
at these lines

Layout

Figure P2.1

2.2 Add two boxes to the layout of Problem 1 (see Fig. P2.2). Using the LASIDRC program with checks 1 and 2, show the design rule violations in the layout.

2.3 Lay out a nominally 250 kΩ resistor using the n-well and the serpentine pattern shown in Fig. P2.3. Assume that the maximum length of a segment is 100 μm. Also design rule check the finished resistor.

2.4 Assuming the n-well depth is 3 μm in the CN20 process, what are the minimum, typical, and maximum values of the n-well resistivity?

2.5 Normally, the scale current of a pn junction is specified in terms of a scale current density, J_s (A/m^2) and the width and length of a junction (i.e., $I_s = J_s \cdot L \cdot W$ neglecting the sidewall component). Estimate the scale current for the diode of Ex. 2.3 if $J_s = 10^{-8}$ A/m^2.

2.6 Repeat Problem 5 including the sidewall component ($I_s = J_s \cdot L \cdot W + J_s \cdot (2L+2W) \cdot$ depth).

2.7 Using the diode of Ex. 2.3 in the circuit of Fig. P2.7, estimate the frequency of the input signal when the AC component of v_{out} is 707 μV (i.e., estimate the 3 dB frequency of the $|v_{out}/v_{in}|$).

2.8 Verify the answer given in problem 7 with SPICE.

2.9 Using SPICE, show that a diode can conduct current from its cathode to its anode when the diode is forward biased.

2.10 Estimate the delay through a 1 MΩ resistor (5 μm by 2,000 μm) made in the CN20 process using the n-well. Verify with SPICE.

2.11 If one end of the resistor in Problem 10 is tied to +5 V and the other end is tied to the substrate that is tied to ground, estimate the depletion capacitance (F/m^2) between the n-well and the substrate in the middle of the resistor. Assume that the resistance does not vary with position along the resistor.

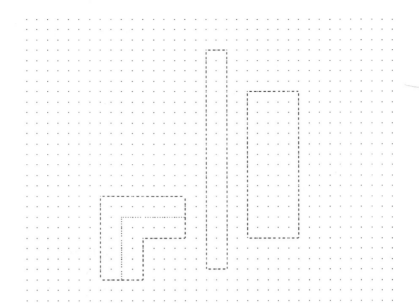

Figure P2.2

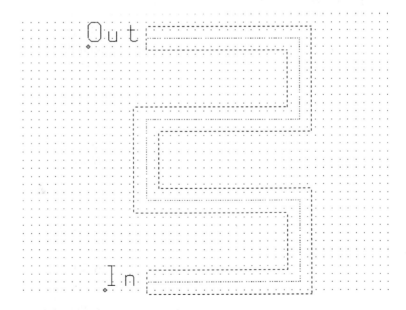

Figure P2.3

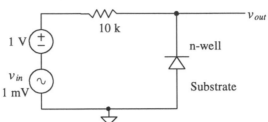

Figure P2.7

2.12 The diode reverse breakdown current, that is, the current that flows when $|V_d| <$ BV (breakdown voltage), is modeled in SPICE by

$$I_D = IBV \cdot e^{-(V_d + BV)/V_T}$$

Assuming 10 μA of current flows when the junction starts to break down at 100 V, simulate, using a SPICE DC sweep, the reverse breakdown characteristics of the diode. (The breakdown voltage, BV, is a positive number. When the diode starts to break down $-BV = V_d$. For this diode, breakdown occurs when $V_d = -100$ V.)

2.13 Repeat Ex. 2.3 if the n-well/p-substrate diode is 50 μm square and the acceptor doping concentration is changed to 10^{15} atoms/cm³.

2.14 Estimate the storage time, that is, the time it takes to remove the stored charge in a diode (see Fig. 2.15), when $\tau_T = 5$ ns, $V_F = 5$ V, $V_R = -5$ V, $C_{j0} = 0.5$ pF, and $R = 1$k. Verify your results with SPICE.

The Metal Layers

Now that we are familiar with the well, let's discuss the metal layers. In particular, for the CN20 process there are two levels of metal. These levels are named metal1 and metal2. The metal layers are used to connect the circuit together. In this chapter, we look at the bonding pad, the design rules for the metal layers, capacitances associated with the metal layers, crosstalk, sheet resistance, and metal migration.

3.1 The Bonding Pad

The bonding pad is at the interface between the die and the package or the outside world. One side of a wire is soldered to the pad, while the other side of the wire is connected to a lead frame. (The lead frame, in part, is the actual pins we see in a packaged integrated circuit.) At this point we will not concern ourselves with electro static discharge (ESD) protection, which is an important design consideration when designing the pad.

3.1.1 Laying out the Pad

The basic size of the bonding pad specified by MOSIS is a square 100 μm x 100 μm. For a probe pad[1], used to probe the circuit with a microprobe station, the size should be greater than 6 μm x 6 μm. A pad that uses metal2 is shown in Fig. 3.1. Notice, in the cross-sectional view, the layers of insulator (SiO_2 in most cases) under and above the metal2. These layers are used for isolation between the other layers in the CMOS process.

Before proceeding any further, we might ask the question, "What is the capacitance from this metal2 box to the substrate?" This is important because we have

---

[1] The minimum size of a probe pad is set by the minimum overglass size of 6 μm. In general, and if possible, probe pads should be a square with a side measuring 75 μm.

to drive this capacitance in order to get a signal off the chip. From the data sheets for the process, the Orbit electrical parameters given in Appendix A, the specification of metal2 to substrate plate capacitance is 13 to 15 aF/μm^2. If the measured plate capacitance is 14 aF/μm^2, then the capacitance of the metal2 bonding pad to substrate is 0.14 pF. This illustrates that when designing high-performance CMOS digital or analog circuits, the design engineer must be aware of the limitations, in this case the capacitances, inherent in the process being used. Layer 12, in the CN20 setups, is used to specify the metal2 layer. The metal layers are sometimes referred to as wires because these layers are used to connect circuits, resistors, MOSFETs, and capacitors, together.

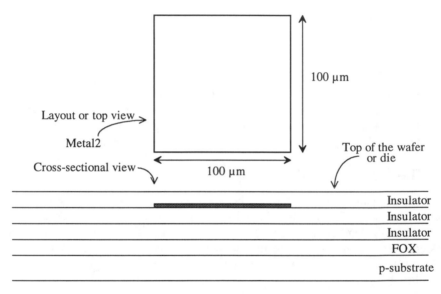

Figure 3.1 Layout of metal2 used for bonding pad with associated cross-sectional view.

Since an insulator is covering the pad, we cannot bond (connect a wire) to the pad. To specify an opening or cut in the glass, we use the PAD or overglass layer, LASI layer 13. The top layer insulator on the chip is also called passivation. The passivation helps protect the chip from contamination. Openings for bonding pads are called cuts in the passivation. Orbit specifies 5 microns between the edge of metal2 and the PAD layer. A complete pad using metal2 with the pad layer is shown in Fig. 3.2.

Often metal1, layer 10, is placed under metal2 when laying out the pads. This is so either metal layer from the circuit can be connected to the pad. Metal2 is connected to metal1 by the layer via, layer 11. On die the only layer that can be connected to metal2 is metal1, the layer directly below metal2. Metal1 can connect to metal2, poly, or active (n+ or p+). We will discuss this in more detail in the next chapter. The via must be at least 3 μm inside metal2. A bonding pad with both metal1 and 2 is shown in Fig. 3.3. Notice how the via has the effect of removing the glass under metal2. When metal2 is deposited on the wafer, it will make physical contact with metal1 where there is a via. Figure 3.4 shows an expanded corner view of a pad using both metals1 and 2 drawn with LASI.

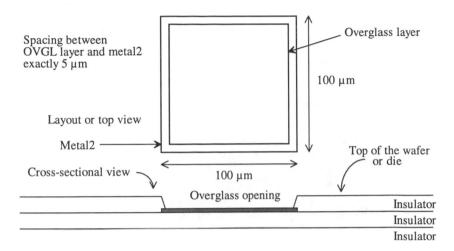

Figure 3.2 Layout of a metal2 pad with pad opening for bonding connection.

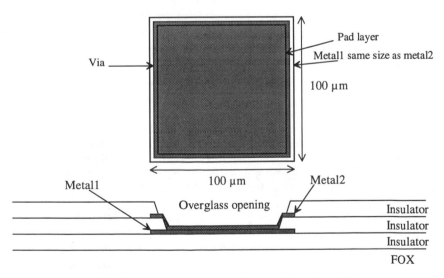

Figure 3.3 A bonding pad using metal1 and metal2.

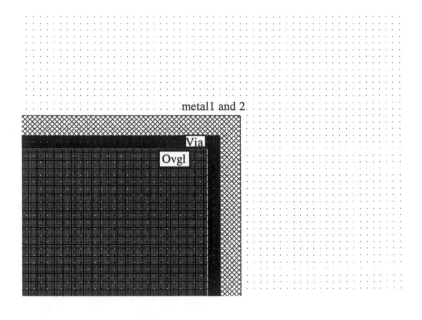

Figure 3.4 Expanded corner view of a pad.

3.1.2 Design Rules for Pads

Figure 3.5 shows the design rules for the bonding pads. The bonding pad size we will use is 100 μm by 100 μm. The ovgl layer (cuts in the passivation) should be a square, 90 μm on a side, centered in the pad, that is, 5 μm from the metal edge. The pad-to-pad spacing must be at least 75 μm. Note that the bonding pad design rules are not checked by LASIDRC.

Example 3.1
Lay out a 40-pad frame for a MOSIS tiny chip.

The size of the tiny chip, from Ch. 1, is 2.2 mm by 2.2 mm. If we assume that ten pads are on each of the four sides and the corners are open, that is, contain no pads (see Fig. 3.6), then we can divide the length of a side by the sum of the number of pads and the corner areas to get the pad cell size, or

$$\text{Cell size} = \frac{2.2 \times 10^{-3}}{12} = 183 \ \mu\text{m} \overset{\text{rounded to}}{\rightarrow} 180 \ \mu\text{m}$$

The chip, or die size, that we will have using a pad cell size of 180 μm is 2160 μm by 2160 μm. Figure 3.7 shows the pad placed in a box, made using the outline layer (layer 58), of 180 μm square. The outline layer has no effect on fabrication. This layer is used to help align the cells when we place them in a higher ranking cell. Figure 3.6 shows the entire padframe. ■

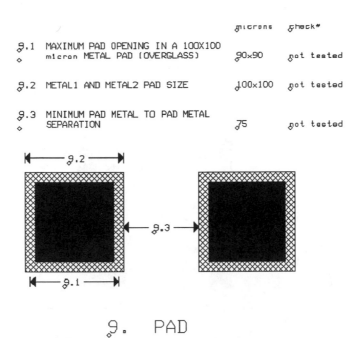

Figure 3.5 Design rules for the bonding pads.

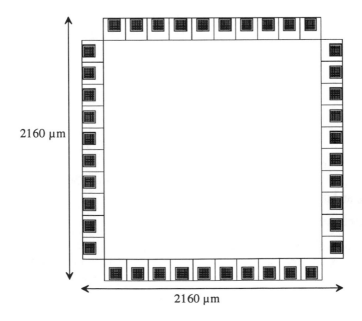

Figure 3.6 Layout of a 40-pad padframe. Final dimensions 2160 µm by 2160 µm.

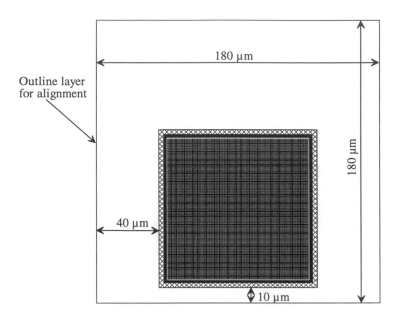

Figure 3.7 Layout of a pad using the outline layer.

3.2 Design and Layout Using the Metal Layers

As mentioned earlier, the metal layers are used to connect the resistors, capacitors, and MOSFETs in a CMOS integrated circuit. We had a taste of the metal1 and 2 layers in the last section. In this section, we begin by discussing the design rules for the metal layers. The parasitic resistances and capacitances of these metal layers are then discussed.

3.2.1 Design Rules for the Metal Layers

The metal and via design rules are shown in Figs. 3.8 through 3.10. The complete set of design rules is given in Appendix A. The minimum width and spacing of metals1 and 2 is 3 µm. Also shown in these figures is the contact layer. This layer connects metal1 to p+, n+ or the poly layers. At this point, we will concern ourselves with metal1, metal2, and the via layers.

3.2.2 Parasitics Associated with the Metal Layers

The basic parasitic resistances and capacitances associated with the metal layers can be calculated from the information given in Appendix A. The main parameters we are interested in are the sheet resistances of the layers, the capacitance between the metal layers and active, poly, substrate, and between one another. Also, there is a finite contact resistance of the via. The following examples illustrate some of the unwanted parasitics associated with these layers.

		microns	check#
6.1	WIDTH	3.0	23
6.2	SPACING	3.0	24
6.3	OVERLAP OF CONTACT	1.0	25
6.4	OVERLAP OF VIA	2.0	26

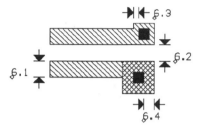

6. METAL 1

Figure 3.8 Metal1 design rules.

		microns	check#
7.1	SPACE TO CONTACT	2.0	27
7.2	SIZE (EXCEPT FOR PADS)	2.0×2.0	28
7.3	SPACING	3.0	29

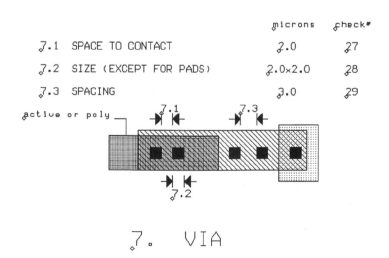

7. VIA

Figure 3.9 Via design rules.

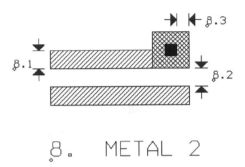

8.1	WIDTH	microns	check#
8.1	WIDTH	3.0	30
8.2	SPACE	3.0	31
8.3	METAL2 OVERLAP OF VIA	2.0	32

8.   METAL 2

Figure 3.10 Metal2 design rules.

Example 3.2

Estimate the resistance of a 1 mm long piece of metal1 with a width of 3 μm. Also estimate the delay through this piece of metal, treating the metal line as an RC transmission line. Assume that no other layers are present under the metal run.

First, we need to calculate the resistance of the metal line. From Appendix A we see that the typical sheet resistance of metal1 is 0.06 Ω/square. The overall resistance of the line is then given by

$$R = 0.06 \cdot \frac{1,000 \ \mu m}{3 \ \mu m} = 20 \ \Omega$$

The capacitance to substrate (we use this value on the data sheets because there are no other layers under metal1 in this example, and we are neglecting the fringing capacitance) is given in the data sheets as 26 aF/μm² (max). The overall capacitance of the line is then

$$C = \frac{26 \times 10^{-18} F}{\mu m^2} \cdot (1,000 \ \mu m) \cdot (3 \ \mu m) = 78 \ fF$$

If we treat the metal line as an RC transmission line, the delay can be estimated as 0.35RC or 0.55 ps, a negligible and unrealistic delay. In general, unloaded metal lines must include inductive effects. An unloaded metal line exhibits a delay of 5.4 ps/mm [2]. ∎

Example 3.3
Estimate the capacitance between a 10 μm square piece of metal1 and an equal-size piece of metal2 placed exactly above the metal1 piece. Sketch the layout and the cross-sectional view. Also sketch the symbol of a capacitor on the cross-sectional view.

The plate capacitance, from Appendix A, between metal1 and metal2 is at most 38 aF/μm², while the fringe capacitance is at most 104 aF/μm. The two layers form a parallel plate capacitor. The capacitance between the plates is given by the sum of the plate capacitance and the fringe capacitance, or

$$C_{12} = \frac{38 \times 10^{-18} \text{F}}{\mu\text{m}^2} \cdot \overbrace{(10 \ \mu\text{m})^2}^{area} + \frac{104 \times 10^{-18} \text{F}}{\mu\text{m}} \cdot \overbrace{(40 \ \mu\text{m})}^{perimeter} = 8 \text{ fF}$$

The layout and cross-sectional view are shown in Fig. 3.11. ∎

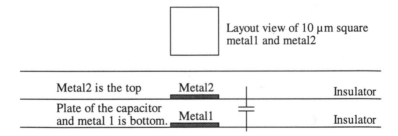

Layout view of 10 μm square metal1 and metal2

| Metal2 is the top | Metal2 | Insulator |
| Plate of the capacitor and metal 1 is bottom. | Metal1 | Insulator |

Figure 3.11 Capacitance between metal1 and metal2.

Example 3.4
In the previous example, estimate the voltage change on metal1 when metal2 changes potential from 0 to 5 V.

The capacitance from metal2 to metal1 was calculated as 8 fF. The capacitance from metal1 to substrate is given by

$$C_{1sub} = \frac{26 \times 10^{-18} F}{\mu\text{m}^2} \cdot \left[\overbrace{10 \ \mu\text{m}}^{area} \right]^2 + \frac{82 \times 10^{-18} F}{\mu\text{m}} \cdot \overbrace{(40 \ \mu\text{m})}^{perimeter} = 5.9 \text{ fF}$$

The equivalent schematic is shown in Fig. 3.12. The voltage on C_{1sub} is given by

$$\Delta V_{metal1} = 5 \cdot \frac{\frac{1}{j\omega C_{1sub}}}{\frac{1}{j\omega C_{1sub}} + \frac{1}{j\omega C_{12}}} = 5 \cdot \frac{C_{12}}{C_{12} + C_{1sub}} = 5 \cdot \frac{8}{8+5.9} = 2.9 \text{ V}$$

A displacement current flows through the capacitors, causing the potential on metal1 to change 2.9 V. This may seem significant at first glance. However, one must remember that most metal lines in a CMOS circuit are being driven from a low-impedance source; that is, the metal is not floating but is being held at some potential. This is not the case in some dynamic circuits or in circuits with high-impedance nodes or long metal runs. ∎

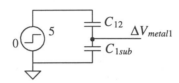

Figure 3.12 Equivalent circuit used to calculate the change in metal1 voltage.

3.2.3 Current Carrying Limitations

Now that we have some familiarity with the metal layers, we need to answer the question, "How much current can we carry on a given width or length of metal?" The factors that limit the amount of current on a metal wire or bus are metal electromigration, and the maximum voltage drop across the wire or buss due to the resistance of the metal layer.

Metal electromigration results from a conductor carrying too much current. This effect is similar to the erosion that occurs when a river carries too much water. The result is a change in the conductor dimensions, causing spots of higher resistance and eventually failure. If the current density is kept below the metal migration threshold current density, J_{Al}, metal migration will not occur. Typically, for aluminum, which is what metal1 and metal2 are made of, the current threshold for migration J_{Al} is $1 \rightarrow 2 \frac{\text{mA}}{\mu\text{m}}$.

Example 3.5

Estimate the maximum current a piece of metal1 3 μm wide can carry. Also estimate the maximum current a bonding pad can receive from a bonding wire.

Assuming that $J_{AL} = 1 \frac{\text{mA}}{\mu\text{m}}$, the maximum current on a 3 μm wide aluminum conductor is given by

$$I_{max} = J_{Al} \cdot W = 10^{-3} \cdot 3 = 3 \text{ mA}$$

The maximum current through a bonding pad is then 100 mA. ∎

Example 3.6

Estimate the voltage drop across the conductor discussed in the previous example when the length of the conductor is 1 cm and the current flowing in the conductor is 3 mA (I_{max}).

The sheet resistance of metal1 is 0.06 Ω/square. The voltage drop across a metal1 wire that is 3 μm wide and 10,000 μm (1 cm) long carrying 3 mA is

$$V_{drop} = (0.06\Omega/\text{square}) \cdot \frac{10,000}{3} \cdot 3 \text{ mA} = 0.6 \text{ V}$$

or a significant voltage drop. If this conductor was used for power, we would want to increase the width significantly; however, if the conductor is used to route data, the size may be fine. ∎

In general, metal2 should be used for power routing. Metal2 is approximately twice as thick as metal1 and therefore has a lower sheet resistance. When routing power the more metal that is used the fewer problems, in general, that will be encountered. If possible, a ground or power plane should be used across the entire die. The more capacitance between the power and ground busses, the harder it is to induce a voltage change on the power plane; that is, the DC voltages will not vary.

3.2.4 Parasitics Associated with the Via

The via layer is used to make a connection between metal1 and metal2. The size of a via is a square with a side measuring exactly 2 μm (except for the bonding pad). The via layer specifies where to remove glass so that when metal2 is deposited it makes contact with metal1. Associated with this connection is a contact resistance. This resistance arises because the metal2 thins as it is deposited in the hole or as a result of the contact potential difference between two different materials. Minimum contact resistance for a via, given in Table A.8 in Appendix A, is 0.05 Ω, while the maximum contact resistance is 0.08 Ω. Also associated with the via is a metal migration limitation of typically 0.4 mA/contact. Because of the thinning of the metal2, the metal migration threshold becomes 0.05 mA/μm of via perimeter.

Example 3.7

Sketch the equivalent schematic for the layout depicted in Fig. 3.13a showing the via contact resistance. Estimate the voltage drop across the contact resistance of the via when 10 mA flows through the via. Comment on the reliability of the via. Suggest a method, by changing the layout, of lowering the effective contact resistance and current density through the via. What is the maximum current that can flow from metal2 to metal1 after the layout change?

The equivalent schematic is in Fig. Ex3.7. If the via contact resistance is 0.08 Ω and 10 mA flows through the via, a voltage drop of 0.8 mV will result. The reliability of the single via is poor. The maximum current that should flow through the via is 0.4 mA. Forcing 10 mA through the via will cause the via to

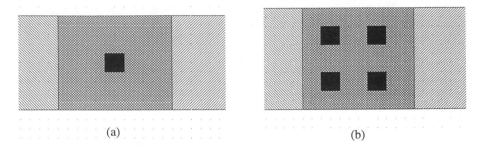

(a)   (b)

Figure 3.13 Layouts used in Example 3.7.

eventually open or become a high-resistance connection. Figure 3.13b shows that adding as many vias as the design rules will allow lowers the contact resistance. The four vias shown in this figure give an effective contact resistance of 0.08/4 or 0.02 Ω due to the contact resistances of each of the vias being in parallel. The maximum current, set by metal electromigration considerations, through the four contacts is 1.6 mA. Increasing the overlap and the number of vias will increase the maximum current to a limitation set by the width of the metal lines. ■

Metal1 ———————ᴡᴡᴡ——— Metal2
                       0.08

Contact resistance

Figure Ex3.7

3.3 Crosstalk and Ground Bounce

Crosstalk is a term used to describe an unwanted interference from one conductor to another. Between two conductors there exists mutual capacitance and inductance, which give rise to signal feedthrough.

Consider the two metal wires shown in Fig. 3.14. A signal voltage propagating on one of the conductors couples current onto the conductor. This current can be estimated using

$$I_m = C_m \frac{dV_A}{dt} \tag{3.1}$$

where C_m is the mutual capacitance, I_M is the coupled current, and V_A is the signal voltage on the source conductor [2]. Treating the capacitance between the two conductors in this simple manner is useful in most cases. If the coupled noise voltage is less than 10 percent of signal voltage, this approximation is accurate to one decimal

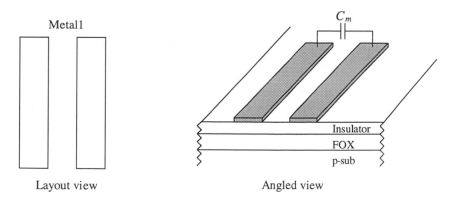

Metal1

Layout view   Angled view

Figure 3.14 Conductors used to illustrate crosstalk.

place [2]. Determining C_m experimentally proceeds by applying a step voltage to one conductor while measuring the coupled voltage on the adjacent conductor. Since we know the capacitance of any conductor to substrate (see Appendix A) we can write

$$\Delta V = V_A \cdot \frac{C_m}{C_m + C_{1s}} \qquad (3.2)$$

where ΔV is the coupled noise voltage to the adjacent conductor and C_{1s} is the capacitance of the adjacent conductor to ground.

The adjacent metal lines shown in Fig. 3.14 also exhibit a mutual inductance. The effect can be thought of as connecting a miniature transformer between the two conductors. A current flowing on one of the conductors induces a voltage on the other conductor. Measuring the mutual inductance begins by injecting a current into one of the conductors. The voltage on the other conductor is measured. The mutual inductance is determined using

$$V_m = L_m \frac{dI_A}{dt} \qquad (3.3)$$

where I_A is the injected current, V_m is the induced voltage, and L_m is the mutual inductance.

At the risk of stating the obvious, crosstalk can be reduced by increasing the distance between adjacent conductors. In many applications (e.g., DRAM), the design engineer has no control over the spacing (pitch) between conductors.

3.3.1 Ground Bounce

Consider the single metal wire shown in Fig. 3.15. The inductance of this wire can be estimated [2], assuming $w > h$, using

$$L \text{ (nH/mm)} = \frac{1.25}{\frac{w}{h} + 1.393 + 0.667 \cdot \ln\left[\frac{w}{h} + 1.44\right]} \text{ (nH/mm)} \qquad (3.4)$$

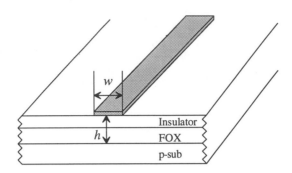

Figure 3.15 Conductors used in calculation of inductance.

Here we are assuming that the thickness of the wire is small compared to the width and that the p-substrate acts like a constant potential or reference ground plane. For the CN20 process, Appendix A, the thickness between metal1 and substrate is 1.5 μm. For a 3 μm wide piece of metal1, the inductance is

$$L \text{ (nH/mm)} = \frac{1.25}{\frac{3}{1.5} + 1.393 + 0.667 \cdot \ln\left(\frac{3}{1.5} + 1.44\right)} = 0.3 \text{ nH/mm}$$

Next consider the circuit shown in Fig. 3.16. A circuit is powered through two wires made using metal1, each 10 mm long and 3 μm wide. Ideally, point A in the figure is held at *VDD*, while point B is fixed at ground potential. Neglecting resistive losses, we find this is true when *I* is constant. However, if *I* changes, the lengths of metal behave as inductors.

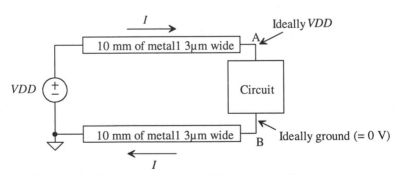

Figure 3.16 Block diagram used to illustrate ground bounce.

Consider a circuit that sources 50 mA in 3 ns (a common requirement in an output driver). The inductance of either metal wire is 3 nH. The change in voltage at point B (keeping in mind point B is supposed to be "ground") is given by

$$V_B = L\frac{dI}{dt} = 3 \text{ nH} \cdot \frac{50 \text{ mA}}{3 \text{ ns}} = 50 \text{ mV}$$

This means the "ground" of the circuit actually jumps 50 mV when the circuit needs this current. This noise or ground bounce generated on the ground conductor can feed into other circuits on the chip and cause problems. What's worse is when ten circuits on chip pull this current through the same conductor at the same time.

Ground bounce can be reduced by increasing the width, and thus decreasing the inductance, of the conductors supplying power to a circuit (the wider the metal the better). Also, increasing the capacitance between the conductor supplying current and the conductor returning current helps reduce ground bounce. The simplest method of increasing the capacitance is to lay the conductors out side by side. An on-chip decoupling capacitor between the circuit *VDD* and ground will help as well.

In the previous analysis, we neglected the resistance of the metal wires. In most practical situations, the inductance of the wire is negligible. The ground bounce is dominated by the resistive drop across the wire. Calculating the ground bounce due to the resistive component in Fig. 3.16 begins by calculating the resistance of a wire. The sheet resistance of metal1 is 0.06 Ω/square. The resistance of the wire is 200 Ω. When $I = 0$, the potential at point B is 0. However, when $I = 5$ mA, the potential at point B is 1 V! Again increasing the width of the conductors, and thus decreasing the resistance, helps to reduce the ground bounce.

3.4 Layout Using Cell Hierarchy

Do not lay out a chip unless the material in this section is understood!

Using cell hierarchy, that is, ranking and nesting of cells, is important to keep the size of the design file (the final GDS file sent to the fab-house) from getting too large. Consider the pad cell shown in Fig. 3.5. Let's assume this cell is called PAD and has a rank of 1. We can generate the layout of Fig. 3.6 in two ways: the right way and the wrong way. Let's consider the wrong way first. Working within the pad cell shown in Fig. 3.17a, use the copy command, **Cpy**, to copy the pad layout Fig. 3.17b, until the number of pads needed is copied into the cell. This is the wrong way to do layout. Each time the cell is copied, boxes or polygons are generated at the new location in the cell, resulting in a significant increase in the size of the design file. In addition, a change in the pad, for example, changing the size of metal2 or adding objects to the cell, must be made to each of the copied pads. This can result in a waste of time. The **Cpy** command should be used as little as possible when laying out a chip.

The correct procedure for laying out the padframe of Fig. 3.6 begins by making a cell called PADFRAME with a rank of 2 using the **Cell** command button. Once in the padframe cell, the object is changed to PAD, the cell we want to use 40 times in the padframe cell using the **Obj** command button. Notice that the PAD cell has a lower ranking than the PADFRAME cell. Figure 3.18 shows use of the **Add** command to add the PAD cell to the PADFRAME cell. Each time the PAD cell is added to the

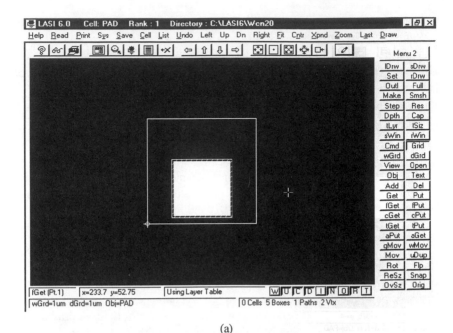

(a)

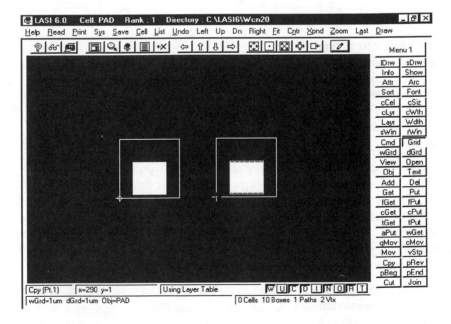

(b)

Figure 3.17 How not to lay out a padframe, (a) basic pad cell with a rank of 1 and (b) using the Cpy command to lay out the padframe of Fig. 3.6.

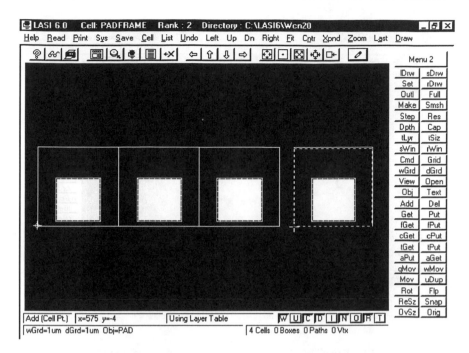

Figure 3.18 Correct method of laying out the PADFRAME cell.

PADFRAME cell a single specifying line, used to reference the added cell and the location, is added to the design file. The result is a significantly smaller design file. Also, if a change is made to the PAD cell, the change is automatically propagated through the hierarchy to the other higher ranking cells using the PAD cell. Figure 3.19 shows the cells of Fig. 3.18 drawn in outlines using the **Outl** command. This can speed up the redraw time.

Cell hierarchy can be extended to every aspect of layout. For example, we know that the via size, used in the CN20 process, is exactly 2 µm by 2 µm square. Instead of drawing a via this size each time we need to make a connection between metal1 and metal2, we can lay out a cell called VIA with a rank of 1 that is a box on the via layer measuring 2 µm by 2 µm square. Now each time we need to make a connection between metal1 and metal2 we simply use the via cell as an object (time is saved since the size of the via is fixed in the VIA cell). The same idea can be used for the active and poly contacts that will be discussed in the next chapter.

When designing VLSI digital circuits, it is common to use minimum-size MOSFETs. A minimum-size MOSFET can be laid out in a cell called NMIN with a rank of one. Each time a MOSFET is needed in a circuit, the NMIN cell can be used. This idea can and should also be used in analog VLSI circuits. Note that an added cell cannot be edited from a higher ranking cell. For example, to edit the PAD cell of Fig. 3.18 we would first have to leave the PADFRAME cell and return to the PAD cell.

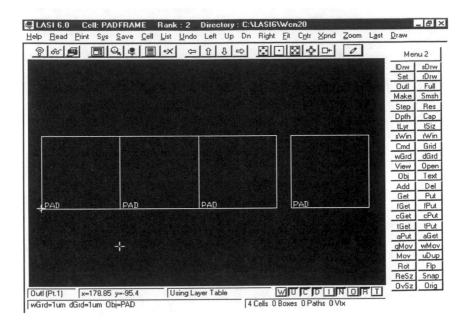

Figure 3.19 Drawing cells in the outline mode can speed up redraw time.

REFERENCES

[1]   W. Tanner, *MOSIS User Manual,* Release 4.0, August 1994.

[2]   H. W. Johnson and M. Graham, *High Speed Digital Design: A Handbook of Black Magic,* Prentice-Hall Publishing Company, 1993. ISBN 0-13-395724-1.

PROBLEMS

3.1   Lay out and DRC the pad given in Fig. 3.5. Also lay out the padframe shown in Fig. 3.6. The pad cell should be a rank of 1. The padframe should have a rank of 2. Use unique names for the cells. Keep these cells backed up. We will add ESD protection in the next chapter. This is basically the beginnings of the chip you will submit to MOSIS.

3.2   Sketch the cross-sectional view of the layout shown in Fig. P3.2.

3.3   Sketch the cross-sectional view across the center of the pad in Fig. 3.5 without the via layer present. Does the outline layer, layer 58, have any fabrication significance? Why would the outline layer be used?

3.4   What would be the result of submitting a chip without the OVGL layer drawn in the pad cell?

3.5   Explain how to label the pin numbers on the padframe cell of Problem 1 using the **Text** command. Use a Text size of at least 20 μm.

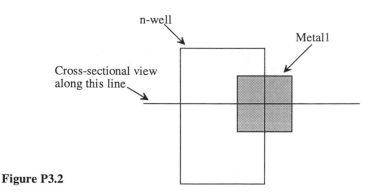

Figure P3.2

3.6   Estimate the change in voltage on metal1 from a change in voltage on metal2 directly above and the same size. Assume the change in metal2 is 1 V. Consider only the plate capacitance, not the fringe capacitance.

3.7   What is the maximum current that a 5 µm wide piece of metal2 can carry? How many vias will be needed to connect this piece of metal2 to metal1 at the maximum current?

3.8   What is the minimum width of a metal2 wire used to supply 20 mA of current to a circuit? How many vias are needed to connect the metal2 wire to a metal1 wire?

3.9   Estimate the inductance of a 4 µm wide piece of metal2 in nH/mm.

3.10 Sketch the cross-sectional views of the layouts shown in Fig. P3.10 at the three locations.

3.11 Estimate the voltage change on metal1, in the layout of Fig. P3.11, if metal2's potential changes from 0 to 5 V.

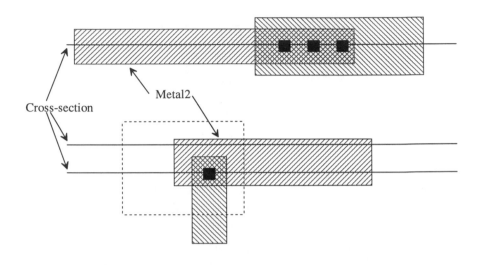

Cross-section

Metal2

Figure P3.10

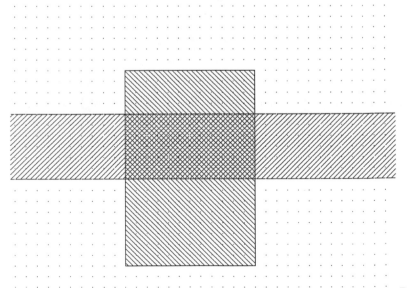

Figure P3.11

The Active and Poly Layers

In this chapter we discuss the active layers, both n+ and p+, and the poly layers. The active layers are used to make the source and drain regions of the MOSFETs. The active layers are also used to connect metal1 to the substrate or the well.

The poly layer is used to form the gate of the MOSFETs. Poly is a short name for polysilicon. Polysilicon is made up of small crystalline regions of silicon. Therefore, in the strictest sense poly is not amorphous silicon,[1] and it is not crystalline silicon such as the wafer.

The poly layer can also be used to connect MOSFETs together. The main limitation when using the poly layer for interconnection is its sheet resistance. As we saw in the last chapter, the sheet resistance of the metal layers is less than 0.1 ohms/square. The sheet resistance of the doped poly can be on the order of 20 Ω/square. The capacitance to substrate is also larger for poly simply because it is closer to the substrate. Therefore, the delay through a poly line can be considerably longer than the delay through a metal line.

The first section of this chapter develops the design rules for the active, poly1, poly2, and contact layers. We use these design rules to lay out a standard cell frame. The standard cell frame allows us to have a fixed area to lay out the n- and p-channel transistors without worrying about power supply routing and substrate/well tie downs. Circuits designed using the standard cell frame can be placed into another cell end to end. This makes layout easier; however, the cost for this is a larger layout. For the most area-efficient use of the silicon, full custom design must be used.

The remaining sections of this chapter discuss the fabrication sequence used to form n+, p+, and poly layers. The chapter ends with a discussion of parasitics.

---

[1] Amorphous silicon is made up of randomly organized Si atoms, while crystalline Si is formed by silicon atoms organized in an orderly fashion.

4.1 Design Rules

Before we start discussing the design rules for the active area, let's discuss an example layout. Consider the actual layout of an n-well resistor shown in Fig. 4.1. When we discussed the layout of the n-well resistor before, we did not discuss how to make contact to the outside world. In Fig. 4.1 we see the basic n-well resistor. We also see an n+ (heavily doped n-type) region in the n-well. This is used to make an "ohmic" contact to metal1. If metal1 is connected directly to the n-well, a rectifying or schottky diode is formed. These metal1 leads can be connected to the outside world through the pads.

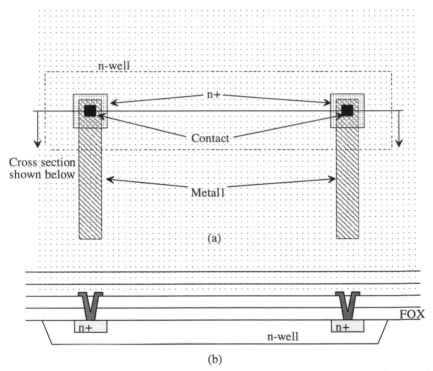

Figure 4.1 Layout of an n-well resistor with active contacts to metal1 (a) Actual layout using LASI, (b) cross-sectional view across the contacts.

4.1.1 Design Rules for the n+/p+ Active Layers

The active design rules apply to both the n+ and p+ active area. A layout interpretation of the design rules is shown in Fig. 4.2. Note that the location of the active area; that is, whether it is in the substrate or the well affects the rule applied to the active area. In general, n+ active in the well is used to keep the potential of the well fixed, except in the case of the resistor when current flows through the n-well, and p+ in the substrate is used to set the potential of the substrate.

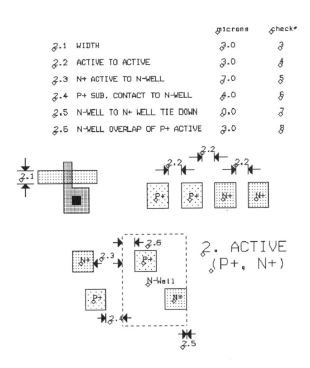

		microns	check*
2.1	WIDTH	3.0	3
2.2	ACTIVE TO ACTIVE	3.0	4
2.3	N+ ACTIVE TO N-WELL	7.0	5
2.4	P+ SUB. CONTACT TO N-WELL	4.0	6
2.5	N-WELL TO N+ WELL TIE DOWN	0.0	7
2.6	N-WELL OVERLAP OF P+ ACTIVE	3.0	8

Figure 4.2 Active area design rules.

Example 4.1

Lay out a diode that is 10 μm by 10 μm using the n-well as the cathode. Sketch a cross-sectional view of the diode. If the substrate is held at ground potential, what is the minimum voltage allowable on the cathode (the n-well) of the diode. Why?

The layout and cross-sectional view are shown in Fig. 4.3. If the substrate is at ground the n-well potential should be greater than –0.6 V to keep the substrate/n-well diode from becoming forward biased. ∎

The contact design rules are shown in Fig. 4.4. The contact layer is used to connect metal1 to n+, p+, and the poly layers. The size of the contact is exactly 2 μm by 2 μm.

One useful function of the diode is in electrostatic discharge protection. Since the input impedance of a MOSFET is capacitive, small amounts of charge placed on a bonding pad, from something or someone external to the chip connected to a gate, can cause the gate oxide of the MOSFET to break down. To avoid this situation, protection schemes are used as shown in Fig. 4.5. These schemes rely on a diode becoming forward biased and providing a low-impedance path to pull the excessive charge away from the MOSFET gate. The scheme shown in Fig. 4.5b uses the n+ as a resistor to

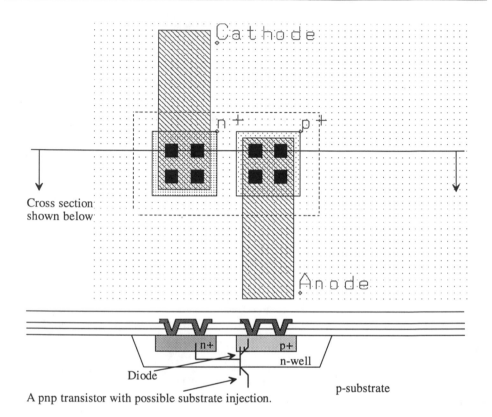

A pnp transistor with possible substrate injection.

Figure 4.3 Layout of a 10 um by 10 um diode using the n-well as the cathode.

		microns	check
5.1	CONTACT SIZE (EXACTLY)	2X2	17
5.2	SPACING	2.0	18
5.3	POLY OVERLAP	2.0	19
5.4	ACTIVE OVERLAP	2.0	20
5.5	POLY CONTACT TO ACTIVE EDGE	3.0	21
5.6	ACTIVE CONTACT TO GATE	3.0	22

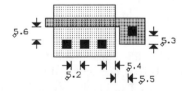

5. CONTACT

Figure 4.4 Contact design rules.

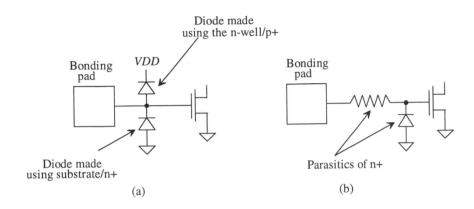

Figure 4.5 Two methods of protecting against ESD damage.

limit the current through the pad and the n+/substrate diode to protect against negative transients and the breakdown voltage of the diode to protect against positive transients. Positive transient protection can also be achieved by adding the p+/n-well diode of Fig. 4.5a. This scheme can be thought of as a distributed resistor/diode, difficult to analyze analytically. Normally, a combination of both schemes are used together with good power and ground busses to provide a low-resistance path for removal of the unwanted charge.

Example 4.2
Modify the pad discussed in Ex. 3.1 to include the diode protection given in Fig. 4.5a. Also include power (*VDD*) and ground (*VSS*) busses on the pad.

The completed pad is shown in Fig. 4.6. Note that *VSS* is connected to the substrate via p+ and that *VDD* is connected to the n-well under part of the pad by way of n+. Also notice that the n+ and p+ areas extend beyond the boundary of the outline of the cell. This is used when the cells are abutted together. The n+ and p+ areas overlap, giving a continuous connection around the die. At least two of the bonding pads must be connected to the *VDD* and *VSS* busses. ∎

4.1.2 Design Rules for Poly1

The design rules for poly1 are shown in Fig. 4.7. Poly1 is used in making the MOSFET gates and with poly2 capacitors. As we will see in Chapter 7, there is an additional layer of poly named poly2. This layer is used for making capacitors with poly1. Poly2 can also be used to form MOSFETs. However, in order to keep the design rule checks to a minimum, we will assume that only poly1 can be used to make MOSFETs and that poly2 use is limited to capacitors.

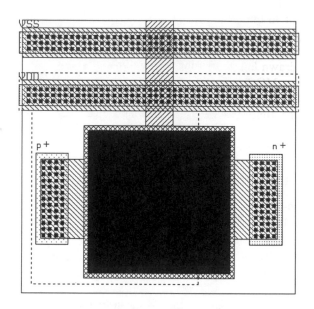

Figure 4.6 Pad with ESD diodes and power busses.

3.1	WIDTH	2.0	9
3.2	SPACE	3.0	10
3.3	GATE OVERLAP OF ACTIVE	2.0	11
3.4	ACTIVE OVERLAP OF GATE	3.0	not tested
3.5	FIELD POLY1 TO ACTIVE	1.0	not tested

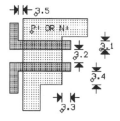

3.   POLY1

Figure 4.7 Design rules for poly1.

Consider the layout shown in Fig. 4.8. Poly1 is drawn over the n+ active area. As we will see later in the chapter, the poly will block n+ diffusion, keeping the area under the poly (gate of the MOSFET) from becoming n+. The width of the poly intersecting with the active area is called the drawn width, while the length of the poly over active is the drawn length. The drawn width and length are the W and L used to specify the size of a MOSFET. Notice that the n+ regions are implanted through the thin gate oxide (GOX). Also note that the poly over the field oxide (FOX) forms a MOSFET called the field MOSFET, an unwanted parasitic. The FOX is the field MOSFETs gate oxide. An implant is used (see Fig. 2.10) to increase the threshold voltage of the field device so that it cannot turn on.

4.2 Layout of a Standard Cell Frame

When laying out n-channel and p-channel transistors, as shown above, it is easy to forget about the bulk connection, that is, the p-substrate for the n-channel transistors and the n-well for the p-channel transistors. Complicating things is the need for *VDD* and *VSS* connections throughout the circuits. Using a standard cell frame can solve both of these layout requirements. The frame provides *VDD* and *VSS* busses in addition to tying down the substrate and well. The drawback of the standard cell frame is increased layout size.

Figure 4.9 shows a standard cell frame that can be used for layout. The width of the outline in this figure is 60 μm, while the height is 120 μm. The metal1 labeled *VSS* is connected to p+ (under metal1), which makes an ohmic contact to the p-substrate. This connection sets the substrate potential to *VSS*. Using the standard cell frame, we won't have to concern ourselves with the substrate connection of the n-channel transistors. The n-well in the top portion of the standard frame cell is connected to *VDD* by the n+ under the metal1 labeled *VDD*. The p-channel transistors are drawn in the n-well, while the n-channel transistors are drawn in the bottom portion of the frame. If, while doing layout, the width of the frame is too narrow, we simply add the frames end to end. Similarly, when laying out a system, we can place the cells end to end, which automatically routes *VDD* and *VSS*.

4.3 Patterning the Active Layers

In Ch. 2 we described how to pattern the n-well. This can be thought of as making the substrate for the p-channel transistors (unless we are using the n-well as a resistor or as part of a diode). The region outside the active regions as defined by the n+ and p+ masks is called the field region. The field region is used to isolate the active regions. The oxide in this region is called Field OXide (FOX) or Recessed OXide (ROX). Using oxide to isolate the active regions is called LOCal Oxidation of Silicon, or LOCOS for short.

Figure 4.10 shows the general patterning steps used in creating the active areas and MOSFETs. We assume that the n-well locations on the die have already been

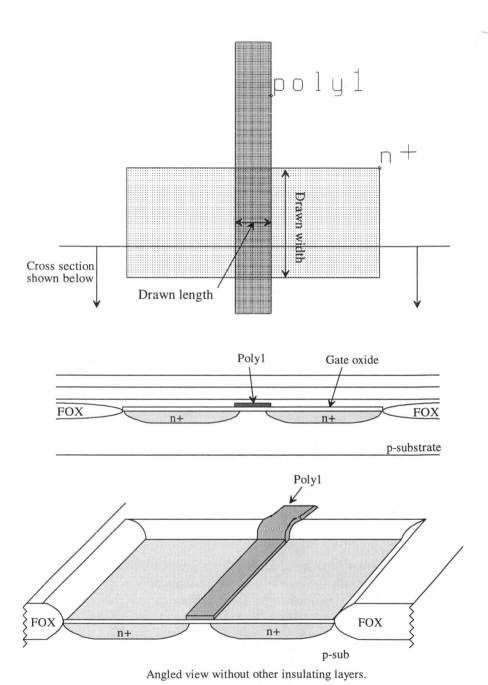

Angled view without other insulating layers.

Figure 4.8 Layout of a MOSFET, cross-sectional and angled view.

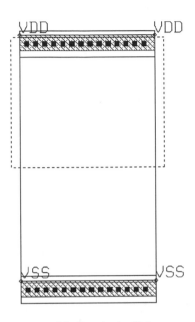

Figure 4.9 Standard cell frame.

processed (Figs. 4.10a and b).   The first step is to cover the active areas in the p-substrate and n-well with a stress relief oxide. Nitride, SiN$_3$, is deposited on top of the stress relief oxide (Figs. 4.10c and d). This thin oxide acts like a cushion for the nitride. The areas not covered by the nitride, that is, the p-substrate areas, are implanted with a p field implant, (Fig. 4.10e), for increasing the threshold voltage of the field devices. This implant is used to increase the threshold voltage of the field MOSFETs, the devices formed by poly over the FOX. Not shown in Fig. 4.10 are the steps involved in implanting the n-type field implant used to raise the threshold voltage of the field MOSFETs in the n-well region.

After the field implants are in place, the FOX is grown (Fig. 4.10f), and the nitride and stress relief oxides are removed.   At this point we should point out an unwanted characteristic of LOCOS, namely, bird's beak. Consider Fig. 4.11a. This figure shows the layout of a 10 μm by 10 μm n+ active area with a poly gate. Figure 4.11b shows the cross-sectional area with field oxide in place. (This is a 90° rotation from the cross-sectional view shown in Fig. 4.8.)

Notice that the oxide has encroached upon the area defined by the n+ active mask. The shape of this encroachment is sometimes referred to as bird's beak. We will refer to this attribute of LOCOS as simply oxide encroachment. This encroachment will affect the width of the MOSFETs. In other words, the width of the MOSFET will be decreased by two times the length of the oxide encroachment. The BSIM SPICE model discussed in Ch. 6 gives the parameter *DW*, or delta width, to describe the oxide

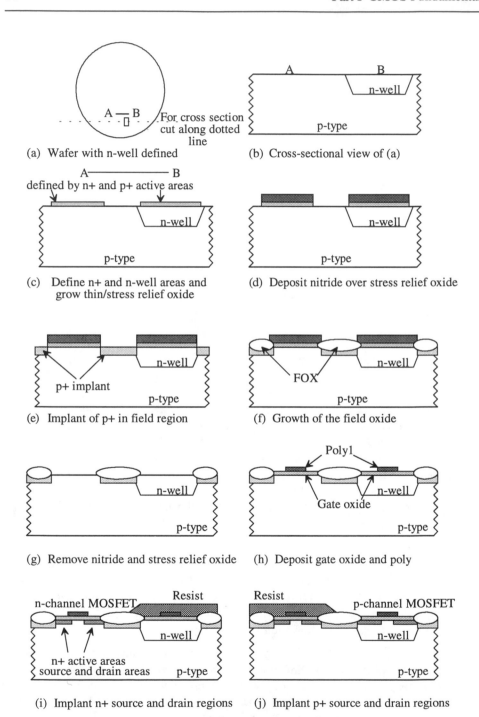

(a) Wafer with n-well defined

(b) Cross-sectional view of (a)

(c) Define n+ and n-well areas and grow thin/stress relief oxide

(d) Deposit nitride over stress relief oxide

(e) Implant of p+ in field region

(f) Growth of the field oxide

(g) Remove nitride and stress relief oxide

(h) Deposit gate oxide and poly

(i) Implant n+ source and drain regions

(j) Implant p+ source and drain regions

Figure 4.10 Sequence of events used in MOSFET formation.

encroachment. If the drawn width of the MOSFET gate is W, then the effective width is given by

$$W_{eff} = W_{drawn} - DW \qquad (4.1)$$

The parameter DW is equal to twice the oxide encroachment. The BSIM SPICE model will automatically subtract DW from the drawn width. (The drawn width is the W specified by the engineer in the SPICE simulation.)

Example 4.3

Assuming the drawn width of the MOSFET in Fig. 4.11 is 10 µm, estimate the effective width after oxide encroachment.

From the BSIM model parameters for the n-channel MOSFET in Appendix A, DW (dw) = 0.1355 µm; therefore, the actual or effective width of the MOSFET shown in Fig. 4.11 is 9.87 µm. Notice that the smaller the drawn width of the device, the larger effect the oxide encroachment has on the effective width. ∎

Let's return to the sequence of events in Fig. 4.10. The next step in the process is to grow the thin gate oxide and deposit the poly (Fig. 4.10h). This is followed by depositing resist, to isolate the n+ areas, and n+ implant. The result is shown in Fig. 4.10i. This is our basic MOSFET structure. From a circuits point of view, several important events occur during these last steps of MOSFET fabrication. When the poly is deposited, it is doped "in situ" to decrease its resistivity. For CN20 the gate is an n+ doping. When the n+ implant is applied to the wafer, the poly, which is used as a mask to keep the n+ from going under the gate region[2], is doped further. The n+ poly gate also acts as a mask for the p-channel transistors. The doping introduced by the p+ implant is not sufficient to significantly change the doping levels in the polysilicon gate. Typically, sheet resistances for poly are on the order of 20 Ω/square. In order to lower the sheet resistance of the poly, a refractory metal and poly mixture called silicide is deposited to the top of the polysilicon. This silicide/poly sandwich is referred to as a polycide gate. Typical sheet resistances of polycide gates are 2-3 Ω/square. The Orbit process (CN20) we are using does not silicide the gate regions.

Another important thing happens when we implant n+ and p+. The dopant atoms diffuse under the gate of the MOSFET. This effect is shown in Fig. 4.12. As can be seen in this figure, the lateral diffusion (*LD*) subtracts from the drawn gate length. The level 2 SPICE model uses *LD* to define the lateral diffusion. The BSIM model defines *DL* (dl), or delta length, to be two times LD. The effective channel length is given by

$$L_{eff} = L_{drawn} - 2 \cdot LD = L_{drawn} - DL \qquad (4.2)$$

---

[2] This is termed a "self-aligned" CMOS process because the gate automatically becomes aligned to the drain and the source regions of the MOSFET.

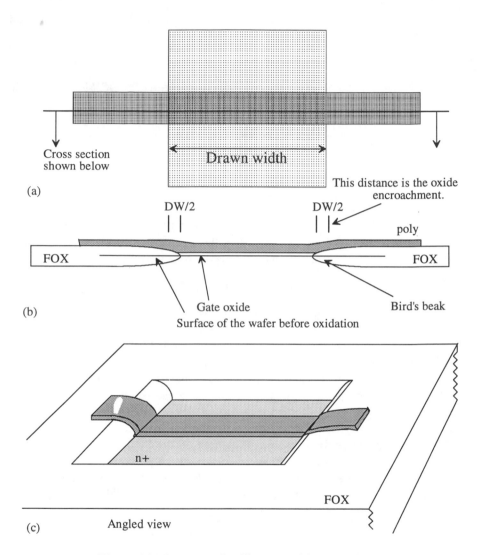

Figure 4.11 Layout used to illustrate oxide encroachment.

Example 4.4

Assuming the drawn length of the MOSFET in Fig. 4.12 is 2 μm, estimate the effective length of the MOSFET after lateral diffusion.

From the BSIM model parameters for the n-channel MOSFET given in Appendix A, $DL = 0.64$ μm and therefore the effective length of the device is 1.36 μm. ■

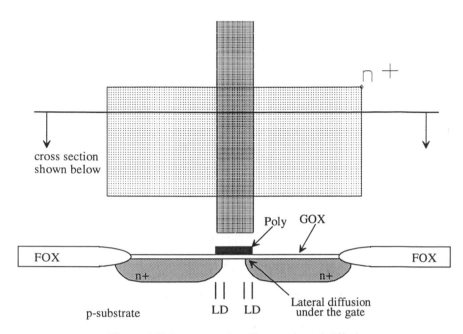

Figure 4.12 Layout used to illustrate lateral diffusion.

The BSIM model will automatically calculate the effective width and length of the MOSFET when given the drawn lengths and widths. The level 2 model will only calculate the effects due to lateral diffusion. (W_{eff} must be given by the designer.)

4.4 Layout of the MOSFET

The layout of the MOSFET without contacts (connections to a circuit) was shown in Figs. 4.8, 4.11, and 4.12. Consider the layout shown in Fig. 4.13 of a MOSFET with connections to the source, drain, and gate. Note that the source and drain of the MOSFET are interchangeable. The substrate connection of the MOSFET is not shown in this figure. Placing the MOSFET in the bottom of the standard cell frame of Fig. 4.9 provides the substrate connection. Using the pad of Fig. 4.6 in a padframe anchors the substrate to *VSS* around the die. P-channel MOSFETs, which are laid out in the n-well, must have a well tie down, that is, an n+ region connected to *VDD* or the source of the MOSFET.

4.4.1 Parasitics Associated with the Active Layers

Associated with the source and drain implants (n+ and p+) is a sheet resistance. We can get the sheet resistance from the tables in Appendix A or directly from the BSIM model parameters. The n-channel transistor has n+ source and drain regions, while the p-channel transistor has p+ source and drain regions. The sheet resistance of the n+ implant is given in the n-channel BSIM model by "rsh" (= 27.9 Ω).

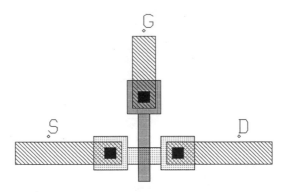

Figure 4.13 Layout of a 2 μm (= L) by 3 μm (= W) n-channel MOSFET with metal1 wire connections. Substrate connection not shown.

Example 4.5
Estimate the resistance of the following layout (3 μm wide and 100 μm long) made using the n+ implant. Neglect corner effects.

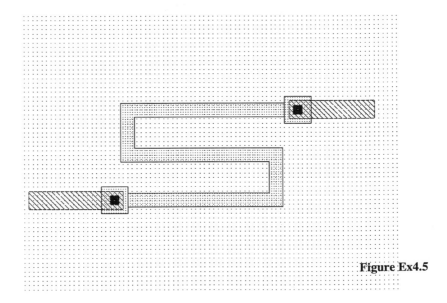

Figure Ex4.5

This layout is an n+ resistor with a value of $R = 27.9 \cdot \dfrac{100}{3} \approx 930 \ \Omega$. ■

The depletion capacitance of the n+ and p+ regions consists of a bottom component and a sidewall component (see Fig. 4.14). The zero-bias depletion capacitances are given in the BSIM model by cj (bottom, F/m^2) and $cjsw$ (sidewall, F/m). The built-in potential of the bottom and the sidewall is pb and $pbsw$ respectively, and the grading coefficients of the bottom and sidewall are mj and $mjsw$, respectively. These parameters are given in the BSIM model. The depletion capacitance can be calculated knowing the area of the drain (or source) and the perimeter, for the n+ to bulk, by

$$C_{jdep} = \frac{cj \cdot A_D}{\left(1 + \frac{V_{DB}}{pb}\right)^{mj}} + \frac{cjsw \cdot P_D}{\left(1 + \frac{V_{DB}}{pbsw}\right)^{mjsw}} \tag{4.3}$$

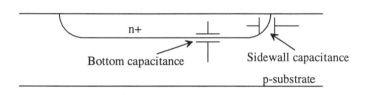

Bottom capacitance   Sidewall capacitance

p-substrate

Figure 4.14 Depletion capacitances of n+ to p-substrate.

Example 4.6

Estimate the depletion capacitance between the drain of the MOSFET shown in Fig. 4.13 and the substrate.

The size of the n+ implant shown in this figure is approximately 6 μm by 6 μm, neglecting the small active area between the poly and the contact. This gives A_D = 36 x 10^{-12} m² (36 μm²) and P_D = 24 μm. The capacitance as a function of V_{DB} (= drain voltage minus the substrate voltage) is given by

$$C_{jdep} = \frac{1.04 \times 10^{-4} \frac{F}{m^2} \cdot 36 \times 10^{-12} m^2}{(1 + \frac{V_{DB}}{0.8})^{.66}} + \frac{2.2 \times 10^{-10} \frac{F}{m} \cdot 24 \times 10^{-6} m}{(1 + \frac{V_{DB}}{0.8})^{0.18}}$$

$$C_{jdep} = \frac{3.75 \text{ fF}}{(1 + \frac{V_{DB}}{0.8})^{.66}} + \frac{5.3 \text{ fF}}{(1 + \frac{V_{DB}}{0.8})^{0.18}}$$

This result shows how the depletion capacitance changes with drain-substrate potential. ■

In many situations, during the course of designing circuits we will neglect the depletion capacitance voltage dependence and simply use the zero-bias capacitance. In the above example, we would use 9 fF as the capacitance between the drain and substrate.

REFERENCES

[1]   G. Massobrio and P. Antognetti, *Semiconductor Device Modeling with SPICE,* 2nd ed., McGraw-Hill, 1993. ISBN 0-07-002469-3.

[2]   D. A. Hodges and H. G. Jackson, *Analysis and Design of Digital Integrated Circuits,* McGraw-Hill Publishing Company, 2nd ed., 1988.   ISBN 0-07-029158-6.

[3]   R. S. Muller and T. I. Kamins, *Device Electronics for Integrated Circuits,* 2nd ed., John Wiley and Sons, 1986. ISBN 0-471-88758-7.

PROBLEMS

4.1   Lay out a nominally 250 kΩ resistor with metal1 wire connections. DRC your layout. What would happen if the layout did not include n+ under the contacts?

4.2   Lay out and DRC the pad of Fig. 4.6. Comment on the layout; that is, describe the operation and layout of the power busses and protection diodes.

4.3   Sketch the cross-sectional views across the *VDD* and *VSS* power busses of the standard cell frame of Fig. 4.9.

4.4   How much does the oxide encroach upon the width of the p-channel MOSFET in the CN20 process? Does the oxide encroachment affect the length of the MOSFET? Sketch the location of the oxide encroachment on the layout (top view) of a MOSFET.

4.5   How much does the lateral diffusion extend under the gate of a p-channel MOSFET in the CN20 process? Does the lateral diffusion affect the width of the MOSFET? Sketch the location of the lateral diffusion on the layout of the MOSFET.

4.6   Lay out an n-channel MOSFET, similar to Fig. 4.13, with an *L* and *W* of 5 μm. Include a p+ connection to the substrate for the fourth terminal of the MOSFET. DRC the layout.

4.7   Using the standard frame layout, both a p- and n-channel MOSFET with *L* = *W* = 5 μm. Is the substrate and well connected to *VSS* and *VDD,* respectively? How?

4.8   What is the maximum depletion capacitance of an n+ implant that measures 10 μm by 10 μm? (*Hint*: The maximum capacitance is simply the zero-bias depletion capacitance, assuming the junction is zero or reverse biased.) If the implant is held at a constant potential and the substrate potential is reduced, what happens to the depletion capacitance?

4.9   Estimate the delay through a 1 mm length, 2 μm wide run of poly1 loaded with 100 *fF* capacitors every 10 μm.

4.10   Sketch the cross-sectional views across the lines shown in Fig. P4.10.

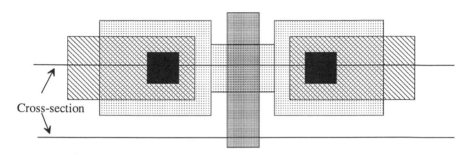

Figure P4.10

4.11   Sketch the cross-sectional views across the lines shown in Fig. P4.11.

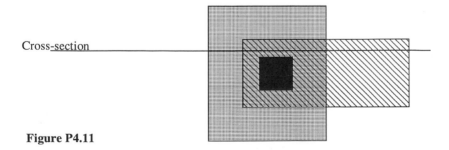

Figure P4.11

4.12   Lay out a diode using the n+ diffusion and the p-substrate. Assume the diode area is a square where the side length is 20 μm. Show connections to the anode and cathode of the diode. Sketch the cross-sectional view of the diode. Specify the SPICE statement and model for the diode (*Hint*: to determine I_s of the diode, use "js" given in the n-channel BSIM model of Appendix A and the area of the diode). Are there any restrictions on the diode anode voltage?

4.13   The layout of a p-channel MOSFET shown in Fig. P4.13 is incorrect. What is the problem?

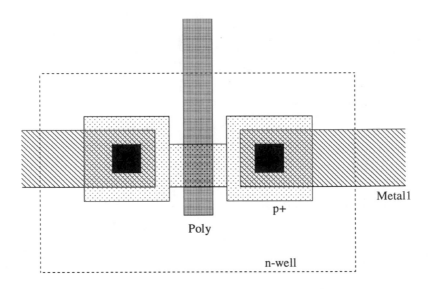

What is the fatal error associated with this layout?

Figure P4.13

Chapter 5

The MOSFET

At this point we should have some appreciation for the parasitics, that is, capacitances and resistances, associated with a CMOS process. In this chapter we discuss the MOSFET operation. To begin let's define the symbols used to denote the n-channel and p-channel MOSFETs (see Fig. 5.1). When the substrate is connected to VSS and the well is tied to VDD, we will use the simplified models shown at the bottom of the figure. It is important to keep in mind that the MOSFET is a four-terminal device and that the source and drain of the MOSFET are interchangeable.

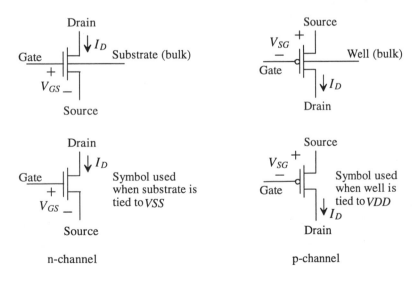

Figure 5.1 Symbols used for n- and p-channel MOSFETs.

5.1 The MOSFET Capacitances

Consider the MOSFET shown in Fig. 5.2 and its associated cross-sectional view. Associated with the drain and source regions to the substrate is a depletion capacitance that was discussed in the previous chapter. In this section we will concentrate on the capacitances associated with the gate electrode, that is, the capacitance from the gate to ground in Fig. 5.2.

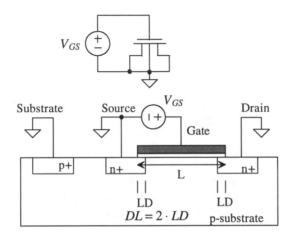

Figure 5.2 Cross-sectional view of MOSFET used to calculate capacitances.

5.1.1 Case I: Accumulation

Let's first consider the case when $V_{GS} < 0$ (Fig. 5.3). Under this condition, mobile holes from the substrate are attracted under the gate oxide. The thickness of the oxide in the SPICE MOSFET model is given by the parameter *TOX*. The capacitance between the *gate* electrode and the *substrate* electrode is given by

$$C_{gb} = \frac{\varepsilon_{ox} \cdot \overbrace{(L - 2 \cdot LD)}^{L_{eff}} \cdot W}{TOX} \tag{5.1}$$

where $\varepsilon_{ox} (= 3.97 \cdot 8.85 \text{ aF/}\mu\text{m})$ is the dielectric constant of the gate oxide, W is the drawn width (neglecting oxide encroachment), and $L - 2 \cdot LD$ is the effective channel length. The capacitance between the gate and drain or source is given by

$$C_{gd,s} = \frac{\varepsilon_{ox} \cdot LD \cdot W}{TOX} = \text{Gate-drain (or source) overlap capacitance} \tag{5.2}$$

neglecting oxide encroachment. The gate-drain overlap capacitance is present in a MOSFET regardless of the biasing conditions. This capacitance is specified in SPICE MOSFET models by the variables *CGDO* and *CGSO* with units of farads/meter. Estimation of C_{gd} or C_{gs} using the measured BSIM model parameters uses

$$C_{gd} = CGDO \cdot W = \frac{\varepsilon_{ox} \cdot LD}{TOX} \cdot W \qquad (5.3)$$

and

$$C_{gs} = CGSO \cdot W \qquad \text{(farads)} \qquad (5.4)$$

The total capacitance, independent of the width and length of the MOSFET, between the gate and ground in the circuit of Fig. 5.2 is the sum of C_{gd}, C_{gs}, and C_{gb} and is given by

$$C'_{ox} = \frac{\varepsilon_{ox}}{TOX} \qquad \text{(farads/meter}^2) \qquad (5.5)$$

The term C'_{ox} is called the oxide capacitance, which for the CN20 process is approximately 800 aF/μm^2. Knowing the width and length of a MOSFET gives a total capacitance from the gate of the MOSFET in Fig. 5.2 to ground of

$$C_{ox} = C'_{ox} \cdot W \cdot L \qquad \text{(farads)} \qquad (5.6)$$

There is a significant resistance in series with C_{gb} in Fig. 5.3 from the resistivity of the p-substrate. The resistivity of the n+ source and drain regions tends to be small enough to neglect in most circuit design applications.

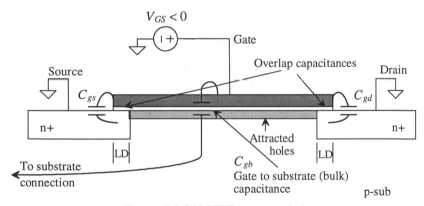

Figure 5.3 MOSFET in accumulation.

5.1.2 Case II: Depletion

Referring again to Fig. 5.2, let's consider the case when V_{GS} is not negative enough to attract a large number of holes under the oxide and not positive enough to attract a large number of electrons. Under these conditions, the surface under the gate is said to be depleted. Consider Fig. 5.4. As V_{GS} is increased from some negative voltage, holes will be displaced under the gate, leaving immobile acceptor ions that contribute a negative charge. We see that as we increase V_{GS} a capacitance between the gate and the induced channel under the oxide exists. Also, a depletion capacitance between the induced channel and the substrate is formed. The capacitance between the gate and the source/drain is simply the overlap capacitance, while the capacitance between the gate and the substrate is the oxide capacitance in series with the depletion capacitance. The depletion layer shown in Fig. 5.4 is formed between the substrate and the induced channel. The MOSFET operated in this region is said to be in weak inversion or the subthreshold region because the surface under the oxide is not heavily n+.

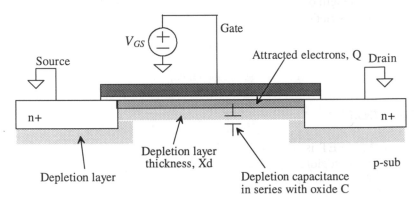

Figure 5.4 MOSFET in depeletion.

5.1.3 Case III: Strong Inversion

When V_{GS} is sufficiently large (> V_{THN}, the threshold voltage of the n-channel MOSFET) so that a large number of electrons are attracted under the gate, the surface is said to be inverted, that is, no longer p-type. Figure 5.5 shows how the capacitance from the gate to ground changes as V_{GS} changes for the MOSFET configuration of Fig. 5.2. This figure can be misleading. Remember that when the MOSFET is in the accumulation region the majority of the capacitance to ground, C_{gb}, runs through the large parasitic

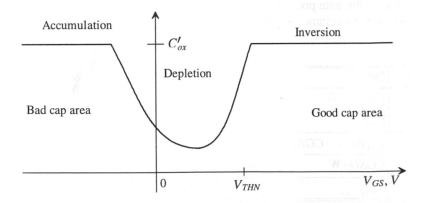

Figure 5.5 Capacitance to ground at the gate terminal of the circuit shown in Fig. 5.2 plotted against the gate-source voltage.

resistance of the substrate. Also note that the MOSFET makes a very good capacitor when $V_{GS} > V_{THN}$ + a few hundred mV. We will make a capacitor in this fashion many times while we are designing circuits.

Example 5.1

Suppose the following MOSFET configuration is to be used as a capacitor. If the width and length of the MOSFET are both 100 μm, estimate the capacitance. Are there any restrictions on the voltages we can use across the capacitor?

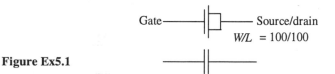

Figure Ex5.1

Since the MOSFET is to be used as a capacitor, we require operation in the strong inversion region, with the gate potential always at V_{THN} + 100 mV above the source/drain potentials.   The capacitance between the gate and the source/drain is then $C_{tot} = C'_{ox} \cdot W \cdot L$. For the CN20 process this results in

$$C_{tot} = (800 \text{ aF/μm}^2)(100 \text{ μm})(100 \text{ μm}) = 8 \text{ pF}$$

Note that we did not concern ourselves with the substrate connection. Since we are assuming strong inversion, the bulk (substrate) connection will only affect the capacitances from the drain/source to substrate (those from the source/drain implant regions). We will see, however, that the connection of the substrate will significantly affect the threshold voltage of the devices. ■

5.1.4 Summary

Figure 5.6 shows our MOSFET symbol with capacitances. The capacitance C_{gb} is associated with the gate poly over the field region. The gate-drain capacitance, C_{gd} and the gate-source capacitance C_{gs} are determined by the region of operation (see Table 5.1).

Name	Off	Triode	Saturation
C_{gd}	$CGDO \cdot W$	$\frac{1}{2} \cdot W \cdot L \cdot C'_{ox}$	$CGDO \cdot W$
C_{db}	C_{jdep}	C_{jdep}	C_{jdep}
C_{gb}	$C'_{ox}WL_{eff} + CGBO \cdot L$	$CGBO \cdot L$	$CGBO \cdot L$
C_{gs}	$CGSO \cdot W$	$\frac{1}{2} \cdot W \cdot L \cdot C'_{ox}$	$\frac{2}{3} \cdot W \cdot L \cdot C'_{ox}$
C_{sb}	C_{jdep}	C_{jdep}	C_{jdep}

Table 5.1 MOSFET capacitances.

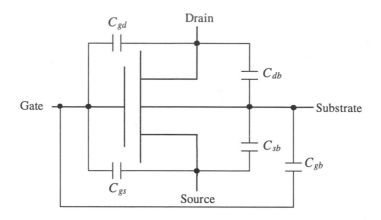

Figure 5.6 MOSFET capacitances.

5.2 The Threshold Voltage

In the last section we said that the semiconductor/oxide surface is inverted when V_{GS} is greater than the threshold voltage V_{THN}. Under these conditions, a depletion region exists between the inverted channel and the substrate. The thickness of the depletion region (Fig. 5.4) is given from pn junction theory by

$$X_d = \sqrt{\frac{2\varepsilon_{si}\phi}{qN_A}} = \sqrt{\frac{2\varepsilon_{si}|\phi_s - \phi_F|}{qN_A}} \tag{5.7}$$

where N_A is the number of acceptor atoms in the substrate, ϕ_s is the electrostatic potential at the oxide-silicon interface, and the electrostatic potential of the p-type substrate is given by

$$\phi_F = -\frac{kT}{q}\ln\frac{N_A}{n_i} \tag{5.8}$$

where n_i is the intrinsic carrier concentration of silicon ($= 14.5 \times 10^9$ atoms/cm^3). The depletion region in the p-type semiconductor is void of mobile holes. The absence of holes in this region leaves a net negative charge due to the immobile acceptor atoms that remain behind. This charge is equal to the charge attracted under the gate. The charge/unit area is given by

$$Q'_b = qN_AX_d = \sqrt{2\varepsilon_{si}qN_A|\phi_s - \phi_F|} \tag{5.9}$$

If the surface electrostatic potential at the oxide interface, ϕ_s, is simply the same as the bulk electrostatic potential ϕ_F (i.e., $\phi_s = \phi_F$ and then $Q'_b = 0$), the MOSFET is operating in the accumulation mode, or the MOSFET is OFF in circuit terms. At this point both ϕ_s and ϕ_F are negative numbers. The number of holes at the oxide-semiconductor surface is N_A, the same concentration as the bulk.

As V_{GS} is increased, the surface potential becomes more positive. When $\phi_s = 0$, the surface under the oxide has become depleted (the carrier concentration is n_i). When $\phi_s = -\phi_F$ (a positive number), the channel is inverted (electrons are pulled under the oxide forming a channel), and the electron concentration at the semiconductor-oxide interface is equal to the substrate doping concentration. The value of V_{GS} when $\phi_s = -\phi_F$ is arbitrarily defined as the threshold voltage, V_{THN}. Note that the surface potential changed a total of $2\phi_F$ between the strong inversion and depletion cases.

For $V_{GS} = V_{THN}$ ($\phi_s = -\phi_F$), the fixed negative charge in the depletion region is given by

$$Q'_{bo} = \sqrt{2qN_A\varepsilon_{si}|-2\phi_F|} \qquad (5.10)$$

with units of coulombs/m². Up to this point we have assumed that the substrate and source were tied together to ground. If the source of the n-channel MOSFET is at a higher potential than the substrate, the potential difference is given by V_{SB}; the negative charge present in the depletion region then becomes

$$Q'_b = \sqrt{2qN_A\varepsilon_{si}|-2\phi_F + V_{SB}|} \qquad (5.11)$$

Example 5.2

For a substrate doping of 10^{15} atoms/cm³, $V_{GS} = V_{THN}$ and $V_{SB} = 0$, estimate the electrostatic potential in the substrate region and at the oxide-semiconductor interface, the depletion layer width, and the charge contained in the depletion region and thus the inverted region under the gate.

The electrostatic potential of the substrate is

$$\phi_F = -\frac{kT}{q}\ln\frac{N_A}{n_i} = -26 \text{ mV} \cdot \ln\frac{10^{15}}{14.5 \times 10^9} = -290 \text{ mV}$$

and therefore the electrostatic potential at the oxide semiconductor interface ($V_{GS} = V_{THN}$), ϕ_s, is 290 mV. The depletion layer width is given by

$$X_d = \sqrt{\frac{2 \cdot 11.7 \cdot (8.85 \times 10^{-18} \text{F/}\mu\text{m})(2 \cdot 0.29\text{V})}{(1.6 \times 10^{-19}\frac{C}{\text{atom}})(10^{15}\frac{atoms}{\text{cm}^3})(\frac{\text{cm}^3}{10^{12}\mu\text{m}^3})}} = 0.866 \ \mu\text{m}$$

and the charge contained in this region, from Eq. (5.10) or (5.9) with $\phi_s = -\phi_F$ by

$$Q'_{bo} = qN_AX_d = \left(1.6 \times 10^{-19}\frac{C}{\text{atom}}\right)\left(10^{15}\frac{\text{atoms}}{\text{cm}^3}\right)\left(\frac{\text{cm}^3}{10^{12}\mu\text{m}^3}\right)(0.866\mu\text{m})$$

$$= 139 \ \frac{\text{aC}}{\mu\text{m}^2}$$

which is the charge contained in the depletion region or oxide-semiconductor interface. Note that this is true only when $V_{GS} = V_{THN}$. ■

Consider the MOSFET shown in Fig. 5.7. We will assume that the applied V_{GS} = V_{THN} so that the preceding discussions and assumptions hold. The potential across the gate-oxide capacitance, C'_{ox}, is simply

$$V_{BC} = \frac{Q'_b}{C'_{ox}}$$ (5.12)

The surface potential *change*, V_C (= $\Delta\phi_s$), from the equilibrium case is $|2\phi_F|$. (The absolute voltage of the channel is 0 V; that is, the source, drain, and channel are at ground.) The potential needed to change the surface potential and overcome the depletion layer charge is given by

$$V_B = \frac{Q'_b}{C'_{ox}} - 2\phi_F$$ (5.13)

An additional positive charge exists at the silicon-oxide interface due to imperfections during growth of the gate oxide or as the result of an ion implantation used to adjust the threshold voltage of the MOSFET. We will call this positive charge Q'_{ss} with units of coulombs/area. Equation (5.13) may be rewritten including these surface charges as

$$V_B = \frac{Q'_b - Q'_{ss}}{C'_{ox}} - 2\phi_F$$ (5.14)

The final component needed to determine the threshold voltage is the contact potential between point D (the bulk) and point A (the gate material) in Fig. 5.7. The potential difference between the gate and bulk (p-substrate) can be determined by summing the difference between the materials in the MOS system shown in Fig. 5.8.

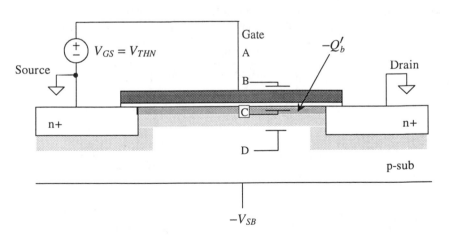

Figure 5.7 Calculation of the threshold voltage.

Adding the contact potentials, we get $(\phi_G - \phi_{ox}) + (\phi_{ox} - \phi_F) = \phi_G - \phi_F$. The contact potential between the bulk and the gate poly, we will assume n+ poly with doping concentration $N_{D, poly}$, is given by

$$\phi_{ms} = \phi_G - \phi_F = \frac{kT}{q} \ln\left[\frac{N_{D, poly}}{n_i}\right] + \frac{kT}{q} \ln \frac{N_A}{n_i} \tag{5.15}$$

The threshold voltage, V_{THN}, is given by

$$V_{THN} = \frac{Q'_b - Q'_{ss}}{C'_{ox}} - 2\phi_F - \phi_{ms} \tag{5.16}$$

$$= -\phi_{ms} - 2\phi_F + \frac{Q'_{bo} - Q'_{ss}}{C'_{ox}} - \frac{Q'_{bo} - Q'_b}{C'_{ox}} \tag{5.17}$$

$$= -\phi_{ms} - 2\phi_F + \frac{Q'_{bo} - Q'_{ss}}{C'_{ox}} + \frac{\sqrt{2q\varepsilon_{si}N_A}}{C'_{ox}}\left[\sqrt{|2\phi_F| + V_{SB}} - \sqrt{|2\phi_F|}\right] \tag{5.18}$$

When the source is shorted to the substrate, $V_{SB} = 0$, we can define a zero substrate bias or simply zero-bias threshold voltage as

$$V_{THN0} = -\phi_{ms} - 2\phi_F + \frac{Q'_{bo} - Q'_{ss}}{C'_{ox}} \tag{5.19}$$

We can define a body effect coefficient or body factor by

$$\gamma = \frac{\sqrt{2q\varepsilon_{si}N_A}}{C'_{ox}} \tag{5.20}$$

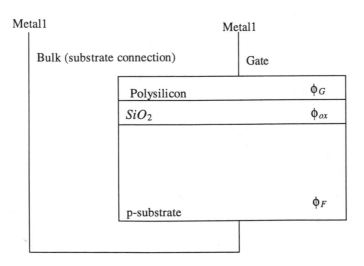

Figure 5.8 Determining the contact potential between poly and substrate.

Equation (5.18) can now be written as

$$V_{THN} = V_{THN0} + \gamma \left(\sqrt{|2\phi_F| + V_{SB}} - \sqrt{|2\phi_F|} \right) \qquad (5.21)$$

It is interesting to note that a voltage, called the flatband voltage V_{FB}, must be applied for the oxide-semiconductor interface surface potential, ϕ_s, to become the same potential as the bulk surface potential, ϕ_F. The flatband voltage is given by

$$V_{FB} = -\phi_{ms} - \frac{Q'_{ss}}{C'_{ox}} \qquad (5.22)$$

The zero-bias threshold voltage may then be written in terms of the flatband voltage as

$$V_{THN0} = V_{FB} - 2\phi_F + \frac{Q'_{bo}}{C'_{ox}} \qquad (5.23)$$

These equations describe how the threshold voltage of the MOSFET is affected by substrate doping, oxide thickness, source/substrate bias, gate material, and surface charge density.

Example 5.3

Assuming $N_A = 10^{15}$ atoms/cm^3 and $C'_{ox} = 800$ aF/μm^2 (CN20), estimate γ (GAMMA). Compare this estimate to the value of γ given in the SPICE level 2 n-channel model presented in Appendix A.

From Eq. (5.20) the calculated γ is

$$\gamma = \frac{\sqrt{2 \cdot 1.6 \times 10^{-19} \frac{col}{atom} \cdot 11.7 \cdot 8.85 \frac{aF}{\mu m} \cdot 10^{15} \frac{atoms}{cm^3} \cdot \frac{cm^3}{10^{12} \mu m^3}}}{800 \frac{aF}{\mu m^2}} = 0.228 \ V^{1/2}$$

while the γ given in the level 2 SPICE model is 0.4179 (= GAMMA). We assume that the doping concentration of the substrate was too low. A more accurate assumption may be 3.4×10^{15} atoms/cm^3. ∎

Example 5.4

Estimate the zero-bias threshold voltage for the MOSFET of Ex. 5.2. Assume that the poly doping level is 10^{20} atoms/cm^3. What happens to the threshold voltage if sodium contamination causes an impurity of 40 aC/μm^2 at the oxide-semiconductor interface?

The electrostatic potential between the gate and substrate is given by

$$-\phi_{ms} = \phi_F - \phi_G = -290 \ mV - 26 \ mV \cdot \ln\frac{10^{20}}{14.5 \times 10^9} = -879 \ mV$$

$$-2\phi_F = 580 \ mV$$

$$\frac{Q'_{bo}}{C'_{ox}} = \frac{139\frac{aC}{\mu m^2}}{800\frac{aF}{\mu m^2}} = 173 \text{ mV}$$

$$\frac{Q'_{ss}}{C'_{ox}} = 50 \text{ mV}$$

The threshold voltage, from Eq. (5.19) without the sodium contamination, is −126 mV; with the sodium contamination the threshold voltage is −176 mV. ∎

These threshold voltages would correspond to depletion devices, that is, MOSFETs that conduct when the $V_{GS} = 0$. In CMOS applications this is highly undesirable. To compensate or adjust the value of the threshold voltage, the channel, the area under the gate poly, can be implanted with p+ ions. This will effectively increase the value of the threshold voltage by Q'_c/C'_{ox} where Q'_c is the charge density/unit area due to the implant. If N_I is the ion implant dose in atoms/unit area, then we can write

$$Q'_c = q \cdot N_I \tag{5.24}$$

and the threshold voltage by

$$V_{THN0} = -\phi_{ms} - 2\phi_F + \frac{Q'_{bo} - Q'_{ss} + Q'_c}{C'_{ox}} \tag{5.25}$$

Example 5.5
Estimate the dose required to change the threshold voltage, in Ex. 5.4 without sodium contamination , to 1 V.

From Eqs. (5.24) and (5.25) and the results of Ex. 5.4

$$V_{THN0} = -126 \text{ mV} + \frac{qN_I}{C'_{ox}} = 1 \text{ V}$$

This gives $N_I = 563 \times 10^9$ atoms/cm². ∎

These calculations lend some insight into how the threshold voltage is affected by the different process parameters. In practice, the results of these calculations do not exactly match the measured threshold voltage. From a circuit design engineer's point of view, the threshold voltage and the body factor are measured in the laboratory. Measurements are performed when the BSIM model parameters (discussed in the next chapter) are extracted. The design engineer then has measured data available in the form of the BSIM model parameters.

5.3 IV Characteristics of MOSFETs

Now that we have some familiarity with the factors influencing the threshold voltage of a MOSFET, let's derive the large-signal IV (current/voltage) characteristics of the MOSFET, namely operation in the triode and the saturation regions. The following derivation is sometimes referred to as the gradual-channel approximation. The charge distribution in the channel is assumed to be essentially constant.

5.3.1 MOSFET Operation in the Triode Region

Consider Fig. 5.9 where $V_{GS} > V_{THN}$, so that the surface under the oxide is inverted and $V_{DS} > 0$, causing a drift current to flow from the drain to the source. In our initial analysis, we will assume that V_{DS} is sufficiently small so that the threshold voltage and the depletion layer width are approximately constant.

Initially, we must find the charge stored on the oxide capacitance C'_{ox}. The voltage, with respect to the source of the MOSFET, of the channel a distance y away from the source is labeled $V(y)$. The potential difference between the gate electrode and the channel is then $V_{GS} - V(y)$. The charge/unit area in the inversion layer is given by

$$Q'_{ch} = C'_{ox} \cdot [V_{GS} - V(y)] \tag{5.26}$$

However, we know that a charge Q'_b is present in the inversion layer from the application of the threshold voltage, V_{THN}, necessary for conduction between the drain and the source. This charge is given by

$$Q'_b = C'_{ox} \cdot V_{THN} \tag{5.27}$$

The total charge available in the channel, for conduction of a current between the drain and the source, is given by the difference in these two equations, or

$$Q'_I(y) = C'_{ox} \cdot (V_{GS} - V(y) - V_{THN}) \tag{5.28}$$

where Q'_I is the charge in the inverted channel.

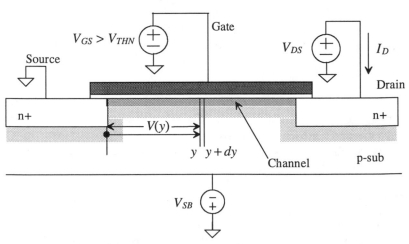

Figure 5.9 Calculation of the large-signal behavior of the MOSFET in the triode (ohmic) region.

The differential resistance of the channel region with a length dy and a width W is given by

$$dR = \overbrace{\frac{1}{\mu_n Q_I'(y)}}^{\text{eff. sheet Res.}} \cdot \frac{dy}{W} \qquad (5.29)$$

where μ_n is the average electron mobility through the channel with units of $cm^2/V\cdot sec$. The mobility is simply a ratio of the electron (or hole) velocity cm/sec to the electric field, V/cm. This parameter is used, in the BSIM model parameters, to fit the IV curves of the SPICE models to the measured results. Also, for short channel devices, the mobility decreases when the velocity of the carriers starts to saturate. This causes the effective sheet resistance in Eq. (5.29) to increase, resulting in a lowering of the drain current.

The differential voltage drop across this differential resistance is given by

$$dV(y) = I_D \cdot dR = \frac{I_D}{W\mu_n Q_I'(y)} \cdot dy \qquad (5.30)$$

or substituting Eq. (5.28) and rearranging

$$I_D \cdot dy = W\mu_n C_{ox}'(V_{GS} - V(y) - V_{THN}) \cdot dV(y) \qquad (5.31)$$

At this point, let's define the transconductance parameter for a MOSFET. For an n-channel MOSFET, this parameter is given by

$$KP_n = \mu_n \cdot C_{ox}' = \mu_n \cdot \frac{\varepsilon_{ox}}{TOX} \qquad (5.32)$$

and for a p-channel MOSFET, it is given by

$$KP_p = \mu_p \cdot C_{ox}' = \mu_p \cdot \frac{\varepsilon_{ox}}{TOX} \qquad (5.33)$$

where μ_p is the mobility of the holes in a PMOS transistor. Typical values of KP in the CN20 process are 50 $\mu A/V^2$ and 17 $\mu A/V^2$ for n- and p-channel transistors, respectively.

The current can be obtained by integrating the left side of Eq. (5.31) from the source to the drain, that is, from 0 to L and the right side from 0 to V_{DS}. This is shown below:

$$I_D \int_0^L dy = W \cdot KP_n \cdot \int_0^{V_{DS}} (V_{GS} - V(y) - V_{THN}) \cdot dV(y) \qquad (5.34)$$

or

$$I_D = KP_n \cdot \frac{W}{L} \cdot \left[(V_{GS} - V_{THN})V_{DS} - \frac{V_{DS}^2}{2} \right] \text{ for } V_{GS} \geq V_{THN} \text{ and } V_{DS} \leq V_{GS} - V_{THN} \quad (5.35)$$

This equation is valid when the MOSFET is operating in the triode (linear) region. This is the case when the induced channel extends from the source to the drain.

Furthermore, we can rewrite Eq. (5.35) by defining the following transconductance parameter:

$$\beta = KP_n \cdot \frac{W}{L} \tag{5.36}$$

or

$$I_D = \beta \cdot \left[(V_{GS} - V_{THN})V_{DS} - \frac{V_{DS}^2}{2} \right] \tag{5.37}$$

The equivalent equation for the p-channel MOSFET operating in the triode region is

$$I_D = KP_p \cdot \frac{W}{L} \cdot \left[(V_{SG} - V_{THP})V_{SD} - \frac{V_{SD}^2}{2} \right] \text{ for } V_{SG} \geq V_{THP} \text{ and } V_{SD} \leq V_{SG} - V_{THP} \tag{5.38}$$

where the threshold voltage of the p-channel MOSFET is assumed to be positive (see Appendix A). In fact, all voltages in Eqs. (5.35) and (5.38) are positive.

5.3.2 The Saturation Region

The voltage $V(L)$ in Eq. (5.28) is simply V_{DS}. In the previous subsection, we said that V_{DS} is always less than $V_{GS} - V_{THN}$ so that at no point along the channel is the inversion charge zero. When $V_{DS} = V_{GS} - V_{THN}$, the inversion charge under the gate at L (the drain-channel junction) is zero. This drain-source voltage is called $V_{DS,sat} (= V_{GS} - V_{THN})$, and it indicates when the channel charge becomes *pinched* off at the drain-channel interface. Further increases in V_{DS} do not cause an increase in the drain current[1]. Figure 5.10 shows that the depletion region between the drain and substrate increases, causing the channel to pinch off. If V_{DS} is increased until the drain-substrate depletion region extends from the drain to the source, the device is said to be *punched* through. Large currents can flow under these conditions, causing device failure. The maximum voltage, for near minimum-size channel lengths, that can be applied between the drain and source of a MOSFET is set by the "punchthrough" voltage (see Appendix A). For long channel lengths, the maximum voltage is set by the breakdown voltage of the drain (n+) implant/substrate diode, also specified in Appendix A.

When a MOSFET is operated with its channel pinched off, that is, $V_{DS} \geq V_{GS} - V_{THN}$ and $V_{GS} \geq V_{THN}$, it is operating in the saturation region. Substitution of $V_{DS,sat}$ into Eq. (5.35) yields

$$I_D = \frac{KP_n}{2} \cdot \frac{W}{L} \cdot (V_{GS} - V_{THN})^2 = \frac{\beta}{2}(V_{GS} - V_{THN})^2$$

$$\text{for } V_{DS} \geq V_{GS} - V_{THN} \text{ and } V_{GS} \geq V_{THN} \tag{5.39}$$

---

[1] We will see that this is not entirely true. An effect called channel length modulation causes the drain current to increase with increasing drain-source voltage.

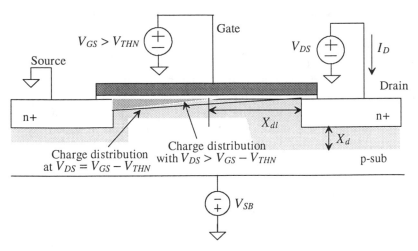

Figure 5.10 The MOSFET in saturation (pinched off).

We can define an electrical channel length of the MOSFET as the difference between the drawn channel length, neglecting lateral diffusion, and the depletion layer width, X_{dl}, between the drain n+ and the channel under the gate oxide by

$$L_{elec} = L_{drawn} - X_{dl} \tag{5.40}$$

Substituting this into Eq. (5.39), we obtain a better representation of the drain current

$$I_D = \frac{KP_n}{2} \cdot \frac{W}{L_{elec}} (V_{GS} - V_{THN})^2 \tag{5.41}$$

Qualitatively, this means that since the depletion layer width increases with increasing V_{DS} the drain current will increase as well. This effect is called *channel length modulation*. To determine the change in output current with drain-source voltage, we simply take the derivative of Eq. (5.41) with respect to V_{DS}, or

$$\frac{\partial I_D}{\partial V_{DS}} = -\frac{KP_n}{2} \frac{W}{L_{elec}^2} (V_{GS} - V_{THN})^2 \cdot \frac{dL_{elec}}{dV_{DS}} = I_D \cdot \left[\frac{1}{L_{elec}} \frac{dX_{dl}}{dV_{DS}} \right] \tag{5.42}$$

where it is common to let

$$\lambda_c = \frac{1}{L_{elec}} \cdot \frac{dX_{dl}}{dV_{DS}} \tag{5.43}$$

Typical values for λ_c, called the channel length modulation parameter, range from greater than 0.1 for short channel devices to 0.01 for long channel devices. Equation 5.41 can be rewritten, taking into account channel length modulation as

$$I_D = \frac{KP_n}{2} \cdot \frac{W}{L} (V_{GS} - V_{THN})^2 [1 + \lambda_c (V_{DS} - V_{DS,sat})] \tag{5.44}$$

Unless otherwise specified for digital applications, we will assume $\lambda_c = 0$. For analog applications, λ_c is normally measured. Also note that in Eq. (5.42) we assumed that the mobility does not vary with V_{DS}. In the next chapter we will find that this is not true. Figure 5.11 shows typical curves for an n-channel MOSFET. Notice how the device *appears to go into saturation earlier* than predicted by $V_{DS} = V_{GS} - V_{THN}$. The actual charge distribution in the channel is not constant but rather a function of V_{DS}. $Q'_I(y)$ decreases as we move away from the source of the MOSFET, causing $Q'_I(L)$ to become zero earlier. This is shown in Fig. 5.10. Another reason, especially for short channel devices, why the MOSFET, becomes pinched off earlier is that the mobility is not a constant; that is, the velocity of the electrons saturates above a certain V_{DS} (related to the electric field and channel length), causing $V_{DS,sat}$ and $I_{DS,sat}$ (the drain current at $V_{DS,sat}$) to decrease. Shown in Appendix A are the curves for both n- and p-channel MOSFETs of varying sizes, together with the threshold voltage and transconductance parameters for the MOSFETs in the CN20 process, which is what we will use for hand calculations throughout the book.

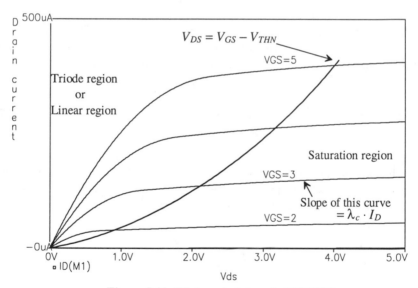

Figure 5.11 IV characteristics of a MOSFET.

C_{gs} Calculation in the Saturation Region

The gate to source capacitance of a MOSFET operating in the saturation region can be determined by solving Eq. 5.28 for the total charge in the inverted channel,

$$Q_I = \int_0^L W \cdot Q'_I(y) \cdot dy = WC'_{ox} \int_0^L (V_{GS} - V(y) - V_{THN})\, dy \qquad (5.45)$$

or solving Eq. 5.31 for dy and substituting yields

$$Q_I = \frac{(W \cdot C'_{ox})^2 \cdot \mu_n}{I_D} \int_0^{V_{GS}-V_{THN}} (V_{GS} - V(y) - V_{THN})^2 \cdot dV(y) \tag{5.46}$$

where the fact that the Q_I goes to zero when $y = L$ occurs when $V_{DS} = V_{GS} - V_{THN}$. Solving this equation using Eqs. 5.32 and 5.39 yields

$$Q_I = \frac{2}{3} \cdot W \cdot L \cdot C'_{ox} \cdot (V_{GS} - V_{THN}) \tag{5.47}$$

We can determine the gate to source capacitance while in the saturation region by,

$$C_{gs} = \frac{\partial Q_I}{\partial V_{GS}} = \frac{2}{3} \cdot W \cdot L \cdot C'_{ox} \tag{5.48}$$

5.4 SPICE Modeling of the MOSFET

In this section we consider the level 1 SPICE model and how it relates to the equations derived in the last section. The level 1 model is a subset of the level 2 and 3 models.

5.4.1 Level 1 Model Parameters Related to V_{THN}

The following SPICE model parameters are related to the calculation of V_{THN},

Symbol	Name	Description	Default	Typ.	Units
V_{THN0}	VTO	Zero-bias threshold voltage	1.0	0.8	Volts
γ	GAMMA	Body-effect parameter	0	0.4	$V^{1/2}$
$2|\phi_F|$	PHI	Surface to bulk potential	0.65	0.58	V
N_A	NSUB	Substrate doping	0	1E15	cm^{-3}
Q'_{ss}/q	NSS	Surface state density	0	1E10	cm^{-2}
	TPG	Type of gate material	1	1	

Using Eq. (5.21), we can calculate the threshold voltage, V_{THN}, given the above parameters. If V_{THN0} or γ is not given, then SPICE will calculate them using the above information and Eqs. (5.20), (5.22), and (5.23). TPG specifies the type of gate material, 1 opposite to substrate, -1 same as substrate, and 0 for aluminum gate.

Example 5.6
Using SPICE and the level 2 model given in Appendix A for the n-channel MOSFET, plot I_D versus V_{GS} for V_{SB} changing from 0 to 5 V and $V_{DS} = 5$ V. Assume the device $W = L = 5$ μm.

The circuit schematic, SPICE output and netlist, are shown in Figs. 5.12 and 5.13. Notice that, as we would predict from Eq. (5.21), the *threshold voltage increases* with increasing source-substrate potential, V_{SB}. The results of this simulation *do not* match well with measured results. The inaccuracy of the level

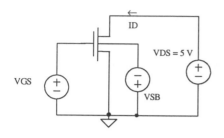

```
*** Top Level Netlist ***
M1_5u_5u 2 3 0 4 CMOSN  L=5u W=5u
VDS 2 0 DC 5
VGS 3 0 DC 0
VSB 0 4 DC 0

.MODEL CMOSN NMOS
(level 2 model in App. A)
.probe
.DC VGS 0 5 .05 VSB 0 5 1
.end
```

Figure 5.12 Schematic and netlist used in Example 5.6.

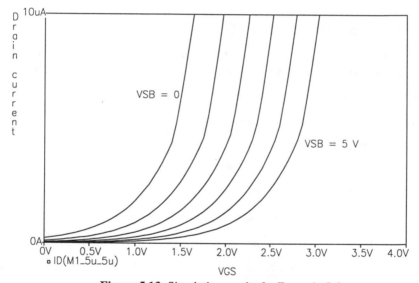

Figure 5.13 Simulation results for Example 5.6.

1, 2, or 3 SPICE models in predicting the threshold voltage makes their use limited to basic functionality tests. The BSIM SPICE model of the next chapter should be used in most simulations (see Fig. 6.2). ∎

5.4.2 Level 1 Model Parameters Related to Transconductance

The following SPICE parameters are related to the calculation of transconductance.

Symbol	Name	Description	Default	Typ.	Units
KP	KP	Transconductance parameter	20E-6	50E-6	A/V^2
t_{ox}	TOX	Gate-oxide thickness	1E-7	40E-9	m
λ	Lambda	Channel-length modulation	0	0.01	V^{-1}
LD	LD	Lateral diffusion	0	2.5E-7	m
$\mu_{n,p}$	UO	Surface mobility	600	580	cm^2/Vs

5.4.3 SPICE Modeling of the Source and Drain Implants

The following SPICE model parameters are related to calculating the parasitics associated with the source/drain implant regions.

Name	Description	Default	Typical	Units
RD	Drain contact resistance	0	40	Ω
RS	Source contact resistance	0	40	Ω
RSH	Source/drain sheet resistance	0	50	Ω/sq.
CGBO	Gate-bulk overlap capacitance	0	4E-10	F/m
CGDO	Gate-drain overlap capacitance	0	4E-10	F/m
CGSO	Gate-source overlap capacitance	0	4E-10	F/m
PB, PBSW	Bottom, sidewall built-in potential	0.8	0.8	V
MJ, MJSW	Bottom, sidewall grading coefficient	0.5	0.5	
CJ	Bottom zero-bias depletion capacitance	0	3E-4	F/m^2
CJSW	Sidewall zero-bias depletion capacitance	0	2.5E-10	F/m
IS	Bulk-junction saturation current density	1E-14	1E-14	A
JS	Bulk-junction saturation current density	0	1E 8	A/m^2
FC	Bulk-junction forward bias coefficient	0.5	0.5	

Example 5.7

Specify the SPICE device statement for the layout shown in Fig. 5.14. Assume that the MOSFET model name is CMOSN.

The width and length of the MOSFET in this figure is 4 and 5 μm respectively. The number of squares in the drain and source regions, NRD and NRS will be calculated next. The number of squares is an approximation, so we neglect the n+ directly under the drain and source contacts. From the figure, the number of squares in the drain ($LNRD/W$) is 6 μm/4 μm or NRD =1.5. For the source NRS = $LNRS/W$ = 16/4 = 4. If the SPICE model specifies a sheet resistance, RSH, then a resistance, NRD·RSH, due to the n+ (or p+) implant sheet resistance is added in series with the drain of the intrinsic MOSFET model. If the sheet resistance is not specified, then the values of RD or RS are added in series with the MOSFET's drain and source, respectively.

One important parameter that is not calculated here is the metal1 to n+ contact resistance. The value of NRD and NRS can be adjusted to account for this contact resistance, or an external resistor can be added. For a large W/L MOSFET NRD and NRS should always be set to zero.

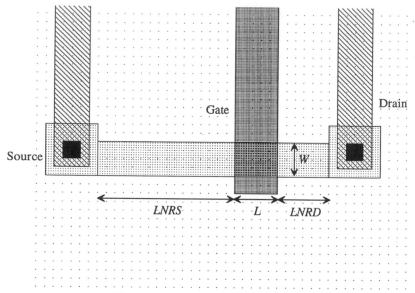

Figure 5.14 Layout of Example 5.7.

In order for SPICE to calculate the depletion capacitances of the drain and source, we need to specify the areas and perimeters of these regions. The area of the drain is 6 μm by 6 μm (36 pm²) plus 24 pm² (*W·LNRD*), while the perimeter is 32 μm (neglecting the portion of the drain implant directly adjacent to the gate poly). The area of the source is 100 pm² and the perimeter is 52 μm. The SPICE statement for a MOSFET, M1 with its drain connected to node 1, gate to node 2, source to node 3, and substrate to node 4 is given by

M1 1 2 3 4 CMOSN L=5u W=4u AD=60p AS=100p PD=32u PS=52u NRD=1.5 NRS=4
■

5.4.4 Layout of the MOSFET

The layouts presented thus far for the MOSFET are for near minimum-size devices. In practice, devices with widths several hundred or even thousand μm are possible. Consider the connection of parallel MOSFETs shown in Fig. 5.15. These MOSFETs

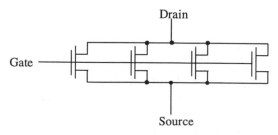

Figure 5.15 Parallel connection of MOSFETs used for layout of a large device.

operate as a single MOSFET with a width equal to the sum of the individual MOSFET's widths, assuming equal lengths. The layout for the circuit of Fig. 5.15 is shown in Fig. 5.16. The drain and source areas are shared between adjacent MOSFETs. Two benefits are achieved with this layout: (1) smaller layout size and (2) reduction of source and drain depletion capacitances. The second benefit is important in analog design or in output driver design where latch-up, discussed in Ch. 11, is a concern. See Ch. 11 for further discussion of layout of large MOSFETs.

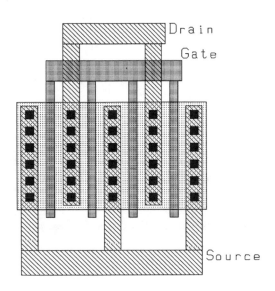

Figure 5.16 Layout of a large (width) MOSFET.

REFERENCES

[1]   D. A. Hodges and H. G. Jackson, *Analysis and Design of Digital Integrated Circuits,* McGraw-Hill Publishing Company, 2nd ed., 1988. ISBN 0-07-029158-6.

[2]   R. S. Muller and T. I. Kamins, *Device Electronics for Integrated Circuits,* 2nd ed., John Wiley and Sons, 1986. ISBN 0-471-88758-7.

[3]   G. Massobrio and P. Antognetti, *Semiconductor Device Modeling with SPICE,* 2nd ed., McGraw-Hill, 1993. ISBN 0-07-002469-3.

Additional texts covering the operation and modeling of the MOSFET.

[4]   Y. P. Tsividis, *Operation and Modeling of the MOS Transistor,* McGraw-Hill, 1987. ISBN 0-07-065381-X.

[5]   R. F. Pierret, *Volume IV in the Modular Series on Solid State Devices-Field Effect Devices,* Addison-Wesley, 1990.

[6]   D. K. Schroder, *Modular Series on Solid State Devices-Advanced MOS Devices,* Addison-Wesley, 1987.

[7]   J. Y. Chen, *CMOS Devices and Technology for VLSI,* Prentice-Hall, 1990. ISBN 0-13-138082-6.

[8]   S. M. Sze, *Physics of Semiconductor Devices,* 2nd ed., John-Wiley and Sons, 1981. ISBN 0-471-05661-8.

[9]   D. A. Neamen, *Semiconductor Physics and Devices-Basic Principles,* Richard D. Irwin, 1992. ISBN 0-256-08405-X.

Texts covering MOSFET fabrication.

[10]   S. Wolf, *Silicon Processing for the VLSI Era-Volume 3: The Submicron MOSFET,* Lattice Press, 1995.

[11]   C. Y. Chang and S. M. Sze, *ULSI technology,* McGraw-Hill, 1996. ISBN 0-07-063062-3.

[12]   R. C. Jaeger, *Modular Series on Solid State Devices-Introduction to Microelectronic Fabrication,* Addison-Wesley, 1989.

PROBLEMS

Unless otherwise stated, use the CN20 process.

5.1   Estimate v_{out} (AC) in the following circuit, Fig. P5.1. Assume that the MOSFET was fabricated using the CN20 process.

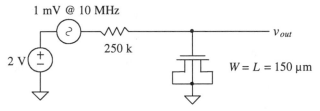

Figure P5.1

5.2   An n-channel MOSFET is known to have $2|\phi_F| = 0.57$ V, $\gamma = 0.45$ V$^{1/2}$, $\mu_n = 550$ cm^2/V·sec, and $V_{THN0} = 0.8$ V. Assume $\lambda = 0$, $n_i = 1.45E10$ atoms/cm^3 and kT/q = 26 mV, compute the value of *KP*. Suppose $W/L = 10/2$. Compute I_D when $V_{GS} = 2$ V, $V_{SB} = 1$ V and $V_{DS} = 1.1$V.

5.3   If a MOSFET is used as a capacitor in the strong inversion region where the gate is one electrode and the source/drain is the other electrode, does the gate overlap of the source/drain change the capacitance? Why? What is the capacitance?

5.4 Repeat Problem 3 when the MOSFET is operating in the accumulation region. Keep in mind that the question is not asking for the capacitance from gate to substrate.

5.5 If the oxide thickness of a MOSFET is 400 Å, what is C'_{ox}?

5.6 Repeat Ex. 5.2 when $V_{SB} = 2$ V.

5.7 Repeat Ex. 5.3 for a p-channel device with a well doping concentration of 10^{16} atoms/cm^3.

5.8 What is the electrostatic potential of the oxide-semiconductor interface when $V_{GS} = V_{THN0}$?

5.9 Repeat Ex. 5.5 to get a threshold voltage of 0.8 V.

5.10 What happens to the threshold voltage in Problem 9 if sodium contamination of 100×10^9 sodium ions/cm^2 is present at the oxide-semiconductor interface?

5.11 How much charge (enhanced electrons) is available under the gate for conducting a drain current at the drain-channel interface when $V_{DS} = V_{GS} - V_{THN}$? Why? Assume the MOSFET is operating in strong inversion, $V_{GS} > V_{THN}$.

5.12 Derive Eq. (5.35) for the p-channel MOSFET following the same procedure presented in the chapter for the n-channel MOSFET.

5.13 Show that the parallel connection of MOSFETs shown in Fig. P5.13 behave as a single MOSFET with a width equal to the sum of each individual MOSFET's width.

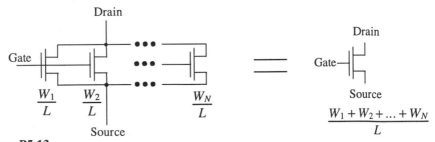

Figure P5.13

5.14 Show that the bottom MOSFET, Fig. P5.14, in a series connection of two MOSFETs cannot operate in the saturation region. Neglect the body effect. [*Hint*: Show that M1 is always in either cutoff ($V_{GS1} < V_{THN}$) or triode ($V_{DS1} < V_{GS1} - V_{THN}$).]

5.15 Show that the series connection of MOSFETs shown in Fig. P5.14 behaves as a single MOSFET with twice the length of the individual MOSFETs. Again neglect the body effect.

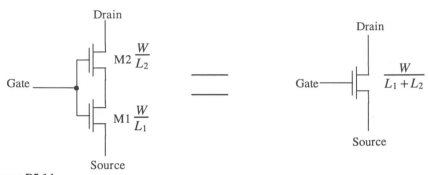

Figure P5.14

5.16   Repeat Ex. 5.6 for a CN20 p-channel MOSFET.

5.17   Lay out and DRC an n-channel MOSFET with a size of 200 μm /2 μm. Use the standard cell frame. Note that the standard cell frames can be placed end to end to achieve a larger frame; that is, the height is constant, while the width of the frame depends on the number used.

5.18   Specify the SPICE statement for the MOSFET of Problem 17.

5.19   The layout of an n-channel MOSFET is shown in Fig. P5.19. What is this device's width and length?

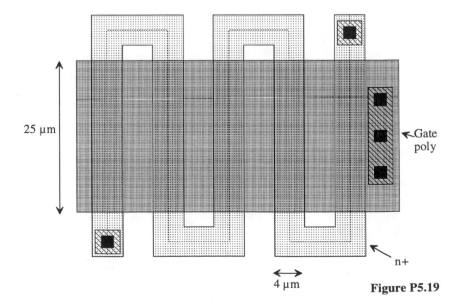

Figure P5.19

The BSIM SPICE Model

The Berkeley Short Channel Igfet (insulated gate field-effect transistor) Model, BSIM, is a SPICE model for both n- and p-channel MOS transistors. Automated parameter extraction and model generation software are used to generate the SPICE model. Most fabrication houses use this model to describe the performance of their devices. The BSIM1 SPICE model [1] is adequate for modeling MOSFETs with channel lengths down to 1 μm [2]. Shorter channel lengths require the use of the BSIM2 [2] or BSIM3 [3] model. This chapter will introduce the BSIM1 and BSIM3 models.

In the past, the level 1, 2, or 3 MOS SPICE models were discussed in texts, and thus courses covering CMOS circuit design. Since most fabricators do not supply a level 1, 2, or 3 SPICE model (MOSIS does supply a level 2 or 3 in addition to the BSIM model), students making the transition from academe to industry were met with learning the BSIM model. Furthermore, students were faced with trying to determine threshold voltage, transconductance parameter, and so on, using the unfamiliar BSIM model. In order to perform hand calculations, we need to be able to extract some of these simple parameters from the BSIM model statement. This is the goal of the chapter.

Figure 6.1 shows a comparison, using Berkeley SPICE3, between the level 2 and the BSIM1 (level 4) models for an NMOS transistor with $W = 3$ μm and $L = 2$ μm. Below is a SPICE netlist used to generate these plots.

```
*** Top Level Netlist ***
M1_3u_2u      3 4 0 0 CMOSN  L=2u W=3u AD=18p AS=18p PD=18u PS=18u
M2_3u_2u      5 4 0 0 CMOSNB L=2u W=3u AD=18p AS=18p PD=18u PS=18u
Vdd      2 0      DC 0
Vgs      4 0      DC 0
VIMTR1           2 3 0V
VIMTR2           2 5 0V

.MODEL CMOSNB NMOS LEVEL=4
```

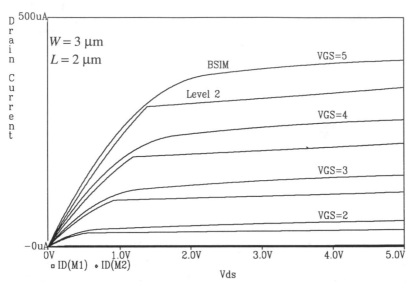

Figure 6.1 A comparison between the Level 2 model and the BSIM1 model in the CN20 process.

```
+vfb=-9.73820E-01, lvfb=3.67458E-01,wvfb=-4.72340E-02
+phi=7.46556E-01,lphi=-1.92454E-24, wphi=8.06093E-24
+k1=1.49134E+00,lk1=-4.98139E-01, wk1=2.78225E-01
+k2=3.15199E-01,lk2=-6.95350E-02,wk2=-1.40057E-01
+eta=-1.19300E-02, leta=5.44713E-02,weta=-2.67784E-02
+muz=5.98328E+02,dl=6.38067E-001,dw=1.35520E-001
+u0=5.27788E-02, lu0=4.85686E-02,wu0=-8.55329E-02
+u1=1.09730E-01, lu1=7.28376E-01,wu1=-4.22283E-01
+x2mz=7.18857E+00,lx2mz=-2.47335E+00, wx2mz=7.12327E+01
+x2e=-3.00000E-03,lx2e=-7.20276E-03,wx2e=-5.57093E-03
+x3e=3.71969E-04,lx3e=-3.16123E-03,wx3e=-3.80806E-03
+x2u0=1.30153E-03, lx2u0=3.81838E-04, wx2u0=2.53131E-02
+x2u1=-2.04836E-02, lx2u1=3.48053E-02, wx2u1=4.44747E-02
+mus=7.79064E+02, lmus=3.62270E+02,wmus=-2.71207E+02
+x2ms=-2.65485E+00, lx2ms=3.68637E+01, wx2ms=1.12899E+02
+x3ms=1.18139E+01, lx3ms=7.24951E+01,wx3ms=-5.25361E+01
+x3u1=2.12924E-02,lx3u1=5.85329E-02,wx3u1=-5.29634E-02
+tox=4.35000E-002, temp=2.70000E+01, vdd=5.00000E+00
+cgdo=3.79886E-010,cgso=3.79886E-010,cgbo=3.78415E-010
+xpart=1.00000E+000
+n0=1.00000E+000 ln0=0.00000E+000 wn0=0.00000E+000
+nb=0.00000E+000 lnb=0.00000E+000 wnb=0.00000E+000
+nd=0.00000E+000 lnd=0.00000E+000 wnd=0.00000E+000
+rsh=27.9   cj=1.037500e-04   cjsw=2.169400e-10  js=1.000000e-08   pb=0.8
+pbsw=0.8   mj=0.66036   mjsw=0.178543   wdf=0   dell=0

.MODEL CMOSN NMOS LEVEL=2 PHI=0.600000 TOX=4.3500E-08 XJ=0.200U TPG=1
+ VTO=0.8756 DELTA=8.5650E+00 LD=2.3950E-07 KP=4.5494E-05
+ UO=573.1 UEXP=1.5920E-01 UCRIT=5.9160E+04 RSH=1.0310E+01
+ GAMMA=0.4179 NSUB=3.3160E+15 NFS=8.1800E+12 VMAX=6.0280E+04
```

```
+ LAMBDA=2.9330E-02 CGDO=2.8518E-10 CGSO=2.8518E-10
+ CGBO=4.0921E-10 CJ=1.0375E-04 MJ=0.6604 CJSW=2.1694E-10
+ MJSW=0.178543 PB=0.800000
.probe
.DC Vds 0 5 .1 Vgs 1 5 1
.end
```

6.1 BSIM1 Model Parameters

All electrical parameters, P', used in the BSIM model are determined using

$$P' = P + \frac{LP}{L - DL} + \frac{WP}{W - DW} \qquad (6.1)$$

L and W are the drawn channel length and width, while DL and DW are the deltalength and deltawidth resulting from lateral diffusion and encroachment discussed earlier. The differences $L - DL$ and $W - DW$ are effective channel length, and widths L_{eff} and W_{eff}. P, LP, and WP are the electrical parameters associated with the electrical parameter P'.

Spice BSIM1 (level 4) parameters in SPICE3.

NAME, units	PARAMETER	W and L dependence
VFB, V	flat-band voltage	LVFB, WVFB
PHI, V	surface inversion potential	LPHI, WPHI
K1, $V^{1/2}$	body effect coefficient	LK1, WK1
K2, none	drain/source depletion charge-sharing coeff.	LK2, WK2
ETA, none	zero-bias drain-induced barrier-lowering coeff.	LETA, WETA
MUZ, cm^2/Vs	zero-bulk-bias mobility	
DL, µm	shortening of channel	
DW, µm	narrowing of channel	
U0, V^{-1}	zero-bias transverse-field mobility degrad. coeff.	LU0, WU0
U1, µm/V	zero-bias velocity saturation coefficient	LU1, WU1
X2MZ, cm^2/Vs	sens. of mobility to substrate bias at $V_{DS} = VDD$	LX2MZ, WX2MZ
X2E, V^{-1}	sens. of DIBL to substrate bias	LX2E, WX2E
X3E, V^{-1}	sens. of DIBL a drain bias at Vds=Vdd	LX3E, WX3E
X2U0, V^{-2}	sens. of trans. field mobility degrad. eff. subs. bias	LX2U0, WX2U0
X2U1, µm/V^2	sens. of velocity saturation effect to substrate bias	LX2U1, WX2U1
MUS, cm^2/Vs	mobility at zero substrate bias and at Vds=Vdd	LMUS, WMUS

X2MS, cm²/Vs	sens. of mobility to substrate bias at Vds=Vdd	LX2MS, WX2MS
X3MS, cm²/Vs	sens. of mobility to drain bias at Vds=Vdd	LX3MS, WX3MS
X3U1, μm/V²	sens. of vel. sat. eff. on drain bias at Vds=Vdd	LX3U1, WX3U1
TOX, μm	gate oxide thickness	
TEMP, C°	temperature at which parameters were measured	
VDD, V	measurement bias range	
CGDO, F/m	gate-drain overlap capacitance per meter channel width	
CGSO, F/m	gate-source overlap capacitance per meter channel width	
CGBO, F/m	gate-bulk overlap capacitance per meter channel length	
XPART	gate-oxide capacitance-charge model flag	
N0	zero-bias sub threshold slope coefficient	LN0, WN0
NB	sens. of sub threshold slope to substrate bias	LNB, WNB
ND	sens. of sub threshold slope to drain bias	LND, WND
RSH, Ω/sq.	drain and source diffusion sheet resistance	
JS, A/m²	source drain junction current density	
PB, V	built in potential of source drain junction	
MJ, none	grading coefficient of source drain junction	
PBSW, V	built in potential of source drain junction sidewall	
MJSW, none	grading coefficient of source drain junction sidewall	
CJ, F/m²	source drain junction capacitance per unit area	
CJSW, F/m	source drain junction sidewall capacitance per unit length	
WDF, m	source drain junction default width	
DELL, m	source drain junction length reduction	

XPART = 0 selects a 40/60 drain/source charge partition in saturation, while XPART=1 selects a 0/100 drain/source charge partition. ND, NG, and NS are the drain, gate, and source nodes, respectively. MNAME is the model name, AREA is the area factor, and OFF indicates an (optional) initial condition on the device for dc analysis. If the area factor is omitted, a value of 1.0 is assumed. The (optional) initial condition specification using I=VDS, VGS is intended for use with the UIC option on the .TRAN control line, when a transient analysis is desired starting from other than the quiescent operating point.

BSIM1 Model Parameters in SPICE2G6

The BSIM model parameters used in SPICE2G6 are slightly different from the BSIM model parameters used in SPICE3.   Many commercial suppliers of SPICE, including Metasoft's HSPICE, have based their models on SPICE2G6.   To change the BSIM models in the file C:\Lasi6\Wcn20\spice.inf from SPICE3 to SPICE2G6, change the following:

LMUS	to	LMS
WMUS	to	WMS
CJSW	to	CJW
JS	to	IJS
PB	to	PJ
PBSW	to	PJW
MJSW	to	MJW
dell	to	DL

Keeping in mind the order of the variables is important.   The BSIM1 model in SPICE3 is known as the level 4 model, whereas in HSPICE it is known as the level 13 model.

6.2 BSIM1 DC Equations

The following subsections give the equations and examples for determining the threshold voltage, drain current, and subthreshold drain current from the BSIM1 model parameters.

6.2.1 The Threshold Voltage

The threshold voltage is calculated in the BSIM1 model using

$$V_{THN} = \text{VFB}' + \text{PHI}' + \text{K1}' \cdot \sqrt{\text{PHI}' + V_{SB}} - \text{K2}' \cdot (\text{PHI}' + V_{SB}) - \text{ETADB}' \cdot V_{DS} \quad (6.2)$$

The voltages V_{SB} and V_{DS} are the substrate-to-source and drain-to-source voltages.   The model parameters are calculated using Eq. (6.1).   ETADB is given by

$$\text{ETADB}' = \text{ETA}' - \text{X2E}' \cdot V_{SB} + \text{X3E}' \cdot (V_{DS} - \text{VDD}) \quad (6.3)$$

It is not practical to calculate the threshold voltage by hand using Eq. (6.2).   For hand calculations, assuming the device we are modeling is large so that the second and third terms in Eq. (6.1) are negligible, a simple method of determining the threshold voltage is

$$V_{THN} = \text{VFB} + \text{PHI} + \text{K1} \cdot \sqrt{\text{PHI} + V_{SB}} - \text{K2} \cdot (\text{PHI} + V_{SB}) \quad (6.4)$$

We arrived at this simplification by neglecting the effect of the device sizes and the lowered drain-induced barrier resulting from nonzero V_{DS}.

Example 6.1
Compare the exact equation (6.2) for V_{THN} with the approximate equation (6.4) when the substrate potential is –5 V, the source is grounded, and the drain is at 2V.   Use the BSIM1 model parameters for the CN20 process with $W/L = 3/2$.

First, let's calculate the simple approximation for the threshold voltage using Eq. (6.4). This is given by

$$V_{THN} = -0.97 + 0.75 + 1.49 \cdot \sqrt{0.75 + 5} - 0.315 \cdot (0.75 + 5) = 1.54$$

Or a generic equation in terms of V_{SB} for this process is

$$V_{THN} = -0.22 + 1.49 \cdot \sqrt{0.75 + V_{SB}} - 0.315 \cdot (0.75 + V_{SB})$$

To calculate the threshold voltage used in the BSIM model, we calculate the primed variables in Eq. (6.2).

$$VFB' = VFB + \frac{LVFB}{L - DL} + \frac{WVFB}{W - DW} = -0.97 + \frac{0.37}{2 - 0.64} + \frac{-0.047}{3 - 0.14} = -0.713 \text{ V}$$

$$PHI' = PHI + \frac{LPHI}{L - DL} + \frac{WPHI}{W - DW} = 0.75 + \frac{0}{2 - 0.64} + \frac{0}{3 - 0.14} = 0.75 \text{ V}$$

$$K1' = K1 + \frac{LK1}{L - DL} + \frac{WK1}{W - DW} = 1.49 + \frac{-0.5}{2 - 0.64} + \frac{0.28}{3 - 0.14} = 1.212 \text{ V}^{\frac{1}{2}}$$

$$K2' = K2 + \frac{LK2}{L - DL} + \frac{WK2}{W - DW} = 0.32 + \frac{-0.07}{2 - 0.64} + \frac{-0.14}{3 - 0.14} = 0.224$$

$$ETA' = ETA + \frac{LETA}{L - DL} + \frac{WETA}{W - DW} = -0.012 + \frac{0.054}{2 - 0.64} + \frac{-0.027}{3 - 0.14} = 0.02$$

$$X2E' = X2E + \frac{LX2E}{L - DL} + \frac{WX2E}{W - DW} = -0.003 + \frac{-0.0072}{2 - 0.64} + \frac{-0.0056}{3 - 0.14} = -0.01$$

$$X3E' = X3E + \frac{LX3E}{L - DL} + \frac{WX3E}{W - DW} = -0.0004 + \frac{-0.003}{2 - 0.64} + \frac{-0.004}{3 - 0.14} = -0.003$$

$$ETADB' = 0.02 + (-0.01)(-5) + (-0.003)(2 - 5) = 0.079$$

Now we substitute these values into Eq. (6.2) to yield,

$$V_{THN} = -0.713 + 0.75 + 1.212 \cdot \sqrt{0.75 + 5} - 0.224 \cdot (0.75 + 5) - 0.079 \cdot 2 = 1.5 \text{ V}$$

We see a difference of 40 mV for this minimum-size device between the exact calculation (6.2) and the hand calculations using Eq. (6.4). The above results should also emphasize that considerably more work is required for performing the full calculations. *These results are not representative of submicron device sizes where the BSIM1 parameters may lose their physical significance.* They may behave like curve-fitting parameters. ■

We will use Eq (6.4) for threshold calculations in this book (see Appendix A). The next example illustrates how a potential difference between the source and substrate (or well) affects the threshold voltage.

Example 6.2
Using the CN20 BSIM1 model parameters, plot the drain current of a minimum-size MOSFET against gate-source voltage when V_{SB} varies from 0 to 5 V and the V_{DS} of the MOSFET is 5 V.

The plot and SPICE netlist are shown below. Notice that the threshold voltage increases as the substrate potential becomes more negative (the body effect). Also notice that the threshold voltage tends to change less as *VSB* is decreased. This fact is used to get better threshold matching in analog CMOS design. The accurate representation of the body effect shown in Fig. 6.2 should be compared to the poor representation given in Fig. 5.13. ■

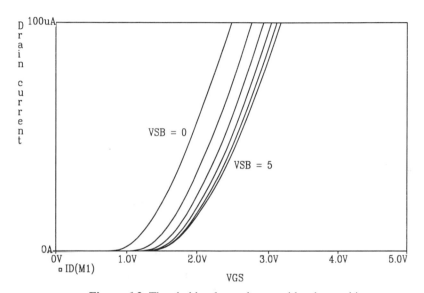

Figure 6.2 Threshold voltage change with substrate bias.

```
*** Top Level Netlist ***
M1_3u_2u        3 4 0 5 CMOSNB  L=2u W=3u
VBB      5 0    DC 0
Vdd      2 0    DC 5
Vgs      4 0    DC 0
VIMTR1          2 3 0V
***** Spice models and macro models *****
.MODEL CMOSNB NMOS LEVEL=4
+vfb=-9.73820E-01, lvfb=3.67458E-01,wvfb=-4.72340E-02
...same model parameters as before
+pbsw=0.8   mj=0.66036   mjsw=0.178543   wdf=0   dell=0
.probe
.DC Vgs 0 2 .05 Vbb 0 -5 -1
.end
```

6.2.2 The Drain Current

When $V_{GS} < V_{THN}$, the MOSFET is in the cutoff region and zero drain current (neglecting the subthreshold current) flows, that is, $I_D = 0$. When $V_{DS} < V_{GS} - V_{THN} > 0$, the MOSFET is in the triode or linear region. The drain current in this region is given by

$$I_{DS} = \frac{MU0'}{[1 + U0Z' \cdot (V_{GS} - V_{THN})]} \cdot \frac{C'_{ox} \cdot \frac{W-DW}{L-DL}}{\left(1 + \frac{U1Z'}{L-DL} \cdot V_{DS}\right)} \cdot \left[(V_{GS} - V_{THN})V_{DS} - \frac{a}{2}V_{DS}^2\right] \quad (6.5)$$

where C'_{ox}, in units of F/μm^2, is given by

$$C'_{ox} = \frac{\varepsilon_{ox}}{TOX} \text{ and } \varepsilon_{ox} = 35.1 \times 10^{-18} \text{ F/}\mu\text{m} \quad (6.6)$$

a is given by

$$a = 1 + \frac{g \cdot K1'}{2\sqrt{PHI' + V_{SB}}} \quad (6.7)$$

and g is given by

$$g = 1 - \frac{1}{1.744 + 0.8364(PHI' + V_{SB})} \quad (6.8)$$

The parameter $MU0'$ is obtained by a quadratic interpolation through three data points: $MU0(@\ V_{DS} = 0)$, $MU0(@\ V_{DS} = VDD)$, and the sensitivity of $MU0$ around the point $V_{DS} = VDD$, or

$$MU0(@V_{DS} = 0) = MUZ' - X2MZ' \cdot V_{SB} \quad (6.9)$$

and

$$MU0(@V_{DS} = VDD) = MUS' - X2MS' \cdot V_{SB} \quad (6.10)$$

The parameters $U0Z'$ and $U1Z'$ account for mobility degradation effects and are given by

$$U0Z' = U0' - X2U0' \cdot V_{SB} \quad (6.11)$$

and

$$U1Z' = U1' - X2U1' \cdot V_{SB} + X3U1' \cdot (V_{DS} - VDD) \quad (6.12)$$

Equation (6.5) is not suitable for hand calculations. We can use the following equation for describing the IV characteristics of a MOSFET in the linear region:

$$I_{DS} = \frac{MUZ \cdot C'_{ox} \cdot W}{L}\left[(V_{GS} - V_{THN}) \cdot V_{DS} - \frac{V_{DS}^2}{2}\right] \quad (6.13)$$

Here we use W and L as the drawn width and length to simplify hand calculations. It should also be pointed out that $MUZ \cdot C'_{ox}$ is approximately the parameter KP used in the level 1-3 models discussed in Ch. 5.

Example 6.3
Compare KP obtained using the BSIM1 model parameters (Eq. [6.13]) for n- and p- MOSFETs, with the value of KP given in the level 2 model.

From the SPICE listings given in the Appendix, we can determine the transconductance parameters. For the n-channel BSIM1, KP is given by

$$KP = MUZ \cdot C'_{ox} = (600 \ cm^2/V \cdot s) \cdot \frac{35.1 \times 10^{-18} F/\mu m}{0.0435 \mu m} \cdot \frac{10^8 \mu m^2}{cm^2} = 48.4 \ \mu A/V^2$$

This is compared with 45 $\mu A/V^2$. For the p-channel, the BSIM1 model gives a KP of 17 $\mu A/V^2$, while the level 2 model supplied gives 15 $\mu A/V^2$. ∎

The drain current of the MOSFET in the saturation region, $V_{DS} > V_{GS} - V_{THN}$, is calculated by

$$I_{DS} = \frac{MU0'}{[1 + U0Z' \cdot (V_{GS} - V_{THN})]} \cdot \frac{C_{ox} \cdot \frac{W-DW}{L-DL}}{2aK} \cdot (V_{GS} - V_{THN})^2 \tag{6.14}$$

where

$$K = \frac{1 + v_c + \sqrt{1 + 2v_c}}{2} \tag{6.15}$$

and

$$v_c = \frac{U1Z'}{L - DL} \cdot \frac{(V_{GS} - V_{THN})}{a} \tag{6.16}$$

Again, Eq. (6.14) is not very useful for hand calculations. The simple solution is

$$I_{DS} = \frac{MUZ \cdot C_{ox} \cdot W}{2 \cdot L} \cdot (V_{GS} - V_{THN})^2 = \frac{KP \cdot W}{2 \cdot L} (V_{GS} - V_{THN})^2 \tag{6.17}$$

This equation will work well for DC hand calculations when V_{DS} is close to $V_{GS} - V_{THN}$. However, this equation shows no V_{DS} dependence. Figure 6.1 showed an increase in drain current with increasing V_{DS} at a constant V_{GS}. We showed earlier that the drain current was dependent on the channel length modulation effect. The mobility in Eq. (6.14) is also dependent on V_{DS}, something we neglected earlier. We can write the mobility, $MU0$, neglecting length and width dependence by using a linear interpolation between the mobility at $V_{DS} = 0$ (MUZ) and the mobility at $V_{DS} = VDD$ (MUS) as

$$MU0 = MUZ + \frac{MUS - MUZ}{VDD} \cdot v_{DS} \tag{6.18}$$

The output resistance of the MOSFET is given by

$$r_o^{-1} = \frac{d}{dv_{DS}}\left(\frac{MU0 \cdot C_{ox} \cdot W}{2 \cdot L}(v_{GS} - V_{THN})^2\right) = \frac{dMU0}{dv_{DS}} \cdot \frac{C_{ox} \cdot W}{2 \cdot L}(v_{GS} - V_{THN})^2 \tag{6.19}$$

where the dependence of L on V_{DS} has been neglected. (Channel length modulation is discussed in Ch. 5.) The output resistance is then given by

$$r_o^{-1} = \left(\frac{MUS - MUZ}{VDD}\right)\frac{C_{ox}W}{2L}(v_{GS} - V_{THN})^2 = \left(\frac{MUS - MUZ}{VDD \cdot MUZ}\right)\frac{MUZ \cdot C_{ox}W}{2L}(v_{GS} - V_{THN})^2$$

$$\tag{6.20}$$

From Eq. (6.17) this is written as

$$r_o^{-1} = \left(\frac{MUS - MUZ}{VDD \cdot MUZ} \right) \cdot I_{DS} = \lambda_m \cdot I_{DS} \tag{6.21}$$

The parameter λ_m is called the mobility modulation parameter. The curve-fitted mobility at $V_{DS} = VDD$ can be less than the mobility at $V_{DS} = 0$. This mobility change has the effect of increasing or decreasing the output resistance of the MOSFET depending on the operating point. When $MUS < MUZ$, Eq. (6.21) gives a negative value for λ_m, which may be close to zero. In this case, we will assume $\lambda_m = 0$, and the channel length modulation will dominate the output resistance. Equation (6.17) may now be written as

$$I_{DS} = \frac{MUZ \cdot C_{ox} \cdot W}{2 \cdot L} \cdot (V_{GS} - V_{THN})^2 [1 + (\lambda_c + \lambda_m)(V_{DS} - V_{DS,sat})] \tag{6.22}$$

where λ_c is the channel length modulation parameter discussed earlier. This equation is very useful in analog applications where the output resistance is important. For digital applications we assume that both modulation parameters are zero.

For short channel lengths, the channel length modulation will dominate the output resistance, whereas for long channel lengths the mobility modulation parameter will dominate. It is also important to note that increasing the channel length will increase the output resistance of the MOSFET to a point. Appendix A shows, for several sizes of MOSFETs, the variation of r_o with V_{DS} and V_{GS}. Notice that r_o for an n-channel MOSFET can change by a factor of 4 in some cases. Hand calculations often give ballpark answers when compared with SPICE. This is understandable given the complexity of the modern MOS device.

Example 6.4
Estimate the λ_m obtained using the BSIM1 model parameters for both the n- and p-channel MOS transistors.

For the n-channel MOSFET BSIM gives $\lambda_m = \dfrac{779 - 598}{5 \cdot 598} = 0.061 \text{ V}^{-1}$ while for the p-channel $\lambda_m = \dfrac{206 - 211}{5 \cdot 211} = -0.005 \text{ V}^{-1}$ which we assume is zero.

Again, it should be emphasized that these equations are useful for hand calculations. They are heavily dependent on the measurement of the BSIM parameters. After hand calculations are performed, we can simulate the circuit using SPICE to get a more accurate representation of performance. ■

6.2.3 The Subthreshold Current

The current that flows when $V_{GS} < V_{THN}$ is called the subthreshold current. When the MOSFET is operating in this region, it is said to be in the weak inversion region. This current is due mainly to diffusion current between the drain and the source, similar to the bipolar junction transistor (BJT). The total drain current $I_{DS,tot}$ is the sum of the weak inversion component and the strong inversion component. From the previous subsection we know that the strong inversion component is zero when $V_{GS} < V_{THN}$. The weak inversion component is modeled by

$$I_{DS,weak} = \frac{I_{exp} \cdot I_{Limit}}{I_{Limit} + I_{exp}} \tag{6.23}$$

where

$$I_{exp} = MU0' \cdot C_{ox} \cdot \frac{W - DW}{L - DL} \left(\frac{kT}{q}\right)^2 e^{1.8} e^{q(V_{GS} - V_{THN})/N'kT} (1 - e^{-qV_{DS}/kT}) \tag{6.24}$$

and

$$I_{Limit} = \frac{MU0' \cdot C_{ox}}{2} \cdot \frac{W - DW}{L - DL} \left(\frac{3kT}{q}\right)^2 \tag{6.25}$$

The subthreshold slope parameter N' is given by

$$N' = N0' - NB' \cdot V_{SB} + ND' \cdot V_{DS} \tag{6.26}$$

An expression useful for hand calculations is

$$I_{exp} = MUZ \cdot C_{ox} \cdot \frac{W}{L} \left(\frac{kT}{q}\right)^2 e^{1.8} e^{q(V_{GS} - V_{THN})/N'kT} (1 - e^{-qV_{DS}/kT}) \tag{6.27}$$

When $V_{DS} > 2kT/q$ ($kT/q = 26$ mV at room temp), Eq. (6.24) can be written as

$$I_{exp} = I_{D0} \cdot \frac{W}{L} e^{q(V_{GS} - V_{THN})/N0 \cdot kT} \tag{6.28}$$

where

$$I_{D0} = MUZ \cdot C_{ox} \cdot \left(\frac{kT}{q}\right)^2 e^{1.8} = KP \cdot \left(\frac{kT}{q}\right)^2 e^{1.8} \tag{6.29}$$

Notice the similarity to the BJT.

Example 6.5
For a CN20 NMOS transistor with $W = L = 5$ μm, use SPICE to plot I_{DS} versus V_{DS} for V_{GS}, changing from 0.65 to 0.75 V. Comment on the usefulness of Eq. (6.28).

SPICE results are shown in Fig. 6.3. Notice that the curves flatten out when V_{DS} is approximately 50 mV. Although Eq. (6.28) eliminates the V_{DS} dependence on the drain current, in practice the drain current increases with increasing V_{DS}. ∎

The output resistance of a MOSFET operating in the subthreshold region (or the strong inversion region) can be characterized using the early voltage parameter, V_A. The output resistance is given by

$$r_o = \frac{V_A + V_{DS}}{I_D} \approx \frac{V_A}{I_D} \tag{6.30}$$

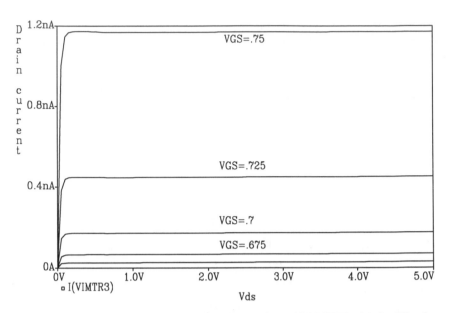

Figure 6.3 Subthreshold characteristics of an n-channel MOSFET with $L = W = 5\ \mu m$.

Figure 6.4 presents a graphical interpretation of this equation.

Subthreshold operation can be very useful for low-power operations. The main problem that plagues circuits designed to operate in the subthreshold region is matching. Since the drain current is exponentially related to the gate-source voltage, any mismatch in these voltages can cause significant differences in the drain current. Another problem relating to the matching of devices is the relative difference between the subthreshold slope parameter, N', of different devices.

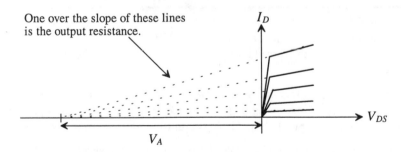

Figure 6.4 Early voltage and its relation to the output resistance of a MOSFET.

The subthreshold region is often characterized using the log I_D plotted against V_{GS} (see Fig. 6.5). We can write the drain current of the MOSFET in the subthreshold region, assuming $I_{exp} \ll I_{Limit}$, by

$$I_D = I_{D0} \cdot \frac{W}{L} \cdot e^{q(V_{GS}-V_{THN})/N' \cdot kT}$$ (6.31)

Taking the log of both sides, we get

$$\log I_D = \log \frac{W}{L} + \log I_{D0} + \frac{q}{kT \cdot N'}(-V_{THN}) \cdot \log e + \overbrace{\left[\frac{q}{kT \cdot N'} \cdot \log e\right]}^{\text{subthreshold slope}} \cdot V_{GS}$$ (6.32)

The reciprocal of the subthreshold slope is given by

$$\text{sub. slope}^{-1} = \frac{kT}{q} \cdot \frac{1}{\log e} \cdot N' = \frac{kT}{q} \cdot \frac{1}{\log e} \cdot (N0' + NB' \cdot V_{BS} + ND' \cdot V_{DS})$$ (6.33)

If $kT/q = 0.026$ V $= V_T$, and $N' = 1$, the reciprocal of the subthreshold slope is 60 mV/decade. From Appendix A, Table A.5, the subthreshold slope is typically 100 mV/decade. The slope parameter N' is used to adjust the model subthreshold slope so that it matches measured data, that is, measured data in the form of Fig. 6.5. Reviewing the BSIM models given in the Appendix reveals that the subthreshold region was not characterized when the models were extracted. The model uses the default 60 mV/decade, which is the result when $N0 = 1$, with the remaining eight subthreshold parameters equal to zero.

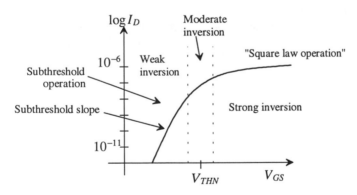

Figure 6.5 Drain current plotted from weak to strong inversion.

6.3 Short Channel MOSFETs

The CN20 process used throughout the text is useful in illustrating the fundamentals of CMOS design. However, modern CMOS transistors have channel lengths that are well below the 2 μm minimum length of the CN20 process. The gradual channel approximation used in the last chapter to develop the current-voltage characteristics of the MOSFET falls apart for modern short channel devices. The electric field under the gate oxide can no longer be treated in a single dimension. In addition, the velocity of the carriers drifting between the source and drain of the MOSFET saturates, an effect called carrier velocity saturation. The result is a reduction in mobility of the electron, μ_n, and thus an increase in the channel's effective sheet resistance [see Eq. (5.29)]. Since the mobility of the electron decreases with increasing temperature,[1] this effect is sometimes referred to as the hot-carrier effect. A cross section of a modern MOS device using a lightly doped drain (LDD) structure, used to reduce the effects of hot carriers, is shown in Fig. 6.6. This structure is formed by implanting a shallow n-, forming the spacer adjacent to the poly, and implanting the n+.

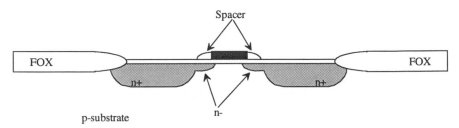

Figure 6.6 LDD MOSFET.

6.3.1 MOSFET Scaling

Reducing the channel length of a MOSFET can be described in terms of scaling theory [5]. A parameter S ($S < 1$) is used to scale the dimensions of a MOSFET. New technologies typically use an S in the neighborhood of 0.7. If a process uses VDD of 5 V, a scaled process would use a VDD' of 3.5 V. In other words

$$VDD' = VDD \cdot S \tag{6.34}$$

The channel length of the scaled process is reduced to

$$L' = L \cdot S \tag{6.35}$$

while the width is reduced to

$$W' = W \cdot S \tag{6.36}$$

Table 6.1 [5] describes how S affects the MOSFET parameters. The main benefits of scaling are (1) smaller device sizes and thus reduced chip size (increased yield and more parts per wafer), (2) lower gate delays, allowing higher frequency operation, and (3) reduction in power dissipation. Associated with these benefits are some unwanted side effects referred to as short channel effects. These unwanted effects are discussed in the next section.

Trends in MOSFET scaling are shown in Fig. 6.7[5]. Notice that the slopes of the curves in this figure are equal to the scaling parameter S.

---

[1] See Ch. 9 for a discussion of the temperature dependence of the mobility.

Parameter	Scaling
Supply voltage (VDD)	S
Channel length	S
Channel width (W)	S
Gate-oxide thickness (tox)	S
Substrate doping (NA)	S^{-1}
Drive current (ID)	S
Gate capacitance (Cox)	S
Gate delay	S
Active power	S^3

Table 6.1 Scaling relationships.

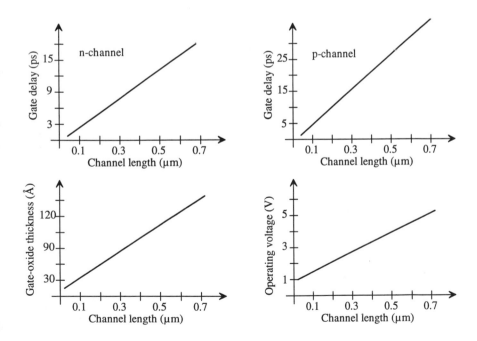

Figure 6.7 Trends in MOS device scaling.

6.3.2 Short Channel Effects

The average drift velocity, v, of an electron plotted against electric field, E^2, is shown in Fig. 6.8. When the electric field reaches a critical value, labeled E_{crit}, the velocity saturates at a value v_{sat}, that is, the velocity ceases to increase with increasing electric field. The ratio of electron drift velocity to applied electric field is the electron mobility, or

$$\mu_n = \frac{v}{E} \tag{6.37}$$

Above the critical electric field the mobility starts to decrease, whereas below E_{crit} the mobility is essentially constant. Rewriting Eq. (5.31) results in

$$I_D = \mu_n \cdot \frac{dV(y)}{dy} \cdot W \cdot C'_{ox}[V_{GS} - V_{THN} - V(y)] \tag{6.38}$$

We are interested in determining how the drain current of a short channel MOSFET changes with V_{GS} when operation lies in the saturation region. (The charge under the gate oxide at the drain channel interface is zero, and the channel is pinched off.) The MOSFET enters the saturation region when $V(L) = V_{DS,sat}$. At high electric fields, the mobility can be approximated by

$$\mu_n = \frac{v_{sat}}{E} = \frac{v_{sat}}{dV(y)/dy} \tag{6.39}$$

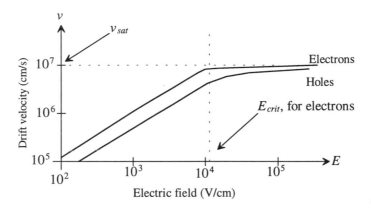

Figure 6.8 Drift velocity plotted against electric field.

---

[2] A simple estimate (ballpark value) of E in a MOSFET is simply V_{DS}/L. This value can be used to determine if short channel effects are present, that is, to determine if the carrier velocity saturates in Fig. 6.8.

so that Eq. (6.38) can be written as [6]

$$I_D = W \cdot v_{sat} \cdot C'_{ox}(V_{GS} - V_{THN} - V_{DS,sat}) \qquad (6.40)$$

This result is important! The drain current of a short channel MOSFET operating in the saturation region increases linearly with V_{GS}. The long channel theory, Eq. (5.39), showed the drain current increasing with the square of the gate-source voltage. This result also presents a practical relative figure of merit for the modern CMOS process, the drive current per width of a MOSFET. The drive current, I_{drive} ($\mu A/\mu m$), is given by

$$I_{drive} = v_{sat} \cdot C'_{ox}(V_{GS} - V_{THN} - V_{DS,sat}) \qquad (6.41)$$

and therefore

$$I_D = I_{drive} \cdot W \qquad (6.42)$$

The drive current can be estimated using these equations; however, it is normally measured with $V_{GS} = VDD$.

Hot Carriers

The low field mobility decreases with increasing temperature (again see Ch. 9). This effect can be used to describe velocity saturation, that is, assume an increase in temperature at the electric fields above E_{crit} . Carriers traveling at v_{sat} are therefore called hot carriers. Note that the drift velocity and temperature are not proportional. *Hot carriers* is simply a term used to describe mobility reduction and thus velocity saturation.

Hot carriers can tunnel through the gate oxide and cause gate current or become trapped in the gate oxide, having the effect of changing the threshold voltage. Hot carriers can also cause impact ionization (avalanche breakdown).

Oxide Breakdown

For reliable device operation, the maximum electric field across a device gate oxide should be limited to 7 MV/cm [7]. This translates into 0.7 V / 10 Å of gate oxide. A device with t_{ox} of 70 Å should limit the applied gate voltages to 4.9 V for reliable long-term operation.

Drain-Induced Barrier Lowering

Drain-induced barrier lowering (DIBL, pronounced dibble) causes a threshold voltage reduction with the application of a drain-source voltage. The positive potential at the drain terminal helps to attract electrons under the gate oxide and thus increase the surface potential ϕ_s. Since V_{THN} decreases with increasing V_{DS}, the result is an increase in drain current and thus a decrease in the MOSFET's output resistance (see Fig. 6.9).

Substrate Current-Induced Body Effect

Substrate current-induced body effect, (SCBE), is the result of hot carriers causing impact ionization and generating a substrate current. This occurs for electric fields

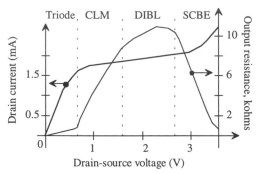

CLM - Channel length modulation dominates output resistance.

DIBL - Drain-induced barrier lowering dominates output resistance.

SCBE - Substrate current -induced body effect dominates output resistance.

Figure 6.9 Characteristics of a MOSFET with W/L = 10 µm / 0.43 µm and TOX = 75 Å [8].

greater than 10^5 V/cm. The substrate current will flow through the resistance of the substrate, increasing the substrate potential and further decreasing the threshold voltage. The result, again, is an increase in drain current and a decrease in the MOSFETs' output resistance.

6.4 The BSIM3 SPICE Model

The BSIM3v3 (BSIM3 version 3) SPICE model is the third generation of BSIM MOSFET model from Berkeley [4].[3] The BSIM3 model was developed for submicron devices. It has approximately 120 possible parameters, each having physical significance such as modeling of noise and temperature effects (which were not modeled in the BSIM1 model). The drain current and its first-order derivative are continuous over all regions of operation, which helps with simulation convergence problems.

Note that here we are using approximations to the actual equations used in the BSIM3v3 model in order to keep the introduction short. For an exact listing, see[4].

The Threshold Voltage Model

The threshold voltage is determined using

$$V_{THN} = V_{THN0} + K_1\left(\sqrt{\phi_s + V_{SB}} - \sqrt{\phi_s} \right) + K_2 V_{SB} + K_1\left(\sqrt{1 + \frac{N_{LX}}{L_{eff}}} - 1 \right)\sqrt{\phi_s} - \Delta V_{THN}$$

$$(6.43)$$

where V_{THN0} is the long channel threshold voltage, ϕ_s (= PHI) is the surface potential, ΔV_{THN} is the threshold voltage reduction due to short channel effects, and K_1, K_2 take into account vertical nonuniform doping effects. The threshold voltage reduction has an exponential dependence on channel length and is given by

$$\Delta V_{THN} = \theta(L) \cdot [2(V_{bi} - \phi_s) + V_{DS}] \qquad (6.44)$$

---

[3] This section is based directly on the information given in [4].

and

$$\theta(L) = D_{VTHN0}[e^{-(L \cdot DVT1)/2l_t} + 2e^{-(L \cdot DVT1)/l_t}] \tag{6.45}$$

where V_{bi} is the built-in potential given by

$$V_{bi} = \frac{kT}{q} \cdot \ln\left(\frac{N_{ch}N_d}{n_i^2}\right) \tag{6.46}$$

and l_t is the characteristic length given by

$$l_t = \sqrt{\frac{\varepsilon_{si}t_{ox}X_d}{\varepsilon_{ox}\eta}} \tag{6.47}$$

The depletion width between the channel and the substrate is X_d, while the average depletion width along the channel is X_d/η. The parameters D_{VTHN0}, $DVT1$, and η can be determined from experimental data.

The Drain Current Model

The operating regions in the BSIM3 model are divided into strong inversion ($V_{GS} > V_{THN}$), weak inversion ($V_{GS} < V_{THN}$), and a transition region ($V_{GS} \approx V_{THN}$). The strong inversion region is divided into the triode region ($V_{DS} < V_{DS,sat}$) and the saturation region ($V_{DS} > V_{DS,sat}$). The drain current in the saturation region is given by

$$I_D = W \cdot v_{sat} \cdot C'_{ox}(V_{GS} - V_{THN} - V_{DS,sat})\left[1 + \frac{V_{DS} - V_{THN} - V_{DS,sat}}{V_A}\right] \tag{6.48}$$

where V_A is the early voltage generated by the MOSFET output generated by the output resistance model. The drain current in the triode region is estimated using

$$I_D = W \cdot \mu_{eff} \cdot C'_{ox} \cdot \frac{W}{L} \cdot \frac{1}{1 + \frac{V_{DS}/L}{E_{crit}}} \left[\frac{V_{GS} - V_{THN} - V_{DS}}{2}\right] \cdot V_{DS} \tag{6.49}$$

where the effective mobility is estimated using

$$\mu_{eff} = 2 \cdot \frac{v_{sat}}{E_{crit}} \tag{6.50}$$

The drain current in the weak inversion or subthreshold region is estimated using

$$I_D = \mu_{eff} \cdot \frac{W}{L} \cdot \left[\frac{kT}{q}\right]^2 \cdot C_d \cdot \left[e^{(V_{GS} - V_{THN} - V_{off})q/nkT}\right][1 - e^{-qV_{DS}/kT}] \tag{6.51}$$

where V_{off} is an offset voltage and n is the subthreshold slope factor.

Model Examples

The BSIM3v3 model is implemented in SPICE3f5 as a level 8 MOSFET model. Metasoft's HSPICE implements their version of the BSIM3v3 model as a level 49 MOSFET model. Below are example model listings [9] for both an n-channel and a p-channel MOSFET model based on BSIM3v3.

```
*model = bsim3v3
*Berkeley Spice Compatibility
* Lmin= .35 Lmax= 20 Wmin= .6 Wmax= 20
.model N1 NMOS
+Level=      8
+Tnom=27.0
+Nch= 2.498E+17 Tox=9E-09 Xj=1.00000E-07
+Lint=9.36e-8 Wint=1.47e-7
+Vth0= .6322    K1= .756  K2= -3.83e-2  K3= -2.612
+Dvt0= 2.812 Dvt1= 0.462  Dvt2=-9.17e-2
+Nlx= 3.52291E-08  W0= 1.163e-6
+K3b= 2.233
+Vsat= 86301.58  Ua= 6.47e-9  Ub= 4.23e-18  Uc=-4.706281E-11
+Rdsw= 650  U0= 388.3203 wr=1
+A0= .3496967 Ags=.1   B0=0.546   B1= 1
+ Dwg = -6.0E-09 Dwb = -3.56E-09 Prwb = -.213
+Keta=-3.605872E-02 A1= 2.778747E-02 A2= .9
+Voff=-6.735529E-02 NFactor= 1.139926 Cit= 1.622527E-04
+Cdsc=-2.147181E-05
+Cdscb= 0  Dvt0w = 0 Dvt1w = 0 Dvt2w = 0
+ Cdscd = 0 Prwg = 0
+Eta0= 1.0281729E-02 Etab=-5.042203E-03
+Dsub= .31871233
+Pclm= 1.114846  Pdiblc1= 2.45357E-03  Pdiblc2= 6.406289E-03
+Drout= .31871233  Pscbe1= 5000000  Pscbe2= 5E-09 Pdiblcb = -.234
+Pvag= 0 delta=0.01
+ Wl =  0 Ww = -1.420242E-09 Wwl =  0
+ Wln =  0 Wwn = .2613948 Ll =  1.300902E-10
+ Lw =  0 Lwl =  0 Lln =  .316394
+ Lwn =  0
+kt1=-.3  kt2=-.051
+At= 22400
+Ute=-1.48
+Ua1= 3.31E-10 Ub1= 2.61E-19 Uc1= -3.42e-10
+Kt1l=0 Prt=764.3

.model P1 PMOS
+Level=      8
+Tnom=27.0
+Nch= 3.533024E+17  Tox=9E-09 Xj=1.00000E-07
+Lint=6.23e-8 Wint=1.22e-7
+Vth0=-.6732829 K1= .8362093 K2=-8.606622E-02 K3= 1.82
+Dvt0= 1.903801 Dvt1= .5333922 Dvt2=-.1862677
+Nlx= 1.28e-8 W0= 2.1e-6
+K3b= -0.24 Prwg=-0.001 Prwb=-0.323
+Vsat= 103503.2 Ua= 1.39995E-09 Ub= 1.e-19 Uc=-2.73e-11
+ Rdsw= 460 U0= 138.7609
+A0= .4716551 Ags=0.12
+Keta=-1.871516E-03 A1= .3417965 A2= 0.83
+Voff=-.074182 NFactor= 1.54389 Cit=-1.015667E-03
+Cdsc= 8.937517E-04
+Cdscb= 1.45e-4 Cdscd=1.04e-4
+ Dvt0w=0.232 Dvt1w=4.5e6 Dvt2w=-0.0023
+Eta0= 6.024776E-02 Etab=-4.64593E-03
```

```
+Dsub= .23222404
+Pclm= .989  Pdiblc1= 2.07418E-02  Pdiblc2= 1.33813E-3
+Drout= .3222404  Pscbe1= 118000  Pscbe2= 1E-09
+Pvag= 0
+kt1= -0.25  kt2= -0.032  prt=64.5
+At= 33000
+Ute= -1.5
+Ua1= 4.312e-9  Ub1= 6.65e-19  Uc1= 0  Kt1l=0
```

6.5 Convergence

A major frustration with using the BSIM SPICE model is the difficulty with convergence. *Assuming* the circuit contains no connection errors, there are basically three parameters that can be adjusted to help convergence. (Other methods of helping convergence for a particular type of analysis are discussed elsewhere in the text.) These three parameters are ABSTOL, VNTOL, and RELTOL [10].

ABSTOL is the absolute current tolerance. Its default value is 1 pA. This means that when a simulated circuit gets within 1 pA of its "actual" value, SPICE assumes that the current has converged and moves onto the next time step or AC/DC value. VNTOL is the node voltage tolerance, default value of 1 μV. RELTOL is the relative tolerance parameter, default value of 0.001 (0.1 percent). RELTOL is used to avoid problems with simulating large and small electrical values in the same circuit. For example, suppose the default value of RELTOL and VNTOL were used in a simulation where the actual node voltage is 1 V. The RELTOL parameter would signify an end to the simulation when the node voltage was within 1 mV of 1 V (1V·RELTOL), while the VNTOL parameter signifies an end when the node voltage is within 1 μV of 1 V. SPICE uses the larger of the two, in this case the RELTOL parameter results, to signify that the node has converged.

Increasing the value of these three parameters helps speed up the simulation and assists with convergence problems at the price of reduced accuracy. To help with convergence, the following statement can be added to a SPICE netlist:

.OPTIONS ABSTOL=1uA VNTOL=1mV RELTOL=0.01

To force convergence, these values can be increased to

.OPTIONS ABSTOL=1mA VNTOL=100mV RELTOL=0.1

The following is a list of common mistakes that can be made when learning to simulate circuits using SPICE.

1.   The first line in a SPICE netlist must be a comment line. SPICE ignores the first line in a netlist file.

2.   One megaohm is specified using 1MEG, not 1M, 1m, or 1 MEG.

3.   One farad is specified by 1, not 1f or 1F. 1F means one femto-farad or 10^{-15} farads.

4.   Areas of MOSFET drains and sources are, in most cases, specified in terms of picometers squared. An area measuring 6 µm on one side and 8 µm on the other side is specified by 48pm or 48E-12.

5.   Voltage source names should always be specified with a first letter of V. Current source names should always start with an I.

6.   Transient simulations display time data; that is, the x-axis in probe is time. A jagged plot such as a sinewave that looks like a triangle wave or is simply not smooth is the result of not specifying a maximum print step size (in the .tran statement) or specifying a maximum print step that is too large. For example, in order to display a 1 kHz sinewave in SPICE, a maximum print step of 10u (ten microeconds) can be specified.

7.   When displaying the results of an AC simulation where the x-axis is frequency, the probe cursor should be displaying either the magnitude or phase of the voltage or current. For example, a cursor that displays "voltage drop at a node" will sum the real and imaginary parts of the voltage at that node and display the somewhat meaningless result. The probe functions vary widely between simulators. (Some simulators are smart enough to start the user in the magnitude mode after performing an AC simulation.)

8.   MOSFET widths and lengths are specified using the letter "u" for microns. A common mistake is to forget to include this letter. For example, a minimum-size MOSFET in the CN20 process is specified by L=2u W=3u not L=2 W=3. The latter means a MOSFET with a 2 meter length and a 3 meter width!

9.   In general, the body connection of a p-channel MOSFET is connected to *VDD*, and the body connection of an n-channel MOSFET is connected to *VSS*. For example, in the CN20 process which is an n-well process, all n-channel bulk connections must be tied to *VSS*. This matter is easily checked in the SPICE netlist.

10.   Convergence in a DC sweep can often be helped by avoiding the power supply boundaries. For example, sweeping a circuit from 0 to 5 V may not converge, but sweeping from 0.1 to 4.9 will.

REFERENCES

[1]   J. R. Pierret, *A MOS Parameter Extraction Program for the BSIM Model*, Electronics Research Laboratory, University of California, Berkeley, Calif. 94720. Memorandum No. UCB/ERL M84/99, November 21, 1984.

[2]   G. Massobrio and P. Antognetti, *Semiconductor Device Modeling with SPICE*, 2nd ed., McGraw-Hill, 1993. ISBN 0-07-002469-3.

[3]   J. H. Huang, Z. H. Liu, M. C. Jeng, P. K. Ko, and C. Hu, *A Robust Physical Predictive Model for Deep-Submicron MOS Circuit Simulator*, Electronics

Research Laboratory, University of California, Berkeley, Calif. 94720. Memorandum No. UCB/ERL M93/57, July 21, 1993.

[4]   D. Foty, *MOSFET Modeling with SPICE: Principles and Practice,* Prentice-Hall, 1997. ISBN 0-13-227935-5.

[5]   M. Bohr, "MOS Transistors: Scaling and Performance Trends," *Semiconductor International,* pp. 75-79, June 1995.

[6]   K. Y. Toh, P. K. Ko, and R. G. Meyer, "An Engineering Model for Short-Channel MOS Devices," *IEEE Journal of Solid State Circuits,* Vol. 23, No. 4, August 1988.

[7]   C. Hu, *ULSI Device Scaling and Reliability,* Research and Development seminar at Micron Semiconductor, Boise, Idaho, 1995.

[8]   J. H. Huang, Z. H. Liu, M. C. Jeng, P. K. Ko, and C. Hu, *A Physical Model for MOSFET Output Resistance,* Electronics Research Laboratory, University of California, Berkeley, Calif. 94720. Memorandum No. UCB/ERL M93/56, July 21, 1993.

[9]   http://www-device.eecs.berkeley.edu/~bsim3/intro.html

[10]   R. Kielkowski, *Inside SPICE: Overcoming the Obstacles of Circuit Simulation,* McGraw-Hill, 1994. ISBN 0-07-911525-X.

PROBLEMS

6.1   Using the CN20 BSIM model parameters for the p-channel MOSFET, calculate, by hand, the zero-bias threshold voltage.

6.2   Calculate KP and C'_{ox} for both the n- and p-channel MOSFETs using the CN20 BSIM model parameters.

6.3   Using SPICE, plot I_D against V_{DS} (V_{SD}) for both the n- and p-channel MOSFETs ($L = 2$ μm and $W = 3$ μm) using the CN20 BSIM model parameters for V_{GS} (V_{SG}) = 1, 2, 3, 4, and 5 V. The SPICE models are located in the file spice.inf in the C:\Lasi6\Wcn20 directory.

6.4   Using SPICE and the BSIM model-generate Fig. 6.5 for an n-channel MOSFET with $L = W = 10$ μm. What is the subthreshold slope?

6.5   Calculate I_{drive} for a short channel MOSFET with $t_{ox} = 75$ Å, $V_{THN} = 0.5$ V, $V_{GS} = VDD = 2.5$ V and $V_{DS,sat} = 1.5$ V. Comment on why $V_{DS,sat}$ is not equal to $V_{GS} - V_{THN}$ for a short channel MOSFET (see Ch. 5).

6.6   Estimate the mobility of the electron and hole at 10^5 V/cm from Fig. 6.8.

6.7   Show that using the level 1 SPICE MOSFET model results in a small-signal output resistance of

$$r_o = \frac{1}{\lambda I_D} = \frac{L}{\lambda \cdot K}$$

where K is essentially constant for small changes in V_{GS}. This equation shows that the output resistance of a MOSFET is linearly dependent on the channel length for a constant drain current. Increasing the channel length of the MOSFET increases the output resistance. This result, from the discussion in this chapter, is *incorrect*. (The level 1 models neglect λ_m.) Consider the results of Problem. 8.

6.8   Using the results of an .OP (operating point analysis), plot the output resistance of a MOSFET versus channel length for the circuit of Fig. P6.8 using the BSIM model for channel lengths from 2 to 10 μm. The SPICE output file should specify the MOSFET small-signal parameters after the simulation has run. The MOSFET output resistance is the reciprocal of the parameter "gds" in this file.

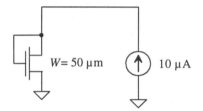

Figure P6.8

6.9   Calculate *KP* and C'_{ox} for both the n- and p-channel MOSFETs using the CMOS14TB BSIM model parameters.

6.10 Using SPICE, plot I_D against V_{DS} (V_{SD}) for both the n- and p-channel MOSFETs ($L = 0.6$ μm and $W = 0.9$ μm) using the CMOS14TB BSIM model parameters for V_{GS} (V_{SG}) = 1, 2, and 3 V. The SPICE models are located in the file spice.inf in the C:\Lasi6\Wcn20 directory.

6.11 Calculate I_{drive} for both the n- and p-channel MOSFETs in the CMOS14TB process.

6.12 Label Field Oxide, the spacer for LDD, gate, drain/source and lateral diffusion in the scanning electron microscope (SEM) picture of Fig. P6.12.

Figure P6.12

Chapter 7

CMOS Passive Elements

In this chapter we discuss the layout of the poly1-poly2 capacitor, as well as the temperature and voltage dependence of the resistor and capacitor, and finally we introduce noise analysis of resistor-capacitor circuits.

7.1 The Second Poly Layer (poly2)

Often, in a CMOS process, a second layer of polysilicon is added for MOSFET and capacitor formation. This poly2 layer follows similar design rules as the poly1 layer. The laycr for contact to poly2 is the "contact" layer ("cont" in the LASI CN20 setups). We can use poly2 for MOSFET formation instead of poly1. The electrical characteristics of the MOSFET made with poly2 are given in the CN20 parametric data available from MOSIS.

7.1.1 Design Rules for Capacitor Formation

The design rules specified here are derived from the Orbit design rules. The design rules for capacitor formation are as follows.

4.1 Width	3 μm
4.2 Space	3 μm
4.3 Poly1 overlap of poly2	2 μm
4.4 Space to active or well edge	2 μm
4.5 Space to contact	3 μm

These rules are illustrated in Fig. 7.1 along with an example cross-sectional view of the capacitor.

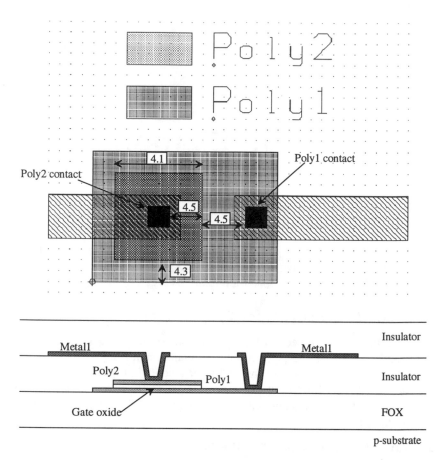

Figure 7.1 Design Rules for Poly1-Poly2 capacitor and associated cross-sectional view.

Example 7.1

Estimate the value of the capacitor shown in Fig. 7.1.

From the Orbit electrical parameters, the capacitance between poly1 and poly2 can vary from a minimum of 443 aF/μm^2 to 557 aF/μm^2. The amount of overlap of poly1 and poly2 shown in the figure is 64 μm^2. So the capacitance can range from 28.4 fF to 35.6 fF. ∎

7.1.2 Parasitics of the Poly Cap

In addition to the capacitance between poly1 and poly2, we also have a capacitance between poly1 and substrate, poly1 (or poly2) and metal1, and possibly metal2 to poly1, poly2, or metal1. These capacitances are shown in Fig. 7.2. When we are laying out capacitors, we should not waste much time trying to analyze each component, but should just try to get an idea for the approximate values.

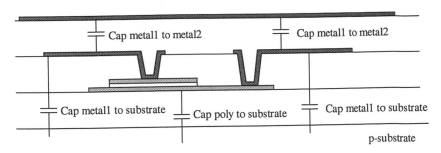

Figure 7.2 Parasitic capacitance associated with poly1-poly2 capacitor.

Another parasitic associated with the capacitor is the contact resistance, metal1 to poly, and the distributed resistance across the plate of the capacitor. To minimize the resistance, contacts should be made everywhere possible on poly. Since we only have access to poly1 on the end, we might want to contact the poly1 down the center or in a finger style similar to laying out a large MOSFET. An example layout is shown in Fig. 7.3. We know that with every contact to poly we have a contact resistance. We think of this as a resistor between metal1 and the plate of the capacitor. Adding another contact halves this resistance because the contact resistances are in parallel. Also, the sheet resistance of poly is on the order of 20 ohms/square. The plate of the capacitor is then

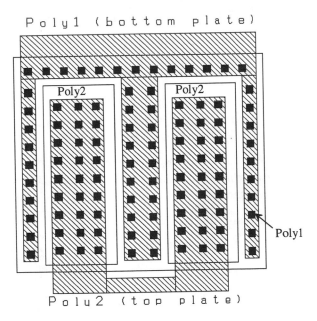

Figure 7.3 Layout of a 1 pF capacitor using multiple contacts and fingers to reduce the series resistance.

actually a distributed RC. The R is from the resistance of the poly, while the capacitance is the actual capacitance we are interested in.

7.1.3 Other Types of Capacitors

As we discussed earlier, we can make a capacitor using metal1 and metal2, a MOSFET with the drain and source shorted with $V_{GS} > V_{THN}$ (operating in the strong inversion region), and we can use the n+ or p+ diffusion regions to substrate or well.

7.2 Temperature and Voltage Dependence of Capacitors and Resistors

The resistor and capacitor in a CMOS process have a value that changes with temperature and voltage. The change is usually listed ppm/°C (parts per million per degree C). The ppm/°C is equivalent to a multiplier of $10^{-6}/°C$.

7.2.1 Resistors

The first-order temperature coefficient of a resistor, TCR, is given by

$$TCR = \frac{1}{R} \cdot \frac{dR}{dT} \tag{7.1}$$

Typical values of TCR for n-well, n+/p+ (active) and poly resistors are 10,000, 2,000, and 1,000ppm/°C respectively. SPICE uses a quadratic, in addition to this first order term, to model the temperature dependence of a resistor

$$R(T) = R_{T0} \cdot [1 + TCR \cdot (T - T_0) + TCR2 \cdot (T - T_0)^2] \tag{7.2}$$

where R_{T0} is the value of the resistor at T_0, typically measured at 27 °C. For hand calculations, we will assume $TCR2$ is 0.

Example 7.2
Estimate the minimum and maximum resistance of an n-well resistor with a length of 100 μm and a width of 10 μm over a temperature range of 0 to 100 °C. Assume that the TCR of the resistor is 10,000 ppm/°C.

The n-well resistor sheet resistance can vary from 2,000 to 3,000 Ω/square. The value of the resistor in this example can change from 20 k to 30 kΩ at 27 °C. The minimum resistance is then determined, using Eq. (7.2) by

$$R_{min} = 20,000 \cdot [1 + 0.01 \cdot (0 - 27)] = 14.6 \text{ k}\Omega$$

and the maximum resistance is determined by

$$R_{max} = 30,000 \cdot [1 + 0.01 \cdot (100 - 27)] = 51.90 \text{ k}\Omega \quad \blacksquare$$

The two resistors discussed in the previous example would correspond to resistors fabricated in two different wafers. Typical matching of n-well resistors on the same wafer is around 1 percent. For a 25 k n-well resistor we might expect another resistor on the same die with the same size to have a value between 24,750 and 25,250 ohms. Again, this is an approximation. Temperature, voltage across, and place on the die all have an effect. If the resistor is placed close to a power amplifier which is dissipating large amounts of power, the resistor will effectively be at a higher temperature than the same-size resistor on the other side of the die.

Another important contributor to a changing resistance is the voltage coefficient given by

$$VCR = \frac{1}{R} \cdot \frac{dR}{dV} \qquad (7.3)$$

where V is the average voltage applied to the resistor, the sum of the voltages on each end of the resistor divided by two. For an n-well resistor, this is the voltage difference between each side of the resistor and the potential of the n-well. The resistance as a function of voltage is then given by

$$R(V) = R_{V0} \cdot (1 + VCR \cdot V) \qquad (7.4)$$

The value R_{V0} is the value of the resistor at the temperature T_0, again typically 27 °C. A typical value of VCR is 200 ppm/V for an n-well resistor. The main contributor to the voltage coefficient is the depletion layer width between the n-well and the p-substrate. The depletion layer extends into the n-well, resulting in an effective increase in the sheet resistance. The thickness of the n-well available to conduct current decreases with increasing potential between the n-well and the substrate.

Example 7.3
Estimate the average resistance of an n-well resistor with a typical value of 50 k at 27 °C, for an average voltage across the resistor of 0, 5, and 10 V.

$$R(0) = 50,000 \cdot (1 + 0.0002 \cdot 0) = 50 \ k\Omega$$

$$R(5) = 50,000 \cdot (1 + 0.0002 \cdot 5) = 50.05 \ k\Omega$$

$$R(10) = 50,000 \cdot (1 + 0.0002 \cdot 10) = 50.10 \ k\Omega$$

This is a small change compared to the change due to temperature. However, as the next example will show, the change due to the applied voltage can have a greater effect on the circuit performance than the temperature. ■

Example 7.4
Compare the change in V_{out} in the following circuit due to VCR with the change due to TCR. Assume that the resistors are fabricated using the n-well.

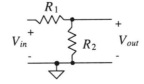

Figure Ex7.4

We start by writing the voltage divider equation and then substituting the temperature or voltage dependence. For the temperature dependence

$$V_{out} = V_{in} \cdot \frac{R2(T)}{R1(T) + R2(T)} = V_{in} \cdot \frac{R2 \cdot [1 + TCR(T - T_0)]}{(R1 + R2) \cdot [1 + TCR(T - T_0)]} = V_{in} \cdot \left[\frac{R2}{R1 + R2} \right]$$

and for the voltage dependence

$$V_{out} = V_{in} \cdot \frac{R2(V)}{R1(V) + R2(V)} = V_{in} \cdot \frac{R2 \cdot (1 + VCR \cdot V_2)}{R1 \cdot (1 + VCR \cdot V_1) + R2 \cdot (1 + VCR \cdot V_2)}$$

where $V_1 = \frac{Vin + Vout}{2}$ and $V_2 = \frac{Vout}{2}$ assuming the substrate is at ground potential. These results show that the voltage divider has no temperature dependence, to a first order, while it does have a voltage dependence. ∎

The preceding example should also show the advantage of ratioing components. For precision design over large temperature ranges, this fact is very important. Usually, a unit resistor is laid out with some nominal resistance. Figure 7.4a shows a unit cell resistor with a nominal resistor value of 20 kΩ. Figure 7.4b shows how the nominal 20 kΩ unit cell resistor is used to implement a nominally 100 kΩ resistor. The errors due to corners and differing perimeters are eliminated with this scheme when the ratio of the resistors is used in a circuit. Amplifier circuits, such as those using op-amps with feedback, are examples of circuits where ratioing is important.

Any precision circuit is susceptible to substrate noise. Substrate noise is the result of adjacent circuits injecting current into one another. The simplest method of

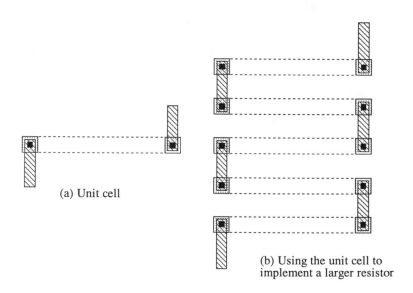

(a) Unit cell

(b) Using the unit cell to implement a larger resistor

Figure 7.4 Using a unit cell in resistor layout.

reducing substrate noise between adjacent circuits is to place a p+ implant (for a p-substrate wafer) between the two circuits that is tied to *VSS*. The diffusion removes the injected carriers and holds the substrate, ideally, at a fixed potential (*VSS*). Figure 7.5 shows the basic idea for a resistor. The p+ implant guards the circuit against minority carrier injection. Since the implants are laid out in a ring, they are often called *guard rings*.

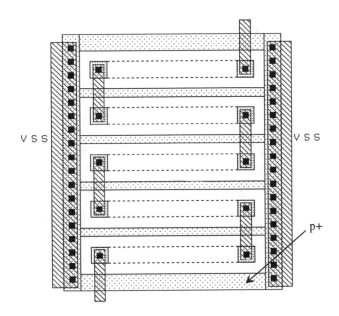

Figure 7.5 Layout of a resistor using guard rings.

Matching between two different resistors can be benefited by using the layout shown in Fig. 7.6. Process gradients, in this case changes in the n-well doping at different places on the die, are spread between the two resistors. Notice that the orientation of the resistors is consistent between unit cells; that is, all cells are laid out either horizontally or vertically. In a practical matched resistor layout, the unit cells in Fig. 7.6 would be surrounded by guard rings. Also, each resistor has essentially the same parasitics.

Common-Centroid Layout

Common-centroid (common center) layout can be used to help improve the matching between two resistors [2] (at the cost of uneven parasitics between the two elements). Consider the arrangement of unit resistors shown in Fig. 7.7a. This arrangement is sometimes called an interdigitated resistor and is the same form as that shown in Fig. 7.6. Furthermore, consider the effects of a linear varying sheet resistance on the overall value of each resistor. If we assign a normalized value to each unit resistor, as shown,

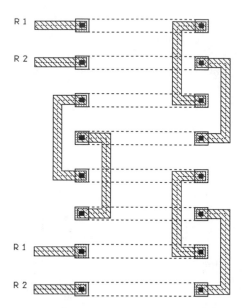

Figure 7.6 Layout of two matched resistors. Note that the parasitic capacitance from metal1 to substrate is essentially the same for each resistor.

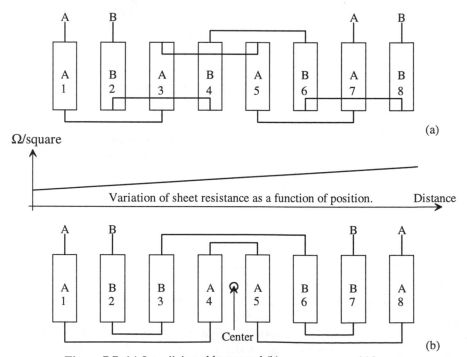

Figure 7.7 (a) Interdigitated layout and (b) common-centroid layout.

then resistor A has a value of 16 and resistor B has a value of 20. Ideally, the resistor values are equal.

Next consider the common-centroid layout shown in Fig. 7.7b, noting that resistors A and B share a common center. The value of either resistor in this figure is 18. In other words, the use of a common center (ABBAABBA) will give better matching than the interdigitized layout (ABABABAB). Figure 7.8 shows the common-centroid layout (two different possibilities) of four matched resistors. Note that common-centroid layout can also improve matching in MOSFETs or capacitors.

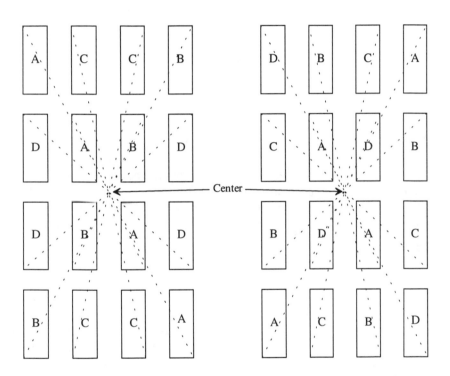

Figure 7.8 Common-centroid layout of four matched resistors (or elements).

Dummy Elements

Another technique useful in improving the matching between two or more elements is the use of dummy elements. Consider the cross-sectional views of n-well shown in Fig. 7.9a. The final amount of diffusion under the resist, on the edges, is different between the outer and the inner unit cells. This is the result of differing dopant concentrations at differing points on the surface (during the diffusion process). This difference will result in a mismatch between unit resistors.

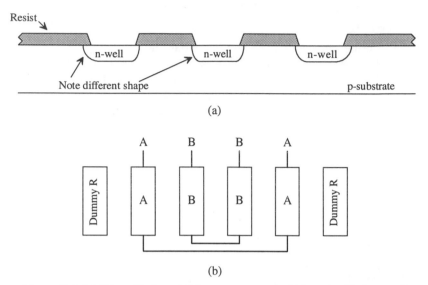

Figure 7.9 (a) Edge effects and (b) a common-centroid layout with dummy R.

To compensate for this effect, dummy elements can be added (see Fig. 7.9b) to an interdigitated or common-centroid layout. The dummy element does nothing electrically. It is simply used to ensure that the unit resistors used in matched resistors see the same adjacent structures.

7.2.2 Capacitors

The first-order temperature coefficient of a capacitor, TCC, is given by

$$TCC = \frac{1}{C} \cdot \frac{dC}{dT} \tag{7.5}$$

A typical value of TCC for a poly1-poly2 capacitor is 20 ppm/°C. Matching of poly1-poly2 capacitors on a die is typically better than 0.1 percent. The capacitance as a function of temperature is given by

$$C(T) = C_{T0} \cdot [1 + TCC \cdot (T-T_0)] \tag{7.6}$$

where C_{T0} is the capacitance at T_0. The voltage coefficient of a capacitor is given by

$$VCC = \frac{1}{C} \cdot \frac{dC}{dV} \tag{7.7}$$

The voltage coefficient of the poly capacitor is in the neighborhood of 10 ppm/V. The capacitance as a function of voltage is given by

$$C(V) = C_{V0} \cdot (1 + VCC \cdot V) \tag{7.8}$$

where C_{V0} is the capacitance between the two poly layers with zero applied voltage, and V is the voltage between the two plates.

The layout of capacitors also uses the unit cell concept. Figure 7.10 shows two nominally 100 fF unit cells (a practical minimum for the CN20 process). The unit cell of Fig. 7.10a allows contacts to poly to be placed directly on top of the thin (or gate) oxide (the oxide between the two poly plates of the capacitor). The bottom plate of the capacitor is made using poly1, while the top plate, the plate whose area determines the capacitance, is made using poly2. A circular disk is used for poly2 (see Fig. 7.10a), so that the amount of poly under etch is constant around the perimeter of the disk. The unit cell of Fig. 7.10b is used when the contact to poly2 must lie over the field region. Sharp corners are avoided in this layout as well. In both unit cells of Fig. 7.10, the contact to poly1 (the bottom plate) is not shown.

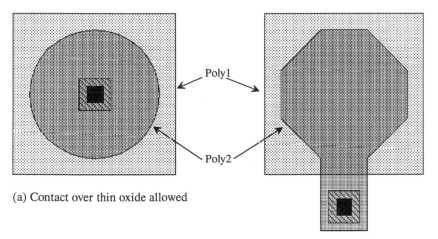

(a) Contact over thin oxide allowed

(b) Contact over thin oxide not allowed

Figure 7.10 Layout of a nominally 100 fF capacitor unit cell.

7.3 Noise in Resistors

Noise is a term used to describe an unwanted signal, voltage, or current present in a circuit. Noise in a resistor is primarily the result of random motion of electrons due to thermal effects. This type of noise is then termed *thermal noise*, or Johnson noise, after J. B. Johnson who performed experimental studies on thermal noise.

Thermal noise in a resistor can be characterized by adding a parallel current generator to the resistor. The root mean squared value of the current generator is given by

$$\sqrt{\overline{i^2}} = \sqrt{\frac{4kT}{R}} \cdot \sqrt{B} \qquad (7.9)$$

where

$$k = \text{Boltzmann's constant} = 1.38 \times 10^{-23} \text{ Watt} \cdot \text{sec/}^\circ\text{K}$$

$$T = \text{temperature in } ^\circ\text{K}$$

$$R = \text{resistance in } \Omega$$

$$B = \text{bandwidth in Hz}$$

Often we do not include the bandwidth over which we calculate the noise until the end of the noise calculation. In this case the units of the RMS noise current are $\text{A}/\sqrt{\text{Hz}}$. An alternative representation is given in terms of the RMS voltage, or

$$\sqrt{\overline{v^2}} = \sqrt{4kTR} = \sqrt{\overline{i^2}} \cdot R \qquad (7.10)$$

Addition of the RMS noise source is shown schematically in Fig. 7.11. Performing a noise analysis on a circuit begins with the addition of the RMS noise sources to the circuit. Then using superposition, we determine the RMS output noise due to each source. We then square each contribution to get the mean squared noise voltage from each noise source. Next, we sum the mean squared components to obtain the total mean squared noise. The total RMS output noise, $\sqrt{\overline{v_{on}^2}}$, is the square root of the integral of the total mean squared output noise, over the bandwidth of interest. The following examples should make the procedures for doing noise analysis clear.

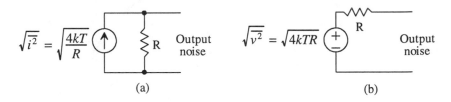

Figure 7.11 Noise sources (a) current and (b) voltage.

Example 7.5

For the circuit shown in Fig. Ex7.5, determine the RMS output and input noise over a bandwidth from DC to 1 kHz. Verify your answer with SPICE.

The RMS noise currents are given by

$$\sqrt{\overline{i_{10k}^2}} = \sqrt{\frac{4 \cdot (1.38 \times 10^{-23}) \cdot (300)}{10,000}} = 1.29 \times 10^{-12} \frac{\text{A}}{\sqrt{\text{Hz}}}$$

and

$$\sqrt{\overline{i_{1k}^2}} = 4.1 \times 10^{-12} \frac{\text{A}}{\sqrt{\text{Hz}}}$$

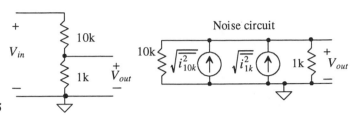

Figure Ex7.5

The root mean squared output noise voltage, $\sqrt{\overline{v_{10k}^2}}$ due to $\sqrt{\overline{i_{10k}^2}}$ is $1.29 \times 10^{-12} \frac{\text{A}}{\sqrt{\text{Hz}}} \cdot \frac{1\text{k} \cdot 10\text{k}}{1\text{k}+10\text{k}}$ or 1.2 nV/$\sqrt{\text{Hz}}$ and the mean squared output noise voltage, $\sqrt{\overline{v_{1k}^2}}$ from the 1k resistor is 3.7 nV/$\sqrt{\text{Hz}}$. The total mean squared output noise voltage over a bandwidth of 1 KHz is given by

$$\overline{v_{on}^2} = \int_{f_L}^{f_H} \overline{v_T^2} \cdot df = \int_0^{1\text{kHz}} (\overline{v_{10k}^2} + \overline{v_{1k}^2}) \cdot df = 15.1 \times 10^{-15} \text{ V}^2$$

Notice that no distinction was made between the noise voltage with and without the bandwidth introduced. The RMS output noise of this circuit is then

$$\sqrt{\overline{v_{on}^2}} = 123 \text{ nV}$$

This noise source voltage source can be added in series with the output of the noiseless circuit. The noise voltage gives the engineer an idea of the minimum voltage that can be measured in a circuit.

Another specification is the effective input noise at the two terminals that are connected to the input source. The effective input noise is a noise voltage in series with the input, which will generate the 123 nV of noise at the output. For this example,

$$\sqrt{\overline{v_{in}^2}} = 123nV \cdot \frac{10\text{k} + 1\text{k}}{1\text{k}} = 1.35 \text{ } \mu\text{V}$$

is the RMS input noise. SPICE3 simulation gives an output mean squared noise of 15.1×10^{-15} V^2 (= 123 nV RMS) and an input noise referenced to 1 V of 1.8×10^{-12} (= 1.35 μV RMS). The SPICE3 netlist is shown below.

```
*** top level netlist ***
.noise   v(2,0)  vin dec 100 1 1k
r1         1 2 10k
r2         2 0 1k
vin        1 0      dc 0 ac 1
.print noise all
.probe
.end
```

```
--------------------------------------------------------------------------------
Index   inoise_total        onoise_total
--------------------------------------------------------------------------------
0         1.821510e-012  1.505380e-014
```

Note that a bandwidth of 1 to 1,000 Hz was used in this simulation rather than DC to 1 kHz. Also note that the reference supply was a voltage, so the units of the SPICE output are in V^2. The input AC source has a magnitude of 1 V, so the input noise SPICE gives is divided by 1V squared. If we had used a 1 mV AC supply, then the inoise_total above would be 1.82×10^{-6} V^2.

The PSPICE netlist is as follows.

```
*** top level netlist ***
.noise v(2,0) vin
.ac dec 100 1 1k
r1 1 2 10k
r2 2 0 1k
vin 1 0 dc 0 ac 1
.probe
.end
```

The magnitude of the input source, in this case "vin" in the netlist, affects the output noise in SPICE3, while in PSPICE the magnitude has no effect on the output noise. The input and output noise voltage density, units $V/\sqrt{Hz}$, is specified in PSPICE by v(inoise) and v(onoise), respectively. The specification for input and output noise current density follows a similar form. Determining the total output noise in PROBE consists of determining the mean squared output voltage, specified by v(onoise)*v(onoise), summing (integrating) each mean squared output voltage over the simulation frequencies, s(v(onoise)*v(onoise)) and taking the square root of the result, sqrt(s(v(onoise)*v(onoise))). The results of the PSPICE simulation are shown in Fig. 7.12. ∎

In summary the procedure for doing a noise analysis is as follows.

1.   Add RMS noise sources to circuit.

2.   Determine RMS circuit output noise from each component using superposition.

3.   Square each component of the output noise to get mean squared component.

4.   Take the sum of each contribution.

For the output noise,

5.   Integrate the sum over the bandwidth.

6.   The square root of this result is the RMS output noise.

and for the input referred noise (noise is always measured on the output of a circuit),

5.   Divide the sum, step 4 above, by the squared magnitude of the network transfer function, that is, $|H(j\omega)|^2 = \left|\frac{V_o(j\omega)}{V_i(j\omega)}\right|^2$, assuming we are interested in noise voltages.

6.   Integrate the result over the bandwidth of interest.

7.   The square root of this result is the RMS input noise.

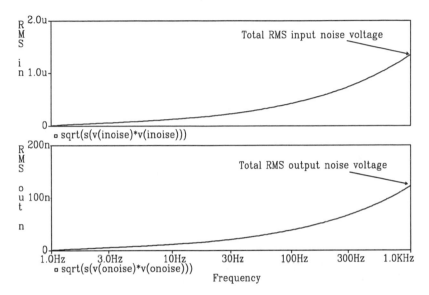

Figure 7.12 PSPICE simulation results for Example 7.5.

Example 7.6
Determine the RMS input and output noise voltages for the following RC circuit over bandwidths from DC to 10 MHz and 100 MHz. Verify your hand calculations with SPICE.

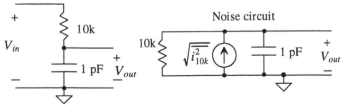

Figure Ex7.6

We begin with step 1, adding the noise sources to the circuit. We know from the previous example that the RMS thermal noise current from the 10 k resistor is

$$\sqrt{\overline{i_{10k}^2}} = 1.29 \times 10^{-12} \frac{A}{\sqrt{Hz}}$$

This generates an RMS noise voltage at the output given by

$$\sqrt{\overline{v_{10k}^2}} = \sqrt{\overline{i_{10k}^2}} \cdot \frac{R}{\sqrt{1 + (2\pi RCf)^2}}$$

which is the end of step 2 given above. Combining steps 3 and 4 gives

$$\overline{v_T^2} = \overline{v_{10k}^2} = \overline{i_{10k}^2} \cdot \frac{R^2}{1 + (2\pi RCf)^2}$$

To determine the output noise using step 5, we get,

$$\overline{v_{on}^2} = \int_{f_L}^{f_H} \overline{i_{10k}^2} \cdot \frac{R^2}{1 + (2\pi RCf)^2} \cdot df$$

This gives the mean squared output noise in units of V², which is what SPICE gives you as well. At this point, a useful relationship is

$$\int \frac{du}{a^2 + u^2} = \frac{1}{a}\tan^{-1}\frac{u}{a} + C \qquad (7.11)$$

Applying this to $\overline{v_{on}^2}$ we get

$$\overline{v_{on}^2} = \frac{\overline{i_{10k}^2} \cdot R^2}{2\pi RC} \cdot [\tan^{-1}(2\pi RC \cdot f_H) - \tan^{-1}(2\pi RC \cdot f_L)]$$

or putting in the numbers for this example with $f_L = 0$ gives

$$\overline{v_{on}^2} = \frac{1.66 \times 10^{-24} \cdot 10k}{2\pi \times 10^{-12}} \cdot \tan^{-1} 2\pi 10k \cdot 10^{-12} \cdot f_H$$

For a bandwidth of 10 MHz, the mean squared output noise voltage is $\overline{v_{on}^2} = 1.48$ nV², while for the bandwidth of 100 MHz $\overline{v_{on}^2} = 3.73$ nV². Both of these results were verified with SPICE. The RMS output noise voltages for the 10 MHz and 100 MHz bandwidths are then 38.5 µV and 61.07 µV, respectively.

To determine the input noise, we must first determine the transfer function, $H(j\omega)$ from step 5 of the input noise calculation, given by

$$H(j\omega) = \frac{1}{1 + j\omega RC} = \frac{V_o(j\omega)}{V_i(j\omega)}$$

Thus

$$\frac{\overline{v_T^2}}{|H(j\omega)|^2} = R^2 \cdot \overline{i_{10k}^2}$$

or a constant independent of frequency. Combining steps 6 and 7 for the input noise calculation gives

$$\sqrt{\overline{v_{in}^2}} = \sqrt{\int_{f_L}^{f_H} \frac{\overline{v_T^2}}{|H(j\omega)|^2} \cdot df} = \sqrt{\int_0^{f_H} R^2 \overline{i_{10k}^2} \cdot df} = R \cdot \sqrt{\overline{i_{10k}^2}} \cdot \sqrt{f_H}$$

The RMS input noise voltage over the bandwidths from DC to 10 MHz and 100 MHz are 40.8 µV and 129 µV, respectively. It is interesting to observe that at large f_H, compared to $\frac{1}{2\pi RC}$, the output noise does not increase, while the input noise continues at a rate determined by the resistor. This is because the capacitor shorts the high-frequency components of the output noise to ground. The following example illustrates the determination of the maximum output noise possible in an *RC* circuit with infinite bandwidth. ∎

Example 7.7
Estimate the output noise of the following *RC* low-pass circuit over the frequency range of DC to infinity.

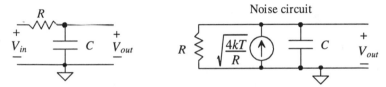

Figure Ex7.7

The total mean squared output noise is given by

$$\overline{v_T^2} = \frac{4kT}{R} \cdot \frac{R^2}{1 + (\omega RC)^2}$$

and the RMS output noise is given by

$$\sqrt{\overline{v_{on}^2}} = \sqrt{\int_0^\infty \overline{v_T^2} \cdot df} = \sqrt{\lim_{f_H \to \infty} \frac{4kT}{R} \cdot \frac{R^2}{2\pi RC} \cdot \tan^{-1}(2\pi RC \cdot f_H)}$$

$$= \sqrt{\frac{kT}{C}} \text{ V} \tag{7.12}$$

This maximum limitation on the output noise is, for obvious reasons, referred to as *kT/C* noise (pronounced "Kay Tee over Cee noise"). The maximum RMS output noise of the *RC* circuit in Ex. 7.6 is given by

$$\sqrt{\overline{v_{on,max}^2}} = \sqrt{\frac{kT}{C}} = \sqrt{\frac{(1.38 \times 10^{-23}) \cdot 300}{10^{-12}}} = 64 \text{ µV}$$

Notice that this maximum RMS output noise is close to what was obtained over the 100 MHz bandwidth, a bandwidth much greater than the 3 dB bandwidth of the circuit. ∎

REFERENCES

[1]   D. J. Allstot and W. C. Black, "Technology Design Considerations for Monolithic MOS Switched-Capacitor Filtering Systems," *Proceedings of the IEEE*, Vol. 71, No. 8, August 1983, pp. 967-986.

[2]   R. A. Pease, J. D. Bruce, H. W. Li, and R. J. Baker, "Comments on Analog Layout Using ALAS!" *IEEE Journal of Solid-State Circuits*, Vol. 31, No. 9, September 1996, pp. 1364-1365.

PROBLEMS

7.1   Lay out and DRC an *RC* circuit with a nominal time constant of 100 ns. Due to process variations, what are the maximum and minimum RC values possible?

7.2   For an n-well resistor fabricated in the CN20 process, plot resistance versus temperature for a 20 kΩ resistor at room temperature (27 °C) over –50 to 100 °C.

7.3   If a 1 pF capacitor at room temperature (27 °C) has a temperature coefficient of 100 ppm/°C, estimate the minimum and maximum capacitances from –40 to 100 °C. Plot the capacitor value versus temperature.

7.4   Lay out and DRC two 50 kΩ matched resistors. Include guard rings in your layout. Using the padframe cell, from Ch. 4, show the connection of the two resistors to four pads and the connection of *VSS* to a *VSS* pad.

7.5   Determine the RMS input and output noise voltages over a bandwidth from DC to 100 kHz for the following circuit.

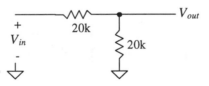

Figure P7.5

7.6   Repeat Problem 5 for the following circuit.

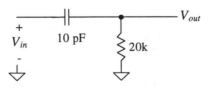

Figure P7.6

Chapter
8

Design Verification with LasiCkt

An important step in the design process is verifying that the layout of an integrated circuit matches the schematics of the integrated circuit. From either the layout or the schematic we should be able to generate a SPICE circuit (a netlist) file and simulate the operation of the IC. A comparison between the netlist and nodelists generated from the layout and schematic of a cell can be used to verify the schematic and layout match. This chapter discusses design verification using the LasiCkt program.

An example chip design, useful in illustrating schematic generation and documentation of layout, is given in the directory C:\Lasi6\W2uchip. It was designed using the MOSIS Scalable CMOS Design Rules (Appendix B) with a λ of 1 μm to give examples of how to generate layouts and schematics for use with LasiCkt. The MOSIS design rules use layers that are slightly different from the layers used with the CN20 setups of Appendix A. For example, to lay out a box of n+ in the CN20 process, we simply select the n+ layer and draw a box, Fig. 8.1a. However, to lay out a box of n+ using the MOSIS design rules, we begin by laying out a box on the "active" layer

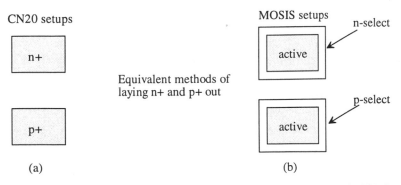

Figure 8.1 Comparison between the layouts of active area using the CN20 and MOSIS setups.

(defines openings in the FOX). To make this active area n+ we draw a box around this area on the n-select (nsel) layer, Fig. 8.1b, (defining an n+ implant). The n-select box must be at least 2 λ bigger than the active box; that is, any edge of active must be at least 2 λ away from any edge of n-select. Similarly, for p+, we draw a box on the active layer, using the MOSIS setups, and surround this box with p-select.

8.1 Fundamentals of LASICKT

The basic drawing of an n-channel MOSFET with connector nodes and labeling is shown in Fig. 8.2. This cell's name is NMOS_SCH and has a rank of 1. (Again the following drawings are available in the C:\Lasi6\W2uchip directory.) The actual symbol of the MOSFET was drawn on the schematic layer (layer 3) using a zero-width path or polygon.

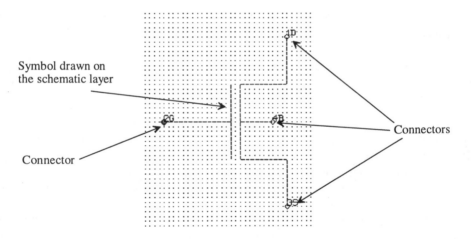

Figure 8.2 Schematic symbol of an n-channel MOSFET.

Specifying Connections to the Symbol

After drawing the symbol, the next step is to label the connections or points on the symbol that will be connected to the wires in the schematic drawing. Text on the connector layer (layer 5) is used to label the connections (sometimes also called pins) to the symbol. The vertex of the text specifies the exact location of the connector. Remember that displaying the vertex can be enabled or disabled by pressing "**t**" on the keyboard followed by executing a **Draw** command.

Layout of the MOSFET

Figure 8.3 shows the layout of two MOSFETs with cell names N3x2 and N15x5 (both with a rank of 1). Again, connector text (on layer 5) is used to specify the connections to the drain, gate, source, and substrate. Notice that the order of the connectors is important; that is, it must correspond to the order of nodes used in SPICE.

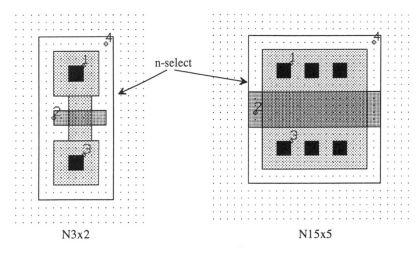

N3x2   N15x5

Figure 8.3 Layout with node connections of an n-channel MOSFET.

8.1.1 The Inverter

To help in understanding the process used to extract a circuit file from a schematic or layout, let's consider the design and layout of a basic inverter. Figure 8.4 shows the placement of NMOS_SCH and PMOS_SCH cells into cell INVERT_SCH (rank 2). Notice that an outline of the cell, enabled by pressing "**i**" on the keyboard, is used to show the actual boundaries of the cells. The next step in drawing the schematic of the inverter is to label the device and parameter information. We will label the NMOS M1 and the PMOS M2 using the device text layer (layer 6), as shown in Fig. 8.5. The layouts in Fig. 8.3 correspond to NMOS devices with width to length ratios of 3/2 and 15/5. Since we have a single symbol for an NMOS transistor, the size of the MOSFET

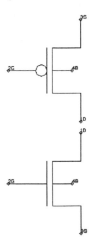

Figure 8.4 Starting the schematic of an inverter.

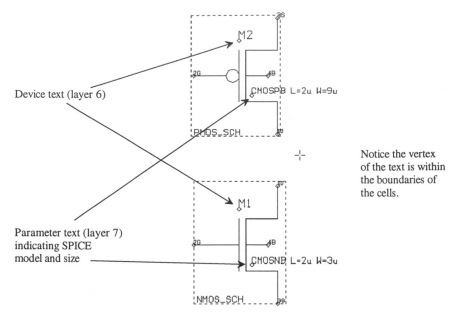

Device text (layer 6)

Notice the vertex
of the text is within
the boundaries of
the cells.

Parameter text (layer 7)
indicating SPICE
model and size

Figure 8.5 Adding device and parameter text to a schematic.

and the model used are specified on the parameter text layers (layer 7). Notice that the vertex of both the parameter (part) text and device text must be placed within the boundaries of the cell.

Labeling the Layout

The layout corresponding to the schematic of Fig. 8.5 is shown in Fig. 8.6. A cell named INVERT with a rank of 2 was created. The cells P9X2, SFRAME, and N3X2 were placed into the INVERT cell. Device and parameter text were used to specify the device name (M1 and M2) and the parameter information (SPICE model, widths, and lengths). It is important that the device name on the schematic correspond directly to a particular device in the layout.

Making Connections

Wires, for use in schematics only, can be drawn on any layer. For the schematics we draw here we will use metal 1 (layer 49). Zero-width paths (which we will call wires) are used to make contact to the device connectors (text on layer 5). We can also label the device connectors directly with a name using node text (layer 4). Figure 8.7a shows how we can label the substrate connection of the NMOS with node 0 (spice ground). A connection is made when a wire crosses the vertex of text written on the connector text layer (layer 5). Figure 8.7b shows a connection between wires. In this figure, the vertex of the wires must coincide for a connection to be made. This enables wires to cross without making a connection. Note that the **fGet** command (around a vertex) was used

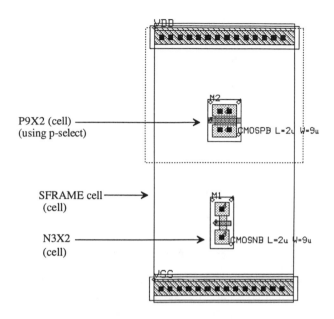

Figure 8.6 Placing the MOSFET cells in the standard cell frame and labeling.

to show the vertices of the wires in this figure. Figure 8.8 shows the schematic of the inverter with wire connections in place and a ground schematic cell added.

The layout of the inverter with interconnections is shown in Fig. 8.9. Poly1, metal1, or metal2 can be used to connect the terminals of the two MOSFETs to *VDD*, ground, the input, and the output. The layer numbers used to specify the interconnect layers, using the MOSIS setups, are 46 (poly1), 49(metal1), and 51(metal2). We might now ask the question, "How does LASICKT know which connections are the inputs,

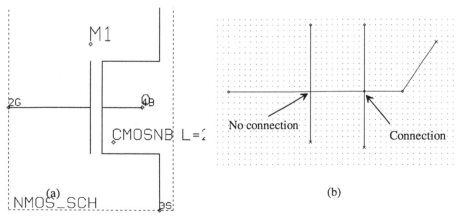

Figure 8.7 How wires make connections.

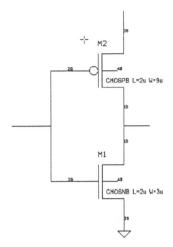

Figure 8.8 Inverter with wire connections.

outputs, and power supplies?" The answer is that we label nodes in the layout and schematic that tell LasiCkt which nodes are the inputs and which are the outputs.

Labeling the Nodes

The next step in creating a schematic or layout is to label the nodes that correspond to inputs/outputs and power supply connections. Figure 8.10 shows that the power supply node is labeled, on the node text layer (layer 4), *VDD*, the input is labeled A, the output

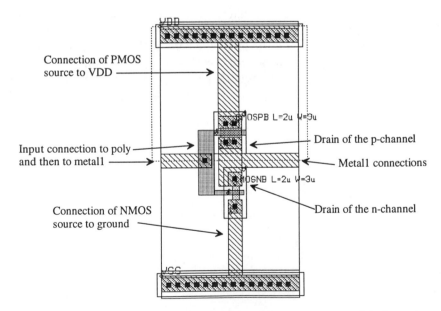

Figure 8.9 Adding connections (poly and metal1) to the layout of the inverter.

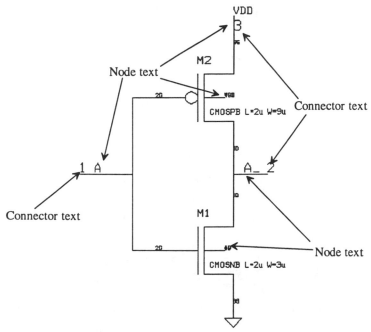

Figure 8.10 Adding node and connector text to the schematic of the inverter.

is labeled A_ and the ground connection is labeled 0 via the ground connector. (Ground is always node 0 in a SPICE circuit file.) It is important to label the input/output nodes, not letting LasiCkt select a name, so that we know how to connect the input voltage or current sources when writing the header file.

In this schematic, we have also labeled the input, output, and *VDD* nodes with connectors (on the connector text layer). Ground was not labeled with a connector since node 0 is universally ground potential. Adding connectors to the inverter allows us to use the inverter as a subcircuit in a higher ranking cell.

The layout of the inverter with labeled nodes and connectors is shown in Fig. 8.11. Notice the direct correlation between the layout and schematic. Both the schematic and layout of the inverter are used in higher ranking cells in the example chip given in the directory C:\Lasi6\W2uchip.

From this layout we can make the following observations.

1. The vertex of the node text must be located on a box or path used for interconnection within a cell. It will have no effect if located on a lower ranking cell. For example, the vertex of the *VDD* node text in Fig. 8.11 must be on the metal1 box connecting the source of the p-channel with the SFRAME (standard frame) cell. Placing it on the metal1 of the underlying SFRAME cell will not label the node.

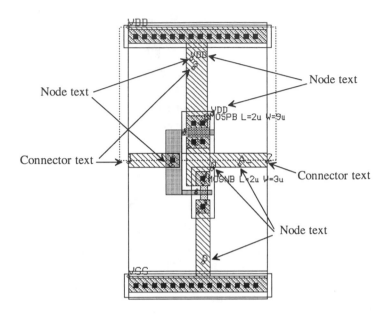

Figure 8.11 Adding node text and connector text to the layout of an inverter.

2. Body connections, substrate or well, must be labeled with a node name separately from the other nodes in the circuit.

3. The vertex of node A in Fig. 8.11 is placed at the intersection of poly1 and metal1. Poly1, metal1, and metal2 are the interconnect layers in a layout. These layers can cross without making contact. Were the layers are connected, using a contact or via, the node text is used to indicate the connection.

4. The vertex of connector text may be placed anywhere in a cell where connection is to be made to the next higher cell. This normally will be where interconnect layers join to the next higher ranked cell.

5. Since connectors are used to specify how cells are connected upward in the cell hierarchy, connectors on a top cell are unnecessary. In fact, they will be flagged as "open connector" errors when LasiCkt compiles. If the cell INVERT was the top cell, there would be no connector text placed on it.

6. There is no node or connector text on the SFRAME cell because it is not an active device. Unless it has an actual SPICE model, it will simply be ignored by LASICKT (unless check for floating cells is enabled).

Adding the additional text to label parameters, device type, nodes, and connectors can clutter a layout or schematic to the point that it becomes unreadable. Using the **View** command button in LASI can limit the number of layers displayed in a drawing.

8.1.2 Running LASICKT

To launch the LasiCkt program, return to the system menu and click on the LasiCkt command button. The screen shown in Fig. 8.12 will appear after selecting the **Setup** button. Filling in the setup information described below, pressing **OK** and then the **Go** button will start **LasiCkt**. If the schematic and layout are drawn using the methods just discussed, LasiCkt will generate SPICE netlist files. The **Comp** command button can be used to compare the schematic and layout nodelists (*.NOD files generated with LasiCkt) to test agreement between the schematic and layout. The SPICE outputs resulting from the simulation of each circuit file can be compared as well.

LASICKT Inputs

Name of Cell - LASICKT will compile a circuit file based on the contents of the cell specified in this field.

Header File - Name of file that LASICKT will insert at the beginning of the circuit file. This file generally contains information such as power supply voltages and input sources.

Footer File - Name of file that LASICKT will append to the end of the circuit file. Generally, this file contains the SPICE models and can be common to all cells in the drawing directory.

Interconnect Layers - Layers used to connect the cells together. For the MOSIS setups used in C:\Lasi6\W2uchip layer 46 (poly1), layer 49 (metal1), layer 51 (metal2),

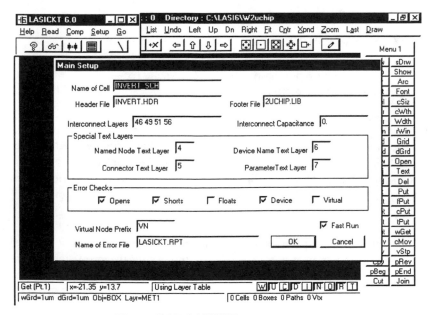

Figure 8.12 LASICKT system screen.

and Layer 56 (poly2) are used. Other layers such as n+ or p+ can be used for connections with the appropriate labeling in the layout.

Interconnect Capacitance - A value that can be used to estimate the capacitance of interconnects based on the area of the interconnection.

Additional information covering the operation or meaning of the inputs used in LasiCkt can be found in the on-line help available by pressing **Help** while at the LasiCkt command menu.

The Header File

The following header files are provided for simulating the operation of the circuits provided in C:\Lasi6\W2uchip.

INVERT	Simple inverter.
COMP1	Comparator with amp and a single buffer output.
COMP2	Comparator with latch and double output buffers.
DFF	D-type flipflop.
NAND	Simple 2-input NAND gate.
OR	Simple 2-input OR gate (NOR+INVERT).
SRFF	SR type flipflop.
TRANGATE	Transmission gate.
WSOTA	Wide-swing operational transconductance amplifier.

The contents of INVERT.HDR are shown below:

```
V1  VDD 0  DC 5V AC 0 0
V2  A 0  DC 0 AC 0 0 PULSE (0 5V 10n 1ns 1ns 50ns 100ns)
.options reltol=0.1 abstol=1u vntol=50mv
.probe
.tran 1ns 150ns
```

The first line in this file specifies the DC voltage source used by the circuit. Note that *VDD* is a node name that corresponds to the labeled nodes in the inverter layout or schematic. The second line is the input pulse to the inverter, discussed in more detail below. The .options statement is used to help with convergence.

.tran 1ns 150ns

This statement is used to specify a transient analysis (the x-axis in probe is time) from 0 to 150ns with a maximum print step of 1ns. Increasing the print step can cause the output to become jagged. Small-print step size can result in a very large output data file (the file used by probe). The full specification for a transient analysis is given by

.tran (print-step) (stop-time) (delay-time) (maximum step size) (UIC)

The delay-time is the time when the data starts being saved to a file. If we wanted to simulate the operation of a circuit from 0 to 100ns but we were only interested in the results from 50ns to 100ns we could use: .tran 1ns 100ns 50ns and only the data from 50 to 100 ns would be saved in the probe data file. The maximum step size will limit the next incremental increase in time as SPICE is simulating the operation of a circuit (and can help with smoothing of SPICE output). Specifying UIC at the end of a .tran statement causes SPICE to use the initial conditions specified in the circuit, for example, the initial voltage across a capacitor.

Pulse Statements

General form:
PULSE(VI V2 TD TR TF PW PER)

Examples:
VIN 3 0 PULSE(-1 1 2n 2n 2n 50n l00n)

PARAMETER	DEFAULT	UNITS
V1 (initial value)		Volts or amps
V2 (pulsed value)		Volts or amps
TD (delay-time)	0.0	Seconds
TR (rise-time)	TSTEP	Seconds
TF (fall-time)	TSTEP	Seconds
PW (pulse width)	TSTOP	Seconds
PER (period)	TSTOP	Seconds

Piecewise Linear Source

General Form:
PWL (T1 V1 <T2 V2 T3 V3 T4 V4 ...>)

Examples:
VCLOCK 7 5 PWL (0 -7 10NS -7 11NS -3 17NS -3 18NS -7 50NS -7)

Extracting a Circuit File from the Inverter Schematic

Pressing **Go** on the LasiCkt system screen shown in Fig. 8.12 will generate a circuit file, shown below, in the file INVERT_SCH.CIR. A circuit file generated from the cell INVERT and placed into the text file INVERT.CIR results in an identical circuit file to INVERT_SCH.CIR.

```
*** SPICE Circuit File of $INVERT***
* START OF INVERT.HDR
V1  VDD 0  DC 5V AC 0 0
V2  A 0  DC 0 AC 0 0 PULSE (0 5V 10n 1ns 1ns 50ns 100ns)
.options reltol=0.1 abstol=1u vntol=50mv
.probe
.tran 1ns 150ns
.plot tran all
.print tran all
* END OF INVERT.HDR
```

```
* MAIN CIRCUIT
M1 A_ A 0 0 CMOSNB L=2u W=3u
M2 A_ A VDD VDD CMOSPB L=2u W=9u

SPICE models not shown

.end
```

SPICE Simulation Results

The simulation results shown in Fig. 8.13 were generated using the circuit file given above and SPICE3. Since the SPICE circuit file generated using the schematic of the inverter and the layout are the same, the SPICE outputs are the same.

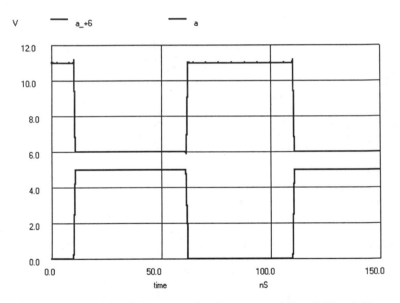

Figure 8.13 Simulation output for the inverter of Figs. 8.10 or 8.11.

8.1.3 Higher-ranking Cells; The OR Gate

The schematic and layout of the inverter can be used in higher ranking cells. To illustrate this, consider the implementaion of an OR gate made using a NOR gate and an inverter. The schematic and layout of the NOR gate with rank 2 (named NOR_SCH and NOR in C:\Lasi6\W2uchip) is shown in Fig. 8.14. These cells were created following the same procedures used to create the inverter cells discussed earlier. SPICE circuit files can be compiled from these cells, keeping in mind that the header file must be changed to NOR.HDR, and the operation of the cells can be simulated.

Creating an OR gate schematic begins with creating a cell, in LASI, named OR_SCH with a rank of three. The INVERT_SCH and NOR_SCH cells are placed into

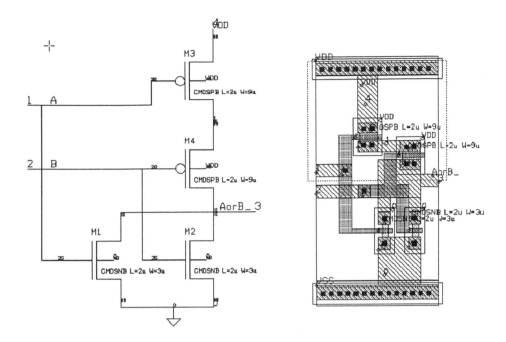

Figure 8.14 Schematic and layout of a NOR gate.

this schematic (Fig. 8.15). Also shown in this figure are the added node and connector names. Figure 8.16 shows this OR gate cell with the NOR and NAND cells drawn as outlines. This figure reveals the location of the text and wires added to the cell. Let's

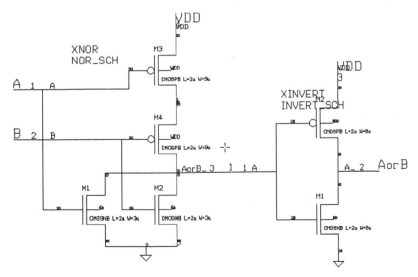

Figure 8.15 Schematic of the OR gate.

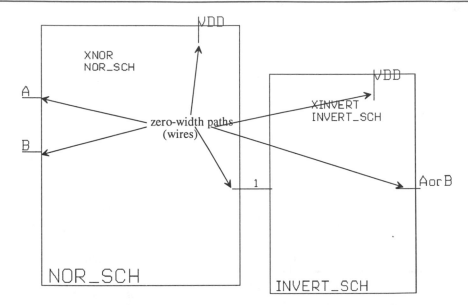

Figure 8.16 The added zero-width paths and text to the OR gate cell.

discuss the labeling shown in this figure, keeping in mind that the first step in generating the OR gate schematic and layout was the addition of the lower ranking cells.

Labeling Devices and Parameters

The "devices" in this cell are labeled XNOR and XINVERT. Since the NOR and the INVERT cells are implemented in SPICE as subcircuits, the first letter in their name must be an X. This is similar to the MOSFET specification that the first letter, when labeling a MOSFET, has to be an M. The vertex of the parameter text is placed within the outline of the cell. The param text is a label telling LASICKT the name of the cell, for this case NOR_SCH and INVERT_SCH.

Labeling Nodes

The most important thing to remember, before labeling any of the nodes in a circuit, is that the vertex of the node text must lie on a box, polygon, or path. *Node text placed on an underlying cell will have no effect unless it is placed exactly on a connector.*

To connect *VDD* to the NOR_SCH and INVERT_SCH cells, we begin by placing a zero-width path on the schematic layer (layer 3) over the vertex of the power connector text. For the NOR_SCH cell schematic, the power supply connection is node 4 on the connector text layer. For the INVERT_SCH cell, the power supply connection is node 3 on the connector text layer. The next step is to place node name text, layer 4, on the added zero-width paths on the schematic layer. As shown in Fig. 8.15, it appears that the source of M3 in the NOR gate cell is labeled twice as *VDD*. This is true. For each level of the schematic in the cell hierarchy, the *VDD* nodes must be labeled.

The next step in the generation of the schematic is to connect the cells together with, and label nodes on, zero-width paths. In this schematic, we have labeled the node connecting the NOR_SCH cell to the INVERT_SCH cell "1." We don't have to label this node. LASICKT will assign a virtual node number if nodetext is not placed on the zero-width wire. Note that the inputs to OR_SCH are labeled A and B, while the output is labeled AorB. Also note that we did not label inputs, output, and power supply with connector text. If we want to use the OR_SCH gate schematic in a higher ranking cell, the connector text must be added.

Labeling the Layout

The layout of the OR gate, with the cells drawn in outline mode, is shown in Fig. 8.17. Notice the added metal1 boxes in the layout. These boxes are placed over the connector text in the lower ranking cells. Note that node 1, in this figure, was not placed on a box; it was placed directly on the connector vertex of both lower ranking cells.

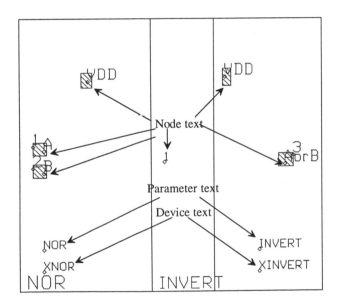

Figure 8.17 Showing the labeling of the OR cell layout.

Simulating the Operation of the OR Gate

LASICKT was used to generate the SPICE circuit file, shown below, using OR.HDR for the header file. The simulation results, using this circuit file, are shown in Fig. 8.18.

```
* START OF OR.HDR
V1  VDD 0  DC 5V AC 0 0
V2  A 0  DC 0 AC 0 0 PULSE (0 5V 5ns 1ns 1ns 50ns 100ns)
V3  B 0  DC 0 AC 0 0 PULSE (0 5V 10ns 1ns 1ns 100ns 200ns)
.options reltol=0.1 abstol=10u vntol=10mv
```

```
.probe
.tran 1ns 200ns
.plot tran all
.print tran all
* END OF OR.HDR

.SUBCKT INVERT A A_ VDD
M1 A_ A 0 0 CMOSNB L=2u W=3u
M2 A_ A VDD VDD CMOSPB L=2u W=9u
.ENDS

.SUBCKT NOR A B AorB_ VDD
M3 1 A VDD VDD CMOSPB L=2u W=9u
M4 AorB_ B 1 VDD CMOSPB L=2u W=9u
M1 AorB_ A 0 0 CMOSNB L=2u W=3u
M2 AorB_ B 0 0 CMOSNB L=2u W=3u
.ENDS

* MAIN CIRCUIT
XINVERT 1 AorB VDD INVERT
XNOR A B 1 VDD NOR

SPICE models not shown
.END
```

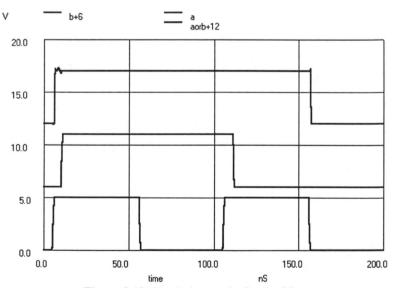

Figure 8.18 Simulation results for the OR gate.

REFERENCE

[1]   Boyce, D.E. *LASICKT Help Manual*, Available by pressing Help while running
        LasiCkt.

Analog MOSFET Models

In this chapter we derive the AC small-signal model of the MOSFET. This model includes the transconductance, output resistance, and capacitances of the MOSFETs. The temperature and noise effects in MOSFETs will also be discussed.

9.1 Low-Frequency MOSFET Model

Before we derive the small-signal models for the MOSFET, it will be useful to review the equations derived in Ch. 6 from the BSIM model parameters. The threshold voltage of an n-channel MOSFET is given by

$$V_{THN} = VFB + PHI + K1 \cdot \sqrt{PHI + V_{SB}} - K2 \cdot (PHI + V_{SB}) \qquad (9.1)$$

The drain current of the n-channel MOSFET in the saturation region, $V_{DS} > V_{GS} - V_{THN}$, is given by

$$I_D = \frac{MUZ \cdot C_{ox} \cdot W}{2 \cdot L}(V_{GS} - V_{THN})^2[1 + (\lambda_c + \lambda_m)(V_{DS} - V_{DS,sat})] \qquad (9.2)$$

where λ_c is the channel length modulation parameter and λ_m is the mobility modulation parameter. We defined

$$\beta = MUZ \cdot C'_{ox} \cdot \frac{W}{L} = KP \cdot \frac{W}{L} \qquad (9.3)$$

so that the drain current in the saturation region can be written as

$$I_D = \frac{\beta}{2}(V_{GS} - V_{THN})^2(1 + \lambda \cdot V_{DS}) \qquad (9.4)$$

where we have assumed $V_{DS,sat}$ is approximately 0 and $\lambda = \lambda_c + \lambda_m$. The drain current of the n-channel MOSFET in the linear or triode region, $V_{DS} < V_{GS} - V_{THN}$, is given by

$$I_D = \beta \cdot \left[(V_{GS} - V_{THN}) \cdot V_{DS} - \frac{V_{DS}^2}{2} \right] \tag{9.5}$$

When the MOSFET is operated in the subthreshold region, $V_{GS} < V_{THN}$, the drain current, assuming $V_{DS} > 100$ mV and $V_{GS} < V_{THN}$ - 100 mV, is given by

$$I_{D,weak} = KP \cdot \frac{W}{L} \cdot \left(\frac{kT}{q} \right)^2 e^{1.8} e^{q(V_{GS}-V_{THN})/N0 \cdot kT} = I_{D0} \cdot \frac{W}{L} \cdot e^{q(V_{GS}-V_{TH})/N0 \cdot kT} \tag{9.6}$$

9.1.1 Small-Signal Model of the MOSFET in Saturation

Consider the circuit shown in Fig. 9.1. The DC sources will be labeled with upper case letters and subscripts (e.g., V_{GS}), and the AC sources will have lower case letters with subscripts (e.g. v_{gs}). The sum of the DC and AC components will be labeled with lower case letters and upper case subscripts (e.g. v_{GS}). What we are trying to get at in Fig. 9.1 is what happens to the drain current when we apply an AC voltage, v_{gs}. We are assuming in the following that $V_{GS} \gg v_{gs}$ or in other words the AC signals are small compared to the DC biasing voltages or currents. Since the MOSFET is in the saturation region, $V_{DS} > V_{GS} - V_{THN}$, the total (AC + DC) drain current is given by

$$i_D = i_d + I_D = \frac{\beta}{2} \overbrace{(V_{GS} + v_{gs}}^{v_{GS}} -V_{THN})^2 (1 + \lambda \cdot V_{DS}) \tag{9.7}$$

The forward transconductance, g_m, of the MOSFET is given by

$$g_m = \left[\frac{\partial i_{DS}}{\partial v_{GS}} \right]_{V_{GS}=const}^{I_{DS}=const} = \beta(V_{GS} + v_{gs} - V_{THN})(1 + (\lambda_c + \lambda_m) \cdot V_{DS}) \tag{9.8}$$

Figure 9.2 shows a circuit model using this result. Notice that the DC sources are removed; that is, this circuit is an AC small-signal model of the MOSFET. We can think of the MOSFET as a voltage-controlled current source or a transconductance amplifier. Before we start making simplifying assumptions to make Eq. (9.8) more user friendly, let's examine how the transconductance will change with signal levels. When

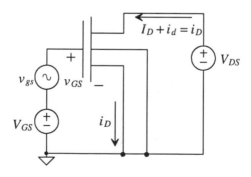

Figure 9.1 Circuit used to determine the forward transconductance.

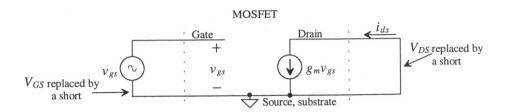

Figure 9.2 Small-signal model of the circuit in Fig. 9.1.

v_{gs} becomes comparable in amplitude (say, within a factor of 10) to V_{GS}, the transconductance variation can be noticeable in the gain of the amplifier. The result for a single-tone (single-frequency sinewave) input to an amplifier is a distorted output. The drain source voltage can modulate the transconductance as well. These problems make "good" (low distortion) large-amplitude voltage amplifiers difficult to design in CMOS. Using feedback can help to reduce the distortion. Another method is to use novel amplifier configurations.

If our MOSFET AC voltages are sufficiently small so that $v_{gs} \ll V_{GS}$ and the product of the AC and DC components of the drain-source voltage with the modulation parameters are less than 1, that is, $1 \gg (\lambda_c + \lambda_m)(v_{ds} + V_{DS})$, then Eq. (9.8) may be written as

$$g_m = \beta(V_{GS} - V_{THN}) = \sqrt{2 \cdot \beta \cdot I_D} \qquad (9.9)$$

From Fig. 9.2 we see that the RMS AC drain current is given by

$$i_d = g_m v_{gs} \qquad (9.10)$$

The transconductance of the MOSFET operating in the weak inversion (subthreshold region) can be arrived at by first rearranging Eq. (9.6), or

$$\overbrace{v_{gs} + V_{GS}}^{v_{GS}} = \frac{kT \cdot N0}{q} \cdot \ln\left(\frac{I_{D,weak}}{I_{D0} \cdot \frac{W}{L}}\right) + V_{THN} \qquad (9.11)$$

The transconductance is then given by

$$g_m^{-1} = \left[\frac{\partial v_{GS}}{\partial i_D}\right]_{I_D = const}^{V_{GS} = const} = \frac{\frac{kT \cdot N0}{q}}{I_{D,weak}} \Rightarrow g_m = \frac{I_D}{V_T \cdot N0} \qquad (9.12)$$

where the designation of weak inversion has been dropped from I_D and $V_T = \frac{kT}{q} = 26$ mV @ $T = 300$ K. The conductance of the MOSFET in the subthreshold region increases linearly with I_D, while when operation is in the strong inversion region the conductance increases as the root of I_D. The model shown in Fig. 9.2 is used for both strong inversion and weak inversion cases.

Example 9.1

Consider the circuit shown below. Determine i_d and an intuitive way to look at the circuit. Use CN20 BSIM model parameters of Ch. 6 or Appendix A.

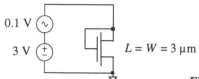

$L = W = 3\,\mu m$

Figure Ex9.1

The substrate connection is not drawn, so we will assume that the substrate is connected to the most negative potential in the circuit, or in this case ground. Since $V_{DS} = V_{GS}$, the transistor is in the saturation region, $V_{DS} > V_{GS} - V_{THN}$. The threshold voltage, from Eq. (6.40), is

$$V_{THN} = -0.22 + 1.49\sqrt{0.75 - 0} - 0.315 \cdot (0.75 - 0) = 0.83 \text{ V}$$

The transconductance, using Eq. (9.9), is

$$g_m = 50\,\frac{\mu A}{V^2} \cdot \frac{3\,\mu m}{3\,\mu m}(3 - 0.83) = 109\,\frac{\mu A}{V}$$

and the RMS drain current is

$$i_d = g_m v_{gs} = (109\,\frac{\mu A}{V}) \cdot (100 \text{ mV}) = 10.9\,\mu A$$

Notice that this drain current flows through both the DC and AC supplies in the circuit. An intuitive way to look at this circuit can be arrived at by observing that the ratio of the AC voltage across the MOSFET (v_{gs}) to the AC current through the MOSFET (i_d) is simply $1/g_m$. For an AC small-signal circuit we can replace the MOSFET with a resistor of value, in this case, of 9.2 kΩ. Note that this intuitive way of looking at the "diode-connected MOSFET" works for both weak and strong inversion. ■

Now let's investigate what happens when the source potential varies and the substrate, gate, and drain are at AC ground. Shown in Fig. 9.3 is the test circuit used to determine how the source-substrate potential variations affect the drain current. In this analysis, we will neglect the forward transconductance effects. These effects arise because $v_{sb} = -v_{gs}$. The body transconductance is given by

$$g_{mb} = -\left[\frac{\partial i_D}{\partial v_{sb}}\right]_{V_{DS}=const}^{I_D=const} = \overbrace{\beta(V_{GS} - V_{THN})(1 + (\lambda_c + \lambda_m)V_{DS})}^{g_m} \cdot \frac{\partial V_{THN}}{\partial v_{sb}} \qquad (9.13)$$

We know that due to the body effect the threshold voltage will not be constant with changing substrate-source potential. Therefore, changes in V_{THN} will modulate the drain current. From Eq. (6.4)

$$\frac{\partial V_{THN}}{\partial v_{sb}} = \frac{K1}{2\sqrt{PHI + V_{SB}}} - K2 = \eta \qquad (9.14)$$

where $K1$ is essentially γ of Ch. 5. Combining Eqs. (9.8), (9.13), and (9.14) gives

$$g_{mb} = g_m \cdot \eta \qquad (9.15)$$

The factor η describes how the threshold voltage changes with reverse body bias. Typical values for η range from 0.6 to possibly a negative number for large source-substrate voltages. For large substrate-source bias voltages we will assume $\eta = 0$. The small-signal model with the effects of the body-substrate current source is shown in Fig. 9.4. Notice that η is a small-signal parameter that is a function of the DC substrate-source potential. Indeed, all of the small-signal parameters derived in this section are a function of DC biasing.

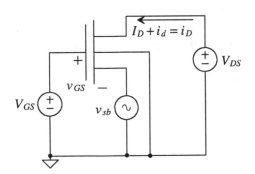

Figure 9.3 Circuit used to determine the body-source transconductance.

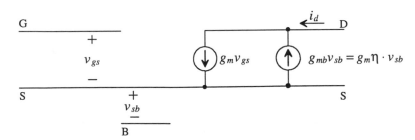

Figure 9.4 Small-signal MOSFET with body-effect current source.

The output resistance of the MOSFET can be determined with the circuit shown in Fig. 9.5. A small AC test voltage, v_{ds}, is applied between the drain and the source. The conductance is determined again by

$$g_{ds} = r_o^{-1} = \left[\frac{\partial i_D}{\partial v_{DS}} \right]_{V_{DS}=const}^{I_D=const} = \frac{\beta}{2}(V_{GS} - V_{THN})^2 (\lambda_c + \lambda_m) = I_D \cdot (\lambda_c + \lambda_m) \qquad (9.16)$$

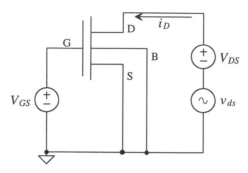

Figure 9.5 Circuit used to determine output resistance of the MOSFET .

This equation combined with Eq. (9.9) gives the maximum gain of a MOSFET stage. Figure 9.6 shows the small-signal model for the MOSFET, including the output resistance. Assuming the source and substrate are connected together (no body effect, $v_{sb} = 0$), we that the maximum voltage gain, in the strong inversion region, is given by

$$\frac{v_{ds}}{v_{gs}} = \frac{-i_d \cdot r_o}{\frac{i_d}{g_m}} = -g_m \cdot r_o = -\frac{\sqrt{2\beta I_D}}{I_D(\lambda_c + \lambda_m)} = -\frac{\sqrt{2\beta}}{\sqrt{I_D} \cdot \lambda} \qquad (9.17)$$

where

$$\lambda = \lambda_c + \lambda_m \qquad (9.18)$$

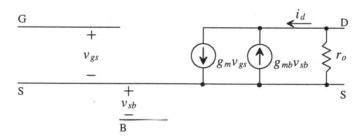

Figure 9.6 Small-signal MOSFET with output resistance.

Example 9.2
Determine the output resistance of an n-channel MOSFET with $L = W = 10$ µm using the BSIM parameters when $V_{GS} = 3$ V. Verify the answer with SPICE.

From Ex. 9.1 $V_{THN} = 0.83$ V, assuming no body effect. The DC drain current is then $I_D = 118$ µA, assuming KP = 50 µA/V². $\lambda \approx \lambda_m = 0.06$ V⁻¹ from Ex. 6.4. The output resistance is $r_o = \frac{1}{0.06 \cdot 118 \times 10^{-6}} \approx 140$ kΩ. Using the BSIM model, SPICE gives, $I_D = 88$ µA and an $r_o = 285$ kΩ at a $V_{DS} = 3$ V. These values are very dependent on the value of V_{DS} used. This example also illustrates how hand calculations give ballpark-type answers. ■

The small-signal output resistance, or channel resistance, of the MOSFET in the triode region is determined by

$$R_{ch}^{-1} = \left[\frac{\partial i_D}{\partial v_{DS}} \right]_{V_{DS}=const}^{I_D=const} = \frac{\partial}{\partial v_{DS}} \beta \left[(V_{GS} - V_{THN})v_{DS} - \frac{v_{DS}^2}{2} \right] \qquad (9.19)$$

and thus,

$$R_{ch} = \frac{1}{\beta(V_{GS} - V_{THN}) - \beta V_{DS}} \approx \frac{1}{\beta(V_{GS} - V_{THN})} \text{ for } V_{DS} \ll V_{GS} - V_{THN} \qquad (9.20)$$

9.2 High-Frequency MOSFET Model

To obtain the high-frequency model of the MOSFET, we will add the MOSFET capacitances to the low-frequency model of Fig. 9.6. The capacitances between the drain and source diffusion regions, discussed in Ch. 5, labeled C_{db} and C_{sb}, and the capacitance of the gate over the field region, C_{gb}, will be added directly to the small-signal model. The capacitance between the gate and the drain is labeled C_{gd} and the capacitance between the gate and source was labeled C_{gs}. Figure 9.7 shows the high-frequency model.

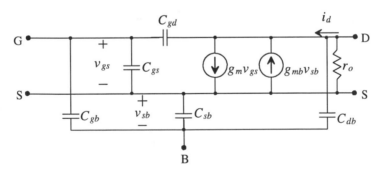

Figure 9.7 Small-signal high-frequency MOSFET model.

Example 9.3
Estimate the capacitances in Fig. 9.7 for the layout (an n-channel MOSFET without substrate connection shown) in Fig. Ex9.3. Use the CN20 model parameters.

Let's start by calculating C_{sb} and C_{db}. The size of the drain and source regions is 6 μm by 14 μm. Since the DC potentials of the drain and source is not specified, we will assume that $V_{DB} = V_{SB} = 0$ V. This will give the maximum possible depletion capacitance of these diffused regions. From the BSIM model parameters and Ch. 5

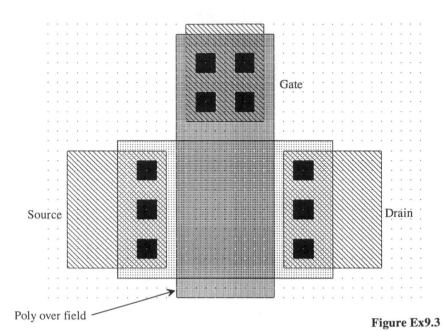

Source   Drain

Poly over field

Figure Ex9.3

$$C_{sb} = C_{sb,bottom} + C_{sb,sidewall}$$

and

$$C_{db} = C_{db,bottom} + C_{db,sidewall}$$

where

$$C_{sb,bottom} = C_{db,bottom} = \frac{CJ \cdot AD}{\left(1 + \frac{V_{DB}}{PB}\right)^{MJ}} = \frac{(1.04 \times 10^{-4}\frac{F}{m^2}) \cdot (84\ \mu m^2)}{\left(1 + \frac{0}{0.8}\right)^{0.66}} = 8.8\ \text{fF}$$

and

$$C_{sb,sidewall} = C_{db,sidewall} = \frac{CJSW \cdot PD}{\left(1 + \frac{V_{DB}}{PBSW}\right)^{MJSW}} = \frac{(2.2 \times 10^{-10}\frac{F}{m}) \cdot (40\ \mu m)}{(1 + \frac{0}{0.8})^{0.18}} = 8.8\ \text{fF}$$

so that $C_{sb} = C_{db} = 17.6$ fF.

The capacitance of the poly over field region, that is, the poly that extends over the n+ above, C_{gb}, is estimated by $CGBO \cdot L = (3.8 \times 10^{-10}\,F/m) \cdot 10\ \mu m = 3.8$ fF. The gate-drain capacitance while in the saturation region is given by $C_{gd} = CGDO \cdot W$, $(3.8 \times 10^{-10}F/m) \cdot 14\ \mu m = 5.3$ fF.   In the triode region, however, $C_{gd} = \frac{1}{2}C'_{ox} \cdot W \cdot L$ or $C_{gd} = \frac{1}{2} \cdot 800 \times 10^{-18}\frac{F}{\mu m^2} \cdot 10\mu m \cdot 14\mu m = 56$ fF.

Finally, the gate-source capacitance of the MOSFET in the saturation region is
$C_{gs} = \frac{2}{3} \cdot C'_{ox} \cdot W \cdot L$, or $C_{gs} = 75$ fF. ∎

9.2.1 Variation of Transconductance with Frequency

Consider the circuit shown in Fig. 9.8. Since the potential across C_{db} and C_{sb} is fixed, we can eliminate them from the following frequency analysis. The small-signal equivalent circuit of Fig. 9.8 is shown in Fig. 9.9. The AC gate-source voltage of the MOSFET is given by

$$v_{gs}(f) = v_s \cdot \frac{\frac{1}{j\omega(C_{gb}+C_{gs}+C_{gd})}}{Z_s + \frac{1}{j\omega(C_{gb}+C_{gs}+C_{gd})}} = v_s \cdot \frac{1}{1+j\omega \cdot Z_s(C_{gb}+C_{gs}+C_{gd})} \quad (9.21)$$

with the transconductance given by, assuming $g_m v_{gs} \gg v_{gs}j\omega C_{gd}$,

$$g_m(f) = \frac{i_d}{v_s} = \frac{g_{m0}}{1+j\omega \cdot Z_s(C_{gb}+C_{gs}+C_{gd})} \quad (9.22)$$

where $g_{m0} = \frac{i_d}{v_{gs}(0)}$ is the low-frequency transconductance.

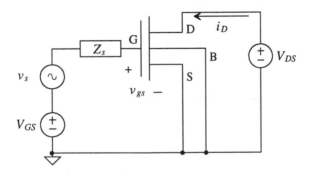

Figure 9.8 Circuit used to determine frequency dependence of the forward transconductance.

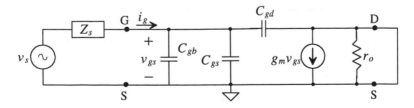

Figure 9.9 Small-signal equivalent circuit of Fig. 9.8.

The current gain of a MOSFET can be defined by

$$\frac{i_d}{i_g} = \frac{g_m \cdot v_{gs}}{v_{gs} \cdot j\omega(C_{gb} + C_{gs} + C_{gd})} = \frac{KP \cdot \frac{W}{L} \cdot (V_{GS} - V_{THN})}{j\omega(C_{gb} + C_{gs} + C_{gd})} \tag{9.23}$$

A useful term used to describe the relative "speed" of a process is the unity current gain transition frequency, f_T, and is defined as the frequency where the magnitude of the MOSFET current gain is unity (making a transition from current gain to attenuation),

$$\left| \frac{i_d}{i_g} \right| = 1 = \frac{KP \cdot \frac{W}{L} \cdot (V_{GS} - V_{THN})}{2\pi f_T \cdot (C_{gb} + C_{gs} + C_{gd})} \tag{9.24}$$

or

$$f_T = \frac{KP \cdot W(V_{GS} - V_{THN})}{2\pi L(C_{gb} + C_{gs} + C_{gd})} \approx \frac{KP \cdot W}{2\pi LC_{gs}}(V_{GS} - V_{THN}) \tag{9.25}$$

This expression neglects the transit time of the carriers through the channel, which in general, for channel lengths within an order of magnitude of minimum length, is a negligible effect.

Example 9.4
Estimate f_T for a minimum-size n- and p-channel device in CN20 with $V_{GS} - V_{THN}$ = 1 V.

For the n-channel MOSFET with $L = 2$ μm and $W = 3$ μm

$$f_T \approx \frac{KP \cdot W}{2\pi L \cdot \frac{2}{3}WLC_{ox}'} = \frac{50\frac{\mu A}{V^2} \cdot 3\mu m}{2\pi \cdot 2\mu m \cdot \frac{2}{3} \cdot 2\mu m \cdot 3\mu m \cdot 800 \times 10^{-18}\frac{F}{\mu m^2}} = 3.74 \text{ GHz}$$

and for the p-channel with $KP = 17\frac{\mu A}{V^2}$, $f_T = 1.27$ GHz. ∎

The most important observation we can make from this example is that the unity current gain frequency is independent of the MOSFET width, or

$$f_T = \frac{3 \cdot KP}{4\pi L^2 C_{ox}'}(V_{GS} - V_{THN}) \tag{9.26}$$

For high-speed design the minimum channel length should be used.

Example 9.5
For the circuit shown in Fig. EX9.5, determine the impedance looking into the source of the MOSFET. Neglect the body effect and assume that over the frequency range of interest $i_d \approx i_s$.

Since $i_d \approx i_s$ we know that $v_s \approx -v_{gs}$ and $i_d \approx -g_m v_s$. The impedance looking into the source is then given by

$$Z_{into \ source} = \frac{-v_s}{i_s} = \frac{1}{g_m} = \frac{1}{g_{m0}} \cdot [1 + j\omega R_s(C_{gb} + C_{gs} + C_{gd})]$$

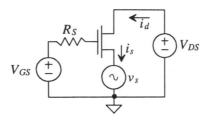

Figure Ex9.5

or

$$Z_{into\ source} = \frac{1}{g_{m0}} + j\omega \frac{R_s(C_{gb} + C_{gs} + C_{gd})}{g_{m0}} \qquad (9.27)$$

or the impedance looking into the source of the MOSFET is a resistor of value $\frac{1}{g_{m0}}$ in series with an inductor of value $\frac{R_s(C_{gb}+C_{gs}+C_{gd})}{g_{m0}}$. ■

The results of this example have several important practical implications. First, the low-frequency resistance looking into the source of a MOSFET is $\frac{1}{g_{m0}}$ regardless of the impedance at the gate. When we are not concerned with the variation of the transconductance with frequency, we will write this resistance as $\frac{1}{g_m} = \frac{1}{\beta(V_{GS} - V_{THN})}$. When the MOSFET is driven with a resistive source impedance, the impedance looking into the source of the MOSFET becomes highly inductive, which will cause ringing in source-follower circuits driving capacitive loads. If the impedance at the gate of the MOSFET is inductive, this occurs in the cascade of two source followers; the output impedance of the second follower can become negative (replace R_s in Eq. [9.27] with the inductor impedance $j\omega L$). A negative impedance is a source of power like a battery. Oscillations and poor step response are probable.

9.3 Temperature Effects in MOSFETs

The threshold voltage and the transconductance parameter change with temperature. The changes in the threshold voltage can be related to the temperature dependence of the surface inversion potential given by

$$\phi_s(T) = PHI(T) = \frac{PHI \cdot T}{T_0} - \frac{3kT}{q} \cdot \ln\left(\frac{T}{T_0}\right) - \frac{E_g(T_0) \cdot T}{T_0} + E_g(T) \qquad (9.28)$$

where E_g, in eV (1 eV = 1.6×10^{-19} J), is the silicon bandgap energy given by

$$E_g(T) = 1.16 - (702 \times 10^{-6}) \cdot \frac{T^2}{T + 1108} \qquad (9.29)$$

The temperature coefficient of the threshold voltage over a range from −100 °C to 100 °C can be estimated as

$$TCV_{THN} = \frac{1}{V_{THN}} \cdot \frac{dV_{THN}}{dT} \approx -3,000 \frac{ppm}{°C} \qquad (9.30)$$

where the threshold voltage is given by

$$V_{THN}(T) = V_{THN}(T_0)[1 + TCV_{THN} \cdot (T - T_0)] \tag{9.31}$$

A PSPICE simulation of a CN20 n-channel MOSFETs threshold voltage variation with temperature is shown in Fig. 9.10. At these low currents, the temperature dependence of the threshold voltage dominates the change in the drain current. We could have also defined an absolute change of $\approx -2.4 \ mV/°C \ (TCV_{THN} \cdot V_{THN})$, dependent on the magnitude of the threshold voltage. Note that SPICE2 or 3 from UC Berkeley does not have temperature- or noise-modeling capability with the BSIM1 model.

```
*** Top Level Netlist ***
M1_10u_10u    3 4 0 0 CMOSNB L=10u W=10u
Vdd      2 0     DC 5
Vgs      4 0     DC 0
VIMTR1 2 3     0V
***** Spice models and macro models *****
.MODEL CMOSNB NMOS LEVEL=4
+VFB=-9.73820E-01, LVFB=3.67458E-01,WVFB=-4.72340E-02
+ ... BSIM model parameters
+pbsw=0.8    mj=0.66036    mjsw=0.178543    wdf=0    dell=0
***** End of spice models and macro models *****
.probe
.DC Vgs .5 1.5 .05  temp -100 101 50
.end
```

The temperature dependence of the transconductance parameter is derived from the temperature dependence of the mobility where T is in kelvin given by

$$\mu(T) = \mu(T_0) \cdot \left(\frac{T}{T_0}\right)^{-1.5} \tag{9.32}$$

or

$$KP(T) = KP(T_0) \cdot \left(\frac{T}{T_0}\right)^{-1.5} \tag{9.33}$$

Figure 9.11 shows that at low-drain currents the effects of the surface potential dominate the changes in the drain current, while at higher currents the mobility effects will dominate. For digital applications, the change in threshold voltage is usually negligible compared to the mobility changes; that is, the mobility changing generally has a much greater impact on the propagation delay than does the threshold voltage. From Fig. 9.11 we can also see that at $V_{GS} = 1.8$ V the drain current is essentially constant. It could be concluded that biasing the device at this gate-source voltage would create temperature- independent circuits. Indeed, this is true. However, variations in the process and a method to accurately bias at this voltage make this result difficult to implement in many cases. (One implementation uses the beta multiplier of Ch. 21.)

Example 9.6
One benefit of using MOSFET transistors in power applications is that the drain current decreases with increasing temperature, keeping the transistor from thermal runaway. Using Fig. 9.11, comment on this assumption.

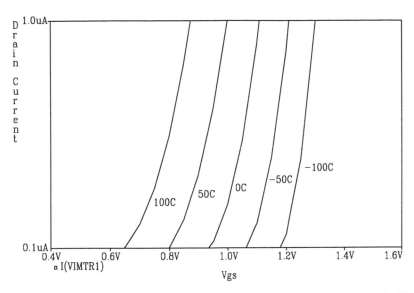

Figure 9.10 Threshold voltage change with temperature for V_{GS} approximately V_{THN}.

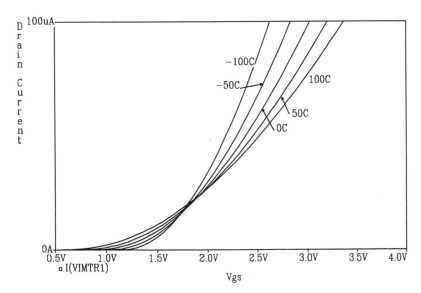

Figure 9.11 Mobility effects when $V_{GS} >> V_{THN}$.

From Fig. 9.11 if we hold the gate-source voltage at 2.5 V, the drain current will decrease with increasing temperature. At $V_{GS} = 2.5$ V, the variation of the mobility with temperature dominates. If we look at $V_{GS} = 1.25$ V, we see that the drain current increases with increasing temperature, due to the surface inversion potential dominating the temperature characteristics. Therefore, unless $V_{GS} >> V_{THN}$, the problem of thermal runaway still exists. ∎

9.4 Noise in MOSFETs

The noise generators in MOSFETs are due to thermal and flicker ($1/f$ noise, pronounced "one over f noise") noise. The RMS thermal noise currents in a MOSFET are generated by the effective channel resistance and by the parasitic drain (R_D), source (R_S), gate (R_G), and substrate (R_B) resistances. The thermal noise generated by the parasitic resistances is given, where the X indicates D, S, G or B terminals, by

$$\sqrt{\overline{i^2_{RX}}} = \sqrt{\frac{4kT}{R_X}} \tag{9.34}$$

where (see Sec. 7.3, $k = 1.38 \times 10^{-23}$ $W \cdot s/°K$) the temperature, T, is in Kelvin, and the measurements are made over a bandwidth, B.

The thermal noise due to the channel resistance, modeled by a resistor of $\frac{3}{2} \cdot \frac{1}{g_m}$ while in the saturation region and R_{CH} in the triode region, is given, in the saturation region, by

$$\sqrt{\overline{i^2_{therm}}} = \sqrt{4kT \cdot \frac{2g_m}{3}} = \sqrt{\frac{8kT}{3}} \cdot \sqrt{2\beta \cdot I_D} \tag{9.35}$$

This is an RMS noise source placed between the drain and source of the MOSFET.

Flicker noise results from trapping of charges at the oxide/semiconductor interface in the inverted channel resulting from imperfections in the silicon crystal structure at the oxide/semiconductor surface. This trapping gives rise to a change in the drain current caused by the carrier recombination/generation taking place via the traps. Since the carrier lifetime in silicon is on the order of tens of microseconds the resulting current fluctuations are concentrated at lower frequencies. The $1/f$ (flicker) noise can be modeled in SPICE (with NLEV = 0, the default) by an RMS noise source given by

$$\sqrt{\overline{i^2_{1/f}}} = \sqrt{\frac{KF \cdot I_D^{AF}}{f \cdot (C'_{ox} \cdot L)^2}} \tag{9.36}$$

where KF is the flicker noise coefficient, a typical value is 10^{-30} $A^{2-AF}(F/m)^2$, I_D is the DC drain current, AF is the flicker noise exponent, a value ranging from 0.5 to 2, and f is the frequency variable we integrate over. A noise model for the MOSFET is shown in Fig. 9.12. All noise sources are RMS. See Sec. 7.3 for a review of noise calculations. Also, note that setting NLEV = 1 in SPICE uses Eq. 9.36 with the L^2 term in the denominator replaced with LW.

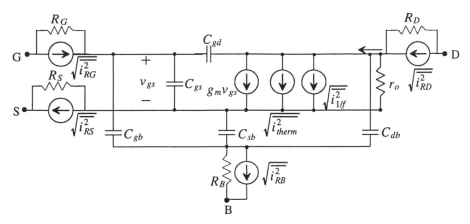

Figure 9.12 Noise model for MOSFET.

For hand calculations, the thermal noise effects due to the parasitic resistances will be neglected. If this is the case, the only noise sources we will evaluate using hand calculations are the channel thermal and flicker noise contributions. Both of these noise sources can be reflected back to the gate of the MOSFET. The RMS magnitudes of the noise voltages in series with the gate terminal are given by

$$\sqrt{\overline{v_{therm}^2}} = \frac{\sqrt{\overline{i_{therm}^2}}}{g_m} \qquad (9.37)$$

and

$$\sqrt{\overline{v_{1/f}^2}} = \frac{\sqrt{\overline{i_{1/f}^2}}}{g_m} \text{ or } \overline{v_{1/f}^2} = \frac{\overline{i_{1/f}^2}}{g_m^2} = \frac{KF \cdot I_D^{AF-1}}{2 \cdot KP \cdot C_{ox}'^2 \cdot W \cdot L \cdot f} \qquad (9.38)$$

The body effect has been neglected. If the source is not connected to the substrate, then g_m should be replaced with $g_m - g_{mb}$ in the above equations. Sometimes Eqs. (9.37) and (9.38) are used to get an intuitive glimpse of how MOSFET noise, on the input to an amplifier, will limit the minimum detectable input signal amplitude.

Example 9.7
Estimate the RMS output noise for the circuit of Fig. Ex9.7 over the bandwidth of 1 to 1 kHz. Use CN20 process parameters with $L = W = 10$ µm and $KF = 10^{-30}$ and $AF = 1.3$. Assume that the parasitic MOSFET resistances generate a negligible amount of thermal noise.

The DC gate voltage is approximately 3.3 V and is determined by solving

$$\frac{5 - V_{GS}}{10k} = \frac{\beta}{2}(V_{GS} - V_{THN})^2 = 25\frac{\mu A}{V^2}(V_{GS} - 0.83)^2$$

The drain current is 153 µA. The noise model for the circuit is shown in Fig. Ex9.7, following step 1 for a noise analysis given in Ch. 7. Using superposition, we can determine the RMS output noise voltage from each noise source. Noting that the gate-source voltage and the output voltage are the same signal, we can write, for the thermal noise current source modeling, the noise generated by the 10 k resistor,

$$10k\left(\sqrt{i^2_{10k}} + g_m v_{gs}\right) = -v_{gs}$$

or

$$v_{gs} = V_{out} = -\sqrt{i^2_{10k}} \cdot \frac{\frac{1}{g_m} \cdot 10k}{10k + \frac{1}{g_m}}$$

This shows that the MOSFET behaves like a resistor of value $\frac{1}{g_m}$ when its gate and drain are tied together. Therefore, to save time and effort we would simply replace the current source above $(g_m v_{gs})$ with this noiseless resistor and perform the noise calculations.

We continue the noise analysis by determining the output RMS noise voltage from each source. Note that you cannot add the noise sources into one noise source, unless you first square each component. The thermal noise from the channel resistance is given by

$$\sqrt{i^2_{therm}} = \sqrt{4 \cdot 1.38 \times 10^{-23} \cdot 300 \cdot \frac{2}{3} \sqrt{4 \cdot 25\mu A/V^2 \cdot 153\mu A}} = 1.2 \frac{pA}{\sqrt{Hz}}$$

and since $\frac{1}{g_m} = 8.1 \text{ k}\Omega$

$$\sqrt{v^2_{therm,out}} = 1.2 \frac{pA}{\sqrt{Hz}} \cdot 10k||8.1k = 5.4 \frac{nV}{\sqrt{Hz}}$$

while the RMS $1/f$ (flicker) noise current is given by

$$\sqrt{i^2_{1/f,d}} = \sqrt{\frac{10^{-30} \cdot (150\mu A)^{1.3}}{f \cdot (10)^2 \cdot (800 \times 10^{-18})^2}} = \frac{408 \text{ pA}}{\sqrt{f}}$$

with RMS output noise voltage, due to $1/f$ noise, given by

$$\sqrt{v^2_{1/f,out}} = 408 \frac{pA}{\sqrt{f}} \cdot 10k||8.1k = 1.82 \frac{\mu V}{\sqrt{f}}$$

The output thermal RMS noise from the 10 kΩ resistor is given by

$$\sqrt{v^2_{10k}} = 10k||8.1k \cdot \sqrt{\frac{4kT}{10k}} = 5.8 \frac{nV}{\sqrt{Hz}}$$

We square each RMS component and integrate over the bandwidth of interest. The total mean squared output noise voltage is given by

$$\overline{v_{on}^2} = \int_{f_L}^{f_H} \left(\overbrace{29.16 \frac{aV^2}{Hz}}^{thermal} + \overbrace{\frac{3.33 \times 10^{-12}V^2}{f}}^{1/f} + \overbrace{33.6 \frac{aV^2}{Hz}}^{res.\ thermal} \right) df$$

or evaluating the integral

$$\overline{v_{on}^2} = (29.16 \frac{aV^2}{Hz} + 33.6 \frac{aV^2}{Hz}) \cdot (f_H - f_L) + \overbrace{3.33\ pV^2 \cdot (\ln f_H - \ln f_L)}^{1/f}$$

The total RMS output noise voltage over the bandwidth of 1 to 1 kHz is given by $\sqrt{\overline{v_{on}^2}} = 4.7\ \mu V$. The RMS output noise is dominated by the $1/f$ noise. From the results above, we observe that the same amount of $1/f$ noise exists in a bandwidth of 1 to 1 kHz as a bandwidth of 1 MHz to 1 GHz, or a bandwidth of once per day to 1,000 times a day. ■

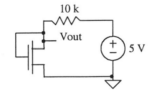

Figure Ex9.7

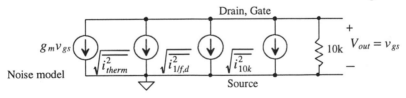

Noise model

REFERENCES

[1]   F. Maloberti, "Layout of Analog and Mixed Analog-Digital Circuits," in J. E. Franca and Y. Tsividis, eds., *Design of Analog-Digital VLSI Circuits for Telecommunications and Signal Processing*, 2nd ed., Prentice-Hall, 1994. ISBN 0-13-203639-8.

[2]   C. D. Motchenbacher and F. C. Fitchen, *Low-Noise Electronic Design,* John Wiley and Sons, 1973. ISBN 0-471-61950-7.

[3]   H. L. Krauss, C. W. Bostian, and F. H. Raab, *Solid State Radio Engineering,* John Wiley and Sons, 1980. ISBN 0-471-03018-X.

PROBLEMS

Unless otherwise stated, use the CN20 process.

9.1   Sketch the p-channel equivalent to Fig. 9.6 using the p-channel variables, v_{sg} and v_{bs}. Are these small-signal voltages positive or negative numbers? Refer to Appendix A.

9.2   Plot the drain current against V_{GS} for the MOSFET shown in Fig. P9.2 using a SPICE DC sweep. What is the approximate slope of a line tangent to this curve at $V_{GS} = 2$ V. What does the reciprocal of this slope correspond to?

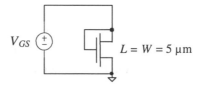

V_{GS} ⊕ $L = W = 5\ \mu m$

Figure P9.2

9.3   Determine the AC drain current for the circuit given in Ex. 9.1 if $L = 2\ \mu m$ and $W = 5\ \mu m$ with $V_{GS} = 1.5$ V and $v_{gs} = 10$ mV. Verify your answer with an AC SPICE simulation.

9.4   Estimate, using hand calculations, the AC drain current that flows in the following circuit. The MOSFET has $W = 15\ \mu m$ and $L = 5\ \mu m$.

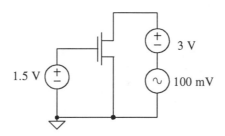

1.5 V ⊕   ⊕ 3 V

~ 100 mV

Figure P9.4

9.5   Verify the answer in Problem 4 using SPICE.

9.6   Repeat Problem 4 with $V_{GS} = 3$ V and $V_{DS} = 1.5$ V (v_{ds} is still 100mV).

9.7   Plot the drain current against V_{SG} for the MOSFET shown in Fig. P9.7 using a SPICE DC sweep.

9.8   Estimate the AC drain current that flows in the circuit of Fig. P9.8. The p-channel MOSFET has $W = 70\ \mu m$ and $L = 5\ \mu m$. How does the AC drain current change if $V_{SD} = 0$, that is, the drain of the p-channel MOSFET is connected to ground?

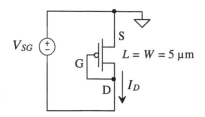

Figure P9.7

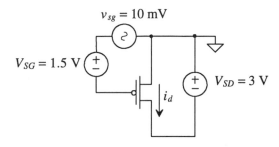

Figure P9.8

9.9   Verify the answer in Problem 8 using SPICE.

9.10   Repeat Problem 8 for the circuit of Fig. P9.10 and verify the answer with SPICE.

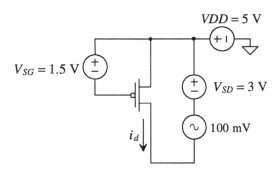

Figure P9.10

9.11   Repeat Ex. 9.3 where the layout shown in Fig. Ex9.3 is a p-channel MOSFET in an n-well.

9.12   Figure P9.12 shows the test setup used to determine a MOSFET f_T. Using this setup and SPICE, verify the results of Ex. 9.4. Remember that the MOSFET must be biased in the saturation region. How are the drain current and gate current related at the frequency f_T and at DC?

9.13   For wide operating range in analog circuits $V_{GS} - V_{THN}$ is made very small. Comment on how this affects the speed of a circuit.

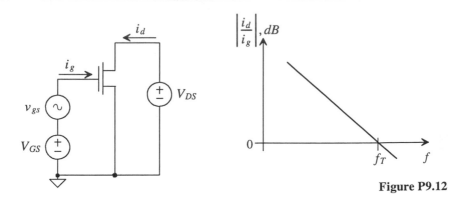

Figure P9.12

9.14 Does the n-channel MOSFET threshold voltage increase or decrease with increasing temperature? Some CMOS processes have MOSFETs with threshold voltages around 0.5 V. Comment on the problems associated with decreasing the threshold voltage in a CMOS process down to a few hundred millivolts. Use drawings to illustrate the problems. (*Hint*: Consider temperature effects and subthreshold current.)

9.15 Verify, with SPICE, the RMS output noise voltage of 4.7 μV given in Ex. 9.7.

9.16 Determine the RMS output noise in the following circuit over the frequency range from 1 to 10 kHz. Verify your answer with SPICE. Neglect parasitic MOSFET resistances and assume that the device capacitances have little effect on the noise performance over this frequency range.

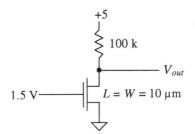

Figure P9.16

Chapter 10

The Digital Model

In this chapter we present the digital model of the MOSFET. At this point, the student should feel relatively comfortable with simulating and laying out CMOS circuits and the parasitics associated with the CMOS process. The transition into digital circuit design should be relatively straightforward.

10.1 The Digital MOSFET Model

Consider the MOSFET circuit shown in Fig. 10.1. Initially, the MOSFET is off, $V_{GS} = 0$, and the drain of the MOSFET is at VDD. If the gate of the MOSFET is taken instantaneously from 0 to VDD, a current given by

$$I_D = \frac{KP_n}{2} \cdot \frac{W}{L} \cdot (VDD - V_{THN})^2 = \frac{\beta}{2} \cdot (VDD - V_{THN})^2 \tag{10.1}$$

initially flows through the MOSFET. Point A in Fig. 10.2 shows the operating point of the MOSFET prior to switching for $VDD = 5$ V. After switching takes place, the operating point moves to point B and follows the curve $V_{GS} = VDD$ down to $I_D \approx 0$ and $V_{DS} = 0$, point C.

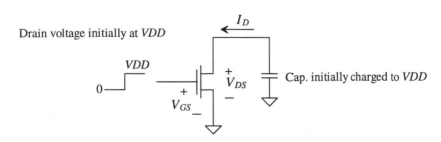

Figure 10.1 MOSFET switching circuit.

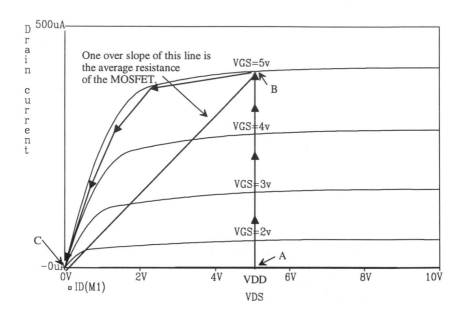

Figure 10.2 Diagram used to determine average resistance of a MOSFET during switching.

An estimate for the resistance between the drain and source of the MOSFET is given by the reciprocal slope of the line BC in Fig. 10.2, or

$$R_n = \frac{VDD}{\frac{KP_n}{2}\frac{W}{L} \cdot (VDD - V_{THN})^2} = R'_n \cdot \frac{L}{W} \tag{10.2}$$

The MOSFET is modeled by the circuit shown in Fig. 10.3. When $V_{GS} > VDD/2$, the switch is closed while V_{GS} values less than $VDD/2$ the switch is open. In the derivation of this model, we assumed that the input step transition occurred in zero time; that is, the risetime was zero, so that the point at which the switch was opened or closed was well defined. In practice, we will not encounter a zero risetime pulse; therefore, the model has limitations. Nevertheless, the model works remarkably well in designing and analyzing digital circuits, giving results that are within a factor of two of simulation or measurement in general applications.

Application of the BSIM model parameters to this digital model can, to a first order, predict the increase in effective resistance for short channel devices. For short channel devices, the drain current increases linearly with V_{GS} rather than as the square. Characterization of these MOSFETs usually results in a lower value of MUZ, accounting for the mobility degradation effects present. The digital model resistance can be written in terms of the BSIM model parameters by

$$R_n = \frac{2L \cdot VDD}{MUZ \cdot C'_{ox} \cdot W \cdot (VDD - V_{THN})^2} = R'_n \cdot \frac{L}{W} \tag{10.3}$$

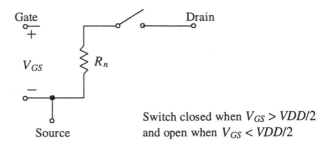

Switch closed when $V_{GS} > VDD/2$
and open when $V_{GS} < VDD/2$

Figure 10.3 Simple digital MOSFET model.

10.1.1 Capacitive Effects

At this point, we need to add the capacitances of the switching MOSFET to our model of Fig. 10.3. Consider the MOSFET shown in Fig. 10.4 with capacitance $C_{ox}/2$ between the gate-drain and the gate-source electrodes. This is the capacitance when the MOSFET is in the triode region. In our development of the digital MOSFET model, we will neglect the depletion capacitances of the source and drain implants to substrate. When the input pulse transitions from 0 to VDD, the output transitions from VDD to 0. The current through C_{gd} ($= C_{ox}/2$), assuming a linear transition, is given by

$$I = C_{gd} \cdot \frac{dV_{gd}}{dt} = \frac{C_{ox}}{2} \cdot \frac{VDD - (-VDD)}{\Delta t} = C_{ox} \cdot \frac{VDD}{\Delta t} = C_{ox} \cdot \frac{dV_{DS}}{dt} \qquad (10.4)$$

The voltage across C_{gd} changes by $2 \cdot VDD$. The current that flows through this capacitance is the drain current of the MOSFET in Fig. 10.4. We can break C_{gd} into a component from the gate to ground and from the drain to ground of value $2C_{gd}$ or C_{ox}. The complete model of a switching MOSFET is shown in Fig. 10.5.

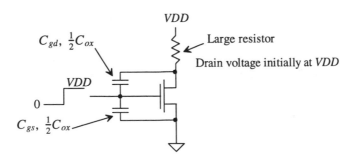

Figure 10.4 MOSFET switching circuit with capacitances.

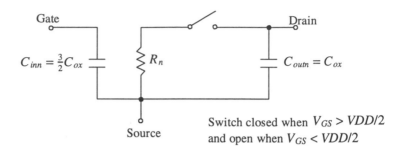

Figure 10.5 Simple digital MOSFET model.

10.1.2 Process Characteristic Time Constant

An important question we can answer at this point is, "What is the intrinsic switching speed of a MOSFET?" Looking at Figs. 10.4 and 10.5, we can see an intrinsic time constant of $R_n C_{ox}$. That is, if the drain is charged to VDD as in Fig. 10.4 and the input switches from 0 to VDD, the output voltage will decay with a time constant of $R_n C_{ox}$. For an n-channel transistor, this is given by

$$\tau_n = R_n C_{ox} = \frac{2L \cdot VDD}{KP_n W(VDD - V_{THN})^2} \cdot C'_{ox} WL = \frac{2L^2 C'_{ox} \cdot VDD}{KP_n \cdot (VDD - V_{THN})^2} \quad (10.5)$$

Notice that the "speed" of a process increases as the square of the channel length and that it is independent of the channel width, W. Also note that the larger VDD, the faster the process. This is very similar to the unity current gain frequency, f_T, we discussed in the last chapter.

Example 10.1
Estimate the process characteristic time constants for CN20, both n- and p-channel devices, using the BSIM model parameters.

We can start the solution of this problem by finding R'_n and R'_p using Eq. (10.3). For the n-channel,

$$R_n = R'_n \cdot \frac{L}{W} = \frac{2 \cdot VDD}{MUZ \cdot C'_{ox}(VDD - V_{THN})^2} \cdot \frac{L}{W} = \frac{2 \cdot 5 \cdot (L/W)}{\left(598 \frac{cm^2}{V \cdot s}\right)\left(800 \frac{aF}{\mu m^2}\right)\left(\frac{10^8 \mu m^2}{cm^2}\right)(5 - 0.83)^2}$$

$$= 12 \text{ k}\Omega \cdot \frac{L}{W}$$

and for the p-channel

$$R_p = R'_p \cdot \frac{L}{W} = \frac{2 \cdot 5 \cdot \frac{L}{W}}{\left(211 \frac{cm^2}{V \cdot s}\right)\left(800 \frac{aF}{cm^2}\right)\left(\frac{10^8 \mu m^2}{cm^2}\right)(5 - 0.92)^2} \approx 36 \text{ k}\Omega \cdot \frac{L}{W}$$

The process characteristic time constants for minimum length devices are given by

$$\tau_n = R_n C_{ox} = 12\text{k} \cdot \frac{2\ \mu\text{m}}{W} \cdot \left(800\frac{\text{aF}}{\mu\text{m}^2}\right) W(2\ \mu\text{m}) = 38\ \text{ps}$$

and

$$\tau_p = R_p C_{ox} = 3\tau_n = 114\ \text{ps} \quad \blacksquare$$

This example has several practical results. The resistance of the n-channel MOSFET is three times smaller than that of the p-channel MOSFET, resulting in a factor of three differences in the time constants. This is because the mobility of the electrons is three times greater than the mobility of holes in the CN20 process. Also note that the effective resistances calculated in this example, for $VDD = 5$V, do not change, so that we can use these in the coming chapters.

10.1.3 Delay- and Transition Times

Before we go any further in the discussion of the digital models, let's define delay and transition times in logic circuits. Consider Fig. 10.6. The top trace represents the input to a logic gate, while the bottom trace represents the output. Note that there was no logic inversion between the input and output; however, the following definitions apply equally well to the case when there is an inversion. The input rise- and falltime are labeled t_r and t_f, respectively. The output rise- and falltime are labeled t_{LH} and t_{HL}, respectively. The delay-time between the 50 percent points of the input and the output are labeled t_{PLH} and t_{PHL} depending on whether the output is changing from a high to a low or from a low to a high. These definitions are extremely important in characterizing the time-domain characteristics of digital circuits.

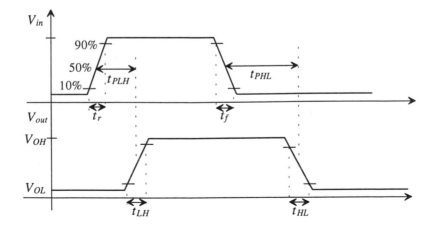

Figure 10.6 Definition of delays and transition times.

For the simple RC circuit shown in Fig. 10.7, the delay-time is given by

$$t_{delay} = 0.7RC \qquad (10.6)$$

and the rise- or falltime is given by

$$t_{rise} = 2.2RC \qquad (10.7)$$

For our simple digital model of Fig. 10.5, we will assume that the propagation delay-time, whether high to low or low to high, is given by one time constant, or

$$t_{PHL}, \; t_{PLH} \approx R_{n,p} \cdot C_{tot} \qquad (10.8)$$

and the output rise- and falltimes are given by

$$t_{HL}, \; t_{LH} \approx 2R_{n,p} \cdot C_{tot} \qquad (10.9)$$

where C_{tot} is the total capacitance from the drain of the MOSFET to ground and $R_{n,p}$ is the effective resistance of the n- or p-channel MOSFET, respectively. These models do not give exact results. The models are useful for determining approximate delay and transition times, usually to within a factor of two.

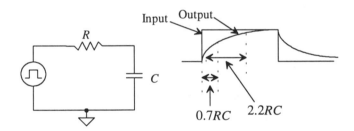

Figure 10.7 Delay- and risetime for a simple RC circuit.

Example 10.2
Using hand calculations, estimate the risetime and delay-time of the following circuits (Fig. 10.8). Compare your results to SPICE simulations.

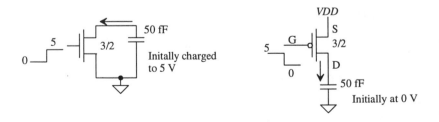

Figure 10.8 Circuits used in Example 10.2.

The effective resistance of the n-channel MOSFET, from Ex. 10.1, is $R_n = 12k$ $\cdot\frac{2\mu m}{3\mu m} = 8$ kΩ and for the p-channel $R_p = 24$ kΩ. C_{ox} is given by $2\mu m \cdot 3\mu m \cdot 800$ aF/$\mu m^2 = 4.8$ fF. The circuit with the digital models drawn is shown in Fig. 10.9. The capacitance, C_{ox}, between the drain and source of the p-channel MOSFET is drawn from the drain to ground rather than from the drain to the source (*VDD*). Electrically, there is no difference in the circuits. Conceptually, it is easier to see that C_{ox} is in parallel with the 50 fF capacitor when the circuit is drawn in this way.

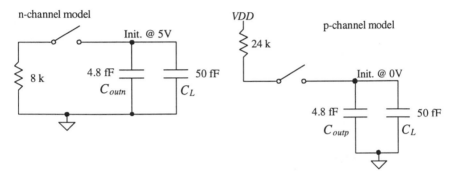

Figure 10.9 Models used to determine switching times in Example 10.2.

The hand calculations of the delay-time for the n-channel transistor, t_{PHL}, is 438 ps, while the falltime, t_{HL}, is 877 ps. For the p-channel $t_{PLH} = 1.3$ ns and $t_{LH} = 2.6$ ns. SPICE simulation results are shown in Fig. 10.10. The PSPICE netlist is shown below. Notice that the .OPTIONS (or .OPTION) statement was used to help with convergence problems (see Ch. 6). ∎

```
*** Top Level Netlist ***
C1      1 0 50f  IC=5
C2      2 0 50f  IC=0
M1      1 3 0 0 CMOSNB  L=2u W=3u
M2      2 4 5 5 CMOSPB  L=2u W=3u
V1      5 0      DC 5 AC 0 0
V2      3 0      DC 0 AC 0 0 PULSE(0 5 1n 1p)
V3      4 0      DC 0 AC 0 0 PULSE(5 0 1n 1p)

.MODEL CMOSNB NMOS LEVEL=4
.... BSIM model parameters of Appendix A
.MODEL CMOSPB PMOS LEVEL=4
.... BSIM model parameters of Appendix A

.OPTION ABSTOL=1u ITL4=100 RELTOL=0.01 VNTOL=.1mv
.probe
.tran 1n 5n 0 .01n uic
.end
```

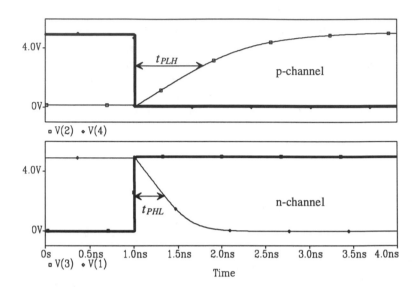

Figure 10.10 Simulation results for Example 10.2.

10.2 Series Connection of MOSFETs

Consider the series connection of MOSFETs shown in Fig. 10.11. The input to this circuit, I, is passed to the output Z when $A = B = C = VDD$ = logic "1." If A, B, or C is at ground (= logic "0"), the output is in the high-impedance state, that is, not a logic 0 or 1. Series connection of MOSFETs occurs frequently in CMOS digital circuit design. In this section, we will analyze the DC and transient behavior of a string of MOSFETs.

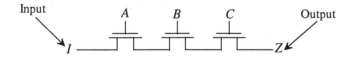

Figure 10.11 Series connection of MOSFETs.

10.2.1 DC Behavior of Series-Connected MOSFETs

To illustrate the DC operation of series-connected MOSFETs, let's use Fig. 10.11 and assume that $I = A = B = C = VDD$ (see Fig. 10.12a). The maximum voltage we can pass from M1 to M2 is $VDD - V_{THN}$ (with body effect). In fact, this is the largest voltage we can pass to the output Z without turning any of the MOSFETs off. Now consider Fig. 10.12b where the input is now a logic low (= 0 V). The output Z can swing all the way down to zero. In other words, the n-channel string passes 0V well and VDD with a threshold drop.

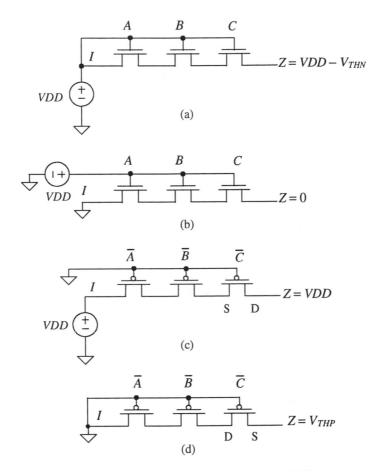

Figure 10.12 DC operation of series-connected MOSFETs.

A p-channel series connection of MOSFETs is shown in Figs. 10.12c and d. Notice that A, B, and C are now active low; that is, the p-channel MOSFETs turn on when $\bar{A} = \bar{B} = \bar{C} = 0$. The p-channel string can pass a logic high without a voltage drop. However, the minimum voltage through a p-channel string is V_{THP} (with body effect).

Example 10.3
Describe the logic function of the circuit shown in Fig. 10.13. What are the minimum and maximum voltages at the output of the circuit?

The logical output of the circuit, Z, is the input I (that is, I is passed to the output) when $A = B = 1$ and $\bar{C} = 0$. Otherwise the output is in the high-impedance state. (At least one of the switches is off.) The output variable Z can vary in voltage from V_{THP} to $VDD - V_{THN}$. For this reason, this circuit is of little practical value. ■

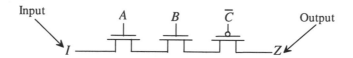

Figure 10.13 Circuit for Example 10.3.

10.2.2 Delay Through Series-Connected MOSFETs

Also of importance in series-connected MOSFETs is the delay. Consider the circuit and its equivalent model shown in Fig. 10.14. In the following analysis we will assume that the capacitance at any internal node is approximated by

$$C_n = C_{inn} + C_{outn} = 2.5C_{ox} \tag{10.10}$$

For a large number of transistors, the series connection of MOSFETs behaves like an RC transmission line with a delay given by Eq. (2.11), or

$$t_d = 0.35C_nR_nl^2 \tag{10.11}$$

where l is the number of MOSFETs in the series connection. Making the appropriate substitutions into this equation gives

$$t_d = 0.35 \cdot 2.5 \cdot C_{ox} \cdot R_n \cdot l^2 \approx C_{ox}R_n \cdot l^2 = \tau_n \cdot l^2 \tag{10.12}$$

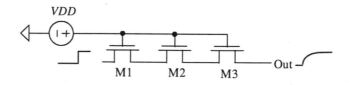

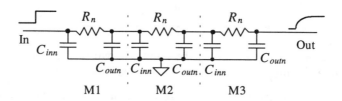

Figure 10.14 Modeling delay through series-connected MOSFETs.

Example 10.4

Estimate and simulate the delay through ten n-channel and ten p-channel MOSFETs. Assume minimum size ($L = 2$ μm and $W = 3$ μm) devices. Use the CN20 parameters.

The digital model resistances of the n- and p-channel MOSFETs are

$$R_n = 12k \cdot \frac{2\ \mu m}{3\ \mu m} = 8\ k\Omega \text{ and } R_p = 36k \cdot \frac{2\ \mu m}{3\ \mu m} = 24\ k\Omega$$

The oxide capacitance of either MOSFET is $C_{ox} = C'_{ox}LW = 800\ aF \cdot 2 \cdot 3 = 4.8$ fF. Using Eq. (10.12), the delay through the series connection of ten n-channel MOSFETs is

$$t_d = C_{ox}R_n l^2 = 4.8\ fF \cdot 8k \cdot (10)^2 = 3.8\ ns$$

while the delay through ten p-channel MOSFETs is

$$t_d = C_{ox}R_p l^2 = 4.8\ fF \cdot 24k \cdot (10)^2 = 11.52\ ns$$

The simulation results are shown in Fig. 10.15. The delay of the n-channel string of MOSFETs is greatest when the string is passing a logic 1, while the delay through the p-channel string is greatest passing a logic 0. Notice that the output of the n-channel string only reaches approximately 3.5 V ($VDD - V_{THN}$), while the output of the p-channel goes down to 1.7 V (V_{THP} with body effect). The delay through the n-channel (p-channel) string is less for an input going from a high to a low (low to a high). Note that the location of the delay-times in Fig. 10.15 is somewhat arbitrary. ∎

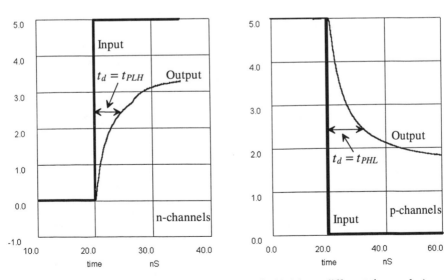

Figure 10.15 Delay simulations for Example 10.4 (note different time scales).

REFERENCE

[1]   R. L. Geiger, P. E. Allen, and N. R. Strader, *VLSI-Design Techniques for Analog and Digital Circuits,* McGraw-Hill Publishing Company, 1990. ISBN 0-07-023253-9.

PROBLEMS

10.1   Repeat Ex. 10.2 for devices with sizes 10/2 (i.e., 10 μm by 2 μm) and a capacitor of value 150 fF. Use the BSIM SPICE models.

10.2   Calculate R'_n and R'_p for the CMOS14TB process. What are the process characteristic time constants?

10.3   Simulate the operation of the circuit shown in Fig. 10.13. Pulse (transient analysis) the input from 0 to 5 V and back to 0. Show and explain the resulting circuit output. Note that the output node cannot be floating. Connect a 100MEG (not 100M) resistor from the output node to ground.

10.4   Estimate the delay through seven n-channel MOSFETs connected in series, similar to Fig. 10.11. The size of the MOSFETs is 10/2.

10.5   Repeat Problem 4 using p-channel MOSFETs.

10.6   Use SPICE to verify the answer to Problem 4.

10.7   Estimate t_{PHL} and t_{PLH} for the circuits shown in Fig. 10.8 when the capacitor is increased to 1 pF. Verify your hand calculations with SPICE.

10.8   The schematic of a standard 10 to 1 scope probe is shown in Fig. P10.8a. The capacitance per foot of the scope cable is approximately 30 pF/ft. The simplified schematic of the probe is shown in Fig. P10.8b. We can use the simplified approximation of the scope probe shown in Fig. 10.8c when calculating probe-loading effects. Repeat Problem 7 using the scope loading of Fig. 10.8c. Note that measuring signals on-chip requires special probes that do not load the MOSFETs on the die. Measuring signals off-chip requires an on-chip buffer to isolate the logic from the off-chip capacitance.

10.9   Repeat Ex. 10.2 using the CMOS14TB process with MOSFETs having a width of 0.9 μm and a length of 0.6 μm.

Possible Student Projects

This section lists some possible student projects for fabrication through the MOSIS service. Generally, two to four student projects should be implemented on one chip. MOSIS will return to the MOSIS liaison (generally, the course instructor) four copies of each chip design submitted.

In addition to the design rule checked designs, in TLC format, the student should turn in (1) one sheet of paper showing the logic level diagrams *and* pin

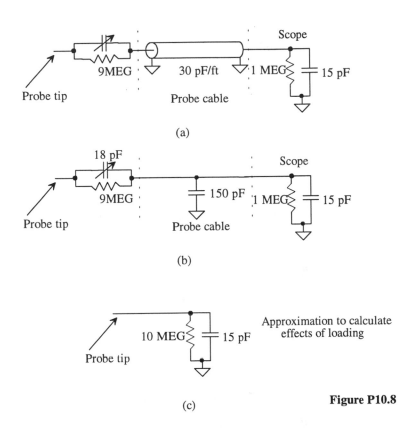

(a)

(b)

(c)   **Figure P10.8**

connections so that whoever is evaluating the chip can quickly determine functionality, and (2) final reports that consist of a block diagram, schematic diagram, layout information, hand calculations, SPICE simulations, and clear explanations of the operation of the circuit.

The report (two copies) is due with the TLC design file on a floppy (the name of the TLC cell to be fabricated should be printed clearly on the floppy disk label). Each group should hand in one disk, while each student should turn in his or her own report. Each designer is solely responsible for his or her designs; it is not a group effort. However, it is the responsibility of each designer to ensure that no fatal problems exist on the chip (e.g., *VDD* shorted to *GND*). Therefore, each student should review the design of the other student's project for his or her own benefit.

1.   Quad 2-input MUX

2.   Clock doubling circuit using exclusive OR gate

3.   Octal buffer with tristate outputs

4.   SR flipflop with tristate outputs

5.   Edge triggered T flipflop

6.   Edge triggered D flipflop

7.   Schmitt trigger

8.   1 of 4 decoder

9.   4-bit dynamic shift register

10.   Bi-CMOS OR gate

11.   Bi-CMOS AND gate

12.   2-bit adder with carryout

13.   Current-starved VCO with center frequency of 20 MHz

14.   2-bit bidirectional transceiver

15.   PE gate to implement $X = \overline{A + BCD + EF}$

16.   One-shot whose output pulse width is determined by external RC

17.   An NMOS super buffer for driving a 20 pF load

18.   An NMOS output driver for driving a 20 pF load

Advanced projects

19.   A 64-bit static RAM which will include storage cell, addressing and decoding circuitry, buffers, write/read enable, and chip select.

20.   Design of a charge pump (voltage generator). The input to the charge pump is VDD (= 5 V) and the output is –3 V. The circuit should be fully simulated. The reference, oscillator, and feedback should be fully simulated and discussed in the final report.

21.   Design of a 64-bit DRAM, which will include storage cell, addressing and decoding circuitry, buffers, write/read enable, and chip select.

22.   A DPLL which will take a 1 MHz input and generate a 4 MHz output. The output should follow the input for frequency changes from 900 kHz to 1.1 MHz. The report should discuss the transient properties of the DPLL, as well as a detailed design of the phase detector, VCO, and loop filter. The entire design should be monolithic; that is, no external components should be used.

CMOS Digital Circuits

Chapter

11

The Inverter

The CMOS inverter is a basic building block for digital circuit design. As Fig. 11.1 shows, the inverter performs the logic operation of A to $\overline{A}$. When the input to the inverter is connected to ground, the output, in accord with the digital models in the last chapter, is pulled to 5 V through the p-channel transistor. When the input terminal is connected to VDD, the output is pulled to ground through the n-channel MOSFET. The CMOS inverter has several important characteristics that are addressed in this chapter: Its output voltage swings from VDD to ground unlike other logic families that never quite reach the supply levels; the static power dissipation of the CMOS inverter is practically zero; the inverter can be sized to give equal sourcing and sinking capabilities; and the logic switching threshold can be set by changing the size of the device.

This chapter concentrates on the DC switching characteristics of the inverter and the transition times associated with driving capacitive loads and RC transmission lines, but it also addresses other types of inverters available in the CMOS process.

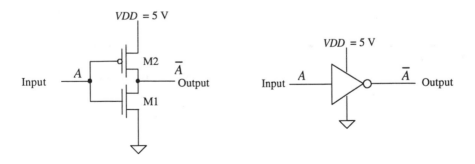

Figure 11.1 The CMOS inverter, schematic, and logic symbol.

11.1 DC Characteristics

Consider the inverter shown in Fig. 11.2 and the associated transfer characteristic plot. In region 1 of the transfer characteristics, the input voltage is sufficiently low (typically not much greater than the threshold voltage of M1), so that M1 is off and M2 is on ($V_{SG} \gg V_{THP}$). As V_{in} is increased, both M2 and M1 turn on (region 2). Increasing V_{in} further causes M2 to turn off and M1 to turn on fully, as shown in region 3.

The maximum output "high" voltage is labeled V_{OH} and the minimum output "low" voltage, V_{OL}. Points A and B on this curve are defined by the slope of the transfer curves equaling -1. Input voltages less than or equal to the voltage V_{IL}, defined by point A, are considered a logic low on the input of the inverter. Input voltages greater than or equal to the voltage V_{IH}, defined by point B, are considered a logic high on the input of the inverter. Input voltages between V_{IL} and V_{IH} do not define a valid logic voltage level. Ideally, the difference in V_{IL} and V_{IH} is zero; however, this is never the case in real logic circuits.

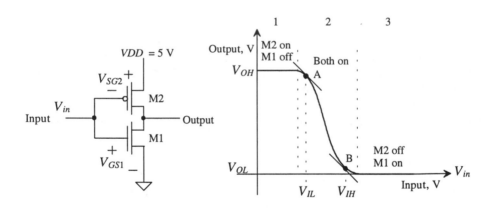

Figure 11.2 The CMOS inverter transfer characteristics.

Example 11.1
Using SPICE, plot the transfer characteristics for the following inverter. From the plot, determine V_{IH}, V_{IL}, V_{OH}, and V_{OL}.

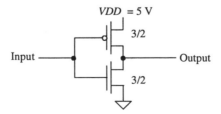

Figure Ex11.1

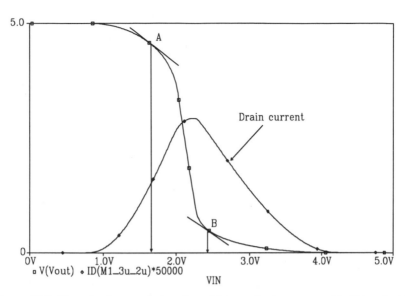

Figure 11.3 Transfer characteristics of a minimum-size inverter used in Example 11.1.

The results of the PSPICE simulation are shown in Fig. 11.3. The netlist for the simulation is shown below. The plot shows that at point A, V_{IL} is approximately 1.7 V and at point B, V_{IH} is approximately 2.4 V. The output voltages, V_{OH} and V_{OL}, are 5 V and 0 V, respectively. Figure 11.3 also shows the scaled drain current of the CMOS inverter. Notice that current flows only when the MOSFETs are switching. ∎

```
*** Top Level Netlist ***
M1_3u_2u Vout 2 0 0 CMOSNB  L=2u W=3u AD=42p AS=42p PD=26u PS=26u
M2_3u_2u Vout 2 Vdd Vdd CMOSPB  L=2u W=3u AD=42p AS=42p PD=26u PS=26u
VDD      Vdd 0   DC 5
VIN       2 0    DC 0

.MODEL CMOSNB NMOS LEVEL=4
+vfb=-9.73820E-01, lvfb=3.67458E-01,wvfb=-4.72340E-02
... see Appendix A for complete listing of BSIM model parameters
.MODEL CMOSPB PMOS LEVEL=4
+ vfb=-2.65334E-01, lvfb=6.50066E-02, wvfb=1.48093E-01
... see Appendix A for complete listing of BSIM model parameters
.probe
.DC Vin 0 5 .01
.end
```

11.1.1 Noise Margins

The noise margins of a digital gate or circuit indicate how well the gate will perform under noisy conditions. The noise margin for the high logic levels is given by

$$NM_H = V_{OH} - V_{IH} \tag{11.1}$$

and the noise margin for the low logic levels is given by

$$NM_L = V_{IL} - V_{OL} \tag{11.2}$$

For $VDD = 5$ V the ideal noise margins are 2.5 V; that is, $NM_L = NM_H = VDD/2$.

Example 11.2

For the minimum-size inverter in Ex. 11.1 determine the noise margins. Comment on making the noise margins closer to ideal.

Using Eqs. (11.1) and (11.2), $NM_H = 5 - 2.4 = 2.6$ and $NM_L = 1.7 - 0 = 1.7$. The high noise margin is almost a whole volt greater than the lower noise margin. This is mainly because the inverter switching point, V_{SP}, is approximately 2.2 V instead of the ideal case of 2.5 V or $VDD/2$. This is discussed further in the next section. ∎

11.1.2 Inverter Switching Point

Consider the transfer characteristics of the basic inverter as shown in Fig. 11.4. Point C corresponds to the point on the curve when the input voltage is equal to the output voltage. At this point, the input (or output) voltage is called the inverter switching point voltage, V_{SP}, and both MOSFETs in the inverter are in the saturation region. Since the drain current in each MOSFET must be equal, the following is true:

$$\frac{\beta_n}{2}(V_{SP} - V_{THN})^2 = \frac{\beta_p}{2}(VDD - V_{SP} - V_{THP})^2 \tag{11.3}$$

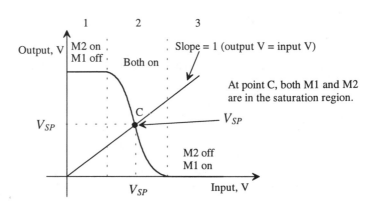

Figure 11.4 Transfer characteristics of the inverter showing the switching point.

Solving for V_{SP} gives

$$V_{SP} = \frac{\sqrt{\frac{\beta_n}{\beta_p}} \cdot V_{THN} + (VDD - V_{THP})}{1 + \sqrt{\frac{\beta_n}{\beta_p}}} \qquad (11.4)$$

Example 11.3

Estimate β_n and β_p so that the switching point voltage of a CMOS inverter is 2.5 V. Assume CN20 parameters and $VDD = 5$ V.

Solving Eq. (11.4) with $V_{SP} = 2.5$ V for the ratio β_n/β_p gives a value of approximately unity. That is,

$$\beta_n = \beta_p = KP_n \frac{W_1}{L_1} = KP_p \frac{W_2}{L_2}$$

Since $KP_n = 3KP_p$, the width of the p-channel transistor must be three times the width of the n-channel, assuming equal-length MOSFETs. For $V_{SP} = 2.5$ V, this requires

$$W_2 = 3W_1$$

which is also the requirement for making $R_n = R_p$. ∎

Example 11.4

Show, using SPICE, transfer curves for the CMOS inverter with transconductance ratios β_n/β_p of 3, 1, and 1/3. Explain what changing the inverter ratio does to the transfer characteristics.

Assuming a channel length of 2 μm for the ratio of 3 $W_1 = W_2 = 3$ μm (one solution), for the ratio of 1, set $W_1 = 3$ μm and $W_2 = 9$ μm; for the ratio of 1/3, $W_1 = 3$ μm and $W_2 = 27$ μm works. Using a simulation similar to that in Ex. 11.1 gives the curves shown in Fig. 11.5. Notice that increasing the transconductance ratio causes V_{SP} to move toward V_{THN}. Inverters are often sized for a specific switching point voltage in digital CMOS circuit design. ∎

11.2 Switching Characteristics

The switching behavior of the inverter can be generalized by examining the parasitic capacitances and resistances associated with the inverter. Consider the inverter shown in Fig. 11.6 with its equivalent digital model. Although the model is shown with both switches open, in practice one of the switches is closed, keeping the output connected to VDD or ground. Notice that the effective input capacitance of the inverter is

$$C_{in} = \frac{3}{2}(C_{ox1} + C_{ox2}) = C_{inn} + C_{inp} \qquad (11.5)$$

The effective output capacitance of the inverter is simply

$$C_{out} = C_{ox1} + C_{ox2} = C_{outn} + C_{outp} \qquad (11.6)$$

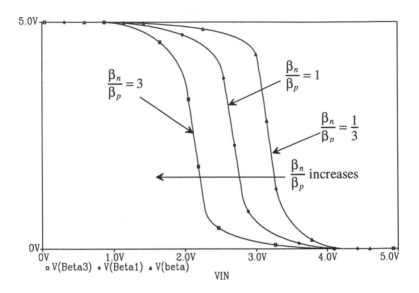

Figure 11.5 Sizing of the CMOS inverter.

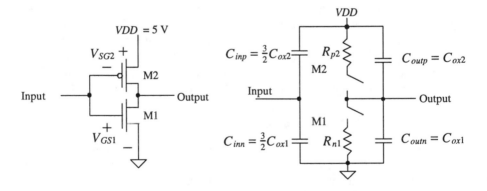

Figure 11.6 The CMOS inverter switching characteristics using the digital model.

The intrinsic propagation delays of the inverter are

$$t_{PLH} = R_{p2} \cdot C_{out} \tag{11.7}$$

$$t_{PHL} = R_{n1} \cdot C_{out} \tag{11.8}$$

Example 11.5
Estimate and simulate the intrinsic propagation delays of the minimum-size inverter.

For the minimum-size inverter $C_{ox1} = C_{ox2} = 3\ \mu m \cdot 2\ \mu m \cdot 800\ aF/\mu m^2 = 4.8\ fF$. $R_{n1} = 12k \cdot 2\ \mu m/3\ \mu m = 8\ k\Omega$, while $R_{p2} = 24\ k\Omega$. The propagation delay times $t_{PHL} = 77\ ps$ and $t_{PLH} = 230\ ps$. The simulation results are shown in Fig. 11.7. ■

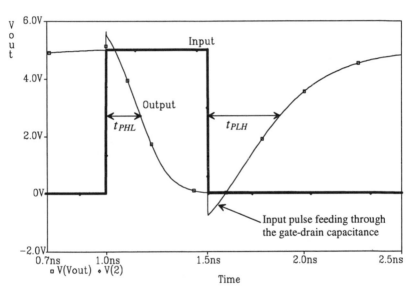

Figure 11.7 Intrinsic inverter delay.

The propagation delays for an inverter driving a capacitive load are

$$t_{PLH} = R_{p2} \cdot C_{tot} = R_{p2} \cdot (C_{out} + C_{load}) \tag{11.9}$$

and

$$t_{PHL} = R_{n1} \cdot C_{tot} = R_{n1} \cdot (C_{out} + C_{load}) \tag{11.10}$$

where C_{tot} is the total capacitance on the output of the inverter, that is, the sum of the output capacitance of the inverter, any capacitance of interconnecting lines, and the input capacitance of the following gate(s).

Example 11.6
Estimate and simulate the propagation delay of a minimum-size inverter driving a 100 fF capacitor.

The schematic of the minimum-size inverter driving a 100 fF load and the logic symbol of the inverter are shown in Fig. 11.8. The sizes adjacent to the inverter correspond to the ratio of the p-channel width to the n-channel width, assuming the lengths of the MOSFETs are the same size. Usually, the lengths are the minimum size available, which for CN20 is 2 μm. The total capacitance, C_{tot}, on the output of the inverter is the sum of C_{out}, the load capacitance and any interconnecting capacitance. In this case C_{tot} = 109.6 fF, assuming no interconnecting capacitance. The propagation delay times are then t_{PHL} = 877 ps and t_{PLH} = 2.63 ns. This can be compared to the simulation results of Fig. 11.9.

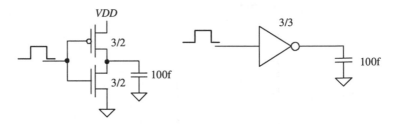

Figure 11.8 Inverter driving a 100 fF load capacitance in Ex. 11.6.

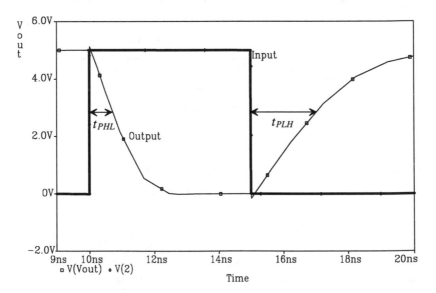

Figure 11.9 Simulation results of minimum-size inverter driving 100 fF.

Notice that in the simulations, a zero risetime input results in delays somewhat less than hand calculations indicate. ■

At this point, the question "How do we size the inverter so that $t_{PHL} = t_{PLH}$?" can be answered. If $R_{n1} = R_{p2}$, the delay times are equal. This is equivalent to making $W_2 = 3W_1$, which was the same requirement used in the previous section for making $V_{SP} = VDD/2$.

11.2.1 The Ring Oscillator

The odd number of inverters of the circuit shown in Fig. 11.10 forms a closed loop with positive feedback and is called a ring oscillator. The oscillation frequency is given by

$$f_{osc} = \frac{1}{n \cdot (t_{PHL} + t_{PLH})} \tag{11.11}$$

assuming the inverters are identical and n is the number (odd) of inverters in the ring oscillator. Since the ring oscillator is self-starting, it is often added to a test portion of a wafer to give an indication of the speed of a particular run.

Consider the case when a minimum-size inverter is used. Under these conditions, C_{tot} is given by

$$C_{tot} = \overbrace{2C_{ox}}^{C_{out}} + \overbrace{3C_{ox}}^{C_{in}} = 5C_{ox} \tag{11.12}$$

where $C_{ox} = 2 \ \mu m \cdot 3 \ \mu m \cdot C'_{ox}$, so that

$$t_{PHL} + t_{PLH} = (R_{n1} + R_{p2})C_{tot} = (12k + 36k)\frac{2}{3} \cdot 5C_{ox} = 160k \cdot C_{ox} \tag{11.13}$$

Also consider the case when the inverters are sized to give equal propagation times. For the delays to be identical, W_2 must equal $3W_1$, which leads to a larger oxide capacitance for M_2, or

$$C_{ox2} = 3C_{ox1} \tag{11.14}$$

Therefore, C_{tot} is given by

$$C_{tot} = \overbrace{4C_{ox}}^{C_{out}} + \overbrace{6C_{ox}}^{C_{in}} = 10C_{ox} \tag{11.15}$$

and the propagation delays are given by

$$t_{PHL} + t_{PLH} = \left(12k\frac{2}{3} + 36k\frac{2}{9}\right)10C_{ox} = 160k \cdot C_{ox} \tag{11.16}$$

which is the same as that given in Eq. (11.13). Although the effective resistance of the p-channel was reduced by a factor of three, the capacitance of M_2 was increased by a factor of three. In general, the ring oscillator frequency is dependent on W, although much less than one would expect. Also note that only five inverters were used. In practice, in order to keep the oscillation frequency in the tens of MHz range, the number of inverters used is 31 (for CN20).

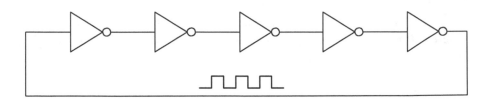

Figure 11.10   A five-stage ring oscillator.

11.2.2 Dynamic Power Dissipation

Consider the CMOS inverter driving a capacitive load shown in Fig. 11.11. Each time the inverter changes states, it must either supply a charge to C_{tot} or sink the charge stored on C_{tot} to ground. If a square pulse is applied to the input of the inverter with a period T and frequency, f_{clk}, the average amount of current that the inverter must pull from *VDD*, recalling that current is being supplied from *VDD* only when the p-channel is on, is

$$I_{avg} = \frac{Q_{Ctot}}{T} = \frac{VDD \cdot C_{tot}}{T} \tag{11.17}$$

The average dynamic power dissipated by the inverter is

$$P_{avg} = VDD \cdot I_{avg} = \frac{C_{tot} \cdot VDD^2}{T} = C_{tot} \cdot VDD^2 \cdot f_{clk} \tag{11.18}$$

Notice that the power dissipation is a function of the clock frequency. A great deal of effort is put into reducing the power dissipation in CMOS circuits. One of the major advantages of dynamic logic (Ch. 14) is its lower power dissipation.

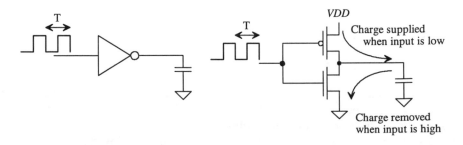

Figure 11.11   Dynamic power dissipation of the CMOS inverter.

To characterize the speed of a digital process, a term called the power delay product (*PDP*) is often used. The *PDP,* measured in joules, is defined by

$$PDP = P_{avg} \cdot (t_{PHL} + t_{PLH}) \tag{11.19}$$

These terms can be determined from the ring oscillator circuit of the previous section. The *PDP* is frequently used to compare different technologies or device sizes; for example, a GaAs process can be compared with a 0.8 μm CMOS process. Although the GaAs process may have a lower propagation delay, the power dissipation may be larger and result in a larger *PDP.*

Example 11.7

Estimate the *PDP* of CN20, using hand analysis of a five-stage ring oscillator with $W_n = W_p = 10$ μm. Simulate the oscillator with SPICE and compare the results to the hand calculations.

The effective resistances of the n- and p-channel MOSFETs are

$$R_{n1} = 12k \cdot \frac{2 \text{ μm}}{10 \text{ μm}} = 2.4 \text{ k}\Omega$$

$$R_{p2} = 36k \cdot \frac{2 \text{ μm}}{10 \text{ μm}} = 7.2 \text{ k}\Omega$$

The input capacitance of any inverter is

$$C_{in} = C_{inn} + C_{inp} = \frac{3}{2}C'_{ox}(W_n L_n + W_p L_p) = \frac{3}{2} \cdot 800 \text{ aF} \cdot (10 \cdot 2 + 10 \cdot 2) = 48 \text{ fF}$$

the output capacitance is

$$C_{out} = C_{outn} + C_{outp} = C'_{ox}(W_n L_n + W_p L_p) = 32 \text{ fF}$$

The total capacitance on the output of any inverter is the sum of its own output capacitance and the input capacitance of the next (identical) stage. This is given by

$$C_{tot} = C_{out} + C_{in} = 80 \text{ fF}$$

thus

$$t_{PHL} + t_{PLH} = (R_{n1} + R_{p2})C_{tot} = (2.4k + 7.2k) \cdot 80 \text{fF} = 768 \text{ ps}$$

The oscillator frequency, from Eq. (11.11), is then

$$f_{osc} = \frac{1}{5 \cdot 768 \text{ ps}} = 260 \text{ MHz}$$

The SPICE simulation results are shown in Fig. 11.12. SPICE gives an f_{osc} of approximately 300 MHz.

Normally, the *PDP* is determined with the minimum-size devices; for CN20 that would be $W_1 = W_2 = 3$ μm. However, for this example, larger than minimum-size devices are specified. The average power dissipated per inverter, using Eq. (11.18), is

$$P_{avg} = 80 \text{ fF} \cdot (5)^2 \cdot (260 \text{ MHz}) = 520 \text{ }\mu W$$

The power delay product, using hand calculations, is 400 fJ (femto-joules). SPICE simulation gives a *PDP* of 330 fJ. ■

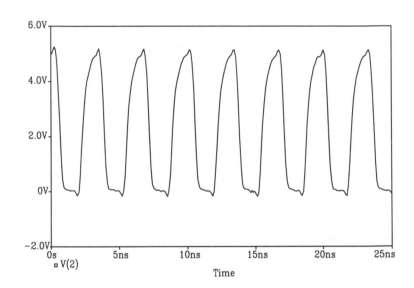

Figure 11.12 SPICE simulation of the five-stage ring oscillator of Ex. 11.7.

11.3 Layout of the Inverter

If care is not taken when laying out CMOS circuits, the parasitic devices present can cause a condition known as latch-up. Once latch-up occurs, the inverter output will not change with the input; that is, the output may be stuck in a logic state. To correct this problem, the power must be removed. Latch-up is especially troubling in output driver circuits. Manufacturers of integrated circuits often use NMOS inverters (discussed later in this chapter) for output drivers, thus eliminating the possibility of latch-up.

11.3.1 Latch-up

Figure 11.13 illustrates two methods of laying out a minimum-size inverter. The cross-sectional view in Fig. 11.14 shows both the n-channel and the p-channel MOSFETs that make up an inverter. Notice first that in Fig. 11.7, the output pulse feeds through the gate-drain capacitance. This causes the output to change in the same direction as the input before the inverter starts to switch. This feedthrough and the parasitic bipolar transistors cause the latch-up.

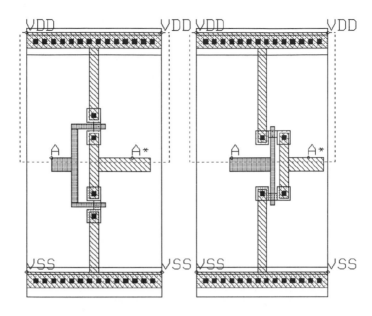

Figure 11.13 Two inverter layout styles.

In Fig. 11.14, the emitter, base, and collector of transistor Q1 are the source of the p-channel, the n-well, and the substrate, respectively. Transistor Q2's collector, base, and emitter are the n-well, substrate, and source of the n-channel transistor. Resistors RW1 and RW2 represent the effects of the resistance of the n-well, and resistors RS1 and RS2 represent the resistance of the substrate. The capacitors C1 and C2 represent the drain implant depletion capacitance, that is, the capacitance between the drains of the transistors and the source and substrate. The parasitic circuit resulting from the inverter layout is shown in Fig. 11.15.

If the output of the inverter switches fast enough, the pulse fed through C2 (for positive going inputs) can cause the base-emitter junction of Q2 to become forward biased. This then causes the current through RW2 and RW1 to increase, causing Q1 to turn on. When Q1 is turned on, the current through RS1 and RS2 increases, causing Q2 to turn on harder. This positive feedback will eventually cause Q2 and Q1 to turn on fully and remain that way until the power is removed and reapplied. A similar argument can be made for negative-going inputs feeding through C1.

Several techniques reduce the latch-up problem. The first technique is to slow the rise- and falltimes of the logic gates, reducing the amount of signal fed through C1 and C2. Reducing the areas of M1 and M2's drains lowers the size of the depletion capacitance and the amount of signal fed through. Probably the best method of reducing latch-up effects is to reduce the parasitic resistances RW1 and RS2. If these resistances are zero, Q1 and Q2 never turn on. The value of these resistances, as seen

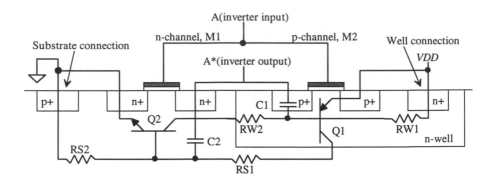

Figure 11.14   Cross-sectional view of an inverter showing parasitic
bipolar transistors and resistors.

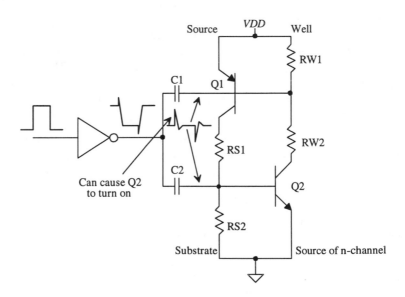

Figure 11.15   Schematic used to describe latch-up.

in Fig. 11.14, is a strong function of the distance between the well and substrate contacts. Simply put, the closer these contacts are to the inverter, the fewer are the chances the inverter will latch up. These contacts should be not only close, but also plentiful. Placing substrate and well contacts between the p- and n-channel MOSFETs provides a low-resistance connection to *VDD* and ground, significantly helping to reduce latchup (see Fig. 11.16 for a simple layout example). Placing n+ and p+ areas between or around circuits reduces the amount of signal reaching a given circuit from another circuit. These diffusions are sometimes called guard rings. Notice that poly cannot be used to connect the gates of the MOSFETs, since poly over the n+ or p+ will be interpreted as a MOSFET. Therefore, metal2 is used to connect the p- and the n-channel MOSFETs together. The cost for reducing the possibility of latch-up is a more complicated layout in a larger area.

Large MOSFETs, required to drive off-chip loads, are especially susceptible to latch-up because of the large drain depletion capacitances. The only way to design latch-up free output drivers is to use only one type of MOSFET, in most cases an n-channel. Eliminating the n-well and the p-channel transistor eliminates the possibility of latch-up. This will be discussed further in the next section.

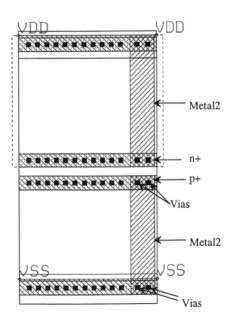

Figure 11.16 Alternative standard-cell frame used for better latch-up protection.

11.4 Sizing for Large Capacitive Loads

Designing a circuit to drive large capacitive loads with minimum delay is important when driving off-chip loads. Consider the inverter string driving a load capacitance, labeled C_{load} and shown in Fig. 11.17. If a single inverter were to drive C_{load}, the delay times would be

$$t_{PHL} + t_{PLH} = (R_n + R_p) \cdot (C_{out} + C_{load}) \tag{11.20}$$

If, moving toward the load, cascading N inverters are used, each inverter larger than the previous by a factor A (that is, the width of each MOSFET is multiplied by A), a minimum delay can be obtained as long as A and N are picked correctly. Each inverter's input capacitance is also larger than the previous inverter's input capacitance by a factor of A. If the load capacitance is equal to the input capacitance of the last inverter multiplied[1] by A, then

$$\text{Input C of final inverter} = C_{in1} \cdot A^N = C_{loaa} \tag{11.21}$$

where C_{in1} is the input capacitance of the first inverter. Rearranging Eq. (11.21) gives

$$A = \left[\frac{C_{load}}{C_{in1}} \right]^{\frac{1}{N}} \tag{11.22}$$

The total delay of the inverter string is given by

$$(t_{PHL} + t_{PLH})_{total} = \underbrace{(R_{n1} + R_{p1})(C_{out1} + A C_{in1})}_{\text{First-stage delay}} + \underbrace{\frac{(R_{n1} + R_{p1})}{A} \cdot (A C_{out1} + A^2 C_{in1})}_{\text{Second-stage delay}} \dots \tag{11.23}$$

where R_{n1} and R_{p1} are the effective resistances of the first inverter and C_{out1} is the output capacitance of the first inverter. As the inverters are increased in size by A, their capacitances, both input and output, increase by A while their resistances decrease by a factor A. The equation (11.23) can be written as

$$(t_{PHL} + t_{PLH})_{total} = \sum_{k=1}^{N} (R_{n1} + R_{p1})(C_{out1} + A C_{in1}) = N(R_{n1} + R_{p1})(C_{out1} + A C_{in1}) \tag{11.24}$$

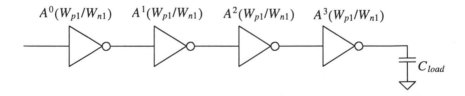

$$A^0(W_{p1}/W_{n1}) \qquad A^1(W_{p1}/W_{n1}) \qquad A^2(W_{p1}/W_{n1}) \qquad A^3(W_{p1}/W_{n1})$$

C_{load}

Figure 11.17   Cascade of inverters used to drive a large load capacitance.

---

[1]   Consider this as if the load capacitance were simulating the input capacitance of the next inverter (if there was another inverter).

or with the help of Eq. (11.22):

$$(t_{PHL} + t_{PLH})_{total} = N(R_{n1} + R_{p1})\left[C_{out1} + \left(\frac{C_{load}}{C_{in1}}\right)^{\frac{1}{N}} \cdot C_{in1} \right] \qquad (11.25)$$

The minimum delay can be found by taking the derivative of this equation with respect to N, setting the result equal to zero, and solving for N. Taking the derivative of Eq. (11.25) with respect to N gives

$$(R_{n1} + R_{p1})C_{out1} + (R_{n1} + R_{p1})C_{in1}\left[\left(\frac{C_{load}}{C_{in1}}\right)^{\frac{1}{N}} + N \cdot \left(\frac{C_{load}}{C_{in1}}\right)^{\frac{1}{N}}\frac{\ln(C_{load}/C_{in1})}{-N^2} \right] = 0 \qquad (11.26)$$

The first term in this equation is the intrinsic delay of the first inverter in our cascade of inverters. If we assume that this delay is small, solving this equation for N gives

$$N = \ln\frac{C_{load}}{C_{in1}} \qquad (11.27)$$

Eqs. (11.27) and (11.22) are used to design a cascade of inverters in order to drive a large capacitance. Note that the larger the first inverter, the fewer the number of inverters needed to drive a given capacitive load. Logic families like the 74HCXX series use fairly large MOSFETs throughout the entire chip. This allows driving large capacitances, say >50 pF, with typically two or three buffer stages. In very-large-scale integration (VLSI) design where the MOSFETs are generally close to minimum size, the number of stages can be greater than this. The following example illustrates output buffer design in its simplest form.

Example 11.8
Estimate $t_{PHL} + t_{PLH}$ for the inverter shown in Fig. Ex11.8 driving a load capacitance of 20 pF. Design a buffer to drive the load capacitance with a minimum delay. Compare the propagation delays of both circuits using SPICE.

The total propagation delay of the unbuffered inverter is given by

$$t_{PHL} + t_{PLH} = \left(12k\frac{2}{3} + 36k\frac{2}{9}\right) \cdot \left(\underbrace{2 \cdot 3 \cdot 800 \text{ aF} + 2 \cdot 9 \cdot 800 \text{ aF}}_{C_{out1}=19.2\text{ fF}} + \underbrace{20 \text{ pF}}_{C_{load}} \right) = 320 \text{ ns !}$$

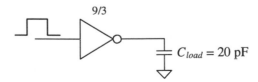

Figure Ex11.8

Designing a buffer begins with determining C_{in1}. For the present case $C_{in1}= \frac{3}{2}C_{out1} = 28.8$ fF. The number of inverters using Eq. (11.27) is

$$N = \ln\left(\frac{20 \text{ pF}}{28.8 \text{ fF}}\right) = 6.54 \rightarrow 7 \text{ stages}$$

In order to maintain the same logic, that is, an inversion of the input signal, we will use seven inverters. In practice, the difference in delay between six and seven inverters is negligible. If we did not want a logic inversion, we would use six stages. The area factor is then

$$A = \left[\frac{20 \text{ pF}}{28.8 \text{ fF}}\right]^{\frac{1}{7}} = 2.55$$

The total delay, using Eq. (11.25), is then

$$(t_{PHL} + t_{PLH})_{total} = 7(16k)(19.2 \text{ fF} + 2.55 \cdot 28.8 \text{ fF}) = 10.4 \text{ ns}$$

or over 30 times faster. Since the p-channel width is three times that of the n-channel width, the propagation delay times, t_{PHL} and t_{PLH}, are equal, or

$$t_{PHL} = t_{PLH} = 5.2 \text{ ns}$$

A schematic of the design is shown in Fig. 11.18. The actual sizes were changed to a number close to that given using the value of A calculated above to make the layout easier. Notice that the first inverter is the same inverter shown above. The SPICE simulation results are shown in Fig. 11.19. Note that the unbuffered inverter does not fully charge the capacitor since the input of the inverter changes back to zero volts 15 ns after it changes to *VDD*. ■

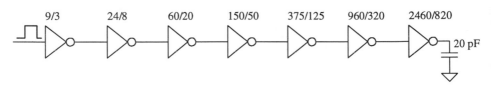

Figure 11.18   Buffer designed in Ex. 11.8.

It should be clear that, although this technique results in the least delay in driving the 20 pF load, the MOSFETs needed are very large. In many applications, the minimum delay through a buffer is not required. A specification that the delay be less than some value is given. Consider the following example.

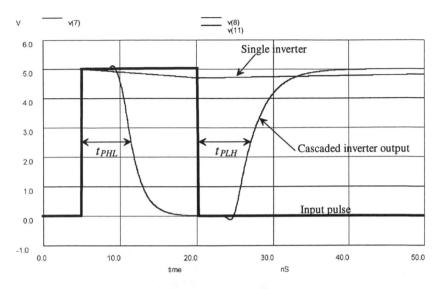

Figure 11.19   Simulation results from Ex. 11.8.

Example 11.9

Redesign the buffer of Ex. 11.8 so that the delay, $t_{PHL} + t_{PLH}$, is less than 15 ns. (The minimum delay was 10.4 ns in Ex. 11.8.)

In order to maintain the logic inversion, either three or five stages should be used. Let's begin by trying three stages. The area factor for three stages is given by

$$A = \left[\frac{20 \text{ pF}}{28.8 \text{ fF}} \right]^{1/3} = 8.86$$

The delay is calculated using Eq. (11.25) and is given by

$$t_{PHL} + t_{PLH} = 3(16k)(19.2 \text{ fF} + 8.86 \cdot 28.8 \text{ fF}) = 13.2 \text{ ns}$$

or

$$t_{PHL} = t_{PLH} = 6.6 \text{ ns}$$

The resulting buffer is shown in Fig. 11.20. The layout size of this buffer is significantly smaller than the buffer designed in Ex. 11.8, while the increase in delay is modest. ∎

Layout of Large MOSFETs

The time and effort it takes to lay out the large MOSFETs used in an output buffer can be greatly reduced using cell hierarchy. As a simple example, let's lay out a 250/2

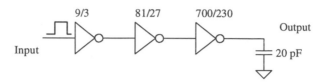

Figure 11.20 Buffer design of Ex. 11.9.

n-channel MOSFET. We begin by creating a cell, called NAA25X2 (n-active-area 25 by 2), with a rank of 1 and shown in Fig. 11.21.

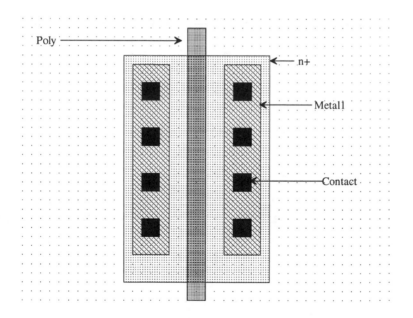

Figure 11.21 Layout of an n-channel MOSFET measuring 25 μm (width) by 2 μm (length).

The next step is to create a cell called N250X2 with a rank of 2. We then set the object (using the **obj** command button in LASI) to the cell NAA25X2 and select the **Add** command button. Figure 11.22 shows four NAA25X2 cells added to the N250X2 cell. The "trick" when placing the NAA25X2 cells is to overlap the source/drain areas. As discussed in Ch. 5, this sharing of areas reduces the depletion capacitance to substrate of the drain/source implants. Figure 11.23a shows the layout of the 250 by 2 n-channel MOSFET with the sources, drains, and gates of each of the individual MOSFETs connected together. The only thing missing from this layout is the connection to the substrate. (Not providing well and substrate connections for the

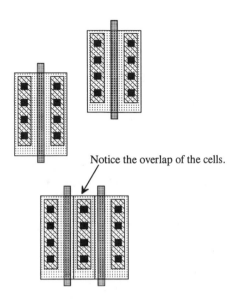

Figure 11.22   Placing basic cells to form a large MOSFET.

MOSFETs is a *fatal* layout error.)   Since the standard-cell frame, SFRAME (with a rank of 1) in the CN20 setups provided with LASI, provides these connections, we could add this frame to our basic layout. The result is shown in Fig. 11.23b.

11.4.1 Distributed Drivers

Consider the driver circuit shown in Fig. 11.24a containing 11 inverters. If all of the inverters shown in the figure are the same size, the delay from the input to the output is

$$t_{PHL} + t_{PLH} = (R_n + R_p)(C_{out} + 10C_{in}) \tag{11.28}$$

Now consider the circuit shown in Fig. 11.24b with 13 inverters. Again, assuming all inverters are the same size, the delay from the input to the output is

$$t_{PHL} + t_{PLH} = (R_n + R_p)[(C_{out} + 2C_{in}) + (C_{out} + 5C_{in})] = (R_n + R_p)[2C_{out} + 7C_{in}] \tag{11.29}$$

which is less delay than the circuit with 11 inverters. Often, distributing the signal into different paths can reduce the propagation delay. At this point we can ask the question, "Why not make the first inverter in the circuits of Fig. 11.24 really large so that it has small effective resistances for driving the ten inverters quickly?" The answer is simply that as we increase the size of an inverter, we also increase its input capacitance. In SPICE simulations, we use ideal voltage sources to drive the first gate in our circuit. In practice, this inverter is driven from another gate somewhere on the chip. Increasing the size will slow the propagation delay-time of the gate driving this inverter.

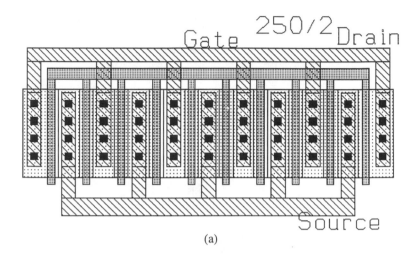

(a)

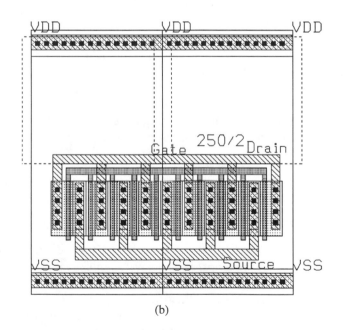

(b)

Figure 11.23 (a) Layout of a 250 by 2 n-channel MOSFET and (b) using a standard-cell frame.

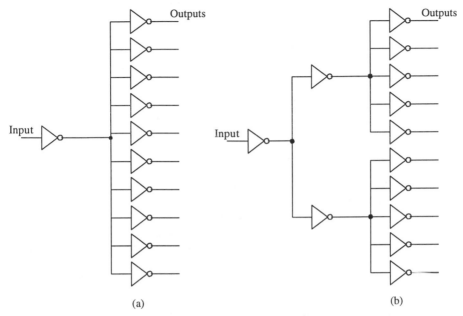

Figure 11.24 Distributed drivers.

11.4.2 Driving Long Lines

Often when designing large systems, a signal may need to be driven across the chip. In some cases, for example, dynamic random-access memory (DRAM), the signal must be transmitted over a line that has a large parasitic resistance and capacitance. We need to develop a method of determining the delay through this line using hand calculations. This will lend insight to the design and help to determine exactly how to design the driver (or drivers).

Consider the driver circuit shown in Fig. 11.25. The inverter is driving an RC transmission line with resistance/unit length, r, capacitance/unit length, c, and unit length, l. We can estimate the delay from the input to voltage across the capacitor by adding the delays. This is given by

$$t_{PHL} + t_{PLH} = (R_n + R_p)(C_{out} + c \cdot l + C_{load}) + 0.35 \cdot rcl^2 + (r \cdot l)(C_{load}) \qquad (11.30)$$

where the first term in this equation is the delay associated with the inverter driving the total capacitance at its output to ground. The second term is the delay through the line, while the last term is an estimate of the delay associated with driving a capacitive load through the line's resistance. The most common method of reducing the delay through the line is to place buffer stages at different locations along the line. This effectively breaks the line up and can lower the overall delay. If C_{load} is a major contributor to the delay, a buffer can be inserted between the RC line and C_{load} to reduce the delay.

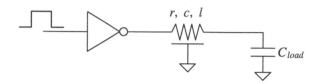

Figure 11.25 Driving an RC transmission line.

11.5 Other Inverter Configurations

Three other inverter configurations are shown in Fig. 11.26. The inverter shown in Fig. 11.26a is an NMOS-only inverter, useful in avoiding latch-up. The inverters shown in Fig. 11.26b and c use a p-channel load, which is, in general, most useful in logic gates with a large number of inputs (more on this in the next chapter). In general, the selection of the MOSFET sizes follows the 4 to 1 rule; that is, the resistance (R_n or R_p) of the load is made four times larger than the resistance of M1. The output logic low will never reach 0 V for these inverters, and thus the noise margins are poorer than the basic CMOS inverter of Fig. 11.1. Also, DC power will be dissipated when the output logic level is a low since a drain current will flow through the inverters. The output high level of the inverter of Fig. 11.26c will reach VDD, while the other inverter's output high level will be a threshold voltage drop below VDD. It might be concluded that the power dissipation of the inverters shown in Fig. 11.26 is greater than the basic CMOS inverter. However, since the input capacitance of these inverters is less than the basic CMOS inverter and the output voltage swing is reduced, the inverter with the greatest power dissipation is determined by the operating frequency. At high operating frequencies, the basic CMOS inverter dissipates the most power.

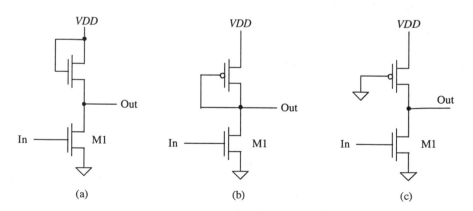

Figure 11.26 Other inverter configurations.

11.5.1 N-Channel Only Output Drivers

Because of the susceptibility of the basic CMOS inverter to latch-up, output drivers consisting of only n-channel MOSFETs are used. Figure 11.27 shows the basic "NMOS super buffer." When the input signal is low, M1 and M4 are off while M2 and M3 are on. The output is pulled to ground through M2. A high on the input to the buffer causes M1 and M4 to turn on pulling the output to $VDD - V_{THN}$, assuming the input high-signal amplitude is VDD.

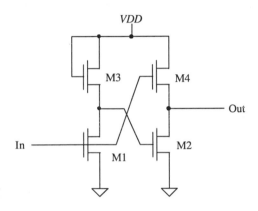

Figure 11.27 NMOS super buffer.

The reduced output voltage of the NMOS-only output buffer can be improved using the circuit of Fig. 11.28. The inverter driving the gate of M2 uses an on-chip generated DC voltage of nominally $VDD + 2$ V. This allows the output signal to reach

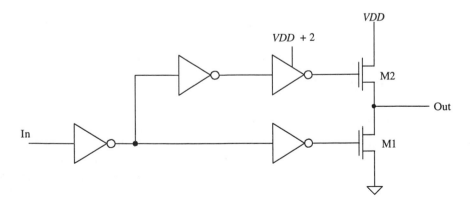

Figure 11.28 Alternative output buffer.

VDD. Thus, the output swings from 0 to *VDD* similar to the CMOS output buffer. Note that with the addition of an enabling logic gate, the gates of M1 and M2 can be held at ground, forcing the output into the high-impedance (Hi-Z) state. This is sometimes referred to as tri-state output since the output can be a 1, 0, or Hi-Z.

11.5.2 Inverters with Tri-State Outputs

Two configurations used in the design of an inverter with tri-state outputs are shown in Fig. 11.29. A high on the *S* input allows the circuit to operate normally, that is, as an inverter. A low on the S input forces the output into the Hi-Z, or high-impedance state. These circuits are useful when data are shared on a communication bus. The logic symbol of the tri-state inverter is also shown in Fig. 11.29.

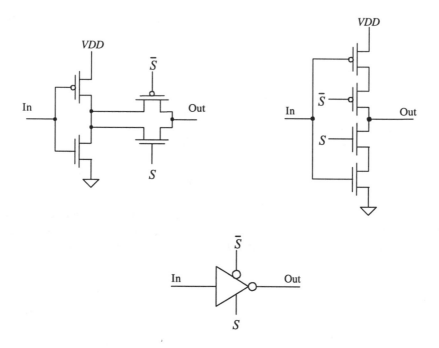

Figure 11.29 Circuits and logic symbol for the tri-state inverter.

11.5.3 The Bootstrapped NMOS Inverter

Consider the modified version of the NMOS inverter of Fig. 11.26 shown in Fig. 11.30. This inverter configuration is called the bootstrapped NMOS inverter. It is used when the output voltage must swing up to *VDD*. To understand the operation, consider first the case when input to the inverter is a logic high. The MOSFET M1 is on, and the output is pulled down to approximately

$$V_{OL} = (VDD - 2V_{THN}) \cdot \frac{R_{n1}}{R_{n1} + R_{n2}} \tag{11.31}$$

For the device sizes given, V_{OL} is approximately 1/2 V. Next consider the case when the input transitions from a high to a low. The MOSFET M4 is used as a capacitor. The idea is to capacitively couple the output pulse to the gate of M2. The result is an increase in the gate potential above VDD, allowing M2 to fully turn on (pulling the output up by its bootstraps). To understand the operation, consider the circuit shown in Fig. 11.31. MOSFET M4 is replaced with a capacitor C_4 $(= W_4 \cdot L_4 \cdot C'_{ox})$. When M1 shuts off, with the input going low, a capacitive voltage divider exists between the output and the gate of M2 given by

$$\text{Change in M2's gate voltage} = (VDD - V_{OL}) \cdot \frac{C_4 + C_{inn2}}{C_4 + C_{inn2} + C_{inn3}} \tag{11.32}$$

Without bootstrapping (via devices M3 and M4), the gate of the M2 is tied to VDD and the output is limited to $VDD - V_{THN}$. If the gate of M2 is bootstrapped up to $VDD + 2$ V, then M2 fully turns on and the output goes to VDD. Therefore, the change in M2's gate potential should be > 2 V. In general, the size ($W \cdot L$) of M4 should be ten times larger than the size of M3. This is equivalent to saying that C_4 should be > $10 \cdot C_{inn3}$. Note that this is a dynamic effect. The gate of M2 under DC conditions is $VDD - V_{THN}$, while the output high is $VDD - 2V_{THN}$ or lower than the nonbootstrapped inverter of Fig. 11.26a.

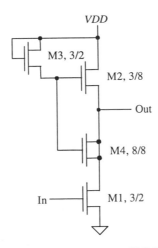

Figure 11.30 Bootstrapped NMOS inverter.

Example 11.10
Simulate the operation of the inverter shown in Fig. 11.30 using SPICE.

The simulation results are shown in Fig. 11.32. Notice how the output doesn't go all the way to ground or VDD. We can decrease the size (W/L) of M2 (increase

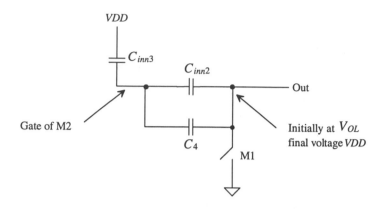

Figure 11.31 Circuit used to illustrate the bootstrapping effect.

the size of its switching resistance) in Fig. 11.30 to make the output go closer to ground. The price we pay for this is an increase in t_{PLH}. Also, the bootstrapped inverter will swing up to *VDD* if we increase the size of the capacitor M4. However, since this capacitor is charged through M3, the resulting increase in charging time will cause the maximum practical operating frequency of the gate to decrease. ■

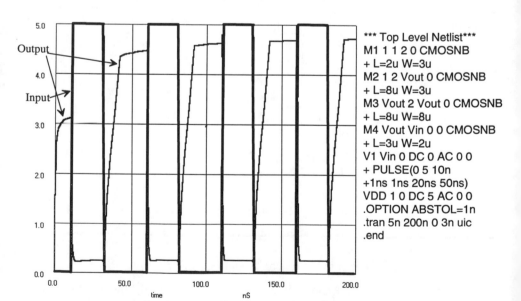

```
*** Top Level Netlist***
M1 1 1 2 0 CMOSNB
+ L=2u W=3u
M2 1 2 Vout 0 CMOSNB
+ L=8u W=3u
M3 Vout 2 Vout 0 CMOSNB
+ L=8u W=8u
M4 Vout Vin 0 0 CMOSNB
+ L=3u W=2u
V1 Vin 0 DC 0 AC 0 0
+ PULSE(0 5 10n
+1ns 1ns 20ns 50ns)
VDD 1 0 DC 5 AC 0 0
.OPTION ABSTOL=1n
.tran 5n 200n 0 3n uic
.end
```

Figure 11.32 Simulation results for Example 11.10.

REFERENCES

[1]   R. L. Geiger, P. E. Allen and N. R. Strader, *VLSI-Design Techniques for Analog and Digital Circuits,* McGraw-Hill Publishing Co., 1990.   ISBN 0-07-023253-9.

[2]   N. H. E. Weste and K. Eshraghian, *Principles of CMOS VLSI Design,* Addison- Wesley, 2nd ed., 1993. ISBN 0-201-53376-6.

PROBLEMS

Use the CN20 process for the following problems unless otherwise stated.

11.1 Design and simulate the DC characteristics of an inverter with V_{SP} approximately equal to V_{THN}. Estimate the resulting noise margins for the design.

11.2 Repeat Ex. 11.6 for MOSFETs with $W = 10$ μm and a load capacitance of 1 pF.

11.3 Estimate the oscillation frequency of a 31-stage ring oscillator using minimum-size inverters.

11.4 Lay out the standard-cell frame of Fig. 11.16. Explain how the added implants help to reduce latch-up.

11.5 Design and simulate the operation of a buffer to drive a 50 pF capacitive load from an inverter with size of 150/50. The $t_{PHL} + t_{PLH}$ should be less than 10 ns.

11.6 Repeat Ex. 11.9, using a maximum delay of 20 ns, where the first inverter in the series is minimum size, that is, 3/2 (p-channel) and 3/2 (n-channel).

11.7 Design and simulate the delay of a minimum-size inverter driving a 1 mm poly line terminated with a 1 pF capacitor.

11.8 Lay out an inverter with a size of 450/150 using the standard-cell frame of Ch. 5.

11.9 Simulate the operation and explain the results for the NMOS super buffer shown in Fig. 11.27.

11.10 Repeat Ex. 11.10 if M4's size is increased to 20/20.

11.11 Repeat Ex. 11.5 using minimum-size (0.9/0.6) MOSFETs in the CMOS14TB process.

11.12 Repeat Ex. 11.6 using minimum-size MOSFETs in CMOS14TB.

11.13 Sketch the cross-sectional views, at the positions indicated, for the layout shown in Fig. P11.13.

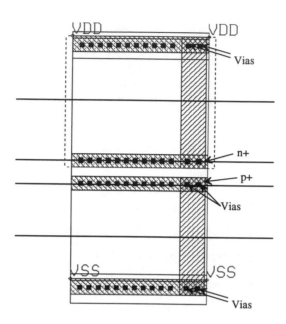

Figure P11.13

Static Logic Gates

In this chapter we discuss the DC characteristics, dynamic behavior, and layout of CMOS static logic gates. Static logic means that the output of the gate is always a logical function of the inputs and always available on the outputs of the gate regardless of time. We begin with the NAND and NOR gates.

12.1 DC Characteristics of the NAND and NOR Gates

The two basic input NAND and NOR gates are shown in Fig. 12.1. Before we get into the operation, notice that each input into the gate is connected to both a p- and an n-channel transistor similar to the inverter of the last chapter. We will make use of the results of Ch. 11 to explain the operation of these gates.

12.1.1 DC Characteristics of the NAND Gate

The NAND gate of Fig. 12.1 requires both inputs to be high before the output will switch low. Let's begin our analysis by determining the voltage transfer curve of a gate with p-channel MOSFETs that have $W = W_p$, $L = L_p$, and n-channel MOSFETs with $W = W_n$, $L = L_n$. If both inputs of the gate are tied together, the gate behaves like an inverter.

To determine the gate switching point voltage, V_{SP}, we must remember that two MOSFETs in parallel behave like a single MOSFET with a width equal to the sum of the individual widths. For the two parallel p-channel MOSFETs in Fig. 12.1, we can write

$$W_3 + W_4 = 2W_p \tag{12.1}$$

again assuming that all p-channel transistors are of the same size. The transconductance parameters can also be combined into the transconductance parameter of a single MOSFET, or

$$\beta_3 + \beta_4 = 2\beta_p \tag{12.2}$$

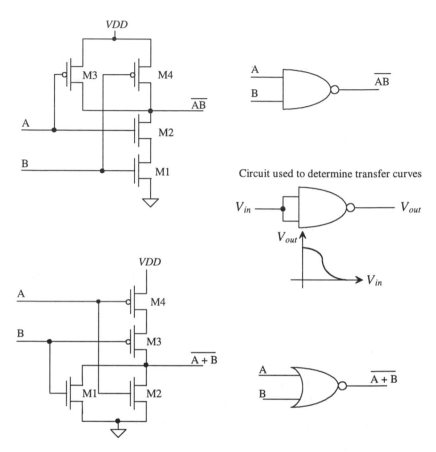

Figure 12.1 NAND and NOR gate circuits and logic symbols.

If we neglect the body effect, then two MOSFETs in series (with their gates tied together) behave like a single MOSFET with a channel length equal to the sum of the individual MOSFET lengths. Referring to the figure above of the NAND, we can write for the n-channel MOSFETs

$$L_1 + L_2 = 2L_n \tag{12.3}$$

and the transconductance of the single MOSFET is given by

$$\beta_1 + \beta_2 = \frac{\beta_n}{2} \tag{12.4}$$

If we model the NAND gate with both inputs tied together as an inverter with an n-channel transistor having a width of W_n and length $2L_n$ and a p-channel MOSFET with a width of $2W_p$ and length L_p, then we can write the transconductance ratio as

$$\text{Transconductance ratio of NAND gate} = \frac{\beta_n}{4\beta_p} \qquad (12.5)$$

The switching point voltage, with the help of Eq. (11.4), of the two-input NAND gate is then given by

$$V_{SP} = \frac{\sqrt{\frac{\beta_n}{4\beta_p}} \cdot V_{THN} + (VDD - V_{THP})}{1 + \sqrt{\frac{\beta_n}{4\beta_p}}} \qquad (12.6)$$

or in general for an n-input NAND gate (see Fig. 12.2), we get

$$V_{SP} = \frac{\sqrt{\frac{\beta_n}{N^2 \cdot \beta_p}} \cdot V_{THN} + (VDD - V_{THP})}{1 + \sqrt{\frac{\beta_n}{N^2 \cdot \beta_p}}} \qquad (12.7)$$

Again, it should be remembered that we have neglected the body effect (an increase in the threshold voltage with increasing V_{SB}). Voltage transfer curves using one input, with the others tied to VDD, will give slightly different results because of this effect.

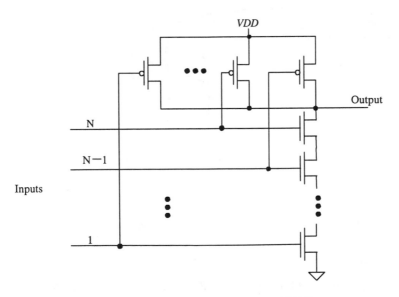

Figure 12.2 Schematic of an n-input NAND gate.

Example 12.1
Determine V_{SP} by hand calculations and compare to a SPICE simulation for a three-input NAND gate using minimum-size devices.

The switching point voltage is determined by calculating the transconductance ratio of the gate, or

$$\sqrt{\frac{\beta_n}{N^2\beta_p}} = \sqrt{\frac{\frac{50\mu A/V^2 \cdot 3\mu m}{2\mu m}}{9 \cdot \frac{17\mu A/V^2 \cdot 3\mu m}{2\mu m}}} = 0.572$$

and then using Eq. (12.7),

$$V_{SP} = \frac{0.572 \cdot (0.83) + (5 - 0.92)}{1.572} = 2.9 \text{ V}$$

The SPICE simulation results are shown in Fig. 12.3. The simulation gives a V_{SP} of approximately 3.1 V. ∎

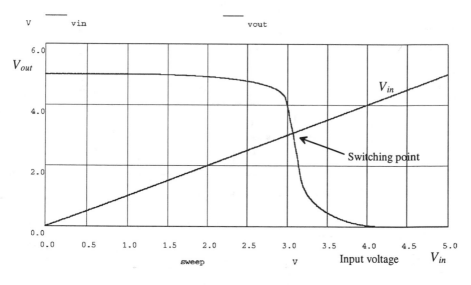

Figure 12.3 Voltage transfer characteristics of the three-input minimum-size NAND gate.

12.1.2 DC Characteristics of the NOR gate

Following a similar analysis for the n-input NOR gate (see Fig. 12.4) gives a switching point voltage of

$$V_{SP} = \frac{\sqrt{\frac{N^2 \cdot \beta_n}{\beta_p}} \cdot V_{THN} + (VDD - V_{THP})}{1 + \sqrt{\frac{N^2 \cdot \beta_n}{\beta_p}}} \qquad (12.8)$$

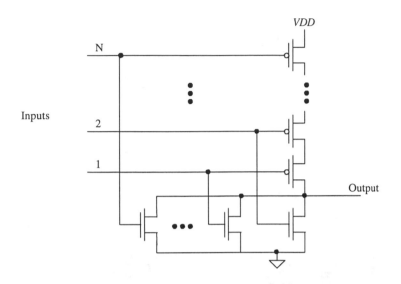

Figure 12.4 Schematic of an n-input NOR gate.

Example 12.2

Compare the switching point voltage of a three-input NOR gate made from minimum-size MOSFETs to that of the three-input NAND gate of Ex. 12.1. Comment on which gate is closer to ideal, that is, $V_{SP} = VDD/2$.

The V_{SP} of the minimum-size three-input NOR gate is 1.35 V, while the V_{SP} of the minimum-size three-input NAND gate was calculated to be 2.9 V. For an ideal gate $V_{SP} = 2.5$ V, so that the NAND gate is closer to ideal than the NOR gate. This arises because the transconductance (actually the mobility) of the n-channel is larger than that of the p-channel. In CMOS digital design, the NAND gate is used most often. This is due partly to the DC characteristics, better noise margins, and the dynamic characteristics. We will also see shortly that the NAND gate has better transient characteristics than the NOR gate. ∎

12.2 Layout of the NOR and NAND Gates

Layout of the three-input minimum-size NOR and NAND gates is shown in Fig. 12.5 using the standard-cell frame. MOSFETs in series, for example, the n-channel MOSFETs in the NAND gate, are laid out using a single-drain and a single-source contact. The active area between the gate poly is shared between two devices. This has the effect of reducing the parasitic drain/source implant capacitances. MOSFETs in parallel, for example, the n-channel MOSFETs in the NOR gate, can share a drain area or a source area. The inputs of the gates are on poly and the outputs are on metal1.

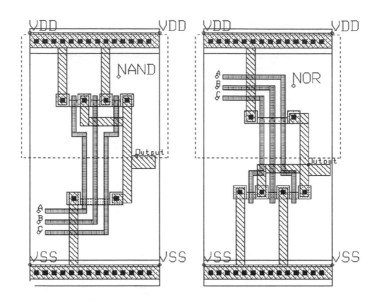

Figure 12.5 Layout of the NAND and NOR gate.

12.3 Switching Characteristics

Consider the parallel connection of identical MOSFETs shown in Fig. 12.6 with their gates tied together. From the equivalent digital models, also shown, we can determine the intrinsic time constant of this chain of N MOSFETs by

$$t_{PLH} = \frac{R_p}{N} \cdot (N \cdot C_{outp}) = R_p C_{outp} \tag{12.9}$$

which for CN20 from Ex. 11.5 is 230 ps. With an external load capacitance, the low to high delay-time becomes

$$t_{PLH} = \frac{R_p}{N}(N \cdot C_{outp} + C_{load}) \tag{12.10}$$

This again assumes that the MOSFET's gates are tied together. For n-channel MOSFETs in parallel, a similar analysis yields

$$t_{PHL} = \frac{R_n}{N}(N \cdot C_{outn} + C_{load}) \tag{12.11}$$

The load capacitance, C_{load}, consists of all capacitances on the output node except the output capacitances of the MOSFETs in parallel.

Consider the series connection of identical n-channel MOSFETs shown in Fig. 12.7. We can estimate the intrinsic switching time of series-connected MOSFETs by

$$t_{PHL} = N \cdot R_n \left(\frac{C_{outn}}{N} \right) + 0.35 \cdot R_n C_{inn}(N-1)^2 \qquad (12.12)$$

The first term in this equation represents the intrinsic switching time of the series connection of MOSFETs, while the second term represents RC[1] delay caused by R_n charging C_{inn}. For the case when $N = 1$, this simply reduces to $R_n C_{outn}$. With an external load capacitance, the high to low delay-time becomes

$$t_{PHL} = N \cdot R_n \cdot \left(\frac{C_{outn}}{N} + C_{load} \right) + 0.35 \cdot R_n C_{inn}(N-1)^2 \qquad (12.13)$$

For p-channel MOSFETs in series, a similar analysis yields

$$t_{PLH} = N \cdot R_p \cdot \left(\frac{C_{outp}}{N} + C_{load} \right) + 0.35 \cdot R_p C_{inp}(N-1)^2 \qquad (12.14)$$

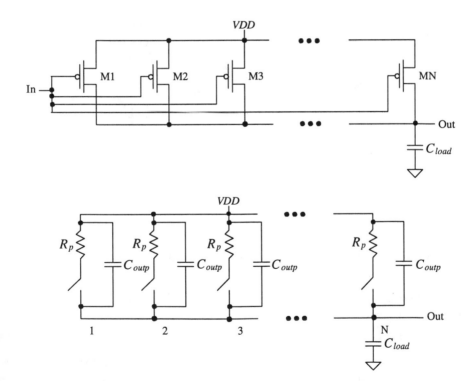

Figure 12.6 Parallel connection of MOSFETs and equivalent digital model.

[1] This effect is similar to the delay through an RC transmission line. The $N - 1$ term is due to the fact that the last MOSFET's source in the string is connected to ground and not to a load.

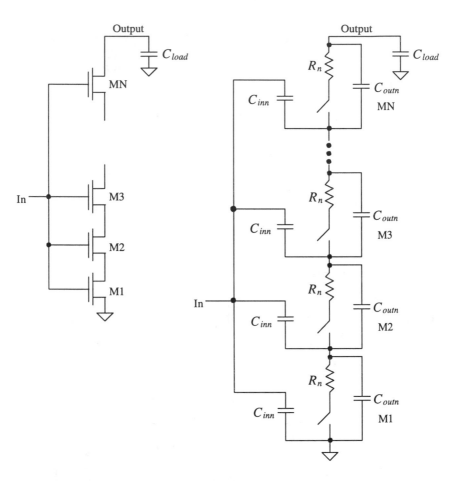

Figure 12.7 Series connection of MOSFETs and equivalent digital model.

These equations are approximations for the propagation delays giving results usually to within a factor of two of the measurements.

12.3.1 NAND Gate

Consider the n-input NAND gate of Fig. 12.8 driving a capacitive load C_{load}. The low-to-high propagation time, using Eq. (12.10), is

$$t_{PLH} = \frac{R_p}{N}\left(N \cdot C_{outp} + \frac{C_{outn}}{N} + C_{load}\right) \tag{12.15}$$

where here C_{load} represents the capacitance external to the gate, whereas in Eq. (12.10) C_{load} represented the capacitance external to the parallel p-channel MOSFETs. If the

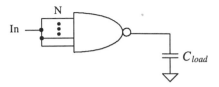

Figure 12.8 An n-input NAND gate driving a load capacitance.

load capacitance is much greater than the output capacitance of the gate, the low to high propagation time can be estimated by

$$t_{PLH} \approx \frac{R_p}{N} \cdot C_{load} \tag{12.16}$$

The high to low propagation time, using Eq. (12.13), is given by

$$t_{PHL} = N \cdot R_n \left[\frac{C_{outn}}{N} + N \cdot C_{outp} + C_{load} \right] + 0.35 \cdot R_n C_{inn} (N-1)^2 \tag{12.17}$$

or if C_{load} is much larger than the output capacitance of the gate

$$t_{PHL} \approx N \cdot R_n \cdot C_{load} \tag{12.18}$$

Example 12.3

Estimate the intrinsic propagation delays, $t_{PHL} + t_{PLH}$, of a three-input NAND gate made using minimum-size transistors. Estimate and simulate the delay when the gate is driving a load capacitance of 100 fF. Assume that inputs are tied together.

For the three-input NAND gate, the three p-channel MOSFETs in parallel are used to pull the output high. Since they are minimum-size

$$R_p = 24 \text{ k}\Omega \text{ and } C_{outp} = 4.8 \text{ fF}$$

this gives a low-to-high propagation time, using Eq. (12.15), with $C_{load} = 0$, of

$$t_{PLH} = \frac{24k}{3} \left[3 \cdot 4.8 \text{ fF} + \frac{4.8 \text{fF}}{3} \right] = 128 \text{ ps}$$

The intrinsic high-to-low propagation time of the three input NAND gate with minimum-size n-channel MOSFETs ($R_n = 8 \text{ k}\Omega$ and $C_{outn} = 4.8$ fF) using Eq. (12.17), with $C_{load} = 0$,

$$t_{PHL} = 3 \cdot 8k \cdot \left[\frac{4.8 \text{fF}}{3} + 3 \cdot 4.8 \text{fF} \right] + 0.35 \cdot 8k \cdot 7.2 \text{fF}(3-1)^2 \approx 500 \text{ ps}$$

With a load of 100 fF, the propagation delays become, $t_{PLH} = 928$ ps and $t_{PHL} = 2.9$ ns. The SPICE simulation results are shown in Fig. 12.9 followed by the SPICE3 netlist. Helping with convergence, the .OPTIONS statement was used, the initially off n-channel MOSFETs were specified, and the input transition times were specified at 0.1 ns instead of the unrealistic conditions of 1 ps. ∎

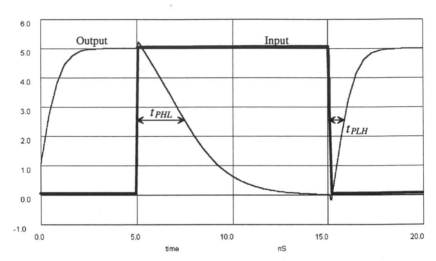

Figure 12.9 Output of the minimum-size NAND gate driving a 100 fF capacitor.

```
*** Top Level Netlist for Example 12.3 ***

C1      5 0 100f
M1      5 1 2 0 CMOSNB  L=2u W=3u AD=36p AS=36p PD=24u PS=24u OFF
M2      2 1 4 0 CMOSNB  L=2u W=3u AD=36p AS=36p PD=24u PS=24u OFF
M3      4 1 0 0 CMOSNB  L=2u W=3u AD=36p AS=36p PD=24u PS=24u OFF
M4      5 1 Vdd Vdd CMOSPB  L=2u W=3u AD=36p AS=36p PD=24u PS=24u
M5      5 1 Vdd Vdd CMOSPB  L=2u W=3u AD=36p AS=36p PD=24u PS=24u
M6      5 1 Vdd Vdd CMOSPB  L=2u W=3u AD=36p AS=36p PD=24u PS=24u
V1      Vdd 0   DC 5
V2      1 0     DC 0 PULSE(0 5 5n .1n .1n 10n )

***** Spice models and macro models *****

.MODEL CMOSNB NMOS LEVEL=4
+VFB=-9.73820E-01, LVFB=3.67458E-01,WVFB=-4.72340E-02
See Appendix A for a complete listing.

.MODEL CMOSPB PMOS LEVEL=4
+ vfb=-2.65334E-01, lvfb=6.50066E-02, wvfb=1.48093E-01
See Appendix A for a complete listing.

.OPTION ABSTOL=1u RELTOL=0.01 VNTOL=1mv ITL4=100
.probe
.tran 100p 20n 0  uic
.plot tran all
.print tran all
.end
```

The delay equations derived in this section are useful in understanding the limitations on the number of MOSFETs used in a NAND gate for high-speed design. However a more useful, though not as precise, method of determining delays can be found by considering the fact that whenever the output changes from VDD to ground the discharge path is through N resistors of value R_n. This is true if all or only one of the inputs to the NAND gate changes, causing the output to change.   Under these circumstances, Eq. (12.18) predicts the high-to-low delay-time, or for *series* connection of N n-channel MOSFETs,

$$t_{PHL} \approx N \cdot R_n \cdot C_{load} \qquad (12.19)$$

The case when the output of the NAND gate changes from a low to a high is somewhat different then the high-to-low case. Referring to Fig. 12.6, we see that if one of the MOSFETs turns on, it can pull the output to VDD independent of the number of MOSFETs in parallel. Under these circumstances, Eq. (12.16) can be used with $N = 1$ to predict the low-to-high delay-time, or for a parallel connection of N p-channel MOSFETs,

$$t_{PLH} \approx R_p \cdot C_{load} \qquad (12.20)$$

We will try to use Eqs. (12.19) and (12.20) as much as possible because of their simplicity. The further simplified digital models of the n- and p-channel MOSFETs are shown in Fig. 12.10. (Input capacitance is not shown.)

Figure 12.10 Further simplification of digital models not showing input capacitance.

Example 12.4
Estimate, using Eqs. (12.19) and (12.20), the propagation delays for the three-input minimum-size NAND gate, with only one input switching driving a 100 fF load capacitance. Compare your results to SPICE.

The propagation delay-times are given by

$$t_{PHL} = 3 \cdot 8k \cdot 100fF = 2.4\,ns$$

and

$$t_{PLH} = 24k \cdot 100\,fF = 2.4\,ns$$

The SPICE simulation results gave $t_{PLH} \approx t_{PHL} \approx 2.3$ ns with one input switching. If we had used Eqs. 12.19 and 12.20 in Ex. 12.3 where all inputs where changing at the same time, the calculated t_{PHL} would have given an underestimate (but not by much), while the calculated t_{PLH} would have overestimated the delay. Also note that since the effective resistance of the p-channel is three times greater than that of the n-channel, the series connection of three NMOS devices gives approximately the same resistance as the single PMOS. The result is equal switching times and the reason the NAND gate is generally preferred over the NOR gate in CMOS circuit design. ■

12.3.2 Number of Inputs

As the number of inputs, N, to a static NAND (or NOR) gate increases, the scheme shown in Fig. 12.2 (Fig. 12.4) becomes difficult to realize. Consider a NOR gate with 100 inputs. This gate requires 100 p-channel MOSFETs in series and a total of 200 MOSFETs ($2N$ MOSFETs). The delay associated with the series p-channel MOSFETs charging of a load capacitance is too long for most practical situations.

Now consider the schematic of an N input NOR gate shown in Fig. 12.11, which uses $N + 1$ MOSFETs. If any input to the NOR gate is high, the output is pulled low through the corresponding n-channel MOSFET to a voltage, when designed properly, of a few hundred millivolts. If all inputs are low, then all n-channel MOSFETs are off and the p-channel MOSFET pulls the output high (to VDD). A simple analysis of the output low voltage, V_{OL}, with *one* input at VDD yields

$$\frac{\beta_p}{2}(VDD - V_{THP})^2 = \beta_n\left[(VDD - V_{THN})V_{OL} - \frac{V_{OL}^2}{2}\right] \qquad (12.21)$$

Assuming that the maximum V_{OL} allowed (the more inputs at VDD the lower V_{OL}) is 500 mV and that the n-channels have $W = 3$ μm and $L = 2$ μm results in a W of 4 μm and an L of 3 μm for the p-channel. In practice, the length of the p-channel can be increased beyond this size to lower V_{OL} further. The static power dissipated by this gate when the output is high, neglecting leakage currents, is zero. When the output is low, a static power is dissipated due to both n- and p-channels conducting. The current that flows under this condition with the above sizes is 150 μA. Decreasing the W/L of the p-channel lowers power draw at the price of increased t_{PLH}.

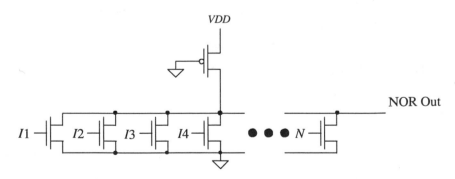

Figure 12.11 NOR configuration used for a large number of inputs.

12.4 Complex CMOS Logic Gates

Implementation of complex logic functions in CMOS uses the basic building blocks shown in Fig. 12.12. We have already used the circuits to implement NAND and NOR gates. In general, any And-Or-Invert (AOI) logic function can be implemented using these techniques. A major benefit of AOI logic is that for a relatively complex logic function the delay can be significantly lower than a logic gate implementation. Consider the following example.

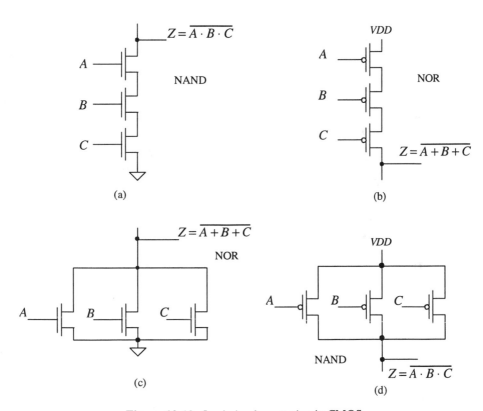

Figure 12.12 Logic implementation in CMOS.

Example 12.5

Using AOI logic, implement the following logic functions:

$$Z = \bar{A} + BC \quad \text{and} \quad Z = A + \bar{B}C + CD$$

The implementation of the first function is shown in Fig. 12.13a. Notice that the p-channel configuration is the dual of the n-channel circuit. The function we obtain is the complement of the desired function, and therefore an inverter is used to obtain Z. Using an inverter is, in general, undesirable if both true and complements of the input variables are available. Applying Boolean algebra to the logic function, we obtain

$$Z = \overline{A} + BC \Rightarrow \overline{Z} = \overline{\overline{A} + BC} = A \cdot (\overline{B} + \overline{C}) \Rightarrow Z = \overline{A \cdot (\overline{B} + \overline{C})}$$

The AOI implementation of the result is shown in Fig. 12.13b. Logically, the circuits of Figs. 12.13a and b are equivalent. However, the circuit of Fig. 12.13b is simpler and thus more desirable. Note that to reduce the output capacitance

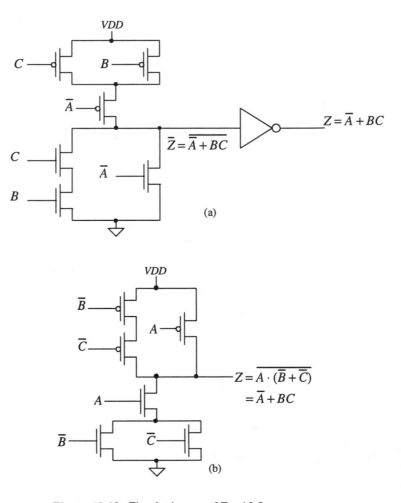

Figure 12.13 First logic gate of Ex. 12.5.

and thus decrease the switching times, the parallel combination of n-channel MOSFETs is placed at the bottom of the logic block.

The second logic function is given by

$$Z = A + \overline{B}C + CD = A + C(\overline{B} + D) \Rightarrow \overline{Z} = \overline{A + C(\overline{B} + D)} = \overline{A} \cdot (\overline{C} + B\overline{D})$$

or

$$Z = \overline{\overline{A} \cdot (\overline{C} + B\overline{D})}$$

The logic implementation is given in Fig. 12.14. ■

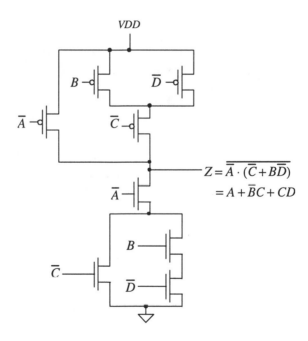

Figure 12.14 Second logic gate of Ex. 12.5.

Example 12.6
Using AOI logic, implement an exclusive OR gate (XOR).

The logic symbol and truth table for an XOR gate are shown in Fig. 12.15. From the truth table, the logic function for the XOR gate is given by

$$Z = A \oplus B = (A + B) \cdot (\overline{A} + \overline{B}) \tag{12.22}$$

or

$$\bar{Z} = \overline{A \oplus B} = \overline{(A + B) \cdot (\bar{A} + \bar{B})} = \bar{A} \cdot \bar{B} + A \cdot B$$

and finally

$$Z = \overline{\bar{A} \cdot \bar{B} + A \cdot B} = A \oplus B \qquad (12.23)$$

The CMOS AOI implementation of an XOR gate is shown in Fig. 12.16. ■

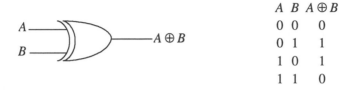

A	B	$A \oplus B$
0	0	0
0	1	1
1	0	1
1	1	0

Figure 12.15 Exclusive OR gate.

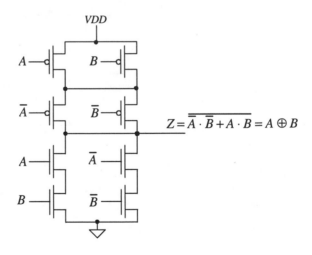

Figure 12.16 CMOS AOI XOR gate.

Example 12.7

Design a CMOS full adder using CMOS AOI logic.

The logic symbol and truth table for a full adder circuit are shown in Fig. 12.17. The logic functions for the sum and carry outputs can be written as

$$S_n = A_n \oplus B_n \oplus C_n$$

$$C_{n+1} = A_n \cdot B_n + C_n(A_n + B_n)$$

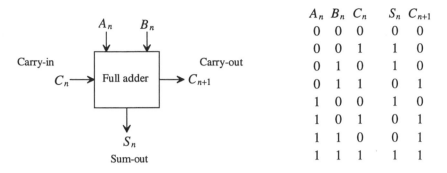

Figure 12.17 Full adder.

The logic expression for the sum can be rewritten as a sum of products

$$S_n = \overline{A}_n \overline{B}_n C_n + \overline{A}_n B_n \overline{C}_n + A_n \overline{B}_n \overline{C}_n + A_n B_n C_n$$

or since

$$\overline{C}_{n+1} = \left(\overline{A}_n + \overline{B}_n \right) \cdot \left(\overline{C}_n + \overline{A}_n \cdot \overline{B}_n \right)$$

the sum of products can be rewritten as

$$S_n = (A_n + B_n + C_n)\overline{C}_{n+1} + A_n B_n C_n$$

The AOI implementation of the full adder is shown in Fig. 12.18. ∎

12.4.1 Cascode Voltage Switch Logic

Cascode voltage switch logic (CVSL) or differential cascode voltage switch logic (DVSL) is a differential output logic that uses positive feedback to speed up the switching times (in some cases). Figure 12.19 shows the basic idea. A gate cross-connected load is used instead of using p-channel switches, as in the AOI logic, to pull the output high. Consider the implementation of $Z = \overline{A} + BC$. (This logic function was implemented in AOI in Fig. 12.13.) N-channel MOSFETs are used to implement Z and $\overline{Z}$ as shown in Fig. 12.20. Figure 12.21a shows the implementation of a two-input XOR/XNOR gate using CVSL, while Fig. 12.21b shows a CVSL three-input XOR/XNOR gate useful in adder design.

12.4.2 Differential Split-Level Logic

Differential split-level logic (DSL logic) is a scheme wherein the load is used to reduce output voltage swing and thus lower gate delays (at the cost of smaller noise margins). The basic idea is shown in Fig. 12.22. The reference voltage V_{ref} is set to $VDD/2 + V_{THN}$. This has the effect of limiting the output voltage swing to a maximum of VDD and a minimum of $VDD/2$. The main drawback of this logic implementation is the increased power dissipation resulting from the continuous power draw through the output leg at a voltage of $VDD/2$. The output leg at VDD draws no DC power.

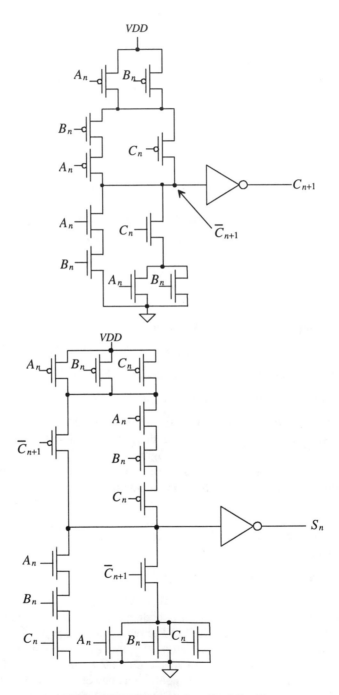

Figure 12.18 AOI implementation of a full adder.

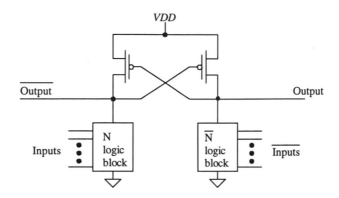

Figure 12.19 CVSL block diagram.

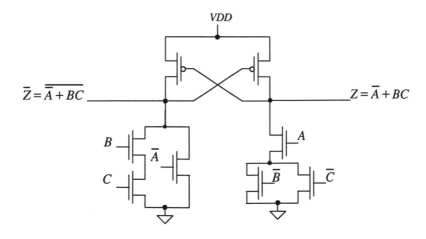

Figure 12.20 CVSL logic gate.

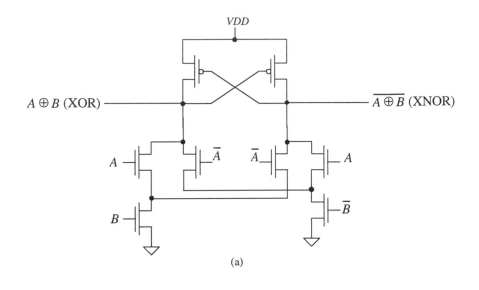

(a)

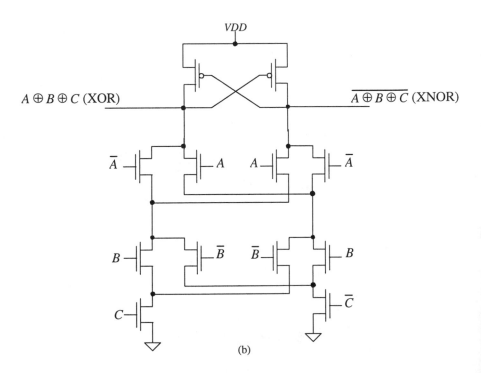

(b)

Figure 12.21 (a) Two-input and (b) three-input XOR/XNOR gates.

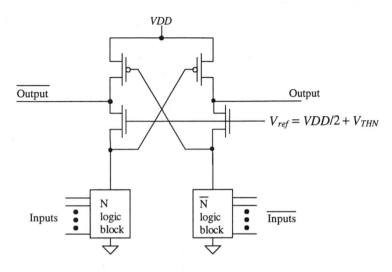

Figure 12.22   DSL block diagram.

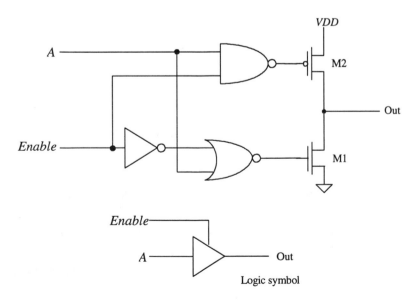

Figure 12.23   Tri-state buffer.

12.4.3 Tri-State Outputs

A final example of a static logic gate, a tri-state buffer, is shown in Fig. 12.23. When the *Enable* input is high, the NAND and NOR gates invert and pass *A* (*VDD* or ground) to the gates of M1 and M2. Under these circumstances, M1 and M2 behave as an inverter. The combination of M1 and M2 with the inversion NAND/NOR gate causes the output to be the same polarity as *A*. When *Enable* is low, the gate of M1 is held at ground and the gate of M2 is held at *VDD*. This turns both M1 and M2 off. Under these circumstances, the output is said to be in the high-impedance or Hi-Z state. This circuit is preferable to the inverter circuits of Fig. 11.29 because only one switch is in series with the output to *VDD* or ground. An inverting buffer configuration is shown in Fig. 12.24.

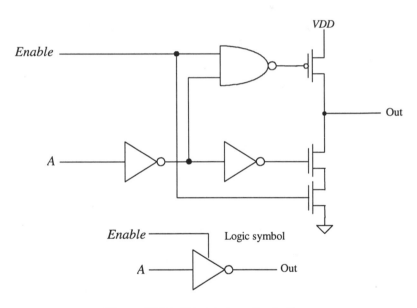

Figure 12.24   Tri-state inverting buffer.

REFERENCES

[1]   M. I. Elmasry, *Digital MOS Integrated Circuits II,* IEEE Press, 1992. ISBN 0-87942-275-0, IEEE order number: PC0269-1.

[2]   J. P. Uyemura, *Circuit Design for Digital CMOS VLSI,* Kluwer Academic Publishers, 1992.

[3]   M. Shoji, *CMOS Digital Circuit Technology,* Prentice-Hall, 1988. ISBN 0-13-138850-9.

PROBLEMS

Use the CN20 process unless otherwise specified.

12.1 Design, lay out, and simulate the operation of a CMOS AND gate with a V_{SP} of approximately 1.5 V. Use the standard-cell frame discussed in Ch. 4 for the layout.

12.2 Design and simulate the operation of a CMOS AOI half adder circuit using static logic gates.

12.3 Repeat Ex. 12.3 for a three-input NOR gate.

12.4 Repeat Ex. 12.4 for a three-input NOR gate.

12.5 Sketch the schematic of an OR gate with 20 inputs. Comment on your design.

12.6 Sketch the schematic of a static logic gate that implements $\left(A + B \cdot \overline{C} \right) \cdot D$. Estimate the worst-case delay through the gate when driving a 50 fF load capacitance.

12.7 Design and simulate the operation of a CSVL OR gate made with minimum-size devices.

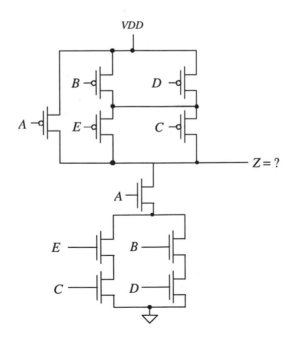

Figure P12.10

12.8 Design a tri-state buffer that has propagation delays under 20 ns when driving a 1 pF load. Assume that the maximum input capacitance of the buffer is 100 fF.

12.9 Sketch the schematic of a three-input XOR gate implemented in AOI logic.

12.10 What logic function does the circuit of Fig. P12.10 implement?

12.11 Calculate the switching point voltage of the gate shown in Fig. P12.11. What logic function does this circuit implement?

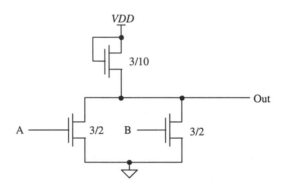

Figure P12.11

12.12 Estimate the minimum and maximum output voltages for the gate of Fig. P12.11.

12.13 The circuit shown in Fig. P12.13 is an edge-triggered one-shot that generates an output pulse, with width t_d, whenever the input makes a transition. Using inverters for delay elements, design and simulate the operation of a one-shot whose output pulse width is 10 ns. Comment on the resulting output if the width of the input pulse is less than t_d.

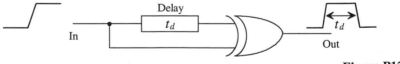

Figure P12.13

The TG and Flip-Flops

The transmission gate (TG) is used in digital CMOS circuit design to pass or not pass a signal. The schematic and logic symbol of the transmission gate (TG) are shown in Fig. 13.1. The gate is made up of the parallel connection of a p- and an n-channel MOSFET. Referring to the figure when S (for select) is high we observe that the transmission gate passes the signal on the input to the output. The resistance between the input and the output can be estimated as $R_n \| R_p$. We begin this chapter with a description of the n- and p-channel pass transistor.

13.1 The Pass Transistor

Consider the single n-channel MOSFET shown in Fig. 13.2a. Assume that the voltage across the capacitor (the output of the pass transistor) is initially 5 V. When the gate (the select line) of the MOSFET is taken to VDD, the MOSFET turns on. In this situation, we can assume that the drain of the MOSFET is connected to the load capacitance and that the source (the input of the pass transistor) is connected to ground, keeping in mind that the drain and source are interchangeable. The delay-time of the capacitor discharging is simply

$$t_{PHL} = R_n C_{load} \tag{13.1}$$

Figure 13.1 The transmission gate.

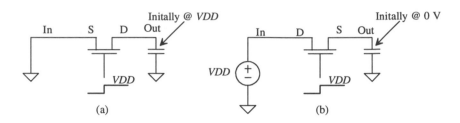

Figure 13.2 An n-channel pass transistor showing transmission of 0 V and VDD.

Now consider Fig. 13.2b where the capacitor is initially at 0 V. In this case, the drain is connected to VDD and the source is connected to the load capacitance. Since the substrate, assumed at VSS = ground, is not at the same potential as the source, we have body effect present causing the threshold voltage to increase. When the gate of this MOSFET is raised to VDD, the load capacitor charges to $VDD - V_{THN}$ where V_{THN}, from Appendix A, is in the neighborhood of 1.5 V. Therefore, the low-to-high delay-time can be estimated by

$$t_{PLH} = R_n C_{load} \text{ for a high voltage of } VDD - V_{THN} \qquad (13.2)$$

In this derivation, we have neglected the parasitic capacitances of the MOSFET. The following example illustrates the switching behavior of the n-channel pass transistor.

Example 13.1
Estimate and simulate the delay through minimum-size n-channel pass transistors using the test setups of Fig. 13.2 driving a 100 fF load capacitance.

We know that for the minimum size ($W = 3$ μm and $L = 2$ μm) the n-channel effective resistance is 8 kΩ. Therefore, the propagation delays $t_{PHL} = t_{PLH} = 800$ ps, remembering that the maximum high voltage is $VDD - V_{THN}$ or approximately 3.5 V. The simulation results are shown in Fig. 13.3. ∎

A similar analysis of the p-channel MOSFET used as a pass transistor gives

$$t_{PHL} = R_p \cdot C_{load} \qquad (13.3)$$

and

$$t_{PLH} = R_p \cdot C_{load} \text{ for a low voltage of } V_{THP} \qquad (13.4)$$

The p-channel pass transistor can pass a logic high without signal loss, while passing a low results in a minimum low voltage of V_{THP} (with body effect). The n-channel transistor can pass a logic low without signal loss, while passing a high results in a maximum high voltage of $VDD - V_{THN}$. One advantage of the p-channel pass transistor is that it can be laid out, in an n-well process, with the well tied to the source, eliminating body effect.

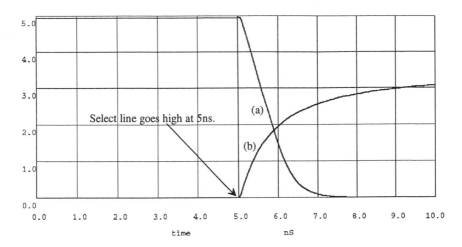

Figure 13.3 Simulation results of an n-channel MOSFET driving a 100 fF load capacitance, (a) with output initially at 5 V passing 0 V, (b) with output initially at 0 V passing 5V.

The intrinsic propagation delays of the n- and p-channel pass transistors (no load capacitance) can be approximated by

$$t_{PHL}, t_{PLH} = R_n C'_{ox} WL = \tau_n \qquad (13.5)$$

and

$$t_{PHL}, t_{PLH} = R_p C'_{ox} WL = \tau_p \qquad (13.6)$$

The pass transistor turning on must discharge the charge stored on the output capacitance of the MOSFET through its own effective resistance.

13.2 The CMOS TG

Since the n-channel passes logic lows well and the p-channel passes logic highs well, putting the two complementary MOSFETs in parallel, as was shown in Fig. 13.1, results in a TG that passes both logic levels well. The CMOS TG requires two control signals, S and $\bar{S}$ (see Fig. 13.4). The propagation delay-times of the CMOS TG are

$$t_{PHL} = t_{PLH} = (R_n \| R_p) \cdot C_{load} \qquad (13.7)$$

The capacitance on the S input of the TG is the input capacitance of the n-channel MOSFET, or C_{inn} ($= 1.5C_{oxn}$). The capacitance on the $\bar{S}$ input of the TG is the input capacitance of the p-channel MOSFET, or C_{inp}. Making the widths of the MOSFETs used in the TG large reduces the propagation delay-times from the input to the output of the TG when driving a specific load capacitance. However, the delay-times in turning the TG on, the select lines going high, increase because of the increase in input capacitance. This should be remembered when simulating. Using a voltage source in

SPICE for the select lines, which can supply infinite current to charge the input capacitance of the TG, gives the designer a false sense that the delay through the TG is limited by R_n and R_p. Often, when simulating logic of any kind, the SPICE-generated control signals are sent through a chain of inverters so that the control signals more closely match what will actually control the logic on die.

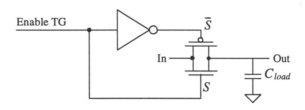

Figure 13.4 The transmission gate with control signals shown.

Example 13.2
Estimate and simulate the delays through the TG shown in Fig. Ex13.2 when minimum-size MOSFETs are used and the load capacitance is 150 fF.

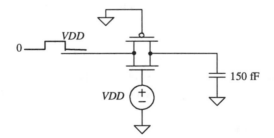

Figure Ex13.2

For minimum-size MOSFETs $R_n\|R_p = 6$ kΩ so that $t_{PLH} = t_{PHL} = 900$ ps. The SPICE simulation results are shown in Fig. 13.5. In this example we have applied the propagation delays defined by Eq. (13.7) somewhat differently. We are estimating how long it takes a change on the input of the TG to reach the output, at the 50 percent points, with the TG enabled. ∎

13.2.1 Layout of the CMOS TG

Figure 13.6 shows the layout of the minimum-size CMOS transmission gate. Often, the inverter used to generate the complementary select signal (see Fig. 13.4) is added to the layout of the cell. The inverter allows the use of a single select signal which can be desirable in systems where both true and complement signals are not available. The n- and p-channel pass transistors can be laid out in a similar manner to the CMOS TG.

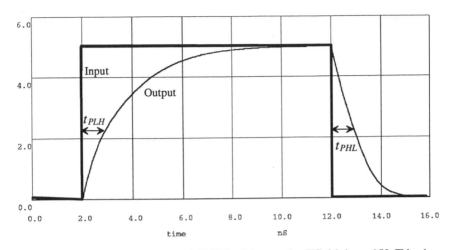

Figure 13.5 Simulation results of CMOS minimum-size TG driving a 150 fF load.

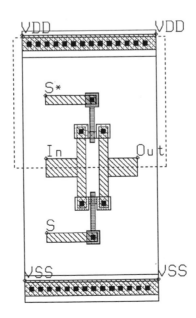

Figure 13.6 Layout of the CMOS transmission gate.

13.2.2 Series Connection of Transmission Gates

Consider the series connection of CMOS transmission gates shown in Fig. 13.7. The equivalent digital model is also depicted in this figure. The output capacitance of the individual MOSFETs is not shown in this figure and will be neglected in the following analysis. The delay through the series connection can be estimated by

$$t_{PHL} = t_{PLH} = N \cdot (R_n \| R_p)(C_{load}) + 0.35 \cdot (R_n \| R_p)(C_{inn} + C_{inp})(N)^2 \quad (13.8)$$

The first term in this equation is simply the time needed to charge C_{load} through the sum of the TG effective resistances, while the second term in the equation describes the RC transmission line effects.

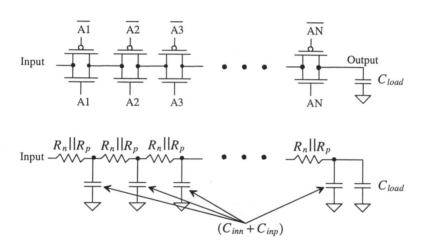

Figure 13.7 Series connection of transmission gates with digital model.

13.3 Applications of the Transmission Gate

In this section, we present some of the applications of the TG [1, 2].

Path Selector

The circuit shown in Fig. 13.8 is a two-input path selector. Logically, the output of the circuit can be written as

$$Z = AS + B\overline{S} \quad (13.9)$$

When the selector signal S is high, A is passed to the output while a low on S passes B to the output.

This same idea can be used to implement multiplexers/demultiplexers (MUX/DEMUX). Consider the block diagrams of a MUX and DEMUX shown in Fig. 13.9. The number of control lines is related to the number of input lines by

$$2^m = n \qquad\qquad (13.10)$$

where n is the number of inputs (outputs) to the MUX (DEMUX) and m is the number of control lines. A 4 to 1 MUX/DEMUX is shown in Fig. 13.10. Note that the MUX is bidirectional; that is, it can be used as a MUX or a DEMUX. The logic equation describing the operation of the MUX is given by

$$Z = A(S1 \cdot S2) + B(S1 \cdot \overline{S2}) + C(\overline{S1} \cdot S2) + D(\overline{S1} \cdot \overline{S2}) \qquad (13.11)$$

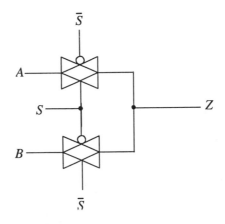

Figure 13.8 Path selector.

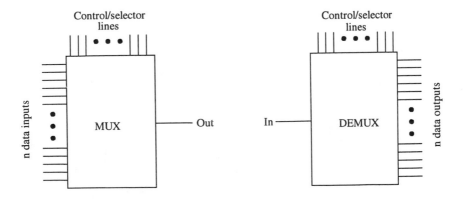

Figure 13.9 Block diagram of MUX/DEMUX.

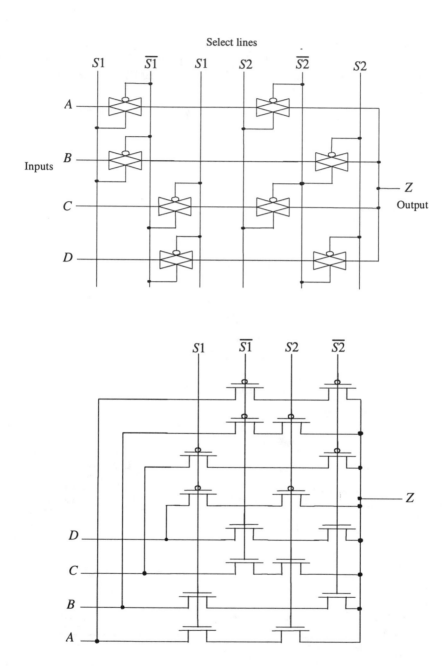

Figuer 13.10 Circuit implemetations of a 4 to 1 MUX/DEMUX.

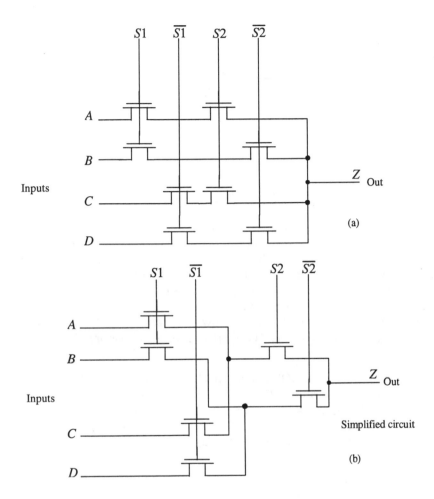

Figure 13.11 MUX/DEMUX using pass transistors.

Figure 13.11a shows a pass transistor implementation of the 4 to 1 MUX. The pass transistor implementation is simpler, using fewer transistors, at the price of a threshold voltage drop from input to output when the input is a high (*VDD*). A simplified version of the circuit of Fig. 13.11a is shown in Fig. 13.11b. Here the MOSFETs connected to *S2* and $\overline{S2}$ are combined to reduce the total number of MOSFETs used. The reduction of the total number of MOSFETs used can be extended to an n-input (output) MUX (DEMUX). Again, it should be remembered that a DEMUX can be formed using the circuits of Figs. 13.10 or 13.11 by switching the inputs with the outputs.

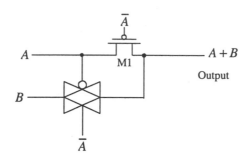

Figure 13.12 TG-based OR gate.

Static Gates

The TG can be used to form static logic gates. Consider the OR gate shown in Fig. 13.12. To understand the operation of the gate, consider the case when both A and B are low. Under these circumstances the pass transistor, M1 is off, and the TG is on. Since the input B is low, a low is passed to the output. If A is high, M1 is on and A is passed to the output. If B is high and A is low, B is passed to the output through the TG. If both A and B are high, the TG is off and M1 is on passing A, a high, to the output.

Figure 13.13 shows an XOR and an XNOR gate made using TGs. Consider the XOR gate with both A and B low. Under these circumstances, the top TG is on and its output is connected to A, a low. If both inputs are high, the bottom TG connects the output to $\overline{A}$, again a low. If A is high and B is low, the top TG is on and the output is connected to A, a high. Similarly, if A is low and B is high, the bottom TG is on and connects the output to $\overline{A}$, a high.

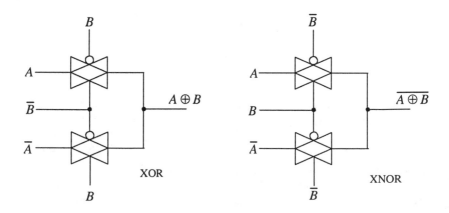

Figure 13.13 TG implementation of XOR/XNOR gate.

13.4 The Flip-Flop

Consider the set-reset flip-flop (SR FF) shown in Fig. 13.14 made using NAND gates. The logic symbol and truth table are also shown in this figure. Consider the case when S is high and R is low. Forcing R low causes Q to go high. Since S is high and Q is high, the $\bar{Q}$ output is low. Now consider the case when both S and R are low. Under these circumstances, the FF outputs are both high. This FF can easily be designed and laid out with the techniques of Ch. 12.

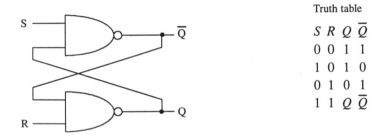

Truth table

S	R	Q	$\bar{Q}$
0	0	1	1
1	0	1	0
0	1	0	1
1	1	Q	$\bar{Q}$

Figure 13.14 Set-reset flip-flop made using NAND gates.

An alternative implementation of the SR flip-flop is shown in Fig. 13.15 using NOR gates. Consider the case when S is high and R is low. For the NOR gate, a high input forces the output of the gate low. Therefore, the $\bar{Q}$ output is low whenever the S input is high. Similarly, whenever the R input is high, the Q output must be low. The case of both inputs being high causes both Q and $\bar{Q}$ to go low, or in other words the outputs of the FF are no longer complements. Figure 13.16 shows the logic symbol of the SR flip-flop. Note that the true and complement locations on the logic symbol are switched with the location of Figs. 13.14 and 13.15.

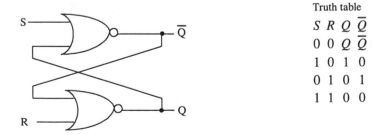

Truth table

S	R	Q	$\bar{Q}$
0	0	Q	$\bar{Q}$
1	0	1	0
0	1	0	1
1	1	0	0

Figure 13.15 Set-reset flip-flop made using NOR gates.

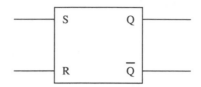

Figure 13.16 Logic symbol of the SR flip-flop.

13.4.1 Clocked Flip-Flops

Clocked flip-flops can be divided into three categories. The first category consists of those in which the clock signal pulse width must be short compared to the propagation delay through the FF. In other words, the clock input should go high and then low before the output of the flip-flop changes state. The second category of clocked FFs consists of those in which the output changes while the clock signal is high. This type of FF is sometimes called a level-sensitive FF. The final category of FF is the edge-triggered type. The output of the FF changes state on the rising or falling edge of the FF.

Short Clock Pulse Widths

The clocked JK FF is shown in Fig. 13.17. The JK FF is constructed using the NAND SR FF and two NAND gates. However, unlike the SR FF, both J and K can be high at the same time without the outputs becoming equal. The operation of the JK FF depends on the previous state of the FF. With the clock signal held low, the inputs and outputs of the SR FF do not change. Holding the clock signal high causes the output of the FF to oscillate between a logic 0 and 1. With application of a short clock pulse and J = K = 0, the outputs of the FF do not change. If K = 1 and J = 0, the output, Q, is 0 after application of the clock pulse, while if, J = 1 and K = 0 the output is set high. If both J and K are high, the output becomes the complement of the previous state.

A toggle or T flip-flop can be constructed using the JK FF by setting J = K = 1, or simply by replacing the three-input NAND gates of Fig. 13.17 with two-input NAND gates. The clock input of the JK FF is used as the T input of the T FF. Application of a short pulse to T input of the T FF causes the output, Q, to toggle states. If the output is a high and the T input goes high, the output changes to a low. Pulsing the T input high again causes the output to change back to a high. The T FF can be used to divide a clock signal in half, keeping in mind the pulse width limit (when using the FF of Fig. 13.17).

Level-Sensitive Flip-Flops

Flip-flops that are clocked with a signal that enables the output to change with the input have no specific pulse width requirements for the clock. To illustrate a level-sensitive

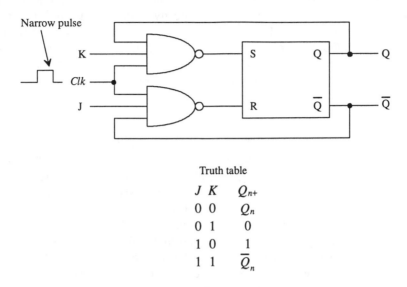

Truth table

J	K	Q_{n+}
0	0	Q_n
0	1	0
1	0	1
1	1	$\overline{Q}_n$

Figure 13.17 Clocked JK FF. Clock pulse width should be short compared to FF delay.

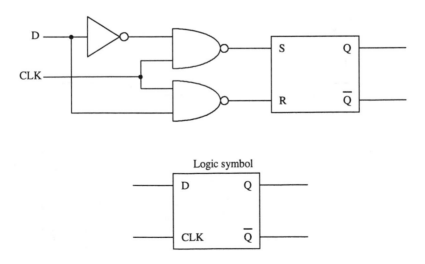

Logic symbol

Figure 13.18 Level-sensitive D flip-flop.

FF, consider the data or D FF shown in Fig. 13.18 with associated logic symbol. When the clock signal is high, the D input can pass directly to the SR FF. If D is a 1 while CLK is high, the output, Q, is a 1, while if D is low the output is a low. If D changes at any time while the CLK input is high, the output will follow. When the CLK signal goes low, the current logic level of D is latched into the SR FF. Note that this FF is not an edge-sensitive FF because the output changes at other times than the edge transition time.

Edge-Triggered Flip-Flops

The JK master-slave FF, shown in Fig. 13.19, is an example of an edge-triggered FF. When the CLK signal goes high, the master JK FF is enabled. Since the slave FF cannot change states when CLK is high, the clock pulse width does not have to be less than the propagation delay of the FF. When CLK goes low, the master data are transferred to the slave. If both J and K are low, the output of the master remains unchanged, and therefore so does the output of the slave. If J = 1 and K= 0 when the CLK pulse goes low, the master output, Q, goes high. When the CLK goes low, the high output of the master is transferred to the slave. The master-slave JK FF behaves just like the JK FF of the previous section except for the fact that the data are not available until CLK goes low and there is no restriction on the pulse width of the clock (i.e., the FF is falling edge triggered). Adding reset or set capability to the FF can be accomplished by adding logic gates between the NAND and the SR FF of Fig. 13.19. The logic gates simply ensure that the SR FF are placed into a certain state upon application of a reset or set signal.

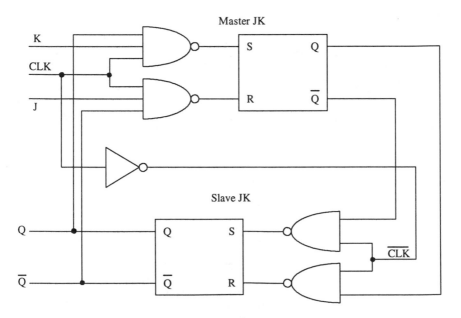

Figure 13.19 Edge-triggered JK master-slave flip-flop.

An implementation of the positive edge-triggered D FF is shown in Fig. 13.20a. The SR FF is made using NAND gates. When the CLK input is low, the outputs of the NAND gates are both high, keeping the SR FF in the "no change mode." When CLK goes high, the logic value on the D input of the FF is transferred to the S input and the complement is transferred to the R input of the SR FF. The $\overline{CLK}$ input of the NAND gates goes low three inverter delays after CLK goes high. This forces both R and S high, putting the flip-flop in the no change mode. The only time that CLK and $\overline{CLK}$ can be high, the condition required to transfer D to the input of the FF, is the time between CLK going high and $\overline{CLK}$ going low. This time is determined by the propagation delay of the inverters. In practice a single inverter, in place of the three, will not provide a sufficient delay to allow the inputs of the SR FF to fully charge to D and $\overline{D}$. Also, there are maximum rise- and falltime requirements on the clock.

Another implementation of the positive edge-triggered D FF using transmission gates is shown in Fig. 13.20b. When the CLK input is low, the logic value at D is setting at node A and $\overline{D}$ is on node B. Transmission gates T2 and T3 are off. The datum on node C is available on the output of the FF and is the result of the previous leading edge transition of the CLK input pulse. When CLK goes high, T1 and T4 turn off, while T2 and T3 turn on and the datum on node is C is transferred, with the appropriate inversion to the outputs. A D FF with set and clear inputs is shown in Fig. 13.20c.

Flip-Flop Timing

The data must be set up or present on the D input of the FF (see Fig. 13.20c) a certain time before we apply the clock signal. This time is defined as the setup time of the FF. To understand the origin of this time, consider the time it takes the signal at D to propagate through T1 and the NAND gate to point B. Before the clock pulse can be applied, the logic level D must be settled on point B. Consider the waveforms of Fig. 13.21. The time between D going high (or low) and the clock rising edge is termed the setup time of the FF and is labeled, t_s.

The wanted D input must be applied t_s before the clock pulse is applied. Now the question becomes "How long does the wanted D input have to remain on the input of the FF after the clock pulse is applied?" This time called the hold time, t_h, is illustrated in Fig. 13.22. Shown in this figure, t_h is a positive number. However, inspection of Fig. 13.20 shows that if the D input is removed slightly before the clock pulse is applied, the point B will remain unchanged because of the propagation delay from D to B. Analysis of this FF would yield a negative hold time. In other words, for the point B to charge to D, a time labeled t_s is needed. Once point B is charged, the D input can be changed as long as the clock signal occurs within t_h.

One final important comment regarding the clock input of a FF is in order. If the clock input risetime is slow, the FF will not function properly. There will not be an abrupt transition between the sets of transmission gates turning on and off. The result will be logic levels at indeterminate states. What is usually done to eliminate this

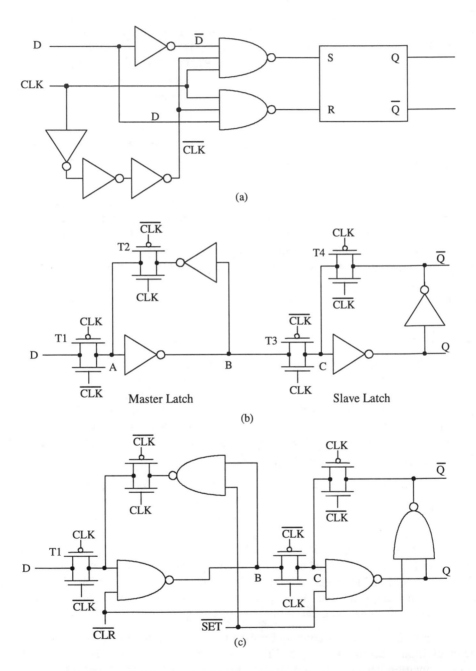

Figure 13.20 Edge-triggered D flip-flops, (a) Gate implementation, (b) TG implementation, and (c) TG implementation with set and clear.

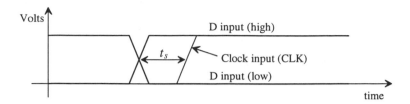

Figure 13.21 Illustrating D FF setup time.

problem is to buffer the clock input through several inverters. This has the effect of speeding up the leading and trailing edges of a slow input pulse and presenting a lower input capacitance on the clock input to whatever is driving the FF. The main disadvantage is the increase in delay times, t_{PHL} and t_{PLH} (defined by clock to output), of the FF. In general, the FFs of Fig 13.20b and c should not be laid out without buffering the clock inputs.

The minimum pulse width of the clock, set, or clear inputs is labeled t_w. The minimum width is determined by the delay through (referring to Fig. 13.20) two NAND gates and a TG. The last timing definition we will consider here is the recovery time, that is, the time between removing the set or clear inputs and a valid clock input. This variable is labeled t_{rec}.

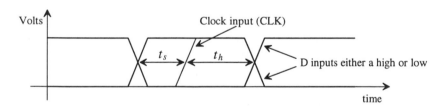

Figure 13.22 Illustrating D FF hold time.

Simple D Flip-Flop

A simple D flip-flop using inverters and TGs is shown in Fig. 13.23. The cross-coupled connection of inverters is sometimes referred to as a latch and is the basis for the static RAM storage cell discussed further in Ch. 17. To understand the operation of this circuit, consider the case when CLK is low. The TGs are off, and the outputs do not change from their previous state. When CLK is high, provided the inverters are sized correctly (more on this shortly), the D input is connected to Q and the $\overline{D}$ input is connected to $\overline{Q}$. Thus, when CLK goes back low, the value of D is remembered and latched. A couple of points should be made regarding this flip-flop: (1) the outputs change with the inputs whenever CLK is high, that is, it is not an edge-triggered flip-flop, and (2) the inputs must supply a current during switching.

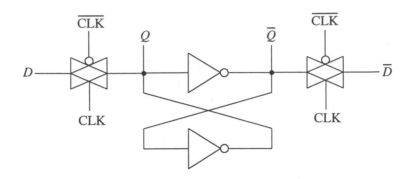

Figure 13.23 Clocked D flip-flop using the basic latch and TGs.

The input DC current comes from the fact that the output of an inverter is connected to each TG. In order to change the voltage at Q and $\overline{Q}$, with the inputs, the effective digital resistances of the inverters should be large compared to the sum of the TG resistance and the driver resistance. (The driver resistance is the effective resistance of whatever gate is driving the TG.) In other words, the R_n and R_p of the inverter should be large. The length of the devices used in the inverters can be longer than the minimum length to reduce the input current.

Metastability in the latch

Consider what happens when the clock inputs in Fig. 13.23 transition low and shut off the TGs just as the inputs, D and $\overline{D}$ are at the switching point voltages of each inverter, V_{SP}. When this happens the inverter based latch is said to be in a *metastable*, or unknown, state since the outputs of the latch are not at well defined logic values. If the input and output of each inverter where exactly at V_{SP} (with the TGs off) then there wouldn't be any imbalance in the circuit and the outputs of the latch would remain unchanged. Over a time, which can be long, noise, together with the positive feedback inherent in the latch, will cause the outputs to become valid logic levels. Metastability can be especially troubling in high-speed digital circuits where the latch must respond to a changing input quickly or the inputs are asynchronous with the clock.

REFERENCES

[1]   J. P. Uyemura, *Circuit Design for Digital CMOS VLSI,* Kluwer Academic Publishers, 1992.

[2]   M. I. Elmasry, *Digital MOS Integrated Circuits II,* IEEE Press, 1992. ISBN 0-87942-275-0, IEEE order number: PC0269-1.

[3]   M. Shoji, *CMOS Digital Circuit Technology,* Prentice-Hall, 1988. ISBN 0-13-138850-9.

PROBLEMS

Unless otherwise stated, use the CN20 process.

13.1 Verify the simulation results shown in Fig. 13.3. If we increase the width of the n-channel pass transistor, what happens to the delay-times? The gate of the pass transistor is driven from some other logic on the chip. What happens to the capacitance seen by this logic when we increase the width of the pass transistor?

13.2 Design and simulate the operation of a half adder circuit using TGs.

13.3 Estimate and simulate the delay through 10 TGs (assume minimum-size) connected to a 100 fF load capacitance.

13.4 Sketch the schematic of an 8 to 1 DEMUX using n-channel pass transistors. Estimate the delay through the DEMUX when the output is connected to a 50 fF load capacitance.

13.5 Verify, using SPICE, that the circuit of Fig. 13.13 operates as an XOR gate.

13.6 Simulate the operation of an SR FF made with NAND gates using minimum-size MOSFETs. Show all four logic transitions possible for the FF.

13.7 Simulate the operation of the clocked D FF of Fig. 13.23b using minimum-size MOSFETs. Comment on any glitches you encounter. Show the FF clocking in a logic 1 and 0. What are the setup and hold times for your design?

13.8 Design and simulate the operation of the FF shown in Fig. P13.8.

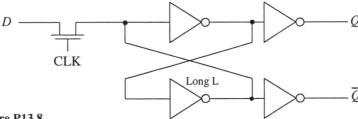

Figure P13.8

13.9 The FF shown in Fig. P13.8 has several practical problems, including not presenting a purely capacitive load at the D-input and large layout size. The FF of Fig. P13.9 is a different implementation of an inverter-based latch which does present a purely capacitive load to the D-input and a (possibly) smaller layout size. Simulate the operation of this FF using the device sizes shown.

13.10 Repeat Ex. 13.1 using minimum-size (0.9/0.6) MOSFETs using the CMOS14TB process.

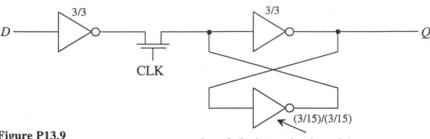

Figure P13.9

Long L (both n- and p-channel) inverter.

13.11 Repeat Ex. 13.2 using the CMOS14TB process.

13.12 Estimate and simulate the delay through 10 TGs (assume minimum-size) connected to a 100 fF load capacitance using the CMOS14TB process.

13.13 Using a SPICE DC sweep, plot the output voltage against the input voltage for the circuit of Fig. P13.13 with the input varying from 0 to 5 V and then from 5 V to 0. Comment on the difference in the plots.

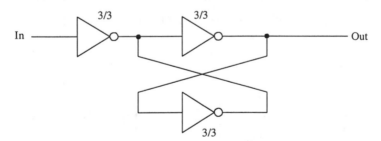

Figure P13.13

Chapter

14

Dynamic Logic Gates

Dynamic or clocked logic gates are used to decrease complexity, increase speed, and lower power dissipation. The basic idea behind dynamic logic is to use the capacitive input of the MOSFET to store a charge and thus remember a logic level for use later. Before we start looking into the design of dynamic logic gates, let's discuss leakage current and the design of clock circuits.

14.1 Fundamentals of Dynamic Logic

Consider the n-channel pass transistor shown in Fig. 14.1 driving an inverter. If we clock the gate of the pass transistor high, the logic level on the input, point A, will be passed to the input of the inverter, point B. If this logic level is a "0," the input of the inverter will be forced to ground while a logic "1" will force the input of the inverter to $VDD - V_{THN}$. When the clock signal goes low, the pass transistor shuts off and the input to the inverter "remembers" the logic level. In other words, when the pass transistor turns on, the input capacitance of the inverter is charged to $VDD - V_{THN}$, or ground, through the pass transistor. As long as this charge is present, the logic value is remembered. What we are concerned with at this point is the leakage mechanisms present which can leak the stored charge off the node. A node, such as the one labeled B in Fig. 14.1, is called a dynamic node or a storage node. Note that this node is a high-impedance node and is easily susceptible to noise (see Ex. 3.4).

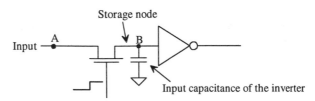

Figure 14.1 Example of a dynamic circuit and associated storage capacitance.

14.1.1 Charge Leakage

Consider the expanded view of the charge storage node shown in Fig. 14.2. Practically, the only leakage path on this node is through the MOSFET's drain (or source since the drain and source are interchangeable) n+ /p-substrate diode. If we consider this node the drain of the MOSFET, the current is given by

$$I_D = I_{leakage} = I_S(e^{-V_B/nV_T} - 1) \tag{14.1}$$

where V_B is the voltage on the storage node to ground, assuming the substrate is at ground potential. From the BSIM model parameters, the scale current is given by

$$I_S = AD \cdot JS \tag{14.2}$$

In order to simplify hand calculations we will assume that the leakage current is equal to the scale current, or

$$I_{leakage} = I_S = AD \cdot JS \tag{14.3}$$

The rate at which the storage node discharges is given by

$$\frac{dV}{dt} = \frac{I_{leakage}}{C_{node}} = \frac{AD \cdot JS}{C_{node}} \tag{14.4}$$

The node capacitance is the sum of the input capacitance of the inverter, the capacitance to ground of the metal or poly line connecting the inverter to the pass transistor, and the capacitance of the drain implant to substrate (the depletion capacitance). For practical applications, we assume that

$$C_{node} \approx C_{in} \text{ of the inverter} \tag{14.5}$$

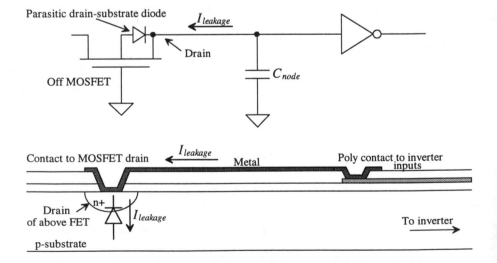

Figure 14.2 Leakage from a storage node through the drain-substrate diode.

Example 14.1

Estimate the discharge rate of the 50 fF capacitor shown below. Assume that the MOSFET drain and source areas measure 6 μm by 6 μm.

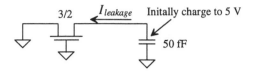

Figure Ex14.1

From the BSIM model parameters, $JS = 10^{-8}$ A/m^2; therefore, the leakage current can be estimated by

$$I_{leakage} = AD \cdot JS = 36p \cdot 10^{-8} = 360 \times 10^{-21} A$$

and the discharge rate is estimated by

$$\frac{dV}{dt} = \frac{360 \times 10^{-21}}{50fF} = 7.2 \ \mu V/s$$

This is a very slow discharge rate. In practice, the MOSFET can have a nonzero gate-source voltage causing a subthreshold current to flow, increasing the discharge rate. Also, the value of current density given in the BSIM model in Appendix A and used above, that is, $JS = 10^{-8}$ A/m^2, is the SPICE default value. This indicates that the leakage current was not measured when generating the SPICE model, indicating another possible source of error. ∎

14.1.2 Simulating Dynamic Circuits

Because of the extremely small leakage currents involved, simulating dynamic circuits can be difficult. First, when SPICE simulates any circuit, it puts a resistor with a conductance value given by the parameter GMIN across every pn junction and MOSFET drain to source. The default value of GMIN is 10^{-12} mhos or a 1 TΩ resistor. A charge storage node at a potential of 5 V has a leakage current, due to GMIN, of 5 pA. Of course, as the node voltage starts to decrease, the leakage current decreases as well. The leakage current calculated in Ex. 14.1 was 360×10^{-21} A, or over a million times smaller than the 5 pA flowing through the default value of GMIN. The value of GMIN can be set using the .OPTIONS command, at the cost of a longer or more difficult convergence time, to a smaller value, say 10^{-15}.

The ABSTOL (current accuracy), RELTOL (relative accuracy), or VNTOL (voltage tolerance) simulation parameters can limit the accuracy of the simulation and give false results. The default value of the current accuracy, ABSTOL, is 1 pA. Since the leakage from the drain-substrate diode can be significantly less than the 1 pA, ABSTOL must be reduced. If we set ABSTOL = 1E-21, the simulation accuracy is helped. However, when simulating, SPICE uses the larger of ABSTOL or the product of RELTOL and the simulation current to determine if convergence has been reached for a given current. Therefore, RELTOL would need to be reduced as well to get SPICE results closer to hand calculations. Note that the charge tolerance, CHGTOL, has nothing to do directly with accuracy unlike VNTOL, ABSTOL and RELTOL.

In practice, we use the default values of SPICE, which give a pessimistic estimate for the discharge time of storage nodes in dynamic circuits. The leakage current, for $VDD = 5$ V, is given by

$$I_{leakage} \approx 5 \; pA = VDD \cdot GMIN \qquad (14.6)$$

and

$$\frac{dV}{dt} = \frac{5 \; pA}{C_{node}} = \frac{VDD \cdot GMIN}{C_{node}} \qquad (14.7)$$

For $C_{node} = 50$ fF, it takes approximately 10 ms for the voltage on the charge storage node to fall 1 V. If 1 V is the most we will allow the node to fall before we apply another clock signal, then the minimum clock frequency is 100 Hz. The following example illustrates the dominance of GMIN in the simulation of a dynamic circuit.

Example 14.2
Simulate the circuit of Ex. 14.1. Estimate the discharge rate of the capacitor due to the default value of GMIN.

The discharge rate from Eq. (14.7) is 1 V per 10 ms for a GMIN of 10^{-12} mhos. The SPICE simulation results are shown in Fig. 14.3. Notice how the leakage drain current is jagged. This is the result of the numerical iteration scheme used by SPICE. The simulation currents will vary by an amount less than ABSTOL, or 1 pA. In most simulations, we do not see the small current variations. ∎

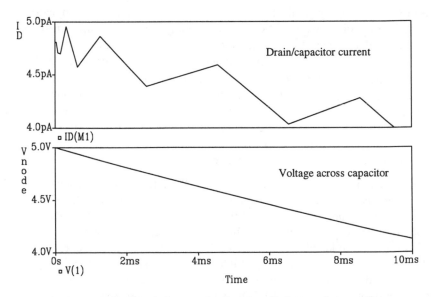

Figure 14.3 Simulation results showing discharge of a capacitor.

14.1.3 Nonoverlapping Clock Generation

Consider the string of pass transistors/inverters shown in Fig. 14.4. This circuit is called a dynamic shift register. When ϕ_1 goes high, the first and third stages of the register are enabled. Data are passed from the input to point A0 and from point A1 to A2. If ϕ_2 is low while ϕ_1 is high, the data cannot pass from A0 to A1 and from A2 to A3. If ϕ_1 goes low and ϕ_2 goes high, data are passed from A0 to A1 and from A2 to A3. If both ϕ_1 and ϕ_2 are high at the same time, the input of the shift register and the output are connected together, which is not desirable in a shift register application. The purpose of the inverter between pass transistors is to restore logic levels, since the n-channel pass transistor passes a high with a threshold voltage drop. Two inverters would be used to eliminate the logic inversion between stages. The clocks used in this dynamic circuit must be nonoverlapping, or logically

$$\phi_1 \cdot \phi_2 = 0 \tag{14.8}$$

There should be a period of dead time between transitions of the clock signals, labeled Δ in Fig. 14.4. The rise- and falltimes of the clock signals should not occur at the same time.

Since the design and layout of the dynamic shift register is straightforward let's concentrate on the generation of clock signals, ϕ_1 and ϕ_2. Note that a simple logic inversion will not generate nonoverlapping clock signals.

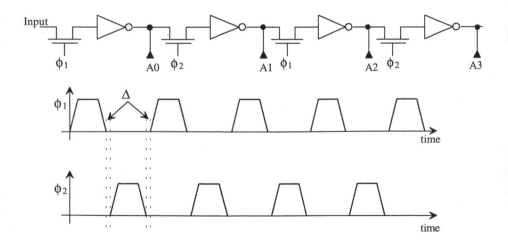

Figure 14.4 Dynamic shift register with associated nonoverlapping clock signals.

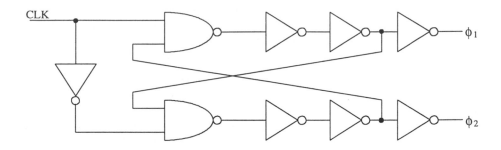

Figure 14.5 Nonoverlapping clock generation circuit.

Consider the schematic of the nonoverlapping clock generator shown in Fig. 14.5. This circuit takes a clock signal and generates a two-phase nonoverlapping clock. The amount of separation is set by the delay through the NAND gate and the two inverters on the NAND gate output. Consider the input clock going high. This forces ϕ_1 high and ϕ_2 low. When the input clock goes low, ϕ_1 goes low. After ϕ_1 goes low, ϕ_2 can go high. When driving long transmission lines such as poly, where the risetime of the signals can be significant, a large number of inverters may need to be used. Line drivers, a string of inverters used to drive a large capacitance, can be used as part of the delay in the nonoverlapping clock generation circuit.

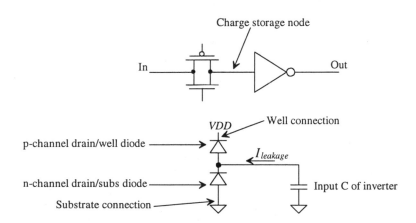

Figure 14.6 CMOS TG used in dynamic logic.

14.1.4 CMOS TG in Dynamic Circuits

The CMOS TG used as a switch to charge or discharge the node capacitance of the charge storage node is shown in Fig. 14.6. Since understanding the charging and discharging of the input capacitance of the inverter follows many of the same analysis and discussions of Ch. 13, we will concentrate here on the charge leakage from the TG.

The leakage of charge off of or onto the input capacitance of the inverter in Fig. 14.6 can be attributed to the drain-well diode of the p-channel MOSFET and the drain-substrate diode of the n-channel MOSFET used in the TG. If these leakage currents were equal, then the leakage of charge off of the storage node would be zero. In general, we use the same hand analysis that was used for the n-channel MOSFET alone; namely, the leakage causes the voltage to change 1 volt in 10 ms. Notice that unlike the n-channel MOSFET, the charge storage node can leak to *VDD* or *VSS* (ground), depending on the size of the drain areas and the leakage currents.

14.2 Clocked CMOS Logic

Clocked CMOS, C²MOS, logic is used to reduce power dissipation and layout size and to increase speed. The standard CMOS static gate requires 2N MOSFETs for an n-input gate. In general, an n-input C²MOS gate requires N + 2 MOSFETs where two MOSFETs are used in the clocking scheme. Additional MOSFETs can be used for buffering or for helping the gate appear more static in operation.

Clocked CMOS Latch

Consider the circuit shown in Fig. 14.7. This circuit performs the dynamic latch operation similar to the circuit of Fig. 14.6. When the clock input ϕ_1 is high, the input is inverted and available on the output of the gate. For low-input clock signals, the

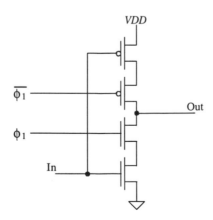

Figure 14.7 A clocked CMOS latch. The clock signals can be generated with an RS FF so that the edges occur essentially at the same moment in time.

output is in the Hi-Z state, or in other words a high-impedance node very susceptible to signal feedthrough. The layout of the C²MOS gate is thus more critical than the static gate. Because of this node, running signal lines above this node in the layout is a definite problem. The output of the gate is not static. When the latch is enabled and ϕ_1 is high, the capacitance on the output node is charged. The same leakage mechanisms present in the CMOS TG latch are present here. This limits the minimum clock frequency to about 100 Hz. Implementing a shift register requires nonoverlapping clocks for adjacent stages. The total number of clock signals needed for a C²MOS shift register is four: the nonoverlapping clocks ϕ_1 and ϕ_2 and their complements.

PE Logic

This section discusses precharge-evaluate logic, or PE logic. Consider the three-input NAND gate shown in Fig. 14.8. The operation of this gate relies on a single clock input. When ϕ_1 is low, the output node capacitance is charged to *VDD* through M5. During the evaluate phase, ϕ_1 is high, M1 is on, and if A0, A1, and A2 are high, the output is pulled low. The logic output is available only when ϕ_1 is high. The output is a logic one when ϕ_1 is low. One disadvantage of PE logic is that the gate logic output is available part of the time and not all of the time as in the static gates.

Several important characteristics of the PE gate should be pointed out. The input capacitance of the PE gate is less than that of the static gate. Each input is connected to a single MOSFET where the static gate inputs are tied to two MOSFETs. Potentially the PE gate is then faster and dissipates less power.

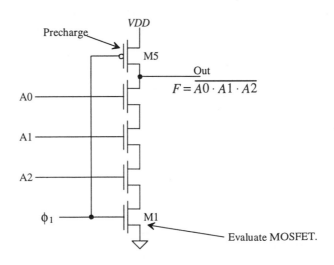

Figure 14.8 Precharge-evaluate three-input NAND gate.

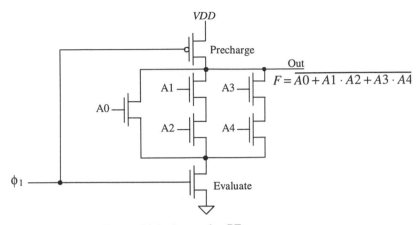

Figure 14.9 A complex PE gate.

The size of the MOSFETs used in a PE gate does not need ratioing for symmetrical switching point voltage. The absence of complementary devices and the fact that the output is pulled high during each half cycle makes the gate V_{SP} meaningless. However, we may need to size the devices to attain a certain speed for a given load capacitance. If the sizes of all NMOS transistors used in Fig. 14.8 are equal, then the t_{PHL} is approximately $4R_n C_{node}$ and the t_{PLH} is $R_p C_{node}$ where C_{node} is the total capacitance on the output node. This may include the interconnecting capacitance and the input capacitance of the next stage. Here we have neglected both the transmission line effects through a series connection of MOSFETs and the intrinsic switching speeds. A more complex logic function, $F = \overline{A0 + A1 \cdot A2 + A3 \cdot A4}$, implemented in PE logic is shown in Fig. 14.9.

Domino Logic

Consider the cascade of PE gates shown in Fig. 14.10. During the precharge phase of the clock, the output of each PE gate is a logic high. This high-level output is connected to the input of the next PE gate. Suppose the logic out of the first PE gate during the evaluate phase is a low. This output will turn off any MOSFETs in the second PE gate. However, during the precharge phase, those same MOSFETs in the second PE gate will be turned on. The delay between the clock pulse going high and the valid output of the first gate will cause the second gates output to glitch or show an invalid logic output. If we can hold the output voltage of the PE gate low, instead of high we can eliminate this race condition. Upon adding an inverter to the PE gate (Fig. 14.11) the condition for glitch-free operation is met. The PE gate with the addition of an inverter is called Domino logic. The name *Domino* comes from the fact that a gate in a series of Domino logic gates cannot change output states until the previous gate changes states. The change in output of the gates occurs similar to a series of falling dominoes. The inverter used in the Domino gate has the added advantage that it can be sized to drive large capacitive loads.

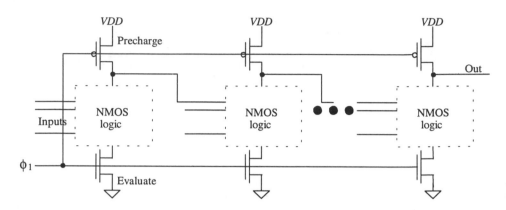

Figure 14.10 Problems with a cascade of PE gates.

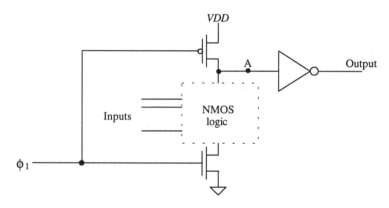

Figure 14.11 Domino logic gate.

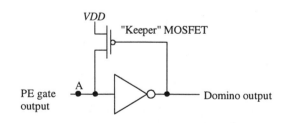

Figure 14.12 Keeper MOSFET used to hold node A in Fig. 14.11
at VDD when PE gate output is high.

One problem does exist with this scheme, however, referring to Fig. 14.11, note that during the precharge phase, node A is charged to *VDD*. If the NMOS logic results in a logic high on node A during the evaluate phase, then that node is at a high impedance with no direct path to *VDD* or ground. The result is charge leakage off of node A when the PE output is a logic high. The circuit of Fig. 14.12 eliminates this problem. A "keeper" p-channel MOSFET is added to help keep node A at *VDD* when the NMOS logic is off. The *W/L* of this MOSFET is small, so that it provides enough current to compensate for the leakage but not so much that the NMOS logic can't drive node A down to ground.

NP Logic (Zipper Logic)

The idea behind implementing a logic function using NP logic is shown in Fig. 14.13. Staggering NMOS and PMOS stages eliminates the need for and delay associated with the inverter used in Domino logic, making higher speed operation possible. A circuit that can easily be implemented in NP logic is the full adder circuit of Fig. 12.18. The NMOS section of the carry circuit is implemented in the first section of the NP logic, while the PMOS section of the sum circuit is implemented in the PMOS section of the NP logic gate.

Pipelining

The NP logic adder just described adds two one-bit words with carry during each clock cycle. Adding two-four bit words can use pipelining [4]; see Fig. 14.14. The bits of the word are delayed, both on the input and output of the adder, so that all bits of the sum reach the output of the adder at the same time. Note, however, that two new four-bit words can be input to the adder at the beginning of each clock cycle and that it takes four clock cycles to finish the addition of the two words. If this circuit were dedicated to continually performing the addition of two words, we could input the words at a very fast rate, around 30 Mwords/s for the CN20 process. However, since performing a single addition requires four clock cycles, applications of pipelining where two numbers are not added continuously can result in longer delay-times.

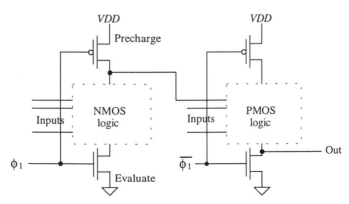

Figure 14.13 NP logic.

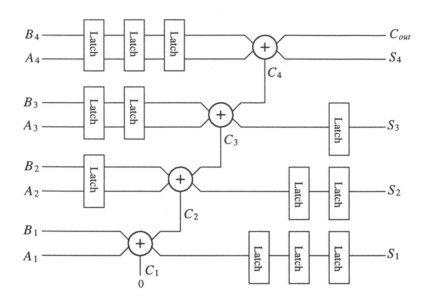

Figure 14.14 A pipelined adder. The latches (clocked) behave as delay elements.

REFERENCES

[1]   R. L. Geiger, P. E. Allen, and N. R. Strader, *VLSI-Design Techniques for Analog and Digital Circuits,* McGraw-Hill Publishing Co., 1990. ISBN 0-07-023253-9.

[2]   M. I. Elmasry, *Digital MOS Integrated Circuits II,* IEEE Press, 1992. ISBN 0-87942-275-0, IEEE order number: PC0269-1.

[3]   J. P. Uyemura, *Circuit Design for Digital CMOS VLSI,* Kluwer Academic Publishers, 1992.

[4]   N. H. E. Weste and K. Eshraghian, *Principles of CMOS VLSI Design,* Addison-Wesley, 2nd ed., 1993. ISBN 0-201-53376-6.

[5]   J. Yuen and C. Svensson, "New Single-Clock CMOS Latches and Flipflops with Improved Speed and Power Savings," *IEEE Journal of Solid-State Circuits,* Vol. 32, No. 1, pp. 62-69, 1997.

PROBLEMS

Unless otherwise stated, use the CN20 process.

14.1   Repeat Ex. 14.2 with a GMIN of 10^{-9} mhos. Use the .OPTIONS to set the value of GMIN.

14.2   Simulate the operation of the nonoverlapping clock generator circuit made using minimum size MOSFETs in Fig. 14.5. Assume that the input clock signal is running at 50 MHz. Show how both ϕ_1 and ϕ_2 are nonoverlapping.

14.3   Design and simulate the operation of a PE gate that will implement the logical function $F = \overline{ABCD} + E$.

14.4   Simulate the operation of the clocked CMOS latch shown in Fig. 14.7. Use minimum-size MOSFETs.

14.5   If the PE gate shown in Fig. 14.9 drives a 50 fF capacitor, estimate the worst-case t_{PHL}.

14.6   Implement an XOR gate using Domino logic. Simulate the operation of the resulting implementation.

14.7   The circuit shown in Fig. P14.7 is the implementation of a high-speed adder cell (1-bit). What type of logic was used to implement this circuit? Using timing diagrams, describe the operation of the circuit.

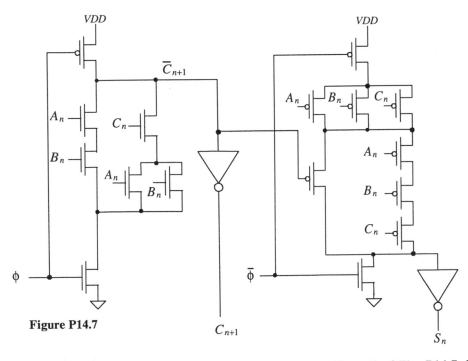

Figure P14.7

14.8   Discuss the design of a two-bit adder using the adder cell of Fig. P14.7. If a clock, running at 20 MHz, is used with the two-bit adder, how long will it take to add two words? How long will it take if the word size is increased to 32 bits?

14.9   Sketch the implementation of an NP logic half adder cell.

14.10   Design (sketch the schematic of) a full adder circuit using PE logic.

14.11   Simulate the operation of the circuit designed in Problem 14.10.

14.12   Figure P14.12 shows one bit of a shift register implemented in the so-called
ratioless NMOS logic.   The term *ratioless* results from the fact that the
MOSFET sizes do not affect the switching point voltages.   Also, this gate can be
laid out in a very small area and the outputs can swing down to ground.   Discuss
and simulate the operation of this circuit.   Keep in mind that ϕ_1 and ϕ_2 are
nonoverlapping clock signals.   What is the maximum output voltage of this
circuit?

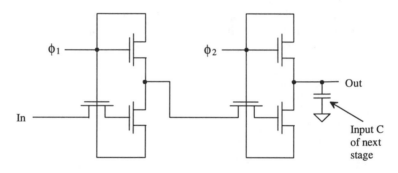

All MOSFETs are minimum size.

Figure P14.12

14.13   Show that the dynamic circuit shown in Fig. P14.13 is an edge-triggered
flip-flop [5].   Note that a single-phase clock signal is used.

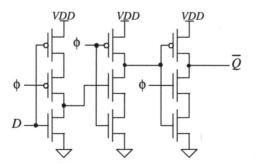

Figure P14.13

Chapter 15

VLSI Layout

The past chapters have concentrated on basic logic-gate design and layout. In this chapter we discuss the implementation of logic functions on a chip where the size and organization of the layouts are of importance. The number of MOSFETs on a chip, depending on the application, can range from tens (an op-amp) to hundreds of millions (a 256 MEG DRAM). Designs where thousands of MOSFETs or more are integrated on a single die are termed *very-large-scale-integration* (VLSI) designs.

To help us understand why chip size is important, examine Fig. 15.1. The dark dots indicate a defect that will lead to a chip which doesn't function properly. Figure 15.1a shows a wafer with nine full die. The partial die around the edge of the wafer are wasted. Five of the nine die do not contain a defect and thus can be packaged and sold. Next consider a reduction in the die size (Fig. 15.1b). We are assuming each die,

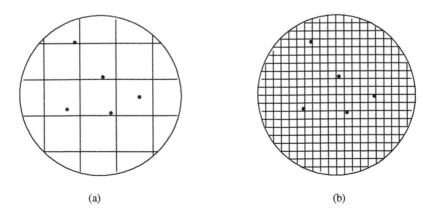

(a)   (b)

Figure 15.1 Defect density effects on yield.

whether discussing the die of Fig. 15.1a or b, performs the same function. This reduction can be the result of having better layout (resulting in a smaller layout area) or fabricating the chips in a process with smaller device dimensions (e.g., going from a 2 μm process to a 0.5 μm process). The total number of die lost (see Fig. 15.1b), due to defects is five; however, the number of good die is significantly larger than the five good die of Fig. 15.1a. The yield (number of good die/total number of die on the wafer) is increased with smaller die size. The result is more die/wafer available for sale. Another benefit of reducing die size comes from the realization that processing costs per wafer are constant and increasing the number of die on a wafer decreases the cost per die.

15.1 Chip Layout

VLSI designs can be implemented using many different techniques including gate-arrays, standard-cells, and full-custom design[1]. Since designs based on gate-arrays are, in general, used where low volume and fast turnaround time are required and the chip designer need know little to nothing[1] about the actual implementation of the CMOS circuits, we will concentrate on full-custom design and design using standard cells.

Regularity

An important consideration when implementing a VLSI chip design is regularity. The layout should be an orderly arrangement of cells. Toward this goal the first step in designing a chip is drawing up a chip (or section of the chip) floor plan. Figure 15.2 shows a simple floor plan for an adder data-path. This floor plan can be added to the floor plan of an overall chip, which includes output buffers, control logic, and memory.

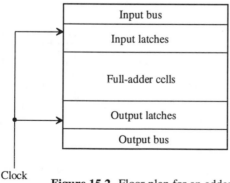

Figure 15.2 Floor plan for an adder.

At this point, we may ask the question, "How do we determine the size of the blocks in Fig. 15.2?" The answer to this question leads us into the design and layout of the cells used to implement each of the logic blocks in Fig. 15.2.

Standard-Cell Examples

Standard cells are layouts of logic elements including gates, flip-flops, and ALU functions that are available in a cell library for use in the design of a chip. *Custom design* refers to the design of cells or standard cells using MOSFETs at the lowest level. *Standard-cell design* refers to design using standard cells; that is, the designer connects wires between standard cells to create a circuit or system. The difference between the two types of design can be illustrated using a printed circuit board-level analogy. A standard-cell design is analogous to designing with packaged parts. The design is accomplished by connecting wires between the pins of the packaged parts. Custom design is analogous to designing the "insides" of the packaged parts themselves.

Figure 15.3 shows an example of an inverter [2]. In addition to keeping the layout size as small as possible, an important consideration, when laying out a standard cell, is the routing of signals. Keeping this in mind, we can state the following general guidelines for standard-cell design:

1. Cell inputs and outputs should be available, at the same relative horizontal distance, on the top and bottom of the cell.

2. Horizontal runs of metal are used to supply power and ground to the cell, a.k.a. power and ground busses. Also, well and substrate tie downs should be under these busses.

3. The height of the cells should be a constant, so that when the standard cells are placed end to end the power and ground busses line up. The width of the cell should be as narrow as the layout will allow. However, the absolute width is not important and can be increased as needed.

4. The layout should be labeled to indicate power, ground, input, and output connections. Also, an outline of the cell, useful in alignment, should be added to the cell layout.

Figure 15.4 illustrates the connection of standard cells to a bus. Note that poly, which runs vertically, can cross the metal1 lines, which run horizontally without making contact. This fact is used to route signals and interconnect standard cells in a VLSI design. Also, in this figure, note how the two inverter standard cells are placed end to end. The result is that power and ground are automatically routed to each cell.

Other examples of static standard cells are shown in Fig. 15.5. A double inverter standard cell is shown in Fig. 15.5a while NAND, NOR, and transmission gate standard cells are shown in Figs. 15.5b, c, and d.

Figure 15.6 shows the layout of a NAND-based SR flip-flop. This layout differs from the others we have discussed. All layouts discussed so far have metal1 and

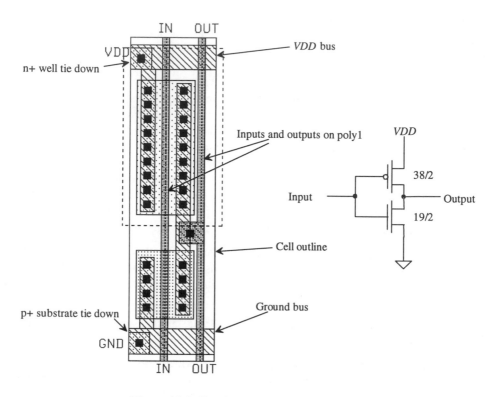

Figure 15.3 Standard-cell layout of an inverter.

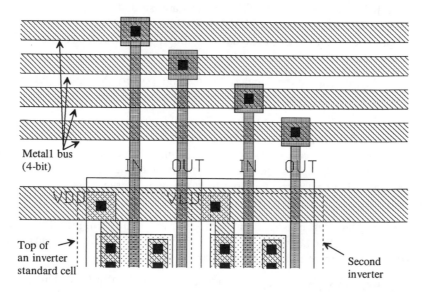

Figure 15.4 Connection of two inverter standard cells to a bus.

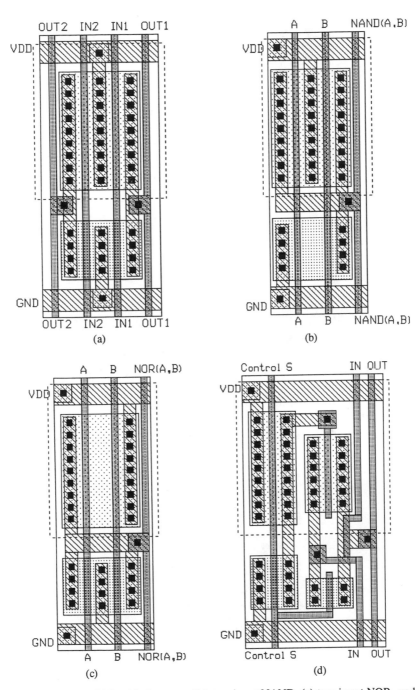

Figure 15.5 (a)Double inverter, (b) two-input NAND, (c) two-input NOR, and (d) transmission gate.

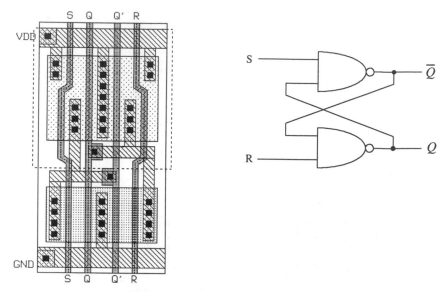

Figure 15.6 SR flip-flop using NAND gates.

contacts adjacent to the gate poly. Also, the gate poly has been laid down without bends. The expanded view of a p-channel MOSFET used in the SR flip-flop is shown in Fig. 15.7. Keeping in mind that whenever poly crosses active (n+ or p+) a MOSFET is formed, we see that the source of the MOSFET is connected to metal through two contacts, while the p+ implant forms a resistive connection to metal1 along the remainder of the device. The layout size, in this case the width of the standard cell, can be reduced using this technique. Because of the bend in the gate, the width of this

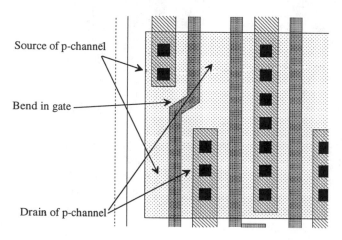

Figure 15.7 Section of the layout shown in Fig. 15.6.

MOSFET is longer than the adjacent MOSFET. This additional width is of little importance and has little effect on the DC and transient properties of the gate. Figure 15.8 shows the NOR implementation of an SR flip-flop.

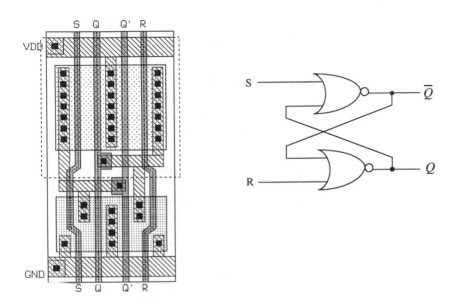

Figure 15.8 SR flip-flop using NOR gates.

Power and Ground Connections

Many of the problems encountered when designing a chip can be related to distribution of power and ground. When power and ground are not distributed properly, noise can be coupled from one circuit onto the power and ground conductors and injected into some other circuit.

Consider the placement of standard cells in a padframe shown in Fig. 15.9a without connections to power and ground shown. Approximately 600 standard cells are shown in this figure. The space between the rows of standard cells is used for the routing of signals. A line drawing of a possible power and ground bussing architecture is shown in Fig. 15.9b. Consider the section of bus shown in Fig. 15.9c. Wire A is used to connect the standard cells in the top row to *VDD*, while wire B is used for connection to ground. Ideally, the current supplied on A (*VDD*) is returned on B (ground). In practice, there exists coupling between conductors B and C, which gives rise to an unwanted signal (noise) on either conductor. This coupling can be reduced by increasing the space between B and C. This reduces the inductive and capacitive coupling between the conductors. Another solution is to increase the capacitance between A and B. A standard-cell decoupling capacitor (Fig. 15.10) can be used toward

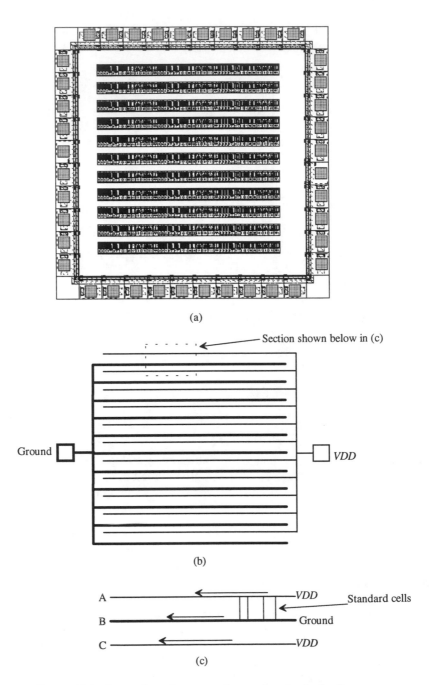

Figure 15.9 Connection of power and ground to standard cells.

this goal. The capacitor is placed in the middle of a standard-cell row. Also, the AC resistive drop effects discussed in Ch. 3 are greatly reduced by inclusion of this capacitor.

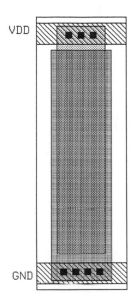

Figure 15.10 Decoupling capacitor.

Coupling is a problem on signal busses as well. Figure 15.11 shows a simple scheme used to reduce coupling. The length of a section, where two wires are adjacent, is reduced by routing the wire to other locations at varying distances along the bus. The inductive or capacitive coupling between two conductors is directly related to the length of the wire runs.

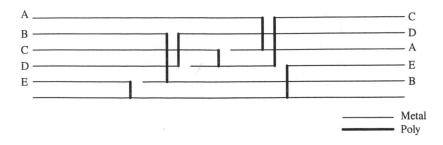

Figure 15.11 Bussing structure used to decrease signal coupling.

An Adder Example

As another example, let's consider the implementation of a four-bit adder. (The floorplan for this adder was shown in Fig. 15.2.) The first components that must be designed are the input and output latches. Figure 15.12a shows the schematic of the D flip-flop used in the latches. This FF is the level-triggered type discussed in the last chapter. When CLK is high, the output, Q, changes states with the input, D. The inverter, I4, is used to provide positive feedback and is sized with a small W/L ratio so that I1 does not need to supply a large amount of DC current to force the latch to change

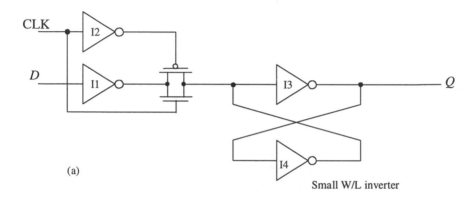

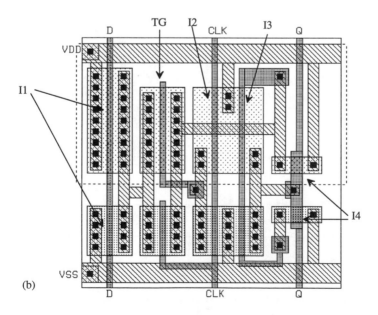

Figure 15.12 Schematic and layout of a D FF.

states. The layout of the FF is shown in Fig. 15.12b. The layout size and the size of the MOSFETs used in these examples are larger than what would be used in practice to make understanding and viewing the layouts easier.

The layout of the static adder is shown in Fig. 15.13. This is the implementation, using near minimum-size MOSFETs, of the AOI static adder of Fig. 12.18. Both the carry-out and sum-out logic functions are implemented in this cell.

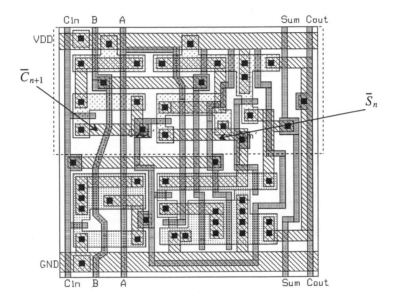

Figure 15.13 Layout of the static adder of Fig. 12.18.

The complete layout of the adder is shown in Fig. 15.14. The two four-bit words, Word-A and Word-B, are input to the adder on the input bus. These data are clocked into the input latch when CLK is high, while the results of the addition are clocked into the output latch when CLK is low. The inverter standard cell of Fig. 15.3 is placed at the end of the output latches and is used to generate $\overline{\text{CLK}}$ for use in the output latches. The inputs and outputs of the adder cells are run on poly because of the short distances involved. The carry-in of the adders is connected to ground, as shown in the figure.

A 4 to 1 MUX/DEMUX

The layout of a 4 to 1 MUX/DEMUX is shown in Fig. 15.15 (based on the circuit schematic of Fig. 13.11). This layout is different from the layouts discussed so far since the circuit does not require power and ground connections and the input/output signals are connected on n+. The select signals are supplied to the circuit on metal1 at the top of the layout. For A to be connected to the output, the signals S1 and S2 should be high. For a large MUX, the propagation delay through the n+ should be considered.

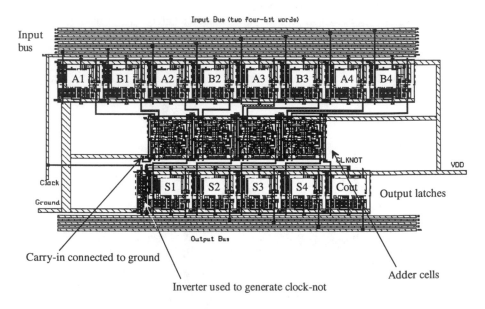

Figure 15.14 Layout of the complete adder.

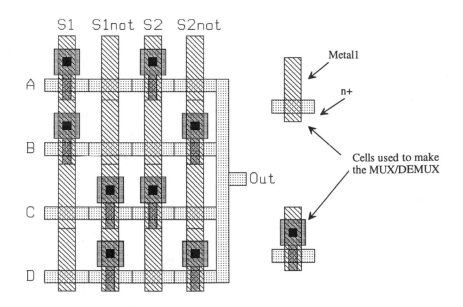

Figure 15.15 Layout of a 4 to 1 MUX/DEMUX.

15.2 Layout Steps *by Dean Moriarty, Crystal Semiconductor*

The steps involved in rendering a schematic diagram into its physical layout are: plan, place, connect, polish, and verify. Let's illustrate each of these steps in some detail through the use of examples.

Planning and Stick Diagrams

The planning steps start with paper and pencil. Colored pencils are useful for distinguishing one object from another. You can use gridded paper to help achieve a sense of proportion in the cell plan but don't get too bogged down in the details of design rules or line widths at this point; we just want to come up with a general plan. A "stick diagram" is a paper and pencil tool that you can use to plan the layout of a cell. The stick diagram resembles the actual layout but uses "sticks" or lines to represent the devices and conductors. When used thoughtfully, it can reveal any special hook-up problems early in the layout, and you can then resolve them without wasting any time.

Figure 15.16a shows the schematic of an inverter. In order to realize the layout of this circuit, it is first necessary to define the direction and metalization of the power supply, ground, input, and output. Since the standard-cell template "sframe" in the CN20 setups does this for us, we'll use it. Power and ground run horizontally in metal1 and are each 7 microns wide. The input and output are accessible from the top or bottom of the cell and will be in metal2 running vertically. Figure 15.16b shows the completed stick diagram. Note the use of "X" and "O" to denote contacts and vias, respectively. The stick diagram should be compared to the resulting layout of Fig. 15.17.

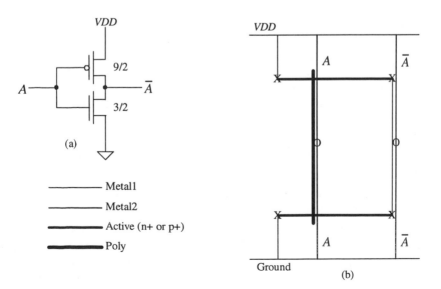

Figure 15.16 (a) Inverter and (b) stick diagram used for layout.

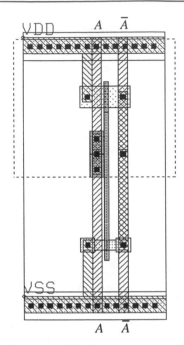

Figure 15.17 Layout of the inverter shown in Fig. 15.16.

Suppose the device sizes of the inverter circuit in Figure 15.16a were quadrupled ($p = 36/2$, $n = 12/2$). Furthermore, let's assume that the maximum recommended poly1 gate width is 20 μm (due to the sheet resistance of the poly) and that exceeding that maximum could introduce significant unwanted RC delays. Let's also suppose that we are to optimize the layout for size and speed (as most digital circuits are). To meet these criteria, it will be necessary to split transistors M1 and M2 in half and lay them out as two parallel "stripes" of 18/2 for the p-channel and two parallel "stripes" of 6/2 for the n-channel. Figures 15.18a-d show the schematics, stick diagram, and layout for this scenario. The output node (drain of M1 and M2) is shared between the stripes so as to minimize the output capacitance. Taking the output in metal2 also helps in this regard. Notice that the stick diagram for this circuit looks like the previous inverter plus its mirror image along the output node. Also observe that the layout of this inverter is mirrored as shown in the stick diagram. This is a common layout technique.

Incidentally, LASICKT will need a schematic similar to Fig. 15.18d to verify the connectivity of the layout. More sophisticated (and much more expensive) CAD software could use the schematic of Fig. 15.18a.

Figure 15.19 shows stick diagrams and layouts for two more common circuits: the two-input NAND and the two-input NOR. Compare the stick diagrams of Figs. 15.19a and c to the layouts of Figs. 15.19b and d. Observe that the output nodes share the active area just as in the previous example. Also note that the spacing between the gates of the series-connected devices is minimum (for the CN20 process).

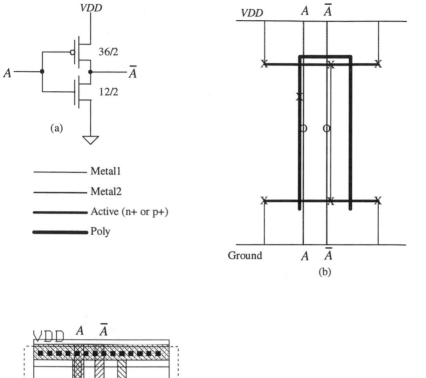

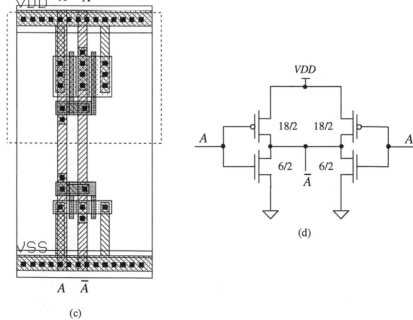

Figure 15.18 (a) Inverter, (b) stick diagram used for layout, (c) layout, and (d) equivalent schematic.

Take another look at the two circuits from a geometrical rather than an electrical viewpoint. Compare the NAND gate layout to the NOR gate layout. Do you see that each can be created from the other by simply "flipping" the metal and poly connections about the x-axis?

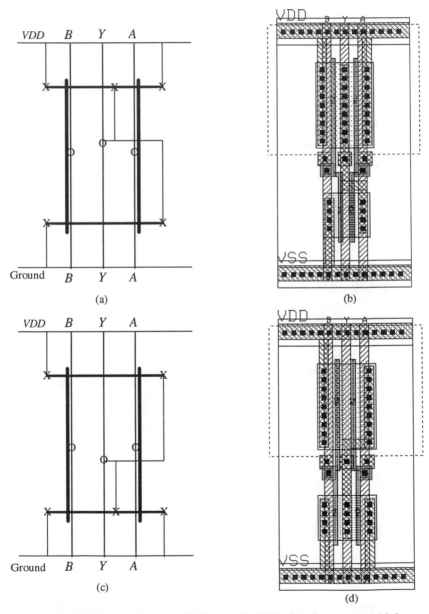

Figure 15.19 (a) NAND stick diagram, (b) layout, (c) NOR stick diagram, and (d) layout.

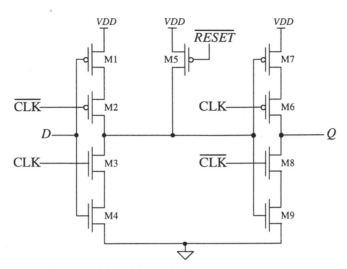

Figure 15.20 Schematic of a dynamic register cell.

Device Placement

Figure 15.20 shows the schematic of a dynamic register cell, while Figs. 15.21 a-c show the stick diagrams and layout for a dynamic register. Compare the schematic of Fig. 15.20 to the stick diagram of Fig. 15.21a. We have labeled this stick diagram "preliminary" for reasons that will soon become apparent. Notice that there is a break or gap in the active area which will form our n-channel devices. Also note that the clock signals CLK and $\overline{\text{CLK}}$ must be "cross connected" from one side of the layout to the other. We don't have to think this through very far to notice that, with this placement of devices, hooking up the clock signals is going to be very difficult. Now look at the stick diagram shown in Fig 15.21b. Notice that we have rearranged the devices so that the active area is a continuous unbroken line. Normally, this "unbroken line" approach to device placement is preferred. It usually results in the most workable device placement. We say "usually" because at times your layout has to fit in an area defined by other blocks around it and you have no control over it. Also observe from Fig. 15.21b that the clock signal hook-up is more straightforward. Compare this stick diagram to the layout of Fig. 15.22c. Obviously, the device sizes used for this circuit are not practical; its purpose is merely to illustrate a layout concept. We can also see that the stick diagram is a useful tool throughout the layout process.

Polish

After your layout is basically finished, it is time to step back and take a look at it from a purely aesthetic point of view. Is it pleasing to the eye? Is the hook-up as straightforward as possible, or is it "busy" and hard to follow? Are the spaces between poly gates and contacts minimum? What about the space between diffusions? Are there

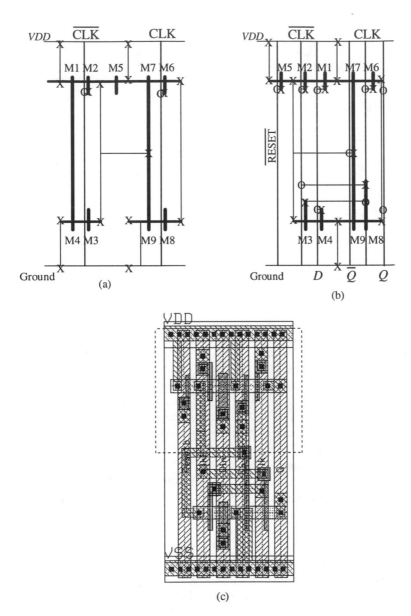

Figure 15.21 Layout of a dynamic register cell.

enough contacts? Did you share all of the source and drain diffusions that can be shared? Are there sufficient well and substrate ties? If you have planned well and followed the plan described here, you shouldn't run into too many problems.

Standard Cells Versus Full-Custom Layout

The standard-cell approach to physical design usually dictates that cell height be fixed and the width be variable when implementing the circuit. Furthermore, standard cells are designed to abut on two sides, usually left and right, and that abutment scheme must be quite regular so that any cell can reside next to any other cell without creating a design rule violation. The standard-cell approach to layout is very useful and is always an excellent place to start. However, in the real world, area on a wafer translates directly into profit and loss (money). Wafer costs are relatively fixed whether they're blank or as tightly packed with circuitry as possible. Therefore, it follows that we want a layout that is as small as possible so that there can be as many die per wafer as possible. These are the economics of the situation. There are also technical advantages to be gained from having as small a layout as possible: interconnecting wires can be as short as possible, thereby reducing parasitic loading and crosstalk effects.

Figure 15.22 shows a typical standard-cell block that has been placed and routed by an automatic tool. Most of the individual cells have been omitted for clarity. Notice the interconnect channels between the rows of standard cells. Power, ground, and clock signal trunks run vertically to both sides of the block by means of a special cell called an "end cap." Cell rows are connected to power and ground through horizontal busses that are part of the standard cells themselves. All remaining connections are made via the routing channels. The standard-cell layouts are designed to accommodate metal2 feedthroughs that run vertically through each cell. The autorouter makes use of this space and adds the feedthroughs as needed in order to connect or pass signals from one routing channel to another. The routing channels and their associated interconnecting wires are the limiting factors for both the density and circuit performance of this type of layout.

Before we continue our discussion of relative layout density, we need to define a metric with which to quantify the matter. It is customary to use the number of transistors per square millimeter of area for this purpose. Because it is a raw number and the common denominator of all circuit layouts, we can use it even when comparing different types of circuitry or even unlike processes.

The density of the standard-cell route shown in Fig. 15.22 is approximately 5,000 transistors per square millimeter. This is fairly representative of the possible density for the channel-based routing approach and the process used (0.8 μm). Figure 15.23 shows a full-custom layout for a digital filter. The circuit area is approximately 2.1 square millimeters. The density is approximately 17,500 transistors per square millimeter, representing a 3.5-fold increase. This circuit, too, is fairly representative of the attainable density of full-custom layout using this particular 0.8 μm process. Both of these circuits were laid out using the same process, and in fact they are from the same die. The device sizes within each block would probably average out to minimum or close to minimum. The main difference affecting density is the interconnect wiring. This overhead associated with interconnect wiring is commonly referred to as the "interconnect burden." The designer must bear this burden in terms of both physical

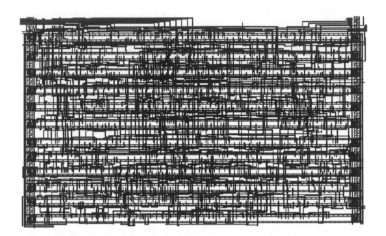

Figure 15.22 Layout based on standard cells.

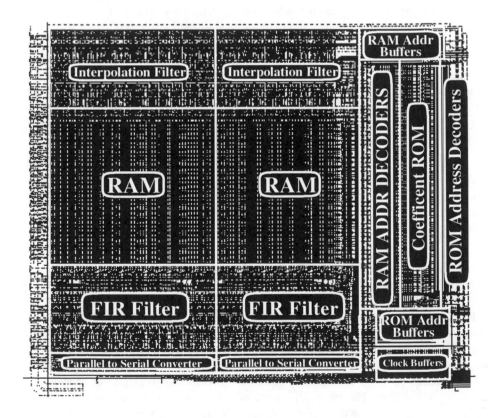

Figure 15.23 Full-custom layout of a digital filter.

(wasted area) and electrical parameters (parasitic loading). Let us examine one method of creating a high-density custom layout that will minimize interconnect burden and circuit area.

Figures 15.24 a-c show a small section of the interpolation filter from Fig. 15.23. In Fig. 15.24a, we see an exploded view of four cells that form part of a data-path: an input data register, a t-gate, a full adder, and an output data register. These are instantiated (placed as a cell) twice, creating a view of eight cells. The two adder cells are slightly different: the carry inputs and outputs are on opposite sides, so that the carry-out can cascade to the carry-in of the next adder by abutting (placing next to one another) the cells. Unlike standard cells, the height and width constraints placed on custom layouts are contextual. In other words, a cell's aspect ratio depends on that of its neighbors. In this case, the width of each cell depended on the maximum allowable width of the widest cell in the group: the data register. Notice the top, bottom, left, and right boundaries of each cell in Fig. 15.24a. Data enter the register cell from the top and are output at the bottom. Clocks, power, ground, and control signals route across all the cells. The adder receives its A and B inputs from the top and outputs their SUM at the bottom. As already mentioned, carry-out and carry-in are available on the left and right edges of the adder, respectively. Figure 15.24b shows a two-bit slice of this data-path with all connections made by cell abutment. Figure 15.24c illustrates how all four edges of each cell join together to complete the hook-up.

We have seen how circuits can be implemented by means of standard cells or custom layout. The time needed to produce a standard-cell route is far less than that of

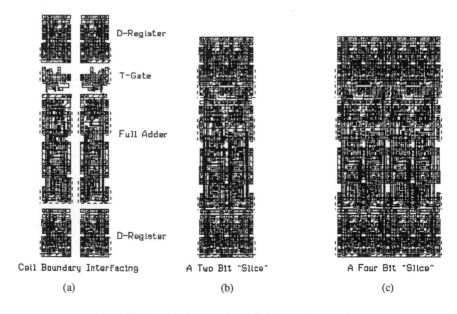

Figure 15.24   Sections of the digital interpolation filter.

a full-custom implementation. The tradeoffs are area and performance. Automatic place and route tools based on routing area rather than routing channels are now coming into use. These promise a compromise solution between the two extremes. The density of their results rivals that of full-custom layout. Perhaps the hand-rendered full-custom layout will someday become a thing of the past. Nevertheless, process technology will continue to advance, circuit designers will continue to design circuits that test the outermost limits of this technology, and the marketplace will still be there demanding ever cheaper, more powerful, and faster products. It is likely then that we will all still have the opportunity to "push a polygon" or two for the forseeable future. There remains no doubt that the future will bring us ever more powerful software tools that will take over the tedious aspects of placing and connecting layouts, leaving to us the more creative aspects of planning and polishing them.

REFERENCES

[1]   N. H. E. Weste and K. Eshraghian, *Principles of CMOS VLSI Design*, Addison-Wesley, 2nd ed., 1993. ISBN 0-201-53376-6.

[2]   D. V. Heinbuch, *CMOS3 Cell Library*, Addison-Wesley, 1988.   ISBN 0-201-11257-4.

[3]   Kerth, Donald A. *"Floorplanning-Lecture Notes"* Crystal Semiconductor, Inc.

[4]   Kerth, Donald A. *"Analog Tricks of the Trade-Lecture Notes."* Crystal Semiconductor, Inc.

[5]   J. Uyemura, *Physical Design of CMOS Integrated Circuits Using L-EDIT*, PWS Publishing Co., 1995. ISBN 0-534-94326-8.

PROBLEMS

15.1   The standard-cell height can be reduced to make standard-cell-based layouts smaller (Fig. P15.1). Using this cell as an approximate height reference, lay out a double inverter.

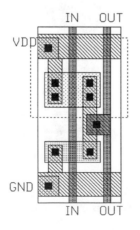

Figure P15.1

15.2 Repeat Problem 15.1 for a two-input NAND gate.

15.3 Repeat Problem 15.1 for a two-input NOR gate.

15.4 Repeat Problem 15.1 for a transmission gate (with the same functionality as Fig. 15.5).

15.5 Repeat Problem 15.1 for a NAND-based SR flip-flop.

15.6 Repeat Problem 15.1 for a NOR-based SR flip-flop.

15.7 Design, lay out, and simulate the operation of a D FF to replace the one described in Fig. 15.12.   Assume that the FF uses mainly minimum-size MOSFETs and that a pass transistor is used for the clocking element.

15.8 Lay out the two-input MUX using TGs shown in Fig. P15.8.

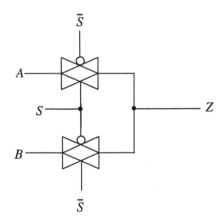

Figure P15.8

15.9 Using the reduced standard-cell frame height described in Problem 15.1, lay out the XOR/XNOR gates of Fig. P15.9. Assume that both the inputs and outputs to the gates are on poly1.

15.10 Lay out a 16-to-1 MUX/DEMUX based on the layout topology given in Fig. 15.15. Lay out another 16 to 1 MUX/DEMUX in as small an area as possible. DRC your final layouts.

15.11 Sketch a stick diagram for the layouts of Fig. 15.5.

15.12 Point out the high-impedance nodes for the schematic of Fig. 15.20. Discuss the concerns one must consider when laying out a circuit with high-impedance nodes.

15.13 List and discuss three reasons to have small layout size.

15.14 Lay out the D FF shown in Fig. P15.14. Show the stick diagram for your layout.

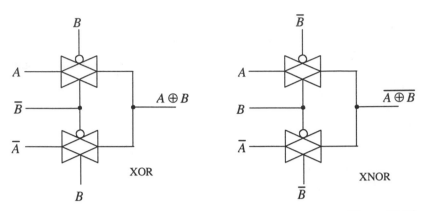

Figure P15.9

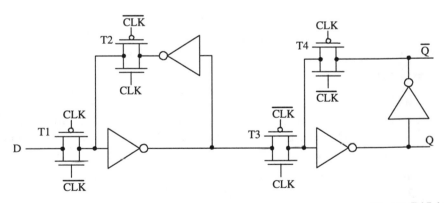

Figure P15.14

BiCMOS Logic Gates

Modern BiCMOS technology began in the early 1980s with high expectations [1]. The name BiCMOS comes from the fact that the logic is made using CMOS and bipolar junction transistors (BJT). The BJTs are used for their high-current capability, while CMOS is used because of its small layout size and ease of implementing logic. With the best of both worlds on a single substrate, high-speed, high-current-driving bipolar transistors and low-power, high-impedance CMOS devices, every major semiconductor foundry now possesses some form of BiCMOS process. Strategies for developing BiCMOS have evolved from the bipolar and the CMOS directions, with advantages and disadvantages associated with each. Bipolar device capabilities have been added to some CMOS processes to improve speed, while CMOS device capabilities have been added to some bipolar processes to minimize power dissipation. A chart comparing CMOS, BiCMOS, and bipolar (with I^2L) technologies can be seen in Fig. 16.1 [2,3]. This chapter focuses on the CMOS process with bipolar capabilities. Although the CN20 process is not a true BiCMOS process, it does contain some BJT options that will allow demonstration of basic digital BiCMOS circuit design. It should be noted that the CMOS14TB process contains no provisions for BJT devices.

Microprocessors are particularly well suited for BiCMOS technology. Typically, three generic categories limit microprocessor performance [1]: (1) Instructions per task, (2) cycles per instruction, and (3) time per cycle. The third category can be greatly improved by increasing the speed critical blocks. A PC microprocessor [4] was developed using a bipolar-based BiCMOS process. Operating at 533 MHz, the microprocessor used high-density CMOS devices that were added to a bipolar process. The floor plan can be seen in Fig. 16.2 [5] in which the speed critical blocks such as the integer and floating point units utilized BJT transistors, while power-consuming cache arrays and I/O cells (for system compatibility) were constructed using CMOS technology.

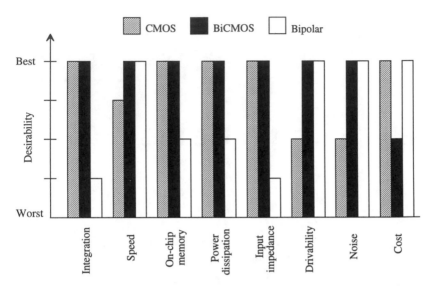

Figure 16.1 Comparison of CMOS, BiCMOS, and bipolar technologies.

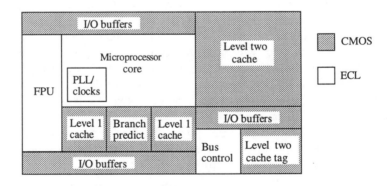

Figure 16.2 Floor plan of a high-performance BiCMOS microprocessor.

16.1 Layout of the Junction-Isolated BJT

We begin this chapter with a discussion of the junction-isolated NPN bipolar junction transistor that is available in CN20 with the addition of a p-diffusion layer called p-base. The layout and operation of the BJT are discussed. Most of the applications of the BJT in digital circuits are in the design of buffer circuits for driving large capacitive loads. The BJT buffer requires less area than the CMOS counterpart.

The vertical NPN BJT in the CN20 process uses the n-well as the collector, the p-base diffusion for the base, and the n+ implant for the emitter. The layout of a 2 x 1

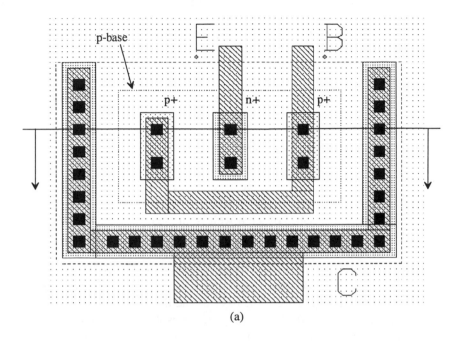

(a)

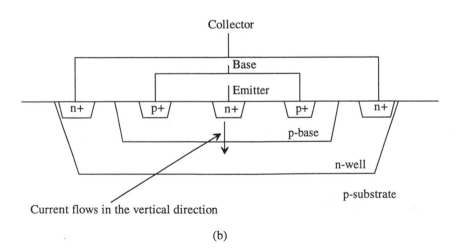

(b)

Figure 16.3 Layout of a 2 x 1 vertical NPN BJT (a) and associated cross-sectional
view (b).

BJT is shown in Fig. 16.3a. The first number in the description of the BJT indicates the number of contacts on the emitter of the BJT, while the second number indicates the number of fingers. The cross-sectional view of the 2 x 1 BJT, is shown in Fig. 16.3b. Note that the metal1 connection to well and p-base require n+ and p+ implants, respectively. The BJT discussed here is called vertical because the majority of the emitter current flows out the bottom of the n+ implant emitter to the n-well collector. In a bipolar only process or a true BiCMOS process, the current flow is also in the vertical direction to a buried n+ sub-collector. Because the CN20 process uses the n-well as the collector, only an NPN transistor can be formed. If the process possessed a p-well and an n-base layer, a PNP transistor could also be formed.

The collector, made with the n-well, of the NPN transistor forms a diode, with the p-type substrate limiting the negative values of collector voltages allowable and isolating the collector from the substrate. (This is very important to keep in mind.) We also know that the sheet resistance of the n-well is on the order of 2,500 Ω/square. For the 2 x 1 BJT shown in Fig. 16.3, the series collector resistance is in the neighborhood of 500 Ω. The U-shaped connection of n+ to the well is used to reduce this resistance and provide better collection of carriers. The p-base diffusion used to make the base of the NPN is also highly resistive. The base contacts on each side of the emitter, instead of on a single side, are used to reduce the base spreading resistance. Typical values of the base spreading resistance, for this small junction-isolated NPN, are 1,000 Ω. The emitter series resistance is tens of ohms and is usually negligible compared to the collector and base series resistances. In short, the problem with using the junction-isolated NPN transistor in a CMOS process to design BiCMOS circuits is the associated large parasitic resistances. The design rules for the p-base layer are shown in Fig. 16.4. These rules correspond to checks 33 through 36 in the CN20.DRC file.

16.2 Modeling the NPN

The junction-isolated NPN bipolar transistor operation is very similar to normal BJT operation, with the exception of large parasitic resistances associated with the base and collector. The symbol for a JI BJT is identical to the normal BJT symbol and is seen in Fig. 16.5.

To develop a digital model for the BJT which is similar to the model we developed for the MOSFET, we will assume that the effective switching resistance between the collector and the emitter is simply R_c. We can define the variable R_{npn} by

$$R_{npn} = R_c \qquad (16.1)$$

The input resistance of the vertical BJT can be estimated by

$$R_{innpn} = R_b \qquad (16.2)$$

A SPICE model for the 2 x 1 BJT in the CN20 process is located on the accompanying disk (or in the files you downloaded) in the file "spice.inf" in the \Wcn20\ directory.

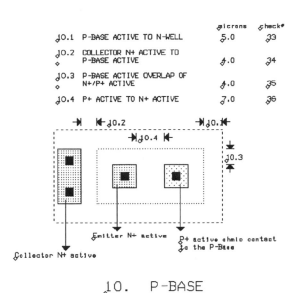

Figure 16.4 Design rules for the p-base layer.

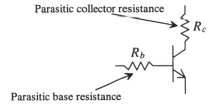

Figure 16.5 Symbol for the vertical BJT showing parasitic resistances.

Example 16.1

Estimate the falltime of the circuit shown in Fig. 16.6. Compare your hand calculations to SPICE. Note that the input 5 V pulse is applied directly to the base of the BJT.

From the SPICE model located in spice.inf R_c = 420 Ω and R_b = 1.2 kΩ. The high-to-low time can be estimated by

$$t_{PHL} = 420 \cdot 5 \text{ pF} = 2.1 \text{ ns}$$

The base current that must be supplied from the input voltage source is 5/1.2 k or approximately 4 mA. The simulation results are shown in Fig. 16.7. ∎

It should be pointed out that the BJT of this example was driven with an ideal voltage source, something that is not available in practice. However, the collector

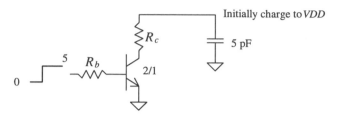

Figure 16.6 Circuit used in Ex. 16.1.

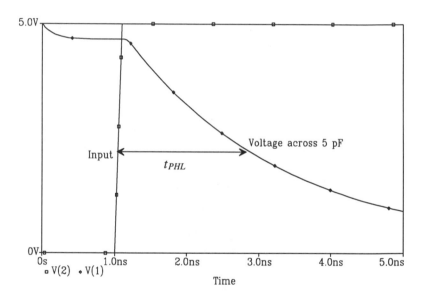

Figure 16.7 Simulation results for Ex. 16.1.

current supplied for an input voltage (the voltage applied to the base of the transistor) of *VDD* is significantly larger than the drain current supplied using the MOSFET switch.

Modeling larger BJTs using the 2 x 1 model can be accomplished using the *A* (area) parameter in SPICE. For example, if you lay out a 2 x 2 BJT, that is, two emitter contacts with two emitter "fingers," we simply set *A* = 2 and use the 2 x 1 model. For a 10 x 3 BJT we set *A* = 15. Here *A* can be thought of as the ratio of the large BJT to the size of the 2 x 1 BJT model. An example of the layout of a 5 x 2 BJT is shown in Fig. 16.8. When using the larger BJT models Eqs., (16.1) and (16.2) are modified by *A*. The resistances now become

$$R_{npn} = \frac{R_c}{A} \tag{16.3}$$

and

$$R_{innpn} = \frac{R_b}{A} \qquad (16.4)$$

Here R_c and R_b are the series resistances specified in the SPICE model used. If A is not specified, SPICE assumes $A = 1$.

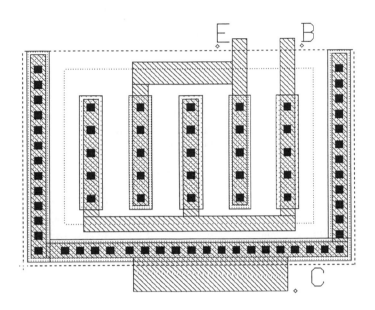

Figure 16.8 Layout of a 5 x 2 junction-isolated NPN transistor.

BJT Capacitances

The BJT capacitances result from the depletion capacitances of the implant regions and from the forward-biased base-emitter junction (the storage capacitance). The base-collector depletion capacitance is estimated either from the SPICE model or from use of the n+ to p-substrate depletion capacitance specified in the n-channel BSIM model. The base-emitter depletion capacitance is estimated using the depletion capacitance from the p+ to n-well specified in the p-channel BSIM model.

More important than these depletion capacitances is a storage capacitance associated with the base-emitter forward-biased diode. If the minority carrier lifetime of the base-emitter junction is called τ_F, then, from Ch. 2, the storage capacitance is given, assuming the emission coefficient, n, is one, by

$$C_{bestor} = \tau_F \cdot \frac{I_E}{V_T} \qquad (16.5)$$

Here I_E is the DC emitter current and V_T is the thermal voltage (kT/q). As the emitter current increases, the storage capacitance increases. The SPICE model supplied in the file "spice.inf" does not specify the τ_F parameter, so simulations will not show charge storage effects (such as the transistor not turning off until the stored charge is removed from the base-emitter junction).

16.3 The BiCMOS Inverter

A BiCMOS inverter is shown in Fig. 16.9. One important aspect of the basic BiCMOS inverter is that the output cannot go to *VDD* or ground, as was the case in the CMOS inverter. This has the effect of lowering the noise margins of the logic. The maximum output voltage is approximately *VDD* – 0.7 V, while the minimum logic output voltage is approximately 0.7 V. The 0.7 V drop for the high and low side comes from the base-emitter voltage drop of Q2 and Q1, respectively. Caution should be exercised when using the output of BiCMOS gates with CMOS logic. The low-output voltage of 0.7 V is very close to the threshold voltage of the n-channel transistor. CMOS gates with switching point voltages close to the threshold voltage are susceptible to noise.

To understand the operation of the basic BiCMOS inverter, consider the case when the input is grounded. MOSFETs M4 and M1 are on, while M2 and M3 are off. The BJT Q1 is off, its base being held at ground potential by M1. The base of Q2 is held at *VDD* by M4. The output voltage is *VDD* – 0.7 V or a logic high. When the input of the BiCMOS inverter is held high, MOSFETs M3 and M2 are on, while M4 and M1 are off. The base of Q2 is held at ground potential, so it is off. Since M2 is on, the output is connected to the base of Q1. This causes Q1 to turn on and pull the output voltage down to 0.7 V.

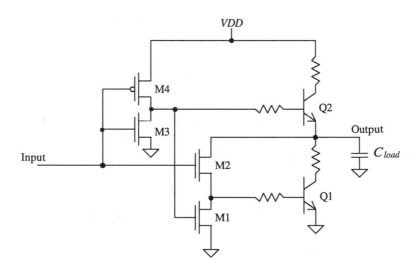

Figure 16.9 Basic BiCMOS inverter showing parasitic collector and base resistances.

Switching Characteristics

The delay associated with the BiCMOS inverter discharging a capacitance, C_{load}, consists of two parts: the delay in Q1 turning on and the delay once Q1 is on discharging C_{load}. The delay associated with discharging C_{load} is given by

$$t_{PHL} = R_{npn} \cdot C_{load} \qquad (16.6)$$

The delay associated with turning Q1 on is comparable to the minority carrier lifetime τ_F, which is in general well under 1 ns. The MOSFET M2 is used to turn Q1 on. A general design rule can be applied to the sizing of M2. We require the effective resistance of M2, R_{n2} to be equal to the base resistance of Q1, R_b. This is a very simple method of sizing the transistors, which gives very good results.

The low-to-high delay-time can be estimated in much the same way as the high-to-low delay. The delay in charging C_{load} is given by

$$t_{PLH} = R_{npn} \cdot C_{load} = t_{PHL} \qquad (16.7)$$

The fact that the propagation delay-times are equal is a side benefit of the BiCMOS inverter. The MOSFET M4 drives the base of Q2. Sizing M4's effective resistance equal to the base resistance of Q2 is a general design rule for the sizing of the MOSFETs. MOSFETs M1 and M3's effective resistance can be made larger than M2 and M4's effective resistance because these MOSFETs simply hold the bases of Q1 and Q2 at ground during the off period. In general, to attain a switching point voltage close to $VDD/2$, M2 and M4 are sized the same as M3.

Example 16.2
Estimate and simulate the delay of the BiCMOS inverter (Fig. 16.10) driving a 10 pF load. Calculate the input capacitance of the inverter.

The propagation delays are simply given by

$$t_{PHL} = t_{PLH} = R_{npn} C_{load} = 420 \cdot 10 \text{ pF} = 4.2 \text{ ns}$$

The SPICE simulation results are shown in Fig. 16.11. Notice how the output of the BiCMOS inverter stays at approximately 0.7 V from the power and ground rails.

The input capacitance is the sum of the input capacitances of the three MOSFETs connected to the input, or

$$C_{in} = \frac{3}{2} \cdot 800 \frac{\text{aF}}{\mu m^2} (2 \cdot 10 + 2 \cdot 10 + 16 \cdot 2) = 86.4 \text{ fF}$$

This input capacitance can easily be driven from logic on chip without excessive delay. The delay of this inverter driving a 10 pF load can be reduced by using larger BJTs with corresponding larger MOSFET drivers.

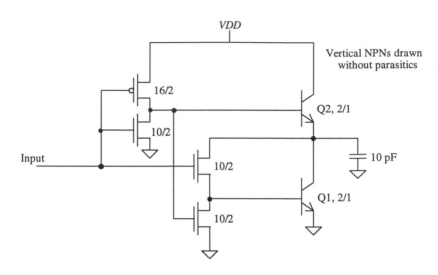

Figure 16.10 Circuit used in Ex. 16.2.

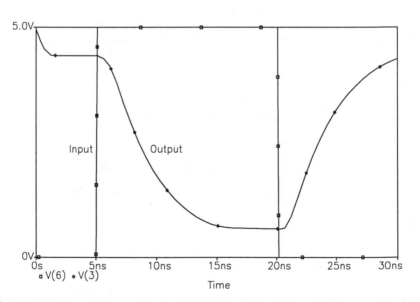

Figure 16.11 Simulation results of a BiCMOS inverter driving a 10 pF capacitive load.

An important characteristic of the BJT has been neglected in the analysis and in the SPICE models. The carrier lifetime, τ_F (resulting in a storage capacitance), was not modeled, and the actual delays can be longer than predicted by SPICE.■

Full-Swing BiCMOS Inverters

Figures 16.12 and 16.13 show two implementations of a full-swing BiCMOS inverter [4, 5]. Consider the first inverter (Fig. 16.12) with its input grounded. MOSFETs M2 and M4 are off while MOSFET M5 is on. MOSFETs M1 and M3 can be thought of as resistors. Since M5 is on, the base of Q2 is pulled to *VDD*. The transistor Q2 is on and pulls the output to *VDD* – 0.7. MOSFET M3, which behaves like a resistor, then pulls the output up to *VDD*. When the input to the inverter is high, M2 and M4 are on and M5 is off. This pulls the base of Q2 to ground, turning it off. At the same time M2 turns on, with the output high, causing Q1 to turn on. Q1 pulls the output down to 0.7 V. From there M1, which behaves like a resistor, pulls the output down to ground. Note that if M1 or M3 does not have a large effective resistance (long L), the circuit will not operate correctly.

The inverter shown in Fig. 16.13 performs better (lower dynamic power dissipation) than the inverter of (a) at the cost of more complicated circuit design. The inverter operation is similar to that given in (a) with the exception that switches M1 and M3 are controlled by the output of the inverter. This topology improves performance by making the MOSFET sizes less critical.

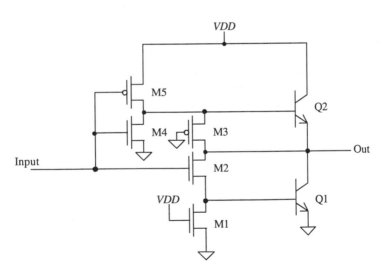

Figure 16.12 A wide-swing BiCMOS inverter.

16.4 Other BiCMOS Logic Gates

The NAND and NOR gates using BiCMOS technology [6, 7] are easily constructed from the standard CMOS static logic gate. As seen in Fig. 16.14, blocks P_1 and P_2 are identical to the all CMOS NAND gate presented in Ch. 12 (Fig. 12.1). Node C

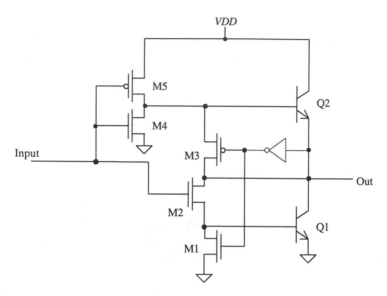

Figure 16.13 A full-swing BiCMOS inverter with improved power dissipation.

represents the output of the all CMOS NAND. Therefore, the output of the BiCMOS gate is identical to the logic state of node C. The transistor pair composed of M5 and M6 (P$_3$) ensures that if the output is low (both inputs are a logic 1), the sinking transistor, Q1, is biased correctly to discharge the load capacitor. The transistor, M7, connects the base of Q1 to *VSS* (or ground) only if the output is high. Examine Table 16.1 for the states of the important nodes and transistors of the NAND gate.

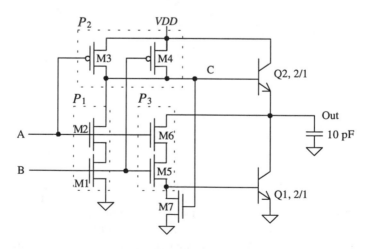

Figure 16.14 A BiCMOS NAND gate.

A B	Node C	M5	M6	M7	Q1	Q2	Out
0 0	1	OFF	OFF	ON	OFF	ON	1
0 1	1	ON	OFF	ON	OFF	ON	1
1 0	1	OFF	ON	ON	OFF	ON	1
1 1	0	ON	ON	OFF	ON	OFF	0

Table 16.1 States of key nodes and components in Fig. 16.14.

The basic BiCMOS NOR gate is seen in Fig. 16.15. Notice again that the blocks P1 and P2 represent the all CMOS NOR presented in Ch. 12 (Fig. 12.1) and that the logic state of node C will again be identical to the output of the BiCMOS gate. Transistor block P_3, composed of M5 and M6, allows Q1 to be biased correctly in order to discharge the output capacitance, while M7 again ensures that the base of Q1 is connected to ground should output be a logic 1. Table 16.2 shows that states of the critical nodes and transistors in Fig. 16.15.

A B	Node C	M5	M6	M7	Q1	Q2	Out
0 0	1	OFF	OFF	ON	OFF	ON	1
0 1	0	ON	OFF	OFF	ON	OFF	0
1 0	0	OFF	ON	OFF	ON	OFF	0
1 1	0	ON	ON	OFF	ON	OFF	0

Table 16.2 States of key nodes and components in Fig. 16.15.

The switching analysis for these gates is very similar to the BiCMOS inverter since the delay-time is dominated by the time necessary to charge and discharge the load capacitance.

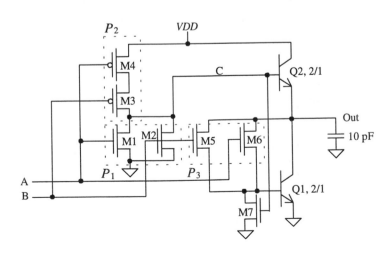

Figure 16.15 A BiCMOS NOR gate.

16.5 CMOS and ECL Conversions Using BiCMOS

One major advantage of using BiCMOS is the ability to mix technologies such as CMOS, ECL, and BiCMOS logic. One advantage of using emitter-coupled logic (ECL) circuits is that the bipolar transistors can double their output current for every 25 mV of change in the base-emitter voltage. This is simply because the collector current, I_C, through a BJT, can be described as

$$I_C = I_S e^{(v_{BE}/V_T)} \tag{16.8}$$

where I_s is the saturation current, V_T is the thermal voltage, and v_{BE} is the instantaneous base-emitter voltage. The expression for the transconductance, which relates the amount of drive current to the input voltage, is

$$g_m(BJT) = \frac{I_C}{V_T} \text{ A/V} \tag{16.9}$$

and is also obviously exponential. As a result, the BJTs can sink or source large amounts of load currents with very small input voltage swings.

Now examine the relationship between gate-to-source voltage and drain current through an MOS device:

$$I_D \approx \frac{\beta}{2}(V_{GS} - V_{THN})^2 \tag{16.10}$$

The transconductance for the MOSFET is

$$g_m(MOS) = \sqrt{I_D 2\beta} = \beta(V_{GS} - V_{THN}) \text{ A/V}$$

which is linear with respect to the input voltage. The amount of input voltage necessary to switch an output from a low to a high or a high to a low is much greater than the BJT case. Typically, the CMOS input signal must overcome the threshold voltage by around 1 V before the output begins to change state. The different switching speeds can be seen in Fig. 16.16 [5].

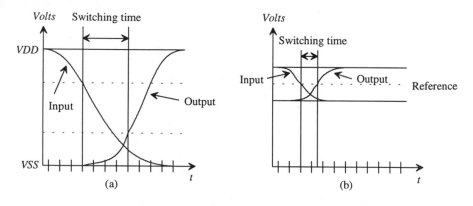

Figure 16.16 Switching times for (a) CMOS logic and (b) ECL.

Notice the differing reference levels shown between the two types of technologies. While CMOS logic typically swings between *VDD* and *VSS*, ECL logic has a much smaller signal swing defining the logic levels. Interface circuits are needed that convert ECL to CMOS logic levels and vice versa.

An ECL to CMOS converter circuit can be seen in Fig. 16.17 [3, 8]. Here, the ECL input signal is level shifted by 2 V_{BE} drops to a current-mode logic (CML) circuit that drives the CMOS output shifter stage. The stepped-down ECL input causes the CML circuit to become imbalanced so that one collector is considered high, while the other collector is considered low. The critical issues in minimizing the delay time are the output swing of the CML stage and the sizes chosen for the CMOS shifter. However, the two specifications are inversely proportional, for increasing the output swing of the CML stage decreases the delay through the CMOS shifter but increases the delay through the CML stage.

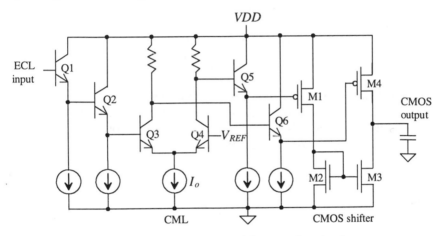

Figure 16.17 ECL to CMOS conversion circuit.

An improved ECL to CMOS translator can be seen in Fig. 16.18 [9]. This circuit has significantly reduced total delay and power dissipation and improved driving capability for capacitive loads over the circuit shown in Fig. 16.17. However, a process allowing both NPN and PNP BJT devices is required. The input can be either complementary or single-ended (with one input tied to a reference voltage). Notice that the two complementary differential input stages drive the output BJT devices. One reason why this circuit has such improved delay-time is that the base of the output transistors, Q5 and Q6, are driven directly from their respective differential input stages. For example, if the ECL inputs were such that ECL1 was high and ECL2 was low, then all the current through M1 would flow through Q4 and into the base of Q6, thus discharging the output node very rapidly. Similarly, if ECL1 was low and ECL2 was high, then all the quiescent current through M2 would flow through Q2, which results in a large amount of current through Q5, thus charging the output node quickly.

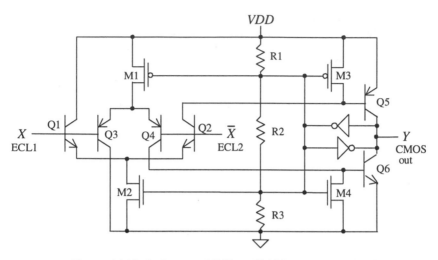

Figure 16.18 An improved ECL to CMOS conversion circuit.

Once the output devices, Q5 and Q6, are turned on, they become saturated until the feedback circuit, composed of inverters I1 and I2, turns off the respective device. For example, if the output node is high, then the output of I1 is low, which in turn pulls the gate of M3 down, thus shorting the base of Q5 to *VDD*. Similarly, if the output node is low, then the output of I1 is high, thus causing M4 to turn on and short the base of Q6 to VSS. Therefore, the output devices are saturated for only a short period of

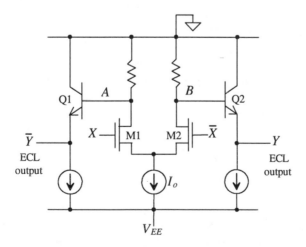

Figure 16.19 CMOS to ECL conversion circuit.

time, while the feedback circuit preserves the logic value at the output. The resistor string is used to keep the base-emitter voltages of the output transistors on the edge of being turned on, thus improving speed even further.

Another conversion circuit is seen in Fig. 16.19 [10]. This circuit translates CMOS signals to ECL logic levels and requires a complemented CMOS input. The input signal causes an imbalance in the source coupled pair, since the current, I_o, is constant. The output swing of the source coupled pair appearing at nodes A and B can be adjusted by wisely choosing the resistance values and input MOS device sizes.

REFERENCES

[1]   A. R. Alvarez, "BiCMOS-Has the Promise Been Fulfilled?," *IEDM 1991*, pp.13.1.1-13.1.4, 1991.

[2]   M. Kubo, I. Masuda, K. Miyata, and K. Ogiue, "Perspective on BiCMOS VLSI's," *IEEE Journal of Solid State Circuits*, vol. 23, no. 1, pp. 5-11, February 1988.

[3]   M. I. Elmasry, "Introduction to BiCMOS Integrated Circuits: A Tutorial," *IEEE BiCMOS Integrated Circuit Design*, IEEE Press, 1994. ISBN 0-7803-0430-6.

[4]   "Exponential Unveils World's Fastest PC Microprocessor at 533 MHz," Press Release by Exponential Technology, Inc., October 21, 1996.

[5]   "Exponential's BiCMOS Technology: Bipolar-Based BiCMOS instead of CMOS-Based BiCMOS," http://www.exp.com/products/x704/bicmos.html, Exponential Technology, Inc., November 1996.

[6]   M. I. Elmasry, *BiCMOS Integrated Circuit Design,* IEEE Press, 1992. ISBN 0-7803-0430-6, IEEE order number: PC0346-7.

[7]   J. P. Uyemura, *Circuit Design for Digital CMOS VLSI,* Kluwer Academic Publishers, 1992.

[8]   S. H. K. Embabi, A. Bellaouar, and M. I. Elmasry, "Analysis and Optimization of BiCMOS Digital Circuit Structures," *IEEE Journal of Solid State Circuits*, vol. 26, no. 4, pp. 676-679, April 1991.

[9]   M. Rau and H. J. Pfleiderer, "An ECL to CMOS Level Converter with Complementary Bipolar Output Stage," *IEEE Journal of Solid-State Circuits*, vol. 30, no. 7, pp. 781-787, July 1995.

[10]   K. Gopalan, *Introduction to Digital Microelectronic Circuits*, Irwin, 1996. ISBN 0-256-12089-7.

PROBLEMS

16.1   Lay out and DRC a junction-isolated 5 x 1 NPN BJT.

16.2   Repeat Ex. 16.1 using a 5 x 1 BJT.

16.3 Verify, using SPICE, the operation of the inverter given in Ex. 16.2.

16.4 Design a full-swing BiCMOS output buffer that has an input capacitance of 100 fF or less and will drive 10 pF with a $t_{PHL} + t_{PLH}$ less than 15 ns.

16.5 Design and simulate the operation of an ECL to CMOS converter based on the circuit topology shown in Fig. P16.5. Assume that the ECL input varies from 4.2 V (a logic high) down to 3.4 V (a logic low).

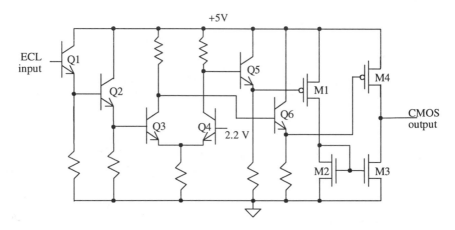

Figure P16.5

Memory Circuits

In this chapter we look into the design of semiconductor memory circuits [1, 2], in particular static random access memory (SRAM) and dynamic random access memory (DRAM). These types of memories are termed random access because any bit of data can be accessed at any time. A block diagram of a RAM is shown in Fig. 17.1.

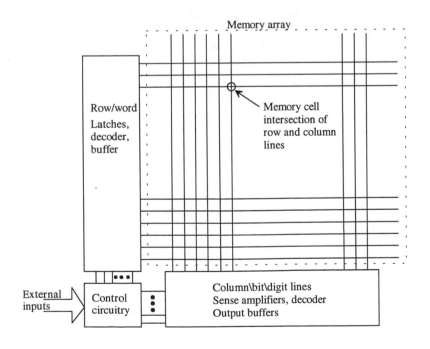

Figure 17.1 Block diagram of random access memory.

External to the memory array are the row and column logic. The row lines are sometimes called word lines, while the column lines are sometimes called bit or digit lines. Referring to the row lines, the row address is latched, decoded, and then buffered. A particular row line will, when selected, go high. This selects the entire row of the array. Since the row line may be long and loaded periodically with the capacitive memory cells, a buffer is needed to drive the line. The address is latched with signals from the control logic. After a particular row line is selected, the column address is used to decode which of the bits from the row are the addressed information. At this point, data can be read into or out of the array through the column decoder. The majority of this chapter will concentrate on the circuits used to implement a RAM.

17.1 RAM Memory Cells

The memory array is at the center of the RAM design. Examples of DRAM and SRAM memory cells are shown in Fig. 17.2. The DRAM memory cell is made up of a pass transistor and a storage capacitor, while the CMOS SRAM memory cell is a cross-coupled connection of inverters. The cross-coupled inverters form a positive feedback circuit, forcing the outputs in opposite directions. The basic operation of the DRAM memory cell was discussed in Ch. 14.

For the SRAM cell shown in Fig. 17.2, consider the case when the row line (word line) is low. Both pass transistors are off, and the datum in the cell is latched, as long as power is applied to the cell. When the row line goes high, the pass transistors turn on. If the pass transistors, width/length ratio is approximately four times that of the SRAM cell transistors' width/length ratio, the data on the bit lines (column lines) are written to the cell. If the widths are comparable in size, the effective resistance of the pass transistor is too large to allow overwriting the cell.

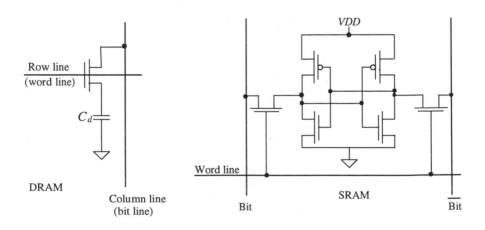

Figure 17.2 Schematic diagrams of RAM memory cells.

17.1.1 The DRAM Cell

Consider the implementation of the DRAM cell shown in Fig. 17.3a. The capacitor is implemented using a MOSFET operating in the inversion region. Schematically, this would be represented by the capacitor in Fig. 17.2 connected to *VDD* instead of ground. The size of the DRAM storage capacitor C_d, is simply the sum of the oxide capacitance of the MOSFET used as a capacitor and the depletion capacitance of the n+ implant to substrate. The row/word line and *VDD* connections to the cell are made using poly1. This greatly simplifies layout.

In order to minimize the layout area of the DRAM array, the scheme shown in Fig. 17.3b is used. In this configuration, the bit line is shared between two cells. When writing to cell0, the word0 line goes high and the datum on the bit line is written to the capacitor M01. This configuration is termed a folded bit line configuration for reasons that are not obvious at this point.

Modern memory processes do not use a MOSFET as the storage capacitor but rather fabricate the storage capacitor using several layers of poly. For the sake of understanding memory circuit design, we will use the CN20 process, for the moment, to describe the implementation of the DRAM.

Consider the layout of the DRAM cell shown in Fig. 17.4. Since the word and *VDD* lines are implemented in poly, the connection to the MOSFETs occurs when the poly runs over the active n+ area. The bit line is implemented in metal1. The DRAM cells are placed adjacent to each other to form the array. A complete DRAM memory array is shown in Fig. 17.5. Notice how adjacent bit lines do not have a word line in common.

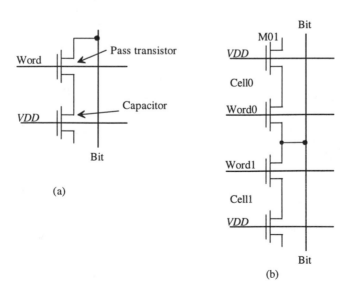

Figure 17.3 Implementation of the DRAM cell.

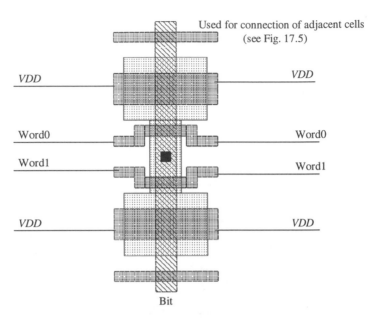

Figure 17.4 Basic layout of a folded bit line DRAM cell.

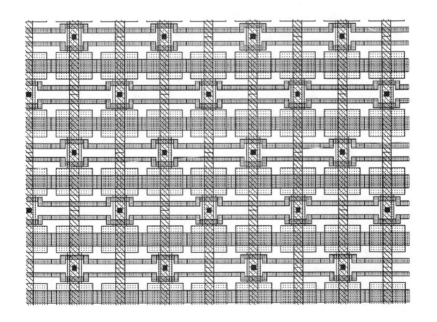

Figure 17.5 Array of DRAM cells.

Example 17.1

This example illustrates the charge sharing principle. Consider the equivalent circuit of the basic DRAM cell shown in Fig. Ex17.1. The unwanted bit line capacitance, C_{bit}, is the sum of the metal1 to substrate capacitance of the bit line and the capacitance of the n+ implant depletion capacitance from each DRAM cell connected to the bit line. Assuming the initial voltage on the DRAM storage capacitance, C_d, is V_d and the initial voltage on the bit line is V_{bit}, calculate the change in voltage on C_{bit} when the word line is taken high.

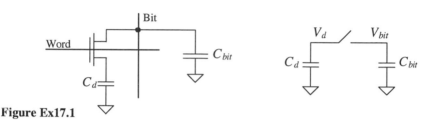

Figure Ex17.1

The initial charge on C_d is $V_d \cdot C_d$, and the initial charge on C_{bit} is $V_{bit} \cdot C_{bit}$. When the word line goes high, the access MOSFET turns on, or in other words the switch closes, connecting the DRAM storage cell capacitance, C_d, directly to the bit line (and its associated capacitance). The voltage across each capacitor must be the same after the switch is closed. This voltage will be called V_{final} and is given by

$$(C_d + C_{bit}) \cdot V_{final} = V_d \cdot C_d + V_{bit} \cdot C_{bit}$$

or

$$V_{final} = \frac{V_d \cdot C_d + V_{bit} \cdot C_{bit}}{C_d + C_{bit}} \tag{17.1}$$

The change in the bit line voltage is simply

$$\Delta V_{bit} = V_{final} - V_{bit} \tag{17.2}$$

Normally, the bit line is precharged to approximately 2.5 V before a read operation, and with typical values of C_d and C_{bit} of 75 fF and 500 fF, respectively, the change in bit line voltage with V_d of 4 V is only 200 mV. The amplifier used to sense this small voltage change is called a sense amplifier or simply sense amp. ∎

Example 17.2

Estimate the delay through a word line that is 2 mm long connected to MOSFET gates measuring 2 µm by 6 µm spaced every 20 µm.

The sheet resistance of poly1, from Appendix A, is 21 Ω/square, while the capacitance of poly1 to substrate is 58 aF/μm^2. The capacitance of each MOSFET gate connected to the word line is

$$C_{inn} = \frac{3}{2} \cdot 6 \cdot 2 \cdot 800 \text{ aF} = 14.4 \text{ fF}$$

If we break the word line into 20 μm increments that are 2 μm wide, the following circuit shown in Fig. Ex17.2 is the result.

Figure Ex17.2

The delay through the entire line can be estimated by

$$t_d = 0.35 rcN^2 = 0.35 \cdot 210 \cdot 16.7f \cdot (100)^2 = 12.5 \text{ ns}$$

As the density of the layout is increased, that is, the number of transistors connected to the word line increases over a given distance, this delay can become even larger. What is normally done in DRAM to circumvent this problem is to break the DRAM into smaller arrays multiplexing the data together. The smaller DRAM array supporting circuitry is easier to design for high-speed operation. ■

Often the substrate is "pumped" to a negative voltage, typically –2 V, in DRAM. This reduces the depletion capacitance between the n+ implant and the substrate, having the effect of lowering the bit line capacitance. Two other benefits of the negative substrate bias are that it increases latch-up immunity and allows the inputs on the chips to go negative without forward biasing the n+/substrate diode. A negative substrate potential has the unwanted effect of increasing the threshold voltage of the access transistor and increasing the junction leakage current. The threshold voltage drop of the access transistor when writing a high level to the storage cell (see Ch. 14) can be compensated for by using a voltage of $VDD + 2$ for the word line drivers. This puts the gate potential of the access transistors above $VDD + V_{THN}$.

Modern DRAM processes do not use a MOSFET in the inversion region for the storage capacitor but rather use a stacked or trenched capacitor (more on this later in the chapter). In the design of these storage capacitors, it is desirable to keep the oxide thickness as thin as possible in order to have the largest capacitance/area. The main limitation, from a circuit designer's point of view, on the oxide thickness is the breakdown voltage. The maximum electric field that should be applied to SiO_2 for reliable operation is 7 MV/cm, or 0.7 V/10 Å (see Ch. 6). For a 100 Å oxide thickness, the maximum potential across the oxide for reliable operation is then limited to 7 V. If the substrate is pumped to –3 V and *VDD* is 5 V, poor reliability is probable when the process t_{ox} is 100 Å. To limit the voltage across the storage capacitor, one plate is held at *VDD*/2 while the other plate, that is, the plate connected to the access transistor, can swing from ground to *VDD*.

17.1.2 The SRAM Cell

The basic SRAM cell shown in Fig. 17.2 is not popular because of its large layout size. The n-well needed for the p-channel transistors is the main cause. The SRAM cell shown in Fig. 17.6 is very popular because of its small layout size. The resistors are made using n+ and p+ polysilicon and are therefore not available in the process described in Appendix A (CN20). The layout of the polysilicon resistor is shown in Fig. 17.7. The resistor can be thought of as a leaky bipolar transistor. Typical resistance

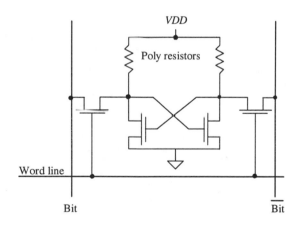

Figure 17.6 SRAM memory cell with poly resistors.

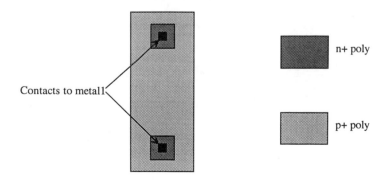

Figure 17.7 Layout of the polysilicon resistor.

values are about 10 MΩ (or more). The CMOS SRAM cell dissipates essentially zero static power, while the power dissipated by the resistor/n-channel SRAM cell is $VDD^2/10$ MΩ, which is 2.5 µW for each cell when VDD is 5 V. An array of 1 million SRAM cells (128 k by 8) dissipates a power of 2.5 W when the resistor/n-channel SRAM cell is used.

17.2 The Sense Amplifier

Consider the portion of a DRAM memory array shown in Fig. 17.8. Two bit lines are shown in this figure. From Ex. 17.1 we found that the bit line may change very little when we are reading a cell. We use the sense amplifier to sense changes on the bit lines and determine whether a 1 or 0 was written to the cell.

Before we illustrate the operation of a sense amplifier, consider what happens to the bit lines when the word line, labeled wordA, in Fig. 17.8 is high. The charge, or absence of charge, on the storage capacitor is used to change the voltage of the bit line. The $\overline{bit}$ line potential remains unchanged and is used as a reference in the sense amplifier. The near proximity of the lines gives excellent noise rejection from coupled signals. Note that the line labeled bit in this figure corresponds to the bit line of the accessed transistor. If the wordB row line were high, the labeling of the bit lines in Fig. 17.8 would be switched. Also note that only one row line can go high, or in other words be selected, at a given time, while all even or odd bit lines intersecting a selected row line can change.

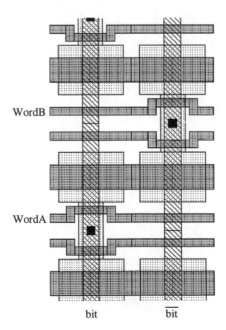

Figure 17.8 Part of a DRAM array showing two bit lines.

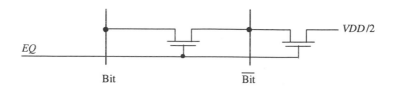

Figure 17.9 Schematic of circuit used to precharge bit lines to *VDD*/2.

The first step in reading out a DRAM cell is to precharge all bit lines to, typically, *VDD*/2. The circuit used to precharge the lines is shown in Fig. 17.9. The control signal *EQ* (equilibrate) is used to turn the two MOSFETs on prior to a read operation, setting both bit lines to *VDD*/2. The MOSFET between the two bit lines is used to ensure that the potential of the lines is equal. This turns out to be very important since the sense amp must discriminate less than 100 mV voltage changes on the bit line.

The n- and p-sense amplifier schematics are shown in Fig. 17.10. The p-sense amplifier in this figure is placed on the top of the array, while the n-sense amplifier is located at the bottom of the array. During precharge of the bit lines, the signals *NLAT* (n-sense amplifier latch) and $\overline{ACT}$ (active pull up) are charged to *VDD*/2. The signals *NSA* and *PSA* are low and high, respectively.

After the digit (bit) lines have been charged to *VDD*/2, the row address decoder selects a word line. If the memory cell at the intersection of the word line and the bit line contains a low and the storage capacitor is discharged, the bit line voltage will fall from the precharged voltage. The $\overline{bit}$ line voltage remains unchanged. Evaluation of the bit line begins when *NSA* is driven high, causing *NLAT* to go low. If the bit line fell from *VDD*/2 (in other words, the accessed bit was a low), the n-sense amplifier would pull the bit line low. If the bit line increased in potential from *VDD*/2, the n-sense amplifier would pull the $\overline{bit}$ to a low. The n-sense amp can only pull a digit or column line low. The final stage in evaluating the digit lines is to fire the p-sense amp. This is done by pulling *PSA* low, causing $\overline{ACT}$ to go high. If the potential of the bit line, after the word line went high, is above *VDD*/2, the p-sense amplifier will pull the bit line high. If the bit line is below *VDD*/2, the p-sense amplifier will pull $\overline{bit}$ up to *VDD*. The p-sense amp can only pull a bit line high.

The sensitivity of the p-sense amplifier is far less critical than that of the n-sense amplifier. After the n-sense amplifier is fired, one of the bit lines is at ground potential while the other is at *VDD*/2. When the p-sense amplifier is fired, it sees this large potential difference. The n-sense amplifier must be capable of discriminating < 100 mV voltage differences, while the p-sense must discriminate *VDD*/2 voltage differences.

Another important attribute of the sense amplifier is its ability to restore or refresh the datum in the memory cell. If the word line stays high during the evaluate phase, the forcing of the bit line high or low by the sense amplifier also forces a high or

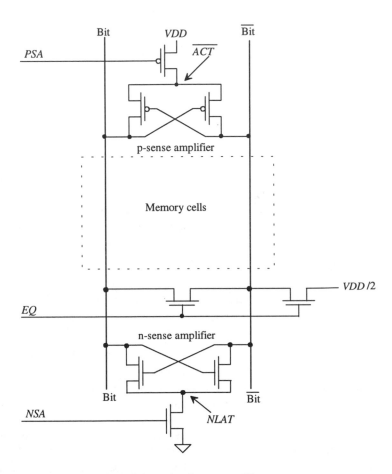

Figure 17.10 Schematic of sense amplifiers.

low potential on the storage capacitor. The change in potential on the $\overline{\text{bit}}$ line has no effect on any of the memory cells since the word line does not access any data on this line. DRAM must be refreshed periodically to keep from losing data. Refreshing is accomplished by sequentially reading each row in the memory array since there is a sense amplifier for each pair of bit lines. The sense amplifier can be shared between different memory arrays on the chip. When the sense amp is shared, isolation transistors are used to select which array the sense amplifier will be used with. Isolation transistors also help to isolate the sense amp from the large bit line capacitance, keeping the sense amp from oscillating.

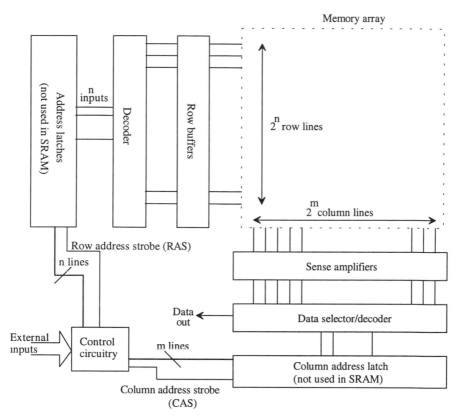

Figure 17.11 Detailed block diagram of a RAM.

17.3 Row/Column Decoders

Consider the more detailed block diagram of a RAM shown in Fig. 17.11. If there are 2^n row lines and 2^m column lines (where a column line is bit and $\overline{bit}$), the total memory size is 2^{n+m} bits of data. In order to minimize the length of the row and column lines, it is desirable to make $n = m$, or in other words, a square array.

Selecting a bit of data begins by latching and decoding the row and column addresses. The row line that goes high as a result of the row decode is buffered to drive the word line across the array (see Ex. 17.2). The column decoder output is fed to a pass transistor. The pass transistor acts like a data selector, allowing the selected bit to be accessed for a read or write operation.

A common row decoder configuration, sometimes called a tree decoder, used in RAM is shown in Fig. 17.12a. The three-bit address A0, A1, and A2 is used to enable the pass transistors used in the decoder. One of the eight outputs is pulled high, that is,

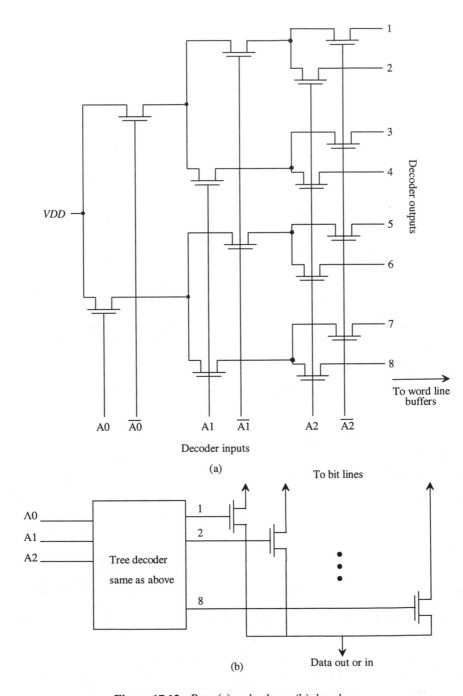

Figure 17.12   Row (a) and column (b) decoders.

$VDD - V_{THN}$, depending on the three-bit address. Placing another pass transistor in series with VDD in this figure allows the decoder to be disabled.

The column address decoder, Fig. 17.12b is similar to the row address decoder with the addition of a pass transistor. The pass transistor is connected to the bit lines. The tree decoder portion of the column decoder is sometimes called a pre-decode. An enable can be used here as well.

Delay through the decoder is an important factor in large DRAM arrays. Consider a ten-bit row address. A total of ten tiers of pass transistors are used with each transistor connected to two others. Since the high output is well below VDD due to the body effect, a buffer circuit must regenerate the logic levels.

Figure 17.13 shows how the output of the decoder is pulled low when it is not selected. A long L MOSFET is used to pull the output of the decoder low when that particular output is not selected. The result is that all decoder outputs are zero except for the output that is selected by the input address. Two inverters are then used to drive the word line capacitance. The switching point voltage of the first inverter is set to compensate for the threshold voltage drop through the pass transistors.

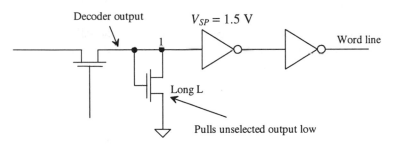

Figure 17.13 Output buffer used in row decoder.

17.4 Timing Requirements for DRAMs

The timing signals used in a normal DRAM read are shown in Fig. 17.14. With the help of Fig. 17.11 these signals will be discussed. The first signal shown, $\overline{RAS}$, is the row address strobe and is active low. The row and column addresses are normally multiplexed together on the die so that one set of address inputs can be used. This reduces the number of pins in the packaged IC. Assuming the row address is set up on the address lines of the DRAM, the falling edge of $\overline{RAS}$ strobes the row address into the row address latches. After the row address has been latched, the address inputs of the chip are changed to the column address. At the falling edge of the column address strobe, $\overline{CAS}$, the column address is strobed into the column address latches. A finite time later, the data are available on the output pins of the DRAM. All internal signals, for example, EQ, $NLAT$, ACT, are derived from $\overline{RAS}$ and $\overline{CAS}$. Not shown in this

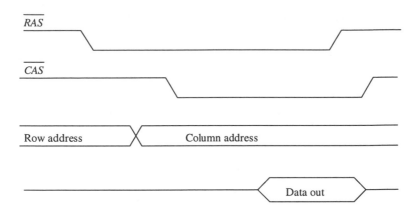

Figure 17.14 Normal read cycle for a DRAM ($\overline{WE}$ = high).

timing diagram is the $\overline{WE}$ signal or write enable signal. This signal is used to put the DRAM in the read or write mode. Normal write operations timing waveforms are shown in Fig. 17.15.

Refreshing the DRAM is accomplished by sequentially addressing the word lines. When a word line is accessed, all bits on this line are refreshed by the sense amplifier as was discussed in Sec. 17.2. Figure 17.16 shows the timing signals for a $\overline{RAS}$ only refresh. The row addresses are sequentially applied to the DRAM during the refresh period.

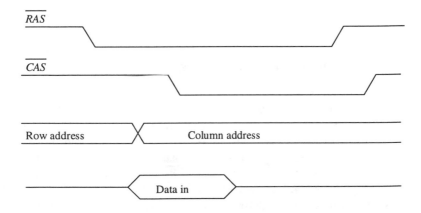

Figure 17.15 Normal write cycle for a DRAM ($\overline{WE}$ = low).

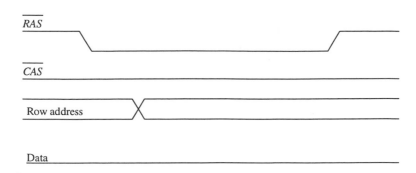

Figure 17.16 $\overline{RAS}$ only refresh cycle.

17.5 Modern DRAM Circuits

The CN20 process is a general-purpose CMOS process. Memory manufactures, however, use CMOS processes specifically designed for memory circuit production. A memory process is specifically engineered to enhance yield. This section will discuss some aspects of modern CMOS memory design.

17.5.1 DRAM Memory Cell Layout

Modern DRAM memory cell layout, often called an mbit (memory bit), is much more involved than the memory cell layout given in Fig. 17.4. Figure 17.17 shows the layout of a modern DRAM memory cell. Again, a shared digit (bit) line contact is used to connect the access transistors to the digit line. The word lines are, again, made using poly1. An important aspect of the modern DRAM cell is the compactness of the mbit. A different set of design rules are used for the mbit layout.

The layout of the capacitor in a process specifically designed for memory is different as well. Figure 17.18 contains a process cross section for the buried capacitor mbit depicted in Fig. 17.17 (a vertical cross section). This type of mbit, employing a buried capacitor structure, places the digit line physically above the storage capacitor [4]. The digit line is constructed from either metal or polycide, while the digit line contact is formed using metal or polysilicon plug technology. The mbit capacitor is formed with polysilicon (poly2) as the bottom plate, an oxide-nitride-oxide (ONO) dielectric, and a sheet of polysilicon (poly3) which forms the common node shared by all mbit capacitors. The capacitor shape can be simple, such as a rectangle, or complex, such as concentric cylinders or stacked discs. Exotic capacitor structures are the topic of many DRAM process papers [5, 6, 7]. The ONO dielectric undergoes optimization to achieve maximum capacitance with minimum leakage. It must also tolerate the maximum DRAM operating voltage without breakdown. A cross-sectional view of an mbit using a trench capacitor is shown in Fig. 17.19 [6, 7].

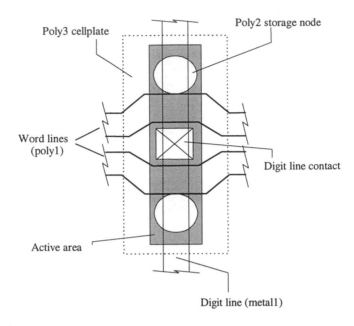

Figure 17.17 Mbit pair layout.

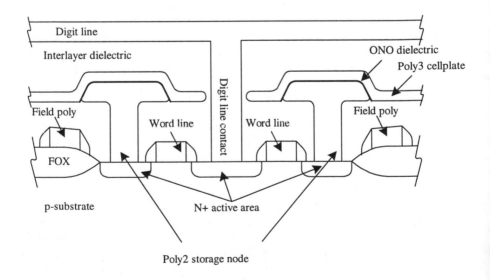

Figure 17.18 Cross-sectional view of a buried capacitor cell, Fig. 17.17, cross section from top to bottom.

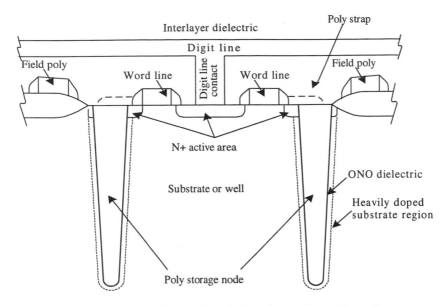

Figure 17.19 Cross-sectional view of a trench capacitor cell.

Mbits organized in an array are shown in Fig. 17.20. A term used often in a periodic layout is *pitch*. The distance between like points in the array is the pitch length. In a memory array, the pitch is usually specified between adjacent bit lines as shown in Fig. 17.20. This figure is also useful to show how the size of a memory cell is specified. If we draw an imaginary box around an individual memory cell (see Fig. 17.20), the area of the box is the cell size. The feature size, F, is normally half the width of the bit line or word line pitch. The area of the memory cell of Figs. 17.17 and 17.20 is $8F^2$.

17.5.2 Folded/Open Architectures

The memory cells discussed thus far are used in the folded bit line architecture shown in Fig. 17.21. The p- and n-sense amplifiers are placed on the top and bottom of the memory array. First-generation DRAM, 64 Kbits and smaller, used the open bit line architecture shown in Fig. 17.22. The sense amplifier's inputs were from two separate memory arrays. The main benefit of the open bit line configuration is the resulting $6F^2$ memory cell (Fig. 17.23). The main drawback (and the reason it is no longer used) is the susceptibility to noise. Noise from the substrate and adjacent word and bit lines does not couple equally into the sense amplifier's inputs. The folded architecture essentially takes the arrays used in the open bit line configuration and "folds" them into a single array. The results are an array where the adjacent bit lines are not used at the same time for reading or writing and a larger array size for a given memory size. (This characteristic is, of course, unwanted.) The noise immunity is increased because of the close proximity. Substrate and coupled noise will ideally feed into each bit line equally.

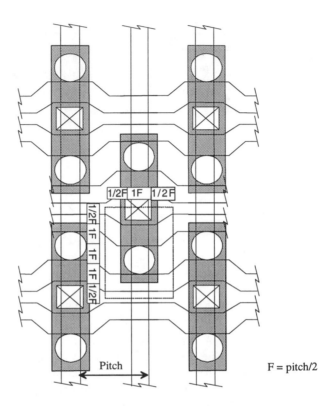

Figure 17.20 How memory cell size is specified (not to scale).

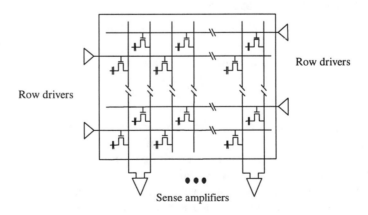

Figure 17.21 Architecture for folded bit lines.

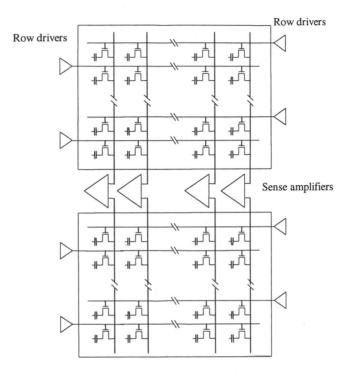

Figure 17.22 Architecture used in an open bit line architecture.

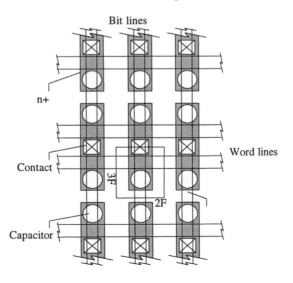

Figure 17.23 Layout of mbit used in an open bit line configuration.

17.6 Other Memory Cells

In addition to the DRAM and SRAM memory cells discussed throughout this chapter, several other memory cells exist [1, 2, 8, 9]. These include read-only-memory (ROM), erasable programmable ROM (EPROM), electrically erasable programmable ROM (EEPROM), and Flash memory.

Read-Only Memory (ROM)

ROM is the simplest semiconductor memory. It is used mainly to store instructions or constants for use in a digital system. The basic operation of a ROM can be explained with the ROM memory shown in Fig. 17.24. Remembering that only one word line (row line) can be high at a time, we see that R_1 going high causes the column lines C_1, C_2, and C_4 to be pulled low. Column lines C_3 and C_5 are pulled high through the long L MOSFET loads at the top of the array. If the information that is to be stored in the ROM memory is not known prior to fabrication, the memory array is fabricated with an n-channel MOSFET at every intersection of a row and column line (Fig. 17.25a). The ROM is programmed (PROM) by cutting (or never making during fabrication) the connection between the drain of the MOSFET and the column line Fig. (17.25b). Because it is not easy to program ROM, it is limited to applications where it is mass produced.

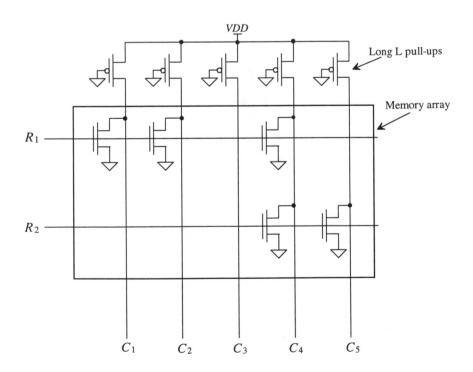

Figure 17.24 A ROM memory array.

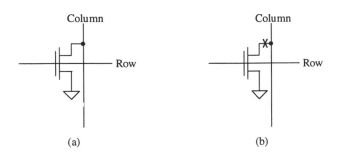

Figure 17.25 (a) n-channel MOSFET at the intersection of every column and row line
and (b) eliminating the connection between the drain and column line
to program the ROM.

Erasable Programmable Read-Only Memory (EPROM)

EPROMs make programming the ROM significantly easier. Consider the cross-sectional view of an EPROM cell shown in Fig. 17.26. This modified n-channel MOSFET is used at the interesection of the column and row lines in the ROM memory array shown in Fig. 17.24. A second layer of polysilicon is added directly above the original polysilicon layer. The original poly layer is floating (i.e., not connected to anything). The second layer (poly2) is now used for connection to the row lines. Note that this is simply a poly2-poly1 capacitor where the bottom plate of the capacitor (poly1) is used in MOSFET formation.

To understand how to program the MOSFET, let's begin by assuming that both gates, poly1 (the bottom) and poly2, are at 0 V. A capacitance exists between poly2 and poly1 as well as between poly1 and the substrate. If the potential of poly2 is increased above 0 V, the voltage between these two capacitors ideally divides evenly since they should be approximately the same value. The result is an increase in the potential on poly1. If the potential on poly2 is raised to approximately $2 \cdot V_{THN}$, then the potential of the poly1 is raised to approximately V_{THN}. Increasing the row line voltage (the voltage

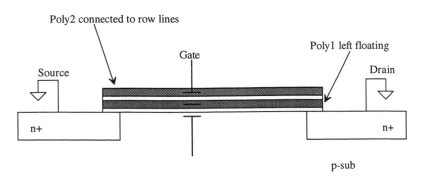

Figure 17.26 EPROM memory cell.

on poly2) to 5 V turns the MOSFET on and pulls the column line low. In other words, the action of the row line going high is enough to turn the MOSFET on and pull the column line low.

To ensure that the MOSFET remains off when the row line goes high, a large voltage, around 25 V, is applied to the row lines to program the MOSFET to remain off during normal operation. The large voltage causes a large current to flow in the MOSFET. In addition, avalance multiplication occurs in the substrate, causing hot-carriers to become trapped on the poly1 [10]. When the large voltage is removed from poly2 (the word line), the voltage on poly1 will drop down to a negative voltage, typically -5 V. This keeps the MOSFET from turning on under normal operation, allowing the column line to stay high. Since both gates are surrounded by SiO_2, a very good dielectric, the charge trapped on poly1 remains trapped for possibly several years. The cells can be reprogrammed by first illuminating the chip with UV light. This causes electron-hole generation in the SiO_2 , having the effect of increasing the insulator's conductivity. This in turn allows the charge trapped on the poly1 to leak off, until equilibrium is reached.

Electrically Erasable Programmable Read-Only Memory (EEPROM)

The inability to reprogram EPROM quickly has led to the use of EEPROM in many applications requiring nonvolatile memory. A voltage generator (see Ch. 18) is used on chip to generate the large voltage needed to program the EEPROM memory cell. The gate-oxide used in EEPROM is thinner than that used in EPROM. The result, when a large voltage, 10 V, is applied to poly2, is a conduction mechanism called Fowler-Nordheim tunneling, [11] between the substrate and poly1. This mechanism, unlike avalance breakdown, can conduct current in both directions between poly1 and substrate. A logic high in an EEPROM is programmed by raising the voltage on poly2 to 10 V, while a logic low is programmed by lowering the voltage to -10 V.

Flash memory is based on both EPROM and EEPROM technologies. Flash memories are programmed in a fashion similar to EPROM where hot electrons are used to accumulate charge on poly2 (see Fig. 17.26). Structurally, the Flash memory cell is the same as the EPROM cell except for the oxide thickness [9]. The oxide thickness is on the order of 100 Å, whereas the oxide thickness used in EPROM is 200-400 Å thick. Unlike EPROM, Flash memories can be erased in a fashion similar to EEPROMs. In other words, the Flash memory is programmed using hot electrons and erased using Fowler-Nordheim tunneling.

REFERENCES

[1]   D. A. Hodges and H. G. Jackson, *Analysis and Design of Digital Integrated Circuits,* McGraw-Hill Publishing Co., 2nd ed., 1988. ISBN 0-07-029158-6.

[2]   R. L. Geiger, P. E. Allen, and N. R. Strader, *VLSI - Design Techniques for Analog and Digital Circuits,* McGraw-Hill Publishing Co., 1990. ISBN 0-07-023253-9.

[3]   B. Keeth, "A Novel Architecture for Advanced High Density Dynamic Random Access Memories," Master's Thesis, University of Idaho, May 1996.

[4]   T. Hamada, "A Split-Level Diagonal Bit-Line (SLDB) Stacked Capacitor Cell for 256Mb DRAMs," *1992 IEDM Technical Digest*, pp. 799-802.

[5]   T. Mohihara et al., "Disk-Shaped Capacitor Cell for 256Mb Dynamic Random-Acess Memory," *Japan Journal of Applied Physics*, Vol. 33, Part 1, No. 8, pp. 4570-4575, August 1994.

[6]   J. H. Ahn et al., "Micro Villus Patterning (MVP) Technology for 256Mb DRAM Stack Cell," *1992 Symposium on VLSI Technical Digest of Technical Papers*, pp. 12-13.

[7]   K. Sagara et al., "Recessed Memory Array Technology for a Double Cylindrical Stacked Capacitor Cell of 256M DRAM," *IEICE Trans. Electron.*, Vol. E75-C, No. 11, pp. 1313-1322, November 1992.

[8]   M. I. Elmasry, *Digital MOS Integrated Circuits II,* IEEE Press, 1992. ISBN 0-87942-275-0, IEEE order number: PC0269-1.

[9]   R. D. Pashley and S. K. Lai, "Flash Memories: The Best of Two Worlds," *IEEE Spectrum,* 1989, pp. 30-33.

[10]   D. Frohman-Bentchkowsky, "FAMOS-A New Semiconductor Charge Storage Device," *Solid-State Electronics,* Vol. 17, pp. 517-529, 1974.

[11]   E. H. Snow, "Fowler-Nordheim Tunneling in SiO_2 Films," *Solid-State Communications,* Vol. 5, pp. 813-815, 1967.

PROBLEMS

17.1   Estimate the capacitance that the pass transistor of Fig. 17.3 contributes to the word line in Fig. 17.3a. Assume that the MOSFET is minimum size, that is, 3 µm by 2 µm.

17.2   Verify the results of Ex. 17.1 using SPICE.

17.3   What layer in the CN20 process is used to make the word lines used in Ex. 17.2? Why can't we use metal1 with contacts to poly for the word lines? If we could, would this improve the delay through the word lines? Why?

17.4   Sketch the cross-sectional view at the location shown for the layout shown in Fig. P17.4.

17.5   Simulate the operation of the SRAM cell shown in Fig. 17.6. Assume that the poly n+/p+ resistors have a value of 10 MEGΩ.

17.6   Explain why it is important to equilibrate the bit lines prior to reading the data from the memory cells. Refer to Fig. 17.10.

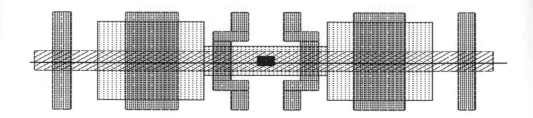

Figure P17.4

17.7   Which DRAM architecture, that is, folded or open, results in the smallest memory bit (mbit) layout size? What is the major advantage of using the folded bit line architecture over the open?

17.8   Suppose it is desirable to fabricate a ROM memory in a process used to make SRAMs. If the n+/p+ poly resistor is available, comment on the concerns that would arise if this resistor were used in place of the p-channel pull-up MOSFET shown in Fig. 17.24. Would there be any advantage to using the n+/p+ pull-up?

17.9   Could we use the CN20 process to fabricate an EEPROM? If so, with the help of the poly2 design rules given in Appendix A, lay out a possible EEPROM cell.

Special-Purpose Digital Circuits

In this chapter we discuss some special-purpose integrated circuits. We begin with the Schmitt trigger, a circuit useful in generating clean pulses from a noisy input signal or in the designing oscillator circuits. Next, we discuss multivibrator circuits, both astable and monostable types. We end this chapter with a discussion of on-chip voltage generators.

18.1 The Schmitt Trigger

The schematic symbol of the Schmitt trigger is shown in Fig. 18.1 along with typical transfer curves [1]. We should note the similarity to the inverter transfer characteristics with the exception of a steeper transition region. Curve A in Fig. 18.1 corresponds to

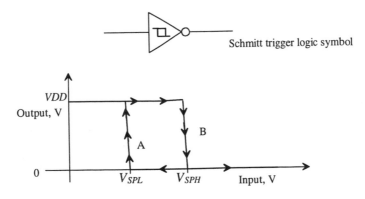

Figure 18.1 Transfer characteristics of a Schmitt trigger.

the output of the Schmitt trigger changing from a low to a high, while curve B corresponds to the output changing from a high to a low. The hysteresis present in the transfer curves is what sets the Schmitt trigger apart from the basic inverter.

Figure 18.2 shows a possible input to a Schmitt trigger and the resulting output. When the output is high and the input exceeds V_{SPH}, the output switches low. However, the input voltage must go below V_{SPL} before the output can switch high again. Note that we get normal inverter operation when $V_{SPH} = V_{SPL}$. The hysteresis of the Schmitt trigger is defined by

$$V_H = V_{SPH} - V_{SPL} \tag{18.1}$$

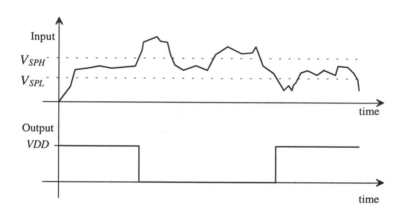

Figure 18.2 Input, top trace, and output of a Schmitt trigger.

18.1.1 Design of the Schmitt Trigger

The basic schematic of the Schmitt trigger is shown in Fig. 18.3. We can divide the circuit into two parts, depending on whether the output is high or low. If the output is low, then M6 is on and M3 is off and we are concerned with the p-channel portion when calculating the switching point voltages, while if the output is high, M3 is on and M6 is off and we are concerned with the n-channel portion. Also, if the output is high, M4 and M5 are on, providing a DC path to VDD.

Let's begin our analysis of this circuit, assuming that the output is high (= VDD) and the input is low (= 0 V). Figure 18.4 shows the bottom portion of the Schmitt trigger used in calculating the upper switching point voltage, V_{SPH}. MOSFETs M1 and M2 are off, with $V_{in} = 0$ V while M3 is on. The source of M3 floats to $VDD - V_{THN}$, or approximately 4 V for VDD = 5 V. We can label this potential V_x as shown in the figure.

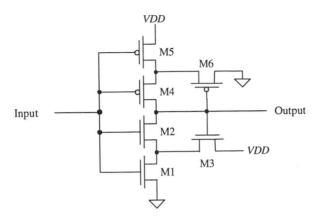

Figure 18.3 Schematic of the Schmitt trigger.

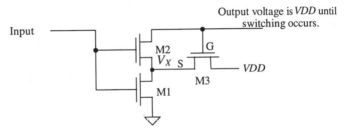

Figure 18.4 Portion of the Schmitt trigger schematic used to calculate upper switching point voltage.

With V_{in} less than the threshold voltage of M1, V_X remains at approximately 4 V. As V_{in} is increased further, M1 begins to turn on and the voltage, V_X, starts to fall toward ground. The high switching point voltage is defined when

$$V_{in} = V_{SPH} = V_{THN2} + V_X \qquad (18.2)$$

or when M2 starts to turn on. As M2 starts to turn on, the output starts to move toward ground, causing M3 to start turning off. This in turn causes V_X to fall further, turning M2 on even more. This continues until M3 is totally off and M2 and M1 are on. This positive feedback causes the switching point voltage to be very well defined.

When Eq. (18.2) is valid, the currents flowing in M1 and M3 are essentially the same. Equating these currents gives

$$\frac{\beta_1}{2}(V_{SPH} - V_{THN})^2 = \frac{\beta_3}{2}(VDD - V_X - V_{THN3})^2 \qquad (18.3)$$

Since the sources of M2 and M3 are tied together, $V_{THN2} = V_{THN3}$ the increase in the threshold voltages from the body effect is the same for each MOSFET. The combination of Eqs. (18.2) and (18.3) yields

$$\frac{\beta_1}{\beta_3} = \left[\frac{VDD - V_{SPH}}{V_{SPH} - V_{THN}}\right]^2 \tag{18.4}$$

The threshold voltage of M1, given by V_{THN} in this equation, is the zero body bias threshold voltage, or 0.83 V for a CN20 n-channel MOSFET. Given a specific upper switching point voltage, the ratio of the MOSFET transconductors is determined by solving this equation. A general design rule for selecting the size of M2, that is, β_2, is to require that

$$\beta_2 \geq 5\beta_1 \text{ and } 5\beta_3 \tag{18.5}$$

Since M2 is used as a switch, we require that it be larger than M1 or M3.

A similar analysis can be used to determine the lower switching point voltage V_{SPL}, resulting in the following design equation:

$$\frac{\beta_5}{\beta_6} = \left[\frac{V_{SPL}}{VDD - V_{SPL} - V_{THP}}\right]^2 \tag{18.6}$$

The following example illustrates the design procedure for a Schmitt trigger.

Example 18.1

Design and simulate a Schmitt trigger with $V_{SPL} = 2$ V and $V_{SPH} = 3$ V using the CN20 process.

We begin by solving Eqs. (18.4) and (18.6) for the transconductance ratios. For the upper switching point voltage,

$$\frac{\beta_1}{\beta_3} = \left[\frac{5-3}{3-0.83}\right]^2 = 0.85 = \frac{KP_N \frac{W_1}{L_1}}{KP_N \frac{W_3}{L_3}} = \frac{W_1 L_3}{W_3 L_1}$$

and for the lower switching point voltage,

$$\frac{\beta_5}{\beta_6} = \left[\frac{2}{5-2-0.91}\right]^2 = 0.92 = \frac{KP_P \frac{W_5}{L_5}}{KP_P \frac{W_6}{L_6}} = \frac{W_5 L_6}{W_6 L_5}$$

An infinite number of solutions exist for these equations. Keeping in mind that we are drawing our MOSFETs on a 1 μm grid[1], one possible set of sizes is

$$W_1 = L_1 = W_5 = L_5 = 3 \ \mu m$$

$$W_3 = 7 \ \mu m \text{ and } L_3 = 6 \ \mu m$$

$$W_6 = 12 \ \mu m \text{ and } L_6 = 11 \ \mu m$$

---

[1] In practice, for the CN20 process, we can draw on a finer grid. A good alternative grid size for precision design in this process is 0.1 μm. Using the **Set** command in LASI, the number of grids can be set as well as the grid spacing.

Selection of the sizes of M2 and M4 follows the design rule given in Eq. (18.5). Noting that the W/L ratios of M1, M3, M5, and M6 are close to unity, we can require that W_2/L_2 and $W_4/L_4 = 5$. Setting $W_2 = W_4 = 10$ μm and $L_2 = L_4 = 2$ μm is a good choice for these device sizes. The simulation results are shown in Fig. 18.5. Two DC simulations were run on the circuit to show both an output high to low and an output low to high transition. ∎

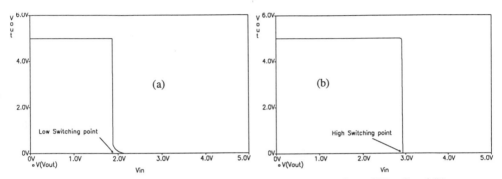

Figure 18.5 Simulation results showing (a) sweeping the input from 5 V to 0 and (b) sweeping the input from 0 to 5 V (showing high switching point).

18.1.2 Switching Characteristics

The propagation delays of the Schmitt trigger can be calculated in much the same way as the inverter of Ch. 11. Defining equivalent digital resistances for M1, M2, M4, and M5 as R_{n1}, R_{n2}, R_{p4}, and R_{p5}, respectively, gives a high to low propagation delay-time, neglecting the Schmitt trigger output capacitance, of

$$t_{PHL} = (R_{n1} + R_{n2}) \cdot C_{load} \qquad (18.7)$$

and

$$t_{PLH} = (R_{p4} + R_{p5}) \cdot C_{load} \qquad (18.8)$$

18.1.3 Applications of the Schmitt Trigger

Consider the waveform shown in Fig. 18.6. A pulse with ringing is a common voltage waveform encountered in busses or lines interconnecting systems. If this voltage is applied directly to a logic gate or inverter input with a V_{SP} of 2.5 V, the output of the gate will vary with the period of the ringing on top of the pulse. Using a Schmitt trigger with properly designed switching points can eliminate this problem.

The Schmitt trigger can also be used as an oscillator (Fig. 18.7). The delay-time in charging and discharging the capacitor is used to set the oscillation frequency. At the moment in time when the output of the Schmitt trigger switches low, the voltage across the capacitor is V_{SPH}. The capacitor will start to discharge toward ground. The voltage across the capacitor is given by

$$V_c(t) = V_{SPH} \cdot e^{-t/RC} \qquad (18.9)$$

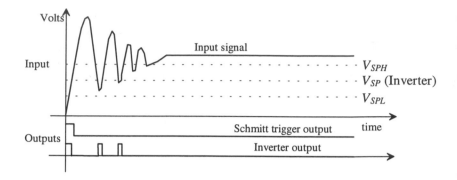

Figure 18.6 Applying a Schmitt trigger to clean up an interconnecting signal.

At the time when $V_c(t) = V_{SPL}$, the output of the Schmitt trigger changes state. This time is given by solving Eq. (18.9) by

$$t_1 = RC \cdot \ln \frac{V_{SPH}}{V_{SPL}} \tag{18.10}$$

A similar analysis for the case when the capacitor is charged from V_{SPL} to V_{SPH} gives

$$V_c(t) = V_{SPL} + (VDD - V_{SPL})\left(1 - e^{\frac{-t}{RC}}\right) \tag{18.11}$$

and

$$t_2 = RC \cdot \ln \frac{VDD - V_{SPL}}{VDD - V_{SPH}} \tag{18.12}$$

The oscillation frequency, neglecting the intrinsic delay of the Schmitt trigger, is given by

$$f_{osc} = \frac{1}{t_1 + t_2} \tag{18.13}$$

The capacitance used in these equations is the sum of the input capacitance of the Schmitt trigger and any external capacitance.

An alternative oscillator using the Schmitt trigger is shown in Fig. 18.8 [2]. Here the MOSFETs M1 and M4 behave as current sources (see Ch. 20) mirroring the

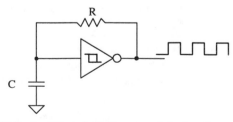

Figure 18.7 Oscillator design using a Schmitt trigger.

current in M5 and M6. When the output of the oscillator is low, M3 is on and M2 is off. This allows the constant current from M4 to charge C. When the voltage across C reaches V_{SPH}, the output of the Schmitt trigger swings low. This causes the output of the oscillator to go high and allows the constant current from M1 to discharge C. When C is discharged down to V_{SPL}, the Schmitt trigger changes states. This series of events continues, generating the square wave output.

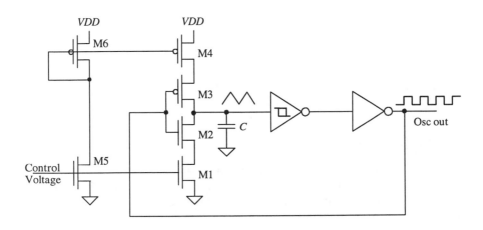

Figure 18.8 Voltage-controlled-oscillator using Schmitt trigger and current sources. MOSFETs M2 and M3 are used as switches.

If we label the drain currents of M1 and M4 as I_{D1} and I_{D4}, we can estimate the time it takes the capacitor to charge from V_{SPL} to V_{SPH} as

$$t_1 = C \cdot \frac{V_{SPH} - V_{SPL}}{I_{D4}} \qquad (18.14)$$

and the time it takes to charge from V_{SPH} to V_{SPL} is

$$t_2 = C \cdot \frac{V_{SPH} - V_{SPL}}{I_{D1}} \qquad (18.15)$$

The period of the oscillation frequency is, as before, the sum of t_1 and t_2.

This type of oscillator is termed a voltage-controlled oscillator (VCO) since the output frequency can be controlled by an external voltage. The currents I_{D1} and I_{D4}, (Fig. 18.8) are directly controlled by the control voltage. As we will see in Ch. 20, the current in M5 is mirrored in M1, M4, and M6, with an appropriate scaling factor dependent on the size of the transistors.

18.1.4 High-Speed Schmitt Trigger

It turns out in practice that the Schmitt trigger of Fig. 18.3 is not easily optimized for high speed. The effective switching resistances of the MOSFETs are difficult to reduce without changing the switching point voltages. An alternative design uses the basic inverter latch, or SRAM storage cell, shown in Fig. 18.9. Although the propagation times of this Schmitt trigger are less than the previous design, the switching point voltages are more difficult to define, which makes it less useful in some cases.

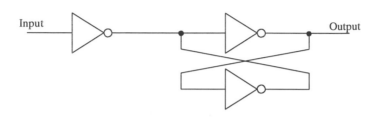

Figure 18.9 Schmitt trigger using the basic latch.

18.2 Multivibrator Circuits

Multivibrator circuits (Fig. 18.10) are circuits that employ positive feedback. There are three types of multivibrators: astable, bistable, and monostable. Astable multivibrator circuits are unstable in either output (high or low) state. The oscillators we have discussed are examples of astable multivibrators. The bistable multivibrator is stable in either the high or low state. Flip-flops and latches are examples of the bistable multivibrator. Monostable multivibrators are stable in a single state. Monostable multivibrators are also called one-shots. In this, section we will discuss the monostable and astable multivibrators.

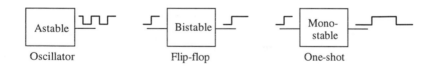

Figure 18.10 Multivibrator circuits.

18.2.1 The Monostable Multivibrator

A CMOS implementation of the monostable multivibrator is shown in Fig. 18.11. Under normal conditions, V_{in} is low and the output of the NOR gate, V_1, is high. The voltage V_2 is pulled high through the resistor, and the output of the inverter, V_3, is a low. Upon application of a trigger pulse, that is, V_{in} going high, both V_1 and V_2 drop to zero volts and the output of the inverter, which is also the output of the monostable, goes high. This output is fed back to the input of the NOR gate holding V_1 at ground potential.

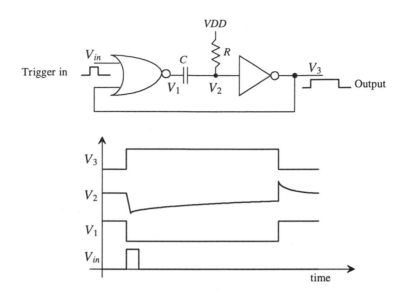

Figure 18.11 Operation of the monostable multivibrator.

After triggering, the potential V_2 will start to increase because C is charged through R. The potential across the capacitor after triggering takes place is given by

$$V_c(t) = V_2(t) - \overbrace{V_1(t)}^{=0} = VDD(1 - e^{\frac{-t}{RC}}) \qquad (18.16)$$

If we assume that the V_{SP} of the inverter is $VDD/2$, then the time it takes for the capacitor to charge to V_{SP} is given by

$$t = RC \cdot \ln \frac{VDD}{VDD - V_{SP}} = RC \cdot \ln(2) \approx 0.7RC \qquad (18.17)$$

This time also defines the output pulse width, neglecting gate delays, since the inverter switches low. The inverter output, V_3, going low causes V_1 to go back to VDD and V_2 to go to $VDD + VDD/2$. If the resistor or capacitor is bonded out (connected to the output pads), the ESD diodes may keep V_2 from going much above $VDD + 0.7$. The time it takes V_2 to decay back down to VDD limits the rate at which the one-shot can be retriggered. Also note that the trigger input can be longer than the output pulse width. Longer output pulse widths may cause V_2 to go as high as 10 V as well as limit the maximum trigger rate.

18.2.2 The Astable Multivibrator

An example of an astable multivibrator is shown in Fig. 18.12. This circuit has no stable state and thus oscillates. To analyze the behavior of this multivibrator, let's begin by assuming that the output, V_3, has just switched high. The output going high causes

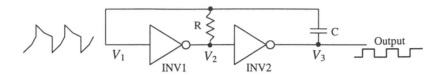

Figure 18.12 An astable multivibrator.

V_1 to go high (actually $VDD + V_{SP1}$), forcing V_2 low. The voltage across the capacitor after this switching takes place is given by

$$V_c(t) = V_1(t) - \overbrace{V_3(t)}^{= VDD} = (VDD + V_{SP1}) \cdot e^{\frac{-t}{RC}} - VDD \qquad (18.18)$$

The output of the astable will go low, $V_3 = 0$, when $V_1 = V_{SP1}$. Substituting this condition into the previous equation gives the time the output is high (or low) and is given by

$$t_1 = RC \cdot \ln \frac{VDD + V_{SP1}}{V_{SP1}} = t_2 \qquad (18.19)$$

If $V_{SP1} = VDD/2$, then

$$t_1 = t_2 = 1.1RC \qquad (18.20)$$

and the frequency of oscillation is

$$f_{osc} = \frac{1}{t_1 + t_2} = \frac{1}{2.2RC} \qquad (18.21)$$

Again, if the resistor and capacitor are bonded out, the ESD diodes on the pads will limit the voltage swing of V_1.

18.3 Voltage Generators

Often when designing CMOS circuits, positive and negative DC voltages are needed that do not lie between VSS and VDD. A simple example of an application where a larger DC voltage source is used was given in Fig. 11.25. In Fig. 11.25 an inverter was powered with a positive voltage supply of $VDD + 2V$. A simple circuit, sometimes called a voltage pump (or more often, a *charge pump*), useful in generating a voltage greater than VDD, is given in Fig. 18.13.

The operation of the voltage pump of Fig. 18.13 can be explained by first realizing that both M1 and M2 operate like a diode. M1 is simply used to pull point A to a voltage of $VDD - V_{THN}$. (The threshold voltage with body effect using the CN20 process with a V_{SB} of 5 V is approximately 1.5 V.) Let's begin the description of the circuit operation by assuming that the output of the inverter is low and point A is at a potential of $VDD - V_{THN}$. When the output of the inverter goes high, the potential at

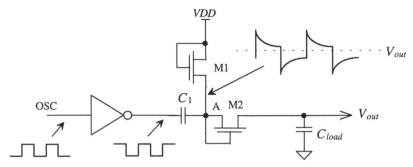

Figure 18.13 Simple pump used to generate a voltage greater than *VDD*.

point A increases to $VDD + (VDD - V_{THN})$. This turns M2 on and charges C_{load} to $2 \cdot (VDD - V_{THN})$ (≈ 7 V for CN20), provided $C_1 \gg C_{load}$ and the oscillator frequency allows the capacitors to fully charge or discharge before changing states. In most practical situations, C_{load} and C_1 are comparable in size, and the output of the pump is loaded with a DC load. The result is an output voltage with a startup time; that is, V_{out} does not immediately rise to $2 \cdot (VDD - V_{THN})$ but requires several oscillator cycles to reach steady state. Also, the output has a ripple dependent on the DC load. Figure 18.14 shows the simulation results for the circuit of Fig. 18.13 using minimum-size MOSFETs with $C_{load} = C_1 = 1$ pF and an oscillator frequency of 10 MHz. Using this simple pump with a DC load, such as a resistor, would require employing larger MOSFETs and capacitors.

The positive voltage pump uses n-channel MOSFETs, while the negative voltage pump, Fig. 18.15, uses p-channel MOSFETs. The reason for this comes from the requirement that the diode formed with the n+ (p+) implant used in the drain/source of the MOSFET combined with the p-substrate (n-well) does not become forward biased.

The capacitors used in Figs. 18.13 and 18.15 can be replaced with MOSFETs. An example of using n-channel MOSFETs in place of the capacitors is shown in Fig. 18.16. The main requirement on a MOSFET used as a capacitor is that its V_{GS} remain greater than V_{THN} at all possible operating conditions. In other words, the MOSFET must remain in the strong inversion region so that its capacitance is a constant $C'_{ox} \cdot W \cdot L$. The capacitor, C_{load}, of Fig. 18.16 must remain in strong inversion since $V_{out} \gg V_{THN}$. Capacitor C_1 remains in strong inversion by realizing that when the inverter output is low, the voltage on the gate of C_1 is $VDD - V_{THN}$ ($= V_{GS}$). When the output of the inverter is high (VDD), the voltage on the gate of C_1 is $2 \cdot VDD - V_{THN}$. In both cases (the output of the inverter high and low), $V_{GS} = VDD - V_{THN}$, and the MOSFET remains in the strong inversion region.

18.3.1 Improving the Efficiency

Figure 18.17 shows a more efficient voltage pump. The improvement in efficiency comes from eliminating the threshold voltage drop at the gate and drain of M7. This allows the output to swing up to $2 \cdot VDD - V_{THN}$.

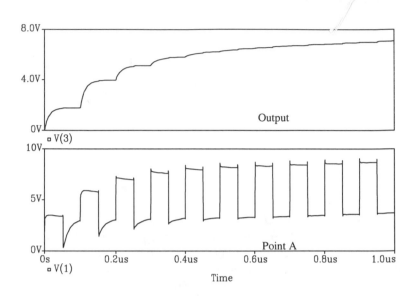

Figure 18.14 Output of simple voltage pump.

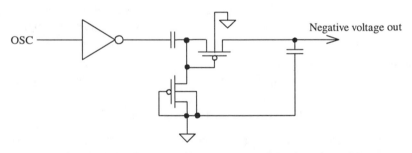

Figure 18.15 Pump used to generate a voltage less than *VSS*.

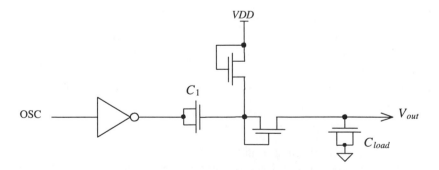

Figure 18.16 Using MOSFETs as capacitors.

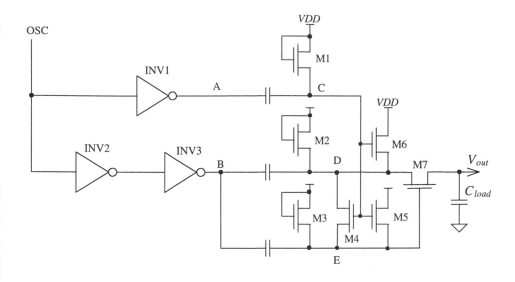

Figure 18.17   Increased efficiency voltage pump.

To understand the operation of the circuit, let's assume that the voltage at point A is low and the voltage at point C is $VDD - V_{THN}$. When the output of INV1 goes high, point A is VDD and the voltage at point C swings up to $2 \cdot VDD - V_{THN}$. This causes M4, M5, and M6 to turn on and pull points D and E to VDD. Now when point B goes high (point A goes back to zero), points D and E swing up to $2 \cdot VDD$ and the output goes to $2 \cdot VDD - V_{THN}$. Note that MOSFETs M2 and M3 are not needed; unless the pump drives a DC load , they never turn on. Also, separating points D and E is unnecessary unless the pump supplies a DC current.

18.3.2 Generating Higher Voltages

The voltage pumps of the last section are limited to voltages less than $2 \cdot VDD$ and greater than $2 \cdot VSS$. Figure 18.18 [3] shows a scheme for generating arbitrarily high voltages (limited by the breakdown voltage of the capacitors or the oxide breakdown of the MOSFETs). Again, the MOSFETs are used as diodes in this configuration.

In the following description of circuit operation, we will assume steady-state operation with no DC load present. When CLK is low, point A is pulled, by M1, to $VDD - V_{THN}$. When CLK goes high, point A swings up to $2 \cdot VDD - V_{THN}$. Diode M2 turns on, and point B charges to $2 \cdot VDD - 2 \cdot V_{THN}$. When CLK goes low, $\overline{CLK}$ goes high and point B swings up to $3 \cdot VDD - 2 \cdot V_{THN}$. When $\overline{CLK}$ goes low, point B swings back down to $2 \cdot VDD - 2 \cdot V_{THN}$. This operation proceeds through the circuit to the last stage of the multiplier. The output of the multiplier swings from $(N+1) \cdot VDD - N \cdot V_{THN}$ down to $(N + 1) \cdot VDD - (N + 1) \cdot V_{THN}$. C_N can be made larger than the other capacitors to reduce this ripple.

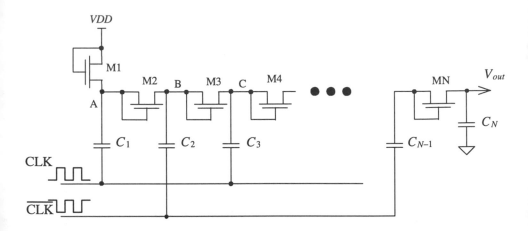

Figure 18.18 N-stage voltage multiplier.

In practice, several problems exist with this circuit. The threshold voltage, due to the body effect, increases for each MOSFET proceeding from M1 to MN. Stray capacitance at each node to ground lowers the efficiency of the multiplier. When the circuit supplies a current charge sharing exists between the capacitors in the circuit. In short, the output impedance of the voltage generator can be very high. This limits the circuit to supplying very small currents. Also, in a practical circuit, the output voltage should have some voltage-limiting circuit on the output of the multiplier to limit the output voltage and protect the circuit from breakdown. A simple limiting circuit is the series connection of several long L n-channel MOSFETs from the output of the circuit to ground.

18.3.3 Example

A common application of a voltage generator in digital circuits is generating a negative substrate bias [4]; that is, instead of tying the substrate to ground, the substrate is held at some negative voltage. Typically, this negative voltage is between −1 and −2 V. "Pumping" the substrate negative is common in virtually all DRAMs. A negative substrate bias has several benefits. It (1) stabilizes n-channel threshold voltages, (2) increases latch-up immunity after power up, (3) prevents forward biasing n+ to p-substrate pn junction, (4) allows chip inputs to go negative without forward biasing a pn junction, (5) prevents substrate from going locally above ground, (6) reduces depletion capacitances associated with the n+ to p-substrate junction and (7) reduces subthreshold leakage current.

A simple substrate pump is shown in Fig. 18.19. In this circuit we can use n-channel MOSFETs to generate a negative potential, unlike Fig. 18.15, since the negative voltage is connected to the substrate. In this situation, the drain/source implants of the n-channel MOSFETs cannot become forward biased. Note the absence of the load capacitance. It turns out that the capacitance of the substrate to everything

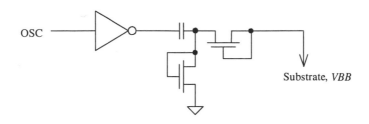

Figure 18.19 Simple substrate pump.

else in the circuit presents a very large capacitance to the pump. In other words, the substrate itself is a very large capacitor.

An example oscillator circuit used to drive the substrate pump is shown in Fig. 18.20. This is a standard ring oscillator with a NAND gate used to enable/disable the oscillator. Capacitors are added at a few points in the middle of the oscillator to increase the delay and lower the frequency of the oscillator in order to allow the pump's capacitors to fully charge.

The final component in a substrate pump is the regulator, a circuit that senses the voltage on the substrate and enables or disables the substrate pump. Using a comparator with hysteresis, a precision voltage reference, and a level shifting circuit, the enable signal can be generated with the circuit of Fig. 18.21. The hysteresis of the comparator determines the amount of ripple on the substrate voltage. Forcing a constant current through the two MOSFETs, M1 and M2, compels their source-gate voltages to remain constant (note how the body effect is eliminated by using p-channel MOSFETs), causing the MOSFETs to behave as if they were batteries. The battery action of M1 and M2 shifts the substrate voltage up so that it lies in the common mode range of the comparator. The number of MOSFETs (in this case two) and the magnitude of the current I determine the substrate voltage.

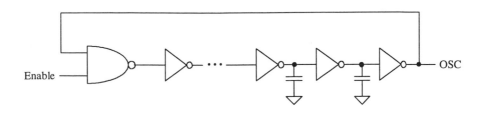

Figure 18.20 Ring oscillator with enable.

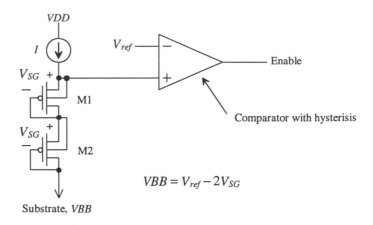

Figure 18.21 Regulator circuit used in substrate pump.

A simpler and less accurate implementation of the regulator is shown in Fig. 18.22. The comparator and voltage reference are implemented with an inverter and M1-M3. MOSFET M3 causes the inverter to have hysteresis, that is, behave like a Schmitt trigger. MOSFETs M2 and M3 form an inverter with a switching point voltage of approximately V_{THN}. The current source of Fig. 18.22 is implemented with the long L MOSFET M4. The level shifting is performed with the n-channel MOSFETs M5 and M6. When point A gets above a potential of V_{THN}, the Enable output goes high, causing the substrate pump to turn on and drive the substrate voltage negative. When point A

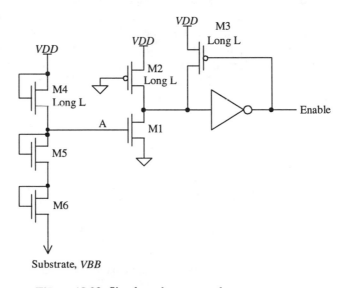

Figure 18.22 Simpler substrate regulator.

gets pulled, via the substrate voltage through M5 and M6, below V_{THN}, the Enable output goes low and the pump shuts off. The substrate voltage generated with this circuit is approximately $-V_{THN}$ with body effect, which is approximately -1 V.

REFERENCES

[1]   D. A. Hodges and H. G. Jackson, *Analysis and Design of Digital Integrated Circuits,* McGraw-Hill Publishing Co., 2nd ed., 1988. ISBN 0-07-029158-6.

[2]   R. Gregorian, K. W. Martin, and G. C. Temes, "Switched-Capacitor Circuit Design," *Proceedings of the IEEE,* Vol. 71, No. 8, August, pp. 941-966, 1983.

[3]   J. F. Dickson, "On-Chip High-Voltage Generation in MNOS (sic) Integrated Circuits Using an Improved Voltage Multiplier Technique," *IEEE Journal of Solid State Circuits*, Vol. SC-11, No. 3, June 1976.

[4]   P. Zagar, *4 MEG DRAM Tutorial-Course Notes,* Micron Semiconductor, Boise, Idaho.

[5]   J. Wu, Y. Chang, and K. Chang, "1.2V CMOS Switched-Capacitor Circuits," *ISSCC*, 1996.

PROBLEMS

18.1   Design a Schmitt trigger with $V_{SPH} = 3$ and $V_{SPL} = 2.5$ V. Perform a SPICE DC sweep on your design, labeling the switching point voltages.

18.2   Estimate t_{PHL} and t_{PLH} for the Schmitt trigger of Ex. 18.1, driving a 100 fF load capacitance.

18.3   Design and simulate the operation of a Schmitt trigger-based oscillator with an output frequency of 10 MHz.

18.4   Estimate the total input capacitance on the control voltage input for the VCO shown in Fig. 18.8.

18.5   Determine, using SPICE simulations, V_{SPH} and V_{SPL} for the Schmitt trigger shown in Fig. 18.9.

18.6   Design an astable multivibrator with an output oscillation frequency of 20 MHz. Use minimum-size inverters in the design. What is the V_{SP} of the inverters?

18.7   Design and simulate the operation of a one-shot that has an output pulse width of 100 ns. Comment on the maximum rate of retrigger and how ESD diodes connected to bonding pads will affect the circuit operation if the resistor and capacitor are bonded out.

18.8   Generate, using SPICE, Fig. 18.14.

18.9   Design a nominally 7 V voltage generator that can supply at most 1 μA of DC current (7 MEG resistor). Simulate the operation of your design with SPICE.

18.10 Design and simulate the operation of a nominally −1 V substrate pump. Comment on the design tradeoffs.

18.11 Figure P18.11 shows a voltage generator [5] useful in increasing the efficiency of the N-stage voltage multiplier. Describe the operation of the circuit. Use timing diagrams to illustrate your understanding.

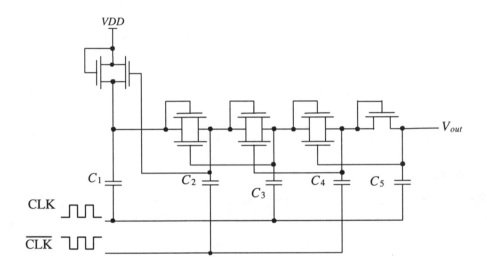

Figure P18.11

Chapter
19

Digital Phase-Locked Loops

The digital phase-locked loop, DPLL, is a circuit that is used frequently in modern integrated circuit design [1]. Consider the waveform and block diagram of a communication system shown in Fig. 19.1. Digital data are loaded into the shift register at the transmitting end. The data are shifted out sequentially to the transmitter output driver. At the receiving end, where the data may be analog (i.e., not have well-defined amplitudes) after passing through the communication channel, the receiver amplifies and changes the data back into digital logic levels. The next logical step in this sequence of events is to shift the data back into a shift register at the receiver and process the received data. However, the absence of a clock signal makes this difficult. The DPLL performs the function of generating a clock signal which is locked or in synchronization with the incoming signal. The generated clock signal is used in the receiver to clock the shift register and thus recover the data. This application of a DPLL is often termed *a clock-recovery circuit* or *bit synchronization circuit*.

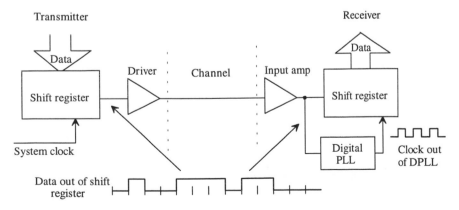

Figure 19.1 Block diagram of a communication system using a DPLL for the generation of a clock signal.

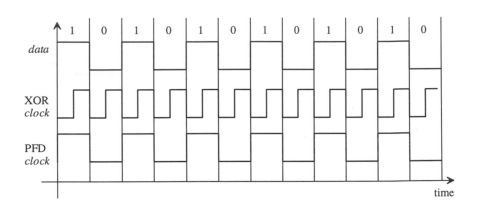

Figure 19.2 Data input to DPLL in lock and possible clock outputs using the XOR phase detector and PFD.

A more detailed picture of the incoming data and possible clock signals out of the DPLL are shown in Fig. 19.2. The possible clock signals are labeled XOR *clock* and PFD (phase frequency detector) *clock* corresponding to the type of phase[1] detector (PD) used. For the XOR PD, the rising edge of the clock occurs in the center of the data, while for the PFD the rising edge occurs at the beginning of the data. The phase of the clock signal is determined by the PD used[2].

A block diagram of a DPLL is shown in Fig. 19.3. The PD generates an output signal proportional to the time difference between the *data in* and the divided down clock, *dclock*. This signal is filtered by a loop filter. The filtered signal, V_{inVCO}, is connected to the input of a voltage controlled oscillator (VCO). Each one of these blocks is discussed in detail in the following sections. Once each block is understood, we will put them together and discuss operation of the DPLL.

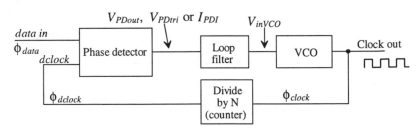

Figure 19.3 Block diagram of a digital phase locked loop.

---

[1] A more correct name for digital applications is time difference detector (TDD).

[2] The PFD is rarely used in clock recovery applications. Figure 19.2 is simply used to illustrate the different phase relations resulting from an XOR PD or PFD in a locked DPLL.

19.1 The Phase Detector

The first component in our DPLL is the phase detector. The two types of phase detectors, an XOR gate and a phase frequency detector (PFD), have significantly different characteristics. This makes understanding the limitations and performance very important. Selection of the type of phase detector is the first step in a DPLL design.

19.1.1 The XOR Phase Detector

The XOR PD is simply an exclusive OR gate. When the output of the XOR is a pulse train with a 50 percent duty cycle (square wave), the DPLL is said to be in lock; or in other words, the clock signal out of the DPLL is synchronized to the incoming data, provided the conditions stated below are met. Consider the XOR PD shown in Fig. 19.4. Let's begin by assuming that the incoming data are a string of zeros and that a divide by two is used in the feedback loop. The output of the phase detector is simply a replica of the *dclock* signal. Since the *dclock* signal has a 50 percent duty cycle, it would appear that the DPLL is in lock. If a logic "1" is suddenly applied, there is no way to know if the clock signal is synchronized (the clock rising edge coincides with the center of the data bit) to the data. This leads to the first characteristic of an XOR PD;

    1. The incoming data must have a minimum number of transitions over a given time interval.

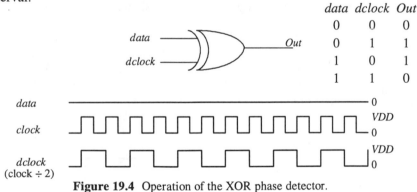

Figure 19.4 Operation of the XOR phase detector.

Now consider the situation when the output of the phase detector, with the data input being a string of zeros, is applied to a simple RC low-pass filter (Fig. 19.5). If RC >> period of the clock signal, the output of the filter is simply *VDD*/2. This leads to the second characteristic of the XOR phase detector.

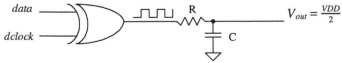

Figure 19.5 How the filtered output of the phase detector becomes VDD/2.

2. With no input data, the filtered output of the phase detector is *VDD*/2.

The voltage out of the loop filter is connected to the input of the VCO. Consider the typical characteristics of a VCO shown in Fig. 19.6. The frequency of the square wave output of the VCO is f_{center} when V_{in} (= V_{center}) is *VDD*/2 (typically). The other two frequencies of interest are the minimum and maximum oscillator frequencies, f_{min} and f_{max} possible, with input voltage V_{min} and V_{max} , respectively. It is important that the VCO continue to oscillate with no input data. Normally, the VCO is designed so that the nominal data input rate and the VCO center frequency are the same. This minimizes the time it takes the DPLL to lock (and is critical for proper operation of the XOR PD).

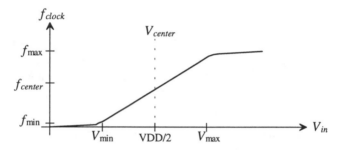

Figure 19.6 Output frequency of VCO versus input control voltage.

Now let's consider an example input to the phase-detector and corresponding output. The data shown in Fig. 19.7 is leading the *dclock* signal. The corresponding output of the phase detector is also shown. If this output is applied to a low-pass filter, the result is an average voltage less than *VDD*/2. This causes the VCO frequency to decrease until the edge of the *dclock* is centered on the data.

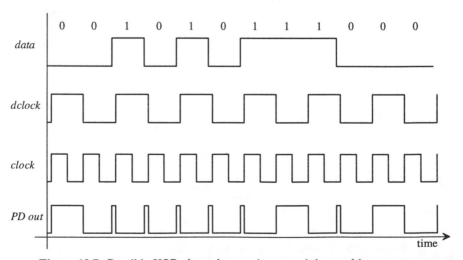

Figure 19.7 Possible XOR phase-detector inputs and the resulting output.

3. The time it takes the loop to lock is dependent on the data pattern input to the DPLL and the loop-filter characteristics.

Since the output of the PD is averaged, or more correctly integrated, noise injected into the data stream (a false bit) can be rejected. A fourth characteristic of this PD is:

4. The XOR DPLL has good noise rejection.

Another important characteristic of a DPLL is whether or not it will lock on a harmonic of the input data. The XOR DPLL will lock on harmonics of the data. To prove that this is indeed the case, consider replacing any of the clock signals of Figs. 19.2, 19.4, or 19.7 with a clock at twice or half the frequency. The average of the waveforms will remain essentially the same. A fifth characteristic of a DPLL using an XOR gate is:

5. The VCO operating frequency range should be limited to frequencies much less than $2f_{clock}$ and much greater than $0.5f_{clock}$ where f_{clock} is the nominal clock frequency for proper lock with a XOR PD.

The loop filter used with this type of PD is a simple RC low-pass filter as shown in Fig. 19.5. Since the output of the PD is oscillating, the output of the filter will show a ripple as well, even when the loop is locked. This modulates the clock frequency, an unwanted characteristic of a DPLL using the XOR PD. This characteristic can be added to our list:

6. A ripple on the output of the loop filter with a frequency equal to the clock frequency will modulate the control voltage of the VCO.

To characterize the phase detector (see Fig. 19.8) we can define the time difference between the rising edge of the *dclock* and the beginning of the *data* as Δt. The phase difference between the *dclock* and *data*, $\phi_{data} - \phi_{dclock}$, is given by

$$\Delta\phi = \phi_{data} - \phi_{dclock} = \frac{\Delta t}{T_{dclock}} \cdot 2\pi \text{ (radians)} \tag{19.1}$$

or, in terms of the DPLL output clock frequency,

$$\Delta\phi = \frac{\Delta t}{2T_{clock}} \cdot 2\pi \tag{19.2}$$

$$f_{clock} = \frac{1}{T_{clock}} = 2f_{dclock} = \frac{2}{T_{dclock}} \tag{19.3}$$

When the loop is locked, the *clock* rising edge is centered on the data; the time difference, Δt, between the *dclock* rising edge and the beginning of the data is simply $T_{clock}/2$ or $T_{dclock}/4$ (see Fig. 19.8c). Therefore, the phase difference between *dclock* and the *data*, under locked conditions, may be written as

$$\Delta\phi = \frac{\pi}{2} \tag{19.4}$$

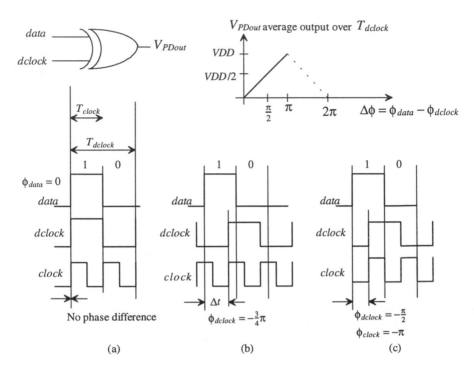

Figure 19.8   XOR PD output for various inputs (assuming input data are
a string of alternating ones and zeros).

Note that the phase difference between *clock* and *data* when in lock is π.

The average voltage out of the phase detector (Fig. 19.8) may be expressed by

$$V_{PDout} = VDD \cdot \frac{\Delta\phi}{\pi} = K_{PD} \cdot \overbrace{\Delta\phi}^{input} \tag{19.5}$$

where the gain of the PD may be written as

$$K_{PD} = \frac{VDD}{\pi} \text{ (V/radians)} \tag{19.6}$$

To aid in the understanding of these equations, consider the diagrams shown in
Fig. 19.8. If the edges of the clock and data are coincident in time (Fig. 19.8a), the
XOR output, V_{PDout}, is 0 V and the phase difference is 0. The loop filter averages the
output of the PD and causes the VCO to lower its output frequency. This causes $\Delta\phi$ to
increase, thus increasing V_{PDout}. Depending on the selection of the loop filter, this
increase could cause the rising edges to increase beyond, or *overshoot*, the desirable
center point as shown in Fig. 19.8b. In this case, the phase difference is $-\frac{3}{4}\pi$, and V_{PDout}
is $\frac{3}{4}VDD$. Figure 19.8c shows the condition when the loop is in lock and the phase
difference is $\pi/2$.

Here we should point out the importance of having the VCO center frequency, f_{center}, equal to the desired clock frequency. If $f_{center} = f_{clock}$, the VCO control voltage, depending on the loop filter, will look similar to Fig. 19.9a during acquisition (the loop trying to lock). If these frequencies are not equal, the control voltage will oscillate, causing the clock to move, in time, around some other point than the center of the data bit (Fig. 19.9b). The actual control voltages will look different from those portrayed in Fig. 19.9 because the VCO control voltage is dependent on the input data pattern as discussed earlier.

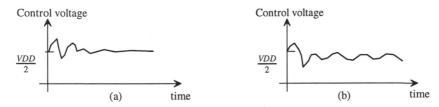

Figure 19.9 Average output voltage of phase detector during acquisition.

To summarize the design criteria of the VCO with the XOR PD, we desire that;

1. The center frequency, f_{center}, should be equal to the clock frequency when the VCO control voltage is *VDD*/2.

2. The maximum and minimum oscillation frequencies, f_{max} and f_{min}, of the VCO should be selected to avoid locking on harmonics of the input data.

3. It is important that the VCO duty cycle be 50 percent. If this is not the case, the DPLL will have problems locking, or once locked the clock will jitter (move around in time).

19.1.2 Phase Frequency Detector

A schematic diagram of the phase frequency detector is shown in Fig. 19.10 [1]. The output of the PFD depends on both the phase and frequency of the inputs. This type of phase detector is also termed a sequential phase detector. It compares the leading edges of the *data* and *dclock*. A *dclock* rising edge cannot be present without a data rising edge. To aid in understanding the PFD, consider the examples shown in Fig. 19.11. The first thing we notice is that the *data* pulse width and the *dclock* pulse width do not matter. If the rising edge of the *data* leads the *dclock* rising edge (Fig. 19.11a), the "up" output of the phase detector goes high while the *Down* output remains low. This causes the *dclock* frequency to increase, having the effect of moving the edges closer together. When the *dclock* signal leads the data (Fig. 19.11b) *Up* remains low while the *Down* goes high a time equal to the phase difference between *dclock* and *data*. Figure 19.11c shows the condition of the loop being locked. Notice that, unlike the XOR PD, the outputs remain low when the loop is locked. Again, several characteristics of the PFD can be described:

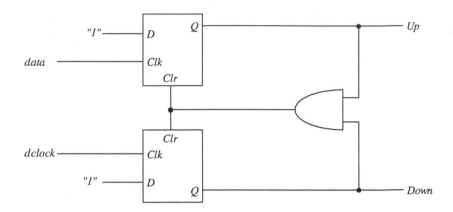

Figure 19.10 Phase frequency detector (PFD).

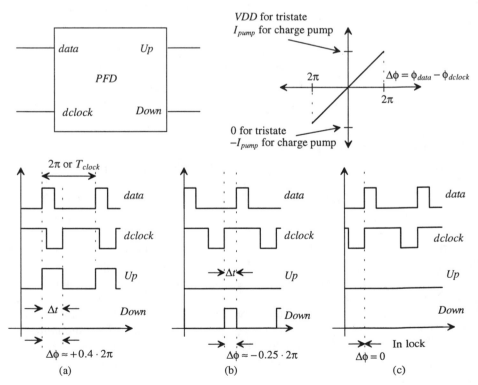

Figure 19.11 PFD phase-detector inputs and outputs.

1. A rising edge from the *dclock* and *data* must be present when doing a phase comparison.

2. The width of the *dclock* and the *data* is irrelevant.

3. The PFD will not lock on a harmonic of the data.

4. The outputs (*Up* and *Down*) of the PFD are both logic low when the loop is in lock, eliminating ripple on the output of the loop filter.

5. This PFD has poor noise rejection; a false edge on either the *data* or the *dclock* inputs will drastically affect the output of the PFD.

The output of the PFD should be combined into a single output for driving the loop filter. There are two methods of doing this, both of which are shown in Fig. 19.12. The first method is called a *tri-state* output. When both signals, *Up* and *Down*, are low, both MOSFETs are off and the output is in a high-impedance state. If the *Up* signal goes high, M2 turns on and pulls the output up to *VDD* while if the *Down* signal is high the output is pulled low through M1. The main problem that exists with this configuration is that the power supply variations can significantly affect the output voltage when M2 is on. The effect is to modulate the VCO control voltage. This wasn't as big a problem when the XOR PFD was used because of the averaging taking place. The second configuration shown in this figure is the so-called *charge pump* [2]. MOS current sources are placed in series with M1 and M2. When the PFD *Up* signal goes high, M2 turns on, connecting the current source to the loop filter. Since the current source can be made insensitive to supply variations, modulation of the VCO control voltage is absent.

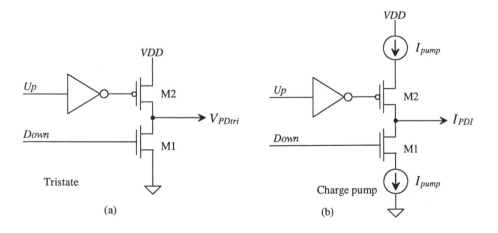

Figure 19.12 (a) Tri-state and (b) charge pump outputs of the PFD.

We can characterize the PFD in much the same manner as we did the XOR PD. We will assume that f_{clock} and f_{dclock} are equal in this analysis. Therefore, the feedback loop, Fig. 19.3, divides by one ($N = 1$). If we again assume that the time difference between the rising edges of the *data* and *dclock* is labeled Δt and the time between leading edges of the *clock* (or the time differences between the leading edges of the *data* since we must have both leading edges present to do a phase comparison) is labeled T_{clock} then we can write the phase ($T_{clock} = T_{dclock}$) as

$$\Delta\phi = \frac{\Delta t}{T_{clock}} \cdot 2\pi \text{ (radians)} \tag{19.7}$$

The phase difference, $\Delta\phi$, is zero when the loop is in lock. The output voltage of the PFD using the tri-state output configuration (see Fig. 19.11) is

$$V_{PDtri} = \frac{VDD - 0}{4\pi} \cdot \Delta\phi = K_{PDtri} \cdot \Delta\phi \tag{19.8}$$

where the gain is,

$$K_{PDtri} = \frac{VDD}{4\pi} \text{ (volts/radian)} \tag{19.9}$$

If the output of the PFD uses the charge-pump configuration, the output current can be written (again see Fig. 19.11) as

$$I_{PDI} = \frac{I_{pump} - (-I_{pump})}{4\pi} \cdot \Delta\phi = K_{PDI} \cdot \Delta\phi \tag{19.10}$$

where

$$K_{PDI} = \frac{I_{pump}}{2\pi} \text{ (amps/radian)} \tag{19.11}$$

The loop filters used with tri-state and the charge-pump outputs are shown in Fig. 19.13. The first loop filter has a transfer function given by

$$V_{inVCO} = \frac{1 + j\omega R_2 C}{1 + j\omega(R_1 + R_2)C} \cdot V_{PDtri} = K_F \cdot V_{PDtri} \tag{19.12}$$

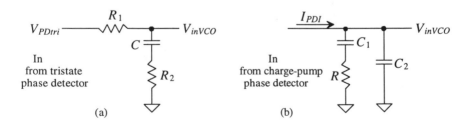

Figure 19.13 Loop filters for (a) tristate output and (b) charge-pump output.

The charge-pump loop-filter transfer function is given (noting the input variable is a current while the output is a voltage) by

$$V_{inVCO} = I_{PDI} \cdot \frac{1 + j\omega R C_1}{j\omega(C_1 + C_2) \cdot \left[1 + j\omega R \frac{C_1 C_2}{C_1 + C_2} \right]} = K_F \cdot I_{PDI} \qquad (19.13)$$

To qualitatively understand how these loop filters work, let's begin by considering the loop filter for use with the tri-state output. For slow variations in the phase difference, the filter acts like an integrator averaging the output of the PD. For fast variations, however, the filter looks like a resistive divider without any integration. This allows the loop filter to track fast variations in the time difference between the rising edges. A similar discussion can be made for the loop filter used with the charge pump. For slow variations in the phase, the current, I_{pump}, linearly charges C_1 and C_2. This gives an averaging effect. For fast variations, the charge pump simply drives the resistor R (assuming C_2 is small), eliminating the averaging and allowing the VCO to track quickly moving variations in the input data.

The design requirements of the VCO used with the PFD can be much more relaxed than those for the XOR PD; therefore, it is preferred over the XOR PD. Since the output of the PD can be a voltage from 0 to *VDD*, the requirement that $f_{center} = f_{clock}$ is not present, although it is still a good idea to design the VCO so that this is true. The oscillation range, $f_{max} - f_{min}$, is not limited by the harmonics of f_{clock} since the PFD will not lock on a harmonic of the clock frequency. The VCO clock duty cycle is irrelevant with the PFD since the PD looks only at rising edges. In high-speed or data communications applications, however, one may be forced to use the XOR PD.

19.2 The Voltage-Controlled Oscillator

The gain of the voltage-controlled oscillator is simply the slope of the curves given in Fig. 19.6. This gain can be written as

$$K_{VCO} = 2\pi \cdot \frac{f_{max} - f_{min}}{V_{max} - V_{min}} \quad \text{(radians/s} \cdot \text{V)} \qquad (19.14)$$

The VCO output frequency, f_{clock}, is related to the VCO input voltage (see Fig. 19.6) by

$$\omega_{clock} = 2\pi \cdot f_{clock} = K_{VCO} \cdot V_{inVCO} + \omega_o \quad \text{(radians/s)} \qquad (19.15)$$

where ω_o is a constant. However, the variable we are feeding back is not frequency but phase. The phase of the VCO clock output is related to f_{clock} by

$$\phi_{clock} = \int \omega_{clock} \cdot dt = \frac{K_{VCO}}{j\omega} \cdot V_{inVCO} \quad \text{(radians)} \qquad (19.16)$$

where this signal can be related to the ϕ_{dclock} by

$$\phi_{dclock} = \frac{1}{N} \cdot \phi_{clock} = \beta \cdot \phi_{clock} \qquad (19.17)$$

where N is the divide by count and β is the feedback factor.

19.2.1 The Current-Starved VCO

The current-starved VCO is shown schematically in Fig. 19.14. Its operation is similar to the ring oscillator discussed in Ch. 11. MOSFETs M2 and M3 operate as an inverter, while MOSFETs M1 and M4 operate as current sources. The current sources, M1 and M4, limit the current available to the inverter, M2 and M3; in other words, the inverter is starved for current. The MOSFETs M5 and M6 drain currents are the same and are set by the input control voltage. The currents in M5 and M6 are mirrored in each inverter/current source stage. An important property of the VCO used in any of the CMOS DPLLs discussed in this chapter is the input impedance. The filter configurations we have discussed rely on the fact that the input resistance of the VCO is practically infinite and the input capacitance is small compared to the capacitances present in the loop filter. Attaining infinite input resistance is usually an easy part of the design. For the charge-pump configuration, the input capacitance of the VCO can be added to C_2.

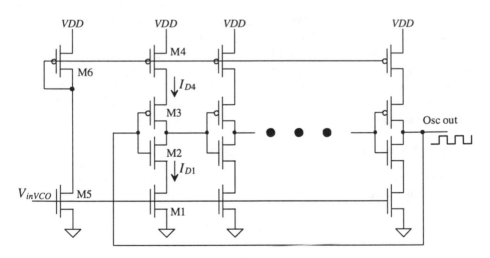

Figure 19.14 Current-starved VCO.

   To determine the design equations for use with the current-starved VCO, consider the simplified schematic of one stage of the VCO shown in Fig. 19.15. The total capacitance on the drains of M2 and M3 is given by

$$C_{tot} = C_{out} + C_{in} = \overbrace{C'_{ox}(W_pL_p + W_nL_n)}^{C_{out}} + \overbrace{\frac{3}{2}C'_{ox}(W_pL_p + W_nL_n)}^{C_{in}} \quad (19.18)$$

which is simply the output and input capacitances of the inverter as discussed in Ch. 11. This equation can be written in a more useful form as

$$C_{tot} = \frac{5}{2} C'_{ox}(W_p L_p + W_n L_n) \tag{19.19}$$

The time it takes to charge C_{tot} from zero to V_{SP} with the constant-current I_{D4} is given by

$$t_1 = C_{tot} \cdot \frac{V_{SP}}{I_{D4}} \tag{19.20}$$

while the time it takes to discharge C_{tot} from VDD to V_{SP} is given by

$$t_2 = C_{tot} \cdot \frac{VDD - V_{SP}}{I_{D1}} \tag{19.21}$$

If we set $I_{D4} = I_{D1} = I_D$ (which we will label $I_{Dcenter}$ when $V_{inVCO} = VDD/2$), then the sum of t_1 and t_2 is simply

$$t_1 + t_2 = \frac{C_{tot} \cdot VDD}{I_D} \tag{19.22}$$

The oscillation frequency of the current-starved VCO for N (an odd number ≥ 5) of stages is

$$f_{osc} = \frac{1}{N(t_1 + t_2)} = \frac{I_D}{N \cdot C_{tot} \cdot VDD} \tag{19.23}$$

which is equal to

$$= f_{center}(@ V_{inVCO} = VDD/2 \text{ and } I_D = I_{Dcenter})$$

Equation (19.23) gives the center frequency of the VCO when $I_D = I_{Dcenter}$. The VCO stops oscillating, neglecting subthreshold currents, when $V_{inVCO} < V_{THN}$. Therefore we can define

$$V_{min} = V_{THN} \text{ and } f_{min} = 0 \tag{19.24}$$

The maximum VCO oscillation frequency, f_{max}, is determined by finding I_D when $V_{inVCO} = VDD$. At the maximum frequency then, $V_{max} = VDD$.

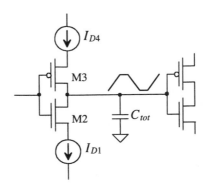

Figure 19.15 Simplified view of a single stage of the current-starved VCO.

The output of the current-starved VCO shown in Fig. 19.14 normally has its output buffered through one or two inverters. Attaching a large load capacitance on the output of the VCO can significantly affect the oscillation frequency or lower the gain of the oscillator enough to kill oscillations altogether.

The average current drawn by the VCO (see Ch. 11) is

$$I_{avg} = N \cdot \frac{VDD \cdot C_{tot}}{T} = N \cdot VDD \cdot C_{tot} \cdot f_{osc} \tag{19.25}$$

or

$$I_{avg} = I_D \tag{19.26}$$

The average power dissipated by the VCO is

$$P_{avg} = VDD \cdot I_{avg} = VDD \cdot I_D \tag{19.27}$$

If we include the power dissipated by the mirror MOSFETs, M5 and M6, the power is doubled from that given by Eq. (19.27), assuming that $I_D = I_{D5} = I_{D6}$. For low-power dissipation we must keep I_D low, which is equivalent to stating that for low-power dissipation we must use a low-oscillation frequency.

Example 19.1

Design a current-starved VCO with f_{center} = 100 MHz using the CN20 process parameters. Estimate the power dissipation. Simulate the design using SPICE.

We begin by calculating the total capacitance, C_{tot}. Using Eq. (19.19) and assuming the inverters, M2 and M3, are sized for equal drive, that is, $L_n = L_p = 2$ μm, $W_n = 3$ μm, and $W_p = 9$ μm, the capacitance is

$$C_{tot} = \frac{5}{2} \cdot 800 \frac{aF}{\mu m^2} \cdot (9 \cdot 2 + 3 \cdot 2) \mu m^2 = 48 \text{ fF}$$

Since the oscillation frequency is relatively large, compare this with the five-stage ring oscillator described in Ch. 11. We will use the minimum number of stages, that is, 5. The center drain current, using Eq. (19.23), is given by

$$I_{Dcenter} = 100 \text{ MHz} \cdot 5 \cdot 48 \text{ fF} \cdot 5 \text{ V} = 120 \text{ μA}$$

To determine the size of M5 and M1 of Fig. 19.14, we solve for W/L in

$$I_{Dcenter} = \frac{\beta_5}{2}(V_{GS} - V_{THN})^2 = \frac{50 \frac{\mu A}{V^2}}{2} \cdot \frac{W_n}{L_n}(2.5 - 0.83)^2 = 120 \text{ μA}$$

giving W_n/L_n = 1.72. We will use W_n = 5 μm and L_n = 3 μm as a close approximation. We use these values to set the size of M5 and M1 (of each stage). Sizing the p-channel we require, $\beta_6 = \beta_5$; this causes M5 to go into the triode region when V_{inVCO} is approximately 3.5 and above. This requirement sets the size of M6 and M4 at W_p = 15 μm and L_p = 3 μm. The power dissipated, using Eq. (19.27), is 1.2 mW. The simulation results are shown in Fig. 19.16.

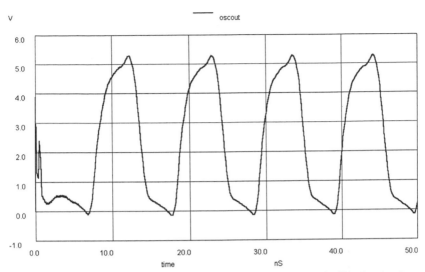

Figure 19.16   Simulation results for the oscillator of Ex. 19.1 (unbuffered output).
Simulations did not include drain and source depletion capacitances, that is,
the perimeters and areas of the drain and source where not specified.

The simulation results did not include a specification of drain and source areas.
Including these capacitances causes the f_{center} to drop to 80 MHz. ∎

19.2.2 Source Coupled VCOs

Another variety of VCO is shown in Fig. 19.17 [3]. These circuits can be designed to
dissipate less power than the current-starved VCO of the last section for a given
frequency. The major disadvantage of these configurations is the need for a capacitor,
something that may not be available in a single-poly pure digital process without using
parasitics for example, a metal1 to metal2 capacitor and the reduced output voltage
swing. However, this configuration is useful when the VCO center frequency is set by
an external capacitor; that is, the capacitor shown in the figure is bonded out.

To understand the operation, let's consider the NMOS source coupled VCO of
Fig. 19.17a. The operation of the CMOS source coupled VCO of Fig. 19.17b is
identical to the NMOS VCO except for the fact that the load MOSFETs M3 and M4
pull the outputs to $VDD - V_{THN}$ for the NMOS VCO and VDD for the CMOS VCO. The
MOSFETs M5 and M6 behave as constant-current sources sinking a current I_D.
MOSFETs M1 and M2 operate as switches. If M1 is off and M2 is on, the drain of M1
is pulled to $VDD - V_{THN}$ by M3. It is very important to remember the body effect in this
discussion. For example, using the CN20 process, the threshold voltage of M3 is
approximately 1.5 V so that the maximum output swing is limited to 3.5 V when the
supply voltage is 5 V. Since the gate of M2 is at $VDD - V_{THN}$, the source and drain (the
Output) of M2 are approximately $VDD - 2 \cdot V_{THN}$. This is the minimum output voltage.
The output voltage swing is limited to V_{THN}, or approximately 1.5 V, a swing from

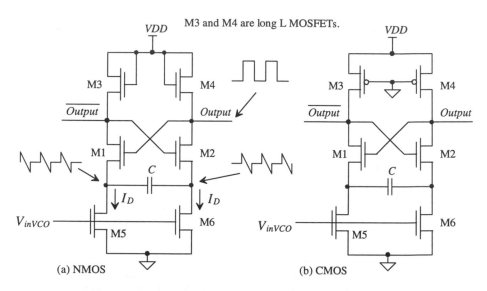

Figure 19.17 Source coupled voltage-controlled oscillators (also known as source coupled multivibrators).

approximately 3.5 V to 2 V. The oscillator would require a buffer, possibly an inverter or self-biased diff-amp circuit to restore CMOS logic levels.

The simplified schematic shown in Fig. 19.18 with M1 off and M2 on will be helpful in determining the oscillator frequency. The *Output*, gate of M1 is approximately $VDD - 2 \cdot V_{THN}$ and is held at this voltage through M2 until M1 turns on and M2 turns off. Initially, at the moment when M1 turns off and M2 turns on, point X is $VDD - V_{THN}$. The current through C, I_D, causes point X to discharge down toward ground. When point X gets down to $VDD - 3 \cdot V_{THN}$, M1 turns on and M2 turns off. In

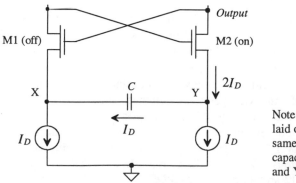

Note: C should be laid out with the same parasitic capacitance at X and Y.

Figure 19.18 Simplified schematic of source coupled oscillator, M1 is on and M2 is off.

other words, the voltage at point X changed a total of $2 \cdot V_{THN}$ before switching took place. The time it takes point X to change $2 \cdot V_{THN}$ is given by

$$\Delta t = C \cdot \frac{2 \cdot V_{THN}}{I_D} \qquad (19.28)$$

Since the circuit is symmetrical, two of these discharge times are needed for each cycle of the oscillator. The frequency of oscillation is given by

$$f_{osc} = \frac{1}{2\Delta t} = \frac{I_D}{4 \cdot C \cdot V_{THN}} \qquad (19.29)$$

The waveforms at the points X, Y, and *Output* are shown in Fig. 19.19 for continuous time operation.

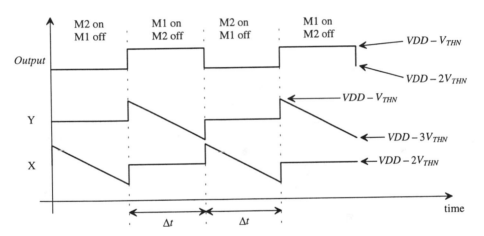

Figure 19.19   Voltage waveforms for the NMOS source coupled VCO.

Example 19.2
Design an NMOS source coupled VCO with a center frequency of 1 MHz. Assume I_D = 10 μA at V_{inVCO} of 2.5 V. (Design of current sources will be discussed in the next chapter.) Simulate the design using SPICE, showing the output of the VCO and points X and Y.

Assuming V_{THN} is approximately 1 V (this approximation is based on M1 and M2's threshold voltage), the value of the capacitor is determined using Eq. (19.29), or

$$C = \frac{10\ \mu A}{4 \cdot 1\ MHz} = 2.5\ pF$$

Since M1 and M2 are used as switches, we want to make their βs large. For the present design we picked W/L = 30/2. MOSFETs M3 and M4 were sized for a

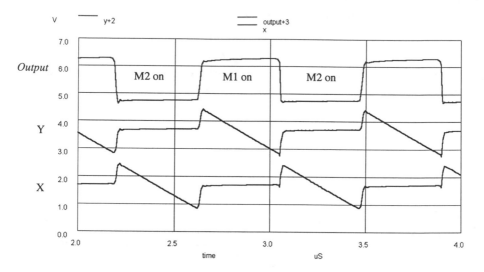

Figure 19.20 SPICE simulation results for Ex. 19.2.

relatively large resistance by setting $W/L = 3/8$. The simulation results are shown in Fig. 19.20. A practical limit on the oscillator frequency using the CN20 process is in the tens of MHz. ∎

Another example of a VCO that is particularly well suited to both high and low frequency operation is shown in Fig 19.21 [4]. The magnitudes of the current sources

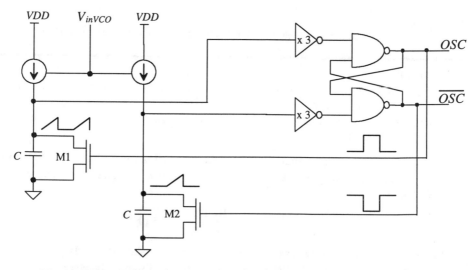

Figure 19.21 VCO with wide operating range (requires two capacitors).

are controlled by an external voltage. The positive feedback around the loop is used to ensure that only one of M1 and M2 are on at a time. The switching points of the inverters combined with the current source determine the oscillator frequency.

19.3 The Loop Filter

The loop filter is the brain of the DPLL. In this section, we discuss how to select the loop-filter values in order to keep the DPLL from oscillating (i.e., keep the V_{inVCO} voltage from oscillating, causing the frequency out of the VCO to wander). If the loop-filter values are not selected correctly, it may take the loop too long to lock, or once locked small variations in the input data may cause the loop to unlock.

In the following discussion, we will be concerned with the *pull-in range* and the *lock range*. The pull-in range, $\pm \Delta \omega_P$, is defined as the range of input frequencies which the DPLL will lock to. The time it takes the loop to lock is labeled T_P and may be a very long time. If the center frequency of the DPLL is 10 MHz and the pull-in range is 1 MHz, the DPLL will lock on an input frequency from 9 to 11 MHz in a time T_P (assuming $N = 1$). The lock range, $\pm \omega_L$, is the range of frequencies in which the DPLL locks within one single beat note between the divided down output (*dclock*) and input (*data*) of the DPLL. *The operating frequency of the DPLL should be limited to the lock range for normal operation.* Once the DPLL is locked, it will remain locked as long as abrupt frequency changes, $\Delta \omega$, in the input frequency (input frequency steps) over a time interval t are much smaller than the natural frequency of the system squared, that is, $\Delta \omega / t < \omega_n^2$.

19.3.1 XOR DPLL

Consider the block diagram of the DPLL using the XOR DPLL shown in Fig. 19.22. The phase transfer function (neglecting the static or DC behavior) is given by

$$H(s) = \frac{\phi_{clock}}{\phi_{data}} = \frac{K_{PD} K_F K_{VCO}}{s + \beta \cdot K_{PD} K_F K_{VCO}} \qquad (19.30)$$

with $s = j\omega$ and the feedback factor, β, is

$$\beta = \frac{1}{N} \qquad (19.31)$$

The transfer function of the loop filter is given by

$$K_F = \frac{1}{1 + j\omega RC} = \frac{1}{1 + sRC} \qquad (19.32)$$

Substituting Eqs. (19.31) and (19.32) into Eq. 19.30 yields

$$H(s) = \frac{\phi_{clock}}{\phi_{data}} = \frac{K_{PD} K_{VCO} \cdot \frac{1}{1 + sRC}}{s + \frac{1}{N} \cdot K_{PD} K_{VCO} \cdot \frac{1}{1 + sRC}} \qquad (19.33)$$

This is a second-order system. $H(s)$ can be rewritten as

$$H(s) = \frac{\frac{K_{PD}K_{VCO}}{RC}}{s^2 + \frac{s}{RC} + \frac{1}{N} \cdot \frac{K_{PD}K_{VCO}}{RC}} = \frac{f_{clock}}{f_{data}} = \frac{(K_{VCO}V_{inVCO})/2\pi + f_o}{f_{data}} \qquad (19.34)$$

since $j\omega \cdot \phi = f$. The natural frequency, ω_n, of this system is given by

$$\omega_n = \sqrt{\frac{K_{PD}K_{VCO}}{N \cdot RC}} \qquad (19.35)$$

and the damping ratio, ζ, is given by

$$\zeta = \frac{1}{2RC\omega_n} = \frac{1}{2} \cdot \sqrt{\frac{N}{K_{PD}K_{VCO} \cdot RC}} \qquad (19.36)$$

The pull-in range is given by

$$\Delta\omega_P = \frac{\pi}{2} \cdot \sqrt{2\zeta\omega_n K_{VCO}K_{PD} - \omega_n^2} \qquad (19.37)$$

while the pull-in time is given by

$$T_P = \frac{4}{\pi^2} \cdot \frac{\Delta\omega_{center}^2}{\zeta\omega_n^3} \qquad (19.38)$$

The lock range of the loop is given by

$$\Delta\omega_L = \pi\zeta\omega_n = \frac{\pi}{2} \cdot \frac{1}{RC} \qquad (19.39)$$

while the lock time is

$$T_L = \frac{2\pi}{\omega_n} \qquad (19.40)$$

Probably the best way to understand how these equations affect the performance of a DPLL is through an example.

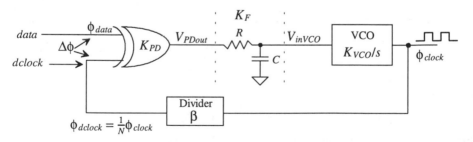

Figure 19.22   Block diagram of a DPLL using a XOR phase detector

Example 19.3
The DPLL of Fig. 19.22 is used to generate a clock for a 9 Mbits/s data stream with data format (non-return to zero level, or simply NRZ-level) shown in Fig.

19.7. Assume that there is a transition at least once every nine bits (an odd-parity bit), that the code for the eight-bit word 255 (11111111) is not allowed, and that some synchronization words are present in the data to help the DPLL lock[3]. Determine R and C of the loop filter, and comment on your selection of VCO.

The width of one bit of *data* is 111 ns (= 1/9 MHz). The frequency of the DPLL output *clock* is 9 MHz. A divide by 2 ($N = 2$) stage is used in the feedback loop to make *dclock* = 4.5 MHz. If the *data* input is an alternating series of ones and zeros as shown in Fig. 19.23, the frequency of the resulting square wave is 4.5 MHz, or one-half of the data rate. This is the major advantage of using NRZ code; that is, the data rate can be twice the channel bandwidth.

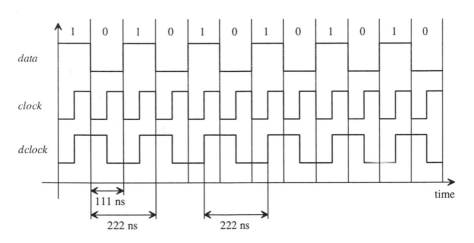

Figure 19.23 Example waveforms for the DPLL of Example 19.3 in lock.

Next let's discuss the VCO. From the previous discussion of Sec. 19.1.1, the maximum f_{clock} can be is 18 MHz and the minimum is 4.5 MHz. When the control voltage is $VDD/2$, the f_{center} of the VCO is 9 MHz. (This is important!) If we try to use the basic current-starved VCO of the previous section, we run into the problem of locking on harmonics of the desired clock frequency. A simple method of setting the maximum current in this VCO, and thus the maximum frequency, is to place a current limiting resistor in series with the drains of M5 and M6 (see Fig. 19.24). The current that flows in M5 and M6 is mirrored by M8 and M9 and sets the variable current in the VCO. To set the minimum current, we add MOSFET M7 to provide a minimum current through M6 when $V_{inVCO} < V_{THN}$. (Note that this current is always present and added to the voltage-variable current through M6.) In all practical situations, the VCO

---

[3] System considerations concerning selection of a data pattern as well as regenerating the clock from NRZ data will be discussed in greater detail later in the chapter.

must be hand tailored to operate over the desired frequency range. Here in itself
is the main problem using the XOR-based DPLL in a pure monolithic form.
Bringing the frequency setting components off chip can help solve this problem.
Precision components can be used to set the operating frequencies. One last
concern when designing the VCO, or any CMOS circuit for that matter, is its
temperature behavior. This subject is covered in more detail in the coming
chapters.

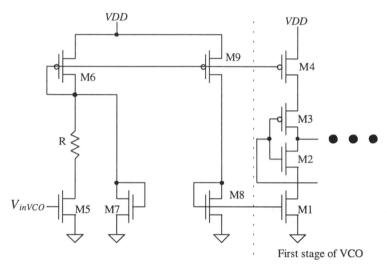

Figure 19.24 Modifications to the current-starved VCO to set minimum and
and maximum frequencies.

If we design the VCO so that when $V_{inVCO} = 4V = V_{max}$, the output oscillator
frequency, f_{max} is 9.1 MHz; when $V_{inVCO} = 1V = V_{min}$, the output frequency, f_{min} is
8.9 MHz then the gain of the oscillator, K_{VCO}, is 2π·67 kHz/V, or

$$K_{VCO} = 418 \times 10^3 \text{ radians/V·s}$$

The only component left in the first-cut design of the loop is selection of RC.
Let's begin by setting the damping ratio ζ to 0.7. Using Eq. (19.36) results in

$$0.7 = \frac{1}{2} \cdot \sqrt{\frac{2}{\frac{5}{\pi} \cdot (418 \times 10^3)RC}} \rightarrow RC = 1.5 \text{ μs}$$

The natural frequency of the second-order system is, from Eq. (19.36),

$$\omega_n = \frac{1}{2 \cdot RC \cdot \zeta} = 476 \times 10^3 \text{ radians/s}$$

The lock range is

$$\Delta\omega_L = \pi\zeta\omega_n = 1 \times 10^6 \text{ radians/s or } \Delta f_L = 167 \text{ kHz}$$

which is outside the VCO operating range. Therefore, the DPLL will lock over the entire VCO operating range, that is, from 8.9 to 9.1 Mbits/s. We do not need to calculate the pull-in range since it is greater than the lock-in range. Again, the design of the oscillator is important in this DPLL. The lock time, T_L, of the DPLL is approximately 13 μs. A simulation of this DPLL using the software provided with [1] is shown in Fig. 19.25. Initially, at $t < 0$, the data rate input is 9.0 Mbits/s, and both the average phase-detector output and the output of the filter are 2.5 V ($VDD/2$ and $\Delta\phi = \pi/2$), point A in Fig. 19.25. At $t = 0$, the data rate jumps to 9.04 Mbits/s. Both the average phase-detector and filter outputs increase in voltage to 3.1 V. The *dclock* is now aligned slightly off center of the *data*, (3.1 V/$K_{PD} = \Delta\phi$ or $\Delta\phi = 0.62\cdot\pi$), keeping the loop locked. ∎

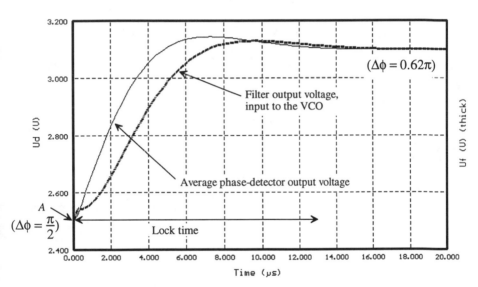

Figure 19.25 Simulation of the DPLL of Ex. 19.3 with a 40 kbit/s data rate step. This corresponds to a frequency step of 20 kHz.

Adding a zero to the simple passive RC loop filter, Fig. 19.26 (called a passive lag loop filter), the loop-filter pole can be made small (and thus the gain of the VCO can be made larger) while at the same time acheiving a reasonable damping factor. The result is an increase in the lock range of a DPLL using the XOR PD and a shorter lock time[1] see Eqs. (19.37) - (19.40). The passive lag loop filter is, in most situations preferred over the simple RC. Again, as Ex. 19.3 showed, if the center frequency of the VCO doesn't match the input frequency the clock will not align at $\pi/2$.

The clock misalignment encountered in a DPLL using an XOR PD and passive filter can be minimized by using the active proportional + integral (active PI) loop filter shown in Fig. 19.27. The transfer function of this filter is given by

$$V_{PDout} \quad R_1 \quad V_{inVCO} \qquad K_F = \frac{V_{inVCO}}{V_{PDout}} = \frac{1+j\omega R_2 C}{1+j\omega(R_1+R_2)C}$$

$$\omega_n = \sqrt{\frac{K_{PD}K_{VCO}}{N(R_1+R_2)C}} \qquad \Delta\omega_L = \pi\zeta\omega_n$$

$$R_2$$

$$C$$

$$\zeta = \frac{\omega_n}{2} \cdot \left(R_2 C + \frac{N}{K_{PD}K_{VCO}}\right)$$

Figure 19.26 Passive lag loop filter used to increase DPLL lock range.

$$K_F = \frac{1+sR_2 C}{sR_1 C} \tag{19.41}$$

The natural frequency of the resulting second-order system is given by

$$\omega_n = \sqrt{\frac{K_{PD}K_{VCO}}{NR_1 C}} \tag{19.42}$$

and the damping ratio is given by

$$\zeta = \frac{\omega_n R_2 C}{2} \tag{19.43}$$

The lock range is

$$\Delta\omega_L = 4\pi\zeta\omega_n \tag{19.44}$$

while the lock time remains $2\pi/\omega_n$. The pull-in range, using the active PI loop filter, is limited by the VCO oscillator frequency. Consider the following example.

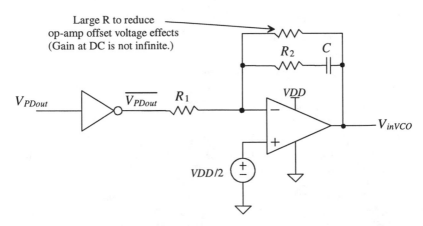

Figure 19.27 Active PI loop filter.

Example 19.4
Repeat Ex. 19.3 using the active PI loop filter.

If the output of the VCO is 10 MHz ($=f_{max}$), when the input voltage is 4 V and 8 MHz ($=f_{min}$) when the input to the VCO is 1 V, then the gain of the VCO is

$$K_{VCO} = 2\pi \cdot \frac{10-8}{4-1} = 4.2 \times 10^6 \text{ radians/V·s}$$

which is much easier to realize in CMOS. If the lock range between *data* and *dclock* is set so that the VCO limits the operating frequency range, say $\Delta f_L = 1$ MHz, then the natural frequency using Eq. (19.44) and assuming $\zeta = 0.7$ is

$$\omega_n = \frac{2\pi \cdot 1 \times 10^6}{4\pi \cdot 0.7} = 713 \times 10^3 \text{ radians/s}$$

Using Eqs. (19.42) and (19.43) with $N = 2$, we can solve for R_1C and R_2C as approximately 6.6 µs and 2 µs, respectively. The lock time of this loop is now 8.8 µs. Simulation results showing a 500 kbit/s data rate step are shown in Fig. 19.28. This corresponds to the data changing from 9 Mbits/s to 9.5 Mbits/s. The DPLL of Ex. 19.3 cannot lock to data input at 9.5 Mbits/s. Note the average PD output is 2.5 V keeping the phase difference between the *dclock* and *data* at $\pi/2$ independent of the VCO center frequency and the DPLL input frequency. ∎

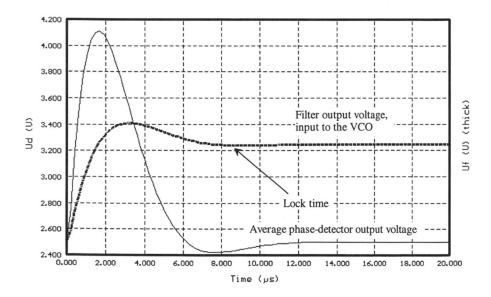

Figure 19.28 Simulation of the DPLL of Ex. 19.4 with a 500 kbit/s input data rate step.

19.3.2 PFD DPLL

A block diagram of a DPLL using the PFD with tri-state output is shown in Fig. 19.29. The phase transfer function is the same as Eq. (19.30)

$$H(s) = \frac{\phi_{clock}}{\phi_{data}} = \frac{K_{PDtri}K_FK_{VCO}}{s + \beta \cdot K_{PDtri}K_FK_{VCO}} \tag{19.45}$$

where

$$K_F = \frac{1 + sR_2C}{1 + s(R_1 + R_2)C} \tag{19.46}$$

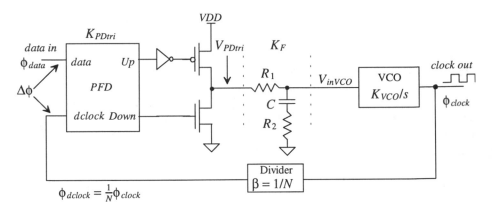

Figure 19.29 Block diagram of a DPLL using a sequential phase detector (PFD).

When this filter is driven with the tri-state output, no current flows in R_1 or R_2 when the output is in the high-impedance state. The voltage across the capacitor remains unchanged. We can think of the filter, tri-state output as an ideal integrator with a transfer function

$$K'_F = \frac{1 + sR_2C}{s(R_1 + R_2)C} \tag{19.47}$$

Substituting this equation into Eq. 19.45 and rearranging results in

$$H(s) = \frac{K_{PDtri}K_{VCO}\frac{1+sR_2C}{(R_1+R_2)C}}{s^2 + s\frac{K_{PDtri}K_{VCO}R_2C}{N(R_1+R_2)C} + \frac{K_{PDtri}K_{VCO}}{N(R_1+R_2)C}} = \frac{\phi_{clock}}{\phi_{data}} = \frac{f_{clock}}{f_{data}} \tag{19.48}$$

From this equation the natural frequency is

$$\omega_n = \sqrt{\frac{K_{PDtri}K_{VCO}}{N(R_1 + R_2)C}} \tag{19.49}$$

and the damping factor is determined by solving

$$2\zeta\omega_n = \frac{K_{PDtri}K_{VCO}R_2C}{N(R_1+R_2)C}$$

(19.50)

which results in a damping factor given by

$$\zeta = \frac{\omega_n}{2} \cdot R_2C$$

(19.51)

The lock range is given by

$$\Delta\omega_L = 4\pi\zeta\omega_n$$

(19.52)

while the lock time, T_L, remains $2\pi/\omega_n$. The pull-in range is limited by the VCO operating frequency. The pull-in time is given by

$$T_P = 2R_1C \cdot \ln\frac{(K_{VCO}/N)\cdot(VDD/2)}{(K_{VCO}/N)(VDD/2)-\Delta\omega}$$

(19.53)

where $\Delta\omega$ is the magnitude of the *input* frequency step.

Example 19.5

Design a DPLL that generates a clock signal at a frequency of 256 kHz from a 1-kHz square wave input. This application of the DPLL, is called *frequency synthesis* (Fig. Ex19.5).

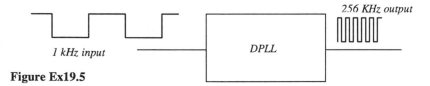

Figure Ex19.5

We will use the DPLL of Fig. 19.29 with a divide by 256 ($N = 256$) in the feedback loop. Furthermore, assume that the VCO oscillates from 500 kHz down to DC for input voltages (V_{inVCO}) from 0 to 5 V. The gain of the VCO is then

$$K_{VCO} = 2\pi \cdot \frac{(500-0)\ KHz}{5-0} = 628\times10^3 \text{ radians/V·s}$$

The lock range, Δf_L will be set to 200 Hz. This means the loop will lock up on an input frequency from 800 to 1200 Hz and generate an output frequency 256 times the input frequency in a time T_L. For example, if 900 Hz is input into the loop, 230 kHz will be output by the DPLL. The natural frequency of the DPLL, again assuming $\zeta = 0.7$, is

$$\omega_n = \frac{2\pi\cdot200}{4\pi\cdot0.7} = 143 \text{ radians/s}$$

Using Eqs. (19.51) and (19.49) gives the loop-filter time constants of approximately, $R_2C = 9.8$ ms and $R_1C = 38$ ms ($K_{PDtri} = VDD/4\pi$). In most cases, the loop filter would be implemented off-chip. The lock time, T_L, for this DPLL is 44 ms. Figure 19.30 shows how the average PD and loop-filter outputs change with a 200 Hz input step in frequency. Initially, at $t < 0$, the input voltage to the VCO is 2.5 V and the input frequency is 1,000 Hz, while the output of the VCO is 256,000 Hz ($N\cdot1,000$ Hz). For times greater than zero, the input frequency jumps to 1,200 Hz, causing the VCO output frequency to jump to 307,000 Hz. Approximately 44 ms after the initial frequency jump, the input voltage to the VCO, V_{inVCO}, settles at approximately 3 V.

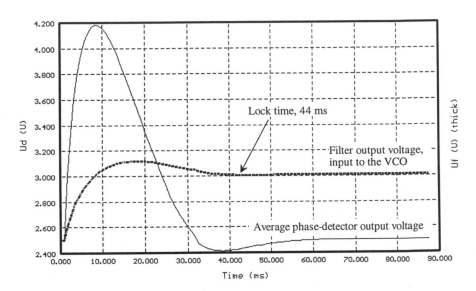

Figure 19.30 Simulation of the DPLL of Ex. 19.5 with an input frequency step of 200 Hz

The maximum VCO output frequency is 500 kHz. This corresponds to an input frequency of 1.95 kHz (divide the VCO output frequency by 256). An 800 Hz frequency step in the input to the DPLL (change the input frequency from 1,000 to 1,800 Hz) results in the waveforms shown in Fig. 19.31. Since the frequency step lies outside the lock range of the DPLL, the VCO input voltage will oscillate before finally locking. The time it takes the loop to pull-in the VCO control voltage, and relock is approximately T_P, or

$$T_P = 2R_1 C \cdot \ln \frac{(628 \times 10^3/256) \cdot (2.5)}{(628 \times 10^3/256)(2.5) - 2\pi \cdot 800} = 130 \text{ ms}$$

Once the loop is locked, it will not lose lock unless the change in the input frequency, $\Delta\omega$, over a time t is greater than the ω^2_n. The relaxed requirements on the VCO operating range make the PFD the phase detector of choice when designing CMOS DPLLs. ∎

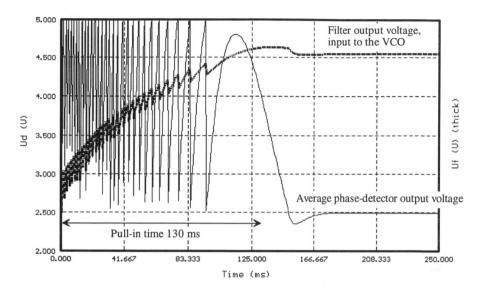

Figure 19.31 Simulation of the DPLL of Ex. 19.5 with an input frequency step of 800 Hz.

The PFD with a charge-pump output is shown in Fig. 19.32. A CMOS implementation of a DPLL using this configuration is, in general, preferred over the tri-state output because of the better immunity to power supply variations. The capacitor C_2 is used to keep $I_{pump} \cdot R$ from causing voltage jumps on the input of the VCO and thus frequency jumps in the DPLL output from occurring. In general, C_2 is set at

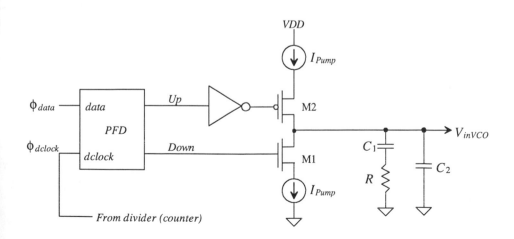

Figure 19.32 PFD using the charge pump.

about one-tenth (or less) of C_1. The loop-filter transfer function neglecting C_2 is given by

$$K_F = \frac{1 + sRC_1}{sC_1} \qquad (19.54)$$

The feedback loop transfer function is given by

$$H(s) = \frac{\phi_{clock}}{\phi_{data}} = \frac{K_{PDI}K_{VCO}(1 + sRC_1)}{s^2 + s(\frac{K_{PDI}K_{VCO} \cdot R}{N}) + \frac{K_{PDI}K_{VCO}}{NC_1}} \qquad (19.55)$$

From the transfer function the natural frequency is given by

$$\omega_n = \sqrt{\frac{K_{PDI}K_{VCO}}{NC_1}} \qquad (19.56)$$

and the damping factor is

$$\zeta = \frac{\omega_n}{2} \cdot RC_1 \qquad (19.57)$$

The lock range and lock time remain the same (using the different values for the natural frequency and damping ratio) as the PFD with the tri-state output. Again, the pull-in range is set by the VCO oscillator frequency range. The pull-in time is given by

$$T_P = 2RC_1 \ln\left[\frac{(K_{VCO}/N) \cdot (I_{pump})}{(K_{VCO}/N) \cdot (I_{pump}) - \Delta\omega} \right] \qquad (19.58)$$

Discussion

When selecting values for the loop filter, we assumed that the output resistance of the phase detector was small (for the XOR and tri-state PD) compared to the impedances used in the loop filter. We also assumed that the input resistance of the VCO was infinite and that the input capacitance of the VCO was small compared to the capacitance used in the loop filter. Considering the parasitics present in the DPLL is an important part of the design.

Examples of CMOS VCOs were given in Sec. 19.2. An example of a CMOS implementation of a PFD is shown in Fig. 19.33[5]. Implementation of the loop filter can use MOSFETs as capacitors and n-well or MOSFETs as resistors.

The center frequency of the VCO was critical for good DPLL performance when using the XOR gate with RC loop filter. If the center frequency, f_{center}, of the VCO (i.e, $V_{inVCO} = VDD/2$) does not match twice the input data rate (when using NRZ code), the DPLL will lock up at a phase different from $\pi/2$. The need for a precision center frequency is eliminated by using the XOR PD with an active PI loop filter or by using the PFD.

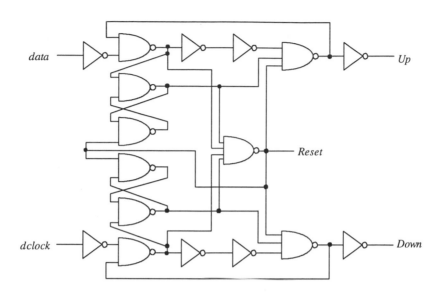

Figure 19.33 CMOS implementation of a PFD.

19.4 System Considerations

System concerns are often the driving force behind the design of a DPLL. Referring to Fig. 19.1, we observe that the data transmitted through the channel should ideally arrive at the receiver with the same shape as it was transmitted with [6]. In reality, the data become distorted. Distortion arises from nonlinearities in the receiver input amplifier and the finite bandwidth of the channel. To understand the conditions for distortionless transmission, consider the block diagram shown in Fig. 19.34. The system has a transfer function in the frequency domain of $H(f)$ and in the time domain $h(t)$. For distortionless transmission, we can relate the input and output of the system by

$$y(t) = K \cdot x(t - t_o) \tag{19.59}$$

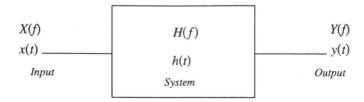

Figure 19.34 Representation of a system with input and output.

where t_o is the time delay through the system and K is a constant. This equation shows that for distortionless transmission through a system the output is simply a scaled, time-delayed version of the input. An interesting observation can be made by taking the Fourier Transform of both sides of this equation,

$$Y(f) = K \cdot X(f)e^{-j2\pi f t_o} \qquad (19.60)$$

The transfer function of a distortionless system can then be written as

$$H(f) = \frac{Y(f)}{X(f)} = Ke^{-j2\pi f t_o} \qquad (19.61)$$

Figure 19.35 shows the magnitude and phase responses of a distortionless system. A system is distortionless when its amplitude response, $|H(f)|$, is a constant, K, and its phase response, $\angle H(f)$, is linear with a slope of $-2\pi t_o$ over all frequencies of interest.

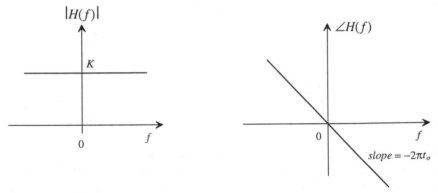

Figure 19.35 Magnitude and phase response of a distortionless system.

The responses shown in Fig. 19.35 are ideal. In practice, the magnitude response of a system may look similar to Fig. 19.36a. At higher frequencies, the magnitude rolls off. To compensate for this roll-off, or other imperfections, a circuit called an equalizer is added in series with the system (Fig. 19.36b). The equalizer has a transfer function in which its magnitude response increases with increasing frequency beyond a point (Fig. 19.36c). If the low-frequency gain of the equalizer is A/K and the low-frequency gain of the system is K, then the resulting gain of the system/equalizer combination is A.

Another source of potential distortion occurs when the receiver input data are regenerated into digital levels. Figure 19.37 shows the basic problem. The analog data generated by the receiver are connected to a comparator. The comparator slices the data, ideally through the middle, and regenerates the digital data. Timing errors occur when the comparator decision level is not correctly set at the middle of the data. Figure 19.37a shows how the analog data should be sliced. While Fig. 19.37b shows the resulting digital data with the decision level too high, the resulting ones are too narrow and the zeros are too wide. What makes slicing the data correctly even more difficult is

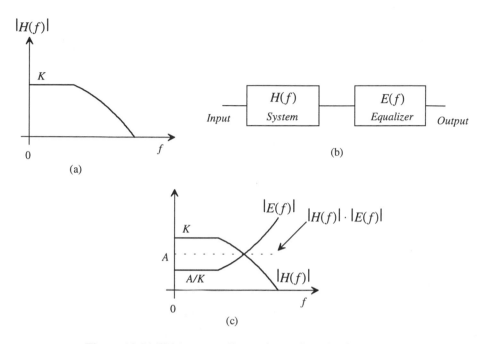

Figure 19.36 Using an equalizer to lower distortion in a system.

the fact that the amplitude response of the channel can change with time and the data pattern can affect the average level of the data. Consider the data and decision-making circuit of Fig. 19.38. The comparator slices the input data around ground. The result, for the waveforms shown in Fig. 19.38, is output logic ones that are too wide (in time). A long string of ones causes the output zeros to be too wide. In both cases, timing errors are possible when clocking the data into the shift register (see Fig. 19.1).

There are two solutions to this problem. The first uses a circuit that will determine the peak positive and negative input analog amplitudes, average the values, and feed back the result to the comparator in the decision-making circuit. The second

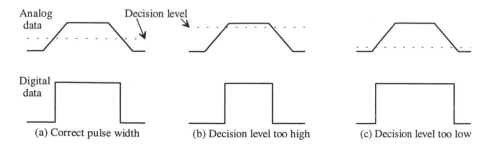

Figure 19.37 Timing errors in regenerating digital data.

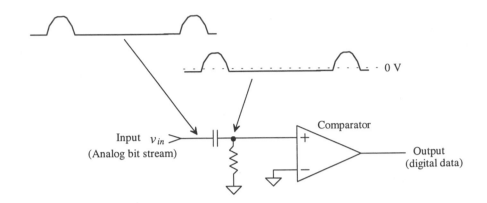

Figure 19.38 The effect of data pattern on the decision-making circuit.

method encodes the digital data so that the duty cycle of the resulting encoded data is 50 percent. The encoding increases the channel bandwidth for a constant data rate.

Consider the peak detector circuit shown in Fig. 19.39. When the input, v_{in}, is greater than the output, v_{peak}, the comparator output goes high, turning the MOSFET on. This causes the capacitor to charge toward VDD. When v_{peak} gets slightly larger than the input voltage, the comparator turns off. This in turn shuts the MOSFET off, leaving the capacitor essentially charged to the peak value of v_{in}.

The peak detector is used in the digital decision circuit shown in Fig. 19.40. Current sources, of a small value, are added to the basic decision circuit to make the peak detectors lossy. This may be necessary so that the decision circuit can follow the slow amplitude and data variations. The buffers shown in the circuit isolate the peak detectors from one another. These buffers can be implemented in some cases using the MOSFET source follower configuration (see Ch. 22). The two resistors perform the averaging function. The average of the analog input is feedback to the inverting input of the comparator. This scheme forces the data to be sliced in the middle, eliminating timing errors.

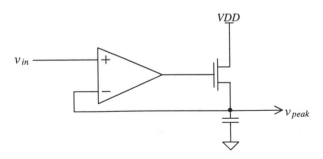

Figure 19.39 CMOS peak detector.

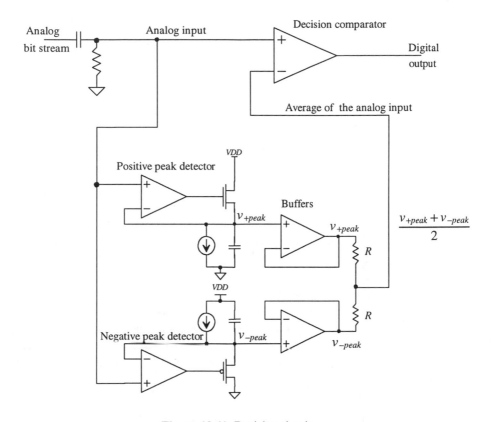

Figure 19.40 Decision circuit.

Encoding the data can eliminate the need for a decision circuit. If the resulting encoded data has a 50 percent duty cycle, it can be passed through a capacitor in the receiver (Fig. 19.38), resulting in an analog signal centered around ground. The comparator can then have its noninverting input connected to ground and slice the data at the correct moments in time (in the middle of the data bit). An example of an encoding scheme is shown in Fig. 19.41. Encoding occurs in the transmitter prior to transmission over the channel. This particular encoding scheme is referred to as the bi-phase format, or more precisely, the bi-phase-level (sometimes called bi-phase-L or Manchester NRZ) format. The cost of using this scheme over the NRZ data format is increased channel bandwidth. Other encoding schemes are shown in Fig. 19.42 [1].

19.4.1 Clock Recovery from NRZ Data

One of the most important steps in the design of a communication system is selection of the transmission format, that is, NRZ, bi-phase, or some other format, together with use of parity, cyclic redundancy code, or some other encoding format.

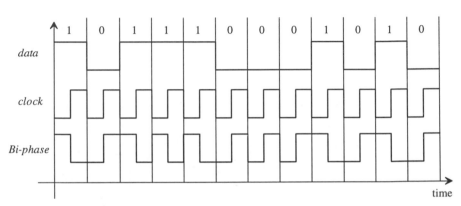

Figure 19.41 Bi-phase data encoding.

In this section, we discuss some of the considerations that go into the design of a clock-recovery DPLL in a system that uses NRZ. The reasons for the selection of the formats used in Ex. 19.3 and the need for a divide by 2 in the feedback loop will be explained.

Let's begin this discussion by considering the NRZ *data* and *clock* shown in Fig. 19.43. Let's further assume that these signals are the inputs to an XOR PD in a DPLL, which is not in lock since the clock is not aligned properly to the data. The resulting output of the XOR PD is shown in this figure as well. If we were to average this output using a loop filter, we would get *VDD*/2. In fact, it is easy to show that shifting the *clock* signal in time has no effect on the average output of the PD. Why? To answer this question, let's use some numbers. Assume that a bit width of *data* is 10 ns (which is also the period of the *clock*). The frequency of the square wave resulting from the alternating strings of ones and zeros is 50 MHz. We know that if we take the Fourier Transform of a square-wave, only the odd harmonics (i.e., 50, 150, and 250 MHz) are present. Since the *clock* signal is at 100 MHz, there is no energy or information common between the *clock* and *data* signals. To remedy this, we divide the clock down in frequency, *dclock*, so that it is at the same frequency as the alternating ones and zeros of the data (divide by 2).

The next problem we encounter, if we use the divide by two in the feedback loop, occurs when we get *data* that is a repeating string of two-ones followed by two-zeroes, Fig. 19.44 (the *dclock* is not locked to the data). Again, there is no common information between the two inputs, and the resulting XOR PD output will always have

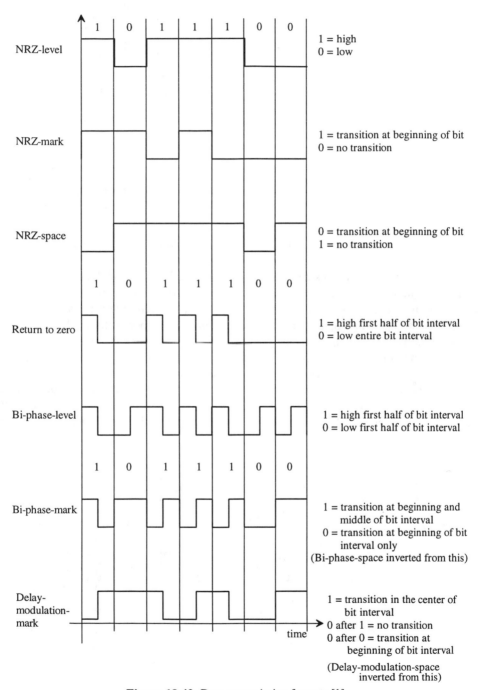

Figure 19.42 Data transmission formats [1].

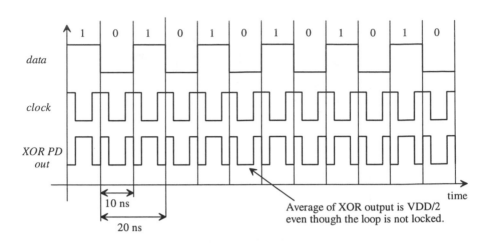

Figure 19.43 The problems of using clock without the divide by 2 to lock on data.

an average of *VDD/2*. In this case, however, the *dclock* is running at 50 MHz and the *data* is a square-wave of 25 MHz. Should we divide the *clock* down further to avoid this situation? The answer is no. However many times we divide the clock down, we can still come up with a data string that will not allow the loop to lock. Also, it is the actual edge transitions (the frequency of the *data* and *dclock*) that is used when the inputs are not pure square-waves. Increasing the width of *dclock* has the effect of removing information and making it more difficult to lock up to the data. One solution to this problem, for an eight-bit word, was given in Ex. 19.3. If we use odd-parity with an eight-bit word (nine bits total) and eliminate, at the transmitting end, the possibility of all eight bits being high, that is, 11111111 or 255, then it is impossible to generate a square-wave.

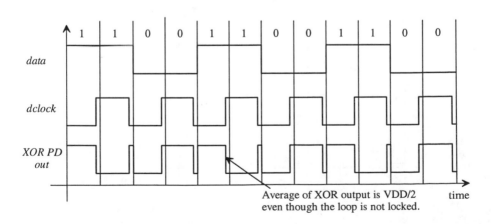

Figure 19.44 Problems trying to lock on a data stream that is one-half the dclock frequency.

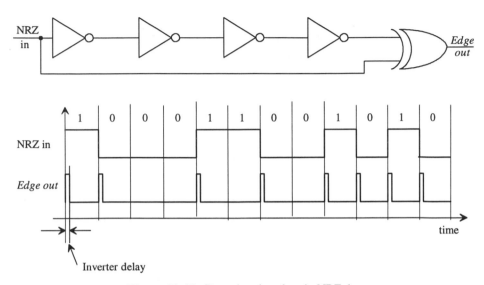

Figure 19.45 Detecting the edges in NRZ data.

The restrictions on the data pattern in a communications system using NRZ data can be reduced by detecting the edges of the input data using an edge detector circuit (Fig. 19.45) [1]. The delay through the inverters sets the width of the output pulses, labeled "*Edge out*" in this circuit. The frequency content of the output pulses will always contain energy at the *clock* frequency and thus the loop can lock up on the data.

As an example, consider the block diagram and data shown in Fig. 19.46. The *Edge output* is connected as the *input* of an XOR based DPLL. The output of the VCO, *clock*, will lock up on the center of the *Edge output*, that is, the rising edge of the *clock* signal will become aligned with the center of *Edge out*. Averaging *PD out*, in this figure, results in *VDD*/2. If the clock is shifted to the left or right in time, the average value of *PD out* will shift down or up, causing the VCO frequency to change, keeping the *clock* aligned to the center of *Edge out*.

Several practical problems exist with this configuration. The delay through the inverters should be constant whether a high-to-low or a low-to-high transition is propagating through the inverters. Also, the delay of the inverters, or whatever element is used for the delay (one common element for high-speed applications is a microstrip line), should be close to one-half of the bit-interval time for best performance. For example, if the bit interval, for the data shown in Fig. 19.46, is 10 ns, then the delay used in the edge detector should be 5 ns. This delay is important since it directly affects the gain of the phase detector and therefore the transient properties of the DPLL.

Not having the *clock* aligned to the center of the data bit can cause problems in high-speed clock-recovery circuits. Simply adding a delay in series with the *clock*

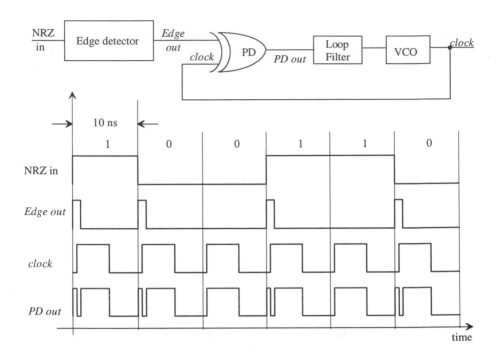

Figure 19.46 Clock-recovery circuit for NRZ using an edge detector.
Note that the DPLL is in lock, when the rising edge of clock
is centered on the edge output pulse.

signal will not solve this problem since the temperature dependence and process
variations of the associated circuit would not guarantee proper alignment. What is
needed is a circuit that is *self-correcting*, causing the clock signal to align to the center
of the data bit independent of the data-rate, temperature, or process variations.

The phase-detector portion of a self-correcting clock-recovery circuit is shown in
Fig. 19.47 [7] along with associated waveforms for a locked DPLL. Nodes A and B are
simply the input NRZ data shifted in time by one-half bit-interval and one bit-interval,
respectively. The outputs of the phase detector are labeled *Increase* and *Decrease*. If
Increase is low more often than *Decrease* the average voltage out of the loop filter and
thus the frequency out of the VCO will decrease. A loop filter that can be used in a
self-correcting DPLL is shown in Fig. 19.48a. This filter is the active-PI loop filter
discussed in Sec. 19.3.1, with an added input to accommodate both outputs of the PD.
Figure 19.48b shows the resulting waveforms in a DPLL where the clock is leading the
center of the NRZ data, and thus *Increase* is high less often than *Decrease*. If the NRZ
data were lagging the center of the data bit *Decrease* would be high less often than
Increase, resulting in an increase in the loop-filter output voltage. Note that in this
discussion we have neglected the propagation delays present in the circuit. For a

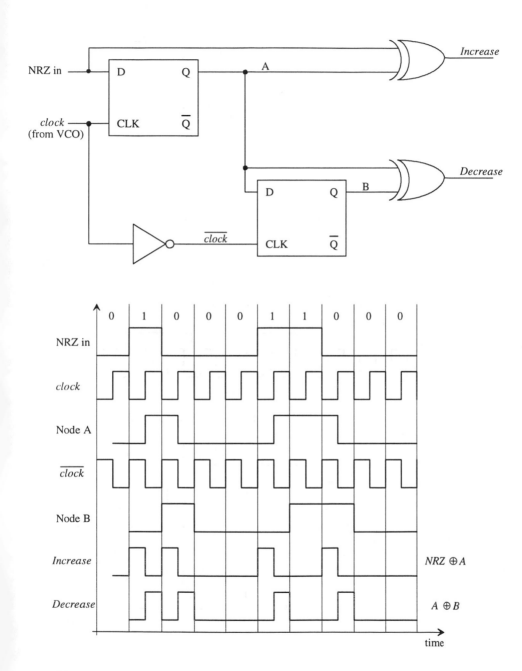

Figure 19.47 The PD portion of a self-correcting clock-recovery circuit in lock [7].

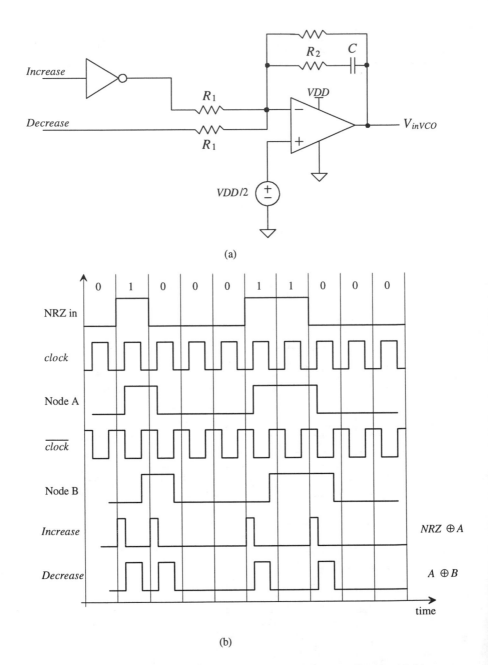

(a)

(b)

Figure 19.48 (a) Possible loop filter used in a self-correcting DPLL and (b)
waveforms when the loop is not in lock and the clock leads the
center of the data.

high-speed self-correcting PD design, we would have to analyze each delay in the PD to determine their effect on the perfomance of the DPLL.

Jitter

Jitter, in the most general sense, for clock-recovery and synchronization circuits, can be defined as the amount of time the regenerated clock varies once the loop is locked. Figure 19.49a shows the idealized case when the *clock* doesn't jitter, while Fig. 19.49b shows the actual situation where the clock-rising edge moves in time (jitters). In these figures, the oscilloscope is triggered by the rising edge of the *data*.

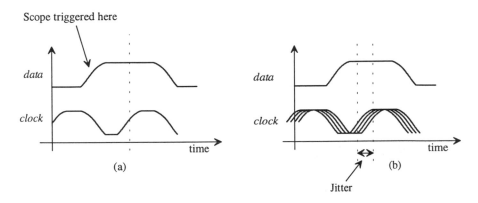

Figure 19.49 (a) Idealized view of clock and data without jitter and (b) with jitter.

In the following discussion, we will neglect power supply and oscillator noise, that is, we will assume that the oscillator frequency is an exact number that is directly related to the VCO input voltage. In the section following this one, we will cover delay-locked loops and will further discuss the limitations of the VCO.

Consider using the charge pump with the self-correcting PD shown in Fig. 19.50. When the loop is locked (from Fig. 19.47), both increase and decrease occur (with the same width) for every transition in the incoming data. Note that unlike a PFD/charge pump combination where the output of the charge pump remains unchanged when the loop is locked, the self-correcting PD/charge-pump combination will generate a voltage ripple on the input of the VCO (similar to the XOR PD with RC or Active PI filter)[4]. Let's assume that this ripple is 10 mV and use the values for VCO gain and frequencies given in Ex. 19.5 to illustrate the resulting jitter introduced into the output clock. The change in output frequency resulting from this ripple is $10\text{mV}\cdot(628 \times 10^3 \text{ radians/V·s})\cdot(1/2\pi)$ or 1 kHz. This means that the output of the

---

[4] Of course, the self-correcting PD should not be used in most frequency synthesis applications. Similarly, the PFD should not be used in most data-recovery applications.

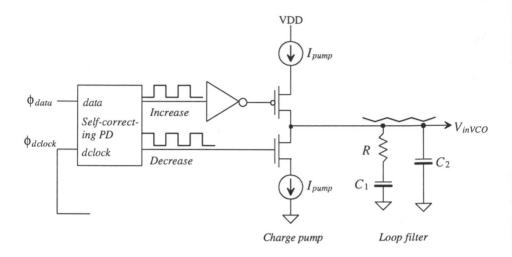

Figure 19.50 Self-correcting PD with charge pump output.

DPLL will vary from 255.5 to 256.5 kHz or, in terms of a jitter specification the clock jitter is 15 ns.

From this example, data dependent jitter can be reduced by

1. *Reducing the gain of the VCO.* Ripple on the input of the VCO has less of an effect on the output frequency. The main disadvantage, in the most general sense, of reducing the gain of the VCO is that the range of frequencies the DPLL can lock up on is reduced. Also, the VCO gain strongly affects the ability to fabricate the VCO without postproduction tuning.

2. *Reducing the bandwidth of the loop filter.* This has the effect of reducing the actual ripple on the input of the VCO. The main drawback here is the increase in chip real estate needed to realize the larger components used in the filter.

3. *Reducing the gain of the PD.* This also has the effect of reducing the ripple on the input of the VCO. This method is, in general, the easiest when using the charge pump since reducing the gain is a simple matter of reducing I_{pump}. Substrate noise can become more of a factor in the design.

Another way of stating the above methods of reducing jitter is simply to say that the forward loop gain of the DPLL, that is, $K_{PD}K_FK_{VCO}$, should be made small. The main problems with using small forward loop gain are the reductions in lock range and pull-in range coupled with the associated increases in pull-in and lock times.

We have not discussed jitter reduction in DPLLs employed in frequency synthesizers or clock synchronizers where the input is a clock signal rather than a bit stream. In these circuits the VCO performance, in general, is the limiting factor.

19.5 Delay-Locked Loops

Problems with PLL output jitter resulting from the VCO output frequency changing (often called oscillator or phase noise) with a constant input voltage (V_{invCO} = constant) has led to the concept of a delay-locked loop (DLL). Figure 19.51 shows the basic block diagram of a DLL. *Assuming* a reference clock is available at exactly the correct frequency, the input data are delayed through a voltage-variable delay line (VVDL) a time t_o until it is synchronized with the reference clock. Jitter is reduced by using an element, the VVDL, that does not generate a signal (like the VCO did). The transfer function ϕ_{clock}/ϕ_{out} is zero[8] (the phase of the reference clock is taken as the reference for the other signals in the DLL, i.e., ϕ_{clock} = 0), so that oscillator noise and the resulting jitter are not factors in DLL design. The jitter considerations discussed in the last section, however, are still a concern since any ripple on the output of the loop filter will cause jitter.

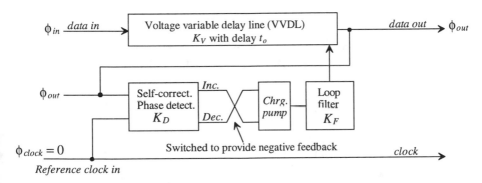

Figure 19.51 Block diagram of a delay-locked loop.

The phase (in radians) of the input data is related to the phase of the output data by

$$\phi_{out} = \phi_{in} + t_o \cdot \frac{2\pi}{T_{clock}} \qquad (19.62)$$

where T_{clock} is the period of the reference clock (or half of the period of the *data in* for a string of alternating ones and zeros). The gain of the VVDL can be written in terms of the delay, t_o, by

$$t_o = K_V \cdot V_{outfilter} \qquad (19.63)$$

where K_V has units of seconds/V and $V_{outfilter}$ is the voltage input to the VVDL from the loop filter. The minimum and maximum delays of the VVDL should, in general, lie between $T_{clock}/2$ and $1.5T_{clock}$ for proper operation. The output of the loop filter can be written as

$$V_{outfilter} = \phi_{out} \cdot K_D \cdot K_F \qquad (19.64)$$

The overall transfer function may now be written as

$$\frac{\phi_{out}}{\phi_{in}} = \frac{1}{1 - K_D K_F K_V \cdot \omega_{clk}} \tag{19.65}$$

where $\omega_{clk} = 2\pi/T_{clock}$. The gain of the self-correcting PD with charge-pump output, with the help of Fig. 19.52 and noting that *Increase* and *Decrease* can occur at the same time, is

$$K_D = -\frac{I_{pump}}{\pi} \text{ (amps/radian)} \tag{19.66}$$

The negative sign is the result of switching the *Increase* and *Decrease* outputs of the self-correcting PD before connection to the charge pump in order to provide negative feedback around the loop. Another benefit of the DLL is that the loop filter can be a simple capacitor, which results in a first-order feedback loop, that is,

$$K_F = \frac{1}{sC_1} \tag{19.67}$$

The transfer function of the DLL relating the input data to the time-shifted output is

$$\frac{\phi_{out}}{\phi_{in}} = \frac{1}{1 + \frac{I_{pump}}{\pi} \cdot \frac{1}{sC_1} \cdot K_V \cdot \omega_{clk}} = \frac{s}{s + K_V \cdot \frac{2I_{pump}}{C_1 T_{clock}}} \tag{19.68}$$

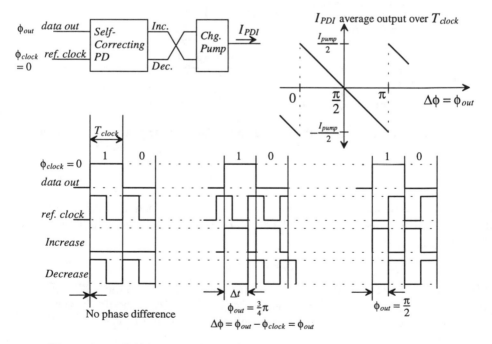

Figure 19.52 Self-correcting PD output for various inputs (assuming input data is a string of alternating ones and zeros).

We know that the frequency of the reference clock must be exactly related to the frequency of the input data. However, there will exist instantaneous changes in the phase of the input data which the output of the DLL should follow. Modeling instantaneous changes in ϕ_{in} by $\Delta\phi_{in}/s$ (a step function with an amplitude of $\Delta\phi_{in}$), we get a change in output phase given by

$$\Delta\phi_{out} = \frac{\Delta\phi_{in}}{s + K_V \cdot \frac{2I_{pump}}{C_1 T_{clock}}} \qquad (19.69)$$

The time it takes the DLL to respond to an input step in phase is simply

$$T_r = 2.2 \cdot \frac{C_1 T_{clock}}{K_V \cdot 2I_{pump}} = \text{number of clock cycles} \cdot T_{clock} \qquad (19.70)$$

This time can be decreased by making C_1/I_{pump} small, which from our discussion in the last section, has the result of increasing the output pulse jitter, that jitter that is independent of the oscillator (i.e., dependent on the input data pattern) in a standard PLL. Decreasing C_1/I_{pump} increases the ripple on the control voltage of the VVDL. Similarly, increasing K_V (the time/volt delay of the VVDL) will increase jitter since a given ripple on the control voltage of the VVDL will have a larger effect on the delay. Again, tradeoffs must be made between response to input variations and output jitter.

Delay Elements

The VVDL is an important component of the DLL. Figure 19.53a shows the basic implementation of a VVDL using adjustable delay inverters. The last two inverters in the VVDL are used to ensure that clean digital signals are output from the VVDL.

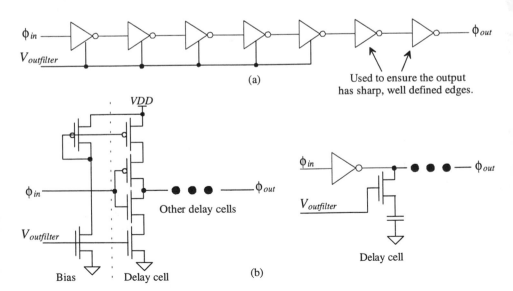

Figure 19.53 (a) VVDL made using inverter delay cells and (b) possible delay cells.

Figure 19.53b shows the circuit schematics for possible delay elements [5, 9-10]. The first delay element should be recognized as a current-starved inverter discussed earlier in the chapter. The second delay element is nothing more than an inverter with a variable load. In practice, these delay elements are rarely used because of the susceptibility to noise and power supply variations but rather fully-differential delay elements are used[5].

Figure 19.54 shows the connection of a fully differential VVDL. Using any number of stages, a fully differential VCO can be implemented with the delay elements (and the proper feedback), while using an even number with the inverting (noninverting) output fed back and connected to the noninverting (inverting) input in-phase (I) and quadrature (Q) signals can be generated [5].

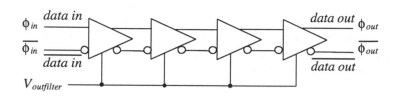

Figure 19.54 Implementation of a fully differential VVDL.

A simple differential delay element is seen in Fig. 19.55a. This circuit is basically a fully differential inverter. The p-channel loads are long L so that the change in $V_{outfilter}$ can more linearly adjust the delay of the element. The signals out of this element are not truly differential because of the unequal high-to-low and low-to-high times so that good power supply rejection is difficult to achieve. The delay element

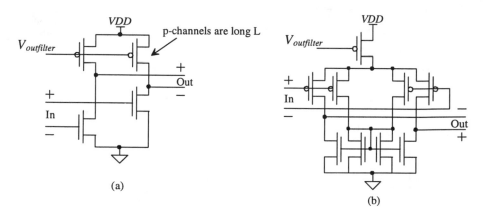

Figure 19.55 (a) A fully differential inverter and (b) a diff-amp-based delay element. Note that the delay element of (a) cannot be used in a differential VCO.

shown in Fig. 19.55b has much better power supply rejection because of the symmetry of the circuit. This circuit will sink and source the same amount of current into a load capacitance (the input capacitance of the next stage) resulting in equal rise and fall times through the delay line. Another example of a delay stage is shown in Fig. 19.56 [10].

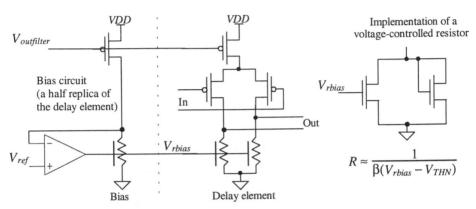

Figure 19.56 A differential delay element based on a voltage-controlled resistor. The bias circuit adjusts the value of the resistors used in the delay elements to sink the current sourced by the p-channel MOSFETs.

Clock Synchronization

Consider the block diagram of the interconnection of a CPU and slave processor [11] shown in Fig. 19.57. The DLL is used in this diagram to synchronize the clock in the master CPU with the clock used in the slave processor. This is very important when common external data/address busses are shared between processors or a processor and memory. Since the "data" input to the DLL is a clock signal that is, it is not data but rather a square-wave, the design of the DLL can be simplified. In particular, we can design the DLL so that it uses only digital components. No analog loop filter or VVDL is needed. Note that simply routing the clock signal around the CPUs (not using a

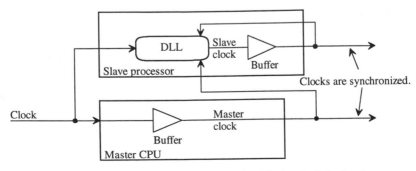

Figure 19.57 Synchronizing the master CPU clock with the clock in the slave processor.

DLL) does not guarantee that the data coming out of the processors are synchronized since the internal delays of the master and slave will not be the same (they are skewed).

The block diagram of a digital-only DLL useful in clock synchronization circuits is shown in Fig. 19.58 [11]. A seven-bit counter is used, in place of an analog loop filter, to select, via the decoder, which tap of the delay line is fed to the output buffers.

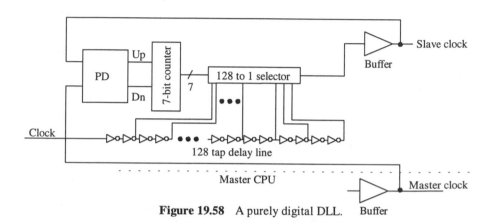

Figure 19.58   A purely digital DLL.   Buffer

REFERENCES

[1]   R. E. Best, *Phase-Locked Loops, Theory, Design and Applications,* McGraw-Hill, 2nd ed., 1993. ISBN 0-07-911386-9.

[2]   F. M. Gardner, "Charge-Pump Phase-Lock Loops," *IEEE Transactions on Communications,* COM-28, No. 11, pp. 1849-1858, November 1980.

[3]   P. R. Gray and R. G. Meyer, *Analysis and Design of Analog Integrated Circuits,* 3rd ed., John Wiley and Sons, 1993. ISBN 0-471-57495-3.

[4]   B. Keeth, R. J. Baker, and H. W. Li, "CMOS Transconductor VCO with Adjustable Operating and Center Frequencies," *Electronics Letters,* Vol. 31, No. 17, pp. 1397-1398, August 1995.

[5]   B. Razavi, *Monolithic Phase-Locked-Loops and Clock Recovery Circuits,* IEEE Press, 1996. ISBN 0-7803-1149-3.

[6]   S. Haykin, *An Introduction to Analog and Digital Communications,* John Wiley and Sons, 1989. ISBN 0-471-85978-8.

[7]   C. R. Hogge, Jr., "A Self Correcting Clock Recovery Circuit," *IEEE Journal of Lightwave Technology,* Vol. LT-3, pp. 1312-1314, December 1985.

[8]   T. H. Lee and J. F. Bulzacchelli, "A 155-MHz Clock Recovery Delay- and Phase-Locked Loop," *IEEE Journal of Solid-State Circuits,* Vol. SC-27, pp. 1736-1746, December 1992.

[9]   M. G. Johnson and E. L. Hudson, "A Variable Delay Line PLL for CPU - Coprocessor Synchronization," *IEEE Journal of Solid-State Circuits*, Vol. SC-23, pp. 1218-1223, October, 1988.

[10]   I. A. Young, J. K. Greason, and K. L. Wong, "A PLL Clock Generator with 5 to 110 MHz of Lock Range for Microprocessors," *IEEE Journal of Solid-State Circuits*, Vol. SC-27, pp. 1599-1607, November 1992.

[11]   A. Efendovich, Y. Afek, C. Sella, and Z. Bikowsky, "Multifrequency Zero-jitter Delay-Locked Loop," *IEEE Journal of Solid-State Circuits*, Vol. 29, pp. 67-70, January, 1994.

PROBLEMS

19.1   Referring to Fig. 19.8, sketch the phase difference between *dclock* and *data* when the width of a data bit is 1 µs and the time between the leading edges of *dclock* and *data* is 0.2 µs. What is the phase difference? What is the average PD output voltage, V_{PDout} for *data* that is an alternating string of ones and zeros.

19.2   Estimate the phase difference and sketch the PFD outputs (both *Up* and *Down*) for the data shown in Fig. P19.2.

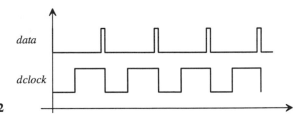

Figure P19.2

19.3   Design and simulate the operation of a 10 MHz VCO (at $V_{inVCO} = VDD/2$) using a current-starved VCO.

19.4   Repeat Problem 19.3 for a source coupled VCO design.

19.5   Derive Eq. (19.30).

19.6   Sketch a simple circuit to implement a divide by 2 in the feedback loop of a DPLL.

19.7   Sketch the phase relationship, in the time domain, between *data* and *dclock* in Ex. 19.3 when the input data rate is 9.04 Mbits/s. Use the results shown in Fig. 19.25.

19.8   Derive Eq. (19.41). State all assumptions.

19.9   Repeat Ex. 19.5 for an input of 1 KHz and an output of 3 kHz. Sketch the logic diagram for the divide by 3 in the feedback loop.

19.10 Verify, in Fig. 19.40, that the resistors average v_{+peak} and v_{-peak}.

19.11 Describe a scheme to implement the return to zero data format shown in Fig. 19.42.

19.12 Using the CN20 CMOS process, design an edge detector that generates outputs with a width of approximately 5 ns.

19.13 What is the optimum delay that should be used in an edge detector? Use the waveforms in Fig. 19.46 to illustrate the resulting PD output when this optimum delay is used.

19.14 Sketch *Increase* and *Decrease*, using the NRZ data in Fig. 19.47, when the rising edge of *clock* is aligned with the beginning of the bit interval.

19.15 Sketch the signals, similar to Fig. 19.52, when the output phase of the DLL is $\pi/4$.

19.16 An important circuit in a receiver is the input comparator or decision circuit. This circuit changes the analog data, with possible amplitudes of a few hundreds of mV, into digital data. Figure P19.16 shows both a single-ended output (see also Figure 26.8) and a differential output receiver useful in high-speed digital design. Using SPICE with v_{in-} fixed at 2.5V determine the minimum voltage swing on v_{in+} (centered around 2.5 V) that will cause the outputs to swing from <1 V to > 4 V. What are the unloaded propagation delays of these circuits?

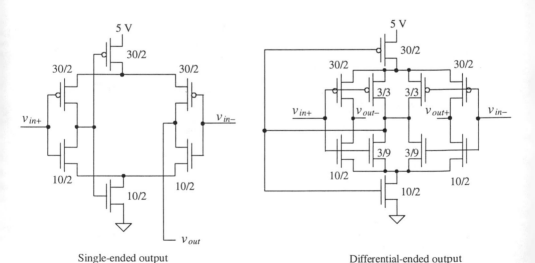

Single-ended output   Differential-ended output

Figure P19.16

Part III

CMOS Analog Circuits

Current Sources and Sinks

The current source/sink is a basic building block in CMOS IC design and is used extensively in analog integrated circuit design. Ideally, the output impedance of a current source/sink should be infinite and capable of generating or drawing a constant current over a wide range of voltages. However, finite values of r_o, and a limited output swing required to keep devices in saturation will ultimately limit the performance of the mirror. The circuits in this chapter will vary in complexity and performance as the design, simulation, and layout issues associated with MOS current sources and sinks are examined.

20.1 The Current Mirror

Figure 20.1 shows the basic current mirror [1-3]. A current flows through M1 corresponding to V_{GS1}. Since $V_{GS1} = V_{GS2}$, ideally the same current, or a multiple of the current in M1, flows through M2. If the MOSFETs are the same size, the same drain current flows in each MOSFET, provided M2 stays in the saturation region. The current I_{D1} is given by

$$I_{D1} = \frac{\beta_1}{2} \cdot (V_{GS1} - V_{THN})^2 \qquad (20.1)$$

while the output current, assuming M2 is in saturation, flowing in M2 is

$$I_{D2} = I_o = \frac{\beta_2}{2}(V_{GS2} - V_{THN})^2 \qquad (20.2)$$

Since $V_{GS1} = V_{GS2}$, the ratio of the drain currents is given by

$$\frac{I_{D2}}{I_{D1}} = \frac{\frac{W_2}{L_2}}{\frac{W_1}{L_1}} = \frac{W_2 L_1}{W_1 L_2} = \frac{\beta_2}{\beta_1} \qquad (20.3)$$

This equation shows how to adjust the W/L ratio of the two devices to achieve the desired output current, I_{D2}. This equation, however, does not show how the output

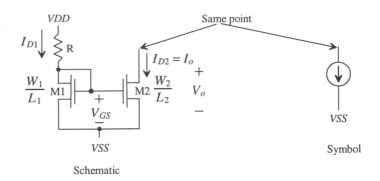

Figure 20.1 The basic current mirror schematic and symbol.

current will change with the voltage across M2, V_o. The reference drain current, I_{D1} in Fig. 20.1, is determined by solving

$$I_{D1} = \frac{VDD - V_{GS} - VSS}{R} = \frac{KP \cdot W_1}{2L_1} \cdot (V_{GS1} - V_{THN})^2 \qquad (20.4)$$

The minimum voltage, V_{min}, across the current sink is set by the requirement that M2 remain in saturation, that is, $V_{min} = V_{DS,SAT} = V_{GS} - V_{THN}$. The output resistance of the current source is simply the output resistance of M2, or

$$r_{o2} = \frac{1}{\lambda I_o} = \frac{1}{\lambda I_{D2}} \qquad (20.5)$$

Looking at Eqs. (20.3) and (20.4), the five variables, L_1, L_2, W_1, W_2, and V_{GS}, are available to the designer to set the currents. Normally, when designing CMOS current mirrors, the values for V_{GS} and L are selected before solving for W to get the desired current. Picking the lengths of all MOSFETs, used in current sources, the same size simplifies Eq. (20.3) to,

$$\frac{I_{D2}}{I_{D1}} = \frac{W_2}{W_1} \qquad (20.6)$$

For digital design we used the minimum length, 2.0 µm for CN20, for our MOSFET switches because we did not care about the output resistance of the MOSFET in the saturation region. For analog design, however, it is extremely important to keep the output resistance as high as possible. It is also desirable to reduce the effects of channel length and mobility modulation discussed in Chs. 5, 6, and 9. These effects are reduced by increasing the channel length of the devices. A general design rule is to set the length of the MOSFETs used in analog applications to two to five times the minimum drawn gate length. Appendix A shows output resistance for different-sized MOSFETs. If we use Eq. (20.5) with this data, we can conclude that a 5 µm device length will give about the same output resistance as a 10 µm device. The sum of the mobility modulation and channel length modulation parameters, λ, for the n- and p-channel MOSFETs is approximately 0.06 V^{-1}. As the plots show, the values of λ, and thus the output resistances, are very dependent on the biasing conditions. From these considerations we can make the following statement: *When designing with the CN20 process, we will assume that the length, L, of the MOSFETs used in analog applications is at least 5 μm.*

The biasing of current mirrors, or any other analog circuit, is greatly simplified if we also assume, and design for, a specific gate-source voltage[1], V_{GS}. Setting V_{GS} (or V_{SG} for the p-channel) close to the threshold voltage, approximately 0.8 V for CN20, results in very large devices, while setting V_{GS} significantly larger than the threshold voltage causes the transistor to enter the triode region too early. An acceptable difference between V_{GS} and V_{THN}, sometimes referred to as the excess gate voltage, ΔV, is several hundred millivolts. Using this general design rule we can make the following assumption: *When designing current sources/sinks with the CN20 process, the gate-source voltage will be initially assumed to be 1.2 V.*

These rules are a starting point in the design. They allow the designer to concentrate on two variables in the basic current mirror (the widths of the devices) rather than five. The V_{GS} is normally adjusted to obtain a desired characteristic (such as minimum voltage across the current source/sink).

Example 20.1
Design a current sink using $VDD = - VSS = 2.5$ V to sink a current of 10 μA. Estimate the minimum voltage across the current source and the output resistance. Simulate the operation of your design with SPICE.

The basic design is shown in Fig. Ex20.1. Here we have selected $V_{GS} = 1.2$ V and the length of the devices as 5 μm. The value of R, assuming $I_{D1} = I_{D2} = 10$ μA, is determined by solving Eq. (20.4), or

$$R = \frac{2.5 - 1.2 - (-2.5)}{10\ \mu A} = 380\ k\Omega$$

We can solve for the width of both M1 and M2 at the same time,

$$I_{D2} = 10\ \mu A = \frac{KP}{2}\frac{W}{L}(V_{GS} - V_{THN})^2 = \frac{50\frac{\mu A}{V^2}}{2}\frac{W}{5\ \mu m}(1.2 - 0.83)^2$$

which gives $W_1 = W_2 = 14.61$ μm which we round up to 15 μm. The requirement for M2 to stay in the saturation region is

$$V_{DS2} \geq V_{GS2} - V_{THN} = 1.2 - 0.83 = 0.37\ V\ = \text{excess gate voltage}$$

As long as the drain of M2 is approximately - 2.13 V or greater, M2 will remain in the saturation region. This result is *important!* When the MOSFETs are

---

[1] This assumption is similar to assuming that the forward voltage drop of a base-emitter junction is 0.7 V. It may be 0.6 or 0.8 V, but we still assume 0.7 V to simplify the biasing of bipolar analog circuits.

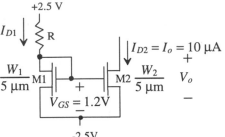

Figure Ex20.1

biased with a V_{GS} of 1.2 V, the minimum voltage from drain to source is 0.37 V (the excess gate voltage) for operation in the saturation region.

The small-signal output resistance of the current source is then approximated using Eq. (20.5), by

$$r_o = \frac{1}{\lambda I_o} = \frac{1}{0.06 \cdot 10\ \mu A} = 1.67\ \text{MEG}$$

For AC small-signal analysis, simply replace the current sources with resistors equivalent to the output impedance connected to ground. The output resistance can also be thought of as the resistance looking into the drain of M2. SPICE simulation results are shown in Fig. 20.2. Notice the effect of the finite output resistance. ■

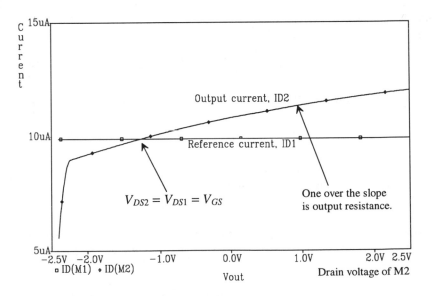

Figure 20.2 Example 20.1 SPICE simulation results.

Example 20.2

Design four current sinks with values 20, 30, 50, and 70 μA. What is the minimum voltage across each current sink? Again assume $VDD = -VSS = 2.5$V.

Following the design procedure of the previous example, we see that this problem is simply a matter of re-sizing the MOSFETs used for sinking current to get the 20, 30, 50, and 70 μA required. The schematic of the design with new sizes is shown in Fig. Ex20.2. Layout methods to avoid problems with oxide encroachment and lateral diffusion are discussed in Sec. 20.1.5. Using a single reference with several current sinks is very common in IC design. Also, note that the 1.2 V we assumed for the value of V_{GS} is not exact; it may be 1.1 or 1.3 V. In this configuration, using the reference MOSFET with its gate shorted to its drain, we observe that the small variations in V_{GS} have little effect on the 10 μA flowing in the reference MOSFET. The minimum voltage across any of the current sinks is, from the previous example, 0.37 V corresponding to a minimum voltage on the drains of –2.13 V. ∎

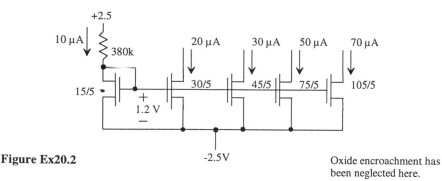

Figure Ex20.2   -2.5V   Oxide encroachment has been neglected here.

Consider the two MOSFETs shown in Fig. 20.3a, each biased with the same current I_1. Since the gate-source voltages of the devices, M1 and M2, and the drain currents are equal, it follows that the drain-source voltages of the devices must be equal, that is, $V_{GS1} = V_{GS2} = V_{DS1} = V_{DS2}$. In Fig. 20.3b, this fact is used to bias a third MOSFET, M3, that is, $V_{GS3} = V_{GS1} = V_{GS2}$. For biasing purposes, we treat the gate of M3 as if it were connected to the gates of M1 and M2. This configuration is very useful in amplifier design where M2 is used as a current source load and M3 is a common source amplifier.

Current sinks are made using n-channel MOSFETs, while current sources are made using p-channel MOSFETs. We can bias the p-channel MOSFETs in the same way we biased the n-channel MOSFETs to form a current source. A p-channel MOSFET mirror is shown in Fig. 20.4 along with its schematic symbol. Sizing the devices for a particular current follows the general design rules as well, that is, $V_{SG} = 1.2$ V and $L = 5$ μm. The following example illustrates the design of a current source.

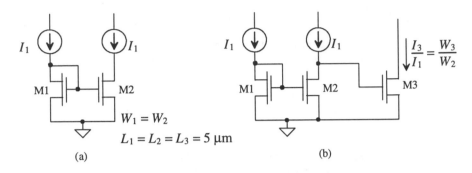

Figure 20.3 (a) The drain of M2 is at the same potential as the gate of M1 or M2 and
(b) using this to bias M3. For biasing purposes, we treat M3 as if its
gate were tied to the gates of M1 and M2.

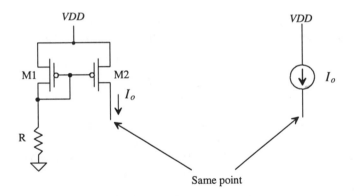

Figure 20.4 The basic current source using a p-channel mirror.

Example 20.3

Using the 10 μA n-channel reference of the previous example, design three
current sources with values of 10, 20, and 50 μA.

The schematic of the design is shown in Fig. Ex20.3. The MOSFETs M1 and
M2 are used to bias M3 at a current of 10 μA. The current that flows in M3 is
mirrored in M4, M5, and M6. To determine the width of M3, we simply solve

$$10 \text{ μA} = \frac{17\frac{\text{μA}}{\text{V}}}{2} \cdot \frac{W_3}{5 \text{ μm}}(1.2 - .91)^2 \rightarrow W_3 \approx 70 \text{ μm}$$

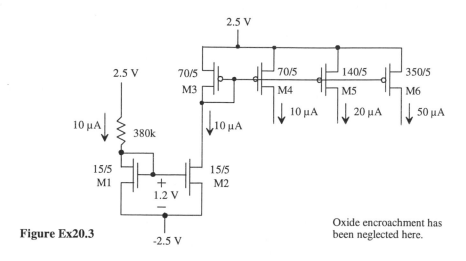

Figure Ex20.3

Oxide encroachment has
been neglected here.

The sizes of M4, M5, and M6 for supplying 10, 20, and 50 μA are 70, 140, and 350 μm, respectively. Because the p-channel transconductance parameter is less than the n-channel transconductance parameter, the widths of the p-channels, for the same current levels and same V_{SG} (V_{GS}), are wider. ■

20.1.1 The Cascode Connection

The cascode connection of basic current mirrors is shown in Fig. 20.5a. This configuration is used to increase the output resistance of a current source or sink. When biasing the cascode configuration, we begin by selecting a V_{GS} for each MOSFET, which will be assumed to be 1.2 V. Calculation of the sizes proceeds in the same manner as with the simple current mirror.

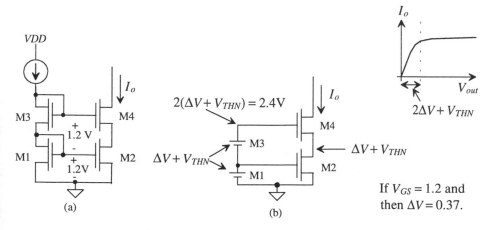

Figure 20.5 The cascode current source.

Calculating the minimum voltage across the current source begins by defining the gate-source biasing voltage in terms of the excess gate-source voltage ΔV as

$$V_{GS} = \Delta V + V_{THN} \tag{20.7}$$

Since $V_{GS} = 1.2$ V, the n-channel excess gate voltage, neglecting the body effect, is $\Delta V = 0.37$ V (see Ex. 20.1) and the p-channel excess gate voltage is 0.290 V. The gate voltage of M4 is $2(\Delta V + V_{THN}) = 2.4$ V, while the source voltage of M4 is $\Delta V + V_{THN} = 1.2$ V. The minimum voltage on the drain of M4, and thus the minimum voltage across the current source, is limited by the requirement that M4 remain in the saturation region, $V_{DS4} \geq V_{GS4} - V_{THN}$ or $V_{D4} \geq \Delta V + (\Delta V + V_{THN}) = 1.57$ V. The minimum voltage across the cascode current mirror is significantly larger than the minimum voltage across the basic current mirror ($\Delta V = 0.37$) calculated in Ex. 20.1.

If the voltage on the gate of M4 can be reduced to $2\Delta V + V_{THN}$, then the voltage on the drain of M2 becomes $\Delta V = 0.37$ V, and the minimum voltage across the current source is reduced to $2\Delta V = 0.74$ V. The circuit shown in Fig. 20.6a illustrates this idea [4]. A battery (M6) is used to drop the potential at the gate of M4 down to $2\Delta V + V_{THN}$. This reduces the voltage on the drain of M4 to $2\Delta V$ before M2 and M4 enter the triode region. Implementation of this current source is shown in Fig. 20.6b. The MOSFET M3 is re-sized to generate $3\Delta V + 2V_{THN}$, that is, $V_{GS3} = 2\Delta V + V_{THN} = 1.57$V, on its gate while M6 is used to drop $\Delta V + V_{THN}$ so that the gate voltage of M4 becomes $2\Delta V + V_{THN}$.
To accomplish this, the width of M3 is made one-fourth the size of the other MOSFETs. Note that a MOSFET with its gate and drain tied together being fed by a constant current (M1 and M3 in Fig. 20.6) behaves as a constant DC potential (a battery).

The output resistance of the cascode current mirror can be determined using the small-signal models derived in Ch. 9 and shown in Fig. 20.7. A test voltage is applied to the drain of M4. Summing the currents at the drain of M4 and neglecting the body effect results in

$$i_t = g_{m4}v_{gs4} + \frac{v_t - (-v_{gs4})}{r_{o4}} \tag{20.8}$$

where

$$v_{gs4} = -i_t \cdot r_{o2} \tag{20.9}$$

The output resistance of the cascode current source is given by

$$R_o = r_{o4}(1 + g_{m4}r_{o2}) + r_{o2} \approx g_{m4}r_o^2 \tag{20.10}$$

where $r_o = r_{o2} = r_{o4}$. This result can be applied to the more general case shown in Fig. 20.8. The small-signal resistance to ground looking into the drain of the MOSFET is

$$R_o = r_o(1 + g_m R) + R \tag{20.11}$$

which is sometimes approximated by

$$R_o \approx r_o(1 + g_m R) \tag{20.12}$$

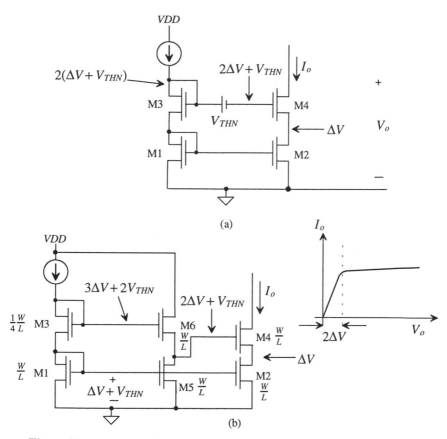

Figure 20.6 Biasing of the cascode current source for lower minimum voltage
across the current source (body effects neglected).

The output resistance of the triple cascode shown in Fig. 20.9 can be written using these
results as

$$
R_o = r_{o6} \left[1 + g_{m6} \overbrace{(r_{o4}(1 + g_{m4}r_{o2}) + r_{o2}))}^{R \text{ looking into drain of M4}} \right] + \overbrace{r_{o4}(1 + g_{m4}r_{o2}) + r_{o2}}^{R \text{ looking into drain of M4}} \approx g_{m6}g_{m4}r_o^3 \quad (20.13)
$$

When performing an AC small-signal analysis, the cascode connection is simply
replaced with a resistor to ground of value given by Eq. (20.10) for the double cascode
or Eq. (20.13) for the triple cascode. Increasing the output resistance of a current
source or sink is simply a matter of stacking up the MOSFETs until the desired output
resistance is reached. The obvious limitation to stacking a large number of cascode
devices is the increase in the minimum voltage across the current source/sink needed to
keep all devices in the saturation region. Other current mirrors that improve the output
impedance using negative feedback will be examined later in the chapter.

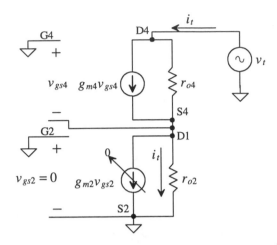

Figure 20.7 Small-signal model of the cascode current source shown in Fig. 20.5.

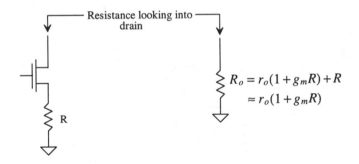

Figure 20.8 Modeling the resistance looking into the drain of a MOSFET.

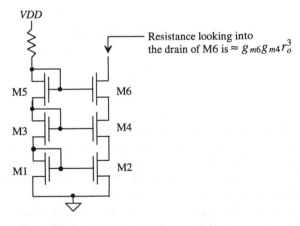

Figure 20.9 The triple cascode current source.

Example 20.4

Repeat Ex. 20.1 using a double cascode current mirror.

The resistor value is calculated by

$$R = \frac{VDD - 2V_{GS} - VSS}{I_{D1}} = \frac{2.5 - 2.4 - (-2.5)}{10 \; \mu A} = 260 \; k\Omega$$

where we have assumed that $I_{D1} = I_{D2} = 10 \; \mu A$ and $V_{GS} = 1.2 \; V$. The sizes of the devices are, from Ex. 20.1, $W = 15 \; \mu m$ and $L = 5 \; \mu m$. Calculation of the output resistance, R_o, begins by calculating the transconductance of the MOSFETs:

$$g_m = \sqrt{2\beta_n I_D} = \sqrt{2 \cdot 50\frac{\mu A}{V^2} \cdot \frac{15}{5} \cdot 10 \; \mu A} = 55 \; \frac{\mu A}{V}$$

and the output resistance of a single MOSFET is $r_o = 1/\lambda I_D = 1/(10\mu A)(0.06) = 1.67 \; M\Omega$. The output resistance of the current source is then

$$R_o \approx g_m r_o^2 = 55\frac{\mu A}{V}(1.67 \; M\Omega)^2 = 152 \; M\Omega$$

The minimum voltage across the current source is $2\Delta V + V_{THN}$ or 1.57 V. SPICE simulation results are shown in Fig. Ex20.4 along with a schematic. Comparing these simulation results to those of Ex. 20.1, we see the increase in the output resistance. Notice that the body effect (increase in the threshold voltage of M3/M4) causes the currents to be less than the 10 μA we designed for. ■

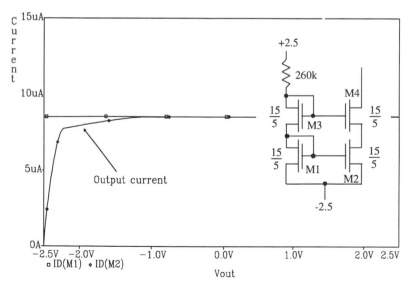

Figure Ex20.4

20.1.2 Sensitivity Analysis

Also of concern when designing current sources and sinks is the sensitivity of the output current to some input, such as the DC power supplies, or the value of a component, such as a resistor. The sensitivity of I_o to VDD, for the simple current mirror in Fig. 20.1, can be defined by

$$S_{VDD}^{I_o} = \lim_{\Delta VDD \to 0} \frac{\frac{\Delta I_o}{I_o}}{\frac{\Delta VDD}{VDD}} = \frac{VDD}{I_o} \cdot \frac{\partial I_o}{\partial VDD} \tag{20.14}$$

The sensitivity of the output current to VDD is given, using Eq. (20.4), and assuming that $I_o = I_{D1}$ and the change in V_{GS} with VDD is small, by

$$S_{VDD}^{I_o} \approx \frac{VDD}{I_o} \cdot \frac{1}{R} \tag{20.15}$$

This equation is useful in answering the question "If VDD changes by 10 percent how much will the output current change?" The percentage change in I_o as a function of the percentage change in VDD is given by

$$\frac{\Delta I_o}{I_o} = S_{VDD}^{I_o} \cdot \frac{\Delta VDD}{VDD} \tag{20.16}$$

Example 20.5
Estimate the variation in I_o, for the current mirror of Ex. 20.1, for VDD changing from 2.4 to 2.6 V. Simulate the sensitivity using PSPICE. Connect the output of the current source, the drain of M2, to a battery of value -1.3 V (approximately the same potential as the drain of M1).

The sensitivity is given by Eq. (20.15) as

$$S_{VDD}^{I_o} = \frac{2.5}{10\mu A} \cdot \frac{1}{380k} = 0.658$$

The percentage change in VDD is 8 percent so that, using Eq. (20.16), the percentage change in the output current is 5.3 percent, that is,

$$\frac{\Delta I_o}{I_o} = 0.658 \cdot \frac{0.2}{2.5} = 0.053 = 5.3 \text{ percent}$$

For an $I_o = 10 \mu A$ the change in the output current is $\Delta I_o = 0.53 \mu A$. In other words, when $VDD = 2.4V$, $I_o = 9.73 \mu A$ and when $VDD = 2.6$ V, $I_o = 10.27 \mu A$. The PSPICE netlist used in this example is as follows.

```
Title
M1        1 1 2 2 CMOSNB  L=5u W=15u
M2        5 1 2 2 CMOSNB  L=5u W=15u
R1        3 1 380k TC1=0.002
VDD       3 0       DC 2.5
Vdrain2 5 0       DC -1.3
VSS       2 0       DC -2.5
```

```
.MODEL CMOSNB NMOS LEVEL=4
BSIM model parameters
.sens I(vdrain2)
.probe
.DC Vdrain2 -2.5 2.5 .01
.end
```

The output sensitivity results are

DC SENSITIVITIES OF OUTPUT I(vdrain2)

ELEMENT NAME	ELEMENT VALUE	ELEMENT SENSITIVITY (AMPS/UNIT)	NORMALIZED SENSITIVITY (AMPS/PERCENT)
R1	3.800E+05	2.441E-11	9.277E-08
VDD	2.500E+00	-2.465E-06	-6.161E-08
Vdrain2	-1.200E+00	-8.302E-07	9.962E-09
VSS	-2.500E+00	3.295E-06	-8.237E-08

The sensitivity, defined by PSPICE, is slightly different than previously discussed. For example, the value of the output current I(vdrain2) given above is:

$$\frac{\partial I_o}{\partial VDD} = \frac{\partial I(vdrain2)}{\partial VDD} = -2.465 \times 10^{-6} \frac{A}{V}$$

which is close to our approximation that

$$\frac{\partial I_o}{\partial VDD} \approx \frac{1}{R} = 2.632 \times 10^{-6} \frac{A}{V}$$

The sign difference is due to the direction of the current flow defined as positive by PSPICE. The normalized sensitivity of the output current to *VDD*, defined by PSPICE and shown in the fourth column of the PSPICE output file, is given by

$$VDD \cdot \frac{\partial I_o}{\partial VDD} \cdot \frac{1}{100} = \frac{Amps}{\%}$$

Performing a sensitivity analysis allows the designer to determine which components have the largest effect on the output variable. The sensitivity of the output current is highly dependent on the voltage across the current source, that is, the potential at the drain of M2. Computer simulations are almost a necessity in determining the sensitivity of an output variable to a circuit component in large circuits. ■

20.1.3 Temperature Analysis

The temperature coefficient of the simple current mirror is given by

$$TC(I_o) = \frac{1}{I_o} \cdot \frac{\partial I_o}{\partial T} = \frac{1}{T} \cdot S_T^{I_o} \tag{20.17}$$

The output current is given by Eqs. (20.3) and (20.4) as

$$I_o = \frac{W_2 L_1}{W_1 L_2} \cdot \frac{VDD - V_{GS} - VSS}{R} \tag{20.18}$$

where we have not assumed that $I_{D1} = I_{D2} = I_o$. If $I_{D1} = I_{D2}$, the ratio of the widths and lengths is unity in this equation (which was assumed for the sensitivity analysis of the previous section). The temperature coefficient is then given by

$$TC(I_o) = -\frac{1}{I_o} \left[\frac{W_2 L_1}{W_1 L_2} \cdot \frac{1}{R} \frac{\partial V_{GS}}{\partial T} + \frac{I_o}{R} \cdot \frac{\partial R}{\partial T} \right] \tag{20.19}$$

From Ch. 7 we know that $\frac{1}{R} \frac{\partial R}{\partial T}$ is the temperature coefficient for a resistor, which, for the n+ resistor, is 2,000 ppm/°C. The gate-source voltage of M2 can be expressed by

$$V_{GS} = V_{THN} + \sqrt{\frac{I_o}{\beta_2/2}} = V_{THN} + \sqrt{\frac{W_2 L_1}{W_1 L_2} \cdot \frac{(VDD - V_{GS} - VSS)}{R \cdot KP(T) \frac{W_2}{2 \cdot L_2}}} \tag{20.20}$$

or, if $VDD - VSS \gg V_{GS}$, then

$$\frac{\partial V_{GS}}{\partial T} = \frac{\partial V_{THN}}{\partial T} + \sqrt{\frac{2L_1}{W_1} \cdot \frac{(VDD - VSS)}{R \cdot KP(T)}} \cdot \left(-\frac{1}{2}\right) \cdot \left[\frac{1}{KP(T)} \frac{\partial KP(T)}{\partial T} + \frac{1}{R} \frac{\partial R}{\partial T} \right] \tag{20.21}$$

The temperature dependence of the threshold voltage was given in Ch. 9. The change in threshold voltage with temperature is given by

$$\frac{\partial V_{THN}}{\partial T} = V_{THN} \cdot TCV_{THN} \tag{20.22}$$

where $TCV_{THN} \approx -3,000 \frac{ppm}{°C}$ or for $V_{THN} = 0.83$ V, the change in threshold voltage with temperature is approximately -2.4 mV/°C. The temperature dependence of the transconductance parameter was modeled, again from Ch. 9, by

$$KP(T) = KP(T_0) \cdot \left(\frac{T}{T_0}\right)^{-1.5} \text{(T is in °K)} \tag{20.23}$$

The change in the transconductance parameter with temperature is given by

$$\frac{\partial KP(T)}{\partial T} = KP(T_0)(-1.5)\left(\frac{T}{T_0}\right)^{-2.5} \cdot \frac{1}{T_0} \rightarrow \frac{1}{KP(T)} \cdot \frac{\partial KP(T)}{\partial T} = \frac{-1.5}{T} \tag{20.24}$$

The change in gate-source voltage, given by Eq. (20.21), can now be written as

$$\frac{\partial V_{GS}}{\partial T} = V_{THN} \cdot TCV_{TH} - \left(\frac{1}{2}\right) \cdot \sqrt{\frac{2L_1}{W_1} \cdot \frac{VDD - VSS}{R \cdot KP(T)}} \left[\frac{1}{R} \frac{\partial R}{\partial T} - \frac{1.5}{T} \right] \tag{20.25}$$

or, substituting into Eq. (20.21),

$$TC(I_o) = -\frac{1}{I_o} \cdot \frac{W_2 L_2}{W_1 L_1} \left[\frac{V_{THN} \cdot TCV_{TH}}{R} - \frac{1}{R} \sqrt{\frac{L_1}{W_1} \frac{VDD - VSS}{2RKP(T)}} \left[\frac{1}{R} \frac{\partial R}{\partial T} - \frac{1.5}{T} \right] \right] - \frac{1}{R} \frac{\partial R}{\partial T}$$

$$\tag{20.26}$$

This result will be applied in the following example.

Example 20.6
Determine the temperature coefficient of the simple current mirror of Ex. 20.1. Compare the temperature coefficient determined with a PSPICE simulation.

The temperature coefficient of the simple current mirror around the temperature 300 °K is determined using Eq. (20.26) by

$$TC(I_o) = \frac{-1}{10\mu} \left[\overbrace{\frac{-0.0024}{380 \times 10^3}}^{-6.3 \times 10^{-9}} - \overbrace{\frac{1}{380 \times 10^3} \sqrt{\frac{5}{6 \cdot 380 \times 10^3 \cdot 50 \times 10^{-6}}} \left(0.002 - \frac{1.5}{300} \right)}^{-1.65 \times 10^{-9}} \right] - .002$$

or the temperature coefficient of the output current at 27 °C is

$$TC(I_o) \approx -1,535 \ ppm/°C = -0.15 \ \%change/°C$$

The output current as a function of temperature is given by

$$I_o(T) = I_o(T_0)(1 + TC(I_o)(T - T_0)) = I(T_0)[1 - 0.0015(T - 27)]$$

The PSPICE simulation results, with TC1 of the resistor set to 0.002, are shown in Fig. 20.10. The hand calculations are close to the simulation results. Note that the temperature coefficient of the current source is not a constant number, but is a function of the operating temperature. This is apparent by the differing distances between the curves of Fig. 20.10. ∎

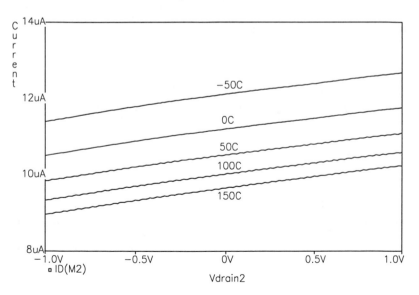

Figure 20.10 Temperature characteristics of the simple current mirror of Ex. 20.1. Notice that the voltage on the drain of M2 is swept from -1 V to + 1 V.

An important question we can ask is "Can we design a simple current mirror with a TC of zero?" Rearranging Eq. (20.26) for a TC(I_o) of zero gives

$$I_o R = \frac{W_2 L_2}{W_1 L_1}\left[-V_{THN}\cdot TCV_{TH} + \sqrt{\frac{L_1}{W_1}\frac{VDD-VSS}{2RKP(T)}}\left[\frac{1}{R}\frac{\partial R}{\partial T}-\frac{1.5}{T}\right]\right]\cdot\frac{1}{\frac{1}{R}\frac{\partial R}{\partial T}} \quad (20.27)$$

This equation relates the output current and the resistor value needed for a current mirror with a TC of zero, given the temperature characteristics of the MOSFET and the resistor.

20.1.4 Transient Response

Another important characteristic of an electronic circuit is how well the circuit performs under AC or varying input signals. Figure 20.11 shows a typical setup for determining the transient or step response of the simple current mirror. The output current of the mirror is given by $I_o = \frac{\beta_2}{2}(V_{GS}-V_{THN})^2$ and is directly dependent on V_{GS}. We are trying to answer the question, "If the voltage across the current source changes abruptly, how does this affect V_{GS} and thus the output current?" The gate-source voltage in this test setup is given by

$$V_{GS} = V_{step}\cdot\frac{Z_{GS}}{Z_{GS}+\frac{1}{j\omega C_{gd2}}} \quad (20.28)$$

where

$$Z_{GS} = \frac{R||\frac{1}{g_{m1}}}{1+j\omega(C_{gs1}+C_{gs2})R||\frac{1}{g_{m1}}} \quad (20.29)$$

Combining these two equations gives the following

$$V_{GS} = V_{step}\cdot\frac{j\omega\left(R||\frac{1}{g_{m1}}\right)C_{gd2}}{1+j\omega(C_{gs1}+C_{gs2}+C_{gd2})R||\frac{1}{g_{m1}}} \quad (20.30)$$

Assuming the step rises very quickly, the peak voltage change, ΔV_{GS}, is given by

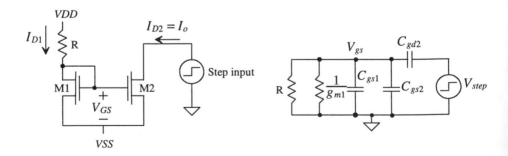

Figure 20.11 Transient response of the simple current mirror.

$$\Delta V_{GS} = V_{step,peak} \cdot \frac{C_{gd2}}{C_{gs1} + C_{gs2} + C_{gd2}} \tag{20.31}$$

and the change in output current is given by

$$\Delta I_o = \frac{\beta_2}{2}(\Delta V_{GS} + V_{GS} - V_{THN})^2 - \frac{\beta_2}{2}(V_{GS} - V_{THN})^2 = \frac{\beta_2}{2}\Delta V_{GS}^2 + g_{m2}\Delta V_{GS} \tag{20.32}$$

The rate at which V_{GS} decays back to its DC value is determined by the time constant, τ. However, the effective resistance needed to charge or discharge the node that defines V_{GS} is dependent on the direction of the pulse generator. If a negative-going pulse is applied to the current mirror, then the effective resistance is simply R. However, if a positive-going pulse is applied to the mirror, then the effective resistance that discharges the change in gate-source voltage is now $1/g_{m1}$ such that

$$\tau = \frac{1}{g_{m1}} \cdot (C_{gs1} + C_{gs2} + C_{gd2}) \tag{20.33}$$

The rate at which the output current decays back to its DC value is approximated by

$$\tau_o \approx 2\tau \tag{20.34}$$

Example 20.7
Determine the peak current deviation and the time it takes the output current to return to its steady-state value for the simple current mirror given in Ex. 20.1 using the test circuit given in Fig. 20.11. Assume the step transitions from 0 to 2.5 V. Compare the hand calculations with PSPICE.

We know that $I_{D1} = I_{D2} = I_o = 10$ μA, $L_1 = L_2 = 5$ μm, and $W_1 = W_2 = 15$ μm. Therefore,

$$C_{gs1} = \frac{2}{3}C'_{ox}W_1L_1 = 40 \text{ fF}$$

$$C_{gs2} = \frac{2}{3}C'_{ox}W_2L_2 = 40 \text{ fF}$$

$$C_{gd2} = CGDO \cdot W_2 = 5.7 \text{ fF}$$

$$\beta_1 = \frac{KP \cdot W_1}{L_1} = 150 \frac{\mu A}{V^2}$$

$$\beta_2 = 150 \frac{\mu A}{V^2}$$

$$g_{m1} = \sqrt{2\beta_1 I_{D1}} = 55 \frac{\mu A}{V}$$

$$g_{m2} = 55 \frac{\mu A}{V}$$

The change in gate-source voltage is given by

$$\Delta V_{GS} = 2.5 \cdot \frac{5.7}{40 + 40 + 5.7} \approx 0.18 \text{ V}$$

and the change in the output current, using Eq. (20.32), is given by

$$\Delta I_o = 75 \times 10^{-6}(0.18)^2 + 55 \times 10^{-6}(0.18) = 12 \ \mu A \ !$$

The time constant for V_{GS} decay using Eqs. (20.33) and (20.34), is given by

$$\tau = \frac{1}{55 \times 10^{-6}} \cdot (40 + 40 + 5.7) \times 10^{-15} \approx 1.5 \ ns$$

and the output current time constant is

$$\tau_o = 3 \ ns$$

The PSPICE simulations are shown in Fig. 20.12. The hand calculations are very close to the simulation results. Note the simulation problems encountered with the output current. Putting a series resistor between the pulse source and the current source helps to "slow" things down enough to eliminate these problems. Of course, the results are affected because the resistor acts to increase the pulse risetime applied to the current source. ∎

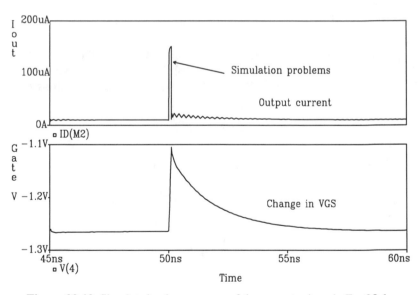

Figure 20.12 Simulated pulse response of the current mirror in Ex. 20.1.

Poor transient response of a current mirror can severely limit the performance of a circuit. In some cases, such as the current sources discussed later in the chapter that use negative feedback, the current source can actually become unstable and oscillate.

20.1.5 Layout of the Simple Current Mirror

When laying out the current mirror, the lateral diffusion and the oxide encroachment on the length and width of the MOSFET must be taken into consideration. The lateral diffusion, *DL,* given in the BSIM model, and the oxide encroachment, *DW,* in the BSIM model will cause errors in the ratio $\frac{W_2 L_1}{W_1 L_2}$. If the requirement that $L_{1drawn} = L_{2drawn}$ is imposed, then the ratio $\frac{L_1}{L_2} = \frac{L_{1drawn} - DL}{L_{2drawn} - DL}$ is unity. Since the widths of the devices determine the relative currents in the mirror, the widths cannot be equal. Figure 20.13a shows a current mirror layout without width compensation. The width ratio, $\frac{W_2}{W_1}$, can be written

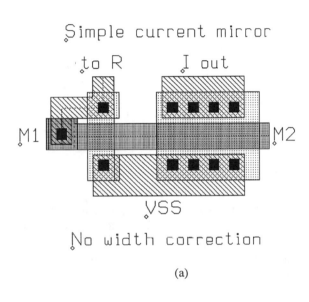

(a)

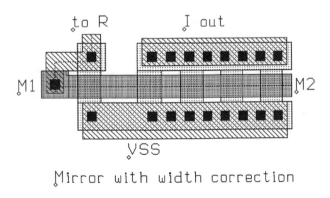

(b)

Figure 20.13 Layout of a MOSFET mirror to eliminate oxide encroachment effects.

as $\frac{W_{2drawn}-DW}{W_{1drawn}-DW}$, which is clearly not an integer if the drawn widths are not equal. Figure 20.13b shows how to layout a current mirror to avoid these problems. The layout of M2 is basically four MOSFETs in parallel. This can be specified in SPICE by adding M = X after the MOSFET statement in the netlist, where X is the number of MOSFETs. The lines in the netlist may look similar to

```
M1    1 1 0 0        CMOSNB        L=5u W=5u
M2    2 1 0 0        CMOSNB        L=5u W=5u M=4
```

20.1.6 Matching in MOSFET Mirrors

Many analog applications are quite susceptible to errors due to layout. In circuits in which devices need to be matched, layout becomes a very critical factor. For example, in the basic current mirror shown in Fig. 20.1, first-order process errors can cause the mirrored current to be significantly different from the reference current. Process parameters such as gate-oxide thickness, lateral diffusion, oxide encroachment and, oxide charge density can drastically affect the performance of a device. Layout methods can be used to minimize the first-order effects of these parameter variations.

Threshold Voltage and Transconductance Parameter

In a given current mirror application, the values for the threshold voltages are critical in determining the overall accuracy of the mirror. Again examine the basic current mirror shown in Fig. 20.1. Since both devices have the same value for V_{GS} and assuming that β for both devices are equal, examine the effect of a mismatch in threshold voltages between the two devices [3]. If it is assumed that the threshold mismatched is distributed across both devices such that

$$V_{THN1} = V_{THN} - 0.5\Delta V_{THN} \tag{20.35}$$

$$V_{THN2} = V_{THN} + 0.5\Delta V_{THN} \tag{20.36}$$

where V_{THN} is the average value of V_{THN1} and V_{THN2} and ΔV_{THN} is the mismatch, then,

$$\frac{I_O}{I_{D1}} = \frac{\beta(V_{GS} - V_{THN} - 0.5\Delta V_{THN})^2}{\beta(V_{GS} - V_{THN} + 0.5\Delta V_{THN})^2} = \frac{\left[1 - \frac{\Delta V_{THN}}{2(V_{GS} - V_{THN})}\right]^2}{\left[1 + \frac{\Delta V_{THN}}{2(V_{GS} - V_{THN})}\right]^2} \tag{20.37}$$

If both expressions are squared and the higher order terms are ignored, then the first-order expression for the ratio of currents becomes

$$\frac{I_O}{I_{D1}} \approx 1 - \frac{2\Delta V_{THN}}{(V_{GS} - V_{THN})} \tag{20.38}$$

Equation (20.38) is quite revealing because it shows that as V_{GS} decreases, the difference in the mirrored currents increases due to threshold voltage mismatch. This is particularly critical for devices that are separated by relatively long distances, since the threshold voltage is quite susceptible to process gradients.

The same analysis can be performed on the transconductance parameter, KP_n. If $KP_{n1} = KP_n - 0.5\Delta KP_n$ and $KP_{n2} = KP_n + 0.5\Delta KP_n$, where KP_n is the average of KP_{n1} and KP_{n2}, then assuming perfect matching on all other parameters, the difference in the currents becomes

$$\frac{I_O}{I_{D1}} \approx 1 + \frac{\Delta KP_n}{KP_n} \tag{20.39}$$

Drain to Source Voltage and Lambda

One aspect of current mirrors that is critical in generating accurate currents is the product of the drain to source voltage and lambda. In the basic current mirror, the ratio of the output current to the reference current is affected by both factors. For example, in Fig. 20.14, assume that the two devices are perfectly matched, except for their values of $\lambda = \lambda_c + \lambda_m$ and V_{DS}. If $V_{DS1} = 2$ V, $V_{DS2} = 4$ V, and $\lambda_1 = 0.05$, $\lambda_2 = 0.04$, then

$$\frac{I_o}{I_{D1}} = \frac{1+(\lambda_c+\lambda_m)V_{DS2}}{1+(\lambda_c+\lambda_m)V_{DS1}} = \frac{1+(0.05)4}{1+(0.04)2} = 1.11 \tag{20.40}$$

resulting in over 11 percent error. In some applications, it may be difficult to ensure that the drain to source voltages are the same for both devices. If possible, effort should be made to design the circuit so that the values of V_{DS} are equal.

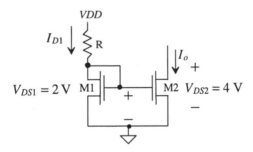

Figure 20.14 Basic current mirror with differing values of drain to source voltages.

Layout Techniques to Improve Matching

Most analog applications require that the length of the gate in an analog design be several times larger than the minimum drawn gate length, since the channel length modulation, λ_c , has less effect on larger devices than smaller ones. As a result, minimum-sized devices are rarely used in analog applications as is found in many digital circuits. However, the larger devices can result in larger parasitics if some layout issues are not considered. Figure 20.15a illustrates a basic MOSFET device with a large W/L. Since the width of the device is quite large, the diffusion resistance of the source and drain can be modeled as shown in Fig. 20.15b. The diffusion resistance can easily be reduced by simply adding as many contacts as possible along the width of both source and drain as seen in Fig. 20.15c. As was discussed in previous chapters, the

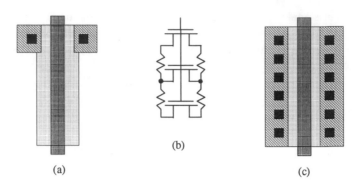

(a)   (b)   (c)

Figure 20.15 (a) Large device with a single contact and (b) its equivalent circuit. (c) Adding more contacts to reduce parasitic resistance.

increase in number of contacts results in lower resistance, more current capability, and a more distributed current load throughout the device. However, as the device size increases, another technique is used which distributes the parasitics (both resistive and capacitive) into smaller contributions.

Examine Fig. 20.16. Here, a single device with a large W/L is split into several parallel devices, each with one-fourth of the original W. One result of splitting the device into several parts is smaller overall parasitic capacitance associated with the reversed-biased diffusion substrate diode (the drain or source depletion capacitance to substrate). Since values of C_{db} and C_{sb} are proportional to W, the split device reduces these parasitics by a factor of $(n + 1)/2n$ where n is the number of parallel devices and is odd. If n is even, then the C_{sb} is reduced by one-half and C_{db} is reduced by $(n + 2)/2n$ [5].

Notice also, that Fig. 20.16 has dummy poly strips on both sides of the device. These strips are used to help minimize the effects of undercutting the poly on the outer

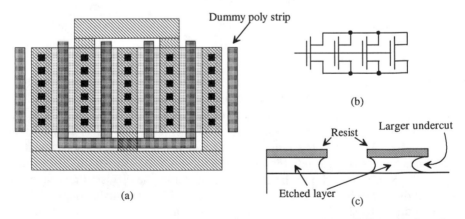

(a)   (b)

                                       (c)

Figure 20.16 (a) A parallel device with dummy strips, (b) the equivalent circuit and (c) undercutting.

edges after patterning. If the dummy strips had not been used, the poly would have been etched out more under the outermost gates, resulting in a mismatch between the four parallel devices.

When matching two devices, it is imperative that the two devices be as symmetrical as possible. Always orient the two devices in the same direction, unlike that illustrated in Fig. 20.17. Splitting the devices into parallel devices and interdigitizing them can distribute process gradients across both devices, thus improving matching. An example of this can be seen in Fig. 20.18. Note the use of dummy poly strips on this layout. A good exercise at this point is to lay out the mirror of Fig. 20.18 in a common-centroid arrangement (see Figs. 7.7 and 7.9).

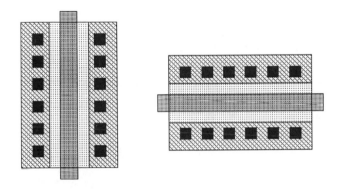

Figure 20.17 Devices with differing orientation.

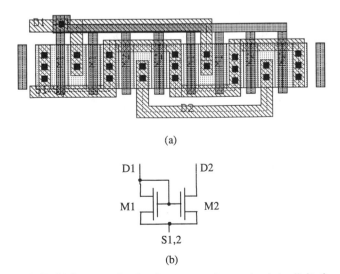

(a)

(b)

Figure 20.18 (a) Layout of a simple current mirror using interdigitation and (b) equivalent circuit.

20.2 Other Current Sources/Sinks

The basic current mirror can be improved significantly with negative feedback. There are two widely used current mirrors of this type, known as a Wilson current mirror and the regulated cascode. Both offer stable current values for wide voltage swings and enhanced output impedance.

The Wilson mirror [1, 3] can be seen in Fig. 20.19. Series sampling is used at the output to increase the output impedance and stabilize the drain current I_{D4}. The feedback mechanism is as follows. Suppose that I_{D1} is a constant and stable reference current and the voltage, V_o, increases, which causes I_{D4} to increase. I_{D2} increases by the same amount while I_{D1} stays constant. As a result, the voltage at node A decreases, thus decreasing V_{GS4} and stabilizing the current through M4.

The output impedance can be found by analyzing the circuit shown in Fig. 20.20. Note that since the gate of M1 is a DC voltage, $v_{gs1} = 0$ and the only contribution to the small-signal circuit from M1 is the resistance, r_o. Source absorption also converts the controlled current source of $g_{m3}v_{gs3}$ into its equivalent resistance of $1/g_{m3}$. From this circuit, we can generate the following equations,

$$v_{sb4} = v_{gs2} \tag{20.41}$$

$$v_{gs2} = i_t(r_{o3}||\frac{1}{g_{m3}}) \tag{20.42}$$

$$v_{gs4} = -v_{gs2}[1 + g_{m2}(r_{o1}||r_{o2})] = -i_t(r_{o3}||\frac{1}{g_{m3}})[1 + g_{m2}(r_{o1}||r_{o2})] \tag{20.43}$$

$$i_t = g_{m4}v_{gs4} - g_{mb4}v_{sb4} + \frac{v_t - v_{gs2}}{r_{o4}} \tag{20.44}$$

Plugging Eqs. (20.41 - 20.43) into Eq. (20.44), yields,

$$R_{out} = \frac{v_t}{i_t} = r_{o4}\left[1 + g_{m4}(r_{o3}||\frac{1}{g_{m3}})(1 + g_{m2}(r_{o1}||r_{o2}))\right] + g_{mb4}\left[(r_{o3}||\frac{1}{g_{m3}}) + \frac{1}{r_{o4}}(r_{o3}||\frac{1}{g_{m3}})\right]$$

$$\tag{20.45}$$

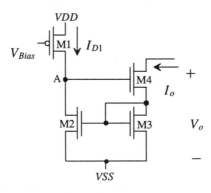

Figure 20.19 Wilson current mirror.

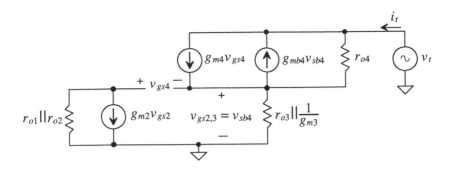

Figure 20.20 Small signal model of the Wilson current mirror used to determine output resistance.

which can be approximated as

$$R_{out} \approx r_{o4}\left[1 + g_{m2}(r_{o1}||r_{o2}) + g_{mb4}(\frac{1}{g_{m3}}) + \frac{1}{r_{o4}g_{m3}}\right] \qquad (20.46)$$

If it is assumed that $r_{o3}||\frac{1}{g_{m3}} \approx \frac{1}{g_{m3}}$ and that $g_{m3} \approx g_{m4}$ then Eq. (20.46) can be further reduced. If the last two terms in the equation are assumed to be negligible and if $r_o = r_{o1} \approx r_{o2} \approx r_{o4}$ then R_{out} becomes,

$$R_{out} \approx r_o + g_{m2}\frac{r_o^2}{2} \qquad (20.47)$$

It now becomes apparent that with the r_o^2 term in Eq. (20.47), the Wilson current mirror has an output impedance similar in size to the cascode current mirror.

One problem associated with this mirror is the minimum voltage required to keep I_o constant. As seen in Fig. 20.19, the minimum output voltage, V_o, necessary to keep both M3 and M4 in saturation is

$$V_o(\text{min}) = V_{GS3} + V_{DS4,sat} = V_{GS3} + V_{GS4} - V_{THN4} \qquad (20.48)$$

which can be written in terms of the drain current, I_o, as

$$V_o(\text{min}) = \sqrt{\frac{2I_o}{\beta_3}} + V_{THN3} + \sqrt{\frac{2I_o}{\beta_4}} = 2\sqrt{\frac{2I_o}{\beta_{3,4}}} + V_{THN3} \qquad (20.49)$$

if $\beta_3 = \beta_4$. As I_o increases, $V_o(\text{min})$ becomes larger by two times the square root of I_o.

Regulated Cascode Current Source/Sink

Another popular type of current mirror is the regulated cascode current mirror [6]. This current mirror also uses negative feedback to stabilize the output current and increase the output impedance to a higher degree than the Wilson mirror. Figure 20.21 illustrates a regulated cascode current mirror. The transistors M2 and M4 make up the negative feedback loop, which stabilizes I_o. If I_o were to increase, then, since a constant

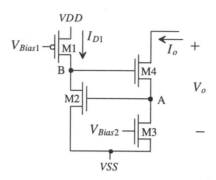

Figure 20.21 Regulated cascode current sink.

current flows through M3, the voltage on node A would rise, which in turn would increase the current through M2. Since the current through M1 is also a constant amount, the voltage at node B would decrease, thereby offsetting the increase in I_o by decreasing V_{GS3}.

This current mirror offers some significant advantages over any of the previously discussed current mirrors. First, the output impedance of this current mirror is significantly higher. The small-signal equivalent circuit is shown in Fig. 20.22. Since M1 and M3 are used in conjunction with DC bias sources, only their resistances, r_{o1} and r_{o3}, respectively, will appear in the small-signal model. However, both M2 and M4 contribute to the model. Notice that there is only a subtle difference in the small-signal model shown for Fig. 20.7 (cascode) and Fig. 20.20 (Wilson). The main difference is the fact that the device, M3, is no longer gate-drain connected, and so the controlled source, $g_{m3}v_{gs}$ is no longer source absorbed. The analysis for calculating the output resistance R_{out} is identical to that performed for the Wilson current mirror, except that the resistor that was represented as, $r_{o3}\|1/g_{m3}$ is now just r_{o3}. The resistance, R_{out} now becomes,

$$R_{out} = \frac{v_t}{i_t} = r_{o4}\left[1 + g_{m4}r_{o3}(1 + g_{m2}(r_{o1}\|r_{o2})) + g_{mb4}r_{o3} + \frac{r_{o3}}{r_{o4}}\right] \quad (20.50)$$

which is dominated by the effective r_o^3 term composed of r_{o3}, r_{o4} and the parallel combination of r_{o1} and r_{o2}. R_{out} can be further reduced to

$$R_{out} = \frac{v_t}{i_t} \approx g_{m2}g_{m4}(r_{o1}\|r_{o2})r_{o3}r_{o4} \approx \frac{g_m^2 r_o^3}{2} \quad (20.51)$$

if all the devices are matched. Assuming typical values of $g_m = 20\ \mu A/V$ and $r_o = 10 M\Omega$, we see that the output of the regulated cascode is approximately 400 GΩ!

Another positive attribute of the regulated cascode is that its minimum value of V_o is lower than most of the other configurations. Since the negative feedback stabilizes

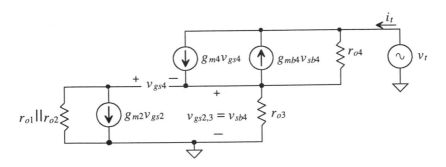

Figure 20.22 Small-signal model of the regulated cascode current mirror used to
determine output resistance.

the voltages at nodes A and B, the output resistance is so large that M4 can become
nonsaturated and still maintain a fairly constant current. The minimum output voltage
is then determined by the fact that I_{D2} starts to change as node A starts to drop and M3
goes into nonsaturation. Therefore, the minimum output voltage is approximately a
single $V_{DS3}(\text{sat})$.

One might question how the current sink shown in Fig. 20.21 can be used as a
current mirror. With some modification, a true current mirror can be formed. Examine
Fig. 20.23. With the addition of transistors M1, M5, M6, and M7, the input current, I_{D1}
can be mirrored quite accurately through M3. If the current through M2 is identical to
I_o, and $\beta_2 = \beta_3$, note that $V_{GS2} = V_{GS3} = V_{DS3}$ and that M3 is effectively always saturated
since $V_{DS3} \geq V_{GS3} - V_{THN3}$. This condition fails, however, if V_{GS2} changes.

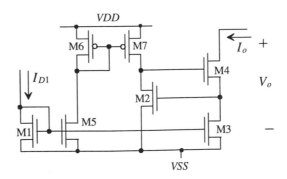

Figure 20.23 Regulated cascode current mirror.

Example 20.8
Estimate and determine, through SPICE simulation, the output resistance of the
current sink shown in Fig. Ex20.8.

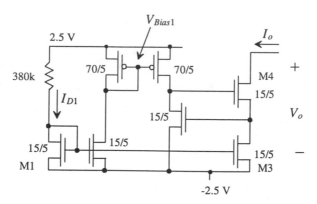

Figure Ex20.8

From Ex. 20.1 the current that flows in all MOSFETs in Fig. Ex20.8 is 10 μA and the output resistance is 1.67 MEG. The transconductance, g_m, from Ex. 20.4 is 55 μA/V. The output resistance can be estimated, using Eq. (20.51), as 6.9 GΩ, or approximately 50 times larger than the basic cascode of Ex. 20.4. Simulation results are shown in Fig. 20.24. The minimum voltage across the current source is approximately 1.3 V (0.3 V) with both M3 and M4 saturated (only M3 saturated). This corresponds to –1.2 V (–2.2 V) on the drain of M4. The current through the 380k resistor is slightly different from the output current when all MOSFETs are in the saturation region. This is the result of differing drain-source voltages between M1 and M3. The regulated cascode

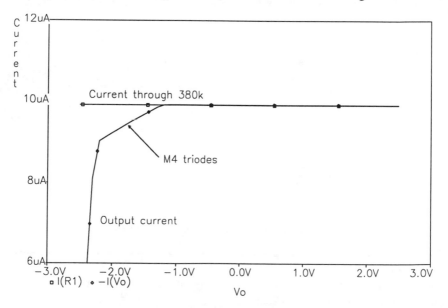

Figure 20.24 Simulation results for Ex. 20.8.

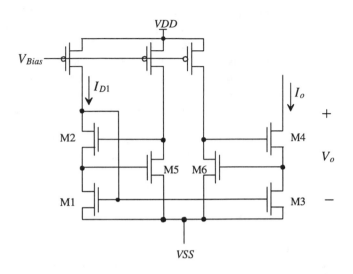

Figure 20.25 Improved regulated cascode current mirror.

current mirror of Fig. 20.25 overcomes this mismatch by forcing the drain-source voltage of these two transistors equal. This configuration is preferred when matching is important, that is, matching between I_{D1} and I_o. ∎

Wide-Swing, Low-Voltage Current Mirror

A wide-swing current mirror is shown in Fig. 20.26 [7, 8]. Wide swing here means that the minimum voltage across the current sink is $2\Delta V$ ($\Delta V = 0.37$ V when $V_{THN} = 0.83$ and

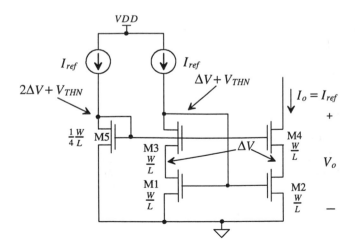

Figure 20.26 A high-swing cascode current mirror.

$V_{GS} = 1.2$ V). This design is based on the wide-swing current mirror of Fig. 20.6 and is very useful in practical circuit design. MOSFETs M1 through M4 form a current mirror and therefore can be used in any of the applications discussed so far. We will see that current mirrors can be used as active loads for differential pairs and in voltage references. The minimum voltage across the wide-swing current source is one threshold voltage drop less than the regular cascode of Fig. 20.5, while the output resistance is the same, that is, $g_m \cdot r_o^2$. An example of a 10 µA current source based on the example values given in this chapter is shown in Fig. 20.27. MOSFETs M6 through M11 are added to supply current to M5. M5 is used to generate the bias voltage on the gates of M3 and M4 for wide-swing operation. Figure 20.28 shows the simulation results for the circuit of Fig. 20.27 when the output of the current source is swept from –2.5 to 2.5 V. This figure should be compared to Fig. Ex20.4.

Note that the body effect has been neglected in this analysis. This will cause V_{GS3} and V_{GS4} to be larger than predicted by $\Delta V + V_{THN}$. The current sink functions properly because $V_{DS,sat}$ (of M1 and M2) is smaller than predicted by $V_{GS} - V_{THN}$ (see Ch. 5). In several practical situations, for example, using this mirror in a folded-cascode op-amp, the size of M5 must be made smaller (perhaps $(1/5) \cdot W/L$).

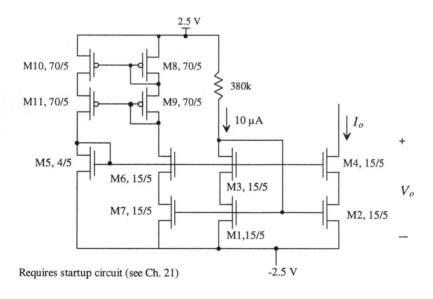

Figure 20.27 A 10 µA wide-swing current sink.

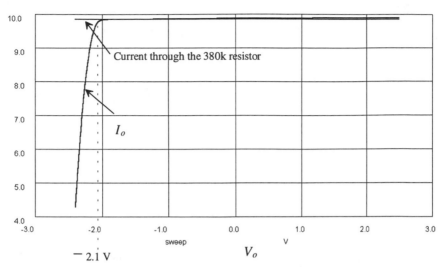

Figure 20.28 Simulation results for the current mirror of Fig. 20.27.

REFERENCES

[1]   P. E. Allen and D. R. Holberg, *CMOS Analog Circuit Design,* Holt, Rinehart and Winston, 1987. ISBN 0-03-006587-9.

[2]   R. L. Geiger, P. E. Allen, and N. R. Strader, *VLSI-Design Techniques for Analog and Digital Circuits,* McGraw-Hill Publishing Co., 1990. ISBN 0-07-023253-9.

[3]   P.R. Gray and R.G. Meyer, *Analysis and Design of Analog Integrated Circuits,* 3rd ed., John Wiley and Sons, 1993. ISBN 0-471-57495-3.

[4]   T. C. Choi, R. T. Kaneshiro, R. Broderson, and P. R. Gray, "High-Frequency CMOS Switched Capacitor Filters for Communication Applications," *IEEE Journal of Solid State Circuits,* Vol. SC-18, pp. 652-664, December 1983.

[5]   U. Gatti, F. Maloberti and V. Liberali, "Full Stacked Layout of Analogue Cells," *Proc. IEEE Int. Symp. on Circuits and Systems,* pp. 1123-1126, 1989.

[6]   E. Säckinger and W. Guggenbühl, "A High-Swing, High-Impedance MOS Cascode Circuit," *IEEE Journal of Solid-State Circuits,* Vol. 25, No. 1, pp. 289-298, February 1990.

[7]   Y. Tsividis and P. Antognetti, *Design of MOS VLSI Circuits for Telecommunications*, Englewood Cliffs, N.J.: Prentice-Hall, 1985, p. 560.

[8]   J. N. Babanezhad and R. Gregorian, "A Programmable Gain/Loss Circuit," *IEEE Journal of Solid State Circuits,* Vol. SC-22, No. 6, pp. 1082-1089, December, 1997.

PROBLEMS

Unless otherwise stated, use CN20 parameters, $VDD = 5$ V and $VSS = 0$ V.

20.1   Repeat Ex. 1 using a current of 1 μA and a V_{GS} of 1 V.

20.2   Design a circuit to bias the MOSFET in Fig. P20.2 so that its drain current is 5-μA. Estimate the minimum voltage across M1 if it is to remain in saturation.

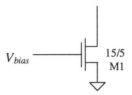

Figure P20.2

20.3   Determine the current that flows in M3 (Fig. P20.3). Neglect oxide encroachment and the finite output resistance of the M3.

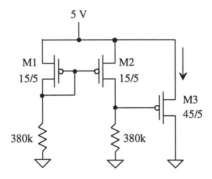

Figure P20.3

20.4   Determine the currents that flow in Fig. Ex20.3 if the size of M2 is increased to 30/5. Again, neglect oxide encroachment. Does V_{GS2} of V_{SG3} change? Why?

20.5   Determine the AC current that flows in v_t in Fig. P20.5. Assume $\lambda = 0.06$.

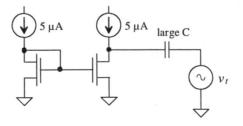

Figure P20.5

20.6 Design a bias circuit so that the current that flows in M1 and M2 of Fig. P20.6 is 1 µA. What are the small-signal resistances looking into the drains of M2 and M1? What is the minimum voltage across M1 and M2 for operation in the saturation region?

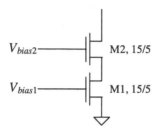

Figure P20.6

20.7 Using a small-signal analysis, verify that Eq. (20.11) is correct.

20.8 Using the circuit topology shown in Fig. P20.8 design current sources with output currents of 10 and 15 µA. Estimate, using hand calculations, the minimum voltage across the current sources and the output resistance. Verify the operation of your design with SPICE. What is the V_{SG} of the 100/5 MOSFETs that source 10 µA of current?

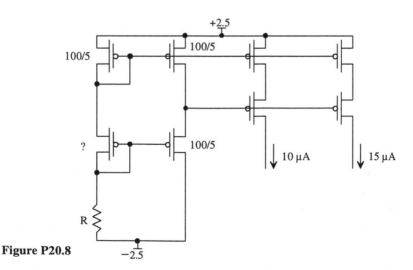

Figure P20.8

20.9 Repeat Ex. 20.5 for *VSS* changing from –2.4 to –2.6 V.

20.10 Verify that the temperature coefficient of the basic current mirror is given by Eq. (20.19).

20.11 Verify that the temperature coefficient of the transconductance parameter is given by

$$\frac{1}{KP(T)} \cdot \frac{\partial KP(T)}{\partial T} = \frac{-1.5}{T}$$

20.12 In order to reduce the power dissipation in current mirrors one might be tempted to design a circuit with a small diode connected MOSFET as shown in Fig. P20.12. The problem with this design is that changes on the drain of M2 can have a large effect on the gate voltage of M1/M2, ultimately being limited by the gate-source capacitance of M2. Using SPICE simulations, show how the output current is affected by an abrupt change across M2 of 1 V and –1 V.

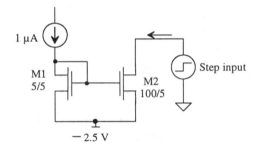

Figure P20.12

20.13 Lay out the current mirror shown in Fig. P20.13, avoiding oxide encroachment.

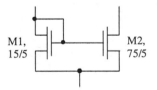

Figure P20.13

20.14 Suppose that a threshold voltage difference of 10 mV exists between the two MOSFETs used in a current mirror made using equal-size MOSFETs. How would you expect the difference in the drain currents to change with the MOSFET V_{GS}? Is matching between the currents better or worse with large V_{GS}? What happens to the minimum voltage across the current mirror if V_{GS} is increased?

20.15 Using the Wilson current mirror shown in Fig. P20.15, and the CN20 process, design a mirror in which $I_o = 150$ μA with $I_{REF} = 50$ μA for $VDD = 5$ V, $VSS = 0$ V. Verify your design with SPICE.

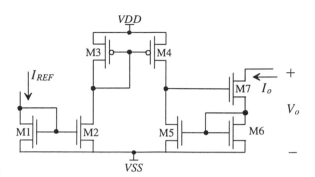

Figure P20.15

20.16 A Wilson current mirror (Fig. P20.15) is designed with $VDD = 5V$, $VSS = 0V$, $I_{REF} = 10\ \mu A$ and all devices sized 10 μ/10μ.

a. Calculate the value of the small-signal output resistance of the Wilson current mirror.

b. Verify part (a) with SPICE.

c. Calculate the minimum output voltage, $V_o(min)$.

d. Construct a plot showing I_o versus $V_o(min)$ for I_o varying from 5 μA to 50 μA in 5 μA increments.

e. Use SPICE to generate I_o versus V_o curves. How does this compare to the plot generated in part (d)?

f. Lay out the devices M4 through M7, assuming that each device is 40/5 using an interditigated layout technique to match devices M5 and M6. Use parallel devices for M4 and M7.

20.17 Repeat Problem 20.16, parts (a-e), for a cascode current mirror made using 15/5 devices.

20.18 Design a regulated cascode current sink of 1 μA. Estimate the output resistance of the current sink. Using SPICE output similar to Fig. 20.24, show the output current change with voltage across the current sink.

20.19 The basic idea used in the regulated cascode current sink is shown in Fig. P20.19. What is the output resistance of this configuration? Show that if the gain A is $g_m(r_o/2)$, then the output resistance of the configuration reduces to that given by Eq. (20.51).

20.20 With respect to the wide-swing current mirror shown in Fig. 20.26, show that, neglecting the body effect, the drains of M1 and M2 are at ΔV (the excess gate voltage) when M5 is one-fourth the size of the other MOSFETs.

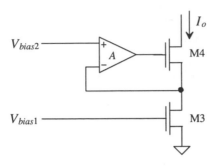

20.21 The basic current mirror finds extensive use as a load in differential amplifiers. Consider the schematic of the modified current mirror shown in Fig. P20.21. This configuration is termed a wide-band configuration since the addition of the source follower helps improve the ability of M1 to drive the large-input capacitance of M2. Estimate the small-signal resistance to ground at both nodes A and B. If a resistor is used to connect node A to *VDD*, write an equation relating the drain currents of M1 and M2 to the resistor value.

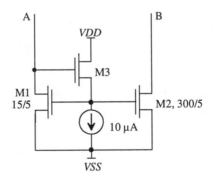

Figure P20.21

Chapter 21

References

This chapter discusses voltage and current references, and in general how we can use voltage references to bias the current sources and sinks discussed in the last chapter. A good reference is insensitive to temperature, process, and supply voltage variations.

21.1 Voltage Dividers

In CMOS integrated circuit design, we can derive reference voltages from the power supplies using the resistor and the MOSFET. Figure 21.1 shows the basic idea. The voltage divider formed with two resistors has the advantage of simplicity, temperature insensitivity (as was shown in Ch. 7), and process insensitivity, that is, changes in the sheet resistance have no effect on the voltage division. The main problem with this circuit is that in order to reduce the power dissipation (i.e., the current through the resistors), the resistors must be made large. Since large resistors require a large area on the die, this voltage divider may not be practical in many cases. The voltage divider formed between the resistor and the MOSFET can be recognized as the same circuit we used for a bias in the current mirror. The final circuit, a voltage divider between an n- and p-MOSFET, has the advantage that the layout can be small. In the following two subsections, we analyze the behavior of these last two voltage dividers.

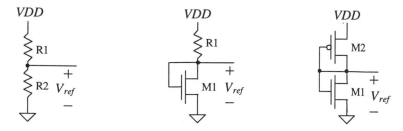

Figure 21.1 Implementation of voltage dividers in CMOS.

21.1.1 The Resistor-MOSFET Divider

The reference voltage used in the resistor-MOSFET divider is equal to the V_{GS} of the MOSFET. We know for this circuit that

$$I_D = \frac{VDD - V_{ref}}{R} = \frac{\beta_1}{2}(V_{ref} - V_{THN})^2 \tag{21.1}$$

or

$$V_{ref} = V_{THN} + \sqrt{\frac{2I_D}{\beta_1}} = V_{THN} + \sqrt{\frac{2(VDD - V_{ref})}{R \cdot \beta_1}} \tag{21.2}$$

The sensitivity of the reference voltage to VDD when $VDD \gg V_{ref}$ is

$$S_{VDD}^{V_{ref}} = \frac{VDD}{V_{ref}} \cdot \frac{\partial V_{ref}}{\partial VDD} \approx \frac{1}{V_{THN} \cdot \sqrt{\frac{2R\beta_1}{VDD}} + 2} \tag{21.3}$$

The temperature coefficient of the resistor-MOSFET voltage divider is given by

$$TC(V_{ref}) = \frac{1}{V_{ref}} \cdot \frac{\partial V_{ref}}{\partial T} \tag{21.4}$$

or with the help of Eq. (20.25) with $VSS = 0$

$$TC(V_{ref}) = \frac{1}{V_{ref}} \left[V_{THN} \cdot TCV_{TH} - \frac{1}{2}\sqrt{\frac{2L_1}{W_1} \cdot \frac{VDD}{R \cdot KP(T)}} \cdot \left[\frac{1}{R}\frac{\partial R}{\partial T} - \frac{1.5}{T} \right] \right] \tag{21.5}$$

A modification of the basic MOSFET divider is shown in Fig. 21.2. The reference voltage in this circuit is given by

$$V_{ref} = V_{GS}\left(\frac{R_1}{R_2} + 1 \right) \tag{21.6}$$

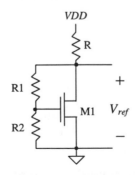

Figure 21.2 Modification of the resistor-MOSFET voltage divider.

21.1.2 The MOSFET-Only Voltage Divider

The MOSFET-only voltage divider shown in Fig. 21.3 generates a reference voltage equal to the voltage on the gates of the MOSFETs with respect to ground. Since $I_{D1} = I_{D2}$ we can write

$$\frac{\beta_1}{2}(V_{ref} - VSS - V_{THN})^2 = \frac{\beta_2}{2}(VDD - V_{ref} - V_{THP})^2 \qquad (21.7)$$

or the reference voltage is given by

$$V_{ref} = \frac{VDD - V_{THP} + \sqrt{\frac{\beta_1}{\beta_2}}\,(VSS + V_{THN})}{\sqrt{\frac{\beta_1}{\beta_2}} + 1} \qquad (21.8)$$

or knowing the reference voltage and the power supply voltages gives

$$\frac{\beta_1}{\beta_2} = \left[\frac{VDD - V_{ref} - V_{THP}}{V_{ref} - VSS - V_{THN}}\right]^2 \qquad (21.9)$$

The sensitivity of V_{ref} with respect to VDD is given by

$$S_{VDD}^{V_{ref}} = \frac{VDD}{VDD - V_{THP} + \sqrt{\frac{\beta_1}{\beta_2}}\,(VSS + V_{THN})} \qquad (21.10)$$

The temperature dependence of the MOSFET-only voltage divider, using Eq. (21.8), and assuming the temperature dependence of the ratio of the transconductance parameters, $\frac{\beta_1}{\beta_2}$, is negligible, is given by

$$TC(V_{ref}) = \frac{1}{V_{ref}} \cdot \frac{\partial V_{ref}}{\partial T} = \frac{1}{V_{ref}} \cdot \frac{1}{\sqrt{\frac{\beta_1}{\beta_2}} + 1} \cdot \left[\frac{\partial(-V_{THP})}{\partial T} + \sqrt{\frac{\beta_1}{\beta_2}}\,\frac{\partial V_{THN}}{\partial T}\right] \qquad (21.11)$$

From Ch. 9 we know that

$$\frac{\partial V_{THN}}{\partial T} = TCV_{THN} \cdot V_{THN} = (-.003°C^{-1})(0.8\ V) = -2.4\ \frac{mV}{°C} \qquad (21.12)$$

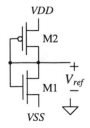

Figure 21.3 MOSFET-only voltage divider used between VDD and VSS.

and

$$-\frac{\partial V_{THP}}{\partial T} = -TCV_{TH} \cdot V_{THP} = -(-0.003°\text{C}^{-1})(0.9\text{V}) = 2.7\frac{\text{mV}}{°\text{C}} \qquad (21.13)$$

To achieve $TC(V_{ref}) = 0$ requires

$$-\frac{\partial V_{THP}}{\partial T} = -\sqrt{\frac{\beta_1}{\beta_2}} \cdot \frac{\partial V_{THN}}{\partial T} \Rightarrow 2.7\frac{\text{mV}}{°\text{C}} = \sqrt{\frac{\beta_1}{\beta_2}} \cdot 2.4\frac{\text{mV}}{°\text{C}} \qquad (21.14)$$

or

$$\sqrt{\frac{\beta_1}{\beta_2}} = 1.125 \qquad (21.15)$$

Zero temperature coefficient, to a first order, can be met by satisfying this equation. However, this ratio is most often set by the desired V_{ref}.

Example 21.1
Design a 3 V reference using the MOSFET-only voltage divider assuming VDD = +5 V and VSS = 0 V. Determine the temperature coefficient of the reference. Verify your hand calculations with SPICE. Compare the power dissipation when $L_1 = L_2 = 5$ μm with $L_1 = L_2 = 50$ μm.

Since we know that $V_{THN} = 0.8$ V and $V_{THP} = 0.9$ V, we can determine the ratio of β_1 to β_2 using Eq. (21.9):

$$\frac{\beta_1}{\beta_2} = \left[\frac{5-3-0.9}{3-0-0.8}\right]^2 = 0.25$$

Setting $L_1 = L_2 = W_1 = 5$ μm gives

$$\frac{KP_n W_1 L_2}{KP_P W_2 L_1} = \frac{\left(50 \times 10^{-6}\frac{A}{V^2}\right)5\mu5\mu}{\left(17 \times 10^{-6}\frac{A}{V^2}\right)W_2 5\mu} = 0.25$$

or solving gives $W_2 = 60$ μm. The temperature coefficient is given by

$$TC(V_{ref}) = \frac{1}{3} \cdot \frac{1}{\sqrt{\frac{50 \cdot 5}{17 \cdot 60}} + 1} \cdot \left[.0027 + \sqrt{\frac{50 \cdot 5}{17 \cdot 60}}(-0.0024)\right] = 337 \text{ ppm/°C}$$

The PSPICE simulation results are shown in Fig. 21.4. The temperature coefficient of this reference calculated from the figure is approximately 400 ppm/°C. The small difference between PSPICE and our hand calculations can be attributed to the temperature coefficients of the threshold voltages. The drain current when $L = 5$ μm is 125 μA, and the power dissipated by the circuit is 600 μW. When $L = 50$ μm, the drain current drops to 12.5 μA and the power dissipation becomes 60 μW. Reducing the power dissipation requires that we use larger devices. Using the threshold voltage drop of an additional device in series with the two of Fig. 21.3 will help with this problem. ∎

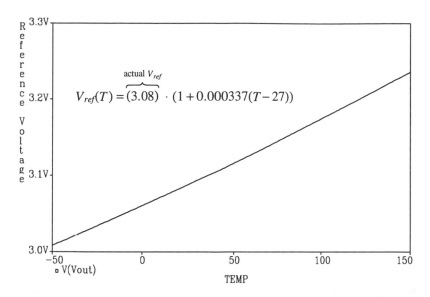

Figure 21.4 Temperature characteristics of the voltage divider designed in Ex. 21.1.

A three-MOSFET voltage divider is shown in Fig. 21.5. Two reference voltages are available in this configuration. The two p-channel MOSFETs are laid out in their own wells to avoid body effect. Designing voltage dividers in modern processes requires the designer to look at the maximum gate-source voltage. Generally, the maximum potential across the gate oxide should be below 7 MV/cm, or 0.7 V/10 Å. Circuits designed with a process that has an oxide thickness of 100 Å should limit the

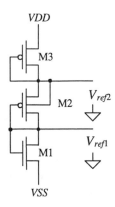

Figure 21.5 Three-MOSFET voltage divider.

oxide-source or oxide-substrate potential to 7 V or less. Reliability in many cases can be related to the maximum voltage or stress across the gate oxide.

The design of this type of divider consists of determining the reference voltages needed, selecting a drain current through the MOSFETs, and solving for the size of the devices.

Example 21.2

Design a 2 V and a 3.5 V voltage reference using the three-MOSFET voltage divider of Fig. 21.5. Assume that $VDD = +5$ V, $VSS = 0$ V, the drain current of the MOSFETs is 10 μA, and that $L_1 = L_2 = L_3 = 20$ μm. Compare the area required for layout of this design with the area used in Ex. 21.1 if $L = 50$ μm.

Since $V_{GS1} = 2$ V, we can solve for W_1 knowing

$$10 \; \mu A = \frac{KP_n}{2} \cdot \frac{W_1}{20 \mu m}(2 - 0.8)^2 \text{ which gives } W_1 \approx 5 \; \mu m$$

The width of M2 is determined in a similar manner, since $V_{SG2} = 1.5$ V, using

$$10 \; \mu A = \frac{KP_p}{2} \cdot \frac{W_2}{20 \mu m}(1.5 - 0.9)^2 \; \Rightarrow W_2 = 65 \; \mu m$$

The width of M3, with $V_{SG3} = 1.5$ V, is also 65 μm.

The area of the reference designed in Ex. 21.1 (using the case when $L = 50$ μm and drain current is 12.5 μA) can be estimated by

$$A_{Ex21.1} = L_1 W_1 + L_2 W_2 = 3250 \; \mu m^2$$

while the area needed in this example is 2700 μm². In addition, the power dissipation is slightly lower, 50 μW, when we use a second MOSFET. ∎

A useful application of the dual-reference voltage divider is shown in Fig. 21.6a. Here the reference is used to bias the cascode current source. The current through the reference, that is, the current through M1, M2, and M3, is selected as equal to or a multiple of the desired current through M4 and M5. Following the discussion of the last chapter, if it is desirable to design a 10 μA current source, we may set the current through M1 to 10 μA and the voltage across M1, V_{GS1} ($= V_{GS4}$), to 1.2 V. The current in M1 is mirrored in M4. The size of M2 can be set equal to the size of M1, while the size of M3 is selected to drop $VDD - 2.4$ V at the bias current.   Sometimes the bias voltages generated by M1 and M2 are simply labeled "Bias" as shown in Fig. 21.6b.   A complicated schematic can be simplified by not showing the bias MOSFETs.

The main drawback of using the references described in this section is that all, that is, the resistor-resistor, resistor-MOSFET, and MOSFET-only dividers are very sensitive to the power supply voltages and temperature. The remainder of this chapter discusses the means of designing voltage references and biasing current sources to reduce the sensitivity to temperature and the power supplies.

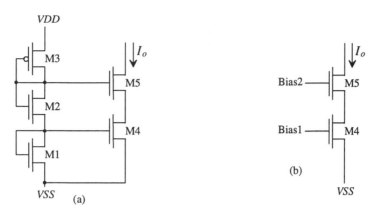

Figure 21.6 Use of the three-MOSFET voltage divider to bias the cascode current source: (a) full schematic and (b) simplified schematic.

21.2 Current Source Self-Biasing

The biasing methods presented so far for current sources have been dependent on the power supply voltages and temperature. Changes in *VDD* or *VSS* directly affect the currents in the circuit. In this section, we discuss three methods of biasing which reduce the effects of power supply variations and possibly temperature on currents present in a current source [1, 2].

21.2.1 Threshold Voltage Referenced Self-Biasing

Consider the biasing circuit shown in Fig. 21.7. MOSFETs M3 and M4 force the same current to flow through M1 and M2. The product of this current and the resistor R is equal to the gate-source voltage of M1 or, neglecting the output resistance of the MOSFETs and the body effect,

$$IR = V_{GS1} = V_{THN} + \sqrt{\frac{2I}{\beta_1}} \qquad (21.16)$$

If β_1 is large, the current, I, is given by

$$I \approx \frac{V_{THN}}{R} \qquad (21.17)$$

In practice, the second term in Eq. (21.16) is not negligible, and the current is given by

$$I = \frac{V_{GS1}}{R} \qquad (21.18)$$

where practical values of V_{GS1} are 1.0 to 1.2 V. These equations predict that the current is independent of the power supply voltages. Indeed, if the output resistance of the MOSFET were infinite, this would be the case. However, the current is dependent, albeit much less than the previous methods discussed, on the supply voltages. Cascoding M3 and M4 helps to make the bias circuit behave more ideal. The accuracy

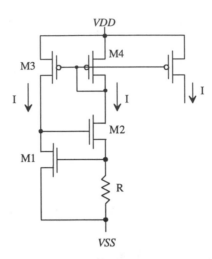

Figure 21.7 Threshold reference self-biasing circuit.

of the current, I, is limited by the threshold voltage accuracy, which may vary by 20 percent, and the n+ resistivity, value which can vary by 20 percent as well.

The temperature performance of the threshold bias circuit can be determined by considering Eq. (21.17). The temperature coefficient of the threshold voltage is (from Ch. 9) approximately −3,000 ppm/°C, while the temperature coefficient of an n+ resistor is (from Ch. 7) 2,000 ppm/°C. The threshold voltage decreases with increasing temperature, while the resistance, R, increases with increasing temperature. The result is a reference current with a large negative temperature coefficient.

The possibility that the current, I, may be zero exists in all self-biased circuits. (Consider what happens when the gates of M3 and M4 are at VDD and the gate of M2 is at VSS.) Figure 21.8a illustrates this graphically. Point A in this figure is the desired operating point, while point B corresponds to $I = 0$. Figure 21.8b shows the reference with a startup circuit used to avoid operation at point B. If the gate of M2 is at or near VSS, M5 turns on and pulls this node upward toward point A. Since the circuit is stable at either point A or B, the gate potential continues increasing until it reaches $2V_{GS}$. This causes M5 to turn off. Once the self-biasing circuit is operating at point A, the startup circuit does not affect the reference operation.

21.2.2 Diode Referenced Self-Biasing

Many biasing circuits use the parasitic pnp transistor formed by the p+ implant (the emitter), the n-well (the base), and the p-type substrate (collector) available in an n-well process in conjunction with the CMOS transistors to generate reference currents or voltages. The layout of the parasitic pnp transistor is shown in Fig. 21.9a with a cross-sectional view shown in Fig. 21.9b. The base of this parasitic transistor is tied to the substrate, which in turn is tied to VSS. The transistor is connected as a diode. This

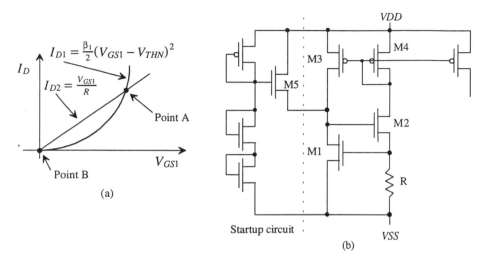

Figure 21.8 (a) Two possible operating points of self-biased circuit. (b) A startup circuit.

configuration reduces the effective resistance in series with the diode and the leakage to substrate.

The emitter size of the pnp is of importance in our applications. The minimum size allowable in the CN20 process is 6 μm square, or an area of 36 μm² (36 × 10⁻¹² m²). Notice how the transistor in Fig. 21.9c is specified by the number "1" indicating minimum size. If we were to specify 8 for the size of the pnp transistor, this would correspond to an emitter area of 8·36 μm² or 288 μm².

It is important to simulate these parasitic transistors using SPICE. Since the base will be tied to the collector, we can use a diode model statement to model the device. The current through a forward-biased diode is given by

$$I_d = I_s \cdot e^{V_d/n \cdot V_T} \tag{21.19}$$

where V_d is the voltage across the diode and V_T is the thermal voltage kT/q or 26 mV at room temperature (see Ch. 2).

The first parameter of interest in Eq. (21.19) is the scale current, I_s. This current can be determined by the BSIM p-channel model parameter, JS, which characterizes the leakage from the p+ source or drain implant to well. For the minimum-size pnp with emitter size A_E the scale current is given by

$$I_s = JS \cdot AE = 10^{-8} \frac{A}{m^2} \cdot 36 \ \mu m^2 = 360 \times 10^{-21} \ A \tag{21.20}$$

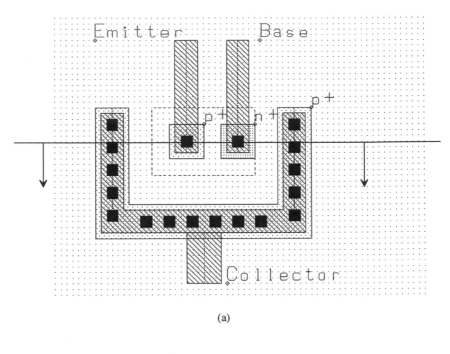

(a)

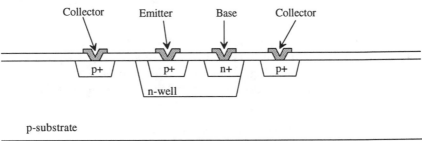

(b)

(c)

Figure 21.9 Layout (a) and cross-sectional view (b) of the parasitic pnp transistor available in an n-well CMOS process and (c) schematic representation of a minimum-size parasitic pnp, that is, emitter area of 6 µm by 6µm.

which is unrealistically low. The default value of the current density, *JS*, was used in the models (this was discussed in greater detail back in Ch. 14), meaning that the current density was not measured. A more realistic value for the saturation current may be

$$I_s = 10^{-15} \text{ A} \tag{21.21}$$

Without measuring the diode characteristics, we cannot determine I_s or *n* (the emission coefficient). If we assume that *n* = 1, we can use the following diode model for the pnp parasitic transistor and get reasonable agreement with experimental results

.MODEL   PNPDIOD   D(is=1E-15 n=1)

The specifications for the diode connected pnp transistors shown in Fig. 21.10 is

D1   1   5   PNPDIOD
D2   2   5   PNPDIOD   8

where the area factor, A (= 8), scales *IS* (i.e., the scale current for D2 is *A·IS*).

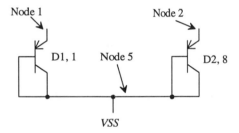

Figure 21.10 Diagram of two parasitic pnp transistors and their implementation in SPICE.

A diode-referenced self-biasing circuit is shown in Fig. 21.11. In this figure the cascode mirrors made with M1 through M8 force the same current, *I*, to flow through D1 and *R*. If the voltage across the diode is V_d, then the current $I (= I_d)$ is given by

$$I = \frac{V_d}{R} = I_s e^{V_d/n \cdot V_T} \tag{21.22}$$

or solving for the resistor value given a current level *I* results in

$$R = \frac{n \cdot V_T}{I} \ln \frac{I}{I_s} \tag{21.23}$$

The main benefit of this circuit over the threshold referenced self-biasing circuit of the previous section is the better matching, from wafer to wafer and on the same die, of the diode voltage over the threshold voltage. The main problem with this biasing circuit, as with the threshold reference biasing scheme, is the temperature dependence. The temperature coefficient of the diode is approximately -2 mV/°C or $-3{,}300$ ppm/°C for a diode with 0.6 V forward bias. This, coupled with the positive temperature coefficient

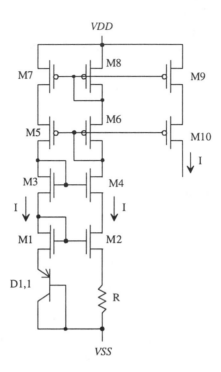

Figure 21.11 Diode referenced self-biasing circuit.

of the n+ resistor of around 2,000 ppm/°C, causes the biasing circuit to have a negative temperature coefficient.

21.2.3 Thermal Voltage Referenced Self-Biasing

A thermal voltage, V_T, referenced self-biasing circuit is shown in Fig. 21.12. Again, the cascode connection of M1 through M8 forces the same current through D1 and D2. D2 has an emitter area K times larger than D1. In this configuration, the voltage across D1 must be equal to the voltage across D2 and R, or

$$V_{d1} = I_{d2}R + V_{d2} \tag{21.24}$$

We know that

$$I_{d1} = I_s e^{V_{d1}/nV_T} \rightarrow V_{d1} = nV_T \cdot \ln\frac{I_{d1}}{I_s} \tag{21.25}$$

and

$$I_{d2} = K \cdot I_s e^{V_{d2}/nV_T} \rightarrow V_{d2} = nV_T \cdot \ln\frac{I_{d2}}{K \cdot I_s} \tag{21.26}$$

Solving for R in terms of the reference current I results in

$$R = \frac{nV_T \cdot \ln K}{I} \text{ or } I = \frac{nk \cdot \ln K}{qR} \cdot T \qquad (21.27)$$

Notice that the current is proportional to the absolute temperature (PTAT). For $K = 8$ and $n = 1$, the voltage drop across R is only 54 mV. Mismatches in the gate-source voltages of M1 and M2 can result in large variations in I when using this scheme. Generally, M1 and M2 are made large for better matching of the IV characteristics. We should also keep in mind that the resistance value R may change up to 20 percent from wafer to wafer. If a precise current is needed, some sort of on-die tuning is required. More often, however, we care more about the relative current levels and temperature changes on a given die than about the die-to-die relative current levels.

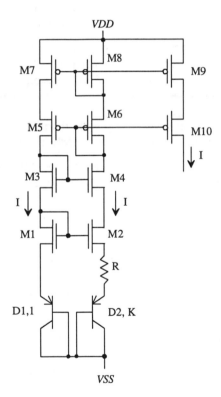

Figure 21.12 Thermal voltage referenced self-biasing circuit.

The main advantage of this current reference is in its temperature characteristics. The temperature dependence of the current I is given by

$$TC_I = \frac{1}{I}\frac{dI}{dT} = \frac{1}{V_T}\frac{\partial V_T}{\partial T} - \frac{1}{R}\cdot\frac{\partial R}{\partial T} \qquad (21.28)$$

The second term in this equation is simply the TC of the resistor, or approximately 2,000 ppm/°C. The first term in this equation is the TC of the thermal voltage, given by

$$TC_{V_T} = \frac{1}{V_T}\frac{\partial V_T}{\partial T} = \frac{q}{kT} \cdot \frac{k}{q} = +3,300 \text{ ppm/°C or ppm/°K} \qquad (21.29)$$

(T in kelvin) where the change in the thermal voltage with temperature is given by

$$\frac{\partial V_T}{\partial T} = \frac{k}{q} = 0.085 \text{ mV/°C} \qquad (21.30)$$

The temperature coefficient of this reference is on the order of +1,000 ppm/°C. The fact that both the temperature coefficient of the thermal voltage and the resistor are positive helps give better temperature characteristics than the diode or threshold voltage references. Practically, the problem with this reference is the matching of the MOSFETs and the susceptibility to externally coupled noise across R.

Note how the temperature coefficient is dependent on a given temperature; that is, we get different temperature coefficients at $T = 300$ °K and $T = 350$ °K. This is true with any temperature coefficient (i.e., TC is a function of temperature).

Example 21.3
Design a 10 μA current source using the thermal voltage self-biased reference. Simulate the temperature characteristics of the design.

Referring to Fig. 21.12, if we require that the current I is 10 μA and that the gate-source voltages are 1.2 V, then the size of the n-channel MOSFETs is 15/5 and the size of the p-channel MOSFETs is 70/5 (see Ex. 20.3). If the emitter area of the diode D2 is eight times larger than D1, then the resistor value is given by

$$R = \frac{26 \text{ mV} \cdot \ln 8}{10 \text{ μA}} = 5.4 \text{ k}\Omega$$

The PSPICE simulation results are shown in Fig. 21.13. The output current is taken as the current through M9 and M10, each sized 70/5, in Fig. 21.12. The netlist is shown below. To aid in convergence of the DC sweep, we used the .NODESET command to set the high-impedance nodes, nodes 2 and 12, and the voltage at the top of the resistor to zero volts. The .NODESET line doesn't force the nodes to zero volts; it just starts the simulation with these initial guesses. Also, the .OPTIONS statement was used to loosen the tolerances and increase the DC and bias point iteration limit, ITL1, and the DC/bias point educated guess iteration limit, ITL2. Notice that the TC of the resistor is specified in the netlist as 2,000 ppm/°C. ■

```
*** Top Level Netlist ***
D1      7 3             pnpdiod
D2      8 3             pnpdiod        8
M1      4 4 7 3         CMOSNB         L=5u W=15u
M2      2 4 11 3        CMOSNB         L=5u W=15u
```

```
M3       1 1 4 3          CMOSNB        L=5u W=15u
M4       9 1 2 3          CMOSNB        L=5u W=15u
M5       1 9 12 6         CMOSPB        L=5u W=70u
M6       9 9 10 6         CMOSPB        L=5u W=70u
M7       12 10 6 6        CMOSPB        L=5u W=70u
M8       10 10 6 6        CMOSPB        L=5u W=70u
M9       13 10 6 6        CMOSPB        L=5u W=70u
M10      14 9 13 6        CMOSPB        L=5u W=70u
R1       8 11     5.4k    TC1=0.002
VDD      6 0      DC 2.5
Vout     14 0     DC 0
VSS      3 0      DC -2.5
.NODESET V(8)=0 V(2)=0 V(12)=0
***** Spice models and macro models *****
.MODEL CMOSNB NMOS LEVEL=4
+...BSIM
.MODEL PNPDIOD D
+IS=1E-15 n=1
.MODEL CMOSPB PMOS LEVEL=4
+ ...BSIM
.OPTION RELTOL=0.01 ABSTOL=1n VNTOL=1m ITL1=100 ITL2=100
.probe
.DC Vout -1 1 .01    temp -25 75 25
.end
```

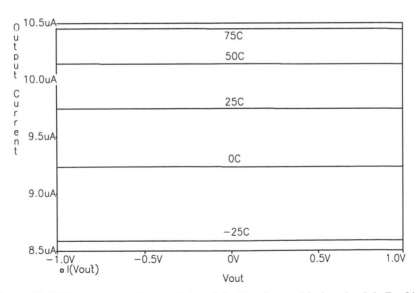

Figure 21.13 Temperature characteristics of the V_T reference biasing circuit in Ex. 21.3.

21.3 Bandgap Voltage References

Bandgap voltage references combine the positive TC of the thermal voltage with the negative TC of the diode forward voltage (the bandgap energy, E_g, of silicon decreases with increasing temperature; refer to Ch. 9) in a circuit to achieve a voltage reference with a zero TC. Once we have a temperature-independent voltage reference, it is a simple matter, with the use of an op-amp, to generate multiples of the reference.

21.3.1 Bandgap Referenced Biasing

The circuit shown in Fig. 21.14 is an example of a bandgap reference. The diode D3 is the same size as D2, while the resistor in series with D3 is L times larger than the resistor in series with D2. The current I in the figure is given by Eq. (21.27), or

$$I = \frac{nV_T \cdot \ln K}{R} \tag{21.31}$$

The reference output voltage with respect to VSS is given by

$$V_{ref} = I \cdot L \cdot R + V_{d3} \tag{21.32}$$

or

$$V_{ref} = (L \cdot n \cdot \ln K)V_T + V_{d3} \tag{21.33}$$

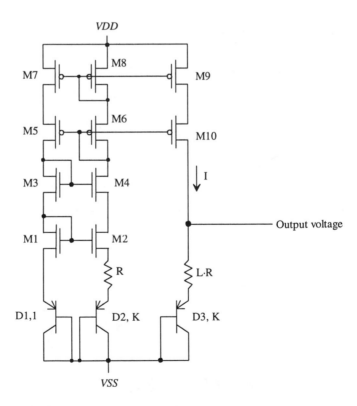

Figure 21.14 A bandgap voltage reference.

The *TC* of the bandgap reference is zero when

$$\frac{\partial V_{ref}}{\partial T} = L \cdot n \cdot \ln K \cdot \overbrace{\frac{\partial V_T}{\partial T}}^{0.085\, mV/^{\circ}C} + \overbrace{\frac{\partial V_{d3}}{\partial T}}^{-2\, mV/^{\circ}C} = 0 \qquad (21.34)$$

This is true when

$$L \cdot n \cdot \ln K = \frac{2}{0.085} = 23.5 \qquad (21.35)$$

For $n = 1$ and $K = 8$ the factor L is 11.3 for a zero *TC*. We will round L up to 12 for the design here. The reference voltage, with respect to *VSS*, may also be written as

$$V_{ref} = (Ln \ln K) \cdot V_T + nV_T \cdot \ln \frac{I}{K \cdot I_s} \qquad (21.36)$$

The reference voltage at 300 °K for $I = 10$ µA, $I_s = 10^{-15}$A, $n = 1$, $K = 8$, and $L = 12$ is 1.25V (i.e., $V_{ref} = 1.25$ if *VSS* = 0 and *VDD* = 5 V). A PSPICE simulation of this bandgap is shown in Fig. 21.15. The current source designed in Ex. 21.3 was used to bias a resistor of value 65 kΩ and a diode with an area of eight times the minimum size. Note: To accurately predict the reference voltage, thorough characterization of the parasitic pnp is required over process and temperature variations.

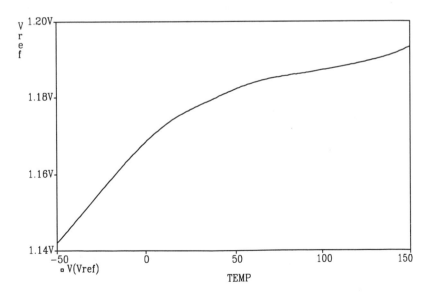

Figure 21.15 PSPICE simulation results of a bandgap voltage reference.

21.4 Beta Multiplier Referenced Self-Biasing

A biasing scheme that deserves special mention is the β multiplier reference of Fig. 21.16 [3]. The width of M2 is made K times larger than the width of M1, so that

$$\beta_2 = K \cdot \beta_1, \text{ assuming } L_1 = L_2 \text{ and } W_2 = K \cdot W_1 \qquad (21.37)$$

and therefore

$$V_{GS1} = V_{GS2} + IR \qquad (21.38)$$

We can write V_{GS} of MOSFETs M1 and M2 in terms of the current I as

$$V_{GS1} = \sqrt{\frac{2I}{\beta_1}} + V_{THN} \qquad (21.39)$$

and neglecting the body effect

$$V_{GS2} = \sqrt{\frac{2I}{K \cdot \beta_1}} + V_{THN} \qquad (21.40)$$

Solving for I using the previous three equations yields

$$I = \frac{2}{R^2 \beta_1} \cdot \left(1 - \sqrt{\frac{1}{K}}\right)^2 \qquad (21.41)$$

This is the basic design equation for this reference. The size parameter K must always be greater than 1. The temperature coefficient of the current reference is given by

$$TC_I = \frac{1}{I}\frac{\partial I}{\partial T} = -2 \cdot \frac{1}{R}\frac{\partial R}{\partial T} - \frac{1}{KP(T)}\frac{\partial KP(T)}{\partial T} = -4,000 \text{ ppm/°C} + \frac{1.5}{T} \qquad (21.42)$$

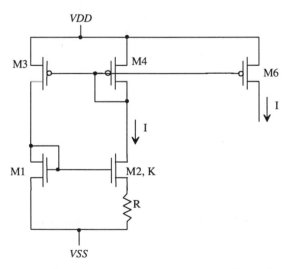

Figure 21.16 β multiplier referenced self-biasing ($\beta_2 = K \cdot \beta_1$).

Example 21.4

Using the β multiplier current reference, design a 10 μA current source. Estimate the *TC* of the reference and assume $VDD = -VSS = 2.5$ V.

If we set $L_1 = L_2 = W_1 = 5$ μm and $K = 4$, then using Eq. (21.41) and solving for R results in

$$R^2 = \frac{2}{10 \text{ μA} \cdot 50\frac{\text{μA}}{\text{V}^2}} \cdot \frac{1}{4} \rightarrow R = 31.6 \text{ k}\Omega$$

The *TC* at 300 °K from Eq. (21.42) is 1,000 ppm/°C. A schematic of the design is shown in Fig. 21.17. Cascode MOSFETs were used to reduce the effects of the finite r_o of the MOSFETs. ■

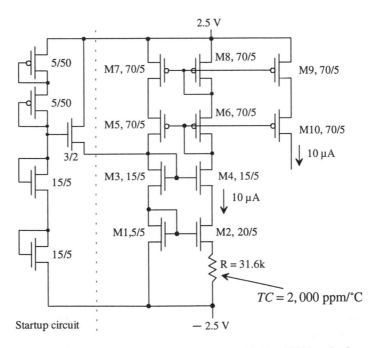

Figure 21.17 A 10 μA reference using a transconductance multiplier self-biased reference.

21.4.1 A Voltage Reference

It is of interest to determine the temperature dependence of V_{GS1} in Fig. 21.16. If the reference is designed correctly, we can make the *TC* of V_{GS1} equal to zero. We define our reference voltage as

$$V_{ref} = V_{GS1} \tag{21.43}$$

Substituting Eq. (21.41) into Eq. (21.38) results in

$$V_{ref} = V_{GS1} = \frac{2}{R\beta_1}\left(1 - \frac{1}{\sqrt{K}}\right) + V_{THN} \qquad (21.44)$$

The change in V_{ref} with temperature is

$$\frac{dV_{ref}}{dT} = \frac{dV_{THN}}{dT} - \frac{2}{R\beta_1}\left(1 - \frac{1}{\sqrt{K}}\right)\left[\frac{1}{R}\frac{\partial R}{\partial T} + \frac{1}{KP(T)}\frac{\partial KP(T)}{\partial T}\right] \qquad (21.45)$$

or

$$\frac{dV_{ref}}{dT} = -2.4 \text{ mV/°C} + \frac{2}{R\beta_1}\left[1 - \frac{1}{\sqrt{K}}\right]\left[-2,000 \text{ ppm/°C} + \frac{1.5}{T}\right] \quad (21.46)$$

At 300 °K Eq. (21.46) is equal to zero when

$$\frac{2}{R\beta_1}\left[1 - \frac{1}{\sqrt{K}}\right] = \frac{2,400}{3,000} = 0.8 \qquad (21.47)$$

Furthermore, if $K = 4$ we require that

$$R = \frac{1}{0.8 \cdot \beta_1} \qquad (21.48)$$

and

$$V_{ref} = 0.8 + V_{THN} = 1.63 \text{ V} \qquad (21.49)$$

There will be an error in the reference voltage associated with the change in the threshold voltage due to process variations and the body effect as well as changes in R, β, and T. The threshold voltage can vary by as much as 20 percent (this is part of the reason why the body effect is neglected so often in our hand calculations). Thus, precision voltages are more difficult to achieve than the bandgap reference of the last section. Often, voltage references (even bandgaps) are adjusted on die using fuses or by trimming a resistor value using a laser. The following example illustrates the design and temperature performance of a zero TC voltage reference using the β multiplier self-biased reference.

Example 21.5
Design a zero TC voltage β multiplier reference at 300 °K with $VSS = 0$ V and $VDD = 5$ V. Simulate the design for changing VDD and temperature.

We will use the general cascode schematic of Fig. 21.17 (with the same sizes) with VDD swept in the simulations and $VSS = 0$. The resistor for this design is

$$R = \frac{1}{0.8 \cdot 50 \frac{\mu A}{V^2}} = 25 \text{ k}\Omega$$

The simulation results are shown in Fig. 21.18. ∎

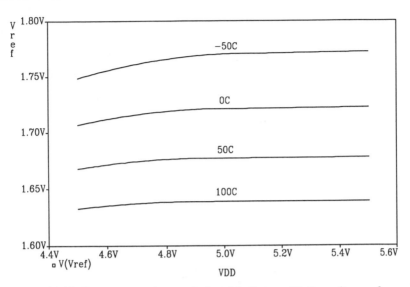

Figure 21.18 Temperature characteristics of the beta multiplier voltage reference.

21.4.2 Operation in the Subthreshold Region

For low-power operation, it may be desirable to operate a current source/sink in the weak inversion, or subthreshold region. Currents in these circuits can be 100 nA or less. Consider the value of the resistor needed for $I_{D1} = 100$ nA in Eq. (20.4). If $VDD - VSS \gg V_{GS}$, then the value of R for the simple current mirror is given by

$$R \approx \frac{VDD - VSS}{I_{D1}} \tag{21.50}$$

If $VDD - VSS = 5$ V, then $R = 50$ MEG! Clearly, the basic current mirror with resistor bias will not be useful in circuits operating in the subthreshold region.

Consider designing the basic β multiplier of Fig. 21.16 so that it operates in the subthreshold region where the current is given, from Ch. 6, by

$$I \approx I_{exp} = I_{DO}\frac{W}{L} \cdot (e^{(V_{GS} - V_{THN})/(n \cdot V_T)}) \tag{21.51}$$

Following the same procedure as in the case of strong inversion

$$V_{GS1} = nV_T \cdot \ln\left[\frac{I \cdot L}{I_{DO} \cdot W}\right] + V_{THN} \tag{21.52}$$

and

$$V_{GS2} = nV_T \cdot \ln\left[\frac{I \cdot L}{I_{DO} \cdot K \cdot W}\right] + V_{THN} \tag{21.53}$$

Solving for the subthreshold current I using Eq. (21.38) and Eqs. (21.52) and (21.53), we get

$$I_D = \frac{n \cdot V_T}{R} \cdot \ln K \tag{21.54}$$

Note the similarity to the thermal voltage referenced self-biasing circuit. Note also that this current reference can be used to generate a bandgap voltage reference as was discussed in Sec. 21.3.1. The following example illustrates the design of a subthreshold current source.

Example 21.6

Design and simulate a 75 nA current source with $VDD = 5$ V and $VSS = 0$.

Using Eq. (21.54) with $K = 8$, $n = 1$, $V_T = 26$ mV @ room temperature (approximately 27 °C) yields

$$R = \frac{26\text{mV}}{75\text{nA}} \cdot \ln 8 = 720 \text{ k}\Omega$$

We will use the general schematic of Fig. 21.17 with W_2, which is now 40 µm.

The resulting SPICE simulation of the output current versus voltage at the drain of M10 is shown in Fig. 21.19. Notice that V_{out} can get very close to V_{DD}, within 100 mV, before M10 begins to shut off. Also notice that the current is almost twice what we designed for. This is due to the body effect of M2. Small variations across R have a large effect on the current. Using the p-channel MOSFETs version of Fig. 21.16 with each p-channel in its own well eliminates the body-effect problem. ∎

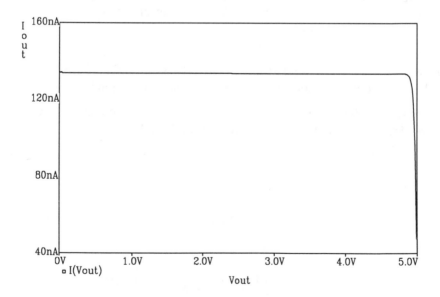

Figure 21.19 Simulation results of Ex. 21.6, the design of a subthreshold current source.

REFERENCES

[1]   P. R. Gray and R. G. Meyer, *Analysis and Design of Analog Integrated Circuits,* 3rd ed., John Wiley and Sons, 1993. ISBN 0-471-57495-3.

[2]   R. L. Geiger, P. E. Allen, and N. R. Strader, *VLSI-Design Techniques for Analog and Digital Circuits,* McGraw-Hill Publishing Co., 1990. ISBN 0-07-023253-9.

[3]   E. Vittoz and J. Fellrath, "CMOS Analog Integrated Circuits Based on Weak Inversion Operation," *IEEE Journal of Solid State Circuits,* Vol. SC-12, No. 3, June 1977, pp. 224-231.

PROBLEMS

21.1   Design a resistor-MOSFET voltage divider, Fig. P21.1, so that the reference voltage, V_{ref}, is 2 V and the current through the resistor is 10 µA. Verify the operation of the circuit using SPICE.

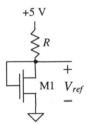

Figure P21.1

21.2   Determine the output voltage and temperature coefficient of the voltage reference in Fig. P21.1 if $W/L = 20/5$ and $R = 400k$. Assume that the *TC* of the resistor is 10,000 ppm/°C. Compare the hand-calculated *TC* of the reference to SPICE simulations.

21.3   Design a MOSFET-MOSFET voltage divider, Fig. P21.3, so that the reference voltage, V_{ref}, is 2 V and the current through each MOSFET is 10 µA. Verify the operation of the circuit using SPICE.

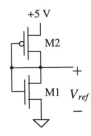

Figure P21.3

21.4   Determine the output voltage and temperature coefficient of the voltage reference in Fig. P21.3 if the *W/L* of both MOSFETs is 20/20. Compare the hand-calculated *TC* of the reference to SPICE simulations.

21.5   Calculate and simulate the sensitivity of the reference described in Problem 4 to *VDD*.

21.6   Select the size of M3 in Fig. P21.6 so that 10 μA flows in all MOSFETs in the circuit. Assume that M5 and M4 are operating in the saturation region.

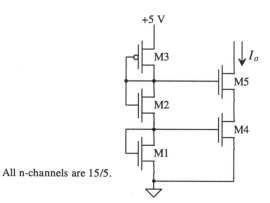

Figure P21.6

21.7   Determine the current *I* in Fig. P21.7. Neglect oxide encroachment and assume that M5 is operating in the saturation region.

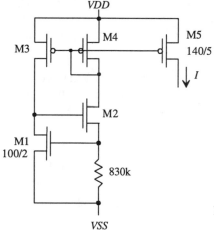

Figure P21.7

21.8   Sketch a startup circuit for the threshold-voltage referenced current source of Fig. P21.7. Assume the maximum current in the startup circuit is 1 μA.

21.9   Lay out a parasitic PNP transistor with an emitter size of 2 (emitter size of 12 μm by 6 μm). What value of IS will be used in the diode model if the area specified for the diode is 1 (which is the same as not specifying an area for the diode)?

21.10 Design a 5 μA current source using the diode referenced self-biasing circuit of Fig. 21.11. Verify the operation of your design using SPICE. (Either include a startup circuit or use the .NODESET statement to keep the circuit from operating at point B in Fig. 21.8.)

21.11 Repeat Ex. 21.3 for a 5 μA current source.

21.12 Verify that Eq. (21.36) is correct. How is the resulting reference voltage affected by using an $L = 12$ instead of 11.3 (Does the reference voltage increase or decrease? Why?)

21.13 The main reason for using the cascode MOSFETs, M1 through M10, shown in Fig. 21.14 is to force the same current through D2 and D3. The main problem with this configuration is the dependence of the reference voltage on changes in the power supply. Using an op-amp to force the same currents through D2 and D3 can improve the sensitivity of the reference voltage to power supply variations (Fig. P21.13). Knowing the reference voltage is given by

$$V_{ref} = V_{d2} + I \cdot L \cdot R$$

show that the current that flows in D1 and D2 is

$$I = \frac{nV_T \cdot \ln K}{R}$$

and therefore

$$V_{ref} = (L \cdot n \cdot \ln K) \cdot V_T + \overbrace{nV_T \cdot \ln \frac{I}{I_S}}^{V_{d2}}$$

Determine the conditions under which the TC of the reference is zero (review Eq. [21.34]).

21.14 Estimate the power dissipated by the startup circuit shown in Fig. 21.17.

21.15 Using the β multiplier voltage reference, design a voltage reference with a TC of +1,000 ppm/°C. Simulate the operation of the reference. Often, in digital systems as the temperature increases, it is desirable to increase the supply voltage used to power the digital logic to compensate for the increase in the MOSFETs' effective resistance and thus the increase in the gate-delay times.

21.16 The β multiplier reference is an example of a circuit that uses positive feedback (the loop gain is less than one). Discuss methods of ensuring a stable operating point for the circuit.

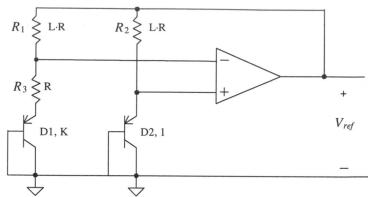

A bandgap reference   **Figure P21.13**

21.17 An important consideration in the design of a voltage or current reference is the design of the startup circuit. Consider the portion of the bandgap reference of Fig. P21.13 shown in Fig. P21.17a. Using SPICE, plot V_{ref} against V_{plus} (the noninverting input of the op-amp) and V_{minus} (the inverting input of the op-amp). The result should look similar to Fig. P21.17b. Design a startup circuit for the bandgap of Fig. P21.13.

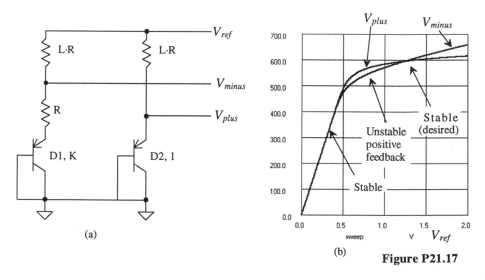

Figure P21.17

21.18 Show that if $K = 1$ and $R_1 \neq R_2$, in Fig. P21.13, the output of the bandgap reference can be written as $V_{ref} = V_{d2} + \dfrac{R_1}{R_3} n V_T \cdot \ln \dfrac{R_1}{R_2}$.

21.19 Repeat Ex. 21.6 for a 0.1 µA current source.

Amplifiers

Single-stage amplifiers are used in virtually every op-amp design. By replacing a passive load resistor with a MOSFET transistor (called an active load), significant amounts of chip area can be saved. An active load can also produce higher values of resistance when compared with a passive resistor, resulting in higher gains.

Several types of active loads are studied in this chapter. The gate-drain connected load consists of a MOSFET with the gate and drain shorted and provides large bandwidths at the cost of reduced gain and low-output impedance. The current source load amplifier has a higher gain and higher output impedance, but usually at the expense of producing a lower bandwidth. Current source load amplifiers are preferred when external feedback is applied to the amplifier to set the gain. The chapter will explore basic single-stage amplifiers with active loads, as well as the corresponding tradeoffs associated with each. The cascode amplifier will also be examined in detail along with several configurations of output stages, including the push-pull amplifier.

22.1 Gate-Drain Connected Loads

The four types of gate-drain connected load amplifiers are shown in Fig. 22.1. All four of these configurations are basic common source voltage amplifiers. In each configuration, M1 and M2 are assumed to be biased in the saturation region.

22.1.1 Common Source Amplifiers

Examine the AC small-signal circuit shown in Fig. 22.1a. Since many MOSFET amplifiers will be analyzed here using small-signal analysis, an intuitive approach will be presented which allows the designer to analyze the circuit very quickly. Begin by replacing M2 with a resistor of value $\frac{1}{g_{m2}}$ and replacing M1 with a current source of value $g_{m1} \cdot v_{in}$, provided $\frac{1}{g_{m2}} \ll r_{o1}$. The equivalent circuit is shown in Fig. 22.2. Note that there is no body effect in either MOSFET and that only the low-frequency model is

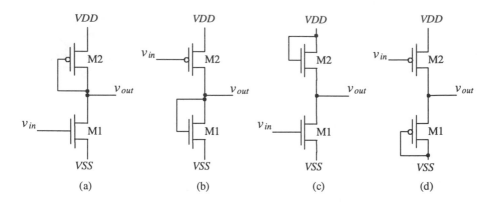

Figure 22.1 The four active load configurations available in CMOS.

shown.   Again, as was seen in Ch. 9, a gate-drain connected MOSFET has a small-signal resistance of value $\frac{1}{g_m}$. The small-signal gain of this circuit is given by

$$\frac{v_{out}}{v_{in}} = \frac{-i_d \cdot \frac{1}{g_{m2}}}{i_d \cdot \frac{1}{g_{m1}}} = -\frac{\frac{1}{g_{m2}}}{\frac{1}{g_{m1}}} = -\frac{\text{resistance in the drain}}{\text{resistance in the source}} = -\frac{g_{m1}}{g_{m2}} \qquad (22.1)$$

This result is very important in the intuitive analysis. It states that the small-signal gain of a common source amplifier is simply the resistance in the drain of M1 divided by the resistance in the source (the resistance looking into the source of M1 added to any resistance from the source of M1 to ground).

The intuitive approach results in an approximation for the gain. The actual value of the resistance in the drain, as will be soon discovered, is $\frac{1}{g_{m2}}||r_{o1}||r_{o2}$. In most cases, however, r_o will be much greater than $\frac{1}{g_m}$, and the approximation is an accurate one.

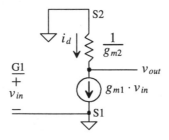

Figure 22.2 Simplified circuit of Fig. 22.1a.

Example 22.1
Determine the small-signal AC gain of the circuit shown in Fig. 22.3 using the intuitive method presented in Eq. (22.1).

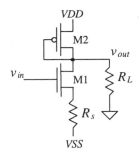

Figure 22.3 Amplifier analyzed in Ex. 22.1.

The small-signal gain of this circuit is given by

$$A_v = \frac{v_{out}}{v_{in}} = - \frac{\text{resistance in the drain}}{\text{resistance in source}}$$

The resistance in the drain is the parallel combination of all resistances connected to the drain of M1, given by

$$\text{Resistance in the drain} = \frac{1}{g_{m2}} \| R_L \text{ for } r_{o1} \gg \frac{1}{g_{m2}}$$

The resistance in the source is the sum of the resistance looking into the source of M1 and the resistance connected from the source of M1 to ground. For the present problem, this resistance is given by

$$R_{\text{in the source}} = \frac{1}{g_{m1}} + R_s$$

The voltage gain is then given by

$$A_v = - \frac{\frac{1}{g_{m2}} \| R_L}{\frac{1}{g_{m1}} + R_s} \text{ which for } R_L \to \infty \text{ and } R_s \to 0 \text{ reduces to } - \frac{g_{m1}}{g_{m2}} \blacksquare$$

We can determine the exact gain of the amplifier of Fig. 22.1a using the full small-signal model shown in Fig. 22.4. Using KCL at the output of the amplifier gives

$$g_{m1} v_{in} + \frac{v_{out}}{r_{o2} \| r_{o1}} = - g_{m2} v_{out} \qquad (22.2)$$

or

$$A_v = \frac{v_{out}}{v_{in}} = - \frac{g_{m1}}{g_{m2} + \frac{1}{r_{o1} \| r_{o2}}} = - \frac{\frac{1}{g_{m2}} \| r_{o1} \| r_{o2}}{\frac{1}{g_{m1}}} = - \frac{\text{resistance in the drain}}{\text{resistance in the source}} \qquad (22.3)$$

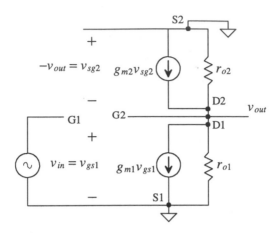

Figure 22.4 Small-signal model of the amplifier shown in Fig. 22.1(a).

For the case when $\frac{1}{g_{m2}} \ll r_{o1} \| r_{o2}$, this reduces to

$$A_v = -\frac{\frac{1}{g_{m2}}}{\frac{1}{gm1}} = -\frac{g_{m1}}{g_{m2}} \tag{22.4}$$

which is the same form as Eq. (22.1).

Now consider the frequency response of the common source amplifier of Fig. 22.1a. The high-frequency equivalent circuit is shown in Fig. 22.5, with the MOSFET capacitances drawn explicitly. A source resistance is added to the circuit to model the effects of the driving source impedance. Miller's Theorem is used to break C_{gd1} into two parts; a capacitance at the gate of M1 to ground and a capacitance from the drain of M1 to ground. Two RC time constants exist in this circuit, one on the input of the circuit and one on the output of the circuit. The input time constant, neglecting C_{gb}, is given by

$$\tau_{in} = R_s(C_{MI} + C_{gs1}) \tag{22.5}$$

where the Miller capacitance at the input, C_{MI} is

$$C_{MI} = C_{gd1}(1 + \frac{g_{m1}}{g_{m2}}) \tag{22.6}$$

When using Miller's Theorem, it should be noted that the capacitance, C_{gd1}, is multiplied by the factor $(1 - K)$, where K is the gain, $-\frac{g_{m1}}{g_{m2}}$.

The output time constant is found in a similar way and is given by

$$\tau_{out} = \frac{1}{g_{m2}} \cdot (C_{gs2} + C_{MO} + C_{db1} + C_{db2}) \tag{22.7}$$

The Miller capacitance which appears at the output, C_{MO}, now becomes

$$C_{MO} = C_{gd1}(1 - \frac{1}{K}) = C_{gd1}(1 + \frac{g_{m2}}{g_{m1}}) \tag{22.8}$$

The frequency response of this amplifier is given by

$$A_v(f) = \frac{-\frac{g_{m1}}{g_{m2}}}{\left(1 + j\frac{f}{f_{in}}\right)\left(1 + j\frac{f}{f_{out}}\right)} \tag{22.9}$$

where

$$f_{in} = \frac{1}{2\pi\tau_{in}} \tag{22.10}$$

and

$$f_{out} = \frac{1}{2\pi\tau_{out}} \tag{22.11}$$

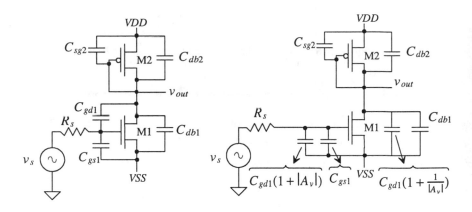

Figure 22.5 Frequency response of the simple common source amplifier with active load.

It should be noted that when using Miller's Theorem, a zero is neglected. For example, in Fig. 22.6, if the input is assumed to be driven from a voltage source so that the input parasitics can be ignored, the transfer function becomes,

$$\frac{v_{out}}{v_{in}} = -g_{m1}R_{Leq}\frac{(1 - s/\omega_{z1})}{(1 + s/\omega_{p1})} \tag{22.12}$$

where ω_{z1} is a right-hand plane zero of value

$$\omega_{z1} = \frac{g_{m1}}{C_{gd1}} \tag{22.13}$$

and ω_{p1} is the pole at the output of the amplifier of value

$$\omega_{p1} = \frac{1}{R_{Leq}(C_{Leq} + C_{gd1})} \qquad (22.14)$$

where $C_{Leq} = C_{sg2} + C_{db1} + C_{db2}$ and $R_{Leq} = r_{o1} \| r_{o2} \| 1/g_{m2}$. The magnitude response of the right-hand plane zero is usually ignored since its value typically places it well beyond the pole frequency. However, the phase response can cause problems since a right-hand plane zero has the same phase as a left-hand plane pole, which could seriously affect stability when the single-stage amplifier is used within an op-amp. These stability issues along with the compensation techniques are presented in Ch. 25.

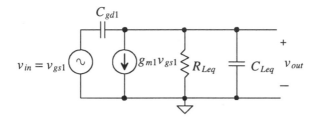

Figure 22.6 Small-signal model used to calculate the RHP zero for amplifiers in Fig. 22.1a-d.

Example 22.2

Determine the gain and bandwidth of the amplifier shown in Fig. 22.7. Determine the output voltage when the input voltage is $10^{-3}\sin 2\pi \cdot 1000t$ volts.

The drain current that flows in M1, and thus M2, is given by

$$I_D = \frac{\beta_1}{2}(V_{GS1} - V_{THN})^2 = \frac{50\mu}{2} \cdot \frac{5\mu}{5\mu}(2.5 - 0.8)^2 = 72 \text{ }\mu A$$

while V_{SG2} is determined by

$$I_D = 72 \text{ }\mu A = \frac{\beta_2}{2}(V_{SG2} - V_{THP})^2 = \frac{17\mu}{2} \cdot \frac{15\mu}{5\mu}(V_{SG2} - 0.9)^2 \Rightarrow V_{SG2} \approx 2.5 \text{ V}$$

The DC portion of the output voltage is approximately 0 V. The AC gain is given by

$$A_v = -\frac{\frac{1}{g_{m2}}}{\frac{1}{g_{m1}}} = -\frac{g_{m1}}{g_{m2}} = -\frac{\sqrt{2\beta_1 I_D}}{\sqrt{2\beta_2 I_D}} = -1$$

Voltage gains in the neighborhood of 1 are common when using gate-drain connected loads. At this point it may appear that this amplifier is almost worthless; however, we will see that when this amplifier is used with additional circuitry (e.g., a current source load for a differential pair) it can be very useful. The output voltage when the input is a 1 mV sine wave at 1 kHz is given by

$$v_{OUT} = \overbrace{0\ V}^{DC} + \overbrace{10^{-3}\sin(2\pi \cdot 1000t + \pi)}^{AC\ component}$$

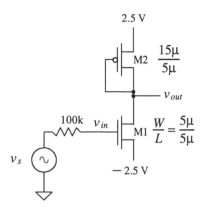

Figure 22.7 Amplifier used in Ex. 22.2.

The bandwidth of the amplifier is calculated by first estimating the capacitances of the MOSFETs. Assuming the drain diffusion to substrate/well is 2.5 V and the n-channel drain area measures 6 μm by 6 μm, while the p-channel drain area measures 15 μm by 6 μm, we find that the capacitances are given by

$$C_{db1} = \frac{CJ \cdot AD}{\left(1 - \frac{V_{db1}}{PB}\right)^{MJ}} + \frac{CJSW \cdot PD}{\left(1 - \frac{V_{db1}}{PBSW}\right)^{MJSW}} = \frac{1.04 \times 10^{-4}\frac{F}{m^2} \cdot 36 \times 10^{-12}m^2}{\left(1 + \frac{2.5}{0.8}\right)^{0.66}} +$$

$$\frac{2.2 \times 10^{-10}\frac{F}{m} \cdot 24 \times 10^{-6}m}{\left(1 + \frac{2.5}{0.8}\right)^{0.18}} = 5.5\ \text{fF}$$

and for the p-channel $C_{db2} = 15.4$ fF. The other capacitances are calculated from

$$C_{gd1} = CGDO \cdot W = 3.8 \times 10^{-10}\frac{F}{m} \cdot 5\mu m = 1.9\ \text{fF}$$

$$C_{gs1} = \frac{2}{3}WLC'_{ox} = \frac{2}{3} \cdot 5\mu m \cdot 5\mu m \cdot 800\frac{aF}{\mu m^2} = 13.3\ \text{fF}$$

$$C_{sg2} = \frac{2}{3}WLC'_{ox} = \frac{2}{3} \cdot 15\mu m \cdot 5\mu m \cdot 800\frac{aF}{\mu m^2} = 39.9\ \text{fF}$$

The input time constant is given by

$$\tau_{in} = 100k \cdot [1.9\text{fF}(1 + 1) + 13.3\text{fF}] = 1.7\ \text{ns}$$

and the output time constant is given by

$$\tau_{out} = \frac{1}{\sqrt{2 \cdot 17\frac{\mu A}{V} \cdot \frac{15\mu}{5\mu} \cdot 72 \, \mu A}} \cdot (39.9 \text{ fF} + 1.9 \text{ fF}(1+1) + 5.5 \text{ fF} + 15.4 \text{ fF}) = 746 \text{ ps}$$

The input time constant dominates. Therefore, the bandwidth of this amplifier can be estimated by

$$f_{3dB} = f_{in} = \frac{1}{2\pi \cdot 1.7\text{ns}} = 93.6 \text{ MHz}$$

Large bandwidth is the main advantage of the active gate/drain-connected load amplifier while low gain is the main drawback. ∎

The previous discussion was centered around common source voltage amplifiers. Consider the circuit shown in Fig. 22.8. This amplifier can be thought of as a transimpedance amplifier, that is, voltage out and current in. The input impedance of this amplifier is given by

$$R_{in} = \frac{1}{g_{m3}} \tag{22.15}$$

The input current, i_{in}, is determined by

$$i_{in} = g_{m3}v_{in} = g_{m3}v_{gs3} = g_{m3}v_{gs1} \tag{22.16}$$

The current through M1, and thus M2, is given by

$$i_d = g_{m1}v_{in} = \frac{g_{m1}}{g_{m3}} \cdot i_{in} \tag{22.17}$$

This equation can be rewritten as

$$\frac{i_d}{i_{in}} = \frac{\beta_1}{\beta_3} = \frac{W_1 L_3}{W_3 L_1} \tag{22.18}$$

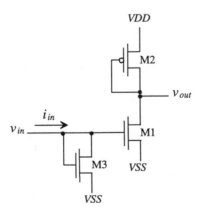

Figure 22.8 An implementation of the transimpedance amplifier.

That is, we can get a current gain simply by adjusting the size of M1 and M3. The transresistance gain of this configuration is given by

$$A_R = \frac{v_{out}}{i_{in}} = -\frac{i_d \frac{1}{g_{m2}}}{i_d \cdot \frac{W_3 L_1}{W_1 L_3}} = -\frac{W_1 L_3}{g_{m2} W_3 L_1} \qquad (22.19)$$

Adding M3 and thus the resistor of value $\frac{1}{g_{m3}}$ to ground lowers the input time constant and thus increases the bandwidth of the amplifier.

Before we proceed any further, let's review, or again verify, how to determine the small-signal, low-frequency resistances looking into the drain and source of a MOSFET. Using this information will greatly aid in our intuitive analysis of CMOS circuits. Consider the test circuit shown in Fig. 22.9. Here the DC voltages ensure that the MOSFET is operating in the saturation region. We are concerned with the AC current that flows through v_t. This test voltage divided by i_d gives us the small-signal AC resistance looking into the drain of the MOSFET. In other words, this is the resistance connected to ground at the drain node.

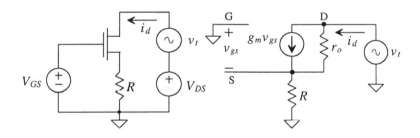

Figure 22.9 Circuit used to determine the resistance looking into the drain of a MOSFET.

Neglecting the body effect, we may write the test voltage as

$$v_t = (i_d - g_m v_{gs})r_o + i_d R \qquad (22.20)$$

where

$$v_{gs} = -i_d R \qquad (22.21)$$

The resistance looking into the drain of the MOSFET is then given by

$$R_o = r_d = \frac{v_t}{i_d} = (1 + g_m R)r_o + R \approx (1 + g_m R)r_o \qquad (22.22)$$

keeping in mind that $r_o \approx \frac{1}{\lambda I_D}$.

Example 22.3

Repeat Ex. 22.1 without neglecting the loading effects of the output resistances of M1 and M2. Use the circuit shown in Fig. 22.3.

The resistance looking into the drain of M2 is $\frac{1}{g_{m2}}\|r_{o2}$, while the resistance looking into the drain of M1 is $r_{o1}(1+g_{m1}R_s)$. The exact gain of the circuit is simply

$$A_v = \frac{v_{out}}{v_{in}} = -\frac{\frac{1}{g_{m2}}\|r_{o2}\|[r_{o1}(1+g_{m1}R_s)]\|R_L}{R_s+\frac{1}{g_{m1}}}$$

or when $r_{o1}\|r_{o2} \gg \frac{1}{g_{m2}}$ this reduces to

$$A_V = \frac{\frac{1}{g_{m2}}\|R_L}{R_s+\frac{1}{g_{m1}}} \quad \blacksquare$$

The resistance at the source of the MOSFET can be found using the circuit shown in Fig. 22.10. Neglecting the body effect and output resistance of the MOSFET gives

$$v_{in} = v_{gs}+i_dR \tag{22.23}$$

Since $g_mv_{gs} = i_d$, this equation can be written as

$$v_{in} = \frac{i_d}{g_m}+i_dR = i_d\left[\frac{1}{g_m}+R\right] \tag{22.24}$$

We can look at this in a simplified manner; that is, $\frac{1}{g_m}$ is the resistance looking into the source of a MOSFET, while R is the resistance connected from the source of the MOSFET to ground. Note that R can be the parallel combination of several resistances.

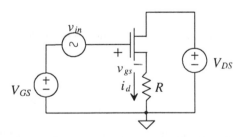

Figure 22.10 Circuit used to determine the resistance looking into the source of a MOSFET.

22.1.2 The Source Follower

Source-follower configurations using active loads are shown in Fig. 22.11. The MOSFET M2 sources current, while M1 sinks current. A source follower implemented in CMOS has an asymmetric drive capability; that is, the ability of the follower to source current is not equal to the ability to sink current for a given bias condition and

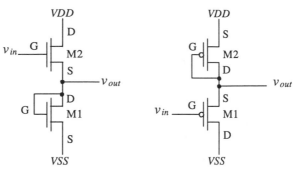

Figure 22.11 Source followers using active loads.

AC input signal. Also, in both configurations, the common drain amplifier exhibits the body effect. Before we analyze the small-signal behavior of this amplifier, let's determine how the body effect comes into play. From Ch. 9

$$g_{mb} = g_m \cdot \eta \tag{22.25}$$

where, using the BSIM SPICE parameters,

$$\eta = \frac{K1}{2\sqrt{PHI + V_{SB}}} - K2 \tag{22.26}$$

From this equation we can see that the larger the source-substrate potential of the MOSFET, the smaller g_{mb}.

Example 22.4

Determine η for the NMOS source follower shown above when the source of M2 is at 0 V potential and $VSS = -2.5$ V using CN20 parameters.

If we use Eq. (22.26) we get

$$\eta = \frac{K1}{2\sqrt{PHI + V_{SB}}} - K2 = \frac{1.49}{2\sqrt{0.75 + 2.5}} - 0.315 = 0.098$$

Therefore, the body transconductance, g_{mb}, is approximately 10 percent of the forward transconductance, g_m. The assumption that the transconductance from the body effect, g_{mb}, is negligible is valid for large V_{SB}. ∎

The small-signal gain of the NMOS source follower shown in Fig. 22.11 is simply determined by a voltage divider between the resistance looking into the source of M2 with the resistance of the gate-drain connected load, M1. The output voltage is given by

$$v_{out} = v_{in} \cdot \frac{\frac{1}{g_{m1}}}{\frac{1}{g_{m1}} + \frac{1}{g_{m2}}} \tag{22.27}$$

or the gain of the source follower with an active load, remembering $g_m = \sqrt{2\beta_1 I_D}$, is

$$A_v = \frac{v_{out}}{v_{in}} = \frac{\frac{1}{g_{m1}}}{\frac{1}{g_{m1}} + \frac{1}{g_{m2}}} = \frac{1}{1 + \frac{g_{m1}}{g_{m2}}} = \frac{1}{1 + \sqrt{\frac{W_1 L_2}{W_2 L_1}}} \qquad (22.28)$$

The output resistance of the source follower shown in this figure is

$$R_{out} = \frac{1}{g_{m1}} \Big|\Big| \frac{1}{g_{m2}} \qquad (22.29)$$

Again, it should be noted that M2 can only source current, while M1 can only sink current and the gain is always less than one.

22.1.3 Common Gate Amplifiers

The common gate amplifier with active load is shown in Fig. 22.12. The input resistance of this amplifier is simply the resistance looking into the source of M1, or

$$R_{in} = \frac{1}{g_{m1}} \qquad (22.30)$$

The gain of this amplifier, again neglecting the body effect, is

$$A_v = \frac{v_{out}}{v_{in}} = \frac{-i_d \cdot \frac{1}{g_{m2}}}{-v_{gs1}} = \frac{-i_d \cdot \frac{1}{g_{m2}}}{-i_d \cdot \frac{1}{g_{m1}}} = \frac{\text{resistance in the drain}}{\text{resistance in the source}} = \frac{\frac{1}{g_{m2}}}{\frac{1}{g_{m1}}} \quad (22.31)$$

or

$$A_v = \frac{g_{m1}}{g_{m2}} = \sqrt{\frac{KP_n}{KP_p} \cdot \frac{W_1 L_2}{W_2 L_1}} \qquad (22.32)$$

which is the same form as the gain for the common source amplifier with active load.

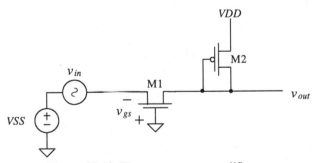

Figure 22.12 The common gate amplifier.

22.2 Current Source Loads

The current source load provides an amplifier with the largest possible load resistance available in a CMOS process.   This section examines single-stage amplifier configurations utilizing current sources (or sinks) as loads.

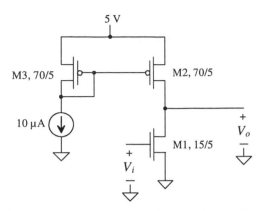

Figure 22.13 Common source amplifier with current source load.

Consider the common source amplifier with current source load shown in Fig. 22.13. The MOSFET M1 is the common source component of the amplifier, while the MOSFET M2 is the current source load. The DC transfer characteristics are shown in Fig. 22.14 for this amplifier. The slope of the line when both transistors are saturated corresponds to the small-signal gain of the amplifier. If we bias M1 and M2 so that they are both saturated, we see that the DC output voltage of the amplifier is very dependent on the DC biasing at the input of the amplifier. This is a common problem in CMOS analog integrated circuit design, that is, determining the exact voltage on the drains of two-series connected p- and n-channel MOSFETs. Feedback is normally employed to set the output voltage of the amplifier at some known value. This results in

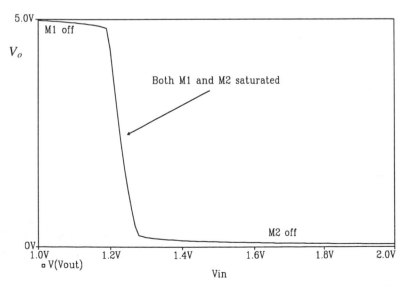

Figure 22.14 DC transfer characteristics for the amplifier shown in Fig. 22.11.

a DC voltage on the input of the amplifier. A common feedback scheme used to bias the single-stage amplifier is to AC couple the input signal to the gate of M1 and to take the output signal from the drain of M2/M1. A large resistor is then placed between the output and the input. When the basic amplifier is designed properly, this forces M1 and M2 into the saturation region.

The resistance looking into the drain of M2 is simply the output resistance given by $r_{o2} \approx 1/\lambda I_D = 1/(0.06 \cdot 10 \ \mu A) = 1.667 \ M\Omega$. This is in parallel with the resistance looking into the drain of M1, which is also $1.667 \ M\Omega$, assuming both λs are $0.06 \ V^{-1}$. If we follow the intuitive method discussed in the last section, the voltage gain of a common source amplifier is the total parallel resistance at the drain of M1 divided by the resistance in the source of M1, or

$$A_v = \frac{v_o}{v_i} = -\frac{r_{o1} \| r_{o2}}{\frac{1}{g_{m1}}} = -\frac{g_{m1}}{g_{o1} + g_{o2}} \tag{22.33}$$

where $r_{o1} = 1/g_{o1}$. It is of interest to determine how the DC bias current affects the small-signal gain. Making the appropriate substitutions into Eq. 22.33 we get

$$A_v = \frac{-\sqrt{2\beta_1 I_D}}{I_D(\lambda_1 + \lambda_2)} = \frac{-\sqrt{2\beta_1}}{(\lambda_1 + \lambda_2) \cdot \sqrt{I_D}} \tag{22.34}$$

This equation shows that the lower the bias drain current, the larger the gain. Figure 22.15 shows the voltage gain of this configuration (Fig. 22.13) plotted against the drain bias current. The gain levels off below about $0.1 \ \mu A$ because the MOSFETs enter the subthreshold region where both the transconductance and output resistance of the MOSFET are linearly dependent on the drain current (see Chs. 6 and 9).

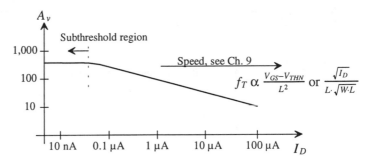

Figure 22.15 Voltage gain of common source amplifier with current source. Note how as the gain decreases the speed of the amplifier increases.

The gain of the amplifier can be increased by using a cascode current load in place of M2. The resistance looking into the drain of the cascode current source/load is much larger than the output resistance of M1. This situation is sometimes referred to as the open circuit gain of a common source amplifier. The open circuit gain of M1 can be written as

$$A_v = -\frac{r_{o1}}{\frac{1}{g_{m1}}} = -g_{m1}r_{o1} = -\frac{\sqrt{2\beta_1 I_D}}{I_D \lambda_1} \qquad (22.35)$$

which increases the gain by a factor of two over the circuit of Fig. 22.13.

The frequency performance of the current source load amplifier is poorer than the gate-drain connected load amplifiers of the last section (also note the gain difference). Referring to Fig. 22.13, the drain of M1 and M2 is termed a high-impedance node, that is, a node with only drain connections. The effective resistance at this node to ground is $r_{o1}\|r_{o2}$. A node connected to the source of a MOSFET or a MOSFET with its drain and gate connected is termed a low-impedance node. The small-signal resistance looking into the source of a MOSFET and the small-signal resistance of the diode connected MOSFET are both $1/g_m$.

Example 22.5
Estimate the bandwidth of the amplifier shown in Fig. 22.13.

We begin by redrawing this schematic with the relevant capacitances included as shown in Fig. 22.16. We assume that the gate of M2 is connected to AC ground

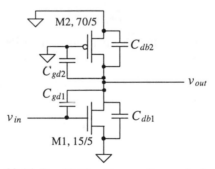

Figure 22.16 Fig. 22.13 with capacitances included.

because of the low small-signal resistance of M3.

The gain of the amplifier is

$$A_v = -g_{m1}(r_{o1}\|r_{o2}) = -\left[\sqrt{2 \cdot 50\frac{\mu A}{V^2} \cdot \frac{15}{5} \cdot 10\ \mu A}\right] \cdot (833k) = -46\ V/V$$

The MOSFET capacitances are now calculated. The gate-drain capacitance of the p-channel MOSFET M2 is given by

$$C_{gd2} = CGDO \cdot W_2 = 5 \times 10^{-10}\frac{F}{m} \cdot 70\ \mu m = 35\ fF$$

while the gate-drain capacitance of M1 is

$$C_{gd1} = CGDO \cdot W_1 = 3.8 \times 10^{-10}\frac{F}{m} \cdot 15\ \mu m = 5.7\ fF$$

To calculate the worst-case depletion capacitances of the drain implants, we can assume zero potential across the junction and areas of 6 µm by the width of the MOSFET. The depletion capacitances are then given by

$$C_{db1} = CJ \cdot \overbrace{6\ \mu\text{m} \cdot W_1}^{\text{Area of drain}} = 1.0 \times 10^{-4}\,\tfrac{\text{F}}{\text{m}^2} \cdot 6\ \mu\text{m} \cdot 15\ \mu\text{m} = 9\ \text{fF}$$

and

$$C_{db2} = 3.25 \times 10^{-4} \cdot 6\ \mu\text{m} \cdot 70\ \mu\text{m} = 137\ \text{fF}$$

Note that because the gain is much higher than the gate-drain connected amplifier, the Miller capacitance that appears at the output, C_{MO}, is,

$$C_{MO} = C_{gd1}\left(1 + \frac{1}{46}\right) \approx C_{gd1}$$

and the output time constant becomes

$$\tau_{out} = (r_{o1} \| r_{o2}) \cdot (C_{gd2} + C_{gd1} + C_{db1} + C_{db2}) = 833\text{k} \cdot 187\ \text{fF} = 155.8\ \text{ns}$$

and the 3 dB frequency of this amplifier is given by

$$f_{out} = \frac{1}{2\pi \cdot \tau_{out}} = 1\ \text{MHz}$$

a significantly lower bandwidth than the amplifier of Ex. 22.2 with a correspondingly higher gain.

If this amplifier was driven from an identical stage with output resistance, $R_{out} = r_{o1} \| r_{o2}$, and capacitance, C_{out}, which is

$$C_{out} = (C_{gd2} + C_{gd1} + C_{db1} + C_{db2}) = 187\ \text{fF}$$

then the input time constant is given by

$$\tau_{in} = R_{out} \cdot [C_{gs1} + (1 + |A_v|)C_{gd1} + C_{out}]$$

where

$$C_{gs1} = \frac{2}{3} \cdot C'_{ox} \cdot W_1 \cdot L_1 = 40\ \text{fF}$$

and

$$\tau_{in} = 866\text{k} \cdot \left(40\ \text{fF} + \overbrace{(1+46)5.7\ \text{fF}}^{\text{Miller capacitance}} + 187\text{fF} \right) = 429\ \text{ns}$$

The input time constant is increased by the Miller effect across the gate-drain capacitance of M1 and the output capacitance of the driving circuit. The methods available to minimize the effect of the Miller capacitance will be explained in the section discussing the cascode connection. ∎

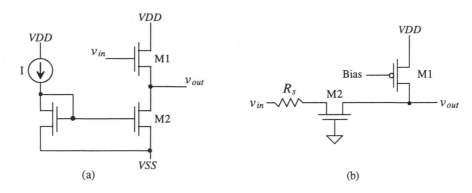

Figure 22.17 (a) Source follower and (b) common gate amplifier.

The source-follower configuration using a current source load is shown in Fig. 22.17a. Neglecting the body effect, we see that the voltage gain of the configuration is given by a voltage divider between the resistance looking into the source of M1 and the resistance looking into the drain of M2, or

$$A_v = \frac{v_o}{v_i} = \frac{r_{o2}}{r_{o2} + \frac{1}{g_{m1}}} = \frac{g_{m1}r_{o2}}{1 + g_{m1}r_{o2}} \approx 1 \qquad (22.36)$$

The output resistance is the parallel combination of the resistance looking into the source of M1 and the resistance looking into the drain of M2, or

$$R_o = \frac{1}{g_{m1}} \| r_{o2} \qquad (22.37)$$

The common gate configuration using a current source load is shown in Fig. 22.17b. The resistance looking into the drain of M1 is r_{o1} while the resistance looking into the drain of M2 is $r_{o2}(1 + g_{m2}R_s)$. The resistance in the source of M2 is $R_s + \frac{1}{g_{m2}}$. Therefore, the gain of the common gate configuration using a current source load is

$$A_v = \frac{[r_{o1} \| (r_{o2}(1 + g_{m2}R_s))]}{R_s + \frac{1}{g_{m2}}} \qquad (22.38)$$

It is important to realize that there are current sourcing/sinking limitations with current source loads. For example, the maximum current the source follower above can sink is the constant current supplied by M2. When the source follower is sourcing current to a load, it must also supply current to M2.

The amplifiers we have discussed so far in this chapter are class A amplifiers; that is, they conduct current over the entire range or cycle of a sinusoidal input signal. Class B amplifiers conduct over half of the cycle, while class AB amplifiers conduct over a range slightly greater than half a cycle. Most often in analog CMOS circuit design we are concerned with class A or AB amplifiers. Class B amplifiers are most often used for the generation of pulses. Usually we make no distinction between the "class" of an amplifier unless we are concerned with power dissipation. Output stages, the stages in a circuit used to drive an external load, are usually where the designer is interested in power dissipation and distortion (a term used to describe how well an amplifier reproduces a scaled version of the input signal). The source-follower circuit of Fig. 22.17(a) is an example of a class A output stage utilized in an op-amp design.

The power conversion efficiency of an amplifier is defined by

$$\% \text{ efficiency} = \frac{\text{Load power, } P_l}{\text{Supply power, } P_s} \times 100 \ \% \tag{22.39}$$

If the source-follower above is driving a resistor, R_L, with an rms voltage v_{out}, then

$$P_l = \frac{(v_{out})^2}{R_L} \tag{22.40}$$

The power supplied to the amplifier can be specified by

$$P_s = I(VDD - VSS) \tag{22.41}$$

If $VDD = -VSS$ and the output voltage can swing up to VDD and down to VSS, then the maximum efficiency of the class A amplifier occurs when

$$v_{out} = \frac{VDD}{\sqrt{2}} = \frac{IR_L}{\sqrt{2}} \tag{22.42}$$

The maximum conversion efficiency of the class A amplifier is 25 percent.

22.2.1 The Cascode Connection

The basic cascode amplifier is shown in Fig. 22.18a. This amplifier has two advantages over the basic common source amplifier with current source load discussed previously; the gain is greater due to the large output resistance, and the absence of the Miller capacitance keeps the input time constant from becoming long. Biasing of the cascode amplifier will be discussed in Ch. 24 in conjunction with diff-amp design.

The small-signal model for the basic amplifier is seen in Fig. 22.18b. The gain from v_{in} to the drain of M1, v_{d1}, can be found by

$$A_{v1} = \frac{v_{d1}}{v_{in}} = \frac{-i_d \cdot (r_{ins2} \| r_{o1})}{i_d \cdot \frac{1}{g_{m1}}} \tag{22.43}$$

where r_{ins2} is the small-signal resistance seen looking into the source of M2. Since the drain of M2 is not on AC ground, it cannot be assumed that $r_{ins2} = 1/g_{m2}$. If a test source is applied as seen in Fig. 22.19, the following equations may be written:

$$i_t = -(g_{m2}v_{gs2} - g_{mb2}v_{sb2} + \frac{v_o - v_t}{r_{o2}}) = g_{m2}v_t - g_{mb2}v_t + \frac{v_t - v_o}{r_{o2}} \tag{22.44}$$

$$v_o = i_t r_{o3} \tag{22.45}$$

Therefore, r_{ins2} is

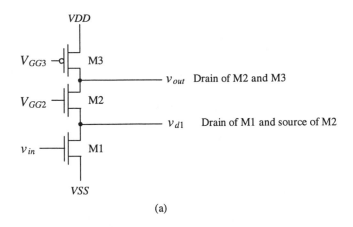

(a)

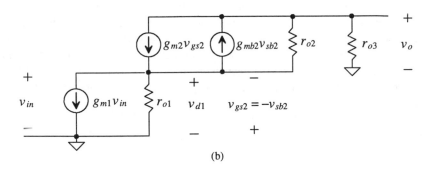

(b)

Figure 22.18 (a) Cascode amplifier and (b) small-signal model.

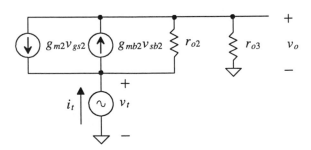

Figure 22.19 Small-signal model used for finding the impedance looking into
the source of M2.

$$r_{ins2} = \frac{v_t}{i_t} = \frac{1 + \frac{r_{o3}}{r_{o2}}}{g_{m2} + g_{mb2} + \frac{1}{r_{o2}}} \approx \frac{1}{g_{m2}}\left(1 + \frac{r_{o3}}{r_{o2}}\right) \tag{22.46}$$

and the gain, A_{v1}, becomes,

$$A_{v1} = \frac{v_{d1}}{v_{in}} \approx -g_{m1}\left[\frac{1}{g_{m2}}\left(1 + \frac{r_{o3}}{r_{o2}}\right)\|r_{o1}\right] \approx -\frac{g_{m1}}{g_{m2}}\left(1 + \frac{r_{o3}}{r_{o2}}\right) \tag{22.47}$$

The gain of the common gate stage, M2, with current source load M1 is

$$A_{v2} = \frac{v_{out}}{v_{d1}} = \frac{-i_d \cdot \overbrace{r_{o2}(1 + g_{m2}r_{o1})}^{\text{R into drain of M2}} \|r_{o3}}{-i_d \cdot \frac{1}{g_{m2}}\left(1 + \frac{r_{o3}}{r_{o2}}\right)} = g_{m2} \cdot \frac{[(r_{o2}(1 + g_{m2}r_{o1}))\|r_{o3}]}{(1 + \frac{r_{o3}}{r_{o2}})} \tag{22.48}$$

The overall gain is the product of A_{v1} and A_{v2} given by

$$A_v = A_{v1} \cdot A_{v2} = -g_{m1} \cdot [(r_{o2}(1 + g_{m2}r_{o1}))\|r_{o3}] \approx -g_{m1}r_{o3} \tag{22.49}$$

which is not much of a gain improvement over the basic common source amplifier with current source load of Fig. 22.13. The output resistance of M3 limits the gain. Figure 22.20 shows a configuration for increasing the resistance looking into the drain of M3. The MOSFETs M3 and M4 form a cascode current source load. The output resistance is the parallel combination of the resistance looking into the drain of M3 with the resistance looking into the drain of M2. This parallel combination can be written as

$$R_o = [r_{o2}(1 + g_{m2}r_{o1})]\|[r_{o3}(1 + g_{m3}r_{o4})] \tag{22.50}$$

and the gain is

$$A_v = -g_{m1} \cdot R_o \tag{22.51}$$

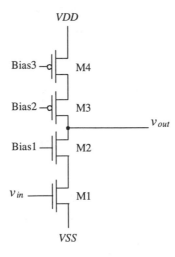

Figure 22.20 Dual cascode amplifier.

The frequency response of the cascode illustrates another advantage of the configuration. Assume that the cascode amplifier is being driven by a circuit with an output resistance and capacitance, R_{L1} and C_{L1}, respectively. Applying Miller's Theorem to the circuit shown in Fig. 22.18a, the high-frequency small-signal model for the cascode can be seen in Fig. 22.21. Note that the capacitances C_2 and C_3 represent the lumped capacitance to ground at those nodes; the values are

$$C_2 = C_{bd1} + C_{bs2} + C_{gs2} + C_{gd1}\left(1 + \frac{1}{|A_{v1}|}\right) \tag{22.52}$$

$$C_3 = C_{gd2} + C_{bd2} + C_{bd3} + C_{gd3} + C_L \tag{22.53}$$

where C_L is the capacitance of any load attached to the output. The input capacitance, C_1, becomes

$$C_1 = C_{gs1} + C_{gd1}(1 + |A_{v1}|) = C_{gs1} + C_{gd1}\left[1 + \frac{g_{m1}}{g_{m2}}\left(1 + \frac{r_{o3}}{r_{o2}}\right)\right] \tag{22.54}$$

Notice that the Miller capacitance that appears at the input is much smaller compared to the simple common source amplifier with a current source load case. This is simply because the gain, v_{d1}/v_{in}, is very small. Therefore, the Miller capacitance no longer dominates the time constant associated with the input. The time constants become

$$\tau_{in} = R_{L1} \cdot (C_{L1} + C_1) \tag{22.55}$$

$$\tau_{out} = r_{o3} \cdot C_3 \tag{22.56}$$

and the time constant associated with the drain of M1 is

$$\tau_{d1} = \left[\frac{1}{g_{m2}}\left(1 + \frac{r_{o3}}{r_{o2}}\right)||r_{o1}\right] \cdot C_2 \tag{22.57}$$

Note that the input time constant, τ_{in}, and the output time constant, τ_{out}, are roughly the same order of magnitude assuming that the cascode is driving another stage with an input capacitance of the same order of magnitude. The time constant associated with

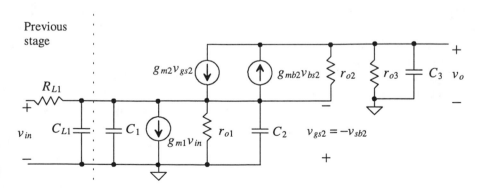

Figure 22.21 High-frequency equivalent circuit for the cascode amplifier.

the drain of M1, τ_{d1}, will be somewhat negligible since it has such a low impedance to ground (from the source of M2) and contributes little to the overall frequency response.

One drawback to using the cascode connection is the reduction in output voltage swing. Since there are more transistors in the circuit, more of the supply voltages will be required to keep the devices in saturation. Again, referring to Fig. 22.18a, the minimum output voltage that will keep all the devices in saturation will be limited by the point at which M2 becomes nonsaturated. This occurs when

$$V_{GS2} - V_{THN2} = V_{DS2} \tag{22.58}$$

or when

$$V_{o(\text{min})} = V_{GG2} - V_{THN2} \tag{22.59}$$

One might wonder why M2 will become nonsaturated before M1. Since the impedance at the output node is much greater than the impedance of node v_{d1}, the voltage at the output will be decreasing at a much faster rate than the voltage at node v_{d1}. This is equivalent to saying that since the gain from the input to the output is much greater than the gain from the input to node v_{d1}, as the input voltage rises, the voltage, v_o, will decrease faster than the voltage at v_{d1}.

The maximum positive voltage that can appear at the output of the cascode while maintaining all transistors in saturation will be limited by the point at which M3 becomes nonsaturated. Applying the same reasoning as before, we find that the maximum positive output voltage occurs when

$$v_{SG} - V_{THP3} = v_{SD} \tag{22.60}$$

or

$$v_o(\text{max}) = V_{GG3} + V_{TH3} \tag{22.61}$$

So the values of the bias voltages become very critical when considering output swing.

22.2.2 The Push-Pull Amplifier

The push-pull amplifier shown in Fig. 22.22 is the inverter discussed in Ch. 11 operating with both M1 and M2 in the saturation region. This configuration, when designed properly, can sink and source equal amounts of current. Both M1 and M2 act as common source amplifiers. The effective resistance at the drains of these transistors to AC ground is the parallel combination of each MOSFET's output resistance. The resistance in the sources of M1 and M2 is the parallel combination of the resistance seen looking into the source of M1 and M2. The gain of this configuration is then

$$A_v = \frac{-i_d \cdot (r_{o1} \| r_{o2})}{i_d \cdot \left(\frac{1}{g_{m1}} \| \frac{1}{g_{m2}} \right)} = -(g_{m1} + g_{m2}) \cdot (r_{o1} \| r_{o2}) \tag{22.62}$$

The push-pull amplifier is popular as an output amplifier because of equal source and sink capability and the ability to pull the output to *VDD* or *VSS*. Figure

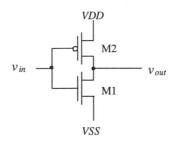

Figure 22.22 The push-pull amplifier.

22.23 shows a configuration that sets the gain of the amplifier to unity. The push-pull configuration is used as a buffer in this schematic; that is, it provides current drive capability. Error amplifiers are used to sense and compare the output voltage and then adjust the gate drive of M1 and M2. Another benefit of this arrangement is that the output resistance of the amplifier is reduced. This is desirable when the amplifier is used as a buffer in op-amp applications. In practice, this configuration has several practical problems, for example, design of the error amplifiers for maximum power conversion efficiency, stability, and crossover distortion when operated class B.

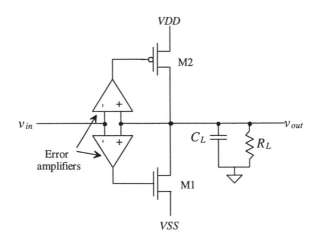

Figure 22.23 Using the push-pull amplifier as a unity gain buffer.

22.3 Noise and Distortion in Amplifiers

Noise

Let's consider the noise performance of the common source amplifier with the current source load shown in Fig. 22.13 and in simplified form in Fig. 22.24a. In our noise analysis, we will neglect the noise contributions of the devices connected to the gate of M2. Figure 22.24b shows the circuit with the noise sources added. Following the procedure of Ch. 7 with the noise models of Ch. 9, the sum of each mean squared contribution is equal to the mean squared output noise given by

$$\overline{v_T^2} = (r_{o1} \| r_{o2} \| \frac{1}{j\omega C_t})^2 \cdot (\overline{i_{1/f,1}^2} + \overline{i_{1/f,2}^2} + \overline{i_{therm,1}^2} + \overline{i_{therm,2}^2}) \qquad (22.63)$$

where C_t is the sum of the capacitances in parallel with r_{o1} and r_{o2}. The RMS output noise over a given bandwidth is

$$\sqrt{\overline{v_{on}^2}} = \left[\int_{f_L}^{f_H} \overline{v_T^2} \cdot df \right]^{1/2} \qquad (22.64)$$

For low frequencies the gain of this amplifier is $-g_{m1} \cdot (r_{o1} \| r_{o2})$, so the squared magnitude of the transfer function is

$$|H(j\omega)|^2 = \frac{(g_{m1}(r_{o1} \| r_{o2}))^2}{\left[1 + \left(\frac{f}{f_{in}} \right)^2 \right] \left[1 + \left(\frac{f}{f_{out}} \right)^2 \right]} \qquad (22.65)$$

while the RMS input noise is

$$\sqrt{\overline{v_{ino}^2}} = \left[\int_{f_L}^{f_H} \frac{\overline{v_T^2}}{|H(j\omega)|^2} \cdot df \right]^{1/2} \qquad (22.66)$$

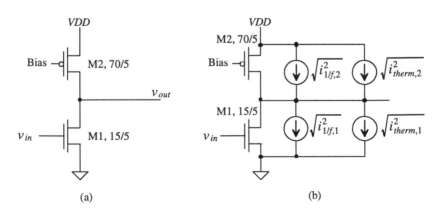

Figure 22.24 (a) Common source amplifier with current source load and (b) with noise sources included.

Distortion

Consider how well this amplifier, Fig. 22.24a, produces a scaled version of the input on its output. Ideally, the output of the amplifier is related to the input by a constant A_v, the voltage gain. The gain for this amplifier is given by

$$|A_v| = g_m(r_{o1}||r_{o2}) \Rightarrow \sqrt{2\beta_1(I_D + i_d)} \cdot \frac{1}{2\lambda(I_D + i_d)} = \frac{\sqrt{2\beta_1}}{2\lambda\sqrt{I_D + i_d}} \quad (22.67)$$

Normally, the AC component of the drain current, i_d, is assumed to be much less than the DC component of the drain current, I_D, and the amplifier gain is essentially constant (small-signal aproximation). If the AC component is comparable to the DC component, noticeable distortion results. The voltage gain for large inputs is dependent on the input signal amplitude.

Characterizing an amplifier begins by applying a pure single-tone sinusoid of the form

$$V_{in}(t) = V_p \sin \omega t \quad (22.68)$$

to the input of the amplifier. The output of the amplifier is a series of tones at an integer multiple of the input tone given by

$$V_{out}(t) = a_1 V_p \sin(\omega t) + a_2 V_p \sin(2\omega t) + \cdots + a_n V_p \sin(n\omega t) \quad (22.69)$$

The magnitude of the fundamental or wanted signal is $a_1 V_p$. Ideally, a_2 through a_n arc zero, and the amplifier is free of distortion. The nth term harmonic distortion is given by

$$HD_n = \frac{a_n}{a_1}, \text{ for } n > 1 \quad (22.70)$$

The total harmonic distortion (THD) is given by

$$\text{THD} = \sqrt{\frac{a_2^2 + a_3^2 + a_4^2 + \cdots + a_n^2}{a_1^2}} \quad (22.71)$$

Output buffers, amplifiers used to drive a large load capacitance or low resistance, are examples of amplifiers where low THD is important. If the output amplifier is biased so that the DC component of the drain current is large compared to the transient or time-varying current, the buffer dissipates too much power for most applications. Therefore, in almost all situations, the biasing current in an output buffer is comparable, or even smaller for class B operation than the time-varying current. We reduce the distortion by employing feedback around the amplifier (see Ch. 23). If the open-loop gain of an op-amp varies from 1,000 to 10,000 depending on the amplitude of the input signal, then the feedback around the amplifier reduces the gain sensitivity. In fact, it is nearly impossible to design a linear output amplifier with low distortion without the use of feedback.

Modeling Distortion with SPICE

SPICE can be used to simulate distortion using a transient analysis and the .FOUR (Fourier) statement [5]. The general form of this statement is .FOUR FREQ OV1 <OV2 OVI . . . where FREQ is the frequency of the fundamental and OV1 . . . are the outputs of the circuit (the voltage or current outputs for which SPICE will calculate distortion). As a simple example, consider the circuit and netlist shown in Fig. 22.25. The input sine wave must have at least one full period for the .FOUR statement to calculate distortion. If more than one period is used, SPICE will use the output over the last full period. Also, the maximum transient step size should be less than the period of the input divide by 100. For the example of Fig. 22.25 with a 1 kHz input (a period of 1 ms), the maximum print size should be 10 µs.

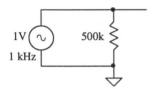

```
*** Top Level Netlist ***
.four 1k V(1)
R1 1 0 500k
V1 1 0 DC 0 AC 0 0 SIN(0 1 1khz 0 0)
.tran 10u 2m 0 10u
.end
```

Figure 22.25 Simple circuit to demonstrate the use of the .FOUR statement.

The resulting simulation output is as follows.

Fourier analysis for v(1):
No. Harmonics: 10, THD: 3.17012e-006 %, Gridsize: 200, Interpolation Degree: 1

Harmonic	Frequency	Magnitude	Phase	Norm. Mag	Norm. Phase
0	0	-5.602e-009	0	0	0
1	1000	0.999632	-9e-6	1	0
2	2000	1.12039e-008	-86.4	1.12e-8	-86.4
3	3000	1.12039e-008	-84.6	1.12e-8	-84.6
4	4000	1.12039e-008	-82.8	1.12e-8	-82.8
5	5000	1.12039e-008	-81	1.12e-8	-81
6	6000	1.12039e-008	-79.2	1.12e-8	-79.2
7	7000	1.12039e-008	-77.4	1.12e-8	-77.4
8	8000	1.12039e-008	-75.6	1.12e-8	-75.6
9	9000	1.12039e-008	-73.8	1.12e-8	-73.8

Notice that, as we would expect, this circuit's output doesn't show any harmonic distortion. SPICE calculates the magnitude and phase of the first nine harmonics and DC. Also note that SPICE automatically calculates the THD for a circuit.

As a more practical example, consider the push-pull amplifier shown in Fig. 26.26. This circuit is biased so that the DC voltage on the gates of the MOSFETs is the same as the DC voltage on the drains (both MOSFETs are in saturation). The 1MEG resistor simulates the effects of a load (and keeps the output coupling capacitor from floating).

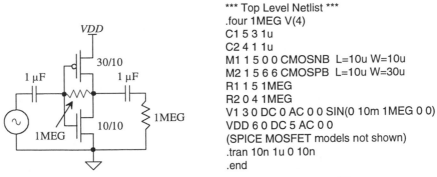

```
*** Top Level Netlist ***
.four 1MEG V(4)
C1 5 3 1u
C2 4 1 1u
M1 1 5 0 0 CMOSNB  L=10u W=10u
M2 1 5 6 6 CMOSPB  L=10u W=30u
R1 1 5 1MEG
R2 0 4 1MEG
V1 3 0 DC 0 AC 0 0 SIN(0 10m 1MEG 0 0)
VDD 6 0 DC 5 AC 0 0
(SPICE MOSFET models not shown)
.tran 10n 1u 0 10n
.end
```

Figure 22.26 Determining distortion in a push-pull amplifier.

The calculated THD will be a strong function of the input signal amplitude. Let's begin by simulating the circuit with a 10 mV input signal. The resulting SPICE output gives a THD of 0.18 percent. If the input amplitude is increased to 100 mV, the THD becomes 1.6 percent.

22.3.1 Modeling Amplifier Noise

Figure 22.27 shows how the input and output RMS noise sources are added to a noiseless amplifier. The signal to noise ratio (*SNR*) of an amplifier is defined by

$$SNR_{in} = 10\log\frac{v_{in}^2}{v_{ino}^2} \qquad (22.72)$$

where v_{in}^2 is the square of the RMS input signal and

$$SNR_{out} = 10\log\frac{v_{out}^2}{v_{on}^2} \qquad (22.73)$$

for the *SNR* on the output of the amplifier. The *SNR* is used to compare signal levels to amplifier noise levels. A *SNR* of 0 dB means that the squared RMS input (or output) signal is equal to the mean squared input-referred (or output) noise of an amplifier. Noise is always measured on the output of a circuit and referred back to the input.

Figure 22.27 Addition of RMS noise voltages to the input or output of an amplifier.

Now consider the practical situation shown in Fig. 22.28 of the amplifier being driven by a voltage source with a source resistance R_s. The source resistance generates

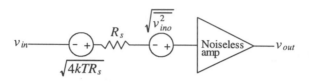

Figure 22.28 Determination of the noise figure of an amplifier.

thermal noise with a mean squared amplitude of $4kTR_s$. We can define the noise figure (*NF*) of an amplifier as the ratio of the output of a noisy amplifier with the source resistor connected over the output of a noise-free amplifier with the noisy source resistor connected. This can be written as

$$NF = 10\log\left[\frac{|H(j\omega)|^2 \cdot \left(4kTR_s + \overline{v_{ino}^2}\right)}{|H(j\omega)|^2 \cdot (4kTR_s)}\right] \tag{22.74}$$

or, where the argument of the logarithm is called the noise factor,

$$NF = 10\log\left[\frac{4kTR_s + \overline{v_{ino}^2}}{4kTR_s}\right] = 10\log\left[\text{noise factor}\right] \tag{22.75}$$

The *NF* is a useful figure to compare the relative noise performance of amplifiers. *NF* varies with source resistance and bandwidth and is usually specified using contours of constant *NF* versus frequency or source resistance. We have assumed in this derivation that the source impedance is much smaller than the input resistance of the amplifier, so that the RMS input signal to the amplifier is v_{in}. A voltage divider between source resistance and amplifier input resistance is used when this is not the case. The input SNR can be calculated, at a given source resistance, by

$$SNR_{in} = 10\log\frac{v_{in}^2}{\overline{v_{ino}^2} + 4kTR_s} = 10\log\frac{v_{in}^2}{4kTR_s} - NF \tag{22.76}$$

Sometimes the noise of an amplifier is specified using the noise temperature, T_n, instead of the NF. Consider the amplifier shown in Fig. 22.29a with the source resistance at $T = 0\ °K$. The amplifier input RMS noise voltage is $\sqrt{\overline{v_{ino}^2}}$. Now consider the circuit shown in Fig. 22.29b where the amplifier is noiseless, $\sqrt{\overline{v_{ino}^2}} = 0$. We increase the temperature of the resistor until the thermal noise from the resistor is equal to $\sqrt{\overline{v_{ino}^2}}$. The temperature of the resistor at this point is labeled T_n for the noise temperature. We can relate the noise temperature to NF by first rewriting Eq. (22.75) as

$$NF = 10\log\left[1 + \frac{\overline{v_{ino}^2}}{4kTR_s}\right] \tag{22.77}$$

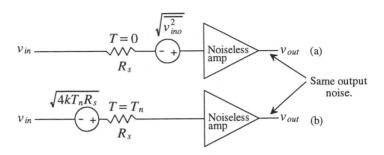

Figure 22.29 Determination of the noise temperature of an amplifier.

The mean squared input noise in Fig. 22.29b is given by

$$\overline{v_{ino}^2} = 4kT_nR_s = 4kTR_s[10^{NF/10} - 1] \qquad (22.78)$$

or the noise temperature in terms of the NF is given by

$$T_n = T(10^{NF/10} - 1) \qquad (22.79)$$

where the noise figure, NF, is in dB and T is the temperature in kelvin at which the NF was measured. The NF can also be written in terms of the noise temperature as

$$NF = 10\log\left(\frac{T_n}{T} + 1\right) \qquad (22.80)$$

22.4 A Class AB Amplifier

Consider the class A common source amplifier shown in Fig. 22.30. We know that the maximum efficiency of this amplifier is 25 percent. In this section, we are interested in the large-signal performance of the amplifier. We know that M1 can only sink current from the load capacitance and the current source I sources current to the load capacitance. The maximum rate at which the load capacitance can be charged is called the slew rate (SR) and is given by

$$SR = \frac{I}{C_L} = \frac{dV}{dt} \qquad (22.81)$$

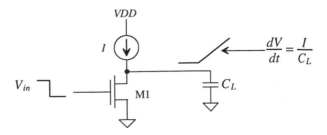

Figure 22.30 Common source amplifier with current source load.

If I is 10 μA and C_L is 1 pF, then the slew rate of the amplifier is 10 V/μs. In many analog or digital applications, this rate of change may be too slow. Usually the capacitance is not a variable available to the design engineer, and increasing the current I results in greater power dissipation. Designing an amplifier that eliminates the current source limitations in charging or discharging the load capacitance is desirable.

Consider the amplifier configuration shown in Fig. 22.31. The value of the batteries determines the class (A, B, or AB) of the amplifier. When the input goes positive, V_{GS2} will increase while V_{SG1} will decrease. This causes the current in M2 to increase and the current in M1 to decrease. The magnitude of the input voltage when M1 turns off is determined by VGG1. This configuration does not exhibit output slew-rate limitations due to the absence of a current source in series with the charging or discharging path. The output voltage swing is limited to a threshold voltage below VDD and above VSS.

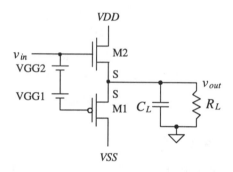

Figure 22.31 Implementation of a class AB (or class B) amplifier.

The floating batteries in Fig. 22.31 can be implemented with the circuit shown in Fig. 22.32. The gate of M6 is biased with a DC voltage so that its DC drain current is equal to the DC current through M3. MOSFET M3 is simply a current source. The gate-source voltages of M4 and M5 are constant (because of the constant current provided by M3) and used as the biasing voltage needed to bias M1 and M2 on. This configuration is useful as an output buffer in an op-amp when the gate of M6 is connected to an active load used in a differential pair and the gate of M3 is connected to the same bias voltage as used in the diff-amp current source.

A class AB amplifier based on a floating current source is shown in Fig. 22.33 [6]. The current source in this figure is used to set the DC current flowing in the MOSFETs. Keeping in mind that the input current, i_{in}, can be negative or positive (i.e., flow into or out of the circuit), an increase in i_{in} causes V_{SG4} to decrease (causing M4 to source less current) and V_{GS1} to increase (causing M1 to sink more current). The output voltage of this circuit can swing from VDD to VSS; thus, this configuration finds extensive use in low-voltage design.

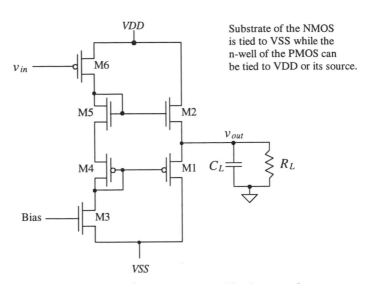

Substrate of the NMOS is tied to VSS while the n-well of the PMOS can be tied to VDD or its source.

Figure 22.32 Practical implementation of a class AB amplifier in a complementary source-follower configuration.

The circuit shown in Fig. 22.34 is another implementation of a class AB amplifier useful in driving capacitive loads (a load resistor kills the gain of this amplifier). The schematic block symbol, an op-amp with the noninverting terminal grounded, is also shown in this figure. Let's begin the analysis of this amplifier by determining the DC operating point with input, v_{in}, grounded. Let's also assume that all n- and p-channel MOSFETs in this figure are the same size, say, 15/5 for the NMOS and 70/5 for the PMOS. If this is true, then the bias voltages generated by M13 through

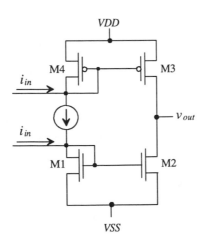

Figure 22.33 Class AB amplifier using floating current sources.

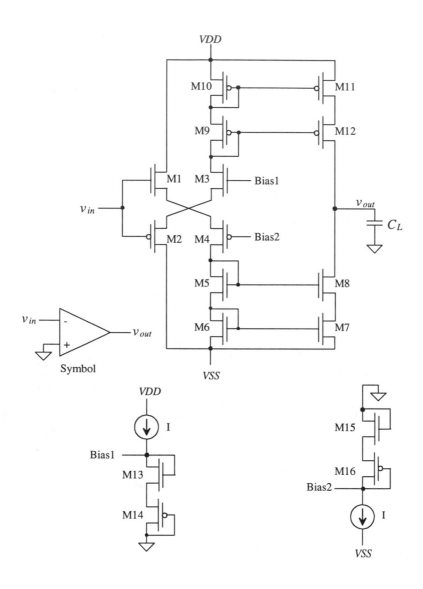

Figure 22.34 Implementation of a class AB amplifier.

M16 force a current I to flow in all remaining MOSFETs. If output driver MOSFETs M7, M8, M11, and M12 are K times larger than M5, M6, M9, and M10, then a current $K \cdot I$ flows in these output MOSFETs. Sizing the output transistors is useful if larger-current drive capability is needed. The open-loop gain of this amplifier is given by

$$A_v = -\frac{g_{m12}r_{o12}r_{o11}\|g_{m7}r_{o7}r_{o8}}{\frac{1}{g_{m1}}+\frac{1}{g_{m4}}} \tag{22.82}$$

This type of amplifier is sometimes called an operational transconductance amplifier (OTA) and finds use in sampled data circuits where analog data are stored on a capacitor.

REFERENCES

[1]   P. E. Allen and D. R. Holberg, *CMOS Analog Circuit Design,* Holt, Rinehart and Winston, 1987. ISBN 0-03-006587-9.

[2]   R. L. Geiger, P. E. Allen, and N. R. Strader, *VLSI - Design Techniques for Analog and Digital Circuits,* McGraw-Hill Publishing Co., 1990. ISBN 0-07-023253-9.

[3]   P. R. Gray and R. G. Meyer, *Analysis and Design of Analog Integrated Circuits,* 3rd ed., John Wiley and Sons, 1993. ISBN 0-471-57495-3.

[4]   P. R. Gray and R. G. Meyer, "MOS Operational Amplifier Design - A Tutorial Overview," *IEEE Journal of Solid State Circuits,* Vol. SC-17, No. 6, pp. 969-982, December 1982.

[5]   P. W. Tuinenga, *SPICE - A Guide to Circuit Simulation and Analysis Using PSPICE,* 3rd ed., Prentice Hall, 1995. ISBN 0-13-436049-4.

[6]   R. Hogervorst, K. J. de Langen, and J. H. Huijsing, *1.1.2 Low-Power Low-Voltage VLSI Operational Amplifier Cells,* 1996 ISCAS Tutorials.

PROBLEMS

22.1   Show that the gate-drain connected MOSFET shown in Fig. P22.1 behaves like a small-signal resistor of value $1/g_m$.

Figure P22.1

22.2   Using small-signal models, verify that the gain given in Ex. 22.1 is correct.

22.3   Verify Miller's Theorem using the amplifier configuration shown in Fig. P22.3a.

22.4   Consider the schematic of a two-MOSFET amplifier shown in Fig. P22.4. Determine the following:

   a. The DC voltages and currents.

   b. The small-signal low-frequency gain.

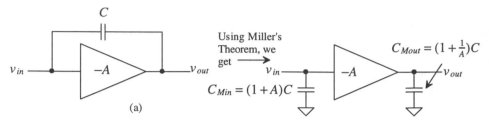

Figure P22.3

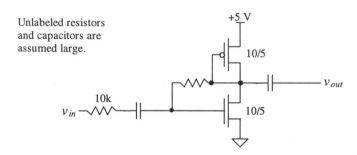

Figure P22.4

c. C_{sg2}, C_{db2}, C_{gd1}, C_{gs1}, and C_{db1}. Assume that the areas of the drains and sources of the MOSFETs measure 6 μm by 10 μm.

d. The frequency response.

e. The location of the zero in the amplifier frequency response.

22.5 This problem investigates how the body effect influences the gain of the source-follower amplifier. Consider the source follower shown in Fig. P22.5. Show that the gain of this configuration, assuming an ideal current sink, is given by

$$\frac{v_{out}}{v_{in}} = \frac{g_m}{g_m + g_{mb}} = \frac{1}{1+\eta}$$

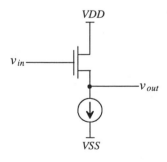

Figure P22.5

22.6   Show that if we include the body effect in the derivation of the resistance looking into the drain of the MOSFET with source resistance shown in Fig. 22.9, the resistance becomes

$$R_o \approx [1 + (1 + \eta)g_m R] \cdot r_o$$

22.7   For the amplifiers shown in Fig. P22.7 derive the small-signal voltage gains using small-signal models. Assume that the MOSFETs used in these amplifiers are biased in the saturation region.

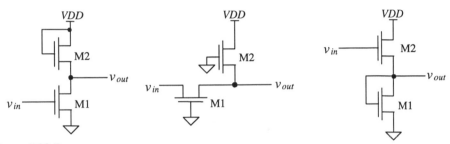

Figure P22.7

22.8   Estimate the small-signal voltage gain of the amplifier shown in Fig. 22.13 assuming that both M1 and M2 are biased in the saturation region.

22.9   Repeat Problem 4 for the amplifier shown in Fig. P22.9.

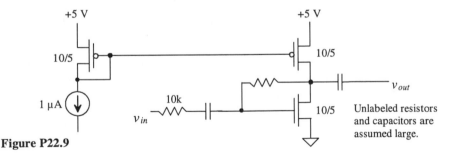

Figure P22.9

22.10 With respect to Fig. 22.17a, what is the small-signal resistance looking into the source of M1? the drain of M2? What is the small-signal resistance from the gate of M2 to ground?

22.11 Verify that Eq. (22.51) is correct. With respect to Fig. 22.20, what is the small-signal resistance looking into the drain of M4 (in terms of g_{mn}, g_{mp}, and r_{on} and r_{op})? the source of M3? the drain of M3? the source of M2? the drain of M1? the drain of M2? the total small-signal resistance at the output (the drains of M2 and M3)?

22.12 Repeat Problem 11 (calculating the actual numbers) if 5 µA of current flows in the MOSFETs of Fig. 22.20. Assume all MOSFETs measure 50/5.

22.13 Using the small-signal models, verify that Eq. (22.62) is correct.

22.14 For the amplifier section shown in Fig. P22.14, estimate the minimum and maximum voltages allowable on the output in order to keep all MOSFETs in the saturation region.

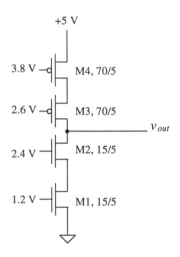

Figure P22.14

22.15 Assume that the amplifier of Fig. P22.9 is driving a 10 pF load capacitor. What is the maximum rate the load capacitance can be charged (what is the slew rate)? Is there a slew-rate limitation for discharging the capacitor?

22.16 For the circuit shown in Fig. P22.16, use SPICE to plot the output voltage against the input voltage. What limits the minimum and maximum output voltage swing? Also plot the current through M1 and M2 against the input voltage.

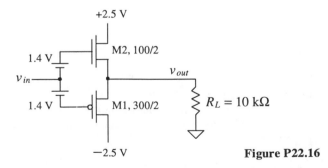

Figure P22.16

Feedback Amplifiers

Feedback is a very powerful concept and has numerous applications. Examples of feedback systems abound in everyday life. For example, the thermostat on an air conditioner unit uses feedback to maintain a comfortable temperature within a room. Similarly, our bodies use feedback to increase antibodies to fight off an infection. By definition, feedback is the process of combining the output of a system with its input. In the air conditioner example, the temperature of the room is considered to be the output of the system, and the temperature set on the thermostat, the system input. The thermostat subtracts the value of the room temperature from the input value. If the temperature in the room is higher than the temperature set on the thermostat, the air conditioner turns on until the temperature in the room is less than or equal to the desired set value.

Thus, it can be said that feedback stabilizes a system. However, not all types of feedback have this property. There are two types of feedback: positive and negative. Negative feedback stabilizes, while positive feedback has the opposite effect. A good example of positive feedback is when a microphone is held too closely to the speaker of a public address (PA) system. More than likely, most people have heard the loud ringing that occurs. The positive feedback causes the system to become unstable, and the undesired effect is clearly recognizable.

Positive feedback occurs when the system output is added to the system input, whereas negative feedback occurs when the system output is subtracted from its input. For our purposes in this chapter, only negative feedback will be considered. While positive feedback can be useful if controlled, its applications will be discussed in a later chapter. However, methods for minimizing positive feedback effects will be discussed here.

23.1 The Feedback Equation

Consider the feedback system shown in Fig. 23.1. The variables used in this diagram are labeled as an x value because they may be either a voltage or a current. The input signal, x_s, and the output signal x_o, may be either a voltage or a current. However, the feedback signal, x_f, must be the same type of signal as the input signal. If the input signal is a voltage, then the feedback signal must also be a voltage.

The block A_{OL} represents an amplifier's open-loop gain and is frequency dependent. The input to the amplifier, x_i, is the difference in the input source signal and the feedback signal, or

$$x_i = x_s - x_f \tag{23.1}$$

and the output is given by

$$x_o = A_{OL}(j\omega) \cdot (x_s - x_f) \tag{23.2}$$

In the following discussion, we will assume that all components of the system are ideal, meaning that the feedback network, β, will not load the amplifier. The specifications for the system can then be defined as follows:

$$A_{OL} = \frac{x_o}{x_i} \tag{23.3}$$

The *feedback factor*, β, is defined as

$$\beta = \frac{x_f}{x_o} \tag{23.4}$$

and the closed-loop gain, A_{CL}, is

$$A_{CL} = \frac{x_o}{x_s} \tag{23.5}$$

Realizing that

$$x_f = \beta \cdot x_o \tag{23.6}$$

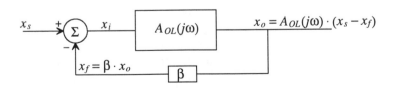

Figure 23.1 Basic block diagram to illustrate the feedback concept.

and plugging Eq. (23.6) into (23.2) and solving for the closed-loop gain, we find that A_{CL} becomes

$$A_{CL} = \frac{x_o}{x_s} = \frac{A_{OL}}{1 + A_{OL}\beta} \tag{23.7}$$

where the frequency dependence of A_{OL} has not been shown.

Note the dependence of the A_{CL} on the value of A_{OL}. If the value of A_{OL} becomes large (A_{OL} approaches infinity), then the value of A_{CL} approaches $\frac{1}{\beta}$. This illustrates the need for having a high-gain amplifier. If A_{OL} is very large, then the closed-loop gain becomes highly dependent on the feedback components.

The term $A_{OL}\beta$ is often referred to as the loop gain and will be used in later sections to determine overall amplifier stability.

23.2 Properties of Negative Feedback on Amplifier Design

When used in the design of amplifiers, feedback can provide a number of advantages. These advantages include desensitizing the gain to process parameter variation, reducing nonlinear distortion, reducing the effects of noise, extending the useful bandwidth of the amplifier, and controlling the input and output impedance levels.

23.2.1 Gain Desensitivity

Since the value of the open-loop gain, A_{OL}, is large, its value may change significantly with temperature, mismatch of devices, and other parameter variations. However, negative feedback desensitizes the closed-loop gain from changes in the open-loop gain. The following derivation illustrates this property.

Differentiating both sides of Eq. (23.7) yields

$$\frac{dA_{CL}}{dA_{OL}} = \frac{1}{(1 + A_{OL}\beta)^2} \text{ or } dA_{CL} = \frac{dA_{OL}}{(1 + A_{OL}\beta)^2} \tag{23.8}$$

Dividing each side of Eq. (23.8) by the corresponding factors in Eq. (23.7),

$$\frac{dA_{CL}}{A_{CL}} = \frac{1}{(1 + A_{OL}\beta)} \cdot \frac{dA_{OL}}{A_{OL}} \tag{23.9}$$

where $\frac{dA_{CL}}{A_{CL}}$ represents the fractional change in A_{CL} for a given fractional change in $\frac{dA_{OL}}{A_{OL}}$.

Using Eq. (23.9), if A_{OL} = 10,000 V/V (assuming a voltage amplifier) and β = 1/10 V/V, it can be seen that if $\frac{dA_{OL}}{A_{OL}}$ = 10%, then the change in the closed-loop gain, $\frac{dA_{CL}}{A_{CL}}$ = 0.01 %! This can easily be verified by using Eq. (23.7) for A_{OL}=10,000 and 9,000, keeping β = 1/10 V/V and solving for A_{CL} for each case. The resulting difference proves the immunity of the closed-loop gain to changes in A_{OL}.

23.2.2 Bandwidth Extension

Negative feedback also increases the usable bandwidth of an amplifier. Assume that an amplifier has the frequency response given by

$$A_{OLH}(s) = \frac{A_{OL}\omega_H}{s + \omega_H} = A_{OL}\frac{1}{\frac{s}{\omega_H} + 1} \qquad (23.10)$$

$A_{OLH}(s)$ is simply the high-frequency dependent version of A_{OL} and is approximated using a first-order pole at ω_H. Plugging Eq. (23.10) into Eq. (23.7) yields the high-frequency dependent closed-loop version of Eq. (23.7)

$$A_{CLH}(s) = \frac{A_{OLH}(s)}{1 + A_{OLH}(s)\beta} = \frac{A_{OL}\omega_H}{s + \omega_H(1 + A_{OL}\beta)} \qquad (23.11)$$

which can be rewritten as

$$A_{CLH}(s) = \frac{A_{OL}}{(1 + A_{OL}\beta)} \cdot \frac{1}{\frac{s}{\omega_H(1+A_{OL}\beta)} + 1} = \frac{A_{OL}}{(1 + A_{OL}\beta)} \cdot \frac{1}{\frac{s}{\omega_{HF}} + 1} \qquad (23.12)$$

Note that the resulting equation is composed of two parts. The first factor should be recognized as Eq. (23.7) and is simply the expression for the closed-loop gain at midband. The second factor is the frequency dependent term. It is interesting to observe that the original –3 dB frequency ω_H in Eq. (23.10) is now multiplied by $(1+A_{OL}\beta)$. Figure 23.2 illustrates the bandwidth extension from using negative feedback. The original open-loop frequency response is drawn in the solid line, while the closed-loop frequency response is illustrated in the dashed line. The closed-loop response shows a decrease in the gain and at the same time an increase in bandwidth.

If one were to decrease the value of β, from Eq. (23.11), it can be seen that two interesting effects occur. First, the closed-loop value of the gain will increase, since the

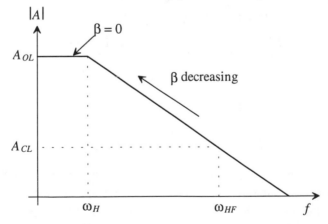

Figure 23.2 Extension of the high-frequency pole by using feedback.

first factor in Fig. 23.12 increases with a decrease in β. Second, the frequency response decreases since ω_H is multiplied by $(1 + A_{OL}\beta)$ and β is decreasing. The resulting frequency response curve is seen in Fig. 23.2. As β decreases, the overall curve follows the original open-loop curve. Naturally, if β decreases to 0, the amplifier no longer contains any feedback and the amplifier is operating in the open-loop configuration. It should be seen that when using feedback, one trades gain for bandwidth.

The same analysis can be applied to the low-frequency response of an amplifier. Equation (23.13) approximates the low-frequency response of an amplifier with a single first-order low-frequency pole.

$$A_{OLL}(s) = A_{OL}\frac{s}{s+\omega_L} \tag{23.13}$$

If we plug Eq. (23.13) into Eq. (23.7), the closed-loop low-frequency response becomes

$$A_{CLL}(s) = \frac{A_{OLL}(s)}{1+A_{OLL}(s)\beta} = \frac{A_{OL}\frac{s}{s+\omega_L}}{1+A_{OL}\frac{s}{s+\omega_L}\beta} = \frac{A_{OL}}{1+A_{OL}\beta}\frac{s}{s+\frac{\omega_L}{1+A_{OL}\beta}} \tag{23.14}$$

The result in Eq. (23.14) is comprised of the standard closed-loop gain at midband and the frequency dependent term. Note however, that with feedback, the original low-frequency pole, ω_L, in Eq. (23.13) is now divided by $1+A_{OL}\beta$ in Eq. (23.14). Figure 23.3 illustrates the effect of the feedback on the low-frequency response of an amplifier.

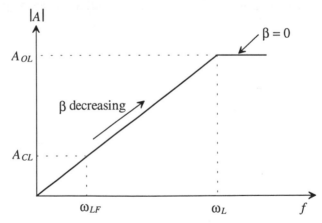

Figure 23.3 Extension of the low-frequency bandwidth by using feedback.

23.2.3 Reduction in Nonlinear Distortion

Negative feedback can also improve the nonlinear behavior of an amplifier. Examine the transfer curve of the voltage amplifier shown in Fig. 23.4 without feedback. Ideally, the amplifier should have a straight line from $-2\ V < V_{in} < 2\ V$. However, nonlinear behavior in the amplifier causes a different slope to occur when $V_{in} > 1V$ and $V_{in} < -1\ V$.

The reasons for nonlinearity will be discussed in a future chapter. Note that the output voltage is limited by the value of the power supply (+ 15 V).

Since the gain of the amplifier can be defined as the slope of the transfer curve, we can deduce that

$$A_{OL} = \frac{\Delta V_O}{\Delta V_{IN}} = 10 \text{ V/V} \quad -10 \text{ V} \leq V_o \leq 10 \text{ V} \tag{23.15}$$

and

$$A_{OL} = \frac{\Delta V_O}{\Delta V_{IN}} = 5 \text{ V/V} \quad V_o > 10 \text{ and } V_o < -10 \text{ V} \tag{23.16}$$

Now suppose that feedback is applied around the amplifier with $\beta = 0.1$ V/V and the transfer curve is redrawn. The gain of the amplifier with feedback becomes

$$A_{CL} = \frac{A_{OL}}{1 + A_{OL}\beta} = \frac{10}{1 + 10(0.1)} = 5 \text{ V/V} \quad -10 \text{ V} < V_o < 10 \text{ V} \tag{23.17}$$

and

$$A_{CL} = \frac{5}{1 + 5(0.1)} = \frac{10}{3} \quad V_o > 10 \text{ V and } V_o < -10 \text{ V} \tag{23.18}$$

The resulting transfer curve is illustrated in Fig. 23.4. Note that with feedback, the overall transfer curve is much more linear, resulting in an amplifier that has less nonlinear distortion than the amplifier shown in Fig. 23.4.

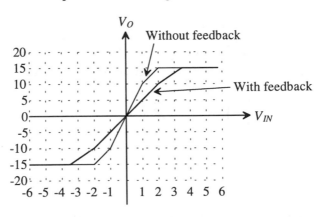

Figure 23.4 Using feedback to improve amplifier linearity.

23.2.4 Input and Output Impedance Control

Input and output impedances of an amplifier can be controlled using negative feedback. As seen in Fig. 23.5, R_i is the small-signal input impedance looking into an amplifier without feedback, and R_{inf} is the input impedance with feedback applied. Similiarly, R_{of}

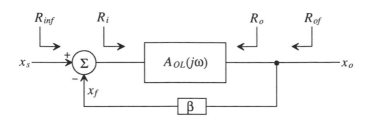

Figure 23.5 Determining the input and output impedances with and without feedback.

and R_o are the output impedances with and without feedback, respectively. Feedback allows us either to increase or decrease both R_{inf} and R_{of} by a factor of $(1 + A_{OL}\beta)$. Although an example of this property is difficult to derive in a general way, proofs of this property will be illustrated with actual circuit examples in a later section. For the time being, it will be sufficient to summarize the impedance control properties of feedback with Table 23.1.   Note that the closed-loop impedances are completely dependent on the type of variable that is used (voltage or current) at the input and the output.

If the input variable, x_s, is a voltage, the closed-loop input resistance, R_{inf}, is equal to the open-loop value of the input resistance, R_{in}, multiplied by the value of $1 + A_{OL}\beta$. If the input variable is a current, R_{inf} is divided by the same factor. Alternatively, if the output variable, x_o, is a voltage, then the closed-loop output impedance is the open-loop output impedance divided by $(1 + A_{OL}\beta)$. And if the output variable is a current, the open-loop output impedance is multiplied by the same factor.

Obviously, this concept is a powerful one. We can adjust both input and output impedances in a larger or smaller direction simply by choosing the type of feedback used in the circuit.

Input Variable, x_s	Output Variable, x_o	R_{inf}	R_{of}
V	V	$R_i \cdot (1 + A_{OL}\beta)$	$R_o / (1 + A_{OL}\beta)$
V	I	$R_i \cdot (1 + A_{OL}\beta)$	$R_o \cdot (1 + A_{OL}\beta)$
I	V	$R_i / (1 + A_{OL}\beta)$	$R_o / (1 + A_{OL}\beta)$
I	I	$R_i / (1 + A_{OL}\beta)$	$R_o \cdot (1 + A_{OL}\beta)$

Table 23.1 Summary of impedances with feedback.

23.3 Recognizing Feedback Topologies

Examine the general single-loop feedback circuit in Fig. 23.6. Remember that x_s, x_i, and x_f must all either be currents or all be voltages, and that the output variable, x_o may be either a voltage or a current. As a result of these restrictions, a total of four types of feedback are possible (Table 23.2).

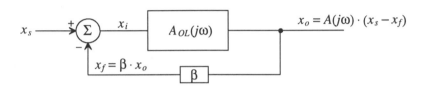

Figure 23.6 Basic block diagram to illustrate the feedback concept.

x_s, x_i, x_f	x_o	Feedback Type (mixing-sampling)
Voltage	Voltage	Series-shunt
Voltage	Current	Series-series
Current	Current	Shunt-series
Current	Voltage	Shunt-shunt

Table 23.2 Feedback type as a function of system variables.

The input summation is often referred to as input mixing. If the input variables, x_s, x_i, and x_f can be written as voltages, the mixing is referred to as series or voltage mixing. If the input variables can be written as currents, it is referred to as shunt or current mixing. The type of variable used at the output determines the second term, known as sampling. If the output variable is a voltage, the sampling is referred to as shunt or voltage sampling. If the output variable is a current, the sampling is known as series or current sampling.

In the analysis of feedback amplifiers, the following terminology will apply. The term *basic amplifier* will correspond to the gain circuit, A_{OL}, in the block level feedback diagrams. When determining the value of A_{OL}, it is very important to determine the loading due to the β network and any source and load resistance. The β network is also known as the feedback network. The overall circuit that includes both A_{OL} and β will be referred to as the *feedback amplifier*.

23.3.1 Input Mixing

Figures 23.7a and b illustrate the inputs to two generic feedback amplifiers. The basic amplifier in 23.7a is used in a series mixing configuration since the equation, $x_i = x_s - x_f$ can be written in terms of the voltages. Note also that the basic amplifier and the β network are in series with each other. In Fig. 23.7b, however, the basic amplifier is used in a shunt mixing configuration. Here, the summation at the input can only be written in terms of currents such that $i_i = i_s - i_f$. The feedback circuit and the basic amplifier circuit are both in parallel, or it can be said that the feedback shunts the basic amplifier.

23.3.2 Output Sampling

Determining the type of variable used at the output is not an obvious endeavor. However, a general model followed by several examples will help clarify the proper procedure. Figures 23.7c and d show the outputs of two generic feedback amplifiers. Shunt sampling is shown in Fig. 23.7c. Note that the feedback circuit is shunting, or in parallel with the basic amplifier circuit. The output variable is considered a voltage because the feedback circuit "senses" or samples the voltage across R_L. Series sampling is seen in Fig. 23.7d. Here, the feedback network is in series with the basic amplifier circuit. The output variable is a current since the feedback circuit now senses the

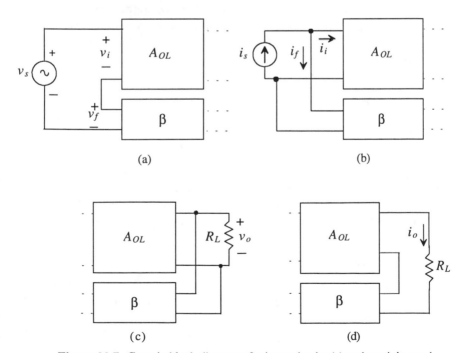

Figure 23.7 Generic block diagrams for input circuits (a) series mixing and (b) shunt mixing and output circuits, (c) shunt sampling, and (d) series sampling.

current through the load resistor. Two rules that help distinguish between the two types of output sampling are:

Rule 1: If one terminal of the output active device (drain or source) is driving the load and the other terminal is attached to the feedback network, the output sampling is series.

Rule 2: If the load and the beta network are connected to the same node, the output sampling is shunt.

23.3.3 The Feedback Network

When analyzing feedback circuits, it would be wise first to recognize the basic amplifier circuit, A_{OL}, and the β network by tracing the small-signal path from the input to the output and back through the feedback. The path from the input through the A_{OL} circuit to the output will be denoted as the forward path and the path from the output back through the β network, as the feedback path. Determining which path the signal takes is not obvious when there are several devices from which to choose. Some rules that will aid in the analysis are as follows:

- The *forward path* through the basic amplifier circuit will always take the path that has the highest gain.

- The AC small-signal will always enter either a gate or a source and will always exit either a drain or a source. The gain from drain to source is very small (at least for linear applications) and is considered to be negligible for most of our applications. A small-signal will never exit through the gate.

- The feedback signal must subtract from the input signal. To ensure that this is the case, one must count the number of inversions around the loop. Every time a signal crosses a gate-to-drain junction (common source amplification), an inversion occurs. Examples of this will be discussed next.

Examine Fig. 23.8a. Note that the signal path in the forward direction progresses through the path with the highest gain—from the gate of M1 to the drain of M1, into the gate of M2 and out the drain of M2. The feedback path consists of the resistors R_1 and R_2. The feedback variable consists of the voltage that appears across R_1.

An Important Assumption

One might attempt to draw the forward path from the gate of M1 to the source of M1, through the feedback resistor, R_2, and to the output. In actuality, this would be a legitimate forward path because in some cases the feedback network is actually bidirectional. However, the signal will have very little gain since the gain from gate to source of any MOS device is a maximum of one (common drain amplification) and the gain from v_f to v_2 in the forward direction will be less than one because of the voltage divider relationship between R_s and R_2. *Thus, for all of the analysis performed in this chapter, the forward path through the amplifier circuit dominates the expression for the total forward gain from input to output, while the forward path through the feedback is assumed to be negligible.* This important assumption greatly simplifies the analysis and is a good one, since the amplifier will have very poor performance if there is not a good deal of gain through the basic amplifier. (Ideally, the value of A_{OL} is infinite.) If an active device is used in the feedback network, the forward gain through the β network is minimized.

Figure 23.8b has the same basic transistor topology, except it uses current as its input. The forward path consists of the source of M1, out the drain of M1 into the gate of M2 and out the source of M2. The feedback path consists of R_s and R_2, and the feedback variable is the current, i_f. Note that the forward path has the highest possible gain path, since M1 acts as a common gate amplifier. The forward path could progress from the input through R_2 and into the drain of M2. However, since the small-signal gain from the drain to the source of M2 is extremely small, it will be neglected.

Counting Inversions Around the Loop

It is also necessary to check the circuit to ensure that the feedback is indeed negative. Counting the number of inversions around the loop in Fig. 23.8a may initially give the wrong impression, because the number of gate-to-drain junctions encountered by the signal around the loop is two. This implies positive feedback since the feedback variable must be inverted with respect to the input signal. However, the final mixing between v_1 and v_f will provide the additional negative behavior needed to ensure the proper feedback. Notice the relationship that exists between v_i and v_f. Since $v_i = v_1 - v_f$ the voltage v_f has a subtractive effect on v_i. Thus, a positive change in the signal entering the gate of M1 will result in a positive change in the source of M1 and a smaller, stable voltage v_i. If v_f had been negative with respect to v_1, then v_i would be, $v_i = v_s + v_f$ and positive feedback would occur.

Now examine Fig. 23.8b. Again, if we count the number of inversions around the loop from input back to the mixing variable, i_f, the number of inversions is odd, which is exactly what is needed, since KCL tells us that $i_i = i_1 - i_f$. The direction of the feedback current from the output is opposite the direction of i_f; however, the signal is inverted, making it equivalent to i_f, as illustrated from Fig. 23.9.

Examples of Recognizing Feedback Topologies

Now that the method for recognizing feedback has been discussed, examples will be presented to solidify the concepts. Again examine Fig. 23.8. Here, two types of feedback are illustrated. Transferring the block diagram concepts to transistor-level circuits can be a difficult task. However, several other rules can be applied which will reveal the type of sampling used in the circuit. The term *input active device* is used to denote the transistor that is being driven by the input source. The term *output active device* is used to denote the transistor used to drive the load.

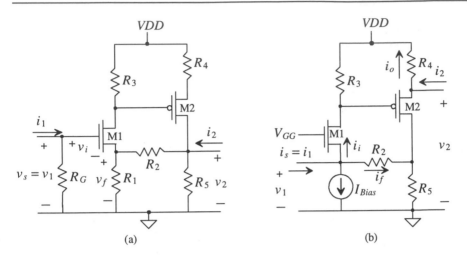

(a)   (b)

Figure 23.8 Feedback topologies (a) series-shunt and (b) shunt-series.

In Fig. 23.8a, the input variables can only be written in terms of voltages such that $v_i = v_s - v_f$. We could attempt to sum currents at the node on the gate. However, the current flowing into the resistor, R_G, will not be the result of any feedback, thus making current mixing impossible. The output sampling is of the shunt type since the β network and the basic amplifier are in parallel (the β network is connected to the same node as the output) and v_o is the voltage being sampled. This amplifier employs series mixing and shunt sampling, and is referred to as a series-shunt feedback amplifier configuration.

Figure 23.8b illustrates shunt-series feedback. The DC current source is viewed as an AC open circuit, and the DC voltage source, V_{GG}, is considered as an AC short circuit. Note how the input variables can only be written in terms of currents forming the expression, $i_i = i_s - i_f$. Series sampling is used at the output since the feedback network is in series with the basic amplifier. This amplifier configuration also follows rule 2, mentioned previously; thus, the proper small-signal output variable is the current, i_o. Note that the direction of the small-signal output current, i_o, is consistent with the direction of the small-signal model, since VDD is considered a small-signal ground.

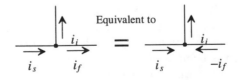

Figure 23.9 Shunt mixing illustrated with an odd number of inversions.

The variables v_1, i_1, v_2, and i_2 may or may not correspond to the defined input and output variables of the feedback circuit. For example, in Fig. 23.8b, the correct output variable is defined as i_o. However, the gain of the feedback amplifier can be ascertained in terms of v_1, i_1, v_2, and i_2, since $v_2 = i_o \cdot R_4$. Also note that v_1/i_1 and v_2/i_2 correspond to the input and output impedances, respectively, of the feedback amplifier.

Series-series feedback is illustrated in Fig. 23.10a. Although the input variables are written in terms of voltages, the output variable is considered to be a current since the feedback network is in series with the amplifier output. Note that the only difference between Figs. 23.10a and 23.8a is in placement of the output terminal. The same holds true for Figs. 23.10b and 23.8b. Lastly, Fig. 23.10b illustrates shunt-shunt feedback. Current summation occurs at the input which typifies shunt or current mixing, and the β network made up of R_s and R_2 shunts the output.

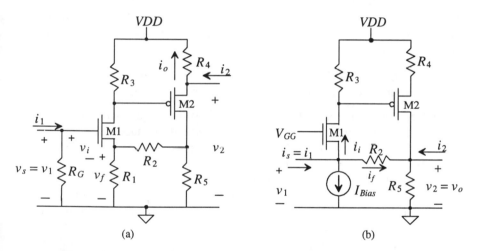

Figure 23.10 Feedback topologies: (a) series-series and (b) shunt-shunt.

23.3.4 Calculating Open-Loop Parameters

Once the feedback topology can be recognized, the analysis of the circuit can begin. We will perform two types of analysis: open-loop and closed-loop. Openloop analysis entails approximately 80 percent of the circuit analysis required to solve a feedback problem. Open-loop values are then used to calculate the closed-loop values, so extreme care must be taken when analyzing the open-loop circuit. We will then "close the loop" and calculate the feedback characteristics of the overall feedback amplifier

The method of analysis for feedback amplifiers can be confusing if we do not distinguish between open- and closed-loop notation. We will denote the open-loop voltage and current variables by use of the * notation. The open-loop parameters include A_{OL}, R_i, R_o, and β and are defined as follows.

$$A_{OL} = \frac{x_o^*}{x_s^*} \text{ and is the open-loop gain of the basic amplifier} \qquad (23.19)$$

$$R_i = \frac{v_1^*}{i_1^*} \text{ and is the open-loop input impedance} \qquad (23.20)$$

$$R_o = \frac{v_2^*}{i_2^*} \text{ and is the open-loop output impedance} \qquad (23.21)$$

$$\beta = \frac{x_f^*}{x_o^*} \text{ and is the gain through the feedback network} \qquad (23.22)$$

In our discussion of general feedback principles in Sec. 23.1, it was assumed that the β network was ideal, meaning that the impedance of the β network did not load the amplifier circuit. However, in real-life applications, the β network can cause loading effects on both the input source and the output of the amplifier circuit. The type of feedback used will determine how the β network loading is calculated. As we progress in the discussion of each feedback topology, the method of determining the β network loading will be presented.

All variables in the following analysis are considered to be small-signal AC voltages or currents. We will perform the open-loop analysis using the following steps:

1. Replace input source with Norton or Thevenin equivalent circuit and calculate open-loop gain. The amplifier circuit will produce a different type of gain for each type of feedback. For example, the open-loop gain for an amplifier used in a series-shunt configuration will have units of V/V. This is a standard voltage amplifier since the output variable is a voltage and the input mixing sums voltages. An amplifier using series-series feedback will have a gain with units of I/V, since the output variable is now a current and the input mixing uses voltages. This type of amplifier is also known as a transconductance amplifier since the units of the gain, I/V, are equivalent to a conductance. Similiarly, an amplifier used in the shunt-series configuration has a gain with units of I/I (a current amplifier), and an amplifier used in a shunt-shunt configuration will have a gain with units of V/I and is known as a transimpedance amplifier.

2. When calculating the open-loop gain of the circuit, consider the β network loading. An equivalent resistance, $R_{\beta i}$ and $R_{\beta o}$, as seen in Fig. 23.11 will represent the total input and output resistance of the β network. The method for calculating $R_{\beta i}$ involves the following steps.

   • If the output sampling is shunt (voltage), short the output node to ground.

   • If the output sampling is series (current), remove the device driving the output load as if it were taken "out-of-socket."

   • Calculate $R_{\beta i}$ as the impedance looking from the input into the β network.

The method for calculating $R_{\beta o}$ involves similar methodology:

- If the input mixing is shunt (current), short the input node to ground.
- If the input mixing is series (voltage), remove the input active device as if it were taken "out-of-socket."
- Calculate $R_{\beta o}$ as the impedance looking from the output into the β network.

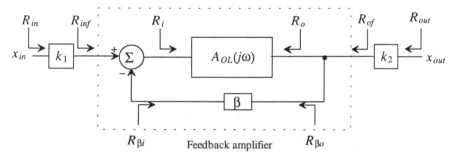

Figure 23.11 Block diagram of a generic feedback amplifier that distinguishes between open- and closed-loop impedances.

The method for determining the loading of the β network is rooted in basic two-port theory where each type of feedback configuration corresponds to one of the four basic two-port topologies. For further information regarding two-port theory, the reader is advised to consult [1]. One method used to easily remember the above rules is to remember that if mixing or sampling is shunt, then "short" the input or output, respectively. If the mixing or sampling is series, then "sever" the input or output device, respectively, by taking it "out-of-socket."

3. Calculate feedback factor, β, for the amplifier. The open-loop circuit is analyzed to determine the gain from the output back to the point where the feedback variable mixes with the input.

4. Determine the open-loop input and output impedances, R_i and R_o. These impedances can be calculated using standard circuit analysis techniques.

23.3.5 Calculating Closed-Loop Parameters

Once all the open-loop parameters are obtained, the closed-loop values are easily calculated. The closed-loop gain is

$$A_{CL} = \frac{x_o}{x_s} = \frac{A_{OL}}{1 + A_{OL} \cdot \beta} \qquad (23.23)$$

for all four topologies. The closed-loop input impedances represent only the input and output impedances of the feedback amplifier. The input impedance from the input source, R_{in}, may correspond to R_{inf} if there is no gain from the source, x_{in}, to the input of the feedback amplifier, x_s. Similiarly, the value of the output resistance, R_{out}, may or may not correspond to R_{of}, depending on the type of sampling used. Figure 23.11 illustrates this distinction. Calculating the input and output impedances of the feedback amplifier is dependent on the type of mixing and sampling circuits used, respectively.

$$\text{For series input mixing, } R_{inf} = R_i(1 + A_{OL}\beta) \tag{23.24}$$

$$\text{For shunt input mixing, } R_{inf} = \frac{R_i}{(1 + A_{OL}\beta)} \tag{23.25}$$

The value of the closed-loop output impedances will be as follows:

$$\text{For series output sampling, } R_{of} = R_o(1 + A_{OL}\beta) \tag{23.26}$$

$$\text{For shunt output sampling, } R_{of} = \frac{R_o}{(1 + A_{OL}\beta)} \tag{23.27}$$

Equations (23.24)-(23.27) will be derived with each specific topology. However, it is important to note the effectiveness of using feedback to control input and output impedances. Note that if series mixing or series sampling is used, the open-loop value of the input or output impedance is multiplied by $(1 + A_{OL}\beta)$; and if shunt mixing or shunt sampling is used, the open-loop values of impedances are divided by the same factor. Now that the general methodology has been described, a detailed examination of the four feedback topologies will be presented. The analysis of the four basic feedback topologies begins at the discrete level and progresses into more complex integrated circuits throughout the section.

23.4 The Voltage Amp (Series-Shunt Feedback)

Consider the ideal voltage feedback amplifier shown in Fig. 23.12, with open-loop values, A_{OL}, β, R_i, and R_o already given. The basic amplifier is a voltage amplifier with gain given V/V. Since this is an ideal feedback amplifier, the β network does not load the basic amplifier, meaning that $R_{\beta o} = \infty$ and $R_{\beta i} = 0$.

To determine what type of feedback is present in the circuit, examine the input variables. Since input variables x_s, x_i, and x_f correspond to the voltages v_s, v_i, and v_f, such that the equation $v_i = v_s - v_f$ can be written the input mixing is series. The output mixing can be determined using the previous rules. Since the output of the amplifier, A_{OL}, and the β network are attached in parallel (both being connected to the load, R_L), the output mixing is considered to be shunt.

From Sec. 23.1, we already know that the closed-loop gain of the amplifier is

$$A_{CL} = \frac{v_o}{v_s} = \frac{A_{OL}}{1 + A_{OL} \cdot \beta} \tag{23.28}$$

Notice that as A_{OL} approaches infinity, Eq. (23.28) approximates to

$$A_{CL} \approx \frac{1}{\beta} \qquad (23.29)$$

Equation (23.29) is important because it illustrates another power feedback concept. The entire gain of the feedback amplifier can be approximated as the inverse of β as the basic amplifier gain increases to higher and higher values.

We can calculate how the feedback affects the input impedance of the amplifier by applying a test voltage directly to the input of the feedback amp shown in Fig. 23.12, so that $v_s = v_{test}$ and $i_i = i_{test}$. Writing a voltage loop at the input and assuming that the feedback network does not load the amplifier gives

$$v_{test} = i_{test} \cdot R_i + \beta \cdot v_o = i_{test} \cdot R_i + \beta \cdot \frac{A_{OL}}{1 + A_{OL}\beta} \cdot v_{test} \qquad (23.30)$$

Equation 23.30 simplifies to

$$R_{inf} = \frac{v_{test}}{i_{test}} = R_i \cdot (1 + A_{OL}\beta) \qquad (23.31)$$

The gain of the open-loop amplifier was decreased by $1 + A_{OL}\beta$, while the input impedance, the impedance the source sees, was increased by $1 + A_{OL}\beta$.

Calculation of the output impedance proceeds in the same way as the calculation of the input resistance except that the test voltage is applied to the output of the amplifier with the input shorted to ground with $v_o = v_{test}$ and $i_o = i_{test}$. Writing a loop equation at the output gives

$$v_{test} = i_{test} \cdot R_o + A_{OL} \cdot v_i = i_{test} \cdot R_o + A_{OL} \cdot (-\beta \cdot v_o) \qquad (23.32)$$

since the input is shorted to ground and $v_i = -v_f = -\beta \cdot v_o$. The output impedance is given by

$$R_{of} = \frac{v_{test}}{i_{test}} = \frac{R_o}{1 + A_{OL}\beta} \qquad (23.33)$$

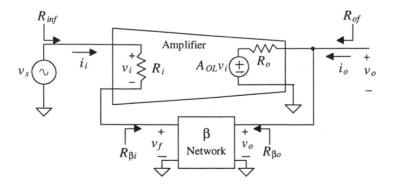

Figure 23.12 An ideal voltage amplifier (series-shunt).

or the output resistance is reduced by $1 + A_{OL}\beta$. Ideally, a voltage amplifier has infinite input resistance and zero output resistance. Adding feedback to a voltage amplifier with finite input resistance and nonzero output resistance help make the amplifier closer to the ideal.

Now that the ideal series-shunt amplifier has been examined from a block level point of view, a discussion of the nonideal series-shunt feedback amplifier at the transistor level with loading effects will be considered. The amplifier seen in Fig. 23.13 was analyzed previously and was determined to use series-shunt feedback.

The small-signal circuit of the feedback circuit is seen in Fig. 23.14. Note that the forward path consists of the following nodes: 1, 2, 3. The feedback path consists of nodes 3 and 4, with the feedback variable, v_f, appearing across R_1. In the previous discussion, it was assumed that the β network did not load the amplifier circuit. However, to accurately calculate the open-loop gain, A_{OL}, the loading of R_1 and R_2 on both the input and the output of the amplifier circuit needs to be considered. Note that the resistor R_S is initially ignored, since it is essentially outside the feedback amplifier.

Since we are analyzing a series-shunt amplifier, we may determine the loading caused by the β network on the input, $R_{\beta i}$ and the output $R_{\beta o}$ in the following way (refer to Fig. 23.15). Looking into the β network from the input, we observe the resistance seen with the output terminal shorted to ground. The equivalent resistance to ground seen is the loading of the β network seen by the input of the amplifier. In this example, R_2 is seen. Therefore, in the open-loop model used to determine A_{OL}, we will include R_2 in parallel with R_1. The loading at the output is found similarly. Since the input mixing is series, we will remove M1 "out-of-socket" and look into the β network from the output. The equivalent resistance seen is then attached to the output of the open-loop model. In this example, the equivalent resistance is $R_2 + R_1$ and is attached to the output of the open-loop model. The resulting open-loop model is seen in Fig. 23.16.

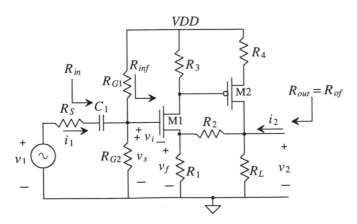

Figure 23.13 Transistor-level series-shunt feedback amplifier.

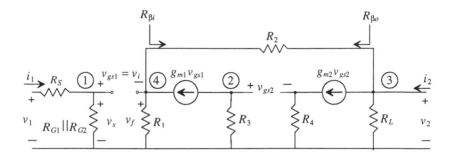

Figure 23.14 Closed-loop small-signal model of Fig. 23.8.

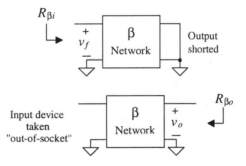

Figure 23.15 Determining the loading due to the feedback network
for a series-shunt amplifier.

We will initially assume that r_o for the MOSFETS is much larger than the discrete resistors and that the bulk and source are tied together ($v_{sb} = 0$). As we progress through the chapter, more difficult circuits will include drain-to-source resistances in our small-signal analysis.

The open-loop model is now ready to be analyzed in order to calculate A_{OL}. Since we are using a series (voltage)-shunt (voltage) feedback amplifier, the units of A_{OL} will be V/V and

$$A_{OL} = \frac{v_2^*}{v_s^*} \tag{23.34}$$

Solving for A_{OL} yields

$$A_{OL} = \frac{v_2^*}{v_s^*} = \left(\frac{v_2^*}{v_{gs2}^*}\right)\left(\frac{v_{gs2}^*}{v_{gs1}^*}\right)\left(\frac{v_{gs1}^*}{v_s^*}\right) = [-g_{m2}R_L||(R_2+R_1)]\left[-\frac{g_{m1}R_3}{1+g_{m2}R_4}\right]\left[\frac{1}{1+g_{m1}(R_1||R_2)}\right]$$

$$\tag{23.35}$$

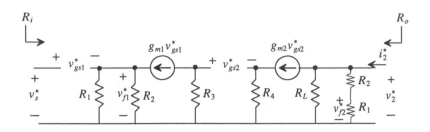

Figure 23.16 Open-loop small-signal model of Fig. 23.8.

Next, the value of β can also be calculated from the open-loop model. Remembering that β is defined as the gain from the output back to the input mixing variable, v_f, we can write

$$\beta = \frac{v_f^*}{v_2^*} = \frac{R_1}{R_1 + R_2} \tag{23.36}$$

since the β network is simply a voltage divider relationship. Notice that the open-loop circuit now contains two values of R_2 and v_f^*. In this example, since r_o was assumed to be infinite, the gain from v_2^* to v_{f1}^* will be zero. If r_o had not been neglected, the gain from v_2^* to v_{f1}^* would have been small but finite. Therefore, it can be said that a reverse path exists through the basic amplifier as well as through the feedback network. However, the gain from v_2^* to v_{f2}^*, though less than one, will be significantly larger than from v_2^* to v_{f1}^*. *Therefore, just as the forward path through the feedback network was neglected, the reverse path through the basic amplifier is assumed to be much smaller than the reverse path through the feedback path. Therefore, the value of β is calculated using the resistor, R_2, closest to the output.*

Next, the value for R_i and R_o will be calculated. These values are determined using the open-loop model generated in Fig. 23.16. Since we are using MOS devices, it should be obvious that the input resistance to the open-loop circuit is ∞. The output resistance, however, can be calculated shorting the gate of M1 to ground and applying a test voltage to the output. Since the input is grounded, $v_{g2} = 0$, and R_o is simply the parallel combination of resistances at the output (assuming that r_{o2} is very large).

$$R_o = \frac{v_o^*}{i_2^*} = R_L \| (R_1 + R_2) \tag{23.37}$$

Once the open-loop values are calculated, the closed-loop values are easily attained. The closed-loop values are

$$A_{CL} = \frac{v_2}{v_s} = \frac{A_{OL}}{1 + A_{OL}\beta} \tag{23.38}$$

$$R_{inf} = \frac{v_s}{i_s} = R_i(1 + A_{OL}\beta) \tag{23.39}$$

$$R_{out} = R_{of} = \frac{v_2}{i_2} = \frac{R_o}{1 + A_{OL}\beta} \tag{23.40}$$

Analysis of this problem has not yet been completed. Notice that we initially neglected the source resistance and the biasing resistors, R_{G1} and R_{G2}, since they played no part in the feedback analysis of the amplifier. However, they will have an effect in the overall gain and need to be considered. The total gain of the entire circuit is

$$\frac{v_2}{v_1} = \frac{v_s}{v_1} \cdot \frac{v_2}{v_s} = \frac{R_{G1} \| R_{G2}}{R_{G1} \| R_{G2} + R_S} \cdot A_{CL} \tag{23.41}$$

and the value of R_{in} as seen by the signal source is

$$R_{in} = \frac{v_1}{i_1} = R_{G1} \| R_{G2} \tag{23.42}$$

and the analysis is now complete. Notice that the value of R_{in} is not the same as R_{inf}.

Example 23.1
For the series-shunt circuit shown in Fig. 23.17a, draw the closed-loop small-signal model and identify the forward and feedback paths, draw the open-loop model, calculate A_{OL}, β, R_i, and R_o, and calculate the closed-loop parameters v_2/v_1, v_1/i_1, and v_2/i_2. You may assume that $r_o = \infty$ for both devices and that DC analysis has already been performed with $g_{m1} = g_{m2} = 1$ mA/V.

The closed-loop small-signal model can be seen in Fig. 23.17b with the forward and feedback paths drawn. The open-loop model with loading effects is seen in Fig. 23.17c. Since the output uses shunt sampling, the value of $R_{\beta i}$ is found by shorting the output node and looking at the load resistance resulting from the feedback network. In this case, $R_{\beta i}$ is equal to R_2. Since the input mixing is series, the input device, M1, is "severed," so that only $R_1 + R_2$ is seen looking into the feedback loop from the output. Thus, $R_{\beta o} = R_1 + R_2$.

The values of r_o are considered to be infinite, so the values of R_i and R_o can be found by inspection, $R_i = \infty$, and $R_o = R_1 + R_2 = 11$ kΩ. Notice that when solving the open-loop circuit (Fig. 23.17c) v_1 and R_G are not included since we are only interested in solving the feedback portion of the circuit.

The open-loop gain, A_{OL}, can be found simply as the gain of two common source amplifiers with source resistance,

$$A_{OL} = \frac{v_2^*}{v_s^*} = \frac{v_2^*}{v_{g2}^*} \cdot \frac{v_{g2}^*}{v_s^*} = \left[\frac{-g_{m2}(R_2 + R_1)}{1 + g_{m2}R_4} \right]\left[\frac{-g_{m1}R_3}{1 + g_{m1}(R_1 \| R_2)} \right] = 109.8 \text{ V/V}$$

The value of β is always the gain from the output to the feedback variable. In this case, a simple voltage divider relationship exists such that

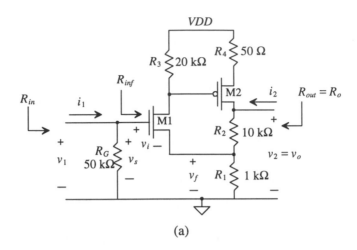

(a)

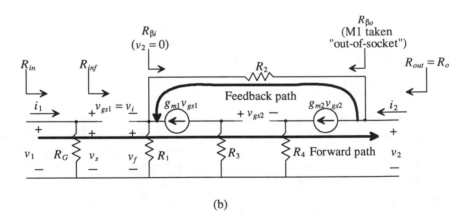

(b)

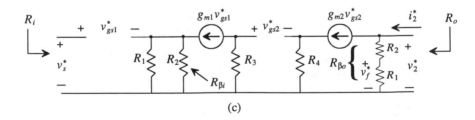

(c)

Figure 23.17 (a) Series-shunt circuit used in Ex. 23.1; (b) its closed-loop
small-signal model; and (c) the resulting open-loop model.

$$\beta = \frac{v_f^*}{v_2^*} = \frac{R_1}{R_1 + R_2} = 0.0909 \text{ V/V}$$

Again, notice that the product of $A_{OL}\beta$ is positive and unitless. Now that the open-loop values are calculated, the closed-loop values can be found by using Eqs. (23.38) - (23.40).

$$A_{CL} = \frac{v_2}{v_s} = \frac{A_{OL}}{1 + A_{OL}\beta} = \frac{109.8}{1 + (109.8 \cdot 0.0909)} = 10 \text{ V/V}$$

A simple check will verify that the solution is correct if $A_{CL} \approx \frac{1}{\beta}$, which is the case.

Notice in Fig. 23.17a and Fig. 23.17b that closed-loop output impedance, R_{of} is equal to the value of R_{out}. However, the value of the closed-loop input impedance, R_{inf}, is not equal to R_{in}, since the feedback amplifier itself excludes the input source and the gate resistor, R_G. The closed-loop values, R_{inf}, and R_{of}, are also easily attained, that is,

$$R_{inf} = \infty \text{ (since } R_i = \infty)$$

and

$$R_{of} = R_{out} = \frac{v_2}{i_2} = \frac{R_o}{(1 + A_{OL}\beta)} = \frac{11 \text{ k}\Omega}{10.98} = 1.002 \text{ }\Omega$$

The last step involves finding R_{in} and v_2/v_1. The value of R_{in} can be found by examining Fig. 23.17a as,

$$R_{in} = \frac{v_1}{i_1} = R_G \| R_{inf} = 50 \text{ k}\Omega$$

The overall gain, v_2/v_1, will be equal to the value of A_{CL} since the input voltage, v_1 is equal to v_s. If we had included a value for a source resistance, then an additional voltage divider relationship would have been needed to formulate v_2/v_1 as seen in Eq. (23.41)

$$\frac{v_2}{v_1} = A_{CL} = 10 \text{ V/V} \quad \blacksquare$$

23.5 The Transimpedance Amp (Shunt-Shunt Feedback)

Shunt-shunt feedback mixes current at its input and samples voltage at the output. Consider the ideal shunt-shunt feedback amplifier shown with open-loop values, A_{OL}, β, R_i, and R_o, in Fig. 23.18. In the ideal case, $R_{\beta i}$ is infinite and $R_{\beta o}$ is zero. Notice that the basic amplifier and the β network are in parallel at both the input and the output. The β network shunts the basic amplifier; therefore, the input variable is a current. Looking into the output of the amplifier, we see that the feedback path is in parallel with or shunts the output signal. Therefore, the output signal is a voltage. Also note that since the input variable is a current and the output variable is a voltage, the basic amplifier

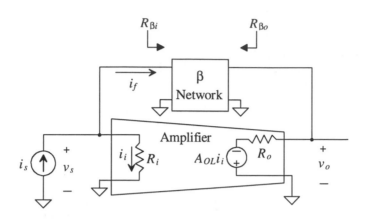

Figure 23.18 An ideal transimpedance (shunt-shunt) amplifier.

has units of V/I, also known as a transimpedance amplifier. Since $A_{OL}\beta$ should always be unitless and positive, β will have units of I/V (mhos).

The closed-loop gain is

$$A_{CL} = \frac{v_o}{i_s} = \frac{A_{OL}}{1+A_{OL}\beta} \tag{23.43}$$

Notice that the basic amplifier circuit is inverting, corresponding to an ideal op-amp circuit. Since A_{OL} is negative, then β must also be negative to ensure that negative feedback exists.

To calculate the input impedance of the transimpedance amplifier we apply a test current source to the input of the amplifier, such that $i_s = i_{test}$ and $v_s = v_{test}$. The test current and the input impedance are determined by

$$i_{test} = \frac{v_{test}}{R_i} + i_f = \frac{v_{test}}{R_i} + \beta v_o = \frac{v_{test}}{R_i} + \beta A_{OL} i_i = \frac{(1+A_{OL}\beta) \cdot v_{test}}{R_i} \tag{23.44}$$

Assuming that the β network does not load the basic amplifier, we note that the closed-loop input impedance becomes

$$R_{inf} = \frac{v_{test}}{i_{test}} = \frac{R_i}{1+A_{OL}\beta} \tag{23.45}$$

A similar analysis of the output impedance of the transimpedance amplifier shows that

$$R_{of} = \frac{R_o}{(1+A_{OL}\beta)} \tag{23.46}$$

The ideal transimpedance amplifier has zero input resistance and zero output resistance. We can see from Eqs. (23.45) and (23.46) that the addition of the feedback helps to make the basic amplifier appear closer to the ideal.

Now examine the transistor level shunt-shunt feedback circuit shown in Fig. 23.19. The closed-loop and open-loop small-signal models are seen in Fig. 23.20 and Fig. 23.21, respectively. The gate is attached to AC ground since it is attached to a DC source and the DC current source I_{SS} is an AC open circuit. Interestingly, since the values of r_{o1} and r_{o2} have been included, there is a feedback path through the basic amplifier. However, this gain is very small compared to the feedback path through the β network and is typically assumed to be negligible.

The open-loop analysis begins by evaluating the effects of the β network on the basic amplifier using the rules stated earlier. The resistance, $R_{\beta i}$, will be the equivalent resistance seen looking into the β network from the input with the output node shorted to ground. The equivalent resistance looking into the β network from the output, $R_{\beta o}$, will be calculated with the input shorted to ground.

Since we are using shunt-shunt feedback, A_{OL} will be defined as

$$A_{OL} = \frac{v_2^*}{i_s^*} = \frac{v_2^*}{v_{g2}^*} \cdot \frac{v_{g2}^*}{v_1^*} \cdot \frac{v_1^*}{i_s^*} \tag{23.47}$$

Solving the first term in Eq. (23.47), can be quite extensive if using standard circuit analysis. A circuit technique based on two-port theory will greatly simplify the analysis.

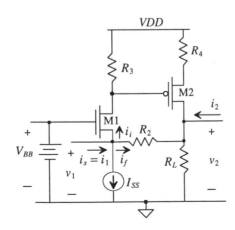

Figure 23.19 Shunt-shunt feedback amplifier.

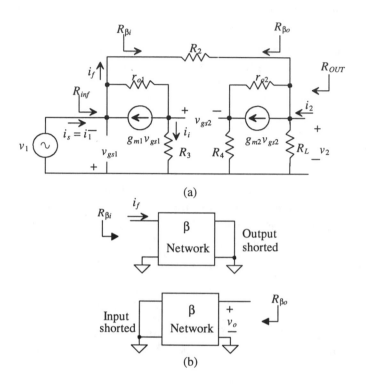

(a)

(b)

Figure 23.20 (a) Closed-loop small-signal model of Fig. 23.19 and
(b) method for determining the feedback network loading.

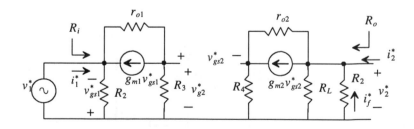

Figure 23.21 Open-loop small-signal model of Fig. 23.19.

For example, the equivalent circuit for the gain, $\frac{v_2^*}{v_{g2}^*}$, can be seen in Fig. 23.22a. However, the circuit seen in Fig. 23.22b shows the equivalent circuit using an equivalent transconductance, G_M, and output resistance, R_{Leq}. The gain of the equivalent circuit, and hence the actual circuit, is

$$\frac{v_2^*}{v_{g2}^*} = G_M R_{Leq} \tag{23.48}$$

The value of R_{Leq} can easily be found as

$$R_{Leq} = R_L \| R_2 \| R_{inD2} \tag{23.49}$$

where R_{inD2} is the resistance seen looking into the drain of M2. From Ch. 20, we know that this resistance is

$$R_{inD2} = [(1 + g_{m2}R_4)r_{o2} + R_4] \tag{23.50}$$

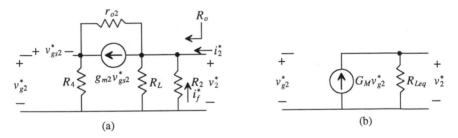

(a)   (b)

Figure 23.22 (a) Solving a portion of Fig. 23.21, including the drain-to-source resistance, and (b) the equivalent transconductance model.

The value of G_M is the short-circuit transconductance [2] and is defined as

$$G_M = \frac{i_o^*}{v_{g2}^*} \; (R_{Leq} = 0) \tag{23.51}$$

which means that the effective transconductance can be found by shorting the equivalent load resistance, in this case $R_L \| R_2$, and finding the gain from the short-circuit current to the input voltage. As seen in Fig. 23.23, the equations used to find G_M are

$$i_o^* = -g_{m2}v_{gs2} + \frac{v_{s2}}{r_{o2}} \tag{23.52}$$

$$v_{s2}^* = -i_o R_4 \tag{23.53}$$

$$v_{s2}^* + v_{gs2}^* = v_{g2}^* \tag{23.54}$$

and solving Eqs. (23.52) - (23.54) yields

$$GM = \frac{i_o^*}{v_{g2}^*} = \frac{-g_{m2}}{1 + g_{m2}R_4 + \frac{R_4}{r_{o2}}} \tag{23.55}$$

the gain, $\frac{v_2^*}{v_{g2}^*}$, becomes

$$\frac{v_2^*}{v_{g2}^*} = \frac{-g_{m2}(R_L||R_2||[(1 + g_{m2}R_4)r_{o2} + R_4)]}{1 + g_{m2}R_4 + \frac{R_4}{r_{o2}}} \tag{23.56}$$

Referring back to Eq. (23.47), the second factor, $\frac{v_{g2}^*}{v_1^*}$, can be found by analyzing Fig. 23.21 as

$$\frac{v_{g2}^*}{v_1^*} = \frac{g_{m1}R_3 + \frac{R_3}{r_{o1}}}{1 + \frac{R_3}{r_{o1}}} \tag{23.57}$$

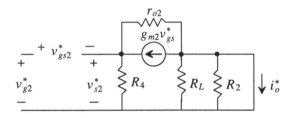

Figure 23.23 Circuit used to determine the equivalent transconductance.

The last term in Eq. (23.47) is simply the input resistance, R_i, of the open-loop circuit shown. Using a test source, we can determine this value to be

$$R_i = \frac{v_1^*}{i_s^*} = \frac{v_t}{i_t}||R_2 = \left(\frac{1 + \frac{R_3}{r_{o1}}}{g_{m1} + \frac{1}{r_{o1}}}\right)||R_2 \tag{23.58}$$

Therefore, the entire expression for the open-loop gain, A_{OL}, becomes

$$A_{OL} = \frac{v_2^*}{i_s^*} = \frac{-g_{m2}(R_L||R_2||((1 + g_{m2}R_4)r_{o2} + R_4))}{1 + g_{m2}R_4 + \frac{R_4}{r_{o2}}} \cdot \frac{g_{m1}R_3 + \frac{R_3}{r_{o1}}}{1 + \frac{R_3}{r_{o1}}} \cdot \left(\frac{1 + \frac{R_3}{r_{o1}}}{g_{m1} + \frac{1}{r_{o1}}}\right)||R_2 \ \Omega \tag{23.59}$$

At first glance, this equation may appear quite daunting. However, notice that if r_{o2} is much greater than R_2, R_4, and R_L and if r_{o1} is much greater than R_3, the open-loop gain simplifies to

$$A_{OL} = [\text{common source amp with source resistance}][\text{common gate amp}][R_i]$$

$$\approx \frac{-g_{m2}(R_L \| R_2)}{1 + g_{m2}R_4} \cdot g_{m1}R_3 \cdot \frac{1}{g_{m1}} \| R_2 \ \Omega \tag{23.60}$$

Equation (23.60) is typically used for discrete designs in which higher values of currents are used, therefore requiring lower resistor values. However, active loads can easily be used, as seen in Fig. 23.24. Here, the resistor, r_{o3} now replaces R_3. Since R_4 is a source degeneration resistor, its value will be small. The active load that replaces R_4 is a gate-drain connected device and equal to $\frac{1}{g_{m4}} \| r_{o4}$. And the resistor, R_L, is now replaced by r_{o5}. Therefore, the open-loop gain of Fig. 23.24 can be written by using Eq. (23.29) and making the proper substitutions:

$$A_{OL} = \frac{-g_{m2}(r_{o5} \| R_2 \| (1 + g_{m2}(\frac{1}{gm4} \| r_{o4}))r_{o2} + \frac{1}{gm4} \| r_{o4}))}{1 + g_{m2}\left(\frac{1}{gm4} \| r_{o4}\right) + \frac{\frac{1}{gm4} \| r_{o4}}{r_{o2}}} \cdot \frac{g_{m1}r_{o3} + \frac{r_{o3}}{r_{o1}}}{1 + \frac{r_{o3}}{r_{o1}}} \left[\frac{1 + \frac{r_{o3}}{r_{o1}}}{g_{m1} + \frac{1}{r_{o1}}} \| R_2 \right]$$

$$\tag{23.61}$$

which can be approximated as

$$A_{OL} = \frac{v_2^*}{i_s^*} \approx \frac{-g_{m2}(R_2)}{1 + g_{m2}\left(\frac{1}{gm4}\right)} \cdot \frac{1 + g_{m1}r_{o3}}{2} \cdot \left(\frac{1}{g_{m1}} \| R_2\right) \Omega \tag{23.62}$$

if it is assumed that $r_{o1} \approx r_{o3}$ and that the discrete resistor, R_2, is relatively small compared to the impedances of the active loads. An active device could have been used to substitute even the resistor, R_2. This will be discussed further later in the chapter.

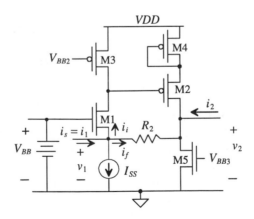

Figure 23.24 Shunt-shunt feedback amplifier using active loads.

Next, the value of β is calculated as

$$\beta = \frac{i_f^*}{v_2^*} = -\frac{1}{R_2} \text{ mhos} \qquad (23.63)$$

Again, note that $A_{OL}\beta$ is unitless and overall positive.

Now that all of the open-loop parameters, A_{OL}, β, R_i, and R_o, have been found, the closed-loop values are easily calculated. The value for the closed-loop gain is

$$A_{CL} = \frac{v_2}{i_s} = \frac{A_{OL}}{1 + A_{OL}\beta} \qquad (23.64)$$

The closed-loop input impedance is

$$R_{inf} = \frac{R_i}{1 + A_{OL}\beta} \qquad (23.65)$$

and the value for the closed-loop output impedance is

$$R_{of} = R_{out} = \frac{R_o}{1 + A_{OL}\beta} \qquad (23.66)$$

We should be able to see a trend in the feedback effects. Shunt mixing and shunt sampling cause the input and output impedances to decrease by the factor $(1 + A_{OL}\beta)$. Similarly, series input mixing and series sampling cause the input and output impedances to increase by the factor $(1 + A_{OL}\beta)$.

In many cases, the overall gain is expressed in terms of a voltage gain, $\frac{v_2}{v_1}$. Since we have calculated the transfer function in terms of the current, i_s, we can easily express this in terms of v_1 by

$$\frac{v_2}{v_1} = \frac{v_2}{i_s(R_{inf})} = A_{CL}(\frac{1}{R_{inf}}) \qquad (23.67)$$

23.5.1 Simple Feedback Using a Gate-Drain Resistor

One of the most popular and simplest examples of shunt-shunt feedback is seen in Fig. 23.25 and consists of a simple inverting amplifier with a resistor connecting the gate and the drain. The feedback resistor, R_2, is usually a large value and serves several important functions. When analyzing the DC characteristics of this circuit, the voltage at the drain will be equal to the voltage on the gate, since there is no DC current flowing through R_2 (when the input is AC coupled). This ensures that the device is always in saturation and provides biasing with no other components needed. We will examine the effects of R_2 on the AC response of the amplifier and soon discover that it has little effect on the gain of the amplifier as well.

Figure 23.26a and b shows the small-signal models for the closed and open-loop circuits, respectively. Since the feedback resistor sums current at the gate of M1, the mixing circuit is shunt. And since the feedback is taken off of the same node as the

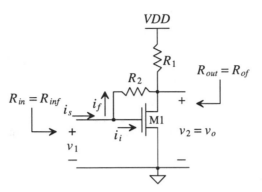

Figure 23.25 Shunt-shunt feedback using a simple gate-drain connected device.

output, the sampling circuit is shunt. Therefore, the value of $R_{\beta i}$ for the open-loop model is found by shorting the output and looking into the feedback loop from the input, and the value of $R_{\beta o}$ is found by shorting the input and looking into the feedback loop from the output.

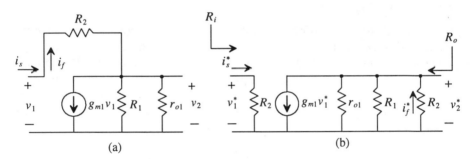

Figure 23.26 Small-signal model of Fig. 23.25: (a) the closed-loop model and (b) the open-loop model.

The open-loop values can be calculated as

$$A_{OL} = \frac{v_2^*}{i_s^*} = \frac{v_2^*}{v_1^*} \cdot \frac{v_1^*}{i_s^*} = [-g_{m1}(R_1 || R_2 || r_{o1})][R2] \text{ V/A} \qquad (23.68)$$

$$R_i = R_2 \qquad (23.69)$$

$$R_o = R_1 || R_2 || r_{o1} \qquad (23.70)$$

and

$$\beta = \frac{i_f^*}{v_2^*} = -\frac{1}{R_2} \qquad (23.71)$$

Using Eqs. (23.43) - (23.46), we find that the closed-loop values become

$$A_{CL} = \frac{v_2}{i_s} = \frac{A_{OL}}{1 + A_{OL}\beta} = \frac{-g_{m1}R_oR_2}{1 + g_{m1}R_oR_2\frac{1}{R_2}} \qquad (23.72)$$

$$R_{inf} = \frac{v_1}{i_s} = \frac{R_2}{1 + g_{m1}R_oR_2\frac{1}{R_2}} \quad \text{and} \quad R_{of} = \frac{R_o}{1 + g_{m1}R_oR_2\frac{1}{R_2}} \qquad (23.73)$$

The value of R_{inf} is also equal to the value of R_{in} since no source resistance is associated with v_1. Notice that the value of the closed-loop gain is dependent on R_2. However, most applications of this amplifier use voltage as the input variable. Therefore, the value of the overall voltage gain becomes

$$\frac{v_2}{v_1} = \frac{v_2}{i_s} \cdot \frac{i_s}{v_1} = A_{CL} \cdot \frac{1}{R_{inf}} = -g_{m1}R_o = -g_{m1}(R_1 \| R_2 \| r_{o1}) \qquad (23.74)$$

and if the value of R_2 is chosen to be much larger than R_1 then its effect on the AC midband gain is minimized.

Example 23.2
Calculate the gain, $\frac{v_o}{v_s}$, of the shunt-shunt amplifier in Fig. 23.27 assuming that $A = 500,000$ V/A, $R_i = 10\ \Omega$, and $R_o = 10\ \Omega$. Notice that the circuit is similar to a simple inverting op-amp.

First, we will source transform the voltage source to a current source, since the input mixing is shunt. The model of the amplifier is in terms of A, not A_{OL}, since we must now include the loading of the feedback resistor, and R_L and R_S in the calculation of A_{OL}. The basic amplifier circuit, a transimpedance amplifier with units of V/I, has an ideal input and output impedance of $0\ \Omega$. Next, we will determine the loading of the β network consisting of R_F on the basic amplifier as seen in Fig. 23.28. A_{OL} can be determined as

$$A_{OL} = \frac{v_o^*}{i_s^*} = -A \cdot \left(\frac{R_F \| R_L}{R_F \| R_L + R_o}\right) \cdot \left(\frac{R_S \| R_F}{R_S \| R_F + R_i}\right) \Omega$$

$$= -500,000 \cdot 0.989 \cdot 0.989 = -489,060 \text{ V/A}$$

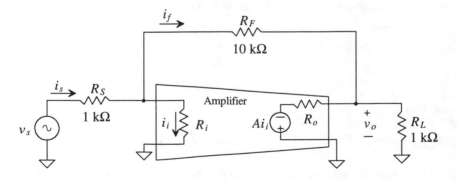

Figure 23.27 A transimpedance amplifier example.

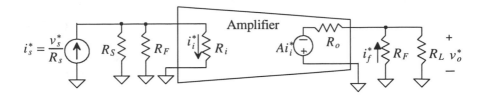

Figure 23.28 Open-loop transimpedance amplifier used in Example 23.2.

The value of β is easily calculated as

$$\beta = \frac{i_f^*}{v_o^*} = -\frac{1}{R_f} = -0.0001 \text{ A/V}$$

And A_{CL} becomes

$$A_{CL} = \frac{v_o}{i_s} = \frac{-489,060}{1 + -489,060 \cdot -0.0001} = -9.8 \text{ k}\Omega \approx R_F$$

Since $v_s = i_s \cdot R_s$, the overall voltage gain is given by

$$\frac{v_o}{v_s} = \frac{v_o}{i_s} \cdot \frac{1}{R_s} = A_{CL} \cdot \frac{1}{R_s} = -\frac{9.8\text{k}}{1\text{k}} = -9.8 \text{ V/V} \approx -\frac{R_F}{R_s}$$

Notice that this is the gain of the standard inverting op-amp configuration. ■

23.6 The Transconductance Amp (Series-Series Feedback)

A series-series feedback amp with open-loop values is shown in Fig. 23.29. Ideally, the values of $R_{\beta i}$ and $R_{\beta o}$ are zero. The feedback resistor R_F is in series with the input and the output of the amplifier. Therefore, the input of the amplifier is a voltage, and the output is a current. The units of A_{OL} will be I/V (a transconductance), and the units of β will be V/I (ohms). Transconductance amplifiers also have high-input and high-output impedance.

The closed-loop gain of the transconductance amplifier is

$$A_{CL} = \frac{i_o}{v_s} = \frac{A_{OL}}{1 + A_{OL}\beta} \text{ A/V} \tag{23.75}$$

The input impedance is again given by applying a test voltage to the input of the feedback amp and calculating the current that flows into the input of the amplifier. Also, the assumption that the feedback network does not load the amplifier will be used. Setting $v_{test} = v_s$, $i_{test} = i_s$ and writing a loop at the input of the amplifier give

$$v_{test} = i_{test} \cdot R_i + v_f = i_{test} \cdot R_i + \beta \cdot A_{OL} \cdot i_{test} \cdot R_i \qquad (23.76)$$

and so the input resistance with feedback is now

$$R_{inf} = \frac{v_{test}}{i_{test}} = R_i \cdot (1 + A_{OL}\beta) \text{ for } R_o \to \infty \qquad (23.77)$$

keeping in mind that $i_{test} \cdot R_i = v_i$ and $A_{OL} \cdot v_i = i_o$. The output resistance is determined by applying a test current at the output with the input source shorted, and is given by

$$v_{test} = (i_{test} - A_{OL}v_i) \cdot R_o = [i_{test} - A_{OL} \cdot (-\beta i_{test})] \cdot R_o \qquad (23.78)$$

since $i_{test} = i_{out}$, $v_i = -v_f$, and $R_{\beta o} = 0$. The output resistance is given by

$$R_{of} = \frac{v_{test}}{i_{test}} = R_o \cdot (1 + A_{OL}\beta) \qquad (23.79)$$

The ideal transconductance amplifier has infinite output and input resistance. Again, feedback helps to make the amplifier appear closer to the ideal.

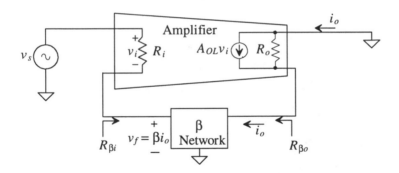

Figure 23.29   An ideal transconductance amplifier.

Now examine the transistor level circuit in Fig. 23.30. Notice the similarity to Fig. 23.13, the circuit used in the series-shunt example; the only difference is the location of the output connection. The feedback loops in both circuits are identical, so the feedback in Fig. 23.30 is known to be negative.

Since the output and the feedback are connected to two separate terminals of the output device, the output variable is a current, sampling i_o. The small-signal model for this circuit is shown in Fig. 23.31 with the open-loop, small-signal model shown in Fig. 23.32. Since the output sampling is a current, loading of the β network will be slightly different from that of the series-shunt example. The input utilizes series mixing; therefore the loading of the β network on the output will be identical to the series-shunt example discussed previously ($R_{\beta o} = R_1 + R_2$). However, since the output sampling is

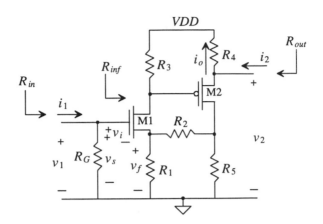

Figure 23.30 Transistor-level series-series feedback amplifier.

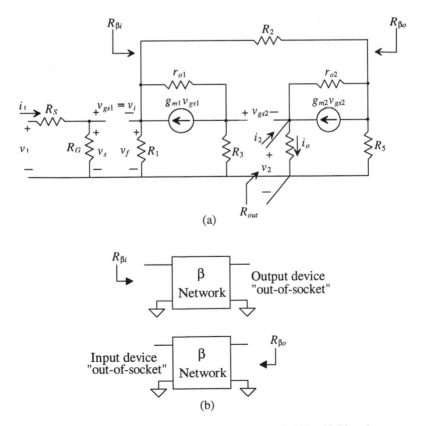

Figure 23.31 (a) Closed-loop small-signal model of Fig. 23.30 and
(b) method for determining feedback loading.

series, the equivalent resistance, $R_{\beta i}$, will be the resistance seen looking into the β network from the input, with the output device taken "out-of-socket" and $R_{\beta i} = R_2 + R_5$.

Once the open-loop model has been constructed, A_{OL} can be calculated as

$$A_{OL} = \frac{i_o^*}{v_s^*} = \frac{i_o^*}{v_{g2}^*}\frac{v_{g2}^*}{v_s^*} \tag{23.80}$$

where the term, $\dfrac{i_o^*}{v_{g2}^*}$, can be determined by using straightforward circuit analysis to solve $\dfrac{v_2^*}{v_{g2}^*}$ and then dividing the result by R_4,

$$\frac{i_o^*}{v_{g2}^*} = \frac{g_{m2}}{1 + g_{m2}R_4 + \dfrac{R_4 + R_5 \| (R_2 + R_1)}{r_{o2}}} \tag{23.81}$$

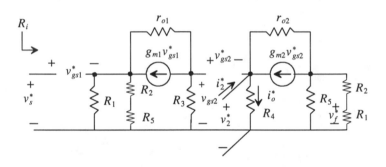

Figure 23.32 Open-loop small-signal model of Fig. 23.30.

The term, $\dfrac{v_{g2}^*}{v_s^*}$, is found by using the G_M method presented in the previous section on shunt-shunt feedback and is

$$\frac{v_{g2}}{v_s^*} = \frac{-g_{m1}(R_3 \| [(1 + g_{m1}R_A)r_{o1} + R_A)]}{1 + g_{m1}R_A + \dfrac{R_A}{r_{o1}}} \quad \text{mhos} \tag{23.82}$$

where $R_A = R_1 \| (R_2 + R_5)$. The feedback factor, β, is

$$\beta = \frac{v_f^*}{i_o^*} \approx \frac{-R_5 R_1}{R_5 + R_1 + R_2} \quad \Omega \tag{23.83}$$

And the closed-loop gain is simply

$$A_{CL} = \frac{i_o}{v_s} = \frac{A_{OL}}{1 + A_{OL}\beta} \quad \text{mhos} \tag{23.84}$$

The value of R_i is obviously infinite, resulting in an identical value of R_{inf}. Therefore, $R_{in} = R_{inf} || R_G = R_G$.

Calculating R_o for a series output requires some explanation. Examine Fig. 23.33. The value of R_o is the value seen looking in series with the load resistor. In this case, the value of R_o becomes

$$R_o = R_4 + \frac{\frac{R_B}{r_{o2}} + 1}{\frac{1}{r_{o2}} + g_{m2}} \approx R_4 + \frac{1}{g_{m2}} \qquad (23.85)$$

where $R_B = R_5 || (R_1 + R_2)$ and the closed-loop value becomes

$$R_{of} = R_o(1 + A_{OL}\beta) \qquad (23.86)$$

Notice, however, that R_{of} is not the same as R_{out}, in this case. Typically, R_{out} is designated as the resistance in parallel with the load. Taking the resistance in series with the load is not a practical specification. Therefore, the resistance R_{out} can be described as seen in Fig. 23.34. In part (a), it can be seen that $R_{of} = R_o(1 + A_{OL}\beta)$ and that $R'_{of} = R_{of} - R_4$. If we want to find a value for R_{out}, using Fig. 23.34b, R_{out} is simply

$$R_{out} = R_4 || R'_{of} = R_4 || (R_{of} - R_4) \qquad (23.87)$$

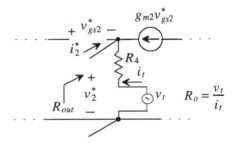

Figure 23.33 Calculation of the output impedance for the circuit in Fig. 23.30.

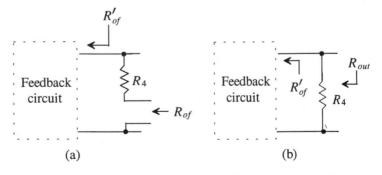

(a)   (b)

Figure 23.34 Determining the output resistance of a series sampling circuit.

23.7 The Current Amplifier (Shunt-Series Feedback)

The last feedback topology to be discussed is the shunt-series feedback amplifier, also known as a current amplifier. As can be expected, both A_{OL} and β have units of I/I, and we can expect the input impedance to be very low and the output impedance very high. Figure 23.35 illustrates the ideal shunt-series amplifier with open-loop values included. Based on past derivations, we can expect that

$$R_{inf} = \frac{R_i}{(1 + A_{OL}\beta)} \tag{23.88}$$

and R_{of} to be

$$R_{of} = R_o(1 + A_{OL}\beta) \tag{23.89}$$

The derivations of this topology will be left to the reader in the Problems section.

The transistor-level circuit shown in Fig. 23.36 is similar to the shunt-shunt topology, except for the placement of the output signal. The reader will also be asked to analyze the shunt-series circuit in the Problems section.

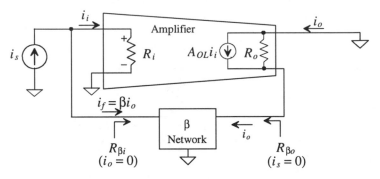

Figure 23.35 An ideal current feedback amplifier.

Example 23.3
Examine the current amplifier seen in Fig. 23.37a. Using feedback analysis, draw open-loop small-signal models and derive values for A_{OL}, β, R_i, R_o, R_{out}, and the overall voltage gain, $\frac{v_2}{v_1}$.

Notice that although this circuit is similiar to the cascode current sink, we can also use it, though unconventionally, as a voltage op-amp. The analysis begins by identifying the feedback circuit. The current summation that occurs at the gate of M4 indicates that shunt mixing is utilized. Since the output and the feedback are taken off separate terminals of the output device, the output sampling is series. The open-loop model is seen in Fig. 23.37b. The only loading on the input due to the feedback network is r_{o2}, since M4 can be taken

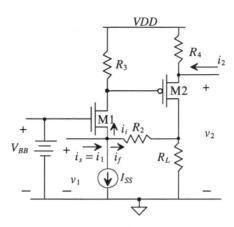

Figure 23.36 A shunt-series feedback amplifier.

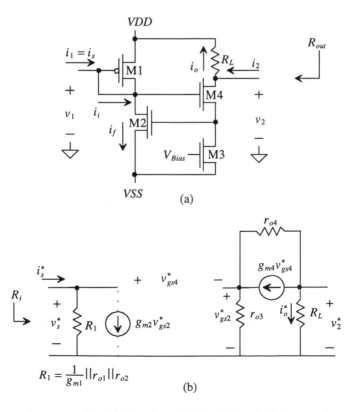

Figure 23.37 (a) Circuit used in Ex. 23.3 and (b) the open-loop model.

"out-of-socket" ($v_{gs2} = 0$). The feedback does not load the output at all. However, we must remember to include the controlled source for determining i_f. The dependent source, $g_{m2}v_{gs}$, will not be included in calculating A_{OL} but will be needed for calculating β. Solving this open-loop circuit for A_{OL} yields (see Problem 23.28):

$$A_{OL} = \frac{i_o^*}{i_s^*} = \frac{v_2^*}{v_s^*} \cdot \frac{R_1}{R_L} = \frac{-g_{m4}(R_L\|((1+g_{m4}r_{o3})r_{o4}+r_{o3}))}{1+g_{m4}r_{o3}+\frac{r_{o3}}{r_{o4}}} \cdot \frac{R_1}{R_L} \approx \frac{-g_{m4}R_1}{1+g_{m4}r_{o3}+\frac{r_{o3}}{r_{o4}}} \text{ A/A}$$

The output, R_o, is found by the same manner presented in the discussion of series-series amplifiers and is

$$R_o = R_L + (1+g_{m4}r_{o3})r_{o4} + r_{o3}$$

Similiarly, the value of R_{out} becomes

$$R_{out} = (R_o(1+A_{OL}\beta) - R_L)\|R_L$$

and R_i is simply R_1, with

$$R_{if} = \frac{R_1}{1+A_{OL}\beta}$$

The feedback variable is $i_f^* \approx g_{m2}v_{gs2}^* = -g_{m2}i_o^*r_{o3}$, and the value of β can be calculated as

$$\beta = \frac{i_f^*}{i_o^*} = -g_{m2}r_{o3} \text{ A/A}$$

The overall voltage gain is

$$\frac{v_2}{v_1} = \frac{i_o \cdot R_L}{i_s \cdot R_{if}} = \frac{A_{OL}}{1+A_{OL}\beta} \cdot \frac{R_L}{R_{if}} \text{ V/V} \quad \blacksquare$$

23.8 Stability

The previous sections illustrated the benefits of feedback and the corresponding tradeoff in gain. However, a critical concern must be examined when applying negative feedback. Some circuits will cause a phase shift in the input signal large enough that the feedback becomes positive (the output adds to the original input), resulting in an unstable system. The occurrence of instability can be minimized with some careful analysis of both the open-loop amplifier, A_{OL}, and the feedback network, β.

The loop gain is defined as

$$T = A_{OL}\beta \tag{23.90}$$

Remember that the product of $A_{OL}\beta$ must itself always be positive. By examining the frequency response of the loop gain, T, the overall stability of the system can be determined. The stability of the system can be summarized with the following rules and is illustrated in Fig. 23.38.

Case 1: If the change in the phase of $A_{OL}\beta$ is equal to 180° and the magnitude is below 0 dB, the system will be stable.

Case 2: If the change in the phase of $A_{OL}\beta$ is equal to 180° and the magnitude equals 0 dB, the system may or may not be stable.

Case 3: If the change in the phase of $A_{OL}\beta$ is equal to 180° and the magnitude is above 0 dB, the system will be unstable.

One might wonder why the 180° phase shift and the magnitude of 0 dB (gain of 1) are such critical factors. The answer may be better understood with a qualitative rather than a quantitative explanation. In order for positive feedback to occur, the output must be added back to its original input signal. The PA system example illustrates this concept. If the loudspeaker, which represents the output of the system, is added back to the input (the microphone), the system becomes unstable because the amplifier is attempting to amplify its own output. The result is a loud, high-pitched, ringing sound, which most people have unfortunately experienced.

In negative feedback applications, the addition of the feedback signal to the original input occurs because of the additional phase shift introduced by the frequency dependent components within A_{OL} and β. Because the feedback signal is already

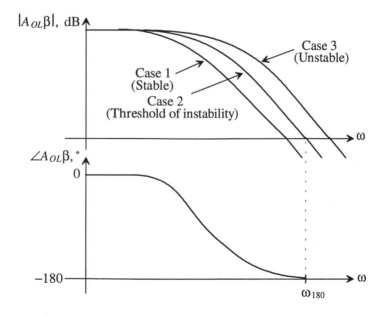

Figure 23.38 Stability analysis using the frequency response of the loop gain.

inverted with respect to the input, an additional 180° of phase shift will cause the feedback signal to be positive with respect to the input. It is this second 180° degrees of phase shift that becomes the important concern. Thus, the frequency response of the loop gain must be examined. The magnitude of 0 dB is also important. If the gain around the loop is less than 1, the output settles to a stable value. However, if the gain around the loop is greater than 1, the amplifier output grows and becomes unstable quickly.

Assume that A_{OL} can be described with the following frequency response:

$$A_{OL}(s) = -\frac{10}{\left(1 + \frac{s}{10}\right)^2} \qquad (23.91)$$

Two poles exist at $\omega = 10$ rad/s with a gain at DC equal to -10 V/V. Also assume that β is frequency dependent and has a single pole at $\omega = 10$ rad/s:

$$\beta = \frac{-1}{\left(\frac{s}{10} + 1\right)} \qquad (23.92)$$

The loop gain then, is

$$A_{OL}\beta = \frac{10}{\left(1 + \frac{s}{10}\right)^2} \cdot \frac{1}{\left(\frac{s}{10} + 1\right)} \qquad (23.93)$$

The Bode plot of the loop gain can be seen in Fig. 23.39. It can be seen that since the phase plot crosses 180° slightly before the magnitude plot crosses 0 dB, the system is unstable.

A more precise method of performing the same analysis will now be described. We can use the exact formulas to solve for the exact frequency which the phase plot crosses 180°. The phase of the loop gain, by definition, is

$$Arg[A_{OL}(j\omega)\beta(j\omega)] = \tan^{-1}\left(\frac{\omega}{z_1}\right) + \tan^{-1}\left(\frac{\omega}{z_2}\right) + ... + \tan^{-1}\left(\frac{\omega}{z_n}\right)$$

$$- \tan^{-1}\left(\frac{\omega}{p_1}\right) - \tan^{-1}\left(\frac{\omega}{p_2}\right) - .... - \tan^{-1}\left(\frac{\omega}{p_n}\right) \qquad (23.94)$$

where $z_1, z_2 \ldots z_n$ are the zeros in the system and $p_1, p_2 \ldots p_n$ are the poles. Since our example has three poles at the same frequency and no zeros, the phase response of the loop gain can be expressed as

$$Arg[A_{OL}(j\omega)\beta(j\omega)] = -3\tan^{-1}\left(\frac{\omega_{180}}{10}\right) = -180 \qquad (23.95)$$

where ω_{180} is the frequency at which the phase response is equal to $-180°$. Solving for ω_{180} yields

$$\omega_{180} = 17.32 \text{ rad/s} \qquad (23.96)$$

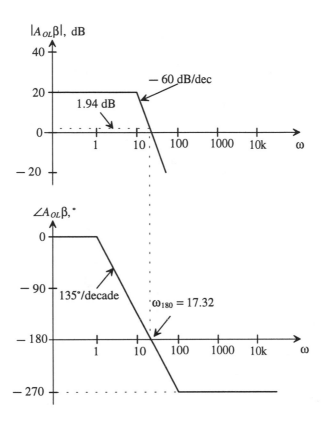

Figure 23.39 Stability analysis using the frequency response of the loop gain.

The magnitude of the loop gain is defined as,

$$20\text{Log}|A_{OL}(j\omega)\beta(j\omega)| = 20\text{Log}(A_o) + 20\text{Log}\sqrt{\left(\tfrac{\omega}{z_1}\right)^2 + 1} + 20\text{Log}\sqrt{\left(\tfrac{\omega}{z_2}\right)^2 + 1} + ..$$

$$+ 20\text{Log}\sqrt{\left(\tfrac{\omega}{z_n}\right)^2 + 1} - 20\text{Log}\sqrt{\left(\tfrac{\omega}{p_1}\right)^2 + 1} - .. - 20\text{Log}\sqrt{\left(\tfrac{\omega}{p_n}\right)^2 + 1} \qquad (23.97)$$

where, z_1, $z_2 \ldots z_n$ are the zeros of the system, p_1, p_2, $\ldots p_n$ are the poles, and A_o is the midband gain. Plugging in the value for ω_{180} into Eq. (23.97), we can solve for the magnitude of the loop gain at the point at which its phase is $-180°$:

$$20\text{Log}|A_{OL}(j\omega)\beta(j\omega)| = 20\text{Log}(10) - 3\left(20\text{Log}\sqrt{\left(\tfrac{17.32}{10}\right)^2 + 1}\right) \qquad (23.98)$$

$$|A_{OL}(j\omega_{180})\beta(j\omega_{180})| = 1.94 \text{ dB} \qquad (23.99)$$

which, using rule 3, verifies that the system is unstable.

Stability is typically measured using two specifications: gain margin and phase margin. Gain margin is defined as the difference between the magnitude of $A_{OL}\beta$ at ω_{180} and unity, whereas phase margin is defined as the difference between the value of the phase at the frequency at which the magnitude of $A_{OL}\beta$ is equal to unity and ω_{180}. Figure 23.40 illustrates both definitions. It should be noted that phase margin is the typical specification used for stability. Amplifiers should be designed to have a phase margin of at least 45°, though 60° phase margin is more acceptable.

In our previous discussion, only the frequency domain analysis was considered. The effect of phase margin in the time domain is related to settling time. As the phase margin increases, less time is required for the signal to settle. As the phase margin approaches 0°, the signal will oscillate indefinitely.

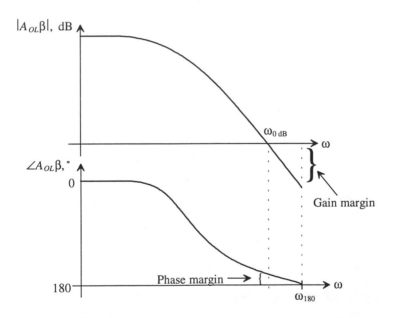

Figure 23.40 Gain margin and phase margin definitions.

23.8.1 The Return Ratio

In some cases, it may be more practical to determine the loop gain from a system point of view. One method used to determine a good approximation of the loop-gain frequency response is to break the loop, input a test signal, and determine the returned value back to the point that the loop was broken. This method is called the return ratio

(RR) method. Consider the block diagram of the single-loop structure in Fig. 23.41 with the loop "open." The gain around the loop is,

$$RR = \frac{x_f}{x_i} = -T = -A_{OL}\beta \tag{23.100}$$

This gain represents the path from the input back around through the feedback network. It should be noted that the RR method may vary quite extensively from the two-port method in finding the value of $A_{OL}\beta$, but for purposes of plotting the loop gain frequency response, the RR should be sufficient [3].

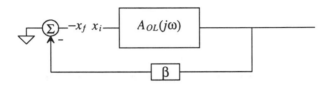

Figure 23.41 Determining the loop gain by opening the feedback loop.

The RR is found by first replacing all independent sources with their ideal impedances. Next, a dependent source is chosen, and the feedback loop is broken between the chosen source and the rest of the circuit. An independent test source is inserted at the node where the dependent source resided and a test signal is injected into the loop. The RR is then the ratio of the returned signal (which now appears across the dependent source) and the test signal, such that

$$RR = -\frac{v_r}{v_t} \tag{23.101}$$

An example will illustrate this method further.

Example 23.4
Determine the RR for the series-shunt op-amp circuit shown in Fig. 23.42a and compare that value to the value $A_{OL}\beta$ using the two-port method. Assume that the op-amp can be modeled as shown in Fig. 23.42b.

Since there is only one dependent source in the circuit, deciding where to break the loop is a simple endeavor. Next, the dependent source is separated from the rest of the circuit, and an independent source, v_t, is put in its original position as seen in Fig. 23.43a. The returned signal is now across the dependent source. By inspection, the value of the RR can be found to be

$$RR = -\frac{v_r}{v_t} = A \cdot \left(\frac{R_I\|R_1}{R_I\|R_1+R_2}\right)\cdot\left(\frac{R_2+R_I\|R_1}{R_2+R_I\|R_1+R_o}\right) = 49,726$$

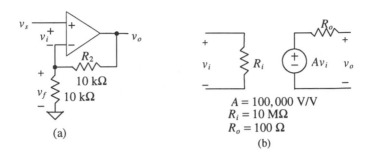

Figure 23.42 (a) A series-shunt amplifier used in Ex. 23.4 and (b) the model
used for the op-amp.

Next, the open-loop model for the original circuit is found using the two-port
analysis presented in Sec. 23.2. The circuit, taking the β network loading effects
into account, can be seen in Fig. 23.43b. The value of A_{OL} is easily calculated as

$$A_{OL} = \frac{v_o^*}{v_s^*} = A \cdot \left(\frac{R_1 + R_2}{R_o + R_1 + R_2}\right) \cdot \left(\frac{R_I}{R_1 \| R_2 + R_I}\right) = 99,497$$

The value of β by inspection is

$$\beta = \frac{v_f^*}{v_o^*} = \frac{R_1}{R_1 + R_2} = 0.5$$

and the value of $A_{OL}\beta$ is

$$A_{OL}\beta = (99,497)(0.5) = 49,748$$

The two methods yielded very similiar answers. ∎

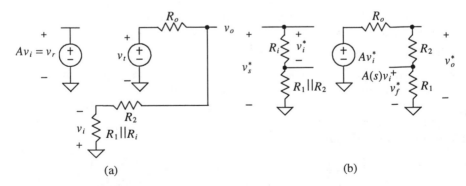

Figure 23.43 (a) The model used to calculate return ratio and (b) the model
used for the two-port analysis of loop gain for Fig. 23.42.

If $A_{OL}\beta$ is frequency dependent, then the phase and gain margin of the circuit can be plotted so as to determine the phase and gain margin. Now suppose that the circuit used in Ex. 23.4 was frequency dependent as seen in Fig. 23.44. Using the same strategy presented in Ex. 23.4 and using the RR method, the value of the loop gain becomes

$$RR = -\frac{v_r}{v_t} = \frac{49,726}{(s/200+1)^3} \tag{23.102}$$

The gain and phase margin can then be analyzed to determine if the system is stable (see Problem 23.34).

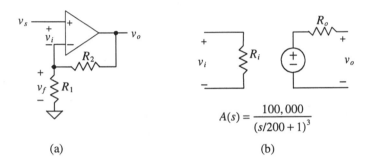

(a)   (b)

Figure 23.44 (a) A series-shunt amplifier used in Ex. 23.4 and (b) the model used for the op-amp.

REFERENCES

[1] A. S. Sedra and K. C. Smith, *Microelectronic Circuits*, Saunders College Publishing, 1991. ISBN 0-03-051648-X.

[2] P. E. Allen and D. R. Holberg, *CMOS Analog Circuit Design*, Holt, Rinehart and Winston, 1987. ISBN 0-03-006587-9.

[3] P. J. Hurst, "Exact Simulation of Feedback Circuit Parameters," *IEEE Transactions on Circuits and Systems*, Vol. 38, No. 11, pp. 1382-1389, November 1991.

PROBLEMS

23.1 An op-amp is designed so that the open-loop gain is guaranteed to be 150,000 ± 10 percent V/V. If the amplifier is to be used in a closed-loop configuration with $\beta = 0.1$ V/V, determine the tolerance of the closed-loop gain.

23.2 What is the maximum possible value of β using resistors in the feedback loop of a noninverting op-amp circuit? Sketch this op-amp circuit when $\beta = 1/2$.

23.3   Examine the feedback loop in Fig. P23.3. A noise source, v_n, is injected in the
system between two amplifier stages. (a) Determine an expression for v_o which
includes both the noise and the input signal, v_s. (b) Repeat (a) for the case where
there is no feedback ($\beta = 0$). (c) If $A_1 = A_2 = 200$, and feedback is again applied
around the circuit, what value of β will be required to reduce the noise by
one-half as compared to the case stated in (b)?

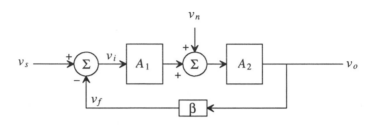

Figure P23.3

23.4   An amplifier can be characterized as follows:

$$A(s) = 10,000 \cdot \frac{100}{s + 100} \text{ V/V}$$

A series of these amplifiers are connected in cascade, and feedback is used
around each amplifier. Determine the number of stages needed to produce an
overall gain of 1,000 with a high-frequency rolloff (at -20 dB/decade) occurring
at 100,000 rad/sec. Assume that the first stage produces the desired
high-frequency pole and that the remaining stages are designed so that their
high-frequency poles are at least a factor of four greater.

23.5   An amplifier can be characterized as follows:

$$A(s) = 1,000 \cdot \frac{s}{s + 100} \text{ V/V}$$

and is connected in a feedback loop with a variable β. Determine the value of β
for which the low-frequency rolloff is 50 rad/sec. What is the value of the
closed-loop gain at that point?

23.6   Make a table summarizing the four feedback topologies according to the
following categories: input variable, output variable, units of A_{OL}, units of β,
method to calculate $R_{\beta i}$ and $R_{\beta o}$, and expressions for A_{CL}, R_{if}, and R_{of}.

23.7   Using the two n-channel common source amplifiers shown in Fig. P23.7 and the
addition of a single resistor, draw (a) a series-shunt feedback amplifier, (b) a
series-series feedback amplifier, (c) a shunt-shunt feedback amplifier, and (d) a
shunt-series amplifier. For each case, identify the forward and feedback paths,
ensure that the feedback is negative by counting the inversions around the loop,

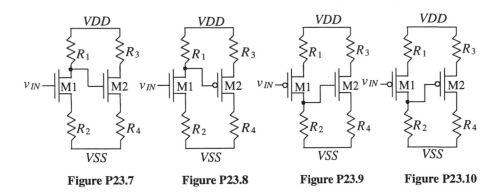

| Figure P23.7 | Figure P23.8 | Figure P23.9 | Figure P23.10 |

and label the input variable, the feedback variable, and the output variable. Assume that the input voltage, v_{IN}, has a DC component that biases M1.

23.8 Repeat Problem 23.7 using the two-transistor circuit shown in Fig. P23.8.

23.9 Repeat Problem 23.7 using Fig. P23.9.

23.10 Repeat Problem 23.7 using Fig. P23.10.

For each of the following feedback analysis problems, assume that the circuit has been properly DC biased and that MOSFETs have been characterized. The n-channel devices have $g_m = 0.06$ A/V and $r_o = 70$ kΩ. The p-channel devices have $g_m = 0.04$ A/V and $r_o = 50$ kΩ.

23.11 Using the series-shunt amplifier shown in Fig. P23.11, (a) identify the feedback topology by labeling the mixing variables and output variable, (b) verify that negative feedback is employed, (c) draw the closed-loop small-signal model, and (d) find the expression for the resistors $R_{\beta i}$ and $R_{\beta o}$.

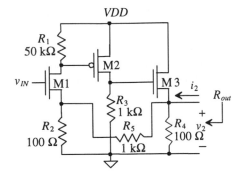

Figure P23.11

23.12 Using Fig. P23.11 and the results from Problem 23.11, (a) draw the small-signal open-loop model for the circuit and (b) find the expressions for the open-loop parameters, A_{OL}, β, R_i, and R_o and (c) the closed-loop parameters, A_{CL} and R_{out}. Note that finding R_{in} is a trivial matter since the signal is input into the gate of M1.

23.13 Using the series-shunt amplifier shown in Fig. P23.13, (a) verify the feedback topology by labeling the mixing variables and the output variable closed-loop small-signal model and (b) find the values of $R_{\beta i}$ and $R_{\beta o}$.

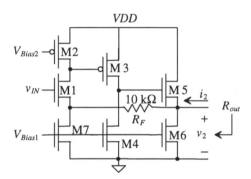

Figure P23.13

23.14 Using the series-shunt amplifier shown in Fig. P23.13 and the results from Problem 23.13, (a) draw the small-signal open-loop model for the circuit and (b) calculate the open-loop parameters, A_{OL}, β, R_i, and R_o and (c) the closed-loop parameters, A_{CL}, and R_{out}. Note that Fig. P23.13 is identical to Fig. P23.11 except the resistors have been replaced with active loads.

23.15 Using the principles of feedback analysis, find the value of the voltage gain, $\frac{v_2}{v_{in}}$ and $\frac{v_2}{i_2}$ for the series-shunt circuit shown in Fig. P23.15.

23.16 A shunt-shunt feedback amplifier is shown in Fig. P23.16. (a) Identify the feedback topology by labeling the input mixing variables and the output variables, (b) verify that negative feedback is employed, (c) draw the closed-loop small-signal model, and (d) find the values of $R_{\beta i}$ and $R_{\beta o}$.

23.17 Using the shunt-shunt amplifier shown in Fig. P23.16 and the results from Problem 23.16, (a) draw the small-signal open-loop model for the circuit and (b) calculate expressions for the open-loop parameters, A_{OL}, β, R_i, and R_o and (c) the closed-loop parameters, A_{CL}, R_{in}, and R_{out}.

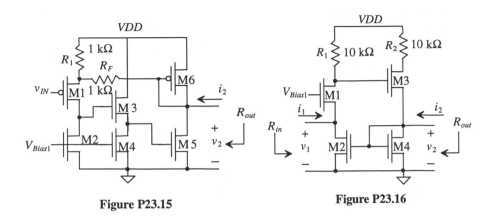

Figure P23.15   **Figure P23.16**

23.18 Using the principles of feedback analysis, find the value of the voltage gain, $\frac{v_2}{v_1}$, $\frac{v_1}{i_1}$, and $\frac{v_2}{i_2}$ for the shunt-shunt circuit shown in Fig. P23.18.

23.19 Using the series-series feedback amplifier shown in Fig. P23.19, (a) identify the feedback topology, (b) verify that negative feedback is employed, (c) draw the closed-loop small-signal model, and (d) find the values of $R_{\beta i}$ and $R_{\beta o}$.

23.20 Using the series-series amplifier shown in Fig. P23.19 and the results from

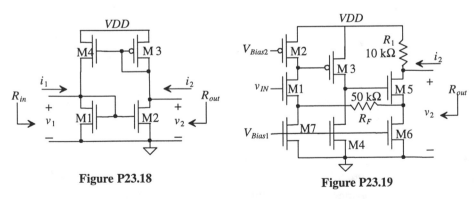

Figure P23.18   **Figure P23.19**

problem 23.19, (a) draw the small-signal open-loop model for the circuit and (b) calculate the open-loop parameters, A_{OL}, β, R_i, and R_o and (c) the closed-loop parameters, A_{CL}, R_{in}, and R_{out}.

23.21 Using the shunt-series amplifier in Fig. 23.35, derive the expressions for A_{OL}, R_{if}, and R_{of}.

23.22 Convert the shunt-shunt amplifier shown in Fig. P23.16 into a shunt-series feedback amplifier without adding any components. (a) Identify the feedback topology, (b) verify that negative feedback is employed, (c) draw the closed-loop small-signal model, and (d) find the values of $R_{\beta i}$ and $R_{\beta o}$.

23.23 Using the shunt-series amplifier from Problem 23.22, (a) draw the small-signal open-loop model for the circuit and (b) calculate the open-loop parameters, A_{OL}, β, R_i, and R_o and (c) the closed-loop parameters, A_{CL}, R_{in}, and R_{out}.

23.24 A feedback amplifier is shown in Fig. P23.24. Identify the feedback topology and determine the value of the voltage gain, $\frac{v_2}{v_1}$, R_{in}, and R_{out}.

23.25 Notice that the amplifier shown in Fig. P23.25 is a simple common source amplifier with source resistance. Explain how this is actually a very simple feedback amplifier and determine the type of feedback used. Determine A_{OL} and β.

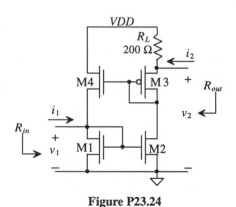

Figure P23.24

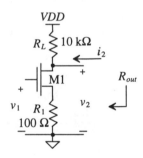

Figure P23.25

23.26 A feedback amplifier is shown in Fig. P23.26. Identify the feedback topology and determine the value of the voltage gain, $\frac{v_2}{v_1}$, R_{in}, and R_{out}.

23.27 A feedback amplifier is shown in Fig. P23.27. Identify the feedback topology and determine the value of the voltage gain, $\frac{v_2}{v_1}$ and R_{out}.

23.28 Prove that the expression for the open-loop gain of Ex. 23.3 is correct.

23.29 Determine if the amplifier seen in Fig. 23.44 is stable.

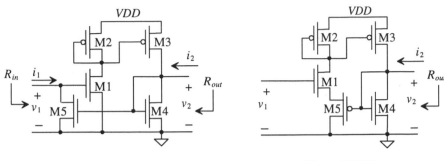

Figure P23.26   **Figure P23.27**

23.30 The op-amp shown in Fig. P23.30a can be modeled with the circuit of Fig. P23.30b. With a feedback factor, $\beta = 1$, determine if the op-amp is stable (and the corresponding phase and gain margins) for the following transfer function and $\omega_2 = 10^5$, 10^6, 10^7, and 5×10^6 rad/sec.

$$A_{OL}(j\omega) = \frac{10,000}{\left(1 + j\frac{\omega}{100}\right)\left(1 + j\frac{\omega}{\omega_2}\right)}$$

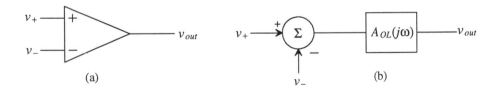

(a)   (b)

Figure P23.30

23.31 The phase plot of an amplifier is shown in Fig. P23.31. The amplifier has a midband gain of $-1,000$ and 3 zeros at $\omega = \infty$ and three other unspecified poles. If the amplifier is configured in a feedback configuration and β is frequency independent, what is the exact value of β that would be necessary to cause the amplifier to oscillate?

23.32 You have just measured the gain of the op-amp circuit shown in Fig. P23.32. You know from basic op-amp theory that the gain of the circuit should be $-R_2/R_1$ V/V. However, your measurements with $R_2 = 10$ kΩ and $R_1 = 1$ kΩ revealed that the gain was only -5 V/V. What is the open-loop gain of the op-amp?

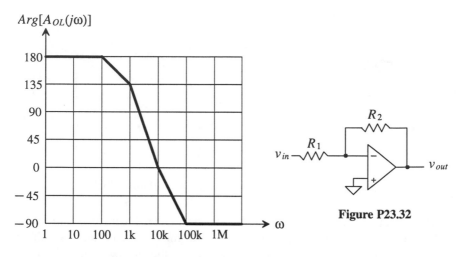

Figure P23.31

Figure P23.32

23.33 Using the circuit shown in Fig. P23.33 and the *RR* method, find a value of R_1 and A_o which will cause the phase margin to equal 45° at $\omega = 8,000$ rad/sec. The amplifier can be modeled as having an infinite input impedance and zero output resistance and has a frequency response of

$$A(s) = \frac{-A_o}{(s/200 + 1)(s/10,000 + 1)}$$

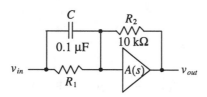

Figure P23.33

23.34 Determine if the system with a return ratio as described by Eq. (23.102) is stable.

Differential Amplifiers

This chapter is concerned with the differential amplifier, an amplifier that amplifies the difference between two signals. The diff-amp is a fundamental building block in CMOS analog integrated circuit design, and an understanding of its operation and design is extremely important. In this chapter, we discuss three basic types of differential amplifiers: the source coupled pair, the source cross-coupled pair, and the current differential amplifier.

24.1 The Source Coupled Pair

The source coupled pair comprised of M1 and M2 is shown in Fig. 24.1 [1]. The MOSFET current mirror made using M5 and M6 is used to provide a source current for the pair of I_{SS}. Throughout this book unless otherwise stated, we will assume that M1 and M2 of the source coupled pair are of equal size, so that $\beta_1 = \beta_2 = \beta$. Summing the AC and DC components of the current at the sources of M1 and M2 yields

$$I_{SS} = i_{D1} + i_{D2} \qquad (24.1)$$

If we label the input voltages at the gates of M1 and M2 as v_{I1} and v_{I2} we can write the difference as

$$v_{DI} = v_{I1} - v_{I2} = v_{GS1} - v_{GS2} \qquad (24.2)$$

or in terms of the AC and DC components of the differential input voltage, v_{DI},

$$v_{DI} = V_{GS1} + v_{gs1} - V_{GS2} - v_{gs2} \qquad (24.3)$$

When the gates of M1 and M2 are grounded

$$I_{D1} = I_{D2} = \frac{I_{SS}}{2} \qquad (24.4)$$

Since we know that a saturated MOSFET follows the relation

$$i_D = \frac{\beta}{2}(v_{GS} - V_{THN})^2 \tag{24.5}$$

the difference in the input voltages may be written as

$$v_{DI} = \sqrt{\frac{2}{\beta}} \left(\sqrt{i_{D1}} - \sqrt{i_{D2}} \right) \tag{24.6}$$

Using Eqs. (24.6) and (24.1), we can write expressions for the drain currents of the MOSFETs in saturation in terms of the difference voltage (assuming $v_{DI} > 0$) as

$$i_{D1} = \frac{I_{SS}}{2} \left| 1 + \sqrt{\left(\frac{\beta v_{DI}^2}{I_{SS}} - \frac{\beta^2 v_{DI}^4}{4I_{SS}^2} \right)} \right| \tag{24.7}$$

and

$$i_{D2} = \frac{I_{SS}}{2} \left| 1 - \sqrt{\left(\frac{\beta v_{DI}^2}{I_{SS}} - \frac{\beta^2 v_{DI}^4}{4I_{SS}^2} \right)} \right| \tag{24.8}$$

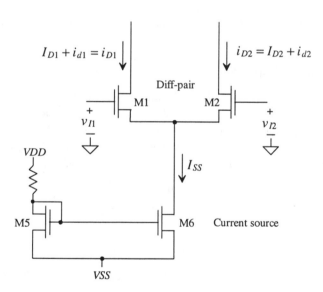

Figure 24.1 Differential amplifier.

In order to understand the operation of this amplifier, let's assume that $v_{I2} = 0$, that is, the gate of M2 is connected to ground, so that $v_{I1} = v_{DI}$. If v_{I1} is held at VDD; then M2 is off and M1 conducts the current I_{SS} (i.e., $i_{D1} = I_{D1} = I_{SS}$). If, for linear operation, M1 and M2 are required to remain in the saturation region, then the point at which the input voltage turns M2 off is the same point at which all of I_{SS} begins to flow in M1. If we label the input voltage at this point v_{DIMAX}, then setting Eq. (24.5) equal to

I_{SS} and realizing at this point that the source potential of the diff-pair is $- V_{THN}$ (due to M2 starting to turn off), we arrive at

$$v_{DIMAX} = \sqrt{\frac{2I_{SS}}{\beta}} \qquad (24.9)$$

This result also holds for the case when v_{I2} is not zero or in general:

$$v_{DIMAX} = v_{I1} - v_{I2} = \sqrt{\frac{2I_{SS}}{\beta}} \qquad (24.10)$$

Example 24.1
Referring to the diff-amp of Fig. 24.1, if $I_{SS} = 10$ µA and $W_2 = W_1 = 15$ µm and $L_2 = L_1 = 5$ µm, determine and compare to simulations v_{DIMAX}.

From Eq. (24.9) we have

$$v_{DIMAX} = \sqrt{\frac{2 \cdot 10 \, \mu}{50\frac{\mu A}{V^2} \cdot \frac{15}{5}}} = 364 \text{ mV}$$

The simulation results are shown in Fig. 24.2. ■

The minimum difference voltage, when all the current flows in M2, can be determined in a similar manner as

$$v_{DIMIN} = -\sqrt{\frac{2I_{SS}}{\beta}} \qquad (24.11)$$

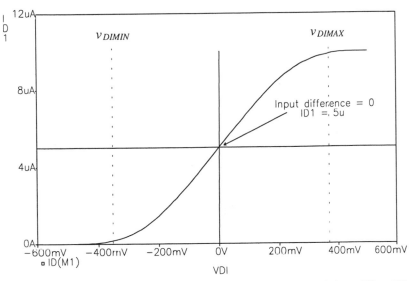

Figure 24.2 Transfer characteristics of the source coupled differential amplifier of Ex. 24.1.

The transconductance gain of the differential amplifier is given by the slope of the transfer characteristics in Fig. 24.2 between v_{DIMIN} and v_{DMAX} (when M1 and M2 are in the saturation region). This large-signal transconductance is given by

$$G_m = \frac{di_{D1}}{dv_{DI}} = \frac{I_{SS}}{2\sqrt{2I_{SS}/\beta}} = \frac{\sqrt{2\beta I_{SS}}}{4} = \frac{g_m}{4} \qquad (24.12)$$

which is of the same form as the small-signal transconductance of a MOSFET. Note that the transconductance can be increased by increasing I_{SS} or making the MOSFETs M1 and M2 larger (wider).

24.1.1 Current Source Load

The CMOS differential amplifier is most often used with a current source load as shown in Fig. 24.3. Consider the case when the gates of M1 and M2 are grounded. Under these circumstances, a current $I_{SS}/2$ flows in M1-M4. The drain voltage of M4 is at the same potential as the gate of M3/M4 (and the drain of M3). Therefore, the potential difference between the drain of M1 and M2 is zero. The fact that the drain of M4 is at the same potential as the gate of M3 is used to bias the next stage at a particular current level. We can select the sizes of the devices similar to the method used in determining the sizes of the current sources and sinks. We begin by selecting a minimum length based on channel length modulation considerations. For the CN20 process, we will use 5 μm for the lengths. The widths are determined by the gate-source voltage of the MOSFETs. In many designs, the W/L ratios of the diff-amp with current source load are made very large. The voltage above V_{THN} may be only 0.1 V in these designs. In general, we will set V_{GS} to 1.2 V in our designs. Note that the body effect when using

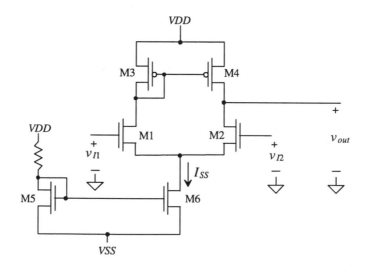

Figure 24.3 Differential amplifier with current source load M3 and M4.

the n-channel MOSFET diff-amp may increase the threshold voltage of M1 and M2 to approximately 1.2 V. This simply means that the circuit we design will have a V_{GS} of 1.5 V or 0.3 volts above V_{THN}. The benefits of using an n-channel diff-pair in an n-well process are better threshold voltage matching because of the body effect and lower capacitance at the sources of the MOSFETs to AC ground. In a practical case one might use a p-channel MOSFET diff-pair (see Fig. 24.4) on the input of the amplifier in order to eliminate the body effect.

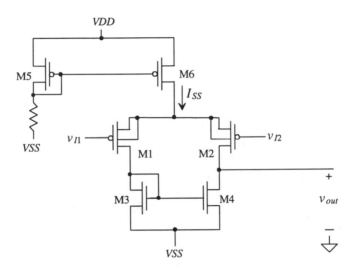

Figure 24.4 P-channel differential amplifier.

Example 24.2
For the differential pair with current source load of Fig. 24.3, plot the transfer characteristics using SPICE when $I_{ss} = 20$ μA and $W_1 = W_2 = 15$ μm, $W_3 = W_4 = 70$ μm, while the length of the devices is 5 μm. Assume $VDD = -VSS = 2.5$ V.

In this simulation, we will ground the gate of M2 and perform a DC sweep of v_{n}. The results are shown in Fig. 24.5. Note that when both inputs are at ground potential the output voltage is 3.8 V. MOSFETs M1 through M4 have 10 μA drain current, and the gate-source voltages are 1.2 V. The slope of this curve around the $v_{n} = 0$ is the voltage gain of the amplifier, which we will determine quantitatively later in the section. ∎

Up to this point, we have only considered the difference between v_{n} and v_{n}, with the common voltage between the two being ground or 0 V. However, another important issue that needs to be discussed is the range of common input voltages the amplifier will continue to operate properly. In other words, if a common signal is input on the gates of M1 and M2, a maximum and minimum voltage exists in which the transistors fail to

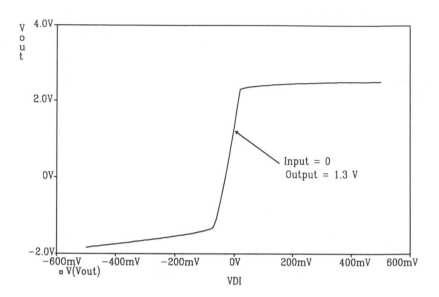

Figure 24.5 DC sweep of the differential amplifier of Ex. 24.2.

stay in the saturation region.   This range of common input signals is also known as the *common-mode range* (CMR).

Consider the schematic shown in Fig. 24.6 of a differential amplifier with the gates of M1 and M2 tied together. If v_I starts at zero and is taken toward *VSS*, we will reach a point where M6 will go into the triode region. The v_I at this point is the minimum input voltage allowed on the input of the diff-amp for linear operation. We will label this input potential as

$$v_{IMIN} = \text{minimum allowable input voltage} \qquad (24.13)$$

This minimum voltage is given by summing the voltages from the input of the diff-amp to *VSS* when the input is at v_{MIN} and M6 is on the verge of going into the triode region or

$$v_{IMIN} = \overbrace{\sqrt{\frac{I_{SS}}{\beta_1}} + V_{THN}}^{V_{GS}\text{ of M1 or M2}} + \overbrace{\sqrt{\frac{2I_{SS}}{\beta_6}}}^{V_{DS6}\,=\,V_{GS6}-V_{THN}} + VSS \qquad (24.14)$$

The maximum allowable input voltage occurs when v_{IMAX} is taken toward *VDD* and M1 and M2 go into the triode region. This occurs when

$$V_{DS1} = V_{GS1} - V_{THN} \rightarrow V_{D1} = V_{G1} - V_{THN} \qquad (24.15)$$

Since $V_{G1} = v_{IMAX}$, we get

$$v_{IMAX} = VDD - \overbrace{\left[\sqrt{\frac{I_{SS}}{\beta_3}} + V_{THP}\right]}^{V_{SG3}} + V_{THN} \tag{24.16}$$

Since V_{THP} is approximately equal to V_{THN}, this reduces to

$$v_{IMAX} \approx VDD - \sqrt{\frac{I_{SS}}{\beta_3}} \tag{24.17}$$

The input common-mode range over which all the MOSFETs in the differential amplifier behave linearly is then,

$$\text{Positive CMR (or limit)} = v_{IMAX} \tag{24.18}$$

and

$$\text{Negative CMR (or limit)} = v_{IMIN} \tag{24.19}$$

We have not included the body effect on the threshold voltages in this derivation. This omission can be justified by considering how we defined a MOSFET in saturation, that is, $V_{DS} \geq V_{GS} - V_{THN}$, back in Ch. 5. We found that the MOSFET entered the saturation region earlier than predicted by this equation because of the uneven charge distribution in the channel. Including the body effect in this derivation would not help us predict the common-mode range with any additional accuracy. Consider the following example.

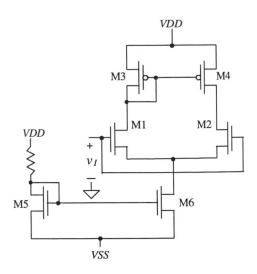

Figure 24.6 Differential amplifier configuration used to determine input common-mode range.

Example 24.3

Estimate and compare with simulations both the positive and negative CMR for the differential amplifier of Ex. 24.2. Assume the MOSFET M6 used in the current mirror has $W_6 = 30$ μm and $L_6 = 5$ μm for an I_{SS} of 20 μA.

We begin by calculating the positive CMR

$$V_{IMAX} = 2.5 - \sqrt{\frac{20 \text{ μA}}{17\frac{\text{μA}}{V^2} \cdot \frac{70}{5}}} = 2.1 \text{ V}$$

and the negative CMR is

$$V_{IMIN} = \sqrt{\frac{20 \text{ μA}}{50\frac{\text{μA}}{V^2} \cdot \frac{15}{5}}} + 0.83 + \sqrt{\frac{2 \cdot 20\text{μA}}{50\frac{\text{μA}}{V^2} \cdot \frac{30}{5}}} - 2.5 = -0.94 \text{ V}$$

Figure 24.7 shows the current through M1 ($I_{SS}/2 = 10$ μA) and the drain-source voltage of M1 plotted against a voltage applied to the gates of M1 (or M2 since they are tied together). The negative CMR is shown in the figure as the point where the drain current of M1 starts to go to zero. The positive CMR is set by the point where M1 goes into the triode region ($V_{DS1} < V_{GS1} - V_{THN}$). Note that the sum of the currents, $I_{D1} + I_{D2}$, is not constant because of the finite output resistance of M6, that is, the current I_{SS} increases as V_{DS6} increases. These results will be used extensively in the design of op-amp circuits in the next chapter. ∎

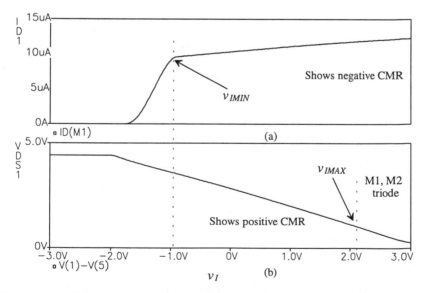

Figure 24.7 (a) Drain current of M1 plotted against input voltage (gates of M1 and M2 tied together) and (b) the drain-source voltage of M1 going into triode.

Now let's determine the small-signal gain of the differential amplifier of Fig. 24.3. Let's begin by assuming that the gate of M2 is at AC ground. The input voltage is given by

$$v_{i1} = v_{gs1} - v_{gs2} = i_{d1} \cdot \frac{1}{g_{m1}} - i_{d2} \cdot \frac{1}{g_{m2}} \tag{24.20}$$

Ideally, zero AC current flows into the drain of M6 so that

$$i_{d1} = -i_{d2} = i_d \text{ and } g_{m1} = g_{m2} = g_m \tag{24.21}$$

and

$$v_{i1} = i_d\left(\frac{2}{g_m}\right) \tag{24.22}$$

The idea that the drain current of M2 flows out of M2's drain rather than into the drain is simply understood by thinking about how the overall AC + DC current changes. Increasing v_{i1} causes i_{D1} to increase and i_{D2} to decrease. A decrease in the overall drain current of M2 is taken to mean that the AC current is flowing out of the drain. This is explained better with the use of Fig. 24.8. The current entering the output node is composed of a current from the current mirror and a current from M2. The MOSFET's M3 drain current is mirrored by M4, and the increase in drain current through M1 is equal to the decrease in drain current through M2. The total current, $2i_d$ is driven into the resistance at the output node. The resistance looking into the drain of M4 is given by

$$r_{o4} = \frac{1}{\lambda I_D} \tag{24.23}$$

while the resistance looking into the drain of M2 is

$$R_{intoD2} = r_{o2}\left(1 + g_{m2} \cdot \frac{1}{g_{m1}}\right) \approx r_{o2} \tag{24.24}$$

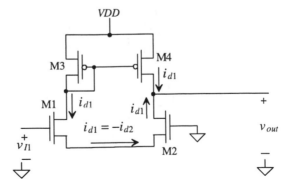

Figure 24.8 Differential amplifier with AC currents shown.

The voltage gain of the differential amplifier is given by

$$A_v = \frac{v_{out}}{v_{i1}} = \frac{v_{out}}{v_{i1} - v_{i2}} = \frac{2i_d(r_{o2}\|r_{o4})}{i_d \cdot \frac{2}{g_m}} = g_m(r_{o2}\|r_{o4}) \qquad (24.25)$$

This gain can also be written

$$A_v = \frac{2\sqrt{\beta}}{(\lambda_2 + \lambda_4)\sqrt{I_{SS}}} \qquad (24.26)$$

In other words, lowering the diff-pair bias current increases the gain at the price of bandwidth and slew rate.

Consider the differential amplifier driving a load capacitance C_L shown in Fig. 24.9. If an input step is applied to the diff-amp that goes from 0 to VSS, then M1, M3, and M4 shut off. MOSFET M2 goes into the triode region, and the current I_{SS} is applied directly to the load capacitance. Under these conditions, the capacitor is discharged at its maximum rate, called the slew rate. This rate of discharge is given, assuming that the load capacitance is large compared to the output capacitance of the differential amplifier, by

$$\frac{dV}{dt} = \frac{I_{SS}}{C_L} \text{ (V/µs)} \qquad (24.27)$$

It can be seen that if we applied a positive step pulse M2 would turn off and the current through M1, M3, and M4 would be I_{SS}. Therefore, the maximum rate at which the load capacitance can be charged is given by this equation as well. When the capacitor is fully charged, M4 cuts off and the current in M1 and M3 remains I_{SS}. In the derivation of the slew-rate, we have neglected the MOSFET capacitances present in the output

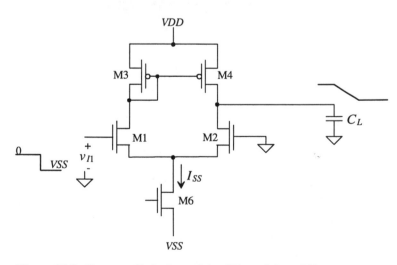

Figure 24.9 Slew-rate limitations of the differential amplifier.

node. In the derivation of the amplifier's small-signal frequency response, we will not neglect these capacitances.

Consider the schematic of the output node shown in Fig. 24.10. We know from the derivation of the small-signal gain that the effective resistance to ground at this node is approximately $r_{o2}||r_{o4}$. The total capacitance at the output node to ground is given by

$$C_{tot} = C_L + C_{db4} + C_{gd4} + C_{db2} + C_{gd2} \quad (24.28)$$

The time constant at the output node is given by

$$\tau_{out} = (r_{o2}||r_{04}) \cdot C_{tot} \quad (24.29)$$

The upper 3 dB frequency is given by

$$f_{out} = \frac{1}{2\pi\tau_{out}} \quad (24.30)$$

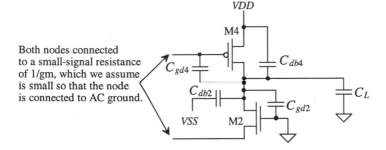

Figure 24.10 Capacitances at the output node used to calculate the frequency response of the differential amplifier.

Example 24.4
For the differential amplifier shown in Fig. 24.11, calculate the slew rate and the small-signal upper 3 dB frequency. Compare your hand calculations with SPICE. Assume that the length of the drain implant regions is 6 μm and that the width of the implant is equal to the width of the MOSFET.

We begin this problem by calculating the MOSFET capacitances. We will approximate the drain depletion capacitances using the bottom zero bias capacitance, or

$$C_{db2} = cj \cdot 6\ \mu m \cdot W_2 = 1.04 \times 10^{-4} \cdot 6\mu \cdot 15\mu = 9.4\ fF$$

and

$$C_{db4} = cj \cdot 6\mu \cdot 70\mu = 137\ fF$$

The drain-gate capacitances of the MOSFETs are given by

$$C_{gd2} = CGDO \cdot W_2 = 3.8 \times 10^{-10} \cdot 15 \ \mu m = 5.7 \ fF$$

and

$$C_{gd4} = CGDO \cdot W_4 = 5 \times 10^{-10} \cdot 70 \ \mu m = 35 \ fF$$

The total capacitance is given by

$$C_{tot} = 2 \ pF + 137 \ fF + 35 \ fF + 9.4 \ fF + 5.7 \ fF = 2.19 \ pF$$

The slew rate is given by

$$SR = \frac{dV}{dt} = \frac{20 \ \mu A}{2.19 \ pF} = 9.1 \frac{V}{\mu s}$$

The output time constant is given by

$$\tau = (r_{02} \| r_{o4}) \cdot C_{tot} = \frac{1}{2\lambda \frac{I_{SS}}{2}} \cdot C_{tot} = \frac{1}{0.06 \cdot 20 \ \mu A} \cdot 2.19 \ pF = 1.8 \ \mu s$$

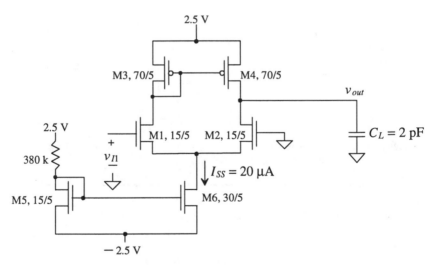

Figure 24.11 Differential amplifier used in Ex. 24.4.

The upper 3 dB frequency is, from Eq. (24.30), 87 kHz. SPICE simulation results are shown in Fig. 24.12. The SR is determined by pulsing the gate of M1 from 0 to − 2.5 V, while the gate of M2 is held at ground potential. The hand calculations were very close to the simulation results. ■

This example presents an important, but not obvious, problem encountered in SPICE simulation. Consider what would happen if, during the SPICE simulation of the slew rate, we pulsed the gate of M1 negative at the beginning or close to the beginning

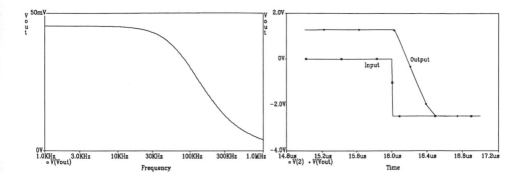

Figure 24.12 SPICE simulation results for Ex. 24.4.

of the simulation. Steady-state conditions would not have been reached in the circuit, and the simulation would likely not converge. However, if we apply the input signal after the circuit has reached steady-state conditions, say approximately 15 μs for the present simulation, the simulation converges (and the frustration level is low). This results in a general rule for simulating analog circuits in SPICE, that is:

> *When performing a transient simulation allow a considerable amount of time for the circuit to reach steady-state conditions before applying the inputs.*

24.1.2 Common-Mode Rejection Ratio

An important aspect of the differential amplifier is its ability to reject a common signal applied to both inputs. Often, in analog systems, signals are transmitted differentially, and the ability of an amplifier to reject coupled noise into each line is very desirable. Consider the amplifier shown in Fig. 24.13. Since the input is no longer a differential signal, the common source node can no longer be considered an AC ground. The current source at the source of M1 and M2 has been replaced with its small-signal output resistance. If we apply an AC signal to the gates of M1/M2, equivalent to saying that a common signal is applied to the input of the differential amplifier, we can calculate the common-mode gain. We begin by writing the AC small-signal common-mode input voltage, v_c, as

$$v_c = v_{gs1,2} + 2i_d r_{o6} \qquad (24.31)$$

The MOSFETs M1 and M2 each source a current, i_d, through the output resistance of the current source, r_{o6}. This equation may be rewritten as

$$v_c = i_d \left(\frac{1}{g_m} + 2r_{o6} \right) \approx i_d \cdot 2r_{o6} \qquad (24.32)$$

The output voltage, because of the symmetry of the circuit, is given by

$$v_{out} = -i_d \cdot \frac{1}{g_{m3}} = -i_d \cdot \frac{1}{g_{m4}} \qquad (24.33)$$

The common-mode gain is

$$A_c = \frac{v_{out}}{v_c} = -\frac{1/g_{m4}}{2r_{o6}} = -\frac{1}{2g_{m4}r_{o6}} \tag{24.34}$$

assuming $g_{m3} = g_{m4}$. The common-mode gain can be decreased (ideally it is zero) by increasing the output resistance of the current source connected to the source coupled pair. The difference-mode gain was given by Eq. (24.25) as $|A_v| = g_{m1}(r_{o2}||r_{o4})$. The common-mode rejection ratio (CMRR), in dB, of the differential amplifier with current source load is given by

$$CMRR = 20\log\left|\frac{A_v}{A_c}\right| = 20\log\left|g_{m1}(r_{o2}||r_{o4}) \cdot 2g_{m4}r_{o6}\right| \tag{24.35}$$

or in generic terms where the output resistance of the current source is R_{out},

$$CMRR = 20\log\left|2g_{m1}g_{m4}(r_{o2}||r_{o4})R_{out}\right| \tag{24.36}$$

Using a cascode current source can greatly increase the CMRR at the price of increased negative common-mode range (CMR). Also, in this analysis we have only discussed low-frequency CMRR. At high frequencies the capacitance at the sources of M1 and M2 to ground (this capacitance is in parallel with the r_{o6} in Fig. 24.13) dominates the output impedance of the current source. The effect causes the CMRR to decrease with increasing frequency.

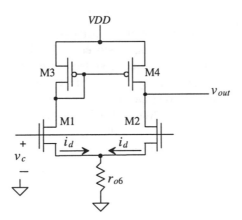

Figure 24.13 Differential amplifier configuration used to determine input CMRR.

24.1.3 Noise

Consider the diff-amp shown in Fig. 24.14, with the noise sources of each MOSFET drawn in the schematic. From Ch. 9 we can write the mean squared noise current generated between each MOSFET's drain and source as

$$\overline{i_n^2} = \overline{i_{therm}^2} + \overline{i_{1/f}^2} = 4kT \cdot \frac{2}{3}g_m + \frac{KF \cdot I_D^{AF}}{f \cdot (C_{ox}' \cdot L)^2}, \quad \left(\frac{A^2}{Hz}\right) \tag{24.37}$$

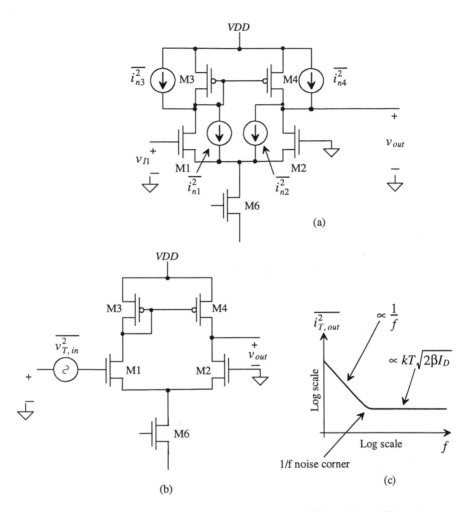

Figure 24.14 (a) Noise sources added into the basic differential amplifier and (b) modeling the amplifier noise on the input and (c) total output noise over a 1-Hz bandwidth.

The noise current generated in M1 (or M3) is mirrored in M4, so that the total mean squared noise current at the output node is

$$\overline{i_{T,\,out}^2} = \overline{i_{n1}^2} + \overline{i_{n2}^2} + \overline{i_{n3}^2} + \overline{i_{n4}^2} \tag{24.38}$$

This is assuming that M6 doesn't generate any noise current and we are interested in a frequency range where the capacitances in the circuit do not come into play. The total mean squared output noise is

$$\overline{v^2_{T, out}} = \left(\overline{i^2_{n1}} + \overline{i^2_{n2}} + \overline{i^2_{n3}} + \overline{i^2_{n4}} \right) (r_{o2} || r_{o4})^2, \quad \left(\frac{V^2}{Hz} \right) \qquad (24.39)$$

while the total mean squared input-referred noise is

$$\overline{v^2_{T, in}} = \frac{\overline{v^2_{T, out}}}{A^2_v} = \frac{1}{g^2_{m1}} \left(\overline{i^2_{n1}} + \overline{i^2_{n2}} + \overline{i^2_{n3}} + \overline{i^2_{n4}} \right) \qquad (24.40)$$

Both the thermal and 1/f input noise contributions can be reduced by reducing the source coupled pair bias current, I_{SS}. Increasing the length of the load MOSFETs M3 and M4, relative to the length of M1 and M2, has the effect of making the input or output 1/f noise contributions determined mainly by M1 and M2.

24.1.4 Matching Considerations

One important aspect of differential amplifiers is the dependence of the performance on the matching of devices. Layout techniques can be used to minimize the first-order effects of mismatch due to oxide gradients and other process variations. Interdigitation was introduced in Ch. 20 when discussing current mirrors. However, a more effective layout technique should be used when matching is critical, as is the case for the

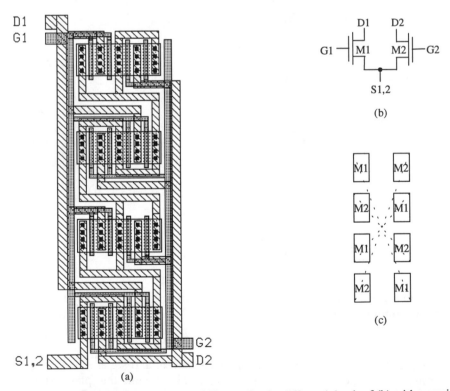

Figure 24.15 (a) Common-centroid layout for the differential pair of (b) with parasitic matching (c) layout of MOSFETs and common center.

differential amplifier. Common-centroid layout, as the name implies, constructs two devices symmetrically about a common center in the layout. This allows the two devices to cancel process gradients in both the x and y directions and exposes both devices to heat sources in an identical manner.

As seen in Fig. 24.15, the source coupled pair is interdigitated in four separate rows and alternately connected in the vertical direction. The common source node (S1,2) intertwines the entire structure. Note how the layout is completely symmetrical about the centroid (center). Although this layout technique is labor intensive, the benefits would warrant the extra time required to minimize the nonideal effects.

The input offset voltage of a diff-pair, resulting from mismatches in threshold voltage, geometrys and load resistances, can be determined with the help of Fig. 24.16 [2]. The value of the offset-voltage is given by

$$V_{OS} = V_{GS1} - V_{GS2} = V_{THN1} + \sqrt{\frac{2I_{D1}}{\beta_1}} - V_{THN2} - \sqrt{\frac{2I_{D2}}{\beta_2}} \quad (24.41)$$

We can define the differences and averages of the threshold voltage, load resistance, and geometry as ΔV_{THN} ($V_{THN1} - V_{THN2}$), V_{THN} (average of V_{THN1} and V_{THN2}), ΔR_L ($R_{L1} - R_{L2}$), R_L (average of R_{L1} and R_{L2}), $\Delta(W/L)$ [$(W_1/L_1) - (W_2/L_2)$] and (W/L) [average of (W_1/L_1) and (W_2/L_2)]. Thus, we can make the appropriate substitutions into Eq. (24.41), requiring $I_{D1}R_{L1} = I_{D2}R_{L2}$, with the result:

$$V_{OS} = \Delta V_{THN} + \frac{V_{GS} - V_{THN}}{2} \cdot \left[\frac{-\Delta R_L}{R_L} - \frac{\Delta(W/L)}{(W/L)} \right] \quad (24.42)$$

The threshold voltage mismatch must be reduced using layout techniques. Mismatches resulting from unequal geometry and differences in the load resistance can be reduced by designing with a small V_{GS} (V_{GS} close to V_{THN}). This should be compared with Eq. (20.38), which showed that using small V_{GS} in a current mirror resulted in a large mirrored current difference.

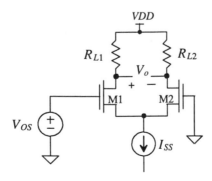

Figure 24.16 Determination of diff-amp input offset voltage.

24.2 The Source Cross-Coupled Pair

The source cross-coupled pair is shown in Fig. 24.17 [3, 4]. This configuration is very useful in practical circuit design. Assuming that both inputs are connected to ground and that all n-channels and all p-channels are the same size, a current I_{ss} flows in all the MOSFETs in the circuit. Normally, we could design the circuit so that $I_{ss} = 10$ μA and the gate-source voltage is 1.2 V. This results in our already familiar sizes of 15/5 for the n-channels and 70/5 for the p-channels.

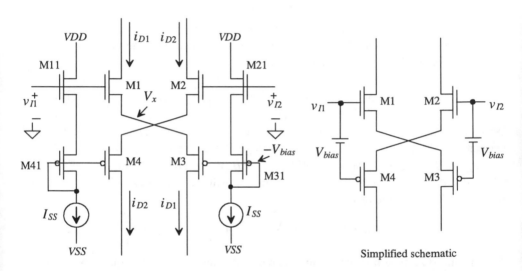

Figure 24.17 Source cross-coupled differential amplifier, n-channel input devices.

With the gate of M2 grounded, the gate of M3 is at a constant potential that we will label $- V_{bias}$ (or approximately –2.4 V with the above sizes). We will label the potential at the sources of M1 and M3 as V_x. The input voltage can be written, assuming that M1 and M4 remain in the saturation region, as

$$v_{I1} = \sqrt{\frac{2i_{D1}}{\beta_1}} + V_x + V_{THN} \tag{24.43}$$

and for M3

$$i_{D1} = \frac{\beta_3}{2}(V_x + V_{bias} - V_{THP})^2 \tag{24.44}$$

Solving Eq. (24.44) for V_x yields

$$V_x = \sqrt{\frac{2i_{D1}}{\beta_3}} - V_{bias} + V_{THP} \tag{24.45}$$

Substituting this equation into Eq. (24.43) results in

$$v_{I1} = \sqrt{2i_{D1}}\left[\frac{1}{\sqrt{\beta_1}} + \frac{1}{\sqrt{\beta_3}}\right] - V_{bias} + V_{THP} + V_{THN} \tag{24.46}$$

When $v_{I1} = 0$ then $i_{D1} = I_{SS}$. This equation can be rewritten as

$$i_{D1} = \frac{1}{2} \cdot (v_{I1} + V_{bias} - V_{THP} - V_{THN})^2 \frac{\beta_1\beta_3}{\left(\sqrt{\beta_1} + \sqrt{\beta_3}\right)^2} \tag{24.47}$$

while a similar analysis for i_{D2}, with v_{I1} grounded, yields

$$i_{D2} = \frac{1}{2} \cdot (v_{I2} + V_{bias} - V_{THP} - V_{THN})^2 \frac{\beta_2\beta_4}{\left(\sqrt{\beta_2} + \sqrt{\beta_4}\right)^2} \tag{24.48}$$

Normally, $\beta_1 = \beta_2$ and $\beta_3 = \beta_4$. Note that the diff-amp output currents, i_{D1} and i_{D2}, are available at both the top and bottom of the amplifier.

Earlier in the chapter we defined the difference in the input voltages as

$$v_{DI} = v_{I1} - v_{I2} \tag{24.49}$$

Equations (24.47) and (24.48) can be written in the more general case by substituting v_{DI} for v_{I1} in Eq. (24.47) and $-v_{DI}$ for v_{I2} in Eq. (24.48).

An important characteristic of the source cross-coupled differential amplifier is that as v_{DI} is increased, i_{D1} continues to increase and i_{D2} shuts off while the opposite is true when v_{DI} decreases. In other words, the amplifier is operating as class AB where neither of the output currents is zero as long as their magnitudes remain less than I_{SS}. The maximum difference in potential between the two inputs so that neither output current is zero is given by

$$|v_{DMIN}| = -V_{bias} + V_{THP} + V_{THN} \tag{24.50}$$

Example 24.5
Referring to the diff-amp of Fig. 24.17, if $I_{SS} = 10$ μA and all NMOSs have a 15/5 size while all PMOSs have a 70/5 size, estimate v_{DMIN} and compare to simulations.

We know that with these sizes and biasing current, the gate-source potential of all MOSFETs is approximately 1.2 V. Knowing this, we get

$$|v_{DMIN}| = -2.4 + 0.91 + 0.83 = -660 \text{ mV}$$

The simulation results are shown in Fig. 24.18 where $v_{I2} = 0$. This figure should be compared with Fig. 24.2 where the drain current flattens out at I_{SS}. It should be obvious from Fig. 24.18 that a source cross-coupled diff-amp does not exhibit slew-rate limitations. The lack of slew-rate limitations results from the absence of a current source in series with M1 or M2. ∎

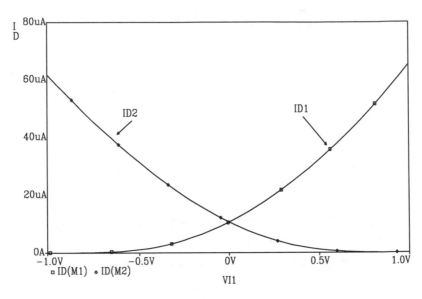

Figure 24.18 Transfer characteristics of the diff-amp of Ex. 24.5.

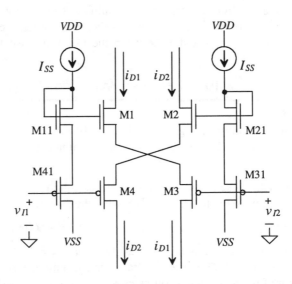

Figure 24.19 Source cross-coupled differential amplifier, p-channel input devices.

The p-channel version of the source cross-coupled differential amplifier is shown in Fig. 24.19. Normally, the version, p-channel or n-channel, of this differential amplifier used is dependent on the common-mode range (CMR). The n-channel version has the best positive CMR, while the p-channel version has the best negative CMR. Calculation of the CMR is presented in the next section.

24.2.1 Current Source Load

The source cross-coupled differential amplifier with active loads is shown in Fig. 24.20. Note that we could have easily added a pair of active loads in series with the drains of M1 and M4 to obtain a differential output amplifier. The small-signal gain of this amplifier can be analyzed intuitively. Superposition will be used to find the gain from the output back to each input. First, determine the small-signal gain, $\frac{v_o}{v_{f1}}$, with v_{f2} grounded. Neglecting the bulk effect, we can see that the voltage gain from the gate of M1 to the source of M1 is simply a common drain amplifier of the form

$$\frac{v_{s1}}{v_{f1}} = \frac{g_{m1} R_{Leq}}{1 + g_{m1} R_{Leq}} \text{ V/V} \tag{24.51}$$

where R_{Leq} is simply the load seen by the source of M1. This can be found by using a test source (as seen in Fig. 24.21) to determine the effective impedance seen looking into the source of M3. This results in

$$R_{Leq} = \frac{v_t}{i_t} = \frac{1 + \frac{1}{g_{m5} r_{o3}}}{g_{m3} + \frac{1}{r_{o3}}} \approx \frac{1}{g_{m3}} \tag{24.52}$$

Therefore, the voltage gain from v_{f1} to the source of M1 becomes

$$\frac{v_{s1}}{v_{f1}} = \frac{g_{m1} \frac{1}{g_{m3}}}{1 + g_{m1} \frac{1}{g_{m3}}} \tag{24.53}$$

The voltage gain from v_{s1} to v_{g9} is the same form as a common gate amplifier. This gain is

$$\frac{v_{g9}}{v_{s1}} = g_{m3} \frac{1}{g_{m5}} \tag{24.54}$$

And finally, the gain from the gate of M9 to the output is simply $- g_{m9}(r_{o9} \| r_{o10})$, and

$$\frac{v_o}{v_{f1}} = -\frac{g_{m1} \frac{1}{g_{m5}}}{1 + g_{m1} \frac{1}{g_{m3}}} g_{m9}(r_{o9} \| r_{o10}) \text{ for } v_{f2} = 0 \tag{24.55}$$

Using the same type of analysis, we find that the expression from v_{f2} to the output is

$$\frac{v_o}{v_{f2}} = \frac{g_{m2} \frac{1}{g_{m6}}}{1 + g_{m2} \frac{1}{g_{m4}}} g_{m10}(r_{o9} \| r_{o10}) \text{ for } v_{f1} = 0$$

$$\tag{24.56}$$

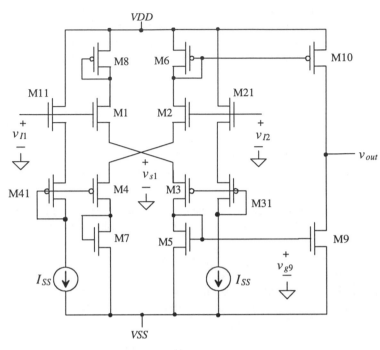

Figure 24.20 Source cross-coupled differential amplifier, with current source loads.

Superposition can now be used to determine the total effect of each input on the output. Therefore, if M1 and M2 are matched, as are M3 and M4, and if M5 and M6 are designed so that their g_m's are equal as are M7 and M8 (respectively), then the differential gain can be expressed as

$$v_o = (v_{I2} - v_{I1}) \frac{2 \cdot g_{m1,2} \frac{1}{g_{m5,6}}}{1 + g_{m1,2} \frac{1}{gm3,4}} g_{m9,10}(r_{o9} \| r_{o10}) \Rightarrow \frac{v_o}{v_d} = \frac{2 \cdot g_{m1,2} \frac{1}{g_{m5,6}}}{1 + g_{m1,2} \frac{1}{gm3,4}} g_{m9,10}(r_{o9} \| r_{o10})$$

$$(24.57)$$

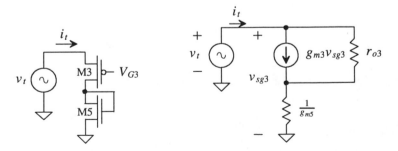

Figure 24.21 (a) Simplified schematic and (b) small-signal model..

The positive common-mode voltage range is determined in exactly the same manner as with the source coupled diff-amp. The positive CMR is given by

$$v_{IMAX} \approx VDD - \sqrt{\frac{2I_{SS}}{\beta_6}} \qquad (24.58)$$

while the negative CMR is given by

$$v_{IMIN} = \overbrace{\sqrt{\frac{2I_{SS}}{\beta_1}} + V_{THN}}^{V_{GS1}} + \overbrace{\sqrt{\frac{2I_{SS}}{\beta_3}}}^{V_{DS3}} + \overbrace{\sqrt{\frac{2I_{SS}}{\beta_5}} + V_{THN}}^{V_{GS5}} + VSS \qquad (24.59)$$

The positive CMR is similar to the basic source coupled differential amplifier. However, the negative CMR is considerably worse. Figure 24.22 shows a biasing scheme that is useful for shifting the CMR of the diff-amp into the center of the supply range. Capacitors between the gates of M1-M4, Fig. 24.22, can eliminate the slew-rate limitations caused by the biasing current sources charging the input capacitances of M1 - M4 when low power operation, that is, small bias currents, is desirable.

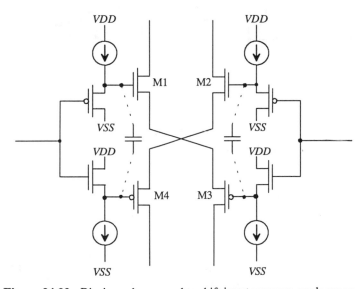

Figure 24.22 Biasing scheme used to shift input common-mode range. Note the body effect will affect the biasing causing the current in M1 - M4 to decrease.

An alternative current source load configuration is shown in Fig 24.23. The gain of this configuration is given by

$$A_v = \frac{v_{out}}{v_{i1} - v_{i2}} = 2 \cdot \frac{r_{o2} \| r_{o6}}{\frac{1}{g_{m1}} + \frac{1}{g_{m3}}} \qquad (24.60)$$

assuming $g_{m1} = g_{m2}$ and $g_{m3} = g_{m4}$.

This differential amplifier is useful as a direct replacement of the source coupled pair, especially when slew-rate limitations are a concern. The bandwidth and CMRR are approximately the same as the source coupled diff-amp, while the circuit complexity and power dissipation are higher.

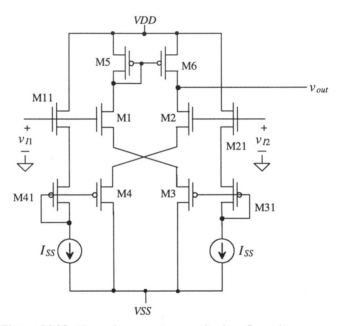

Figure 24.23 Alternative current source load configuration.

24.3 Cascode Loads

The differential amplifier with current source load of Fig. 24.3 gave a voltage gain of $g_m(r_{o2}\|r_{o4})$. For many applications this gain may be too low. Using a cascode current source load in place of M3/M4 results in a gain of approximately $g_m r_{o2}$ (review Sec. 22.2.1). This modest improvement in gain can be increased by cascoding M1 and M2 as well. Figure 24.24 shows the resulting circuit configuration. The MOSFET MC6 is selected so that its V_{GS} keeps M1/M2 and MC1/MC2 in the saturation region. The gain of this configuration is given by

$$A_v = g_{m1}(R_{into\ DC2}\|R_{into\ DC4})$$ (24.61)

The resistance looking into the drain of MC2, assuming M2 and MC2 are the same size, is given by

$$R_{into\ DC2} \approx g_{m2} \cdot r_{o2}^2$$ (24.62)

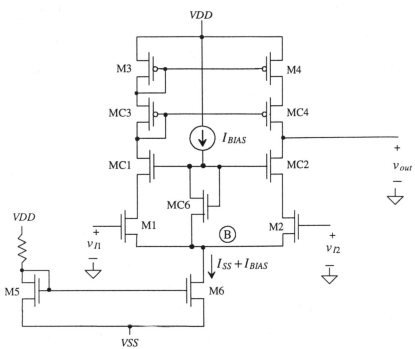

Figure 24.24 Cascode differential amplifier.

and the resistance looking into the drain of MC4, again assuming MC4 and M4 are the same size, is given by

$$R_{into\,DC4} \approx g_{m4} \cdot r_{o4}^2 \qquad (24.63)$$

The gain of the cascode differential amplifier may be written as

$$A_v = g_{m1}\left(g_{m2}r_{o2}^2 \| g_{m4}r_{o4}^2\right) \qquad (24.64)$$

The main drawback of using the cascode differential amplifier is the major reduction in positive common-mode range. In the negative direction, v_{IMIN} is identical to the source coupled diff-amp discussed in Sec. 24.1. The minimum common-mode voltage is limited by M6 becoming nonsaturated. Hence,

$$v_{IMIN} = \sqrt{\frac{2(I_{SS} + I_{BIAS})}{\beta_6}} + \sqrt{\frac{I_{SS}}{\beta_{1,2}}} + V_{THN1,2} + V_{SS} \qquad (24.65)$$

The positive common-mode signal is limited by the additional circuitry required by the cascode loads. Notice that v_{IMAX} in this case will be limited by the amount of voltage necessary to keep M1, MC1, MC3, and M3 in their respective saturation regions. However, an interesting fact concerning this circuit is the DC feedback caused by MC6. Since the current through MC6 is constant, v_{SGC6} will be constant. Therefore, as the common-mode voltage on the gates of M1 and M2 begins to increase, the drains of M1 and M2 will normally decrease. However, since the current through M1 and M2 is also constant, the common source node, marked as node B, also begins to increase. Thus, node B (the gates of MC1 and MC2) also increases. The currents through MC1 and MC2 are also constant; thus, the value of $v_{GSC1,2}$ will attempt to stay constant, pulling the drains of M1 and M2 up. This feedback action attempts to keep M1 and M2 from going into nonsaturation. Therefore, the maximum common-mode voltage, v_{IMAX}, will be limited by MC1 and MC2 going into nonsaturation. This occurs when

$$v_{DSC1} = v_{GSC1} - v_{THC1} \tag{24.66}$$

where the voltage on drain of MC1 (and MC2) is

$$v_{DC1} = v_{DD} - v_{SG3} - v_{SGC3} \tag{24.67}$$

and the voltage on the gate of MC1 (MC2) is

$$v_{GC1} = v_{INMAX} - v_{GS1} + v_{GSC6} \tag{24.68}$$

Plugging into Eq. (24.66) and substituting for the values of V_{SG},

$$v_{INMAX} = V_{DD} - \sqrt{\frac{I_{SS}}{\beta_1}} - \sqrt{\frac{I_{SS}}{\beta_{C3}}} - \sqrt{\frac{I_{SS}}{\beta_3}} + \sqrt{\frac{2I_{BIAS}}{\beta_{C6}}} - 2V_{THP} + V_{THN} \tag{24.69}$$

This assumes that the threshold voltages for all the n- and p-type devices are approximately the same. Example 24.6 will illustrate the degraded CMR for the cascode loaded diff-amp.

Example 24.6
Determine the size of M6, MC6, and MB1-MB3 in Fig. 24.25 so that M1 and M2 remain in saturation. Assume I_{SS} is 20 μA, and estimate the gain of the configuration and the common-mode range.

The gate-source voltage of M1/M2 or MC1/MC2 is determined by solving

$$\frac{I_{SS}}{2} = \frac{\beta_{C1}}{2}(V_{GS} - V_{THN})^2 \rightarrow V_{GS} = 1.2 \text{ V}$$

Therefore, the minimum voltage from the gate of MC1 to the source of M1 (in other words V_{GSC6}) is the sum of $V_{GSC6} + V_{DS,sat}$, or approximately 1.5 V. If we design for $I_{BIAS} = 10$ μA, we can solve for the size of MC6 with

$$10 \text{ μA} = \frac{50\frac{\mu A}{V^2}}{2} \frac{W_{C6}}{5 \text{ μm}}(1.5 - 0.83)^2 \rightarrow W_{C6} \approx 5 \text{ μm}$$

Since a current of 10 μA flows in M5, we set MB3 to the same size, that is, 15/5, so that its drain current is 10 μA as well. MOSFETs MB2 and MB1 form a current mirror. The drain current of MB3 is mirrored in MB1. We can set the sizes of MB2 and MB3 to 70/5. The only remaining MOSFET needing size specifications is M6. The total current that flows in M6 is $I_{SS} + I_{BIAS} = 30$ μA, so

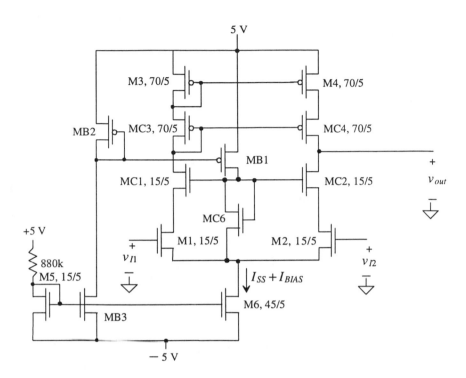

Figure 24.25 Cascode differential amplifier used in Ex. 24.6.

its size is set to 45/5. The gain of the cascode differential amplifier is given, with the help of Eq. (24.64), by

$$A_v = 55\frac{\mu A}{V^2}\left[55\frac{\mu A}{V^2}(833k)^2 \| 70\frac{\mu A}{V^2}(833k)^2\right] = 1,175 \text{ V/V}$$

A noncascode configuration has a gain in the neighborhood of 50 V/V.

The common-mode range will be calculated from Eq. (24.65) and Eq. (24.69). Here the value of v_{IMIN} becomes

$$v_{IMIN} = .365 + .365 + .83 - 5 = -3.44 \text{ V}$$

and the value for v_{IMAX} becomes

$$v_{IMAX} = 5 - .365 - .289 - .289 + .632 - 2(.91) + .83 = 3.77 \text{ V} \quad \blacksquare$$

24.4 Wide-Swing Differential Amplifiers

An increasingly important concern with the advent of low-voltage design is the need for improved common-mode range. A technique used to extend the allowable input swing of a diff-amp is to use two complementary diff-amp stages in parallel [5, 6] as shown in

Fig. 24.26. To understand the operation of this circuit, consider the case when the DC component of the input signal, v_{in}, is such that both diff-amps are on (and the AC component is small compared to the DC). The current through the diff-pairs M1/M2 and M9/M10 is I, while the current through the summing MOSFETs M4 and M12 is $2I$. If M5 is the same size as M4 and if M7 is the same size as M12, then a current $2I$ flows in the output transistors. The small-signal voltage gain, assuming $g_{m1} = g_{m2}$ and $g_{m9} = g_{m10}$, is given by

$$A_v = (g_{m1} + g_{m9})[r_{o7}(2I)\|r_{o5}(2I)] = \frac{(g_{m1} + g_{m9})}{\lambda_7 2I + \lambda_5 2I} = \frac{\sqrt{2\beta_1 I} + \sqrt{2\beta_9 I}}{2I(\lambda_7 + \lambda_5)} \qquad (24.70)$$

If the input is such that the p-channel diff-amp is on and the n-channel diff-amp is off, then a current I flows in the summing MOSFETs M4 and M12 and zero current flows in M1, M2, M3, and M6. The current that flows in M5 and M7 is now I. The small-signal voltage gain is

$$A_v = g_{m9}[r_{o7}(I)\|r_{o5}(I)] = \frac{\sqrt{2\beta_9 I}}{I(\lambda_7 + \lambda_5)} \qquad (24.71)$$

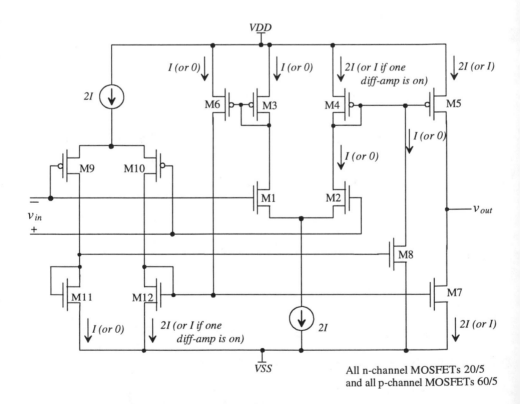

Figure 24.26 Two parallel differential amplifiers used to increase input swing.

The small-signal gain when the p-channel diff-amp is off and the n-channel diff-amp on is given by

$$A_v = g_{m1}[r_{o7}(I)||r_{o5}(I)] = \frac{\sqrt{2\beta_1 I}}{I(\lambda_7 + \lambda_5)} \tag{24.72}$$

It is desirable to have a smooth transition from having a single diff-amp on to having both diff-amps on. If we require

$$\beta_1 = \beta_9 = \beta \Rightarrow G_m = \sqrt{2\beta I} \tag{24.73}$$

then Eqs. (24.70) through (24.72) can be rewritten as

$$A_v = G_m \cdot [r_{o7}(I)||r_{o5}(I)] \tag{24.74}$$

For low distortion a constant gain is important. Also, for a stable amplifier it is important that the amplifier can be compensated (more on this in Ch. 25) to ensure stability. An amplifier with a gain dependent on the input signal amplitude is difficult to compensate.

The amplifier of Fig. 24.26 is called an operational transconductance amplifier (OTA). OTAs are very useful in CMOS analog circuit design. If we set the size of the n-channel MOSFETs in Fig. 24.26 to 20/5 and the p-channel MOSFETs to 60/5 (to satisfy Eq. [24.73] with $KP_n = 50$ μA/V^2 and $KP_p = 17$ μA/V^2) and the current $I = 1$ μA, then the small-signal gain of the circuit, using Eq. (24.72) with $\lambda = 0.06$, is 166 V/V (= A_v). SPICE simulations (using the BSIM model) with $VDD = |VSS| = 2.5$ V resulted in a gain of 240 with common-mode input voltages of 2, 0, and -2 V. Cascoding M5 and M7 can increase the gain of the OTA.

24.4.1 Current Differential Amplifier

Another wide-swing differential amplifier is the current differencing amplifier. The current diff-amp is shown schematically in Fig. 24.27. In the following discussion, let's assume that M1 through M4 are the same size. If both i_1 and i_2 are zero, then a current I_{SS} flows in all MOSFETs in the circuit. Now assume that i_1 is increased above zero, but less than I_{SS}. This causes the current in both M1 and M2 to increase. As a result, the drain current in M3 decreases in order to keep $i_{D2} + i_{D3} = 2I_{SS}$. The decrease in M3 drain current is mirrored in M4, forcing the current i_1 out of the diff-amp. A similar argument can be made for increasing i_2; that is, it causes the output of the differential amplifier to sink a current equal to i_2. Ratioing the sizes of M1-M4 can be used to give the diff-amp a gain or to scale the input currents.

The input impedance of the current differential amplifier is simply the small-signal resistance of a diode connected MOSFET, or

$$R_{in} = \frac{1}{g_m} \tag{24.75}$$

This configuration finds applications in both low-power and high-speed circuit design.

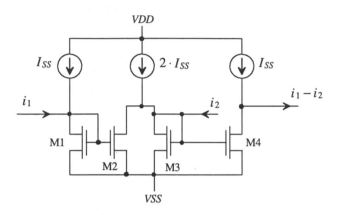

Figure 24.27 Current differential amplifier.

24.4.2 Constant Transconductance Diff-Amp

A rail-to-rail differential amplifier made using n- and p-channel diff-amps is shown in Fig. 24.28 [7,8]. It is desirable, for distortion and compensation reasons, that the overall transconductance of the diff-amp remain constant independent of the region of operation, that is, operation with both n- and p-channel diff-amps on or only a single diff-amp on. A constant g_m is guaranteed over the input range if

$$g_m = g_{mn} + g_{mp} = \sqrt{2\beta_n I_n} + \sqrt{2\beta_p I_p} = \text{constant} \qquad (24.76)$$

where g_{mn} and g_{mp} are the transconductances of the n- and p-channel diff-amps and g_m is the overall transconductance of the input stage. Since β_n and β_p are constant and can be made equal, Eq. (24.76) can be rewritten as (assuming both diff-amps are operating),

$$\sqrt{I_n} + \sqrt{I_p} = \text{constant} \qquad (24.77)$$

This equation always holds if both differential amplifiers are on. The problem with nonconstant transconductance occurs if only one diff-amp is on. If, for example, the common-mode input (common input voltage on the + and − inputs of the differential amplifier) is large enough to shut the p-channel diff-amp off, then $I_p = 0$ and the transconductance of the overall input diff-amp changes.

The solution to this problem begins by making

$$I_n = I_p = I_o \qquad (24.78)$$

When both diff-amps are on (keeping in mind we have already set the requirement that $\beta_n = \beta_p$), Eq. (24.77) reduces to

$$2\sqrt{I_o} = \text{constant} \qquad (24.79)$$

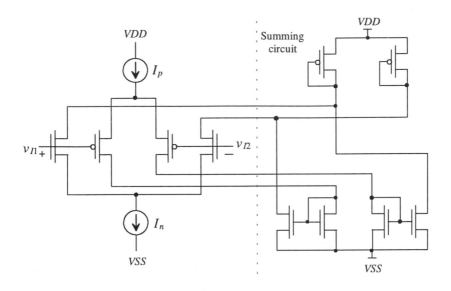

Figure 24.28 Rail-to-rail differential amplifier.

If we add a current of $3I_o$ to I_n (or to I_p) when the p-channel diff-pair (n-channel diff-pair) is off, the transconductance of the pair is constant, or

$$\overbrace{2\sqrt{I_o}}^{\text{both on}} = \overbrace{\sqrt{3I_o+I_n}}^{\text{n diff-amp on}} = \overbrace{\sqrt{3I_o+I_p}}^{\text{p diff-amp on}} \quad \text{if } I_o = I_p = I_n \qquad (24.80)$$

An example of a constant-g_m rail-to-rail input stage is shown in Fig. 24.29. The summing circuit of Fig. 24.28 is not shown in this figure. MOSFETs M1-M4 make up the p- and n-channel diff-pairs, while MP1 and MN1 source the constant current I_o when both diff-pairs are on. (MP1 and MN1 represent the constant current sources shown in Fig. 24.28.) When both diff-pairs are on, the current out of each current diff-amp is approximately 0. (The drain current of M6 and M7 is approximately 0.) If the common-mode input voltage becomes large enough to shut the p-channel diff-pair off, then MOSFETs MS1 and MS2 are off as well. This causes the current in M5 to become I_o. The current in M6 mirrors the current in M5. Since M6 is three times larger than M5, the current in M6 becomes $3I_o$.

Discussion

In general, the constant-g_m diff-amp is used in an op-amp to keep from overcompensating the op-amp and to avoid distortion when the op-amp is used with large signals where the diff-amps, on the input of the op-amp, are turning on and off with variations in the input signal. Several practical problems exist with these uses of the constant-g_m diff-amp. Since the value of g_m may vary with process variations by 20

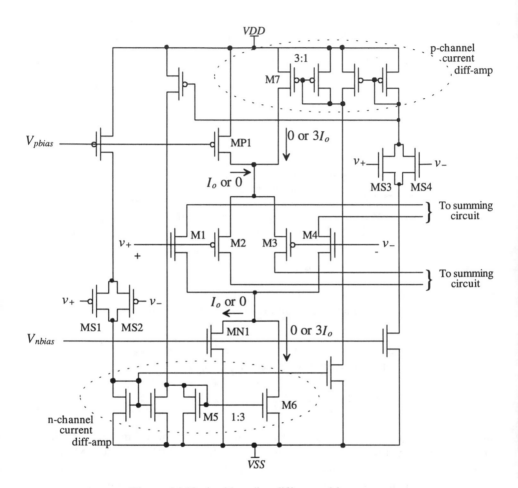

Figure 24.29 A wide-swing diff-amp with constant gm.

percent we may still overcompensate an op-amp design. Using the constant-g_m stage to avoid overcompensation is useful, but since we may over- (or under-) compensate the op-amp with process variations, the added complexity and power draw may not yield improvements justifying the additional circuitry. Using the constant-g_m stage to avoid distortion is based on keeping the unity gain frequency of an op-amp, f_u, a constant independent of the common-mode input voltage. In practice, mismatches (threshold voltage and geometry) in the input diff-pair and changes in DC biasing conditions, resulting from the diff-amps turning off and on, can cause distortion. In some cases, this distortion can be worse than the distortion resulting from the nonconstant f_u (which we get with the nonconstant g_m diff-amp). The distortion resulting from mismatches can be modeled as an offset-voltage that is dependent on the input common-mode voltage. Since the DC currents sourced or sinked from a constant-g_m diff-amp are not constant, the DC operating point of the circuit summing the currents changes with the input signals' amplitude. The result is a change in the low-frequency gain and added distortion.

REFERENCES

[1]   P. E. Allen and D. R. Holberg, *CMOS Analog Circuit Design*, Holt, Rinehart and Winston, 1987. ISBN 0-03-006587-9.

[2]   P. R. Gray and R. G. Meyer, *Analysis and Design of Analog Integrated Circuits*, 3rd ed., John Wiley and Sons, 1993. ISBN 0-471-57495-3.

[3]   R. Castello and P. R. Gray, "A High-Performance Micropower Switched-Capacitor Filter," *IEEE Journal of Solid State Circuits*, Vol. SC-20, No. 6, pp. 1122-1132, December, 1985.

[4]   E. Seevinck and R. F. Wassenaar, "A Versatile CMOS Linear Transconductor/Square-Law Function Circuit," *IEEE Journal of Solid State Circuits*, Vol. SC-22, No. 3, pp. 366-377, June 1987.

[5]   M. Steyaert and W. Sansen, "A High-Dynamic-Range CMOS Op Amp with Low-Distortion Output Structure," *IEEE Journal of Solid State Circuits*, Vol. SC-22, No. 6, pp. 1204-1207, December 1987.

[6]   T. S. Fiez, H. C. Yang, J. J. Yang, C. Yu, and D. J. Allstot, "A Family of High-Swing CMOS Operational Amplifiers," *IEEE Journal of Solid State Circuits*, Vol. 24, No. 6, pp. 1683-1687, December 1989.

[7]   J. H. Botma, R. F. Wassenaar, and R. J. Wiegerink, "A Low-Voltage CMOS Op Amp with a Rail-to-Rail Constant-g_m Input Stage and a Class AB Rail-to-Rail Output Stage," *Proceedings of the 1993 IEEE ISCAS*, p. 1314.

[8]   A. L. Coban, P. E. Allen, and X. Shi, "Low-Voltage Analog IC Design in CMOS Technology," *IEEE Transactions on Circuits and Systems*, Vol. 42, No.11, November 1995.

PROBLEMS

24.1 For the diff-amp shown in Fig. P24.1 determine the drain current of M1 as a function of the input voltage, $v_{I1} - v_{I2}$. Show your work and neglect body effect.

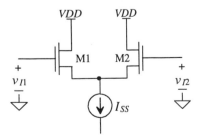

Figure P24.1

24.2   Repeat Ex. 24.1 if the widths of M1 and M2 are increased to 100 μm. Determine the transconductance of the diff-amp. Write i_{d2} as a product of g_m (= $g_{m1} = g_{m2}$) and v_{i1} (with v_{i2} = AC ground), v_{i2} (with v_{i1} = AC ground), and $v_{i1} - v_{i2}$.

24.3   Using the diff-amp topology shown in Fig. P24.3, design a circuit that will change a 1-V square wave into a 0 to 5 V square wave at 1 kHz. Note that many of the MOSFETs in this circuit are operating in the cutoff, saturation, or triode regions.

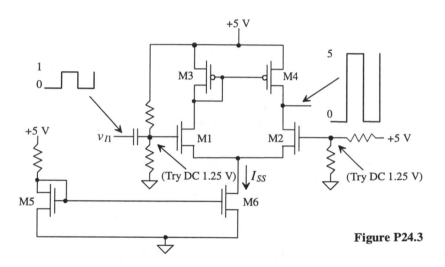

Figure P24.3

24.4   Simulate the operation of the circuit designed in Problem 24.3.

24.5   Determine the small-signal gain and the input common-mode-range (CMR) for the diff-amps shown in Fig. P24.5. All n-channels are 15/5, all p-channels are 70/5, and the resistors are 3.8 MEG.

24.6   Verify the results of Problem 24.5 using SPICE.

24.7   Estimate the slew-rate limitations in charging and discharging a 1 pF capacitor tied to the outputs of the diff-amps shown in Fig. P24.5.

24.8   For the n-channel diff-pair shown in Fig. P24.8, show that if the body effect is included in the analysis of the transconductance, the following relationships are valid.

$$i_{d1} = \frac{g_m}{2}\left[v_{i1}\left(2 - \frac{g_m}{g_m + g_{mb}}\right) - v_{i2} \cdot \frac{g_m}{g_m + g_{mb}}\right] - \frac{g_m \cdot g_{mb}}{g_m + g_{mb}} \cdot \frac{[v_{i1} + v_{i2}]}{2}$$

and

$$i_{d2} = \frac{g_m}{2}\left[v_{i2}\left(2 - \frac{g_m}{g_m + g_{mb}}\right) - v_{i1} \cdot \frac{g_m}{g_m + g_{mb}}\right] - \frac{g_m \cdot g_{mb}}{g_m + g_{mb}} \cdot \frac{[v_{i1} + v_{i2}]}{2}$$

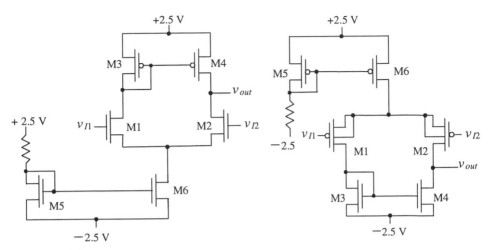

Figure P24.5

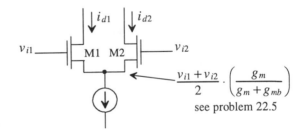

Figure P24.8

24.9   The diff-amp configuration shown in Fig. P24.9 is useful in situations where a truly differential output signal is needed. Determine the following:

(a)   The transconductance of the diff-amp.

(b)   The drain currents of all MOSFETs in terms of the input voltages and g_{mn} (the transconductance of an n-channel MOSFET).

(c)   The small-signal voltage gain, $(v_{o1} - v_{o2})/(v_{i1} - v_{i2})$.

24.10 Determine the CMRR of the diff-amp shown in Fig. 24.13 if a cascode current source is used for the tail current, that is, I_{ss}.

24.11 Develop an expression for the CMRR of the diff-amp of Fig. 24.13 as a function of frequency, assuming the current source output impedance can be modeled by a simple capacitor. Assume that the parasitic capacitance on the drain of M4 is negligible.

24.12 Determine the RMS input noise voltage for the diff-amp of Ex. 24.2 over a bandwidth of 1 to 1 kHz. Assume $KF = 10^{-30}$ and $AF = 1.3$.

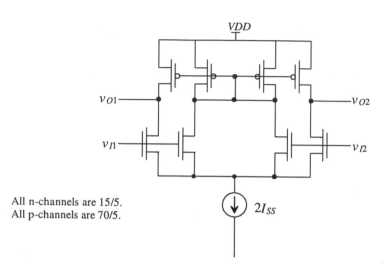

All n-channels are 15/5.
All p-channels are 70/5.

Figure P24.9

24.13 Verify the calculated noise voltage determined in Problem 24.12 with SPICE simulation results.

24.14 Repeat Ex. 24.5 if the MOSFETs (both n- and p-channels) used in the diff-amp are 100/5.

24.15 Determine the small-signal gain of the amplifier shown in Fig. P24.15. Also determine the input CMR of the amplifier if the minimum voltage across a current source/sink is 0.3 V.

24.16 Repeat Problem 24.15 if the active current source loads M6/M8 and M5/M7 are replaced with wide-swing current mirrors, Fig. P24.16. Note that the CMR of the amplifier remains unchanged if V_{bias} is selected correctly; see the discussion in Sec. 20.2.

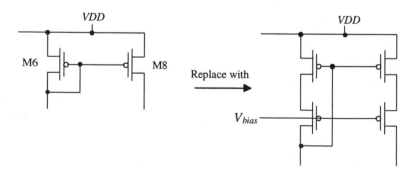

Figure P24.16

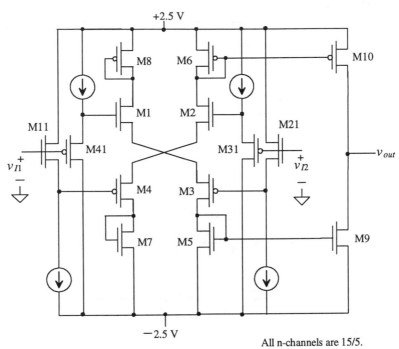

+2.5 V

M8 M6 M10

M1 M2

M11

M41 M31 v_{I2}

v_{I1}

v_{out}

M4 M3

M7 M5 M9

−2.5 V

Figure P24.15

All n-channels are 15/5.
All p-channels are 70/5.
The current sources/sinks are 10 μA.

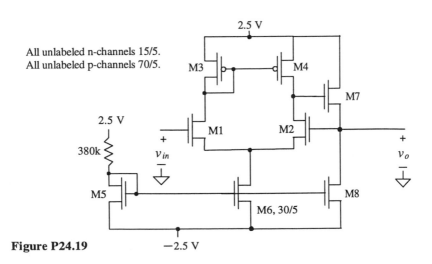

All unlabeled n-channels 15/5.
All unlabeled p-channels 70/5.

2.5 V

M3 M4

M7

2.5 V

380k

v_{in}

M1 M2

v_o

M5

M6, 30/5

M8

Figure P24.19 −2.5 V

24.17 Repeat Ex. 24.6 if I_{ss} is 2 µA.

24.18 If $VDD = -VSS = 2.5$ V, $I_{ss} = 10$ µA, and an n-channel MOSFET size of 15/5, estimate the input CMR (in current) for the current diff-amp of Fig. 24.25.

24.19 Determine v_o/v_{in} for the circuit of Fig. P24.19. This circuit is a voltage follower.

24.20 What is the purpose of MS1/MS2 and MS3/MS4 in the diff-amp of Fig. 24.29?

24.21 Show that the output current of the circuit shown in Fig. P24.21 is related to the square of the difference of the input voltages, v_{I1} and v_{I2}.

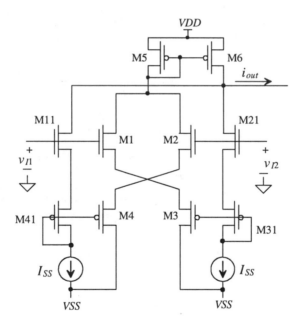

Figure P24.21

Chapter

25

Operational Amplifiers

The operational amplifier (op-amp) is a fundamental building block in analog integrated circuit design [1-3]. A block diagram of the two-stage op-amp with output buffer is shown in Fig. 25.1. The first stage of an op-amp is a differential amplifier. This is followed by another gain stage, such as a common source stage, and finally an output buffer. If the op-amp is intended to drive a small purely capacitive load, which is the case in many switched capacitor or data conversion applications, the output buffer is not used. If the op-amp is used to drive a resistive load or a large capacitive load (or a combination of both), the output buffer is used.

Since it is difficult to characterize a packaged op-amp that cannot drive a resistive load, we will begin the chapter with a discussion of the two-stage op-amp with output buffer. Design of the op-amp consists of determining the specifications, selecting device sizes and biasing conditions, compensating the op-amp for stability, simulating and characterizing the op-amp A_{OL} (open-loop gain), CMR (common-mode range on the input), CMRR (common-mode rejection ratio), PSRR (power supply rejection ratio), output voltage range, current sourcing/sinking capability, and power dissipation.

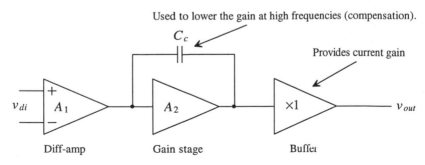

Figure 25.1 Block diagram of two-stage op-amp with output buffer.

25.1 Basic CMOS Op-Amp Design

One possible op-amp design is shown in Fig. 25.2. We will use this configuration to illustrate the design of a CMOS op-amp.

Selection of the Differential Amplifier Bias Current, I_{ss}

Selection of the current I_{ss} is determined by gain, CMR, CMRR, power dissipation, noise, matching considerations and slew rate. (If slew rate is a concern, the source cross-coupled diff-amp should be considered for the first stage of the op-amp.) The small-signal gain of the differential amplifier is given in Eq. (24.25), or

$$A_1 = g_{m1}(r_{o2}\|r_{o4}) = \frac{2\sqrt{\beta}}{(\lambda_2 + \lambda_4)\sqrt{I_{SS}}} = \frac{2}{(\lambda_2 + \lambda_4)(V_{GS} - V_{THN})} \tag{25.1}$$

In addition to selecting I_{ss}, we must determine a V_{GS} of the MOSFETs. Often, selection of the bias current is an iterative process: that is, we select a bias current and verify that the above characteristics of most importance are met. In general, the larger (increasing W while holding L constant) we make the MOSFETs in the differential pair at a given bias current, the lower V_{GS}, CMR is increased, input-referred noise is lowered, matching is better, and gain is increased (from the increase in β). The main drawbacks are increased layout area and parasitic capacitances (and thus lower speed).

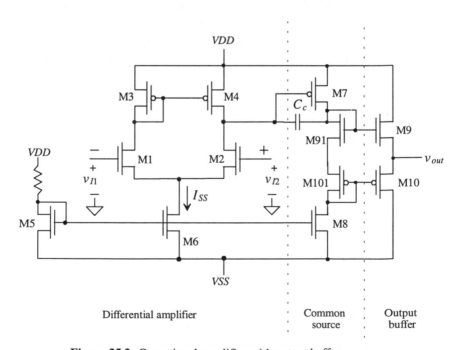

Figure 25.2 Operational amplifier with output buffer.

We may start by selecting $I_{SS} = 20$ μA and $V_{GS} = 1.2$ V, which results in the differential amplifier of Exs. 24.2, 24.3, and 24.4 and Fig. 24.11. When designing an op-amp for off-chip use, the MOSFETs used in the differential input stage normally have widths in the several hundred μm range. For the same I_{SS} this increases the open-loop gain of the op-amp while decreasing V_{GS} (a desirable effect to improve op-amp common-mode range).

Selection of Device Sizes

The next step in the design of the op-amp is selection of the second-stage biasing current. The same considerations that applied to selection of the differential amplifier biasing current apply to the selection of the second-stage biasing current. It would be helpful at this point to review Fig. 20.3. With both inputs to the op-amp, (i.e., the gates of M1 and M2), at the same potential, the same current flows in M3 and M4. The result is that the drain of M4 is at the same potential as its gate. If we want to set the current that flows in M7 at 10 μA, we simply size it the same size as M4. If we want 5 μA to flow, we size M7 so that its W/L is half of M4s, that is, 35/5. We then size M8 so that its drain current, 10 μA in the present case, is equal to the current in M7. The gain of the second stage is given, assuming that $1/g_{m91} + 1/g_{m101}$ is small compared to $r_{o7} \| r_{o8}$, by

$$A_2 = -g_{m7} \cdot (r_{o7} \| r_{o8}) = \frac{-\sqrt{2\beta_7 I_{D7}}}{(\lambda_7 + \lambda_8) I_{D7}} = \frac{-2}{(\lambda_7 + \lambda_8)(V_{SG} - V_{THP})} \quad (25.2)$$

The open-loop gain of the op-amp is given by

$$A_{OL} = A_1 \cdot A_2 = g_{m1}(r_{o2} \| r_{o4}) \cdot [-g_{m7}(r_{o7} \| r_{o8})] \quad (25.3)$$

If we use the sizes of Fig. 25.3, the open-loop gain of the op-amp is

$$|A_{OL}| = \sqrt{2 \cdot 50 \frac{\mu A}{V^2} \cdot \frac{15}{5} \cdot 10 \mu A} \cdot \left[\frac{1}{(0.06 + 0.06)10\mu A} \right]^2 \cdot \sqrt{2 \cdot 17 \frac{\mu A}{V^2} \cdot \frac{70}{5} \cdot 10 \mu A}$$

where the braces indicate g_{m1}, $(r_{o2} \| r_{o4}) \cdot (r_{o7} \| r_{o8})$, and g_{m7}.

$$= 2,600 \frac{V}{V}$$

Decreasing the bias currents to 1 μA will increase the open-loop gain by a factor of 10.

The output stage is a class AB type discussed in Ch. 22. The MOSFETs M91 and M101 are used to bias the source-follower buffers M9 and M10. The gate of M9 can swing up to VDD. If we want no more than 1.2 V across the gate-source potentials of M9 or M10, then the maximum current we can source, neglecting the body effect, is given by

$$I_{OUTMAX} = \frac{\beta_9}{2}(V_{GS9} - V_{THN})^2 = 50 \frac{\mu A}{V^2} \cdot \frac{150}{2}(1.2 - 0.83)^2 = 500 \text{ μA} \quad (25.4)$$

while the minimum current is given by

$$I_{OUTMIN} = \frac{\beta_{10}}{2}(V_{SG10} - V_{THN})^2 = 17\frac{\mu A}{V^2} \cdot \frac{700}{2}(1.2 - 0.91)^2 = -500\ \mu A \quad (25.5)$$

All device lengths are 5 µm
except output buffer, which
uses L = 2 µm since the
output resistance is less
important.

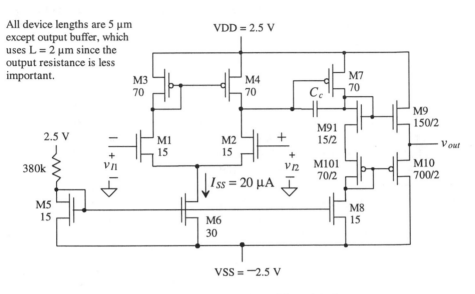

Figure 25.3 Operational amplifier with sizes.

Under these circumstances (load currents), the maximum output swing is given by

$$V_{OUTMAX} = VDD - 1.2 \qquad (25.6)$$

and

$$V_{OUTMIN} = VSS + 1.2 \qquad (25.7)$$

The maximum output current is highly dependent on the output voltage of the amplifier. For example, if $V_{out} = 0$ V, the op-amp, with 2.5 V supplies, can source a current of

$$I_{OUTMAX} = 50\frac{\mu A}{V^2} \cdot \frac{150}{2}(2.5 - 0.83)^2 = 10\ mA \qquad (25.8)$$

The power dissipated by M9 is $2.5 \cdot 0.010 = 25$ mW! After reviewing the temperature characteristics of MOSFETs from Ch. 9, we can see that V_{GS9} will increase (until M9 shuts off) keeping the device from thermal runaway. Although the device and op-amp should not be harmed by this excessive power draw, it still may be important to avoid this situation. Implementation of a maximum current detect or short-circuit protection circuit on the output of the op-amp is needed. Consider the short-circuit protection circuit of Fig. 25.4. When the output current exceeds a value set by

$$I_{OUTMAX} = \frac{V_{THN}}{R} \qquad (25.9)$$

for an op-amp sourcing current or

$$I_{OUTMIN} = \frac{V_{THP}}{R} \quad (25.10)$$

for an op-amp sinking current, the gate drive is removed from M9 or M10. This keeps the MOSFETs from sourcing or sinking too much current. Typical values of the resistor for the present circuit are 1 kΩ.

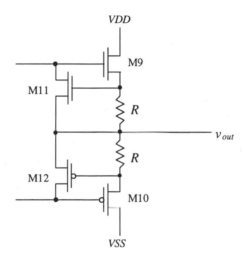

Figure 25.4 Addition of M11, M12, and the resistors to output stage for short-circuit protection.

Compensating the Op-amp

The last step in our first cut of the op-amp design is to select the compensation network. The closed-loop gain of an op-amp can be written in terms of the feedback equation by

$$A_{CL} = \frac{A_{OL}}{1 + A_{OL}\beta} \quad (25.11)$$

We know from our discussion in Ch. 23 that the feedback amplifier becomes unstable when the loop gain $A_{ol}\beta = -1$. This is equivalent to the requirement that

$$|A_{OL}\beta| = 1 \text{ and } \angle A_{OL}\beta = \pm 180° \quad (25.12)$$

We know that β represents the amount of the output signal that is fed back and subtracted from the input of the amplifier. Therefore, the largest value of β possible without an amplifier in the feedback loop occurs when all of the output is fed back to the input, or in other words β = 1. The voltage follower configuration shown in Fig. 25.5 is an example of a closed-loop amplifier with β = 1. Under these circumstances, we can rewrite Eq. (25.12) as

$$|A_{OL}| = 1 \text{ and } \angle A_{OL} = \pm 180° \quad (25.13)$$

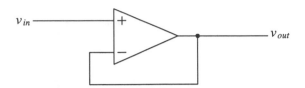

Figure 25.5 Voltage follower configuration, an example of a closed-loop amplifier with unity feedback factor.

We will use this result (i.e., we will consider the open-loop gain of an amplifier) in the following discussion on compensating the operational amplifier.

The small-signal model of the two-stage op-amp, considering the high-impedance nodes that give rise to the dominant poles, is shown in Fig. 25.6. The resistance to ground at the output of the differential amplifier is given by

$$R_1 = r_{o2}||r_{o4} \qquad (25.14)$$

while the capacitance, see Eq. (24.28), at this node is given by

$$C_1 = C_{gs7} + C_{gd7}(1+|A_2|) + C_{db4} + C_{gd4} + C_{db2} + C_{gd2} \qquad (25.15)$$

For the present amplifier

$$R_1 = 833 \text{ k}\Omega$$

With the help of Ex. 24.4 and knowing

$$C_{gs7} = \frac{2}{3} \cdot C'_{ox} \cdot W_7 L_7 = 187 \text{ fF}$$

and

$$C_{gd7} = W \cdot CGDO = 70 \text{ μm} \cdot 5 \times 10^{-10} \frac{\text{F}}{\text{m}} = 35 \text{ fF}$$

The gain of stage two, A_2, from Eq. (25.2) is -58 V/V. The capacitance C_1 is given, with the results of Ex. 24.4, by

$$C_1 = 187 + 35(1+58) + 190 = 2.44 \text{ pF}$$

The pole resulting from this time constant for the circuit shown in Fig. 25.3 is given by

$$f_1 = \frac{1}{2\pi R_1 C_1} = \frac{1}{2\pi \cdot 833 \text{ k} \cdot 2.44 \text{ pF}} = 78 \text{ kHz}$$

The pole resulting from the high-impedance node at the drain of M7 is characterized by R_2 and C_2, or

$$R_2 = r_{o7}||r_{o8} \qquad (25.16)$$

and

$$C_2 = C_{gd7}\left[1 + \frac{1}{|A_2|}\right] + C_{db7} + C_{db8} + C_{gd8} + C_{gd9} + C_{gd10} \qquad (25.17)$$

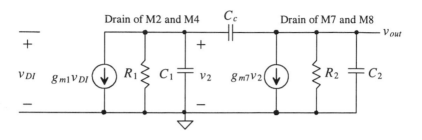

Figure 25.6 Small-signal model of two-stage op-amp.

The value of R_2 is the same as R_1 because the bias currents of the first and second stages are identical. The value of C_2 is the sum of $C_{gd7} = 35$ fF, $C_{db7} = 137$ fF, $C_{db8} = 9.4$ fF, $C_{gd8} = 5.7$ fF, and

$$C_{gd9} = 150\ \mu m \cdot 3.8 \times 10^{-10} = 57\ \text{fF}$$

and

$$C_{gd10} = 700\ \mu m \cdot 5 \times 10^{-10} = 350\ \text{fF}$$

or $C_2 = 600$ fF. The pole resulting from the high-impedance node at the drains of M9 and M10 is located at a frequency given by

$$f_2 = \frac{1}{2\pi R_2 C_2} = 318\ \text{kHz}$$

So far we have not considered the effect of adding a compensation capacitor C_c. Before we try to compensate the amplifier to ensure that it remains stable, let's discuss simulation of the open-loop gain of the amplifier. The simulation of the open-loop gain will be another indication of the dominant pole locations and can be compared to the hand calculations given above. A circuit useful in simulating the open-loop gain of an

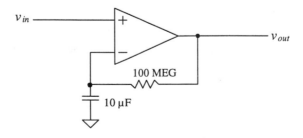

Figure 25.7 Circuit configuration used to measure open-loop gain and frequency response.

op-amp with or without compensation is shown in Fig. 25.7. The resistive feedback ensures a stable DC operating condition, while the capacitor/resistor combination eliminates output AC signals fed back from the output of the op-amp. The SPICE simulation results of the amplifier described above without compensation and given schematically in Fig. 25.3 are shown in Fig. 25.8. The point at which the open-loop gain is unity (0 dB) corresponds to a phase shift of 160°, or a − 20 ° phase margin. In other words, highly unstable. Normally, we desire no less than a + 45 ° phase margin when compensating an op-amp.

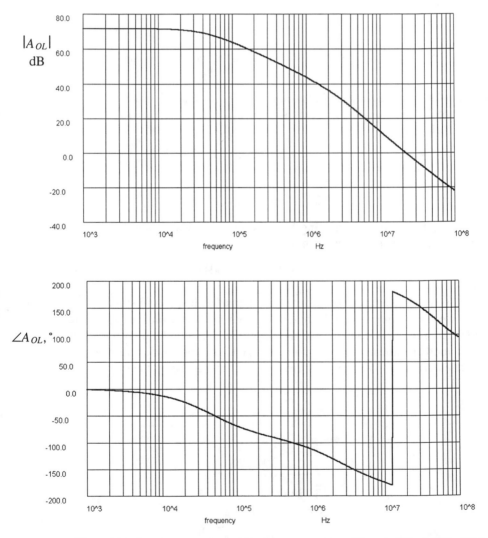

Figure 25.8 Open-loop uncompensated frequency response of the amplifier of Fig. 25.3.

The compensation capacitor causes the gain to roll off long before the second pole takes effect. If C_c is much larger than C_1 ($v_2 \approx 0$ and all of $g_{m1}v_{DI}$ flows through C_c), then we can write the output voltage in terms of the input voltage as

$$v_{out} = \overbrace{g_{m1}v_{DI}}^{\text{I out of 1st stage}} \cdot \frac{1}{j\omega C_c} \tag{25.18}$$

$$A_{OL} = \frac{v_{out}}{v_{DI}} = \frac{g_{m1}}{j\omega C_c} \tag{25.19}$$

If we label the frequency when the magnitude of the open-loop gain is unity as f_u, then we can select the compensation capacitor based on

$$|A_{OL}| = \frac{g_{m1}}{2\pi f C_c} \tag{25.20}$$

or when $|A_{OL}|$ is unity

$$C_c = \frac{g_{m1}}{2\pi f_u} \tag{25.21}$$

This equation is useful for general-purpose op-amp design especially where bandwidth of the amplifier is not critical.

Compensation of the op-amp of Fig. 25.3 begins by inspecting the open-loop gain magnitude and phase plots of Fig. 25.8. Our hand calculations showed $f_1 = 78$ kHz and $f_2 = 318$ kHz, while SPICE simulation results give $f_1 = 50$ kHz and $f_2 = 1.5$ MHz. If we set the unity gain frequency at the location of the second pole, the phase of A_{OL} is approximately 135°, corresponding to a phase margin of 45°. To give a little more margin, we will set f_u to 1 MHz. The compensation capacitor is then

$$C_c = \frac{55\frac{\mu A}{V^2}}{2\pi 10^6} = 8.75 \text{ pF}$$

The compensated op-amp phase and magnitude plots are shown in Fig. 25.9. The phase margin is approximately 45°. Practically speaking, it may be wise to increase the compensation capacitor to 10 pF for a little extra head room.

At this point we may ask the question, "Why do we need a phase margin anyway?" The answer to this question involves the realization that a large load capacitance can increase the phase shift through the op-amp and the feedback network normally does not have zero propagation delay. Consider using an n-well resistor in the feedback loop. The delay through the n-well resistor translates into a phase shift at a given frequency given by

$$\theta = t_{delay} \cdot f \cdot 360° \tag{25.22}$$

Another important characteristic of adding a compensation capacitor is that the compensation capacitor combined with the diff-pair bias current sets the slew rate out of the op-amp. The maximum rate the output of an op-amp can change is approximated by I_{ss} charging C_c or

$$SR = \frac{dV_{out}}{dt} = \frac{I_{SS}}{C_c} \qquad (25.23)$$

This is an important design factor in some situations. Replacing the source coupled pair with the source cross-coupled pair can eliminate slew-rate limitations in the basic two-stage op-amp at the price of a more complicated circuit and high-power dissipation.

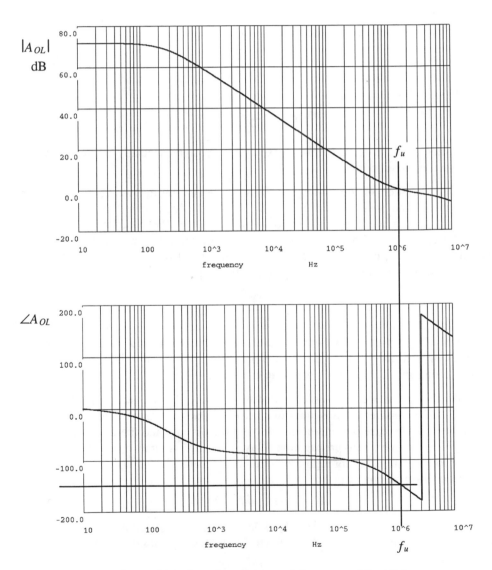

Figure 25.9 Open-loop compensated op-amp of Fig. 25.3.

In the derivation of this compensation scheme, we assumed that C_c was much greater than C_1. This approximation has several practical problems. Reviewing Fig. 25.9, we see that A_{OL} has a zero at approximately f_u. The effect of the zero is to cause f_u to be less well defined. A more exact analysis of the two-stage op-amp *with* compensation capacitor of Fig. 25.6 gives a transfer function of

$$\frac{v_{out}}{v_{DI}} = \frac{g_{m1}g_{m7}R_1R_2\left(1 - \frac{sC_c}{g_{m7}}\right)}{s^2R_1R_2[C_1C_2 + C_c(C_1 + C_2)] + s[R_1(C_1 + C_c) + R_2(C_2 + C_c) + g_{m7}R_1R_2C_c] + 1}$$

(25.24)

The poles of this equation can be approximated by

$$p_1 \approx \frac{-1}{(1 + g_{m7}R_2)C_cR_1}$$

(25.25)

and

$$p_2 \approx \frac{-g_{m7}C_c}{C_2C_1 + C_2C_c + C_cC_1}$$

(25.26)

A right-hand plane zero is also present and is given by

$$z = \frac{g_{m7}}{C_c}$$

(25.27)

Comparing this result with Eq. (25.20), we see that if the transconductance of the first and second stages is comparable, the unity gain frequency of the op-amp and the location of the zero occur close together. Since the zero is in the right-half plane, the phase margin is further degraded. Intuitively, this effect can be understood by realizing that the output of the first stage is fed directly to the output of the second-stage, without inversion, through the compensation capacitor effectively degrading the phase margin.

To circumvent this feedthrough problem, we can insert a resistor in series with the compensation capacitor as shown in Fig. 25.10. Addition of the resistor R_z shifts the zero to a location given by

$$z = \frac{1}{C_c\left(\frac{1}{g_{m7}} - R_z\right)}$$

(25.28)

If we select R_z so that it is equal to $1/g_{m7}$ the zero disappears. Making R_z larger than $1/g_{m7}$ causes the zero to add to the open-loop phase response, increasing the phase margin. For this reason, this type of compensation is sometimes referred to as lead compensation.

For the op-amp of Fig. 25.3 with $g_{m7} = 70$ μA/V^2, the value of the zero canceling resistor can be determined by

$$R_z = \frac{1}{g_{m7}} = 14.5 \text{ k}\Omega$$

(25.29)

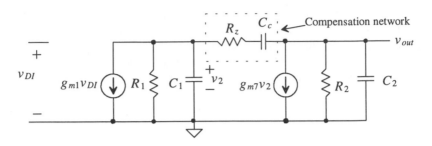

Figure 25.10 Practical compensation of a two-stage op-amp.

The selection of the compensation capacitor C_c is unchanged from our previous discussion since R_z has no effect on the low-frequency pole. A complete schematic[1] of the op-amp is shown in Fig. 25.11 with the compensation network in place. The open-loop gain magnitude and phase with this compensation network are shown in Fig. 25.12. The phase margin of the op-amp with the R_z / C_c combination is greatly improved over the C_c alone. *Any practical two-stage op-amp design should use the zero nulling resistor in the compensation network.*

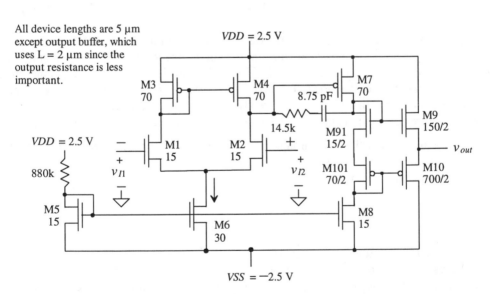

Figure 25.11 Op-amp schematic with compensation.

---

[1] We could bias the op-amp with a current source instead of the 880k resistor shown in Fig. 25.3. We chose to use the resistor to keep the schematic simple. In practice, any of the current references of Ch. 21 can provide the current through M15 and thus bias the op-amp.

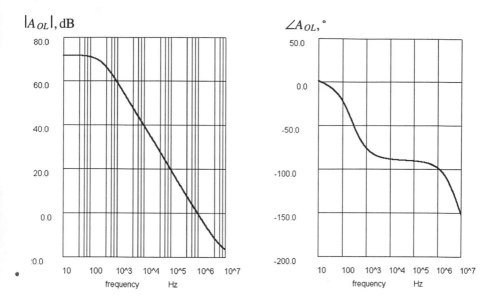

Figure 25.12 Open-loop gain and phase response of the amplifier of Fig. 25.11 with zero nulling resistor. Phase margin is approximately 80°.

The resistor, R_z, can also be implemented using a MOSFET (M13 in Fig. 25.13). This implementation is more desirable in general because the effective channel resistance of M13 can be selected to track the inverse of g_{m7} over temperature and process variations. A bias circuit consisting of M14, M15, and M16 generates a gate voltage for M13 (the MOSFET used as R_z). The channel resistance of a p-channel MOSFET in the triode region, from Ch. 9, is given by

$$R_{ch} = \frac{1}{\beta(V_{SG} - V_{THP})} = R_z \qquad (25.30)$$

or setting this in terms of g_{m7} we arrive at

$$g_{m7} = \beta_7(V_{SG7} - V_{THP}) \text{ and } R_{ch}^{-1} = \beta_{13}(V_{SG13} - V_{THP}) \qquad (25.31)$$

Using these two equations to determine the product of g_{m7} and R_{ch}, assuming $\beta_7 = \beta_{14} = \beta_{15}$ and $\beta_8 = \beta_{16}$, gives

$$g_{m7}R_{ch} = \frac{\beta_7}{\beta_{13}} = g_{m7}R_z \qquad (25.32)$$

To make the zero vanish, from Eq. (25.28), we set the product equal to unity. For our op-amp of Fig. 25.11, we simply size M13 of Fig. 25.13 to 70/5, which is the same size as M7. For lead compensation (i.e., to increase the phase margin further), we set the product to a number greater than zero. The price we pay for reducing the op-amp sensitivity to process variations is a more complicated circuit and higher power dissipation.

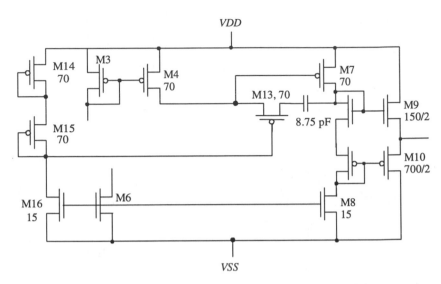

Figure 25.13 Partial schematic of an op-amp showing the use of M13 as a process-insensitive resistor used for compensation.

25.1.1 Characterizing the Op-Amp

Now that we have a rough op-amp design, we need to characterize the op-amp to determine its characteristics.

Input Offset Voltage

An op-amp parameter of great importance is the input offset voltage. Ideally, if we ground both inputs to the op-amp, the output voltage is zero. Realistically, the output voltage is nonzero because of mismatches in the input stage, called *random offsets*, and because the potential on the output of the second stage (for the example op-amp of this chapter, on the drains of M7 and M8) is not well defined, leading to a *systematic offset*. The random offsets in current mirrors and differential amplifiers due to differences in geometry and threshold voltages have already been discussed. The systematic offset results in a worst-case offset voltage of

$$V_{os,sys} = \frac{VDD}{A_{OL}} \text{ or } \frac{VSS}{A_{OL}} \qquad (25.33)$$

The offset voltage of an op-amp is modeled as a DC voltage source in series with the noninverting input of an offset free op-amp. This is shown in Fig. 25.14a. The systematic offset is simulated by grounding the inverting input of the op-amp and sweeping the noninverting input. Simulation results for the op-amp of Fig. 25.11 are shown in Fig. 25.14b. The offset voltage is equal to the distance between the curve and the origin along the x-axis and in this case is approximately 0.1 mV. The slope of the transfer characteristics corresponds to the open-loop gain of the op-amp.

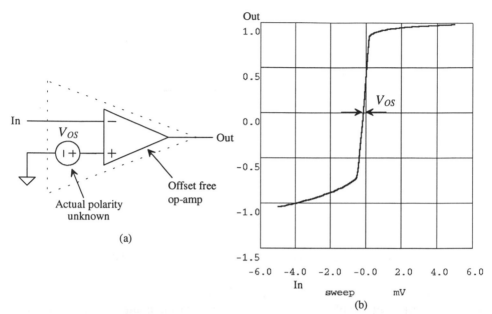

Figure 25.14 Simulation results of op-amp shown in Fig. 25.11 showing systematic offset.

Characterizing the Output Voltage Swing

The output voltage swing can be determined using Eqs. (25.4) through (25.7) for the class AB output stage of the op-amp shown in Fig. 25.11. The simulation results of 25.14b show that the output swing is approximately +/– 0.8 V. We are interested here in the range of outputs that are linearly related to the inputs. From this figure we see that input voltages of approximately +/– 0.5 mV on the input (neglecting the offset) result in an output voltage of $A_{OL} \cdot (v_{p} - v_{n})$.

Common Mode Rejection Ratio

The CMRR of an op-amp is calculated by multiplying the common-mode gain, A_c, of the input differential amplifier with the gain of the second stage and dividing the result into the open-loop gain A_{OL}. The CMRR of an op-amp in dB is given by

$$\text{CMRR} = 20 \log \left(\frac{A_{OL}}{A_c A_2} \right) = 20 \log \left(\frac{A_1}{A_c} \right) \qquad (25.34)$$

which shows that the op-amp CMRR is determined by the differential stage. Simulation of the CMRR can be performed with the circuits shown in Fig. 25.15. The CMRR is determined using Eq. (25.34) with the circuit outputs A_{OL} and $A_c \cdot A_2$. For the op-amp of Fig. 25.11, the CMRR is approximately 75 dB [from Eq. (24.35)] and the A_{OL}, from Fig. 25.12, is 72 dB. Knowing the CMRR and A_{OL}, we give the common-mode gain by

$$A_{cm} = (A_2 A_c)(\text{dB}) = A_{OL}(\text{dB}) - \text{CMRR}(\text{dB}) = 72 - 75 = -3 \text{ dB} = 0.707 \text{ V/V} \qquad (25.35)$$

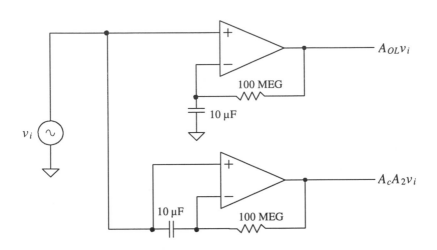

Figure 25.15 Circuit configuration used to simulate CMRR.

We can think of A_{cm} as the open-loop gain of the op-amp for common-mode signals. If the applied differential voltage is zero and we change the common-mode voltage by ΔV_c, then the output voltage will change by $\Delta V_o = A_{cm} \cdot \Delta V_c$. To compensate for the change in the output voltage, a nonzero input differential voltage develops on the input of the op-amp (an offset voltage that is a function of the common-mode voltage). This offset voltage can be estimated by

$$\Delta V_{OS} = \frac{\Delta V_o}{A_{OL}} = \frac{A_{cm}}{A_{OL}} \cdot \Delta V_c = \frac{\Delta V_c}{\text{CMRR}} \qquad (25.36)$$

If we apply a 1 V common-mode signal to both inputs of the op-amp of Fig. 25.11, the offset voltage will change by 175 μV.

Power Dissipation

The power dissipated by an op-amp, P, is simply the product of the sum of the currents flowing in the current sources or sinks with the power supply voltages. For the op-amp of Fig. 25.11, $P = (VDD - VSS) \cdot (I_{D5} + I_{D6} + I_{D8} + I_{D10}) = 700\ \mu\text{W}$.

Power Supply Rejection Ratio

The power supply rejection ratio (PSRR) is a term used to describe how well an amplifier rejects noise or changes on the VDD and VSS power buses. This parameter can be extremely important in precision analog design. Consider the test setup shown in Fig. 25.16. The positive PSRR is defined by

$$\text{PSRR}^+ = \frac{A_{OL}}{v_{out}/v^+} \qquad (25.37)$$

while the negative PSRR is defined by

$$PSRR^- = \frac{A_{OL}}{v_{out}/v^-} \qquad (25.38)$$

Let's begin by considering how the output voltage of the op-amp of Fig. 25.11 changes with changes on VDD (i.e., v^+). At relatively low frequencies, the compensation capacitor starts to effectively short the gate of M7 to its drain. At this point, we can think of M7 as a small-signal resistance of value $1/g_{m7}$. Since the resistance looking down into the drain of M91 is dominated by r_{o8} and $r_{o8} \gg 1/g_{m7}$ almost all of v^+ is coupled to the output buffer and thus to the output of the op-amp. The gain from v^+ to v_{out} is plotted in Fig. 25.16c. For the negative PSRR, the effect of the compensation capacitor shorting the gate of M7 to its drain is to effectively short the drain of M8 to VDD. This causes the negative PSRR to increase with increasing frequency (see Fig. 25.16d). Because of the compensation scheme used, the PSRR of two-stage op-amps is poor. Discussed later in the chapter are a group of op-amps referred to as operational transconductance amplifiers (OTAs). These op-amps are one stage and are compensated by a load capacitance. In general, OTAs show much better PSRR than the two-stage variety discussed so far.

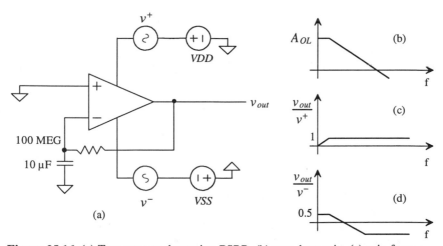

Figure 25.16 (a) Test setup to determine PSRR, (b) open-loop gain, (c) gain from AC signal on VDD to output, and (d) gain from AC signal on VSS to output.

Slew Rate

The slew-rate limitations of an op-amp built with the source coupled diff-amp is given by Eq. (25.23). The configuration shown in Fig. 25.17 is useful in measuring or simulating the slew-rate of an op-amp. For the op-amp of Fig. 25.11, the SR is 20 μA/8.75 pF or 2.25 V/μs. Normally, the slew-rate, overshoot, and settling time are specified for different loads. We will assume in the simulation of Fig. 25.11 that the op-amp is driving a 10 pF load (equivalent to a standard scope probe capacitance). The simulation results are shown in Fig. 25.18 for a 0 to 1 V step input into the follower. The simulation results and hand calculations are close.

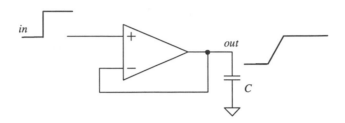

Figure 25.17 Circuit used to measure the slew rate of an op-amp.

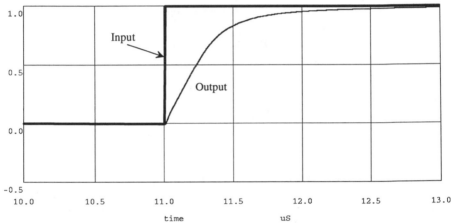

Figure 25.18 Slew-rate simulation results of the op-amp of Fig. 25.11 driving a 10 pF load.

If the SR is important in a particular application, the source cross-coupled configuration should be considered. An op-amp based on the sizes of Fig. 25.11 using the source cross-coupled pair is shown in Fig. 25.19. Again, as we discussed with the source coupled pair, the widths of M1/M2 and M3/M4 should be increased beyond 15 μm in practice. Increasing the widths has the effect of increasing the transconductance of the differential amplifier and thus increasing the gain of the op-amp.

25.1.2 Compensating the Op-Amp Without Buffer

The two-stage CMOS op-amp without output buffer is shown in Fig. 25.20. This configuration is useful as a comparator (without the compensation network included) or when the load is purely capacitive, a situation that occurs frequently when the op-amp is used on-chip in sampled data circuits. When compensating the two-stage op-amp, the load capacitance should be included with the calculation of C_2. Since the load capacitance affects the location of the second pole, it is possible that the compensated amplifier/load capacitance combination is unstable.

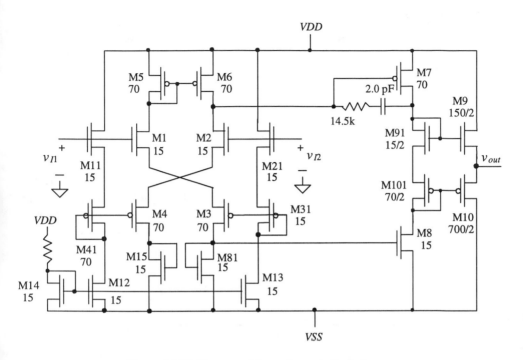

Figure 25.19 Op-amp without slew-rate limitations.

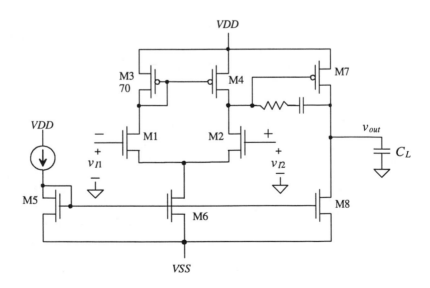

Figure 25.20 Two-stage op-amp without buffer.

25.1.3 The Cascode Input Op-Amp

An unbuffered op-amp based on the cascode differential input stage is shown in Fig. 25.21. The cascode input stage is used to increase the overall open-loop gain of the op-amp. A source-follower stage consisting of M13 and M12 is used to provide a DC level shift between the output of the diff-amp and the output of the op-amp. This allows the output to swing up to *VDD* and down to *VSS*. The location of the first pole in this configuration is shifted lower in frequency due to the higher output resistance of the cascode configuration. Although the op-amp can have significantly higher gain than the basic current mirror configuration, the input common-mode range and CMRR are poorer. Because of the large gain of the first stage, this op-amp can be used without an output buffer. We must keep in mind that the output can source a current via M14 but is limited to sinking a current set by the size and biasing conditions of M15. One application of this amplifier is in a voltage regulator circuit where the op-amp is continuously sourcing a current to the load (keeping in mind the PSRR limitations).

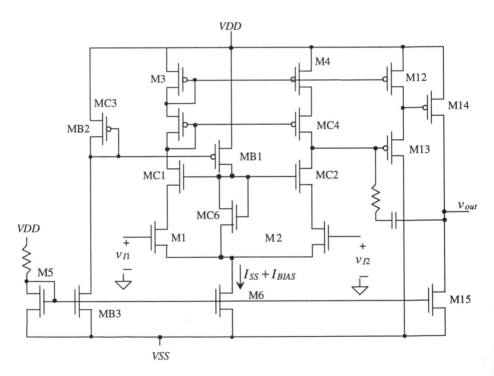

Figure 25.21 Unbuffered op-amp using a cascode differential amplifier.

25.2 Operational Transconductance Amplifiers

The operational transconductance amplifier (OTA) is basically an op-amp without an output buffer. *An OTA without buffer can only drive capacitive loads.* An OTA can be defined as an amplifier where all nodes are low impedance except the input and output nodes. An example OTA is shown in Fig. 25.22. Note that the basic op-amp of Fig. 25.20 is not considered an OTA because the drain of M4 is a high-impedance node and not the input or output of the amplifier. Assuming that $\beta_1 = \beta_2$, $\beta_{31} = \beta_{41}$, we observe that the current i_{d31} or i_{d41} is given by

$$- i_{d31} = i_{d41} = \frac{g_{m1}}{2}(v_{I2} - v_{I1}) = i_d \qquad (25.39)$$

Furthermore, if $\beta_4 = K \cdot \beta_{41} = K \cdot \beta_{31} = K \cdot \beta_3$ and $K \cdot \beta_{51} = \beta_5$, then $i_{d4} = - i_{d5} = K \cdot i_{d41} = - K \cdot i_{d31}$. If the impedance of the capacitor is large compared to $r_{o4} \| r_{o5}$, then the output voltage of the OTA is given by

$$v_{out} = 2K i_d(r_{o4} \| r_{o5}) \qquad (25.40)$$

and the voltage gain is

$$A_v = \frac{v_{out}}{v_{I2} - v_{I1}} = K g_m(r_{o4} \| r_{o5}) \qquad (25.41)$$

where we have assigned the noninverting input of the OTA as the gate of M2. However, as the name states, we are interested in the transconductance (current out, voltage in) of the amplifier. If the impedance of the load capacitor or the resistance of

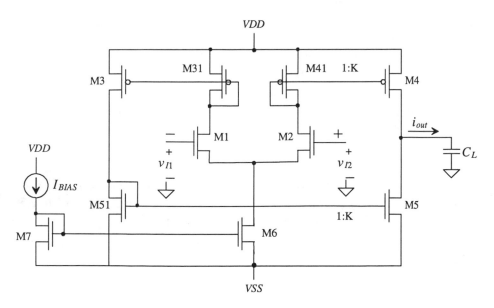

Figure 25.22 Basic configuration of an OTA.

an external load is small compared to the output resistance, $r_{o4}\|r_{o5}$, then the output current flows mainly in the external load. Under these circumstances, we can write the output current as

$$i_{out} = i_{d4} - i_{d5} = 2Ki_d \tag{25.42}$$

The transconductance of the OTA is given by

$$G_m = \frac{i_{out}}{v_{I2} - v_{I1}} = g_m K \tag{25.43}$$

In the following discussion, we will assume $K = 1$ so that $G_m = g_m$. In other words, the transconductance of the OTA is set by the transconductance of the input differential amplifier. A useful feature of the OTA is that its transconductance can be adjusted by the bias current. Filters made using the OTA can be tuned by changing the bias current labeled I_{BIAS} in Fig. 25.22. The symbol for the OTA is shown in Fig. 25.23.

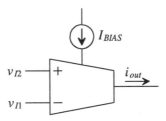

Figure 25.23 Schematic symbol of an OTA.

Example 25.1

Consider the OTA shown in Fig. 25.24.   Determine I_{BIAS} and the transconductance of the OTA, g_m, as a function of $V_{control}$.

The bias current can be determined by realizing that the drain current of M10 and the drain current (and thus I_{BIAS}) of M8 are equal. The bias current can be written in terms of $V_{control}$ as

$$I_{BIAS} = \frac{V_{control} - V_{GS10}}{100 \text{ k}\Omega}$$

Since the W/L ratio of M10 is so large, we can assume $V_{GS10} \approx V_{THN} \approx 1$ V. Using this result, I_{BIAS} varies from 0 to 15 μA for $V_{control}$, changing from 1 to 2.5 V, respectively. If $V_{control}$ is set to 2 V, then I_{BIAS} is 10 μA.

The transconductance of the OTA is given, keeping in mind I_{SS} (the drain current of M6) of the differential amplifier is $2I_{BIAS}$, by

$$g_m = \sqrt{2\beta I_{BIAS}} \tag{25.44}$$

or in terms of $V_{control}$

$$g_m = \left[\frac{2\beta_1(V_{control} - 1)}{100 \text{ k}\Omega} \right]^{1/2} = \left[\frac{2 \cdot 50\frac{\mu A}{V^2} \cdot 15(V_{control} - 1)}{100 \text{ k}\Omega \cdot 5} \right]^{1/2} \quad (25.45).$$

For $V_{control} = 2$ V, we get an OTA gain of $G_m = g_m = 55$ µA/V. ∎

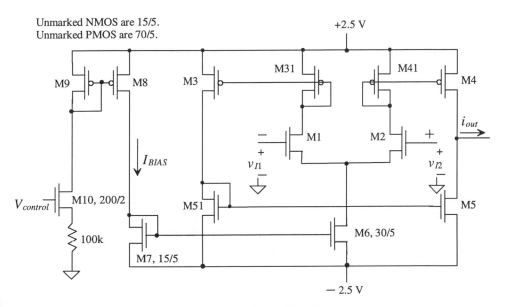

Figure 25.24 OTA of Ex. 25.1.

Example 25.2
Using the OTA of Fig. 25.24 with $V_{control}$ equal to 2 V ($I_{BIAS} = 10$ µA), determine the transfer function, v_{out}/v_{in}, of the circuit shown in Fig. 25.25. Compare the hand calculations to a SPICE simulation.

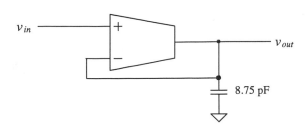

Figure 25.25 OTA circuit for Ex. 25.2.

The current out of the OTA, i_{out}, is $g_m(v_{in} - v_{out})$. The output voltage is given by

$$v_{out} = i_{out} \cdot \frac{1}{j\omega C} = \frac{g_m}{j\omega C_L} \cdot (v_{in} - v_{out})$$

where C_L is the load capacitance, 8.75 pF in this case. The transfer function of the circuit is given by

$$\frac{v_{out}}{v_{in}} = \frac{1}{1 + j\omega\left(C_L \cdot \frac{1}{g_m}\right)} \tag{25.46}$$

The circuit of Fig. 25.25 performs the same function as a simple low-pass single-time constant filter with a capacitor value of C_L and a resistor value of $1/g_m$. The major advantage of this circuit is that the source, v_{in}, sees the input impedance of the OTA, which is practically infinite. The present circuit has a pole at a frequency of

$$f = \frac{1}{2\pi \cdot 8.75p \cdot \frac{1}{55\mu}} = 1 \text{ MHz}$$

The simulation results are shown in Fig. 25.26. An *important observation* we should make at this point is that increasing the load capacitance increases the phase margin of the OTA circuit (the dominant pole moves downward). ∎

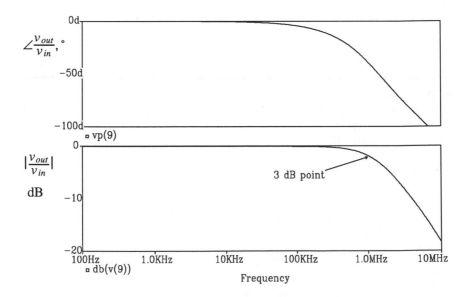

Figure 25.26 Phase/magnitude response of the OTA circuit of Fig. 25.25.

A high-pass OTA circuit [4] is shown in Fig. 25.27. The transfer function of the circuit is given by

$$\frac{v_{out}}{v_{in}} = \frac{j\omega\left(C \cdot \frac{1}{g_m}\right)}{1 + j\omega\left(C \cdot \frac{1}{g_m}\right)} \tag{25.47}$$

Another name for the OTA-capacitor filter circuits of Figs. 25.25 and 25.27 is transconductor-C filters.

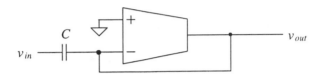

Figure 25.27 OTA circuit implementing a high-pass filter.

Two practical concerns when designing an OTA for filter applications are the input signal amplitude and the parasitic input/output capacitances. Large signals cause the OTA gain to become nonlinear. (An AC SPICE analysis does not reveal large-signal distortion.) The external capacitance should be large compared to the input/output parasitics of the OTA. This limits the maximum frequency of a filter built with an OTA and causes amplitude or phase errors. The errors can usually be tuned out with proper selection of I_{BIAS}.

Figure 25.28 shows a biquadratic (biquad for short) useful as a low-pass, high-pass, band-pass, or band-reject filter [4]. The input conditions and filter type are given in Table 25.1. If we assume that the transconductances of each stage are equal, then the natural frequency of the filters is given by

$$\omega_o = \frac{g_m}{\sqrt{C_1 C_2}} \tag{25.48}$$

while the Q of the filter is

$$Q = \sqrt{\frac{C_2}{C_1}} \tag{25.49}$$

An ideal transconductance amplifier has infinite input and output resistances. The OTA of Fig. 25.22 has a modest output resistance. The output resistance of the OTA causes an error in the transfer function in filter applications. Increasing the output resistance is accomplished with the cascode OTA of Fig. 25.29. The transconductance of the amplifier is unchanged from the noncascode case. Adding the cascode MOSFETs has the effect of increasing the output resistance of the OTA while

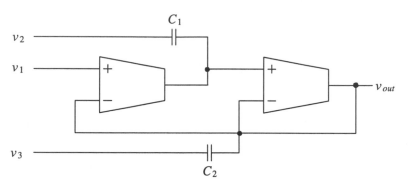

Figure 25.28 Biquad filter implementation using OTAs.

reducing the output swing. The MOSFET M9 is used to keep the drain of M3 at the same potential as the drain of M4 and thus eliminating errors due to channel length modulation. The bias voltages can be selected so that M4 and M5 are kept in the saturation region.

Filter Type	Input Conditions	Transfer Function
Low-pass	$v_{in} = v_1$, with v_2 and v_3 grounded	$\dfrac{g_m^2}{s^2 C_1 C_2 + s C_1 g_m + g_m^2}$
High-pass	$v_{in} = v_3$, with v_1 and v_2 grounded	$\dfrac{s^2 C_1 C_2}{s^2 C_1 C_2 + s C_1 g_m + g_m^2}$
Band-pass	$v_{in} = v_2$, with v_1 and v_3 grounded	$\dfrac{s C_1 g_m}{s^2 C_1 C_2 + s C_1 g_m + g_m^2}$
Band-reject	$v_{in} = v_1 = v_3$ with v_2 grounded	$\dfrac{s^2 C_1 C_2 + g_m^2}{s^2 C_1 C_2 + s C_1 g_m + g_m^2}$

Table 25.1 Design equations for the Biquad filter.

25.2.1 Wide-Swing OTA

A wide-swing OTA first introduced in Ch. 24 is shown in Fig. 25.30 [5]. Wide swing here means that the input CMR is close to the supply voltages. Simulation results using the OTA are shown in Fig. 25.31 for an input voltage swing from zero up to +5 V. The output voltage swing of the OTA is limited by the supply voltages. The transconductance of the OTA (from Sec. 24.4) is $g_m = \sqrt{2\beta I} = \sqrt{2 \cdot \frac{20}{5} \cdot 50\frac{\mu A}{V^2} \cdot 1\,\mu A}$ $= 20\,\frac{\mu A}{V}$ for one diff-amp conducting and $40\,\frac{\mu A}{V}$ when both diff-amps are conducting. The unloaded low-frequency voltage gain of the OTA, again from Sec. 24.4, is 166 V/V independent of the number of diff-amps conducting current.

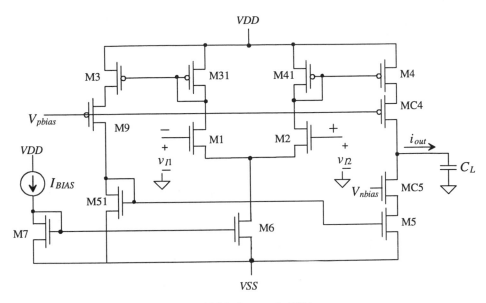

Figure 25.29 A cascode OTA.

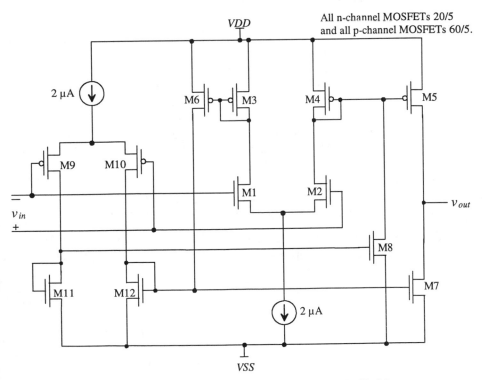

Figure 25.30 Wide-swing OTA first introduced in Ch. 24.

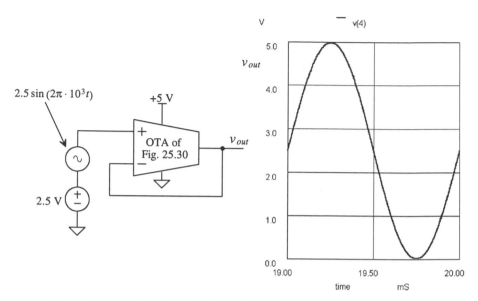

Figure 25.31 Simulation results (using BSIM model) of the OTA of Fig. 25.30 with VDD = +5 V and VSS = 0.

A circuit that operates in triode, saturation, or cutoff, such as the OTA of Fig. 25.30, is very difficult to simulate using SPICE. Using the level 2 model helps with convergence at the price of accuracy. For example, simulating the OTA of Fig. 25.30 using the level 2 model resulted in an open-loop gain of approximately 18, while using the BSIM model resulted in a voltage gain of 240. This result makes it clear that the level 2 model is useful only in functional testing, verifying that the circuit is connected correctly in the simulation. In order to achieve convergence using the BSIM model, the simulation included the .OPTIONS statement with RELTOL = 0.1, ABSTOL = 0.1u, and VNTOL = 50 mV. Simulating the circuit using AC analysis and achieving convergence are considerably easier than a DC sweep or transient analysis. An AC analysis begins by calculating the operating point. (This is the difficult part of the AC simulation.) This is followed by replacing the active devices with small-signal models.

Before we discuss the AC analysis of the OTA circuit of Fig. 25.30 to determine stability, we should discuss compensating OTAs. The closed-loop gain of an OTA circuit is given by

$$A_{CL} = \frac{A_{OL}}{1 + A_{OL} \cdot \beta}$$

(25.50)

The maximum value of the feedback factor, β, without an amplifier in the feedback loop is unity, that is, all of the output voltage is fed back and subtracted from the input. Setting the feedback factor to unity follows the same procedure used in Sec. 25.1 [see Eqs. (25.12) and (25.13)] to determine the gain and phase margins of the op-amp. An

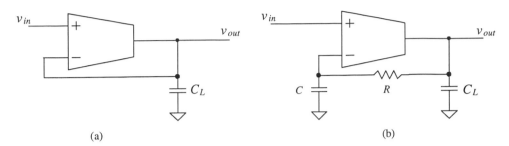

Figure 25.32 OTA circuit with (a) unity feedback factor and (b) DC stability.

OTA with a unity feedback factor is shown in Fig. 25.32a while Fig. 25.32b shows the configuration used to determine the open-loop gain of the OTA. The resistor/capacitor combination helps stabilize the DC output voltage of the OTA, ensuring that the MOSFETs used in the OTA operate in the saturation region. The feedback resistor should be much larger than the output resistance of the OTA, and the RC product should be much longer than the reciprocal of the lowest frequency of interest. This is equivalent to stating that R and C of Fig. 25.32b should have no effect on the AC performance of the circuit. If we assume that the impedance of the load capacitor is small compared to the output impedance of the OTA, then we can write

$$v_{out} = i_{out} \cdot \frac{1}{j\omega C_L} = g_m v_{in} \cdot \frac{1}{j\omega C_L} \qquad (25.51)$$

or

$$\frac{v_{out}}{v_{in}} = \frac{g_m}{j\omega C_L} \Rightarrow \left| \frac{v_{out}}{v_{in}} \right| = \frac{g_m}{2\pi f C_L} \qquad (25.52)$$

When determining the phase margin, we are interested in the frequency at which the magnitude of v_{out}/v_{in} is unity. Using Eq. (25.52) and labeling the frequency at which the magnitude of v_{out}/v_{in} is unity, f_u, we get

$$f_u = \frac{g_m}{2\pi C_L} \text{ or } \frac{K \cdot g_m}{2\pi C_L} \text{ if } K \neq 1 \text{ (see Eq. 25.43)} \qquad (25.53)$$

It would appear that the OTA (for feedback factors of unity or less) is always stable. Indeed, this is the case, provided the impedance of C_L is small compared to the output impedance of the OTA and the parasitic poles are at a much higher frequency than f_u.

The open-loop gain and phase responses of the OTA of Fig. 25.30 can be simulated using the test circuit shown in Fig. 25.33 (with $C_L = 0$). Figure 25.34 shows the simulation results with $V_{CM} = 2.5$ V. The unity gain frequency, f_u, is 60 MHz, while the phase at f_u is 150°, a 30° phase margin. An interesting observation can be made if V_{CM} is set to 0.5 V and then 4.5 V. These values of V_{CM} correspond to situations where the n- and p-channel diff-amps are not conducting. Under these conditions, the small-signal transconductance of the OTA is half of the value when both diff-amps are

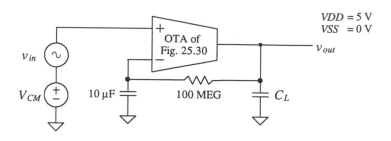

Figure 25.33 OTA circuit used to determine open-loop characteristics of OTA of Fig. 25.30.

on.   The result is a shift in the unity gain frequency of the OTA, while the low-frequency gain of the OTA is unchanged, as discussed in Ch. 24. The unity gain frequency, f_u, with V_{CM} set at 0.5 or 4.5, is approximately 30 MHz; see Fig. 25.35 for the case $V_{CM} = 0.5$ V.

Compensating the OTA is simply a matter of selecting a minimum load capacitance. Increasing the load capacitance actually increases the phase margin of the OTA. Looking at the phase characteristics of the OTA in Fig. 25.34 (without a load), we see that adding and increasing the load capacitance shifts the point where the phase is −45° to a lower frequency. (Increasing the load capacitance lowers the first pole in the transfer function.)   The point where the phase is −100° remains essentially

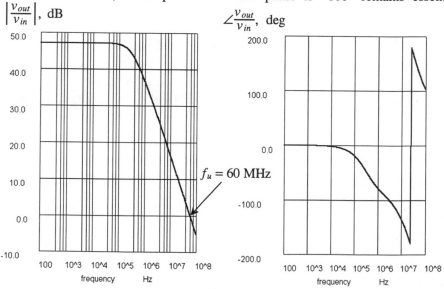

Figure 25.34 Open-loop simulation results of the OTA in Fig. 25.30 using the setup of Fig. 25.33 with $C_L = 0$ and $V_{CM} = 2.5$ V (both diff-amps on).

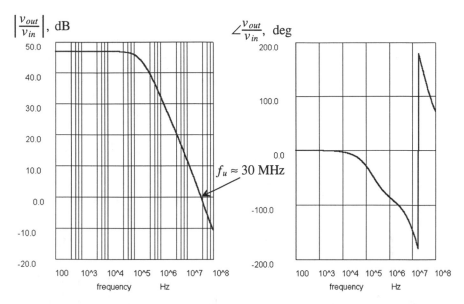

Figure 25.35 Open-loop simulation results of the OTA in Fig. 25.30 using the setup of Fig. 25.33 with $C_L = 0$ and $V_{CM} = 0.5$ V (p-channel diff-amp on).

unchanged since the other parasitic poles are not affected by the load capacitance. The frequency at a phase of $-100°$ is approximately 3 MHz. If we set $f_u = 3$ MHz, then the phase margin is $80°$ (assuming no interaction from the other parasitic poles present in the circuit). From the beginning of the section, the transconductance of the OTA is 40 $\mu A/V^2$. Using Eq. (25.53) results in a minimum-load capacitance of 2.1 pF. Simulation results using $C_L = 2.1$ pF are shown in Fig. 25.36. The simulation results show $f_u = 2.4$ MHz and a phase margin of $70°$. The difference between the hand calculations and the simulation results can be attributed to the parasitic poles present in the OTA. The parasitic poles present in any CMOS analog amplifier ultimately limit the complexity of the circuit. In practice, a minimum-load capacitance greater than 2.1 pF would be used. A possible practical load capacitance minimum of 5 pF would be appropriate in this design.

The OTA with Buffer

An op-amp can be made using an OTA with buffer circuit. The buffer is used to isolate the OTA from the load. The load can be resistive, capacitive, or a combination of the two. Compensating the OTA/buffer is somewhat different from compensating the OTA alone. Consider the OTA/buffer shown in Fig. 25.37. The buffer used in this figure is a source follower. To begin, let's try to compensate the op-amp by putting a load capacitance on the output of the source follower. This is similar to what we did for the OTA alone. Now, however, the output of the amplifier is a low impedance, that is, the output impedance of the source follower. Therefore, the load capacitance should have little effect on the phase margin of the amplifier. For the cases of very large load

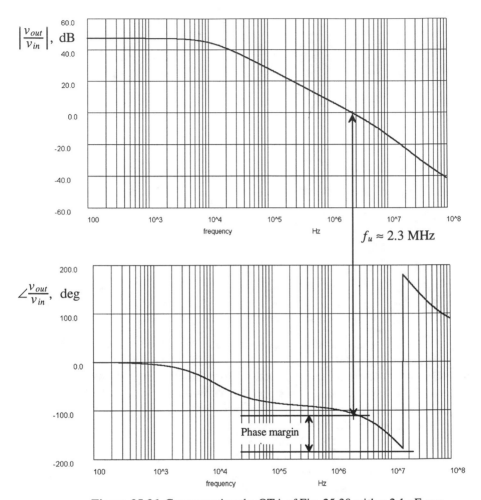

Figure 25.36 Compensating the OTA of Fig. 25.30 with a 2.1 pF cap.

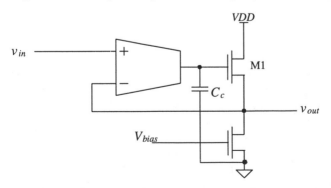

Figure 25.37 OTA with source-follower buffer (an op-amp).

capacitances or a very small source follower, the load capacitance may indeed affect the phase margin of the amplifier. Normally, however, the buffer is designed so that it can easily source and sink (by making the value of the current source load used in the source follower large) the current needed to drive the load capacitance.

Next, we could try to compensate the amplifier using the input capacitance of the source follower. One might try to use the gate-source capacitance of M1 as the compensation capacitance on the output of the OTA. Ideally, however, the output voltage of the OTA and the output voltage of the source follower are the same, that is, the source follower has a voltage gain of +1. This means that the voltage on each side of the gate-source capacitance changes at the same rate. In other words, the AC voltage between the gate and source of M1, v_{gs1}, is zero (ideally). This means that zero displacement current flows through C_{gs1}, and therefore it cannot be used for compensation. The gate-drain capacitance of M1 can be used for compensation since it is connected directly to AC ground (*VDD*). Using C_{gd1} would require, for a practical value of compensation capacitance, M1 to be too large for most situations.

This discussion shows that the op-amp of Fig. 25.37 should be compensated with a capacitance from the output of the OTA to ground. The value of the compensation capacitor C_c is selected following the procedures of the last section. Notice that even though there is an amplifier in the feedback loop, the maximum value of the feedback factor, β, is still 1. This is because the amplifier has a voltage gain of unity.

Next consider the OTA/buffer shown in Fig. 25.38. This configuration is useful in voltage regulator design since the output can swing all the way up to *VDD*. With the use of feedback resistors, the gain of the circuit can be set to a multiple of the input voltage. Also, in this configuration β can be larger than one since there is an amplifier, M1, in the feedback loop. Since the amplifier is inverting (M1), the signal fed back is connected to the "+" input of the OTA.

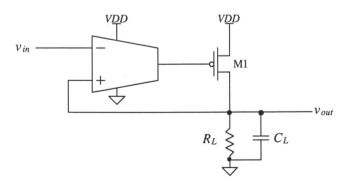

Figure 25.38 Op-amp made using the OTA/common-source amplifier.

This circuit is basically a two-stage op-amp and can be compensated in the same way as the op-amp of Fig. 25.11. A problem exists, however, if the load capacitance is too large. Consider the following example.

Example 25.3
The circuit shown in Fig. 25.39 is used as a voltage regulator that generates an output voltage of 2.5 V with a maximum output current of 10 mA (into R_L) over VDD changes of 4 to 6 V. Design the circuit (select R_z, R_{min}, and C_c) so that the phase margin is approximately 90° when $C_L = 0$. Estimate the maximum C_L allowed so that the phase margin of the circuit remains greater than 45°.

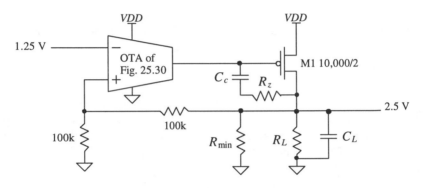

Figure 25.39 Voltage regulator of Ex. 25.3. Note to simulate a circuit with a large MOSFET set NRD and NRS to zero.

The solution to this example begins by following the compensation procedure given in Sec. 25.1. The value of R_1 (which is the output resistance of the OTA) is given by

$$R_1 = r_{o5} \| r_{o7} = 4.2 \text{ MEG for } \lambda = 0.06 \text{ V}^{-1} \text{ and } I = 2 \text{ μA}$$

The value of C_1, since M1 is large compared to output capacitance of the OTA, is given by

$$C_1 = C_{gs1} + C_{gd1}(1 + |A_{v2}|) =$$

$$\frac{2}{3} \cdot (10,000)(2)(800 \text{ aF}) + (10,000 \text{ μm})\left(5 \times 10^{-10} \frac{\text{F}}{\text{m}}\right)(1 + |A_{v2}|) =$$

$$C_1 = 10.67 \text{ pF} + 5 \text{ pF} \cdot (1 + |A_{v2}|)$$

The location of the pole on the output of the OTA, before adding the compensation capacitor, is given by

$$f_1 = \frac{1}{2\pi \cdot R_1 C_1} = \frac{1}{2\pi \cdot 4.2 \text{ MEG} \cdot [10.67 + 5(1 + |A_{v2}|)] \text{pF}}$$

The location of the pole on the output of M1, prior to adding the compensation network, is given by

$$f_2 = \frac{1}{2\pi \cdot R_2 C_2} = \frac{1}{2\pi \cdot (r_{o1} \| R_L \| R_{min}) \left[C_L + 5 \text{ pF}(1 + \frac{1}{|A_{v2}|}) \right]}$$

The reason for adding R_{min} to the circuit will now be explained. If the regulator output current starts to decrease toward zero (i.e., R_L and r_{o1} move toward ∞), then M1 will start to turn off. This in turn pushes f_2 down in frequency toward f_1 (f_1 decreases also up to a point since A_{v2} increases with decreasing output current, see Fig. 22.15). At very low output currents, f_2 can be less than f_1. The result is an unstable feedback loop. Adding R_{min} to the regulator ensures that M1 is always sourcing at least some minimum current. For the present design, with an output voltage of 2.5 V, R_{min} will be set to 10 kΩ.[2] This forces M1 to source a minimum of 250 μA at any given time. If the regulator output current, I_{D1}, increases, then f_2 increases for a constant C_L, increasing the phase margin of the circuit. For this reason, in the remainder of the discussion, we will assume $I_{D1} = 250 \ \mu$A (the worst-case situation) so that

$$r_{o1} = \frac{1}{0.8 \cdot 250 \ \mu A} = 5 \text{ k}\Omega \text{ and } R_L = \infty$$

where $\lambda = 0.8$ for M1 with a channel length of 2 μm. The second stage gain is then

$$|A_{v2}| = g_{m2} \cdot (r_{o1} \| 10 \text{ k}\Omega) = \sqrt{2 \cdot 17\frac{\mu A}{V^2} \cdot \frac{10,000}{2} \cdot 250\mu A} \cdot (5 \text{ k} \| 10\text{k})$$

$$= 21.7 \text{ V/V}$$

where g_{m2} is the transconductance of the second-stage (M1's transconductance). The location of f_1, using this gain, is 300 Hz, while f_2 is given by

$$f_2 = \frac{1}{2\pi \cdot (5 \text{ k} \| 10 \text{ k}) \left[C_L + 5 \text{ pF}(1 + \frac{1}{|21.7|}) \right]} = \frac{1}{2\pi \cdot (3.3 \text{ k}\Omega)(C_L + 5 \text{ pF})}$$

Adding the Compensation Capacitor

We cannot use Eq. (25.18) to determine the unity gain frequency of the loop since C_c will not be much larger than C_1. We can, however, write the open-loop gain (with the compensation capacitor) of the circuit (where g_{m1} is the transconductance of the OTA) as

$$\frac{V_{out}}{V_{in}} = \frac{\overbrace{g_{m1} \cdot R_1}^{\text{OTA gain}} \cdot \overbrace{21.7}^{\text{M1's gain}}}{\left(1 + j\frac{f}{f_{1C}}\right)\left(1 + j\frac{f}{f_{2C}}\right)}$$

---

[2] This value of resistor may be too small for most practical situations.

where f_{1C} and f_{2C} are the frequencies of the two poles with the compensation capacitor added. Since f_{1C} (the dominant pole) is much less than f_{2C}, we can rewrite this equation for frequencies less than f_{2C} as

$$\left| \frac{v_{out}}{v_{in}} \right| \approx \frac{g_{m1} R_1 \cdot 21.7}{\sqrt{1 + \left(\frac{f}{f_{1C}} \right)^2}} = \frac{3,650}{\sqrt{1 + \left(\frac{f}{f_{1C}} \right)^2}}$$

By definition, when $f = f_u$, $|v_{out}/v_{in}| = 1$. Since f_u is much larger than f_{1C}, this equation can be rewritten as

$$f_u = f_{1C} \cdot 3,650 = f_{1C} \cdot A_{v1} \cdot A_{v2}$$

The location of the poles with compensation capacitor are given by

$$f_{1C} = \frac{1}{2\pi \cdot 4.2MEG \cdot [10.7 + (5 + C_c)(1 + 21.7)]} = \frac{1}{2\pi \cdot 4.2MEG \cdot [125 \text{ pF} + C_c(1 + 21.7)]}$$

and

$$f_{2C} = \frac{1}{2\pi \cdot 3.3 \text{ k}\Omega \cdot (C_L + C_c + 5 \text{ pF})} \qquad (25.54)$$

From the design requirements given earlier, the phase margin should be approximately 90° when $C_L = 0$. This can be met if f_{2C} (with $C_L = 0$) is ten times f_u, that is, $f_{2C} > 10 \cdot f_u$. If we set $f_u = 100$ kHz then $C_c = 55$ pF and $f_{2C} = 800$ kHz. The value of f_{2C} is only eight times larger than f_u. However, this value is very dependent on the value of λ when I_{D1} is small. Since our selection of λ, for M1 (0.8), may not be exact, we will simulate the circuit and verify that the design meets the given criteria.

The value of R_z can be picked based on the maximum g_m (g_{m2}) of M1. This occurs when $I_{D1} = 10$ mA or $g_{m2} = 41.23$ mA/V. If we set $R_z = 1/g_{m2}$ then $R_z = 25$ Ω. In order to increase the phase margin of the regulator, we will set $R_z = 100$ Ω in this design. Note that R_z has little effect on the phase margin when the regulator is sourcing small amounts of current. Open-loop simulation results with $C_c = 55$ pF, $R_{min} = 10$ kΩ, $R_L = \infty$, and $C_L = 0$ are shown in Fig. 25.40. The common mode input voltage was set to 2.5 V in this open-loop simulation, and DC stability circuitry was used (see Fig. 25.33).

The unity gain frequency shown in this simulation matches the hand calculations. However, the second pole is about five times larger than the 800 kHz we calculated (this is good). Also, the right-hand plane zero appears to be located very close to the second pole. This causes the phase to decrease twice as fast. The phase margin under these circumstances meets the design criteria (i.e. 90°), when $C_L = 0$. A phase margin of 45° occurs when $f_{1C} = f_u$. Using Eq. (25.54) with $f_{1C} = 100$ kHz results in $C_L = 420$ pF. Simulation results showed that a capacitor five times this value, or 2100 pF, caused the phase margin to become 45° (with $I_{D1} = 250$ μA). Increasing I_{D1} increases the phase margin. ∎

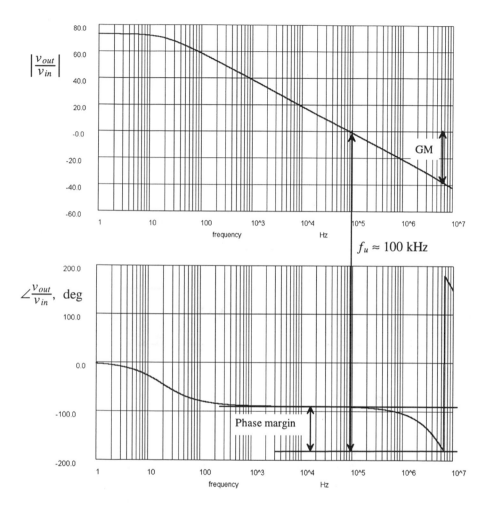

Figure 25.40 Open-loop frequency response for Ex. 25.3 (see text).

A final example of a wide-swing OTA with buffer is shown in Fig. 25.41. This is the OTA of Fig. 25.30, with four MOSFETs added to form an output stage. MOSFETs M13 and M14 can be thought of as "zero canceling resistors" used to shift the right-hand plane zero into the left-hand plane. M13 and M14 also help to reduce the current flowing in M15 and M16. We know that the current through M13 and M14 is well defined. The current through M15 and M16 is less well defined. Increasing the lengths of M13 and M14 helps to reduce the quiescent current in M15 and M16 and push the zero further into the left-half plane.

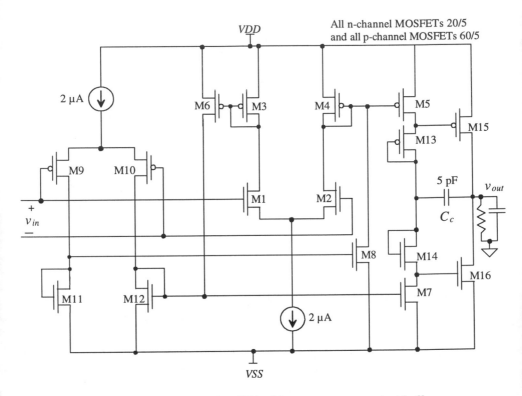

Figure 25.41 Wide-swing OTA with common-source output buffer.

The unity gain frequency of an OTA, f_u, is a strong function of the load capacitance. For the op-amp of Fig. 25.41, the unity gain frequency and the DC gain of the op-amp are a strong function of the load resistance and capacitance. This is due to the fact that the output stage has a gain greater than one. The resistor on the output of the op-amp affects the second-stage gain. The compensation capacitor is selected in the same manner as the OTA without output buffer, assuming the second stage has a gain of one. If the second-stage has a gain greater than one, then the Miller effect multiplies C_c by the second stage gain. This effectively increases the capacitance seen by the first stage, keeping the op-amp compensated. Again, the op-amp can become unstable, as we saw in Ex. 25.3, when the load capacitance becomes too large. Note that because the output buffer is inverting, the "+" and "−" inputs of the op-amp are switched.

25.2.2 The Folded-Cascode OTA

The folded-cascode OTA is shown in Fig. 25.42 [6-8]. The name "folded-cascode" comes from folding down p-channel cascode active loads of a diff-pair and changing the MOSFETs to n-channels. This OTA, like all OTAs, has good PSRR compared to the two-stage op-amp since the OTA is compensated with the load capacitance.

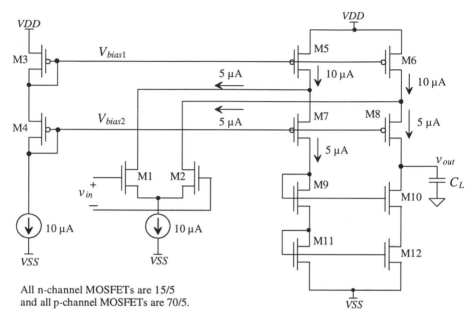

Figure 25.42 Folded-cascode OTA.

To understand the operation of the folded-cascode OTA, consider Fig. 25.42 without the diff-amp (M1/M2) in the circuit. Without the diff-amp present in the circuit, 10 µA flows in all MOSFETs. MOSFETs M3 and M4 provide the DC bias voltages to M5-M8. Note that the cascoded MOSFETs (M5-M12) are not biased for wide-swing operation. A wide-swing biasing circuit could replace M3/M4, and M9-M12 could be replaced with a wide-swing current mirror. Biasing for wide-swing operation increases the output voltage swing. When the diff-amp is added back into the circuit, it steals 5 µA from M7-M12, reducing their drain currents to 5 µA.

Applying an AC input voltage, v_{in}, causes the diff-amp differential drain current to become $g_m v_{in}$ (g_m is the transconductance of the diff-amp). This AC differential drain current is mirrored in the cascoded MOSFETs M7 through M12. The output voltage of the OTA is then

$$v_{out} = g_m v_{in} \cdot R_o \qquad (25.55)$$

where

$$R_o = (\text{R looking into drain of M10}) \| (\text{R looking into drain of M8})$$

$$= [r_{o10}(1 + g_{m10} r_{o12})] \| [r_{o8}(1 + g_{m8} r_{o6})] \qquad (25.56)$$

The gain of the folded-cascode OTA is given by

$$\frac{v_{out}}{v_{in}} = g_m R_o \tag{25.57}$$

Using the values shown in Fig. 25.42, $g_m = 38.7$ μA/V, $r_{o6} = 1.67$ MΩ, $r_{o8} = r_{o10} = r_{o12} = 3.33$ MΩ, $g_{m8} = 50$ μA/V, $g_{m10} = 38.7$ μA/V, we find that the value of R_o is 170 MΩ and $|v_{out}/v_{in}|$ is 6,500 V/V.

The dominant pole of the OTA is located at $1/2\pi R_o C_L$. Parasitic poles exist at the sources of M7/M8 and M9/M10. These parasitic poles should be larger than the unity gain frequency ($f_u = g_m/2\pi C_L$) of the OTA. For the op-amp of Fig. 25.42, setting the unity gain frequency to 1 MHz results in $C_L = 6.1$ pF.

An op-amp can be made using the folded-cascode OTA and source-follower output buffer (Fig. 25.43). The value of the current source, I_{SF}, used in the source-follower depends on the amount of current sinking capability required of the op-amp. For example, if the minimum-load resistance the op-amp will see is 10 kΩ and $VSS = -2.5$ V, then the current source should be at least 2.5/10 kΩ or 250 μA. In other words, the op-amp can sink 250 μA of current. When the op-amp is sourcing current, M15 must supply a current to the current source, I_{SF}, and a current to the load. The source-follower will not degrade the frequency performance of the OTA. However, the maximum output voltage will be limited to less than $VDD - V_{THN15}$.

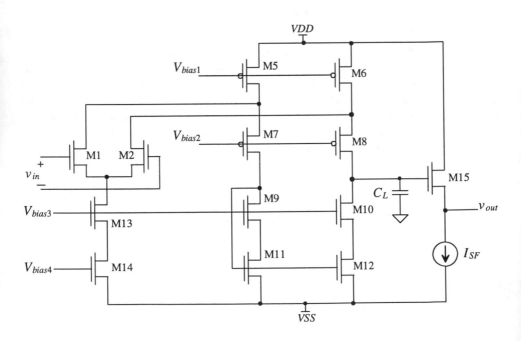

Figure 25.43 Folded-cascode op-amp.

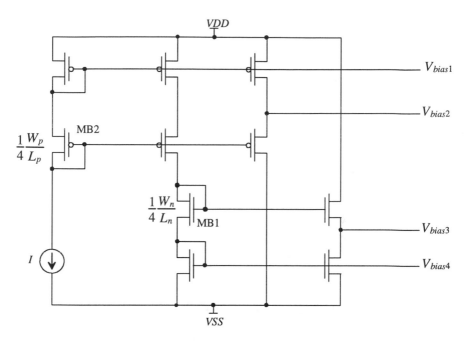

Figure 25.44 Biasing circuit for the op-amp of Fig. 25.43.

The bias voltages can be generated using the circuit of Fig. 25.44. This circuit is basically two wide-swing current mirrors based on the configuration given in Fig. 20.6. MB1 and MB2 are sized one-fourth the size of the other MOSFETs, as was discussed in Ch. 20. If, in Figs. 25.43 and 25.44, $I = 10$ μA and the unmarked n-channels are 15/5 and the unmarked p-channels are 70/5, then MB1 can be sized 30/40 and MB2 can be sized 35/10. This biasing scheme causes 10 μA to flow in MOSFETs M13, M14, M5, M6, and all MOSFETs in the biasing circuit. MOSFETs M1, M2, and M7-M12 have 5 μA of drain current. Note that this biasing scheme gives the maximum voltage swing on the output of the OTA (i.e., the drains of M8 and M10).

Wide-Swing Folded-Cascode OTA

A major advantage of the folded-cascode OTA of Fig. 25.42 is that the positive CMR extends beyond *VDD*, while the negative CMR is limited by the minimum voltage across the input diff-amp. Figure 25.45 shows a wide-swing folded-cascode OTA where the input CMR extends beyond the power supply rails (however, the output does not). The input diff-amp is the wide-swing type discussed in Ch. 24 (Fig. 24.27). If we use our familiar sizes of 15/5 and 70/5 for the n- and p-channel MOSFETs together with the bias circuit of Fig. 25.44 (with 10 μA biasing current), then a current of 10 μA flows in M5, M6, M11-M14, M17, and M18 (under DC conditions and assuming both diff-amps are on). The remaining MOSFETs in the circuit conduct 5 μA. SPICE simulation

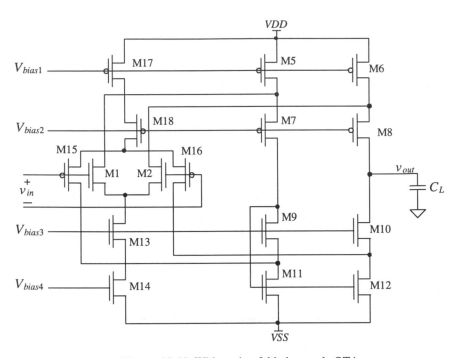

Figure 25.45 Wide-swing folded-cascode OTA.

results using this OTA in a unity gain configuration with $VDD = 5$ and $VSS = 0$ are shown in Fig. 25.46. The input voltage in this figure was swept from -1 to 6 V. Note how the output stage of the OTA limits the maximum and minimum output voltages.

Compensation of the wide-swing folded-cascode OTA follows the procedure used to compensate the wide-swing OTA of Fig. 25.30. The problem of having one or both diff-amps on exists with this configuration as well. An additional problem exists with this configuration, that is, the low-frequency gain is not constant. Nevertheless, this configuration finds extensive use in analog applications. With both diff-amps conducting and a load capacitance of 1 pF, there is a 10 MHz unity gain frequency, with a 70° phase margin and an open-loop gain of 6500 (76 dB). The bias voltages or the size of the MOSFETs M5-M12 should be adjusted to compensate for the larger DC currents when one diff-amp is off. In other words, it is desirable to have wide-swing operation where the drain-source voltages of M5/M6 or M11/M12 are small (equal to the excess gate voltage). But if these voltages are too small, then M5/M6 or M11/M12 will approach the triode region and the gain will decrease.

The constant g_m stage discussed in Ch. 24 can be used as the input diff-amp stage for the wide-swing folded cascode. This solves the problem of varying transconductance for compensating the OTA, but the problem of nonconstant low-frequency gain remains.

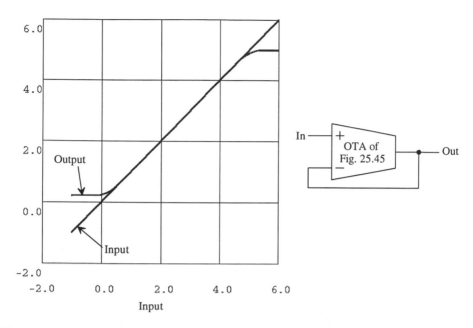

Figure 25.46 DC characteristics of wide-swing OTA of Fig. 25.45 in the unity gain configuration.

Folded-Cascode OTA with Output Buffer

In many applications, the OTA must drive a resistive/capacitive load or a large capacitive load. In these situations, a buffer between the load and OTA is used to ensure that the gain of the OTA is not degraded. Figure 25.47 shows one example of an output buffer that employs source followers. The output MOSFETs M03 and M04 are source followers used to drive the load. MOSFET M01 is simply a current source, while M02 is a p-channel source follower. The source-gate voltage of M02 is selected to bias both M03 and M04 on, that is, $V_{SG2} > V_{THN} + V_{THP}$. The gain through the buffer is approximately unity so that the compensation capacitance C_c does not have to be changed when the buffer is added to the OTA. The main drawback of this configuration is the limited output swing. The maximum output voltage is $VDD - V_{THN}$ (with body effect), while the minimum output voltage is $VSS + V_{THP}$.

A second example of a folded-cascode output buffer is seen in Fig. 25.48. This output buffer configuration is essentially the same as that shown in Fig. 25.41. The gate-drain connected MOSFETs M07 and M08 can be thought of as zero-canceling resistors (i.e., $R_z = 1/g_{m07} \| 1/g_{m08}$). Compensation follows the same procedure used for the two-stage op-amp discussed in the beginning of the chapter. Normally, since the gain of the output buffer can be low, the output MOSFETs, M05 and M06, can be sized with minimum or near-minimum lengths. This has the effect of increasing the output drive capability. This buffer has the added advantage that the outputs can approach the supply voltages.

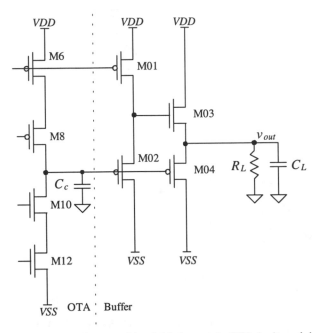

Figure 25.47 Output buffer for use with a folded-cascode OTA (unity gain).

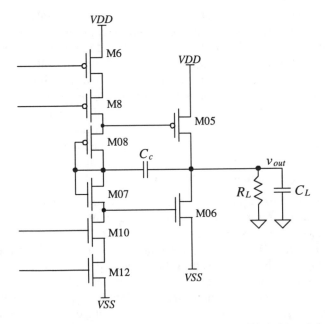

Figure 25.48 Output buffer for use with a folded-cascode OTA (with gain).

Figure 25.49 shows a wide-swing folded-cascode OTA, with output buffer based on floating current sources [7]. This configuration is very useful in low-supply voltage op-amp design. To illustrate the operation and biasing of this op-amp, let's assume that the biasing circuit of Fig. 25.44 is used with $I = 10$ µA, 15/5 n-channels, 70/5 p-channels. MB2 is sized 15/20 and MB1 is 35/10. The generation of the bias voltages V_{biasp} and V_{biasn} will be discussed momentarily. With these sizes, and assuming both diff-amps are conducting current, 10 µA flows in M17/M18, M13/M14, M5/M6, and M11/12. A current of 5 µA flows in M1/M2, M15/M16, M7/M8, and M9/M10. A current of 100 µA flows in MO1/MO2, and a current of 2.5 µA flows in MC1-MC4. The current that flows in M5/M6 and M11/M12 is mirrored in MO1 and MO2.

The floating current sources MC1/MC2 and MC3/MC4 form a positive feedback loop with a gain of 1. The resistance looking into the sources of these parallel MOSFETs is the output resistance of the MOSFETs in series with the output resistance of the cascode MOSFETs of the folded-cascode OTA.

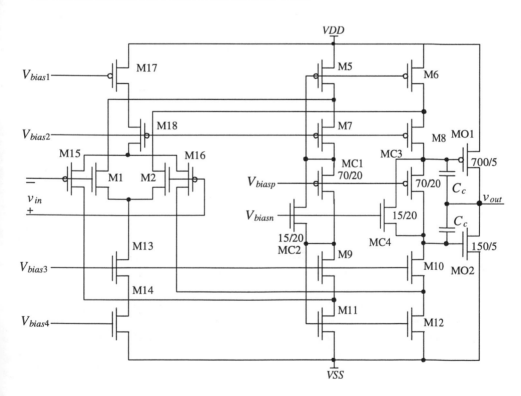

All unlabeled n-channels are 15/5.
All unlabeled p-channels are 70/5.

Figure 25.49 Wide-swing folded-cascode OTA with output buffer (an op-amp).

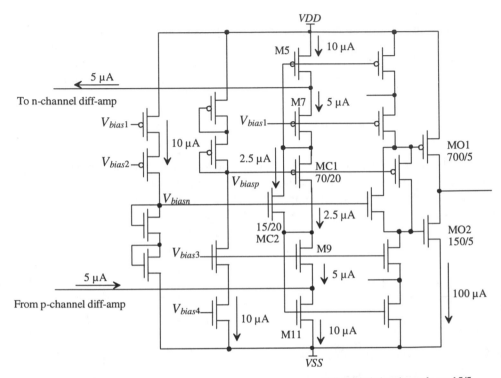

All unlabeled n-channels are 15/5.
All unlabeled p-channels are 70/5.

Figure 25.50 Simplified schematic of biasing circuitry for the op-amp of Fig. 25.49.

The schematic of the biasing circuit is shown in Fig. 25.50. The reason for sizing MC1-MC4 one-fourth the other MOSFETs now becomes obvious: they conduct 2.5 μA of current. An interesting situation occurs when one of the diff-amps on the input of the op-amp shuts off. If the p-channel diff-amp were to turn off, as a result of the input signal going outside the p-diff-pairs CMR, then the 5 μA of current flowing into the drain of M11 from the p-channel diff-amp would become zero. This would cause the current flowing in M01 and M02 to become unbalanced (i.e., 100 μA of current flowing in MO1 and 50 μA of current flowing in MO2). However, without a DC load current, we know that these currents must be equal. The result is an offset voltage on the input of the op-amp that will result in balanced currents in MO1 and MO2. Because of the very large gain of this topology, this offset voltage is usually negligible in comparison to the offsets resulting from device mismatches. Device matching is critical for low-distortion amplifier design, and in many cases the device matching will restrict the use of more elegant circuit structures (e.g., the constant transconductance diff-amp).

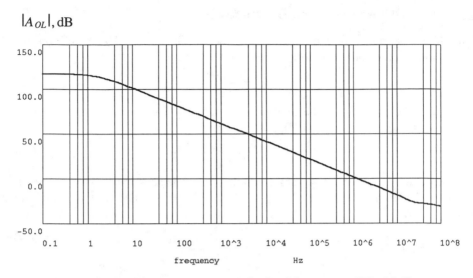

Figure 25.51 Open-loop-compensated gain of the op-amp of Fig. 25.49.

The open-loop compensated gain of the op-amp of Fig. 25.49 is shown in Fig. 25.51. The op-amp was compensated with two 5 pF capacitors. The open-loop DC gain of this configuration is 700,000 V/V. Simulating the characteristics of an op-amp with such a large gain requires that the RELTOL, VNTOL, and ABSTOL remain at their default values or less. Also, the DC stability scheme of Fig. 25.7 may not work well for high-gain op-amps because of the presence of feedback in the circuit. The scheme shown in Fig. 25.52 can be used to simulate the open-loop gain of an op-amp. The value of V_{bias} is changed until all MOSFETs in the op-amp are biased into the saturation region. For the simulation of the op-amp of Fig. 25.49, a 1 V (AC)/2.5 V (DC) voltage source was connected to the positive, or noninverting, input of the op-amp. After several simulation runs, it was found that a 2.49995 V bias on the minus, or inverting, input of the op-amp caused all MOSFETs to operate in the saturation region.

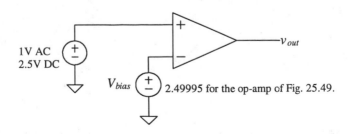

Figure 25.52 Simulating the open-loop response of a high-gain op-amp.

25.3 The Differential Output Op-Amp

The differential output op-amp symbol is shown in Fig. 25.53. The open-loop gain of the op-amp is given in terms of the inputs and outputs by

$$A_{OL} = \frac{v_{o+} - v_{o-}}{v_+ - v_-} \qquad (25.58)$$

This should be compared to the single-ended output op-amp (the op-amp we have been discussing up to this point), which has a gain given by

$$A_{OL} = \frac{v_{o+}}{v_+ - v_-} \qquad (25.59)$$

If we ignore v_{o-}, then the differential output op-amp behaves just like a single-ended op-amp. For linear applications the op-amp is used with a feedback network, the inputs are related by

$$v_+ \approx v_- \qquad (25.60)$$

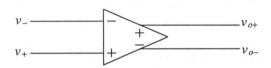

Figure 25.53 Differential output op-amp.

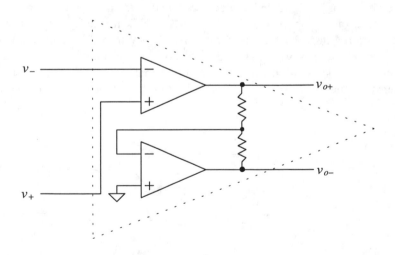

Figure 25.54 Formation of a differential output op-amp using two single-ended op-amps.

while the outputs are related by

$$v_{o+} \approx -v_{o-} \tag{25.61}$$

We can use two single-ended op-amps to form a differential output op-amp (Fig. 25.54). This implementation works well for low frequencies. However, because the op-amps are in different topologies, the phase response of each op-amp is different. This means Eq. (25.61) will not hold for higher (> 1 kHz) frequencies because the time delays through each configuration are sufficiently different.

The importance of using differential output op-amps can be explained with the help of Fig. 25.55. This figure shows a cascade of differential output op-amps, without feedback network. The stray capacitance between the interconnecting metal lines (the metal lines that carry the signals between op-amps) and the substrate or any other noise source are shown. If the metal lines are run close to one another, then the noise voltage will couple even amounts (ideally) of noise into each signal wire. Since the diff-amp, on the input of the op-amp, rejects common signals (signals that are present on both inputs), the coupled noise is not passed to the next op-amp in the string. Variations on the power supply rails are rejected as well. If the differential op-amp is symmetric, then changes on the power supply couple evenly into both outputs, having little effect on the difference, or desired signal, coming out of the op-amp. For these reasons, that is, good coupled noise rejection and PSRR, the differential op-amp is a necessity in any mixed-signal integrated circuit. (Mixed-signal means that both analog and digital circuits are present on the chip - digital logic is great at generating noise.)

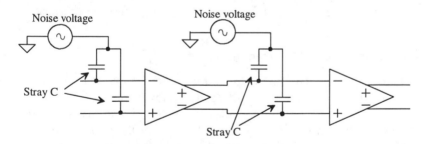

Figure 25.55 Differential output op-amps showing parasitic capacitance and noise.

Normally, the average of the op-amp outputs is called the common-mode output voltage and in terms of the supply voltages, is given by

$$V_{CM} = \frac{VDD + VSS}{2} \tag{25.62}$$

For example, if $VDD = 2.5$ V and $VSS = -2.5$ V, then both outputs are referenced to the common-mode voltage 0 V (ground). If $VDD = 5$ V and $VSS = 0$, then $V_{CM} = 2.5$ V. Figure 25.56a shows a simple differential output op-amp gain configuration (inverting or noninverting, depending on which output is used as the positive output). Since the input voltages to the circuit are equal, the output voltages of the op-amp should remain at V_{CM}. However, if $v_{o+} = v_{o-} = VDD$ or $v_{o+} = v_{o-} = VSS$ or $v_{o+} = v_{o-} =$ anything, then $v_+ = v_-$ (this is a problem). Equation (25.61) can be written to include the common-mode output voltage:

$$v_{o+} + V_{CM} \approx -v_{o-} + V_{CM} \qquad (25.63)$$

To hold the common-mode output voltage of the op-amp at a known voltage, that is, V_{CM}, the configuration of Fig. 25.56b is used. A circuit called "common-mode feedback" (CMFB) is used to sense the average value of the op-amp outputs (that is, V_{CM}). The output of the CMFB circuit is fed back into the op-amp to adjust V_{CM} to the correct value, that is, the value given by Eq. (25.62). The op-amp of Fig. 25.54 has its V_{CM} set by the connection of the "+" input of the bottom op-amp. In this figure, $V_{CM} = 0$ V, and so it can be concluded that this configuration works with $VDD = - VSS$.

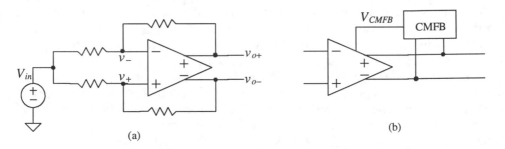

(a)

(b)

Figure 25.56 (a) Simple gain configuration using differential output op-amp and (b) the use of a common mode feedback circuit to adjust the common-mode output voltage.

25.3.1 Fully Differential Folded-Cascode OTA

The fully differential (both inputs and outputs) folded-cascode OTA is shown in Fig. 25.57. The main difference between this OTA and the single-ended OTA of Fig. 25.43 is that the gate of M11 is now tied to V_{bias4} where, as before, it was tied to the drain of M9. The drains of M9 and M7 are now the "−" output of the OTA. Although this change is subtle, it drastically affects the DC biasing of the OTA.

The currents through M7-M12 are now 10 µA. The current through M5 and M6 is 15 µA. The common-mode feedback circuit should adjust the gate potential of M5 and M6, so that they provide the 15 µA of current to M1/M2 and M7-M12. For reasons that will be clear when we discuss the CMFB circuit, the sizes of M5 and M6 are increased to 1.5 times the size of the other p-channel MOSFETs used in the circuit. The biasing circuit of Fig. 25.44 can be used if V_{bias2} is tied to the gate of MB2, and MB2 is resized to the W/L of the other p-channel MOSFETs in the circuit. The change lowers the value of V_{bias2} and the maximum output swing of the OTA. Lowering V_{bias2} allows the voltage on the drains of M5 and M6 to vary without letting them go into the triode region. This is important for proper operation of the CMFB circuit.

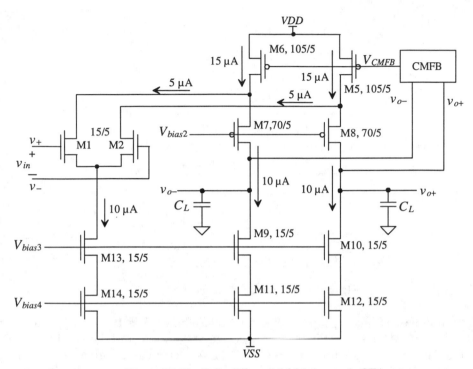

Figure 25.57 Fully differential folded-cascode OTA.

Compensation of the OTA is accomplished in the same manner as the single-ended OTA except now two load capacitances are required, one for each output.

The CMFB Circuit

A common-mode feedback circuit is shown in Fig. 25.58. The inputs to the circuit are v_{o+} and v_{o-} of the OTA, while the output is V_{CMFB}. The unique aspect of this circuit is its rejection of a difference-mode signal on its inputs and amplification of the common-mode signal. This is exactly the opposite of what a single diff-amp does. To be exact, the CMFB circuit amplifies the difference between the average of the outputs, $(v_{o+} + v_{o-})/2$ and V_{CM}. Through the use of feedback, through the OTA, the averages of the outputs and V_{CM} are made equal. This is similar to how feedback is used to set the inputs of an op-amp, v_+ and v_-, equal. We will cover the feedback mechanism further during the discussion of compensating the CMFB circuit. The MOSFETs MF1/MF2 and MF3/MF4 are simply current sources (10 µA for the sizes and currents shown in Fig. 25.57). In the following discussion, we will assume $VSS = 0$, $VDD = +5$ V, and $V_{CM} = 2.5$ V.

To understand the operation of the circuit, let's first consider the DC case when both v_{o+} and v_{o-} are 2.5 V. (This is what we ultimately want, that is, both outputs' DC component equal to the common-mode voltage.) The gates of MF5-MF8 are at 2.5 V,

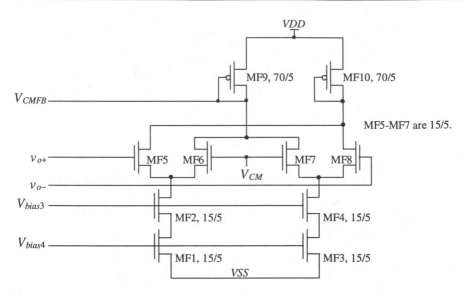

Figure 25.58 Common-mode feedback circuit.

and their drain currents are 5 μA. The current through MF9 is 10 μA. This current sets the V_{SG} of MF9. Since MOSFETs M5 and M6 of the OTA are 1.5 times the size of MF9, a current of 15 μA flows in these MOSFETs. This is the current required by M1/M2 and M7-M12, and so the CMFB circuit output doesn't change. Mismatches in the MOSFETs used in either the CMFB circuit or the OTA will result in an offset voltage.

If v_{o+} and v_{o-} go above V_{CM}, then the drain currents of MF6 and MF7 start to decrease. This causes V_{CMFB} to increase toward *VDD*. The increase in V_{CMFB} causes the drain currents of M5 and M6 to decrease. Since the current through M9-M12 is constant, the result is a decrease in the average output voltages. A similar argument can be made for the case when v_{o+} and v_{o-} are below V_{CM}; that is, the average output voltage increases because the current flowing in MF9 increases. Next consider the correct possible outputs v_{o+} = 3.5 V and v_{o-} = 1.5 V. These voltages lie outside the CMR range of the CMFB diff-amps. An incremental change in the inputs to the OTA can cause the outputs to become unbalanced. For example, if v_{o+} = 3.0 V and v_{o-} = 1.5 V then V_{CMFB} doesn't change (assuming the diff-pair CMR is a few-hundred mV). However, through the use of feedback around the OTA, either resistive (for an OTA with buffer) or capacitive feedback, the outputs will change until the OTA inputs are equal (i.e., v_{+} = v_{-}), which can occur even if the outputs are unbalanced (i.e., $v_{o+} - v_{o-}$ is correct, but v_{o+} does not equal $-v_{o-}$).

One solution to increasing the CMFB circuit's CMR is to increase the lengths of MF5 through MF8. This decreases the transconductance of the diff-amp and thus lowers the gain of the CMFB circuit. Low CMFB circuit gain makes balancing the outputs more difficult. Also the circuit's 3-dB frequency decreases, which can affect the compensation of the CMFB/OTA combination.

Figure 25.59 shows an alternative CMFB circuit [8], which operates over wide output voltage swings. In some cases, the buffers can be implemented using source-followers. The average of v_{o+} and v_{o-} is generated with the help of the resistor/capacitor combination and applied to the gate of M1. This scheme ensures fully balanced outputs over a voltage range limited by the CMR of the buffers.

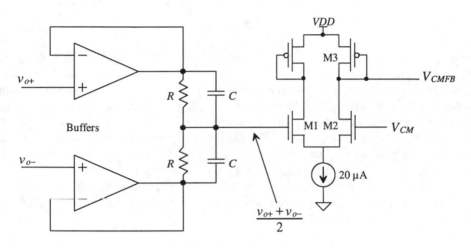

Figure 25.59 Improved range CMFB circuit.

A simple method of changing a single-ended input signal referenced to ground into a signal usable in a differential system is shown in Fig. 25.60. The p-channel MOSFETs perform two functions: (1) DC level shifting (adding V_{SG} of the p-channel

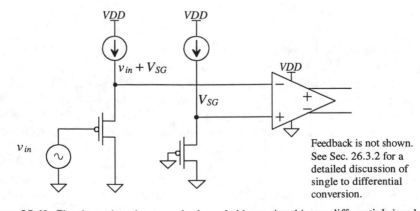

Figure 25.60 Circuit used to change a single-ended input signal into a differential signal.

MOSFETs to the input signal) and (2) buffering the input signal. Note that the value of V_{SG} is not critical - it simply moves the input signal into the CMR of the op-amp. The output DC level of the op-amp is determined by the CMFB circuit. (V_{SG} should be close to V_{CM} if resistive feedback is used in order to reduce the amplifier input DC current.)

Compensating the CMFB Circuit

The CMFB circuit combined with the folded-cascode OTA forms a feedback loop. As with any feedback loop, we must ensure stability. If we apply a common AC signal (let's call it v_{oc}) to both inputs of the CMFB amplifier of Fig. 25.58, the resulting gain is

$$\frac{v_{CMFB}}{v_{oc}} = \frac{g_{m5}}{g_{m9}} \approx 1 \tag{25.64}$$

where g_{m5} is the small-signal transconductance of MF5, MF6, MF7, or MF8 and g_{m9} is the small-signal transconductance of MF9. The bandwidth of this configuration is very wide. This is important since we do not want any phase shift through the CMFB circuit to degrade the phase-margin of the feedback loop. To actually understand the feedback mechanism present, consider the simplified drawing of the OTA/CMFB circuit shown in Fig. 25.61. One side of the folded-cascode amplifier is shown, and the CMFB circuit is drawn as a gain block. The gain from the gate of M5 to the output (the drain of M8) is $g_{m5} \cdot R_o$, where R_o is output resistance of the cascode configuration given by Eq. (25.56). If the bandwidth of the CMFB amp is wide compared to the OTA and the gain is approximately one, then the same load capacitance used to compensate the OTA can be used to compensate the CMFB amp.

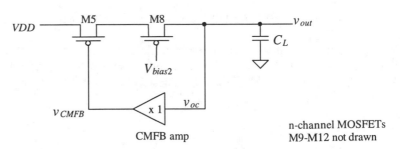

Figure 25.61 Feedback circuit resulting from addition of the CMFB circuit.

Simulation results for the fully differential folded-cascode OTA of Fig. 25.57, using a C_L of 6.1 pF ($f_u = 1$ MHz) on each output, are shown in Fig. 25.62. For these simulations, $VDD = 5$ V, $VSS = 0$, and $v_+ = 2.5$ V (DC) + AC, while $v_- = 2.5$ V + AC@180° phase. Interdigitated layout was employed in the folded-cascode section to increase matching. The folded-cascode OTA with common-source output buffers is shown, drawn slightly different than Fig. 25.57, in Fig. 25.63. The CMFB circuit inputs can now be connected to the outputs of the source followers. Layout of the fully-differential folded-cascode OTA is shown in Fig. 25.64. The bias voltages are supplied on poly1.

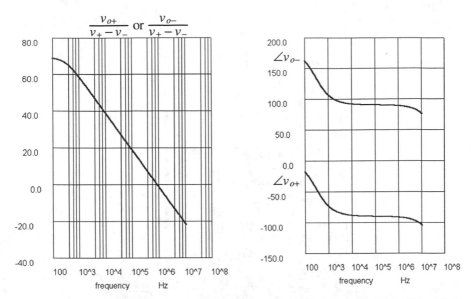

Figure 25.62 Magnitude and phase responses of the folded-cascode OTA of Fig. 25.57.

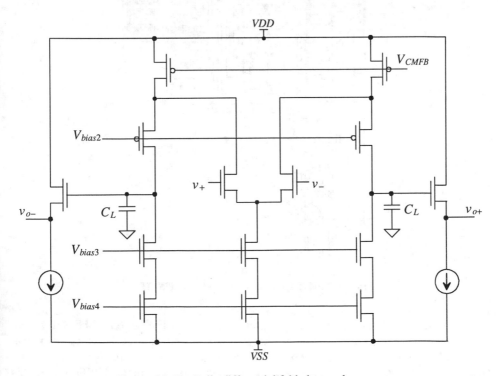

Figure 25.63 Fully differential folded-cascode op-amp.

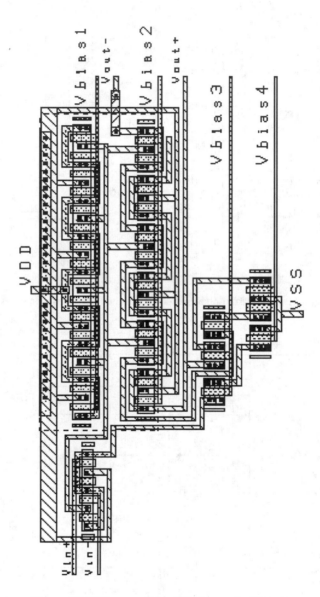

Figure 25.64 Layout of the folded-cascode OTA.

A buffered differential op-amp topology is seen in Fig. 25.65 without the input diff-amp drawn. This topology uses the floating current sources discussed in Sec. 25.2.2. The middle string of MOSFETs shown in this figure are used for biasing. However, if the diff-amps source or sink current into the outer MOSFETs, in this section of the op-amp, the biasing can be affected to the point where the op-amp doesn't

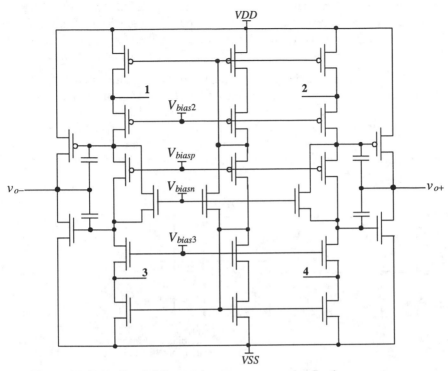

Figure 25.65 Buffered differential output op-amp using floating current sources. Note that the output buffers will not operate class AB when one of the diff-amps is off.

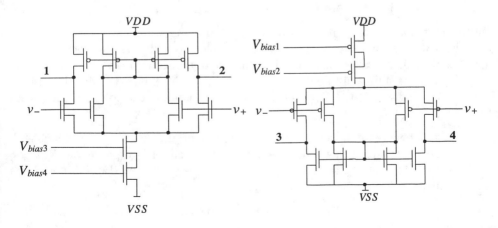

Figure 25.66 Input diff-amps for the op-amp of Fig. 25.65.

function. For this reason, the diff-amps shown in Fig. 25.66 are used. The node numbers in this figure correspond to the labeled nodes in Fig. 25.65. These diff-amps do not source or sink a current; for DC biasing purposes, therefore the diff-amps can be connected directly to the folded-cascode section of Fig. 25.65 without any interaction.

The final component required in the design of this op-amp is the CMFB circuit Fig. 25.67. Since the outputs of the op-amp are buffered, we can connect the outputs directly to the averaging resistors. When the outputs are balanced, the CMFB circuit does not source or sink a current and thus does not affect the folded-cascode section. However, if the average of the outputs is above V_{CM}, then the outputs source a current, shifting the average output voltage lower until the outputs are balanced (i.e., $v_{o+} = -v_{o-}$).

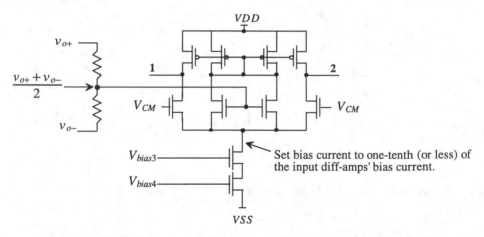

Figure 25.67 Common-mode feedback circuit for the op-amp of Fig. 25.65.

25.3.2 Gain Enhancement

Up to this point, the only method we have had to increase the gain of an op-amp is decreasing the biasing currents. The problem with lowering the bias currents is that our unity gain frequency drops (more on this at the end of the section). In addition, the gain of an op-amp is ultimately limited to the gain of the MOSFET in the subthreshold region (see Fig. 22.15). Gain-enhancement techniques help to overcome these limitations.

Gain-enhancement can be demonstrated with the help of Fig. 25.68 [9]. Without the OTA present, the gain of this cascode amplifier is $g_{m1} \cdot R_o$ where R_o is the resistance looking into the drain of M2, that is, $r_{o2}(1 + g_{m2} \cdot r_{o1})$; see Eq. (20.12). With the OTA in place, the output resistance becomes

$$R_{oGE} = r_{o2}[1 + (1 + A_{OTA})g_{m2}r_{o1}] \tag{25.65}$$

where A_{OTA} is the voltage gain of the OTA. The gain of the cascode configuration with gain enhancement is

$$A_{vGE} \approx A_{OTA} \cdot (g_{m1} \cdot R_o) \qquad (25.66)$$

Adding the OTA to the circuit increased the gain of the configuration by A_{OTA}. The OTA attempts to hold the drain of M1 at V_{bias}. We can apply this technique to increase the gain of the fully differential folded-cascode OTA of the last section but not to a single-ended output op-amp. Because of the lack of CMFB in single-ended output op-amps, the DC biasing becomes too difficult to implement in most cases.

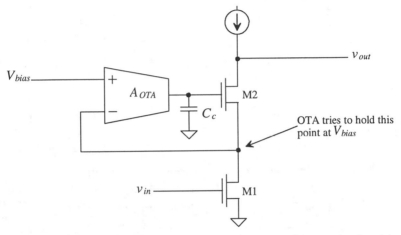

Figure 25.68 Use of gain enhancement to increase cascode open-circuit gain.

Figure 25.69 shows the op-amp of Fig. 25.57 (or Fig. 25.63 without output buffers) implemented using gain-enhancement techniques. To keep things simple, we have used the same DC biasing voltages, those generated with the circuit of Fig. 25.44. In practice, the biasing voltages would be changed since now the sources of M7/M8 are held at V_{bias2}. The wide-swing OTA of Fig. 25.30 was used in this amplifier simply because it has already been characterized in the chapter. The gain of the OTA is 240 V/V (47 dB). The cascode amplifier without gain enhancement had an open-loop gain, from Fig. 25.62, of 68 dB (single-ended output) and 74 dB (double-ended output). We can estimate the open-loop gain of the OTA of Fig. 25.62 as 68 dB + 47 dB, or 115 dB. Note that since we have not changed the transconductance of the input diff-amp, we can still compensate the OTA with a 6.1 pF for a unity gain frequency of 1 MHz. The gain enhancement simply shifts the open-loop 3 dB frequency of the OTA downward. Simulation results are shown in Fig. 25.70.

High-Speed CMOS Op-Amp Design

Up to this point we have not been concerned about designing an op-amp for high-speed operation. The unity gain frequency of the OTAs we have discussed thus far is given by $g_m/2\pi C_L$ where g_m is the transconductance of the input diff-amp and C_L is the load capacitance used to compensate the OTA. We might conclude that simply by increasing

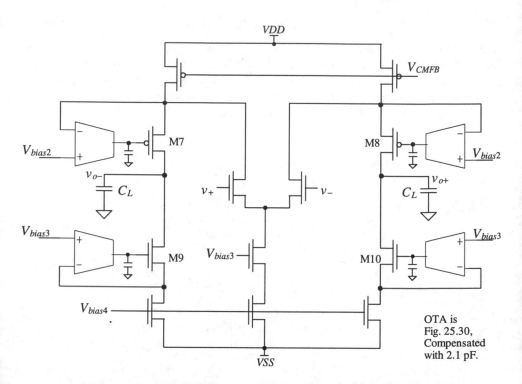

Figure 25.69 Fully differential folded-cascode OTA with gain enhancement.

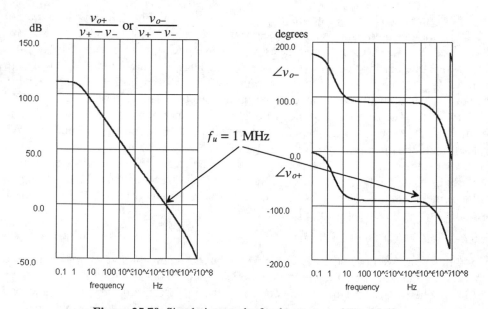

Figure 25.70 Simulation results for the op-amp of Fig. 25.69.

the width of the input diff-pair, possibly to several thousand μm, we can increase the unity gain frequency. While this is true to a point, the intrinsic speed of the MOSFET limits the practical unity gain frequency. (Every MOSFET introduces a parasitic pole into the overall transfer function of the op-amp.) In Ch. 9 we discussed the unity gain frequency of a MOSFET, f_T. This frequency f_T is determined by the point where the gate current and drain current of the MOSFET are equal and is given by

$$f_T = \frac{3 \cdot KP}{4\pi \cdot C'_{ox}L^2} \cdot (V_{GS} - V_{THN}) \qquad (25.67)$$

The channel length, L, and V_{GS}, can be used to increase f_T. For high-speed operation, the minimum channel length should be used. In the CN20 process, this means that all MOSFETs used in an amplifier should have a channel length of 2 μm. The gate-source voltage of a MOSFET should also be large. However, as we know, increasing the gate-source voltage of a MOSFET decreases the input CMR and output voltage swing. A reasonable value of $V_{GS} - V_{THN}$ for high-speed operation is 0.5 V.

The folded-cascode OTA of Fig. 25.57 can be implemented using 1000/2 MOSFETs and increasing the biasing current to 1 mA. MOSFETs M5 and M6 are 1500/2 because they supply 1.5 mA. Note that the CMFB circuit can be scaled down to reduce power dissipation; for example, the MOSFETs used in this circuit can be 100/2. With a 5 pF compensation capacitor, the simulation results are shown in Fig. 25.71. The open-loop gain is 48 dB (250), while the unity gain frequency is 70 MHz. The low value of open-loop gain is the result of using a 2 μm device with a λ of approximately 0.8 V^{-1} and a large biasing current (1 mA).

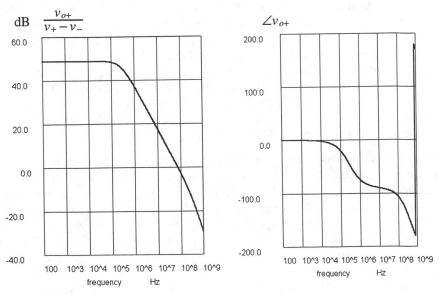

Figure 25.71 Simulation results of OTA shown in Fig. 25.57 using 1000/2 devices.

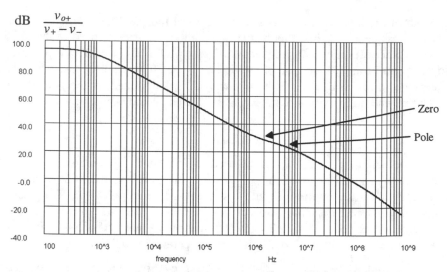

Figure 25.72 Simulation of uncompensated OTA shown in Fig. 25.69 using 1000/2 devices. (This OTA is stable without a compensation capacitor.)

The gain of this OTA can be increased using the gain-enhancement technique shown in Fig. 25.69. With gain enhancement we would expect the low-frequency gain to become 48 dB + 47 dB (the gain of the OTA of Fig. 25.30), or 95 dB. The open-loop, uncompensated gain of the gain-enhanced OTA is shown in Fig. 25.72.

Several characteristics of this plot deserve comment. First, the OTA used to enhance the gain, Fig. 25.30, does not need to be a wide-bandwidth stage. In fact, as long as its unity gain frequency is greater than the nonenhanced 3-dB bandwidth of the main OTA (simulation results shown in Fig. 25.71), the added OTA will not affect the gain. In other words, the OTA of Fig. 25.30 with 2.1 pF compensation capacitor has an f_u of 2.3 MHz. As long as the 3-dB bandwidth of the folded-cascode OTA without enhancement is smaller than 2.3 MHz, the added OTA will not affect the low-frequency gain. From Fig. 25.71 the 3-dB frequency of the nonenhanced OTA is approximately 200 kHz, which is less than the 2.3 MHz (f_u) of the added OTA.

Next we should notice that a pole and zero occur at almost the same frequency. This is called a doublet [10]. Although the phase margin is not degraded from the presence of this doublet, the settling time can be affected. With the use of gain enhancement, the settling time is improved (shortened). The feedback loop formed with the added OTA helps to decrease the time it takes the error signal to propagate back through the overall amplifier. Since the added OTAs form localized feedback loops on the outputs of the amplifiers, the error signal can get fed back with less delay. Adding the compensation capacitor of 5 pF to the gain-enhanced OTA gives the simulation results shown in Fig. 25.73. Keeping in mind that this op-amp was designed with a 2 μm process, we see that the results are impressive: a low-frequency gain of 94 dB (50,000) and a unity gain frequency of 30 MHz.

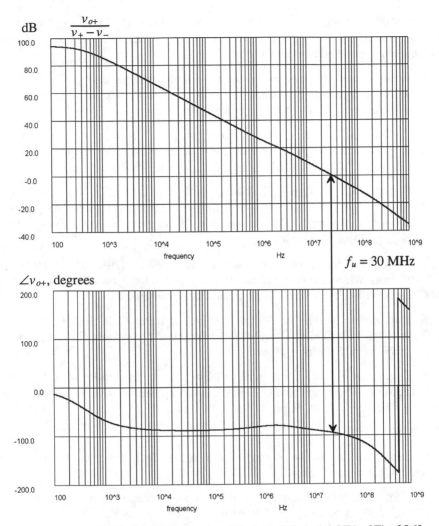

Figure 25.73 Simulation results of the fully differential OTA of Fig. 25.69
(see text for description).

REFERENCES

[1]   P. E. Allen and D. R. Holberg, *CMOS Analog Circuit Design,* Holt, Rinehart
       and Winston, 1987. ISBN 0-03-006587-9.

[2]   P. R. Gray and R. G. Meyer, *Analysis and Design of Analog Integrated Circuits,*
       3rd ed., John Wiley and Sons, 1993. ISBN 0-471-57495-3.

[3]   P. R. Gray and R. G. Meyer, "MOS Operational Amplifier Design - A Tutorial
       Overview," *IEEE Journal of Solid State Circuits,* Vol. SC-17, pp. 969-982,
       December 1982.

[4]   R. L. Geiger and E. Sánchez-Sinencio, "Active Filter Design Using Operational Transconductance Amplifiers: A Tutorial," *IEEE Circuits and Devices Magazine*, pp. 20-32, March 1985.

[5]   M. Steyaert and W. Sansen, "A High-Dynamic-Range CMOS Op-Amp with Low-Distortion Output Structure," *IEEE Journal of Solid-State Circuits,* Vol. SC-22, No. 6, pp. 1204-1207, December 1987.

[6]   T. C. Choi, R. T. Kaneshiro, R. Broderson, and P. R. Gray, "High-Frequency CMOS Switched Capacitor Filters for Communication Applications," *IEEE Journal of Solid State Circuits,* Vol. SC-18, pp. 652-664, December 1983.

[7]   R. Hogervorst, J. P. Tero, R. G. H. Eschauzier, and J. H. Huijsing, "A Compact Power-Efficient 3 V CMOS Rail-to-Rail Input/Output Operational Amplifier for VLSI Cell Libraries," *IEEE Journal of Solid State Circuits,* Vol. 29, pp. 1505-1513, December 1994.

[8]   M. Banu, J. M. Khoury, and Y. Tsividis, "Fully Differential Operational Amplifiers with Accurate Output Balancing," *IEEE Journal of Solid State Circuits,* Vol. 23, No. 6, pp. 1410-1414, December 1988.

[9]   K. Bult and G. J. G. M. Geelen, "A Fast-Settling CMOS Op-Amp for SC Circuits with 90-dB DC Gain," *IEEE Journal of Solid State Circuits,* Vol. 25, pp. 1379-1384, December 1990.

[10]   B. Y. Kamath, R. G. Meyer, and P. R. Gray, "Relationship Between Frequency Response and Settling Time of Operational Amplifiers," *IEEE Journal of Solid State Circuits,* Vol. SC-9, pp. 347-352, December 1974.

PROBLEMS

25.1   Calculate the gain of the op-amp shown in Fig. 25.3 if the resistor is replaced with a 1 μA current source.

25.2   Determine the maximum output source and sink currents for the op-amp of Fig. 25.11 for a maximum V_{GS} (of M9) or V_{SG} (of M10) of 1.5 V.

25.3   Referring to the op-amp of Fig. 25.11, determine the small-signal resistance looking into

(a) the sources of M1 and M2 (neglect body effect)

(b) the drain of M6

(c) the source of M7

(d) the drains of M1 and M2

(e) the sources of M9 and M10 (neglect body effect).

25.4   What is the small-signal resistance to ground at the drains of M1/M3, M4/M2, and M7/M91 in Fig. 25.11?

25.5 If the drain current of M6, in Fig. 25.11, is reduced to 5 μA, what are the drain currents in M3, M4, M7, and M9?

25.6 If width of M7, in Fig. 25.11, is increased to 140 μm, how would M8 be resized? With these sizes, what quiescent current would flow in M9 and M10?

25.7 Derive Eq. (25.24).

25.8 Assume that an op-amp is needed that will always be used with feedback to get a gain of ten. Suggest a modification to Eq. (25.13) for compensating this op-amp. Suggest a circuit configuration, based on Fig. 25.7, that will allow simulating the stability of the op-amp.

25.9 Simulate the operation of the op-amp of Fig. 25.11 with SPICE and verify the results shown in Fig. 25.12.

25.10 Simulate the operation of the op-amp of Fig. 25.11 and determine the input offset voltage, the output voltage swing, the CMRR, the power dissipation, PSRR, and the slew-rate limitations.

25.11 If the OTA of Fig. 25.22 is made using 15/5 n-channels, 70/5 p-channels, and biased with a 10 μA current source, determine the small-signal, low-frequency gains $i_{out}/(v_{l2} - v_{l1})$ and $v_{out}/(v_{l2} - v_{l1})$.

25.12 Repeat Ex. 25.1 if the bottom of the 100k resistor in Fig. 25.24 is connected to –2.5 V.

25.13 Repeat Ex. 25.2 using the high-pass transconductor-C filter shown in Fig. 25.27. The capacitor value remains 8.75 pF.

25.14 If the OTA of Fig. 25.29 is made using 15/5 n-channels, 70/5 p-channels, and biased with a 10 μA current source, determine the small-signal, low-frequency gains $i_{out}/(v_{l2} - v_{l1})$ and $v_{out}/(v_{l2} - v_{l1})$.

25.15 If the minimum voltage across the current sources in Fig. 25.30 is 0.3 V, estimate the input CMR if $VDD = 5$ and $VSS = 0$.

25.16 Discuss, in your own words, the problems with compensating the OTA of Fig. 25.30.

25.17 Determine, using hand-calculations, the small-signal, low-frequency voltage gains of the OTA shown in Fig. 25.30 with $V_{CM} = 0.8$ V, 2.5 V, and 4.2V.

25.18 Redesign the regulator of Ex. 25.3 so that the output voltage is 2.0 V. Show all hand calculations, a schematic for your design, and estimate the phase margin.

25.19 If the 10 μA bias current of Fig. 25.42 is changed to 1 μA, what happens to the currents in the folded-cascode OTA? What is the small-signal voltage gain with 1 μA bias current?

25.20 How can the biasing circuit of Fig. 25.44 be simplified? (Hint: Two of the MOSFETs are not necessary.)

25.21 Label the DC biasing currents on the schematic of the wide-swing folded-cascode OTA of Fig. 25.45 if the biasing circuit of Fig. 25.44 is used with $I = 1$ μA, MB1 is sized 15/20, MB2 is sized 35/10, and all other MOSFETs are sized 70/5 (p-channels) and 15/5 (n-channels).

25.22 Estimate the small-signal gains, i_{out}/v_{in} and v_{out}/v_{in}, of the OTA described in Problem 25.21. What size of load capacitance would be needed to compensate the op-amp for an f_u of 1 MHz?

25.23 Design an output buffer for the OTA of Problem 21 that can source or sink 100 μA. Assume, for your design, that $VDD = -VSS = -2.5$ V and that the output voltage swing is +/– 1 V.

25.24 Verify the operation of the op-amp design of Problem 25.21 with SPICE. Show a plot similar to Fig. 25.46 with the op-amp driving a 10 pF load. Using SPICE, show the phase margin of your design with and without the 10 pF load.

25.25 Figure P25.25 shows a floating current source made using MC1 and MC2. Determine the size of MC1/MC2 so that 30 μA flows in M5 and M9.

25.26 If the minimum voltage between the drain and source of any MOSFET in Fig. 25.49 is 0.4 V, estimate the minimum VDD possible for the op-amp assuming $VSS = 0$. Comment on the problems of operating the op-amp at this voltage.

25.27 Discuss the purpose of the CMFB circuit in differential input/output op-amps.

25.28 Determine the CMR of the CMFB circuit of Fig. 25.58. For the CMFB circuit of Fig. 25.59, does the CMR of the diff-amp affect the operation of the circuit? Why or why not?

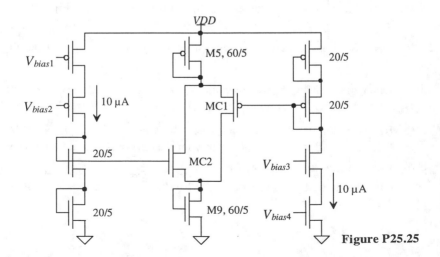

Figure P25.25

Mixed-Signal Circuits

Chapter

26

Nonlinear Analog Circuits

The past six chapters have presented the design of linear analog circuits, circuits whose input signals are linearly related to the circuits' output signals. In this chapter, we discuss several circuits that are not purely analog or digital. We term these circuits *nonlinear analog circuits* (the inputs are not linearly related to the outputs). In particular, we will discuss voltage comparator analysis and design, adaptive biasing, and analog multiplier design.

26.1 Basic CMOS Comparator Design

The schematic symbol and basic operation of a voltage comparator are shown in Fig. 26.1. The comparator can be thought of as a decision-making circuit. If the +, v_+, input of the comparator is at a greater potential than the –, v_-, input, the output of the comparator is a logic 1, whereas if the + input is at a potential less than the – input, the output of the comparator is at a logic 0. Although the basic op-amp of the last chapter can be used as a voltage comparator, in some less demanding low-frequency or speed applications, we will not consider the op-amp as a comparator. Instead, we will discuss practical comparator design and analysis where propagation delay and sensitivity are important.

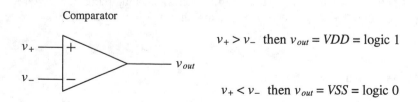

Comparator

$v_+ > v_-$ then $v_{out} = VDD = $ logic 1

$v_+ < v_-$ then $v_{out} = VSS = $ logic 0

Figure 26.1 Comparator operation.

A block diagram of a high-performance comparator is shown in Fig. 26.2. The comparator consists of three stages; the input preamplifier, a positive feedback or decision stage, and an output buffer. The preamp stage (or stages) amplifies the input signal to improve the comparator sensitivity (i.e., increases the minimum input signal with which the comparator can make a decision) and isolates the input of the comparator from switching noise (often called kickback noise) coming from the positive feedback stage (*this is important*). The positive feedback stage is used to determine which of the input signals is larger. The output buffer amplifies this information and outputs a digital signal. Designing a comparator can begin with considering input common-mode range, power dissipation, propagation delay, and comparator gain. We will develop a basic comparator design using a procedure similar to the last chapter's development of the basic op-amp.

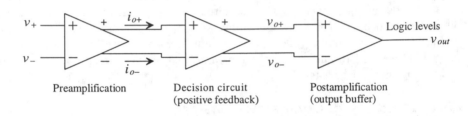

Figure 26.2 Block diagram of a voltage comparator.

Preamplification

For the preamplification stage, we chose the circuit of Fig. 26.3. This circuit is a differential amplifier with active loads. The sizes of M1 and M2 are set by considering the diff-amp transconductance and the input capacitance. The transconductance sets the gain of the stage, while the input capacitance of the comparator is determined by the size of M1 and M2. We will concentrate on speed in this design, and therefore we will set the channel lengths of the MOSFETs to 2 μm. (Channel length modulation gives rise to an unwanted offset voltage.) Notice that there are no high-impedance nodes in this circuit, other than the input and output nodes. Using the sizes given in the schematic, we can relate the input voltages to the output currents by

$$i_{o+} = \frac{g_m}{2}(v_+ - v_-) + \frac{I_{SS}}{2} = I_{SS} - i_{o-} = 20\ \mu A - i_{o-} \qquad (26.1)$$

where, for the present case,

$$g_m = g_{m1} = g_{m2} = \sqrt{2\frac{10}{2} \cdot 50\frac{\mu A}{V^2} \cdot 10\mu A} = 71\ \frac{\mu A}{V} \qquad (26.2)$$

In other words, if v_+ is 10 mV greater than v_-, the output currents, i_{o+} and i_{o-}, are 10.35 μA and 9.65 μA, respectively. To further increase the gain of the first stage, we can size up the widths of MOSFETs M3 and M4 relative to the widths of M31 and M41.

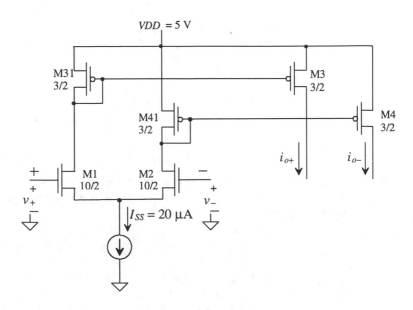

Figure 26.3 Preamplification stage of comparator.

Decision Circuit

The decision circuit is the heart of the comparator and should be capable of discriminating mV level signals. We should also be able to design the circuit with some hysteresis (see Ch. 18) for use in rejecting noise on a signal. The circuit used in the present comparator is shown in Fig. 26.4 [1]. The circuit uses positive feedback from the cross-gate connection of M6 and M7 to increase the gain of the decision element.

Let's begin by assuming that i_{o+} is much larger than i_{o-} so that M5 and M7 are on and M6 and M8 are off. We will also assume that $\beta_5 = \beta_8 = \beta_A$ and $\beta_6 = \beta_7 = \beta_B$. Under these circumstances, v_{o-} is approximately 0 V and v_{o+} is

$$v_{o+} = \sqrt{\frac{2i_{o+}}{\beta_A}} + V_{THN} \qquad (26.3)$$

If we start to increase i_{o-} and decrease i_{o+}, switching takes place when the drain-source voltage of M7 is equal to V_{THN} of M6. At this point, M6 starts to take current away from M5. This decreases the drain-source voltage of M5 and thus starts to turn M7 off. If we assume that the maximum value of v_{o+} or v_{o-} is equal to $2V_{THN}$, then M6 and M7 operate,

$$\beta_A = \beta_5 = \beta_8$$

$$\beta_B = \beta_6 = \beta_7$$

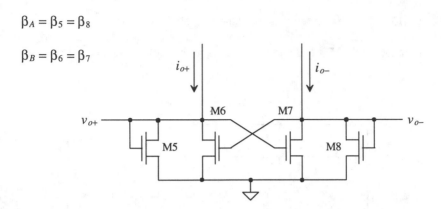

Figure 26.4 Positive feedback decision circuit.

under steady-state conditions, in either cutoff or the triode regions. Under these circumstances, the voltage across M7 reaches V_{THN}, and thus M7 enters the saturation region, when the current through M7 is

$$i_{o-} = \frac{\beta_B}{2}(v_{o+} - V_{THN})^2 = \frac{\beta_B}{\beta_A} \cdot i_{o+} \tag{26.4}$$

This is the point at which switching takes place; that is, M7 shuts off and M6 turns on. If $\beta_A = \beta_B$, then switching takes place when the currents, i_{o+} and i_{o-}, are equal. Unequal βs cause the comparator to exhibit hysteresis. A similar analysis for increasing i_{o+} and decreasing i_{o-} yields a switching point of

$$i_{o+} = \frac{\beta_B}{\beta_A} \cdot i_{o-} \tag{26.5}$$

Relating these equations to Eq. (26.1) yields the switching point voltages (review Ch. 18), or

$$V_{SPH} = v_+ - v_- = \frac{I_{SS}}{g_m} \cdot \frac{\frac{\beta_B}{\beta_A} - 1}{\frac{\beta_B}{\beta_A} + 1} \text{ for } \beta_B \geq \beta_A \tag{26.6}$$

and

$$V_{SPL} = -V_{SPH} \tag{26.7}$$

For the present design under development, we will use 3 μm and 2 μm device widths and lengths (no hysteresis and the offset voltage is not important).

Example 26.1
For the circuit shown in Fig. 26.5, estimate and simulate the switching point voltages for two designs: (1) $W_5 = W_6 = W_7 = W_8 = 3$ μm with $L = 2$ μm and (2) $W_5 = W_8 = 3$ μm and $W_6 = W_7 = 4$ μm with $L = 2$ μm.

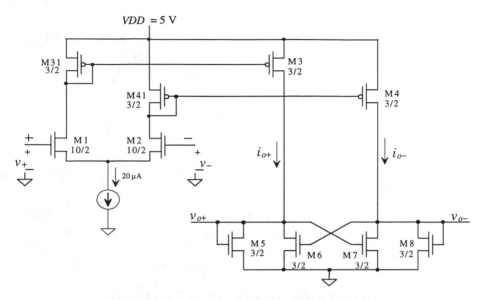

Figure 26.5 Schematic of preamp and decision circuit.

For the first case

$$\beta_A = \beta_B = 50 \frac{\mu A}{V^2} \cdot \frac{3 \, \mu m}{2 \, \mu m}$$

so that $V_{SPH} = V_{SPL} = 0$.   In other words, the comparator does not exhibit hysteresis.

Simulation results are shown in Fig. 26.6. The input v_- is set to 2.5 V while the input v_+ is swept from 2.0 to 3.0 V. Note that the amplitudes of v_{o+} and v_{o-} are limited to approximately $2V_{THN}$.

For the second case

$$\beta_A = 50 \frac{\mu A}{V^2} \cdot \frac{3 \, \mu m}{2 \, \mu m} \text{ and } \beta_B = 50 \frac{\mu A}{V^2} \cdot \frac{4 \, \mu m}{2 \, \mu m}$$

or $\beta_B = 1.33\beta_A$. Using Eqs. (26.6) and (26.7) with $I_{SS} = 20 \, \mu A$ and $g_m = 71$ μA/V, we get

$$V_{SPH} = - V_{SPL} = 40 \text{ mV}$$

The simulation results are shown in Fig. 26.7. Fig. 26.7a shows a sweep of v_+ from 2.4 to 2.6 V, with v_- held at 2.5 V. Since V_{SPH} is 40 mV the decision circuit switches states when v_+ is 40 mV above v_- (i.e. when $v_+ = 2.54$ V). The case of v_+ being swept from 2.6 to 2.4 V is shown in part (b) of the figure. Switching occurs in this situation when $v_+ = 2.46$ V, or 40 mV less than 2.5 V.

■

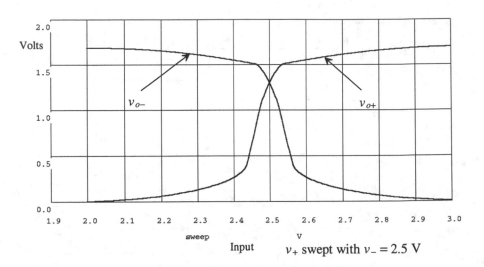

Figure 26.6 Simulation results for case 1 of Ex. 26.1.

Output Buffer

The final component in our comparator design is the output buffer or postamplifier. The main purpose of the output buffer is to convert the output of the decision circuit into a logic signal (i.e., 0 or 5 V). The output buffer should accept a differential input signal and not have slew-rate limitations.

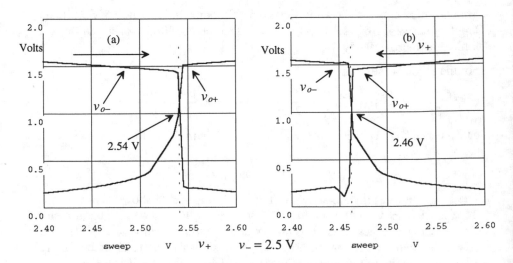

Figure 26.7 Simulation results for case 2 of Ex. 26.2.

The circuit used as an output buffer in our basic comparator design is shown in Fig. 26.8. This circuit is a self-biasing differential amplifier [2]. An inverter was added on the output of the amplifier as an additional gain stage and to isolate any load capacitance from the self-biasing differential amplifier. A DC sweep of this amplifier configuration is shown in Fig. 26.9. The input v_{o+} is swept from 1 to 4 V, while v_{o-} is held at 1 to 3.5 V in 0.5 V increments. It is apparent that the inputs, v_{o+} and v_{o-}, should lie within 1.5 and 3 V for linear operation of the output buffer. Comparing this result with the output of the positive feedback circuit, which varies from 0 to 1.5 V, we see a problem in connecting the decision circuit directly to the output buffer. To shift the output of the decision circuit up approximately 1 V, the circuit of Fig. 26.10 is used. The MOSFET M17 is added in series with the decision circuit to increase the average voltage out of the decision circuit. The size of the MOSFET is somewhat arbitrary. We will set $W_{17}/L_{17} = 100$ μm/2 μm so that the output of the decision circuit is increased by approximately V_{THN}. The complete schematic of the comparator is shown in Fig. 26.11. Unlabeled MOSFETs are 3 μm/ 2 μm.

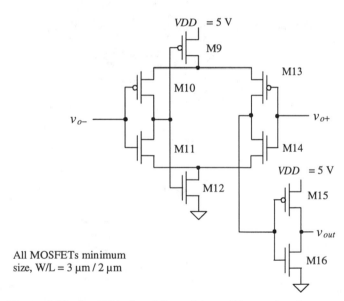

Figure 26.8 A self-biasing differential amplifier used as the comparator output buffer.

26.1.1 Characterizing the Comparator

Comparator Gain and Offset

With v_- held at 2.5 V, the v_+ input of the comparator is swept from 2.49 to 2.51 V, with the result shown in Fig. 26.12. An important parameter of a comparator is its offset voltage. From this figure we can see that the systematic offset voltage is approximately 1 mV. The next chapter will discuss dynamic methods of eliminating the offset voltage.

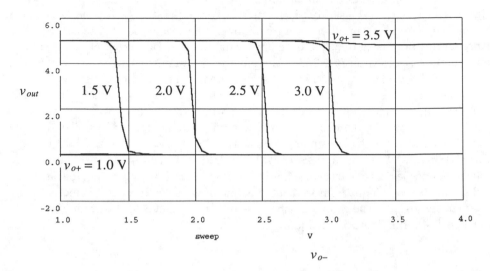

Figure 26.9 DC sweep of the self-biasing amplifier of Fig. 26.8.

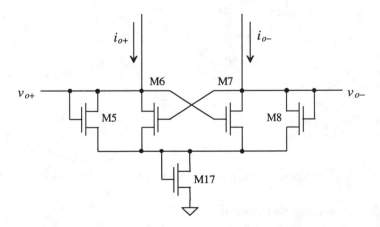

Figure 26.10 Use of a large MOSFET, M17, to level shift the output of the decision circuit.

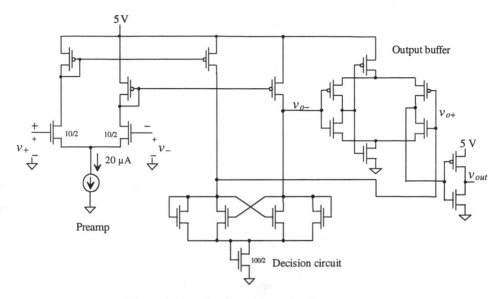

Figure 26.11 Complete schematic of comparator.

If we take the derivative of the transfer curves, the gain of the comparator and thus the smallest difference that can be discriminated between v_+ and v_- become known. Figure 26.13 shows the gain of the comparator at about 2,000 so that approximately 2.5 mV is needed to make the comparator output change logic levels. The comparator gain can be increased by increasing the transconductance of the preamp, that is, by increasing the widths of M1 and M2 or by adding additional inverters. Adding another gain stage to the preamp will also increase the comparator overall gain.

Transient Response

The transient response of a comparator can be significantly more difficult to characterize than the DC characteristics. Let's begin by considering $v_- = 2.5$ V DC, with the v_+ input to the comparator a 10 ns wide pulse and an amplitude varying from 2.49 to 2.51 V. This is termed a narrow pulse with a 10 mV overdrive; we are driving the + input of the comparator 10 mV over the negative input. The simulation results for these inputs are shown in Fig. 26.14. If the pulse amplitude or width is reduced much beyond this, the comparator does not make a full transition. Notice how the simulation starts at 200 ns.

Although these results are interesting, they are not practically useful in some situations. In few cases will the comparator discriminate a signal similar to that of Fig. 26.14a. A better indication of comparator performance is to apply a signal that starts at a voltage where one side of the differential amplifier is cut off and to finish at a voltage slightly larger than the reference voltage (2.5 V in the above example). Applying a 0 to 2.51 V pulse to the v_+ input will cause M1, M3, and M31 to be initially off and then

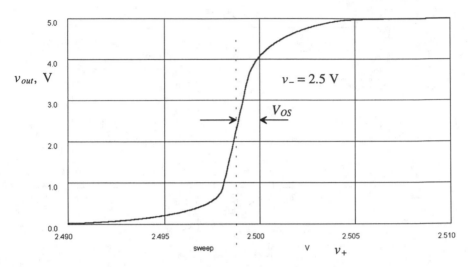

Figure 26.12 DC characteristics of the comparator of Fig. 26.11 with −input connected to 2.5 V.

turn on. The comparator nodes must be charged over a voltage range of 0 to 2.51 V input, wider than the 2.49 to 2.51 V example of Fig. 26.13. In order to keep one side of the preamp from turning off, the circuit of Fig. 26.15 can be used. The diode connected MOSFETs ensure that the voltage between the drains of M31 and M41 are always within a V_{THN} of one another. This configuration is sometimes referred to as a clamped input stage.

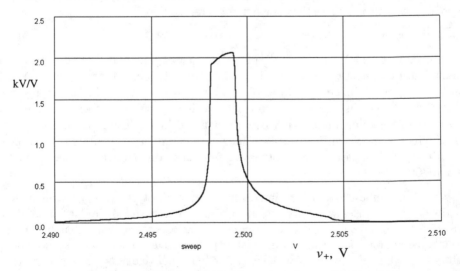

Figure 26.13 Comparator gain as a function of input voltage.

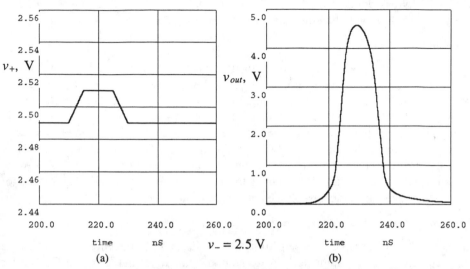

Figure 26.14 Comparator input (a) and corresponding output (b) Inverting input held at 2.5 V potential.

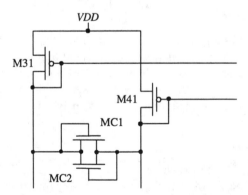

Figure 26.15 Using MOSFETs to keep active loads of preamp from shuting off.

Propagation Delay

Ideally, the propagation delay (the time difference between the input, v_+, crossing the reference voltage, v_-, and the output changing logic states) is zero. For the comparator of Fig. 26.11 with simulation results shown in Fig. 26.14, the delay of the comparator is approximately 10 ns (or less with an overdrive). This should be compared with the delay of an op-amp used as a comparator, which may be several hundred nanoseconds.

Another interesting fact about comparator design is that the delay of a comparator can be reduced by cascading gain stages. In other words, the delay of a single high-gain stage is in general longer than the delay of several low-gain stages.

Minimum Input Slew Rate

The last characteristic we will discuss is the minimum input slew rate of a comparator. If the input signals to the comparator vary at a slow rate (e.g., a sine wave generated from the AC line), the output of the comparator may very well oscillate resulting in an output with a metastable (unknown) state. If a comparator is to be used with slowly varying signals, or in a noisy environment, the decision circuit should have hysteresis. The minimum input slew rate is difficult to simulate with SPICE because of the slow and fast varying signals present at the same instant of time in the circuit. This same situation (metastability) can also occur if the input overdrive is small. The result, for this case, however, is an increase in the delay time of the comparator.

Improvements in Sensitivity and Speed

Improvements in comparator sensitivity and speed can be directly related to improvements in the preamplifier. A derivation (similar to that given in Ch. 11 for minimum delay through an inverter string driving a load capacitance, C_L) of the number of stages, N, needed in a preamplifier to reach minimum delay for a given load capacitance and input capacitance results in

$$N = \ln \frac{C_L}{C_{in}} \tag{26.8}$$

In the derivation of this equation, it was assumed that the driving resistance of any stage is $1/g_{mn}$ (the transconductance of the nth differential amplifier stage). In practice, Eq. (26.8) is of little use because one side of the differential amplifier can be off. For this reason (and others such as size and power draw), comparators with more than two or three preamp stages are not common. Figure 26.16 shows a two-stage preamp with decision circuit.

Clocked Comparators

Figures 26.17 [3] and 26.18 [4] show two clocked comparators used in analog-to-digital converter design. The use of a clock pulse can greatly improve comparator performance and speed.

The clocked comparator shown in Fig. 26.17 is a clocked version of the comparator of Fig. 26.11. The use of a clock eliminates the need for an output buffer used in level shifting the output of the decision circuit. When $\overline{LATCH}$ is high, the comparator behaves like the preamp and decision circuit of Fig. 26.11. The output of the decision circuit is dependent on the input signals. When $\overline{LATCH}$ is taken low, ML1 shuts off, effectively latching the outputs at a logic high or low. That is, the comparator stops comparing and remembers the status of the inputs at the instant $\overline{LATCH}$ is switched low.

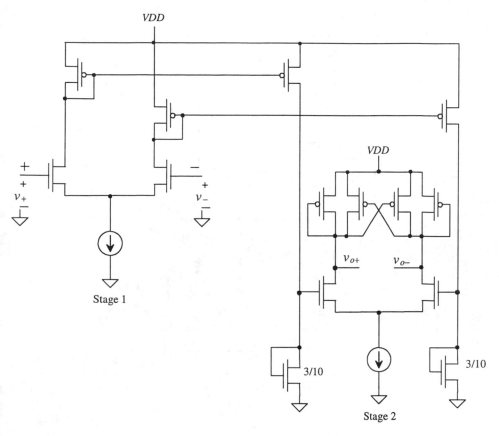

Figure 26.16 Two-stage preamp with decision circuit.

This comparator has the unwanted characteristic of continuously drawing power. Adding a MOSFET switch, similar to ML1 in series with the decision circuit, in series with the preamp current source can significantly reduce power draw. Under DC conditions, the comparator doesn't draw power when $\overline{LATCH}$ is low with this switch added.

The second clocked comparator design shown in Fig. 26.18 is based on the regenerative latch. Again, the comparator is made up of three stages: preamp, positive feedback decision circuit, and output buffers. The regenerative latch (cross-coupled inverters) is made up of M1 through M4. When *Latch* is low, the p-channels, M1 and M2, of the latch are isolated from the n-channels, M3 and M4. In addition, the outputs of the decision circuit are pulled high, that is, the outputs of the comparator are low, $Q = \overline{Q} = 0$. When *Latch* signal transitions high, the regenerative action of the latch combined with the preamplifier causes an imbalance in the decision circuit, forcing the outputs into a state determined by v_+ and v_-.

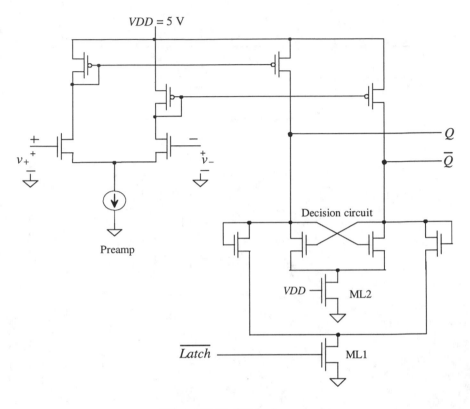

Figure 26.17 Clocked comparator.

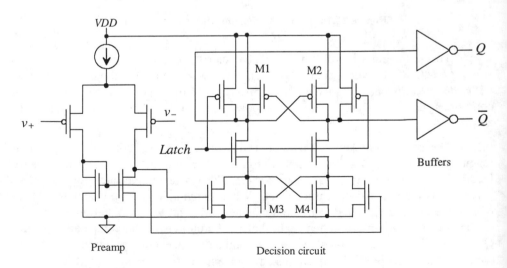

Figure 26.18 Clocked comparator based on the basic latch.

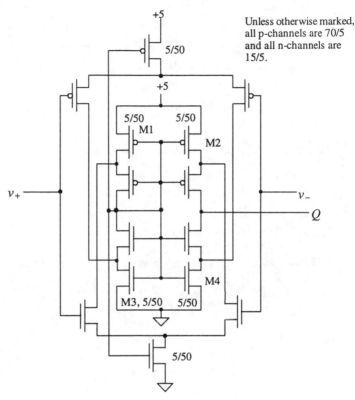

Figure 26.19 Self-biased comparator.

Self-Biasing Comparator

For the final example of a comparator, consider the self-biased comparator shown in Fig. 26.19 [2]. This circuit operates similarly to the self-biased differential amplifier of Fig. 26.8 but with a larger gain and wider input common-mode range. The DC characteristics of the comparator are shown in Fig. 26.20. Note that because of the high gain of this comparator configuration, the delay tends to be longer (hundreds of ns) than the other configurations we discussed. The delay can be reduced, at the price of gain and more power dissipation, by decreasing the widths of M1 through M4 (operating M1 through M4 in the triode region) and by using minimum channel lengths.

26.2 Adaptive Biasing

Adaptive biasing can reduce power dissipation in an amplifier while at the same time increasing output current drive capability [5,6]. Figure 26.21 can be used to help illustrate the idea. When v_{I1} and v_{I2} are equal, the current sources I_{SS1} and I_{SS2} are zero (an open). The diff-amp DC tail current is simply I_{SS}, the same as an ordinarily biased

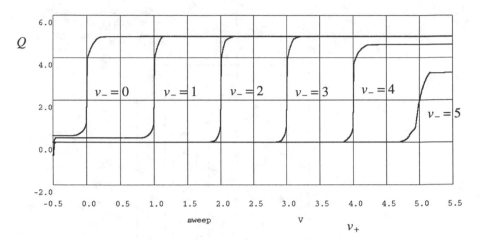

Figure 26.20 DC sweep for the self-biased comparator of Fig. 26.19.

If $v_{I1} = v_{I2}$ then $I_{SS1} = I_{SS2} = 0$.

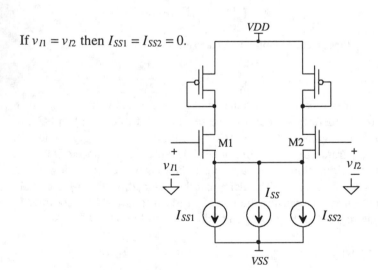

Figure 26.21 Adaptively biased diff-amp.

diff-amp. If v_n becomes larger than v_{n2}, the current source I_{SS1} increases above zero, effectively increasing the diff-amp DC bias current. Similarly, if v_{n2} becomes larger than v_n, the current source I_{SS2} increases above zero. The diff-amp output current is normally limited to I_{SS} when one side of the diff-amp shuts off. However, now the maximum output current is limited to either $I_{SS} + I_{SS1}$ or $I_{SS} + I_{SS2}$. Power dissipation can be reduced using an adaptive bias, and slew-rate problems can be eliminated.

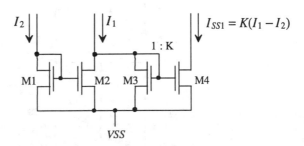

Figure 26.22 Current diff-amp used in adaptive biasing.

The current diff-amp of Fig. 26.22 can be used to implement the current source I_{SS1} or I_{SS2}. If the currents I_1 and I_2 are equal, then zero current flows in M3 and M4. Also, if I_2 is greater than I_1, zero current flows in M3 and M4. If I_1 is larger than I_2, the difference between these two currents $(I_1 - I_2)$ flows in M3. Since M4 is K times wider than M3, a current of $K(I_1 - I_2)$ flows in M4 (normally $K < 1$). Two of these diff-amps are needed to implement the adaptive biasing of the diff-amp of Fig. 26.21.

Figure 26.23 shows the implementation of adaptive biasing into the diff-amp of Fig. 26.21. P-channel MOSFETs are added adjacent to M3 and M4 to mirror the currents through M1 and M2 (I_1 and I_2). The maximum total current available through M1 occurs when M2 is off. Positive feedback exists through the loop M1, M3, M5-M7. Initially, when M2 shuts off, the current in M1 and M3 is I_{SS}. This is mirrored in M5 and M6, and thus I_{SS1} becomes $K \cdot I_{SS}$. At this particular instance in time, the tail current, which flows through M1, is now $I_{SS} + K \cdot I_{SS}$. However, provided the MOSFETs M1, M3 - M7 remain in saturation, this current circles back around the positive feedback loop and increases by K. This continues, resulting in a final or total tail current of

$$I_{tot} = I_{SS} \cdot (1 + K + K^2 + K^3 + ...) \qquad (26.9)$$

If $K < 1$, this geometric series can be written as

$$I_{tot} = \frac{I_{SS}}{1 - K} \qquad (26.10)$$

Setting $K = 0$ (MOSFET M7 doesn't exist) results in no adaptive biasing and a total tail current of I_{SS}. Setting $K = 1/2$ (M7 half the size of M6) results in a total available tail current of $2 \cdot I_{SS}$. Since the tail current limits the slew rate, when the diff-amp is driving

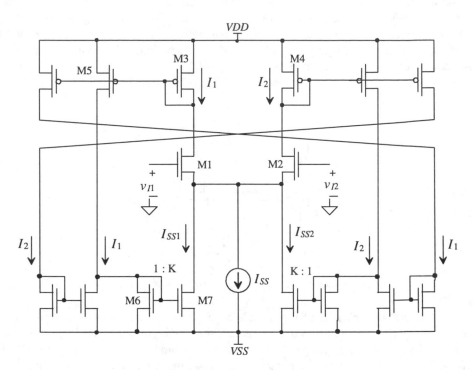

Figure 26.23 Adaptively biased diff-amp.

a capacitive load, making K equal to one eliminates slew-rate limitations while at the same time not increasing static power dissipation. In practice, shuting off one side of the diff-pair, M1/M2, is difficult since the adaptive biasing has the effect of lowering the source potentials of the diff-pair, keeping both MOSFETs on. Adaptive biasing can be used in a comparator where M1 or M2 can shut off. However, the static power dissipation will be large, and therefore there is no benefit over the comparators discussed earlier in the chapter. When applying adaptive biasing to an OTA design, the value of K should be unity or less. [Using a K of 1 or 2 can still result in a finite I_{tot} since the MOSFETs have a finite output resistance and since the MOSFETs in the diff-pair will not shut off, as was assumed in the derivation of Eq. (26.10)]. An adaptively biased OTA is shown in Fig. 26.24 [5].

A final example of an adaptive voltage-follower amplifier is shown in Fig. 26.25 [6]. This amplifier can only source current to a load. If v_{in} and v_{out} are equal, the current that flows in M1 and M2 is $I_{ss} + I_{D6}$. If v_{in} is increased, the current in M1 and M3 increases. This causes the currents in M4 - M6 to increase, effectively increasing the tail current of the diff-pair. The result is a large current available to drive the load. Note that M7 can be sized larger than the other MOSFETs to increase maximum output current.

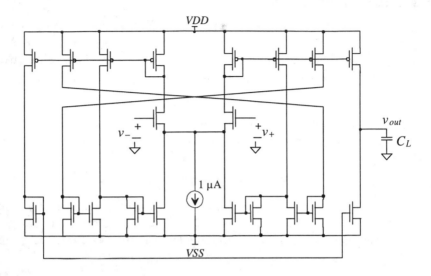

Figure 26.24   Adaptively biased OTA [5].

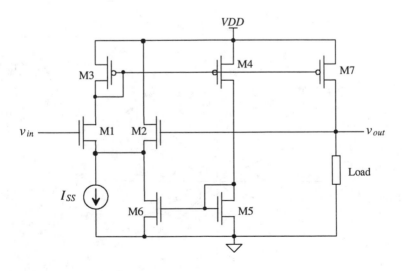

Figure 26.25 Adaptive voltage follower [6].

26.3 Analog Multipliers

Analog multipliers find extensive use in communication systems. Figure 26.26 shows the voltage characteristics of a four-quadrant multiplier [7]. This multiplier is termed a four-quadrant multiplier because both inputs can be either positive or negative. The ideal output of the multiplier is related to the inputs by

$$v_{out} = K_m \cdot v_x v_y \qquad (26.11)$$

where K_m is the multiplier gain with units of V^{-1}. In reality, imperfections exist in the multiplier gain, resulting in offsets and nonlinearities. The output of the multiplier can be written as [7]

$$v_{out} = K_m(v_x + V_{OSx})(v_y + V_{OSy}) + V_{OSout} + v_x^n + v_y^m \qquad (26.12)$$

where V_{OSx}, V_{OSy}, and V_{OSout} are the offset voltages associated with the x-, y-inputs, and the output, respectively. The terms v_x^n and v_y^m represent nonlinearities in the multiplier. Normally, these nonlinearities are specified in terms of the total harmonic-distortion or

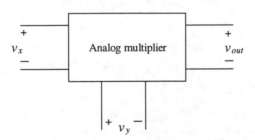

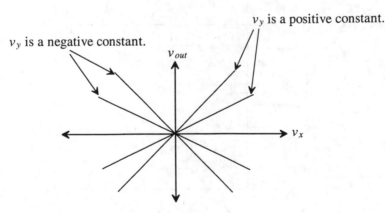

Figure 26.26 Operation of a four-quadrant analog multiplier.

by specifying the maximum deviation in percentages between a straight line and the actual characteristic curves shown in Fig. 26.26 over some range of input voltages.

Although many different techniques exist for implementing analog multipliers in CMOS [8], we will concentrate on a technique useful in high- and low-frequency multiplication [9, 10].

26.3.1 The Multiplying Quad

A CMOS multiplier employing a multiplying quad (M1 - M4) is shown in Fig. 26.27. The multiplying quad operates in the triode region, and thus MOSFETs M1-M4 can be thought of as resistors. For the moment we will not consider the biasing of the quad. The negative output voltage of the multiplier is given by

$$v_{o-} = -R \cdot (i_{D1} + i_{D2}) \tag{26.13}$$

while the positive output voltage is

$$v_{o+} = -R \cdot (i_{D3} + i_{D4}) \tag{26.14}$$

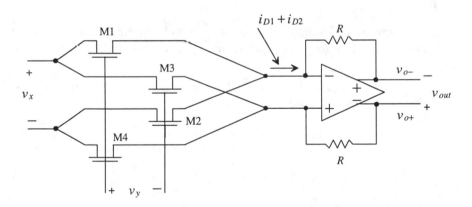

Figure 26.27 CMOS analog multiplier.

The output voltage of the multiplier is

$$v_{out} = v_{o+} - v_{o-} = R \cdot (i_{D1} + i_{D2} - i_{D3} - i_{D4}) \tag{26.15}$$

A simplified schematic of the multiplying quad with biasing is shown in Fig. 26.28. The op-amp inputs are at an AC virtual ground and at a DC voltage of V_{CM} (the op-amp output common-mode voltage). In order to minimize the DC input current on the x-axis inputs, the common-mode DC voltage on this input is set to V_{CM}. The DC biasing voltage on the y-input is set to a value large enough to keep the quad in triode. The input signals have been broken into two parts (e.g., $v_x/2$ and $-v_x/2$) to maintain generality. In practice, the minus inputs can be connected directly to the bias voltages

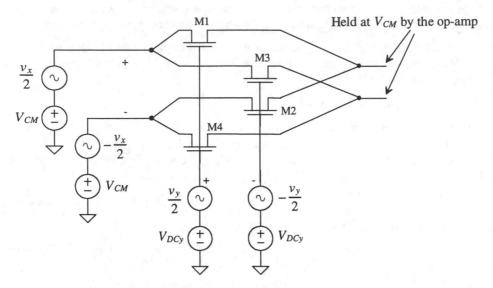

Figure 26.28 Biasing of the multiplying quad.

at the cost of large-signal linearity. (The input is not truly differential in this situation.) However, as discussed earlier, a fully differential system has much better coupled noise immunity.

Using Eq. (A.5) and noticing that the DC gate-source voltage of all MOSFETs is the same, the drain currents can be written as

$$i_{D1} = \beta_1 \left[\left(V_{GS} + \frac{v_y}{2} - V_{THN1} \right) \left(\frac{v_x}{2} \right) - \frac{1}{2} \left(\frac{v_x}{2} \right)^2 \right] \qquad (26.16)$$

$$i_{D2} = \beta_2 \left[\left(V_{GS} - \frac{v_y}{2} - V_{THN2} \right) \left(-\frac{v_x}{2} \right) - \frac{1}{2} \left(-\frac{v_x}{2} \right)^2 \right] \qquad (26.17)$$

$$i_{D3} = \beta_3 \left[\left(V_{GS} - \frac{v_y}{2} - V_{THN3} \right) \left(\frac{v_x}{2} \right) - \frac{1}{2} \left(\frac{v_x}{2} \right)^2 \right] \qquad (26.18)$$

$$i_{D4} = \beta_4 \left[\left(V_{GS} + \frac{v_y}{2} - V_{THN4} \right) \left(-\frac{v_x}{2} \right) - \frac{1}{2} \left(-\frac{v_x}{2} \right)^2 \right] \qquad (26.19)$$

We can design so that $\beta = \beta_1 = \beta_2 = \beta_3 = \beta_4$. We can use Eq. (26.15) together with Eqs. (26.16) - (26.19) to rewrite the output voltage of the multiplier as

$$v_{out} = R\beta \cdot \left(\frac{v_x}{2} \right) \left[\frac{v_y}{2} - V_{THN1} + \frac{v_y}{2} + V_{THN2} + \frac{v_y}{2} + V_{THN3} + \frac{v_y}{2} - V_{THN4} \right] \qquad (26.20)$$

We can see that if $V_{THN1} = (V_{THN2}$ or $V_{THN3})$ and $V_{THN4} = (V_{THN3}$ or $V_{THN2})$, this equation can be rewritten as

$$v_{out} = R\beta \cdot v_x v_y \qquad (26.21)$$

The source of a MOSFET (the terminal we label "source" depends on which way current flows in the MOSFET) in the multiplying quad is connected either to the op-amp or to the x inputs. When the sources of the MOSFETs are connected to the op-amp, all the MOSFETs in the multiplying quad have the same threshold voltage. (Since the source of each MOSFET is tied to the same potential, the body effect changes each MOSFET's threshold voltage by the same amount.) If the + x-input is sinking a current, then the sources of M1 and M3 are the "+" x-input and thus $V_{THN1} = V_{THN3}$. In any case, the threshold voltages of the MOSFETs cancel and Eq. (26.21) holds. Comparing Eqs. (26.21) and (26.11) results in defining the gain of this multiplier as

$$K_m = R \cdot \beta \qquad (26.22)$$

Simulating the Operation of the Multiplier

Simulating the performance and understanding the analog multiplier operation is an important step in the design process. The design of a multiplier consists of designing the op-amp, selecting the sizes of the multiplying quad, and designing the biasing network. Since we covered the design of differential input/output op-amps in the last chapter, it will not be covered here. In order to simulate the performance of a multiplier in SPICE without including the limitations of the op-amp, the simple model shown in Fig. 26.29 will be used. The sum of the multiplying factors associated with the voltage-controlled voltage sources, E1 and E2, is the open-loop gain of the op-amp. A typical SPICE statement for these VCVS (voltage-controlled voltage source) where the op-amp open loop gain is 20,000 is

```
E1      Voplus 8 4 3    1E4
E2      8 Vominus 4 3  1E4
```

where the nodes correspond to those labeled in Fig. 26.29.

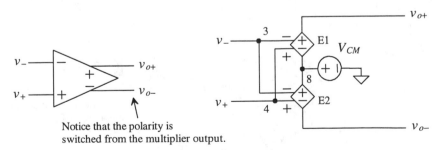

Notice that the polarity is
switched from the multiplier output.

E1 and E2 are voltage-controlled voltage sources.

Figure 26.29 SPICE modeling a differential input/output op-amp with common-mode voltage.

The next problem we encounter in simulating the operation of the multiplier is implementing the differential voltages (e.g. $\pm v_x/2$), in addition to the DC biasing voltages. The setup shown in Fig. 26.20 will be used to implement the biasing and the differential voltage sources. The op-amp common-mode output voltage, V_{CM}, and the x input voltage are set to 1.5 V. The lower this voltage, the easier it is to bias the multiplying quad into the triode region. On the other hand, a reduction in the value of V_{CM} limits the op-amp output voltage swing and thus the multiplier output range. The size of the multiplying quad was set to 10/2. The larger the W/L ratio of the MOSFETs used in this quad, the easier it is to keep the quad in the triode region. On the other hand, using a large W/L increases the required input current. The channel length can be increased to ensure the device operates as a long channel device and follows Eq. (A.5). Since the quad is part of the feedback around the op-amp, long channel devices do not affect the speed. The DC voltage at the y-inputs was set to 3.5 V as a compromise between keeping the multiplying quad in triode and the y-input voltage range. The gain of the multiplier in Fig. 26.30 is, from Eq. (26.22),

$$K_m = 20k \cdot 50\frac{\mu A}{V^2} \cdot \frac{10}{2} = 5 \text{ V}^{-1}$$

A DC sweep showing the operation of the multiplier is shown in Fig. 26.31. The x-input, v_x, was swept from –1 to +1, while at the same time the y-input was stepped from –1 to 1 V in 0.5 V increments. Keeping in mind that the output of the multiplier is $v_{o+} - v_{o-}$, we can understand the data presented in Fig. 26.31 by considering

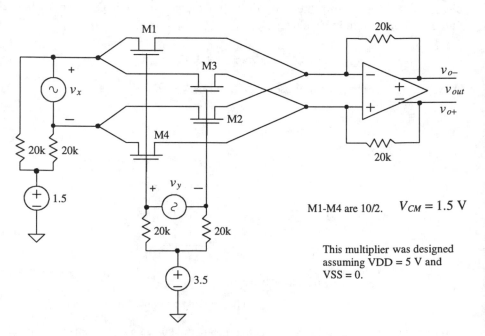

Figure 26.30 SPICE simulation schematic.

points A and B. At point A the y-input is 1 V while the x-input is 0.2 V. The output
voltage of the multiplier is the product of the multiplier gain and these two voltages
(i.e., $5 \cdot 1 \cdot 0.2 = 1$ V). The output voltage at point B is $5 \cdot (-0.5) \cdot (-0.6) = 1.5$ V. Note that
this figure was generated using an almost ideal op-amp. The characteristics do not
show the limitations of the op-amp. In particular, the limited output swing. Below is
the SPICE netlist used to generate this plot.

```
*** Top Level Netlist ***
E1        Voplus 8 4 3    1E4
E2        8 Vominus 4 3  1E4
M1        7 5 4 0 CMOSNB  L=2u W=10u
M2        11 6 4 0 CMOSNB  L=2u W=10u
M3        7 6 3 0 CMOSNB  L=2u W=10u
M4        11 5 3 0 CMOSNB  L=2u W=10u
R10       10 7 20k
R7        10 11 20k
R8        9 5 20k
R9        9 6 20k
Rfn       Vominus 4 20k
Rfp       Voplus 3 20k
VCM       8 0      DC 1.5 AC 0 0
VCMx      10 0     DC 1.5 AC 0 0
VDCy      9 0      DC 3.5 AC 0 0
Vx        7 11     DC 0 AC 1 0
Vy        5 6      DC 1 AC 0 0

.MODEL CMOSNB NMOS LEVEL=4
+VFB=-9.73820E-01, LVFB=3.67458E-01,WVFB=-4.72340E-02
see Appendix A for a complete BSIM listing

***** End of spice models and macro models *****
.OPTION ABSTOL=1u RELTOL=0.01 VNTOL=1mv
```

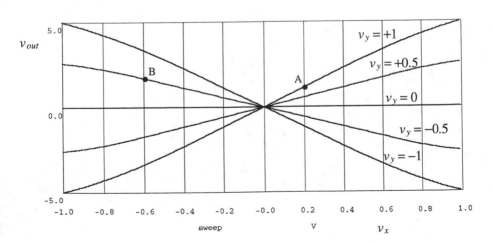

Figure 26.31 DC characteristics of the multiplier of Fig. 26.30.

```
.probe
.DC Vx -1 1 .01 Vy -1 1 .5
.plot dc all
.print dc all
.end
```

26.3.2 Level Shifting

Now that we have discussed the basic operation of the multiplier, the next step is to design the level-shifting stages. Level-shifting stages are used to implement the biasing batteries shown in Fig. 26.28 for both the x- and y-inputs. Before we complete the initial design of our basic multiplier, let's discuss general-level shifting and single-ended to differential conversion since they have many applications in single-supply chip design.

Consider the basic p-channel source-follower circuits shown in Fig. 26.32 without the body effect. The source-gate voltages of the p-channel MOSFETs are used to shift the input signals, which are referenced to ground, upward. This circuit could be useful in implementing the x-input level shifter in our analog multiplier. The x-inputs can actually go negative by V_{THP} before M1 or M2 go into the triode region. The main drawback of using this circuit is the lack of common-mode rejection on the inputs. Also, if we try to drive the circuit single-ended, that is, the minus input connected to ground, the outputs of the circuit are not truly differential. This problem is encountered in any single-ended to differential-ended conversion, including those circuits made using diff-amps. Normally, the best method of designing a single-ended to differential-ended converter is to use a high-gain differential input/output op-amp with feedback. This configuration keeps the signal levels across the input of the op-amp small. Figure 26.33 shows two op-amp configurations, which are useful in single- to

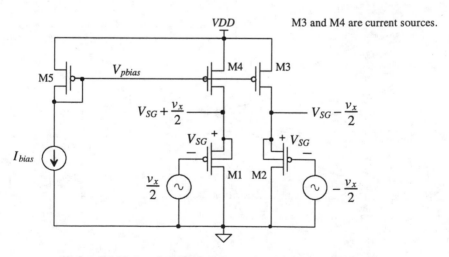

Figure 26.32 Level shifting using p-channel source followers.

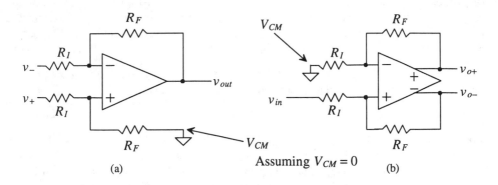

Assuming $V_{CM} = 0$

(a)   (b)

Figure 26.33 (a) Differential to singled ended converter and
(b) single to differential-ended converter.

differential-ended (and vice-versa) conversion. The input/output relationship of the
differential- to single-ended amplifier of Fig. 26.33a is

$$v_{out} = \frac{R_F}{R_I}(v_+ - v_-)$$ (26.23)

The relationship between the input and outputs of the single-ended to differential
amplifier of Fig. 26.33b is

$$v_{o+} - v_{o-} = \frac{R_F}{R_I} \cdot v_{in}$$ (26.24)

In this figure, we assumed a common-mode voltage of zero (ground). Figure 26.34
shows how the single- to differential-ended converter would be connected if the
amplifier was used in a system with a single supply. It is important that all
common-mode voltages used in the circuit have the same temperature coefficient. This
avoids input DC current and keeps the output of the op-amp from moving toward *VDD*

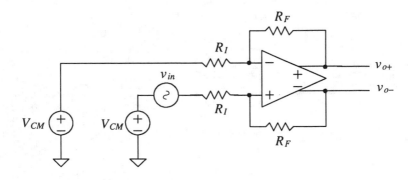

Figure 26.34 Implementation of a single- to differential-ended converter in single-supply system.

or ground. For example, if we were to use p-channel source-followers to generate the V_{CM} on the inputs of the amplifier, then we would also use a source follower to generate the op-amp V_{CM}. This idea will be explained in greater detail when we discuss the design of the x-input level-shifting network in our basic multiplier.

Another level-shifting circuit (also known as a floating battery) is shown in Fig. 26.35 [11]. The battery voltage generated by this circuit is given by

$$V_B = V_{SG1} + V_{SG2} \qquad (26.25)$$

The small-signal resistance of the battery (i.e., the resistance between the drain and source of M2) is $1/g_{m2}$. Normally, to keep $1/g_{m2}$ small, β_2 is made large. This results in a V_{SG2} of approximately V_{THP}, and therefore the size of M1 is adjusted to achieve the desired battery voltage. A potential problem exists with this configuration in its current state. Current can only flow from the source of M2 to its drain. This means that an input signal connected to the drain of M2 cannot source a current. Adding the current source, I_2, as shown in Fig. 26.36a helps solve this problem at the cost of an input current - the current I_2 flows in v_{in} under steady-state conditions. Adding the source-follower, M3, shown in Fig. 26.36b eliminates input current. Using the source-follower shifts the DC output voltage up by V_{SG3}. The major benefit of these configurations is that the input voltage can go negative by V_{THP} and M1 or M3 will remain in the saturation region. If we assume that β_2 and β_3 are large, then $V_{SG3} = V_{SG2} = V_{THP}$. The DC output voltage, for the circuit of Fig. 26.36b, used to shift an input signal upward is

$$V_{DC} = V_{SG1} + 2V_{THP} \qquad (26.26)$$

Input Level Shifting for the Multiplier of Figure 26.27 and 26.28

In the following discussion, we are assuming that the multiplier has to be DC coupled, the inputs are referenced to ground, and only a single supply voltage is used. The biasing design can be greatly simplified if these conditions do not hold. For example, if the inputs can be AC coupled, a simple resistive or MOSFET divider can be used to set

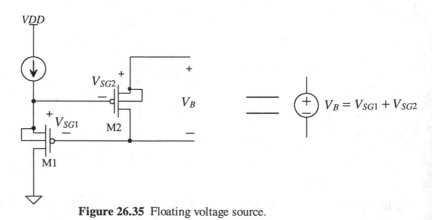

Figure 26.35 Floating voltage source.

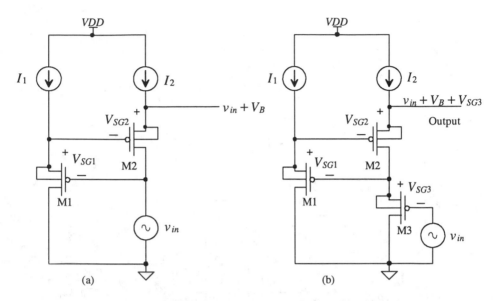

Figure 26.36 Adding input voltage source to battery: (a) with input current and (b) without input current.

the DC operating conditions while the input AC signals can be connected through a capacitor. If the inputs are referenced to, say V_{CM} (which is what would occur if the inputs signals are supplied by a differential input/output op-amp), then diff-amps can be used to generate the bias voltages since the input signals now lie within the CMR of a diff-amp.

The complete analog multiplier with level-shifting stages is shown in Fig. 26.37. Current sources M7 and M8 are biased to source 500 µA of current. The nominal V_{SG} of MOSFETs M5 and M6 is 1.5 V. A channel length of 2 µm was used in all MOSFETs used in the level-shifting networks for high-speed operation. MOSFETs M9 and M10 are used to generate the V_{CM} for the op-amp (the voltage used at the gates of MF6 and MF7 in Fig. 25.50). This guarantees that the DC reference on the x-inputs and the output of the op-amp are the same voltage and track with temperature. This eliminates DC input current through the multiplying quad and keeps the op-amp output from railing at VDD or ground. Using the same circuit (in this case a source follower) for the DC reference on the inputs and the outputs is very *important*. Note that in a dual supply system, where ground is the common-mode voltage, temperature drift and op-amp output wander are not a concern.

The y-input level-shifting network shifts the y-inputs, which are referenced to ground up to nominally 3.6 V. MOSFETs M11 through M20 are biased at 50 µA with a corresponding source-gate voltage of 1.2 V. This level-shifting configuration is wide-band since all MOSFETs are operated in the source-follower configuration. Also,

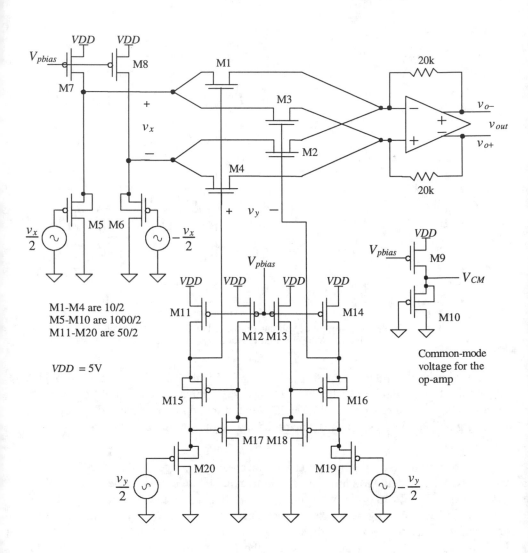

Figure 26.37 Complete analog multiplier.

in both level-shifting networks, the body effect is eliminated by laying out each MOSFET in its own well.

The multiplier output is shown in Fig. 26.38. Note that because the gain of the source-followers is less than one, the overall gain of the multiplier is less than five, which is what we predicted earlier with the simple battery implementation of Fig. 26.30.

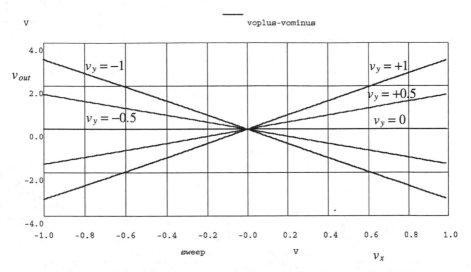

Figure 26.38 Simulation results for the multiplier shown in Fig. 26.37.

26.3.3 Multiplier Design Using Squaring Circuits

An analog multiplier can be designed based on the difference between the sum of two voltages squared and the difference of two voltages squared [12], or

$$V_o = (V_1 + V_2)^2 - (V_1 - V_2)^2 = 4V_1 V_2 \qquad (26.27)$$

The basic sum-squaring and difference-squaring circuits are shown in Fig. 26.39. MOSFETs M1 and M4 are source-followers, while MOSFETs M2 and M4 are called squaring MOSFETs. This circuit is designed so that $\beta_1 = \beta_4 = \beta_{14}$, $\beta_2 = \beta_3 = \beta_{23}$ and $\beta_{14} \gg \beta_{23}$. This makes almost all of the DC bias currents, I_{S12} and I_{S34}, flow in the MOSFETs M1 and M4, respectively. The squaring current, I_{SQ}, assuming that zero current flows through the resistor when both inputs are zero volts (or whatever the common-mode voltage when a single supply is used), is given by

$$I_{SQ(a)} = \frac{\beta_{23}}{4}(V_1 + V_2)^2 \qquad (26.28)$$

Similarly, the squaring current in the difference-square circuit of Fig. 26.39b is given by

$$I_{SQ(b)} = \frac{\beta_{23}}{4}(V_1 - V_2)^2 \qquad (26.29)$$

The output voltage of the sum-square circuit is given by

$$V_{o-} = VDD - I_{SQ(a)}R \qquad (26.30)$$

while the output voltage of the difference-square circuit is given by

$$V_{o+} = VDD - I_{SQ(b)}R \tag{26.31}$$

A multiplier is formed by taking the difference between these voltages. The output voltage of the multiplier of Fig. 26.39 is given by

$$V_{out} = V_{o+} - V_{o-} = R\frac{\beta_{23}}{4}\left[(V_1 + V_2)^2 - (V_1 - V_2)^2\right] \tag{26.32}$$

or, using Eq. (26.27)

$$V_{out} = R\beta_{23} \cdot V_1 V_2 \tag{26.33}$$

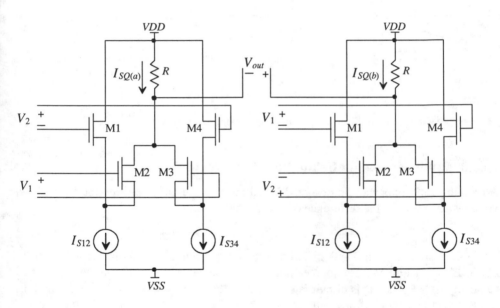

Figure 26.39 (a) Sum-squaring circuit and (b) difference squaring circuit.

REFERENCES

[1]   D. J. Allstot, "A Precision Variable-Supply CMOS Comparator," *IEEE Journal of Solid-State Circuits,* Vol. SC-17, No. 6, pp. 1080-1087, December 1982.

[2]   M. Bazes, "Two Novel Full Complementary Self-Biased CMOS Differential Amplifiers," *IEEE Journal of Solid-State Circuits*, Vol. 26, No. 2, pp. 165-168, February 1991.

[3]   B-S. Song, S. Lee, and M. F. Tompsett, "A 10-b 15 MHz CMOS Recycling Two-Step A/D Converter," *IEEE Journal of Solid-State Circuits*, Vol. 25, No. 6, pp. 1328-1338, December 1990.

[4]   A. Yukawa, "A CMOS 8-bit High-Speed A/D Converter IC," *IEEE Journal of Solid-State Circuits*, Vol. SC-20, No. 3, pp. 775-779, June 1985.

[5]   M. G. Degrauwe, J. Rijmenants, E. A. Vittoz, and H. J. DeMan, "Adaptive Biasing CMOS Amplifiers," *IEEE Journal of Solid-State Circuits*, Vol. SC-17, No. 3, pp. 522-528, June 1982.

[6]   E. A. Vittoz, "Micropower Techniques," Chapter 3 in J. E. Franca and Y. Tsividis (eds.) *Design of Analog-Digital VLSI Circuits for Telecommunications and Signal Processing,* 2nd ed., Prentice Hall, 1994. ISBN 0-13-203639-8.

[7]   S. Soclof, *Applications of Analog Integrated Circuits*, Prentice Hall, 1985. ISBN 0-13-039173-5.

[8]   M. Ismail, S-C. Huang, and S. Sakurai, "Continuous-Time Signal Processing," Chapter 3 in M. Ismail and T. Fiez (eds.), *Analog VLSI: Signal and Information Processing*, McGraw Hill, 1994. ISBN 0-07-032386-0.

[9]   B-S. Song, "CMOS RF Circuits for Data Communications Applications," *IEEE Journal of Solid-State Circuits*, Vol. SC-21, No. 2, pp. 310-317, April 1986.

[10]   J. Crols and M. S. J. Steyaert, "A 1.5 GHz Highly Linear CMOS Downconversion Mixer," *IEEE Journal of Solid-State Circuits*, Vol. 30, No. 7, pp. 736-742, July 1995.

[11]   R. Gregorian and G. C. Temes, *Analog MOS Integrated Circuits for Signal Processing*, John Wiley and Sons, 1986. ISBN 0-471-09797-7.

[12]   H-J. Song and C-K. Kim, "An MOS Four-Quadrant Analog Multiplier Using Simple Two-Input Squaring Circuits with Source Followers," *IEEE Journal of Solid-State Circuits*, Vol. 25, No. 3, pp. 841-848, June 1990.

PROBLEMS

26.1   A very important component of a comparator is its offset voltage. The offset voltage of a comparator can be modeled as a DC voltage source in series with the gate of the MOSFET used in the input diff-pair (Fig. P26.1). Regenerate Fig. 26.6, including an offset voltage of 10 mV. Comment on the effects of an offset voltage on this plot.

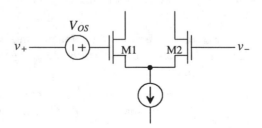

Figure P26.1

26.2   What is the input capacitance of the comparator shown in Fig. 26.11? How is the gain of the comparator affected by using a decision circuit with hysteresis?

26.3   Design and simulate the operation of a comparator with 100 mV hysteresis.

26.4   To determine the sensitivity of a comparator, the amplitude of a wide-input pulse is reduced until the output of the comparator does not make a full logic transition. Determine the minimum sensitivity of the comparator of Fig. 26.11 using SPICE. Note: In practice, making sensitivity measurements is very difficult. The comparator will oscillate, and coupled noise will affect the measurement.

26.5   Simulate the operation of the comparator shown in Fig. 26.18.

26.6   Discuss the benefits and problems of using the output buffer shown in Fig. P26.6 in conjunction with the preamp and decision circuit of Fig. 26.5.

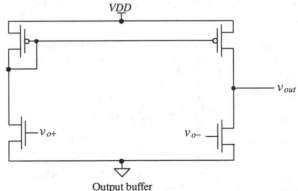

Output buffer

Figure P26.6

26.7   Can the self-biased comparator be used as a wide-swing op-amp? If so, how would the op-amp be compensated?

26.8   Sketch the schematic of an adaptive voltage follower that can source or sink current.

26.9   Discuss the benefits and problems of using long channel lengths in the multiplying quad of Fig. 26.30.

26.10 Verify, using SPICE and the differential input/output model shown in Fig. 26.29, the operation of the converters shown in Fig. 26.33. Apply sine waves to the inputs and perform a SPICE transient analysis.

26.11 Design and simulate the operation of a level-shifting circuit that will shift input signals referenced at ground up nominally 1.5 V. Comment on the frequency response of your design.

Chapter

27

Dynamic Analog Circuits

Chapter 14 discussed dynamic digital circuits, which are useful in reducing power dissipation and the number of MOSFETs used to perform a given circuit operation. Dynamic analog circuits exploit the fact that information can be stored on a capacitor or gate capacitance of a MOSFET for a period of time. In this chapter, we will discuss analog circuits such as sample and holds, current mirrors, amplifiers, and filters using dynamic techniques.

27.1 The MOSFET Switch

A fundamental component of any dynamic circuit (analog or digital) is the switch (Fig. 27.1). An important attribute of the switch, in CMOS, is that under DC conditions the gate of the MOSFET does not draw a current. Therefore, neglecting capacitances from the gate to the drain/source, we find that the gate control signal does not interfere with information being passed through the switch. Figure 27.2 shows the small-signal resistance of the switches of Fig. 27.1 plotted against drain-source voltage. The benefits of using the CMOS transmission gate are seen from this figure, namely, lower overall resistance. Another benefit of using the CMOS TG is that it can pass a logic high or a

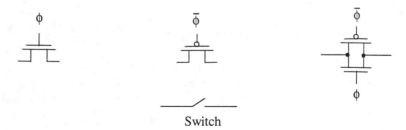

Switch

Figure 27.1 MOSFETs used as switches.

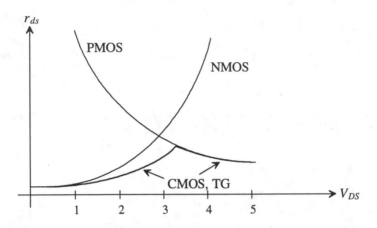

Figure 27.2 Small-signal on-resistance of MOSFET switches [1].

logic low without a threshold voltage drop. The largest voltage an n-channel switch can pass is $VDD - V_{THN}$, while the lowest voltage a p-channel switch can pass is V_{THP}.

While MOS switches may offer substantial benefits, they are not without some detraction. Two nonideal effects typically associated with these switches may ultimately limit the use of MOS switches in some applications (particularly sampled-data circuits such as data converters). These two effects are known as charge injection and clock feedthrough.

Charge Injection

Charge injection can be understood with the help of Fig. 27.3. When the MOSFET switch is on and V_{DS} is small, the charge under the gate oxide resulting from the inverted channel is (from Ch. 5) Q'_{ch}. When the MOSFET turns off, this charge is injected onto the capacitor and into v_{in}. Since v_{in} is assumed to be a low-impedance, source-driven node, the injected charge has no effect on this node. However, the charge injected onto C_{load} results in a change in voltage across it.

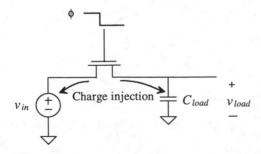

Figure 27.3 Simple configuration using an NMOS switch to show charge injection.

Although the charge injection mechanism is itself a complex one, many studies have sought to characterize and minimize its effects [2-4]. It has been shown that if the clock signal turns off fast, the channel charge distributes fairly equally between the adjacent nodes. Thus, half of the channel charge is distributed onto C_{load}. From Ch. 5, the charge/unit area of an inverted channel can be approximated as

$$Q_I'(y) = C_{ox}' \cdot (V_{GS} - V_{THN}) \tag{27.1}$$

The total charge in the channel must then be multiplied by the area of the channel resulting in

$$Q_I(y) = C_{ox}' \cdot W \cdot L \cdot (V_{GS} - V_{THN}) \tag{27.2}$$

Therefore, the change in voltage across C_{load} (if an NMOS switch is used) is

$$\Delta v_{load} = -\frac{C_{ox}' \cdot W \cdot L \cdot (V_{GS} - V_{THN})}{2C_{load}} \tag{27.3}$$

which can be written as

$$\Delta v_{load} = -\frac{C_{ox}' \cdot W \cdot L \cdot (VDD - v_{in} - V_{THN})}{2C_{load}} \tag{27.4}$$

if it is assumed that the clock swings between V_{DD} and V_{SS}. The threshold voltage can also be substituted into Eq. (27.4) to form

$$\Delta v_{load} = -\frac{C_{ox}' \cdot W \cdot L \cdot (VDD - v_{in} - [V_{THN0} + \gamma(\sqrt{|2\phi_F| + v_{in} - V_{SS}} - \sqrt{|2\phi_F|})])}{2C_{load}} \tag{27.5}$$

Note that Eq. 27.5 illustrates the problem associated with charge injection. The change in voltage across C_{load} is nonlinear with respect to v_{in} due to the threshold voltage. Thus, it can be said that if the charge injection is signal dependent, harmonic distortion results. In sampled-data systems, charge injection will result in nonlinearity errors. In the case where the charge injection is signal independent, a simple offset occurs which is much easier to manage than harmonic distortion. These will be discussed in more detail in Chs. 28 and 29, but it should be obvious here that charge injection effects should be minimized as much as possible.

Capacitive Feedthrough

Consider the schematic of the NMOS switch shown in Fig. 27.4. Here the capacitances between the gate/drain and gate/source of the MOSFET are modeled with the assumption that the MOSFET is operating in the triode region. When the gate clock signal, ϕ, goes high, the clock signal feeds through the gate/drain and gate/source capacitances. However, as the switch turns on, the input signal, v_{in}, is connected to the load capacitor through the NMOS switch. The result is that C_{load} is charged to v_{in} and the capacitive feedthrough has no effect on the final value of v_{out}. However, now consider what happens when the clock signal makes the transition low, that is, the n-channel MOSFET turns off. A capacitive voltage divider exists between the gate-drain (source) capacitance and the load capacitance. As a result, a portion of the clock signal, ϕ, appears across C_{load} as

$$\Delta v_{load} = \frac{C_{overlap} \cdot (V_{DD} - V_{SS})}{C_{overlap} + C_{load}}$$ (27.6)

where $C_{overlap}$ is the overlap capacitance value,

$$C_{overlap} = C'_{ox} \cdot W \cdot LD$$ (27.7)

and LD is the length of the gate that overlaps the drain/source.

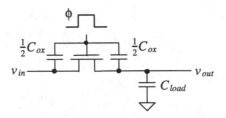

Figure 27.4 Illustration of capacitive feedthrough.

Reduction of Charge Injection and Clock Feedthrough

Many methods have been reported that reduce the effects of charge injection and capacitive feedthrough. One of the most widely used is the dummy switch [4, 5], as seen in Fig. 27.5. Here, a switch with its drain and source shorted is placed in series with the desired switch M1. Notice that the clock signal controlling the dummy switch is the complement of the signal controlling M1, and in addition, should also be slightly delayed.

When M1 turns off, half the channel charge is injected toward the dummy switch, thus explaining why the size of M2 is one-half that of M1. Although M2 is effectively shorted, a channel can still be induced by applying a voltage on the gate. Therefore, the charge injected by M1 is essentially matched by the charge induced by M2, and the overall charge injection is canceled. Note what happens when M2 turns

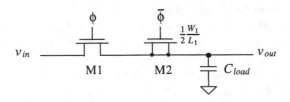

Figure 27.5 Dummy switch circuit used to minimize charge injection.

off. It will inject half of its charge in both directions. However, since the drain and source are shorted and M1 is on, all the charge from M2 will be injected onto the low-impedance, voltage-driven source which is also charging C_{load}. Therefore, M2 charge injection will not affect the value of voltage on C_{load}.

Another method used to counteract charge injection and clock feedthrough is to replace the switch with a CMOS transmission gate (TG). This will result in lower changes in v_{out} because the complementary signals used will act to cancel each other. However, this approach requires precise control on the complementary clocks (the clocks must be switched at exactly the same time) and assumes that the input signal, v_{in}, will be small, since the symmetry of the turn-on and turn-off waveforms are dependent on the input signal.

Fully differential circuit topologies are used to cancel these effects to a first order, as seen in Fig. 27.6. Since the nonideal charge injection and clock feedthrough effects appear as a common-mode signal to the amplifier, they will be reduced by the CMRR of the amplifier. However, the second-order effects resulting from the input signal amplitude dependence will ultimately limit the dynamic range of operation, neglecting coupled and inherent noise in the dynamic circuits. This subject will be discussed in more detail in the next sections.

kT/C Noise

In Ch. 7 we saw that the maximum RMS output noise generated from a simple RC circuit was $\sqrt{kT/C}$ (see Ex. 7.7). If we think of the MOSFET in Fig. 27.7 as a resistor (when the MOSFET is on), then we can add this RMS noise source in series with the output of the capacitor. The noise can be regarded as a sampled voltage onto the capacitor each time the switch is turned on. The RMS noise generated when using a 1 pF capacitor is 64 µV, while a 100 fF capacitor results in a noise voltage of 200 µV. In other words, the larger the capacitor, the smaller the noise voltage sampled on to the storage capacitor. For high-speed systems, it is desirable to use small capacitors since they take less time to charge. When designing a high-speed/low-noise circuit tradeoffs must be made when selecting the capacitor size.

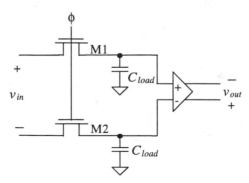

Figure 27.6 Use of a fully differential circuit to minimize charge injection and clock feedthrough.

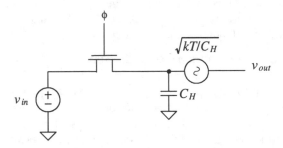

Figure 27.7 How kT/C noise adds to a sampled signal.

Sample-and-Hold Circuits

An important application of the switch is in the sample-and-hold circuit. The sample-and-hold circuit finds extensive use in data converter applications as a sampling gate. A variety of topologies exist, each with their own benefits. The simplest is shown in Fig. 27.8. A narrow pulse is applied to the gate of the MOSFET, enabling v_{in} to charge the hold capacitor, C_H. The width of the strobing gate pulse should allow the capacitor to fully charge before being removed. The op-amp simply acts as a unity gain buffer, isolating the hold capacitor from any external load. This circuit suffers from the clock feedthrough and charge injection problems mentioned in the previous discussion.

Figure 27.9 shows a fully differential sample-and-hold circuit and associated clock waveforms that eliminate clock feedthrough and charge injection to a first order [6]. The switches in this figure are closed when their controlling clock signals are high. The basic operation can be understood by considering the state of the circuit at t_0. At this time, the input signals charge the sampling capacitors. The bottom plates of the capacitors (poly1) are tied directly to the input signals, for reasons that will be explained below. The op-amp is operating in a unity-follower configuration in which both inputs of the op-amp are held at V_{CM}. At this particular instance in time, prior to t_1, the amplifier is said to be operating in the sample mode of operation.

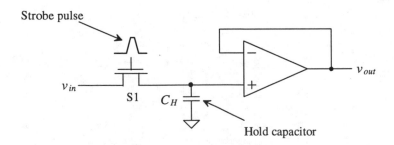

Figure 27.8 A basic sample-and-hold circuit.

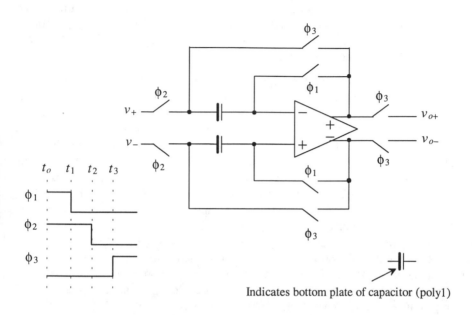

Figure 27.9 Sample-and-hold using differential topology [6].

At t_1 the ϕ_1 switches turn off. The resulting charge injection and clock feedthrough appear as a common-mode signal on the inputs of the op-amp and are ideally rejected. Since the top plates of the hold capacitors (the inputs to the op-amp) are always at V_{CM}, at this point in time the charge injection and clock feedthrough are independent of the input signals. The result is an increase in the dynamic range of the sample-and-hold (the minimum measurable input signal decreases). The voltage on the inputs of the op-amp (the top plate of the capacitor) between t_1 and t_2 is $V_{OFF1} + V_{CM}$, a constant voltage. Note that the op-amp is operating open loop at this time so that the time between t_1 and t_3 should be short.

At t_2 the ϕ_2 switches turn off. At this point in time, the voltages on the bottom plates of the sampling capacitors are v_+ and v_- for the $+$ and $-$ inputs of the circuit, respectively. The voltages on the top plates of the capacitors are $V_{OFF1} + V_{OFF2} + V_{CM}$ (assuming the storage capacitors are much larger than the input capacitance of the op-amp). The term V_{OFF2} is ideally a constant that results from the charge injection and capacitive feedthrough from the ϕ_2 switches turning off. The time between t_1 and t_2 should be short compared to variations in the input signals.

At time t_3 the ϕ_3 switches turn on and the op-amp behaves like a voltage follower, and the circuit is said to be in the hold mode of operation. The charge injection and clock feedthrough resulting from the ϕ_3 switches turn on cause the top plate of the capacitor to become $V_{OFF1} + V_{OFF2} + V_{OFF3} + V_{CM}$, again assuming that the storage capacitors are much larger than the input capacitance of the op-amp. The outputs of the sample-and-hold are v_+ and v_-, assuming infinite op-amp gain since these offsets appear as a common-mode voltage on the input of the op-amp. Note that the terms V_{OFF2} and V_{OFF3} are dependent on the input signals.

The reason for connecting the input signals to the bottom plate of the capacitor can be explained with the help of Fig. 27.10. This figure is a simplified, single-ended version of Fig. 27.9 where the capacitance, C_p, is the parasitic capacitance from the bottom plate to the substrate. With regard to Fig. 27.10a, coupled noise from the substrate sees either the input voltage from the op-amp driving the sample-and-hold (in the sample mode) or the output voltage of the op-amp used in the sample-and-hold itself (in the hold mode). Since the op-amps in either mode directly set this voltage, the substrate noise will, ideally, have little effect on the circuit operation. In Fig. 27.10b, coupled substrate noise feeds directly into the input of the op-amp and can thus drastically affect the output of the sample-and-hold. Another more subtle problem occurs in the circuit of Fig. 27.10b. When the circuit makes the transition to the hold mode at t_3, the output of the op-amp should quickly change to the voltage sampled on the input capacitors. The time it takes the output of the op-amp to change and settle to this final voltage is called the *settling time*. The parasitic capacitance on the input of the op-amp reduces the feedback factor from unity to $C_H /(C_p + C_H)$. This slows the settling time and causes a gain error in the circuit's transfer function. For these reasons, the parasitic capacitance on the top plate of the capacitor should be small. Nothing should be laid out over or in near proximity of poly2.

Another improvement of the basic S/H circuit can be seen in Fig. 27.11 [7, 8]. Here, two amplifiers buffer the input and the output. Notice that switch S_2 ensures that amplifier A1 is stable while in hold mode. If the switch were not present, A1 would be open loop during hold mode and would swing to one of the rails. During the next

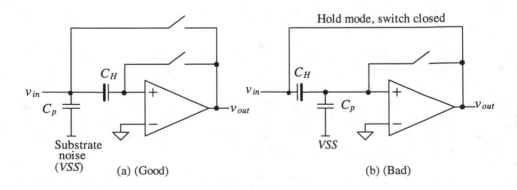

Figure 27.10 Explanation for connecting the bottom plate of the capacitor to the input.

sample mode, it would then be slew limited while going from the supply to the value of v_{in}. However, with the addition of S_2, the output of A1 tracks v_{in} even while in hold mode. The switch S_3 also disconnects A1 from the output during hold mode.

This S/H has its disadvantages, however. The capacitor is still subjected to charge injection and clock-feedthrough problems. In addition, during sample mode, the circuit may become unstable since there are now two amplifiers in the single-loop feedback structure. Although compensation capacitors can be added to stabilize its performance, the size and placement of the capacitors are purely dependent on the type and characteristics of the op-amps.

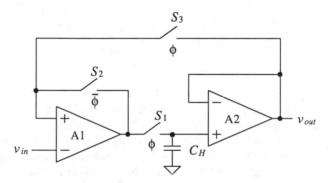

Figure 27.11 A closed-loop S/H circuit.

Another S/H circuit is seen in Fig. 27.12 [7,9]. Here, a transconductance amplifier is used to charge the hold capacitor. A control signal turns the amplifier, A1, on or off digitally, thus eliminating the need for the switches S_2 and S_3 (from Fig. 27.11). Since CMOS op-amps are well suited for high-output impedance applications, this configuration would seem to be a popular one. However, the speed of this topology is dictated by the maximum current output of the transconductance amplifier and the size of the hold capacitor.

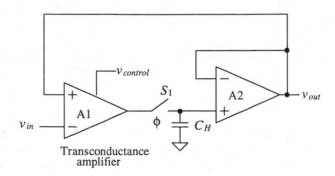

Figure 27.12 A closed-loop S/H circuit using a transconductance amplifier.

An even better S/H circuit can be seen in Fig. 27.13 [7, 8]. The advantage of this circuit may not be completely obvious and warrants further explanation. First, notice that the hold capacitor is actually in the feedback path of the amplifier, A2, with one side connected to the output of the amplifier and the other connected to a virtual ground. When switch S_1 turns off, any charge injected onto the hold capacitor will result in a slight change in the output voltage. However, now that one side of the switch is at virtual ground, the change in voltage is no longer dependent on the threshold voltage of the switch itself. Therefore, the charge injection will be independent of the input signal and will result as a simple offset at the output. An offset error is much easier to tolerate than a nonlinearity error, as will be seen in Chs. 28 and 29.

When sampling, S_1 is closed and S_2 is open, and the equivalent circuit is simply a low-pass filter with a buffered input. The overall transfer function becomes

$$\frac{v_{out}}{v_{in}} = -\frac{R_2}{R_1} \cdot \frac{1}{(sR_2C_H+1)} \tag{27.8}$$

Therefore, this circuit performs a low-pass filter function while sampling. The buffer A1 can be eliminated when we desire a low-input impedance. Once hold mode commences, the output will stay constant at a value equal to v_{in}, while the switch S_2 isolates the input from the hold capacitor. One important issue to note here is that A2 will need to be a buffered CMOS amplifier because of the resistive load attached at v_{out} during hold mode. Notice also that during both sample mode and hold mode, there is only one op-amp in each feedback loop, so this S/H topology is much more stable than the closed-loop structure introduced in Fig. 27.12.

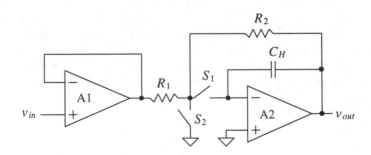

Figure 27.13   A closed-loop S/H circuit using a transconductance amplifier.

27.2 Switched-Capacitor Circuits

Consider the circuit shown in Fig. 27.14a [10, 11]. This dynamic circuit, named a switched-capacitor resistor, is useful in simulating a large value resistor, generally > 1 MΩ. The clock signals ϕ_1 and ϕ_2 form two phases of a nonoverlapping clock signal with frequency f_{clk} and period T (see Ch. 14). Let's begin by considering the case when

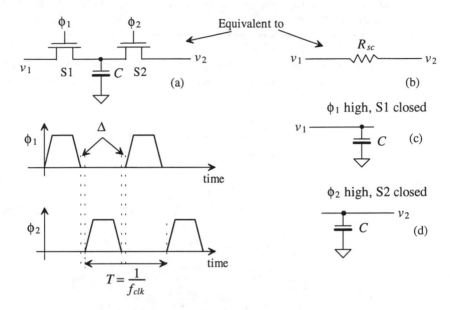

Figure 27.14 Switched-capacitor resistor (a) and associated waveforms, and
(b, c, d) the equivalent circuits.

S1 is closed. When ϕ_1 is high, the capacitor C is charged to v_1. The charge, q_1, stored
on the capacitor during this interval, Fig. 27.14c, is

$$q_1 = Cv_1 \tag{27.9}$$

while if S2 is closed, the charge stored on the capacitor is

$$q_2 = Cv_2 \tag{27.10}$$

If v_1 and v_2 are not equal, keeping in mind that S1 and S2 cannot be closed at the same
time due to the nonoverlapping clock signals, then a charge equal to the difference
between q_1 and q_2 is transferred between v_1 and v_2 during each interval T. The
difference in the charge is given by

$$q_1 - q_2 = C(v_1 - v_2) \tag{27.11}$$

If v_1 and v_2 vary slowly compared to f_{clk}, then the average current transferred in an
interval T is given by

$$I_{avg} = \frac{C(v_1 - v_2)}{T} = \frac{v_1 - v_2}{R_{sc}} \tag{27.12}$$

where the resistance of the switched-capacitor circuit is given by

$$R_{sc} = \frac{T}{C} = \frac{1}{Cf_{clk}} \tag{27.13}$$

In general, the signals v_1 and v_2 should be bandlimited to a frequency at least ten times less than f_{clk} (more on this later).

Example 27.1

Using switched-capacitor techniques, implement the circuit shown in Fig. 27.15a so that the product of RC is 1 ms, that is, the 3-dB frequency of $|v_{out}/v_{in}|$ is 159 Hz.

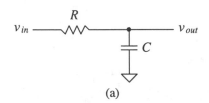

(a)

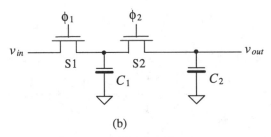

(b)

Figure 27.15   (a) Circuit used in Ex. 27.1 and (b) its implementation using a SC resistor.

The switched-capacitor implementation of this circuit is shown in Fig. 27.15b. The product of RC may now be written in terms of Eq. (27.13) as

$$RC_2 = \frac{C_2}{C_1} \cdot \frac{1}{f_{clk}} \qquad (27.14)$$

This result is important! The product RC_2 is determined by f_{clk}, which may be an accurate frequency derived from a crystal oscillator and the ratio of C_2 to C_1, which will be within 1 percent on a chip. This means that even if the values of the capacitors change by 20 percent from wafer to wafer, the ratio of the capacitors relative to one another, on the same wafer, will remain constant within 1 percent.

It is also desirable to keep C_1 larger than the associated parasitics present in the circuit (e.g., depletion capacitances of the source/drain implants and the stray capacitances to substrate). For the present example, we will set C_1 to 1 pF. The selection of f_{clk} is usually determined by what is available. For the present design, a value of 100 kHz will be used. This selection assumes that the energy present in v_{in} at frequencies above 10 kHz is negligible. The value of C_2 is determined solving Eq. (27.14) and is 100 pF.   Note that the value of the switched-capacitor resistor is 10 MΩ. Implementing this resistor in the CN20 process using n-well would require 4,000 squares! The resulting delay through the n-well resistor, because of the capacitance to substrate, may cause a significant phase error in the transfer function as well. ∎

27.2.1 Switched-Capacitor Integrator

Because the switched-capacitor resistor of Fig. 27.14a is sensitive to parasitic capacitances, it finds little use in many switched-capacitor circuits. Consider the circuit of Fig. 27.16a. This circuit is a switched-capacitor integrator and is the heart of the circuits we will be discussing in the remainder of the section [11]. The portion of the circuit consisting of switches S1 through S4 and C_I forms a switched-capacitor resistor with a value given by

$$R_{sc} = \frac{1}{C_I f_{clk}} \qquad (27.15)$$

The equivalent continuous time circuit for the switched-capacitor integrator is shown in Fig. 27.16b. Notice that v_{in} is now negative. It may be helpful in the following discussion to remember that the combination of switches and C_I of the switched-capacitor integrator can be thought of as a simple resistor. The transfer function of the switched-capacitor integrator is given by

$$\frac{v_{out}}{v_{in}} = \frac{1/j\omega C_F}{R_{sc}} = \frac{1}{j\omega\left(\frac{C_F}{C_I} \cdot \frac{1}{f_{clk}}\right)} \qquad (27.16)$$

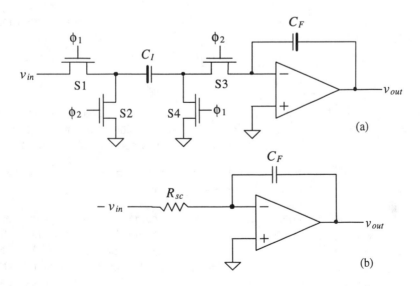

Figure 27.16 (a) A stray insensitive switched-capacitor integrator (noninverting) and (b) the equivalent continuous time circuit.

Again, the ratio of capacitors is present, allowing the designer to precisely set the gain of the amplifier and the integration time constant.

As we mentioned, the switched-capacitor integrator is not sensitive to parasitic or stray capacitances. This can be understood with the use of Fig. 27.17. First, if we realize that C_{p2} (the parasitic capacitance from the right side of C_I to ground) is always connected to ground either through S4 or by the virtual ground on the inverting input of the op-amp, it cannot store a charge (both sides of C_{p2} are connected to ground). The capacitance C_{p1} is charged to v_{in} when S1 is closed and then discharged to ground when S2 closes. Since none of the charge stored on C_{p1} when S1 is closed is transferred to C_I, it does not affect the integrating function. A practical minimum for C_I is 100 fF in the CN20 process.

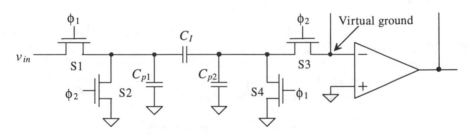

Figure 27.17 Parasitic capacitances associated with a switched-capacitor resistor.

The inverting integrator configuration is shown in Fig. 27.18. Note the simple change of clock signals (S1 and S2) from Fig. 27.16 (the noninverting configuration). The gain of this configuration is given by

$$\frac{v_{out}}{v_{in}} = -\frac{1}{j\omega\left(\frac{C_F}{C_I} \cdot \frac{1}{f_{clk}}\right)} \tag{27.17}$$

An example of a switched-capacitor integrator circuit that combines input signals is shown in Fig. 27.19a. Remembering that each switched-capacitor section can be thought of as a resistor, we note that the relationship between the inputs and output is

$$v_{out} = \frac{v_1}{j\omega\left(\frac{C_F}{C_1}\frac{1}{f_{clk}}\right)} + \frac{v_2}{j\omega\left(\frac{C_F}{C_2}\frac{1}{f_{clk}}\right)} - \frac{v_3}{j\omega\left(\frac{C_F}{C_3}\frac{1}{f_{clk}}\right)} \tag{27.18}$$

Figure 27.19b shows how redundant switches can be combined to reduce the number of devices used. Used alone, the basic integrator has the practical problem of integrating not only the input signal but also the offset voltage of the op-amp. In many applications, a reset switch or resistor is placed across the feedback capacitor (Fig. 27.19b). An example of a lossy integrator circuit useful in first-order filter design is shown in Fig. 27.20. The transfer function of this circuit is given by

$$\frac{v_{out}}{v_{in}} = \frac{R_4}{R_3}\left(\frac{1+j\omega R_3 C_1}{1+j\omega R_4 C_2}\right) = \frac{C_3}{C_4}\left(\frac{1+j\omega\left(\frac{C_1}{C_3}\cdot\frac{1}{f_{clk}}\right)}{1+j\omega\left(\frac{C_2}{C_4}\cdot\frac{1}{f_{clk}}\right)}\right) \qquad (27.19)$$

For low-frequency input signals (low frequencies compared to the pole and zero given in Eq. (27.19), the gain of the lossy integrator is simply

$$\frac{v_{out}}{v_{in}} = \frac{C_3}{C_4} \qquad (27.20)$$

which is again a precise number due to the ratio of the capacitors. Also note that the switched-capacitor resistor in the feedback loop is stray insensitive. The left side of C_4 is always connected to 0 V, while the right side is either connected to ground or to the output of the op-amp.

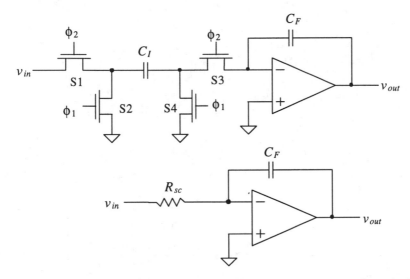

Figure 27.18 A stray insensitive switched-capacitor integrator (inverting).

Example 27.2
Design a switched-capacitor filter with the transfer characteristics shown in Fig. Ex27.2.

We can see that this transfer function has a pole at 500 Hz and a zero at 5 kHz. The lossy integrator of Fig. 27.20 will be used to realize this filter. The low-frequency gain of this circuit is 10 (20 dB). Using Eq. (27.20), we have

$$\frac{C_3}{C_4} = 10$$

while the pole and zero locations are given by

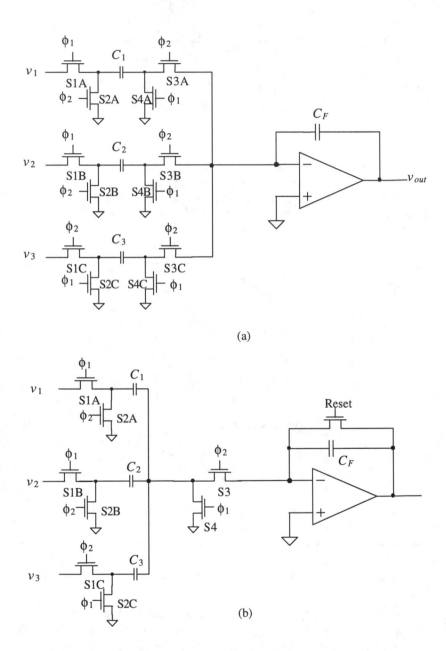

Figure 27.19 (a) Switched-capacitor implementation of a summing integrator and (b) practical implementation of the circuit combining switches and adding reset.

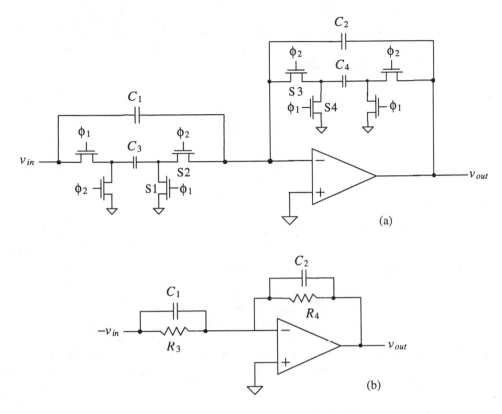

Figure 27.20 Lossy integrator (a) switched-capacitor implementation and (b) continuous time circuit.

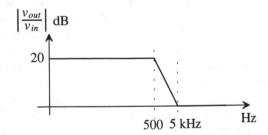

$$f_p = \frac{1}{2\pi\left(\frac{C_2}{C_4} \cdot \frac{1}{f_{clk}}\right)} = 500 \text{ and } f_z = \frac{1}{2\pi\left(\frac{C_1}{C_3} \cdot \frac{1}{f_{clk}}\right)} = 5 \text{ kHz}$$

If we set f_{clk} to 100 kHz and C_4 to 100 fF, then $C_3 = 1.0$ pF, $C_2 = 3.2$ pF, and C_1 = 3.2 pF. ∎

We will now develop an exact relationship between the switching frequency, f_{clk}, and the signal frequency ω [12,13]. Referring to Fig. 27.21, we can write the output of the integrator as the sum of the previous output voltage, $v_{out(n)}$, at a time nT and the contribution from the current sample as

$$v_{out(n+1)} = v_{out(n)} + \frac{C_I}{C_F} \cdot v_{in(n)} \tag{27.21}$$

Since a delay in the time domain of T corresponds to a phase shift of ωT in the frequency domain, we can take the Fourier transform of this equation and get

$$e^{j\omega T} v_{out}(j\omega) = v_{out}(j\omega) + \frac{C_I}{C_F} \cdot v_{in}(j\omega) \tag{27.22}$$

Solving this equation for v_{out}/v_{in} gives

$$\frac{v_{out}}{v_{in}}(j\omega) = \frac{C_I}{C_F}\left(\frac{1}{e^{j\omega T}-1}\right) = \frac{C_I}{C_F}\left(\frac{e^{-j\omega T/2}}{e^{j\omega T/2}-e^{-j\omega T/2}}\right) = \frac{C_I}{C_F}\left[\frac{1}{z-1}\right] \tag{27.23}$$

where $z = e^{j\omega T}$. Remembering $f_{clk} = 1/T$ and $\omega = 2\pi f$, we get

$$\frac{v_{out}}{v_{in}}(j\omega) = \frac{1}{j\omega\left(\frac{C_F}{C_I}\cdot\frac{1}{f_{clk}}\right)}\left(\frac{\frac{\pi \cdot f}{f_{clk}}}{\sin\frac{\pi \cdot f}{f_{clk}}}\cdot e^{-j\pi \cdot f/f_{clk}}\right) \tag{27.24}$$

Ideally, the term on the right in parentheses is unity. This occurs when f is much less than f_{clk}. This equation describes how the magnitude and phase of the integrator are affected by finite f_{clk}.

Capacitor Layout

An important step in the implementation of any switched-capacitor design is the layout of the capacitors. Normally, a unit size capacitor is laid out and then replicated to the desired capacitance. For the previous example, the unit size capacitance would be nominally 100-fF; see Fig. 27.22 (Fig. 7.10)[1]. Again, the absolute value of the capacitors isn't important; rather, the important value is the ratio. A total of 32 of these unit size capacitors would be used to achieve the larger nominally 3.2 pF capacitors in the previous example (see Fig. 27.23). Using this approach eliminates errors due to uneven patterning of poly to a first order. A p+ guard ring is placed around the capacitor to help reduce coupled substrate noise. Substrate noise can also be reduced by laying the capacitor out over an n-well that is tied to *VDD*. Injected minority carriers are collected either by the p+ or the n-well (or a combination of both). If matching of the capacitors is critical, schemes that use a common-centroid layout can be used. Also, as was discussed in Ch. 20, dummy poly strips or capacitors can be placed around the array of capacitors so that the edge differences from underetching poly are eliminated.

---

[1] Note that here we are assuming that a circle can be accurately reproduced on the reticle and patterned on the wafer. In practice, effects such as the finite e-beam size (and the granularity of the grid) used to make the reticle can make this assumption questionable.

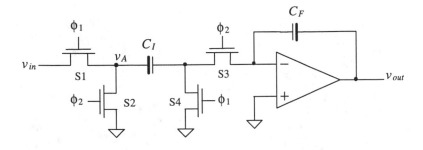

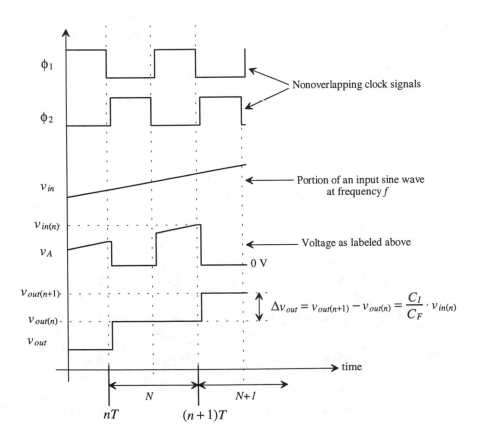

Figure 27.21 Switched-capacitor integrator used to determine the relationship between input frequency and switch clock frequency.

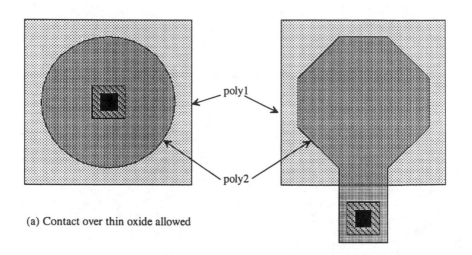

(a) Contact over thin oxide allowed

(b) Contact over thin oxide not allowed

Figure 27.22 Layout of a nominally 100 fF capacitor unit cell.

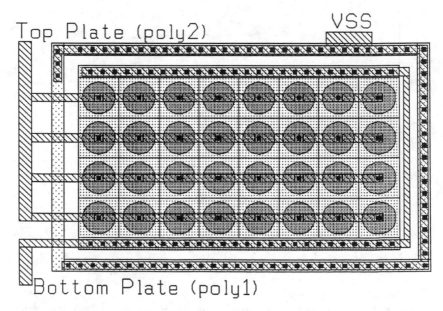

Figure 27.23 Layout of a 3.2 pF capacitor using a 100 fF unit cell.

Implementing Ladder Filters

Consider the ladder low-pass filter shown in Fig. 27.24. The doubly terminated structure is used to reduce sensitivity to component variations [14]. The resistors R_1 and R_5 will be implemented using switched-capacitor techniques along with the inductor and capacitors shown in the schematic. With regard to the notation shown in the figure, the voltage across L_3 and the current through L_3 are labeled v_3 ($= v_2 - v_4$) and i_3, respectively. Also, we will define the product of the current through the inductor with a scaling resistance, R_s (normally $R_s = 1\ \Omega$) as the inductor scaled voltage. This can be written for L_3 as

$$v_3' = i_3 \cdot R_s \qquad (27.25)$$

For the capacitor C_2 in Fig. 27.24, we can write

$$v_2 = \frac{i_2}{j\omega C_2} = \frac{1}{j\omega C_2}(i_1 - i_3) = \frac{v_s}{j\omega C_2 R_1} - \frac{v_2}{j\omega C_2 R_1} - \frac{v_3'}{j\omega C_2 R_s} \qquad (27.26)$$

which results in a summing integrator, that is, C_2 is replaced with a switched-capacitor summing integrator (see Fig. 27.25a). The value of the capacitors in this figure is determined using

$$C_2 R_1 = \frac{C_{F2}}{C_{21}} \cdot \frac{1}{f_{clk}} \qquad (27.27)$$

$$C_2 R_s = \frac{C_{F2}}{C_{22}} \cdot \frac{1}{f_{clk}} \qquad (27.28)$$

For the inductor L_3 we can write

$$v_3' = R_s \cdot i_3 = \frac{v_2}{j\omega L_3/R_s} - \frac{v_4}{j\omega L_3/R_s} \qquad (27.29)$$

This is again a summing integrator (see Fig. 27.25b). The value of the components in this figure are determined by solving

$$\frac{L_3}{R_s} = \frac{C_{F3}}{C_{31}} \cdot \frac{1}{f_{clk}} \qquad (27.30)$$

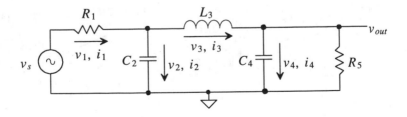

Figure 27.24 Doubly terminated low-pass ladder filter.

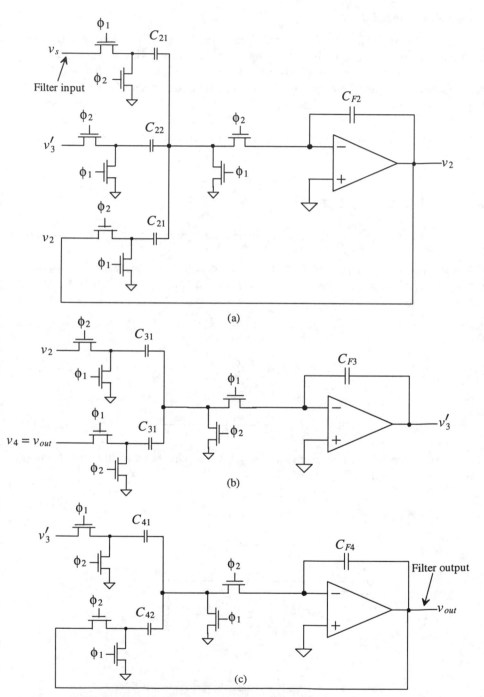

Figure 27.25 (a) Switched-capacitor implementation of R1 and C2 in Fig. 27.24, (b) Switched-capacitor implementation of L3 in Fig. 27.24, and (c) implementation of C4, R5.

Notice that we have switched the phase of the clock signals, ϕ_1 and ϕ_2, in Fig. 27.25b from Fig. 27.25a (look at the switches directly adjacent to the op-amp). This makes the output of the circuit in (a) immediately available to the input of (b) without a time delay, $T (= 1/f_{clk})$. For the voltage $v_4 (= v_{out})$, we can write

$$v_4 = \frac{i_3 - i_5}{j\omega C_4} = \frac{v_3'}{j\omega C_4 R_s} - \frac{v_{out}}{j\omega C_4 R_5} = v_{out} \qquad (27.31)$$

where

$$C_4 R_s = \frac{C_{F4}}{C_{41}} \cdot \frac{1}{f_{clk}} \qquad (27.32)$$

and

$$C_4 R_5 = \frac{C_{F4}}{C_{42}} \cdot \frac{1}{f_{clk}} \qquad (27.33)$$

The implementation of this section of the ladder filter is shown in Fig. 27.25c. Again, notice that the clock phases have been switched from the previous section of the filter. The complete filter of Fig. 27.24 is realized by combining the three sections (one section for each reactive component) shown in Fig. 27.25.

Op-Amp Settling Time

Figure 27.26 shows an op-amp configuration in which the op-amp can source or sink current to a switched-capacitor and a feedback capacitor. The time it takes to charge and discharge these capacitors is important since it directly affects the maximum switched capacitor clocking frequency, f_{clk}. The slew-rate limitations of the op-amp have been discussed in detail already. Let's now consider the limitations due to op-amp finite bandwidth. The closed-loop gain of the op-amp is given by

$$A_{CL} = \frac{A_{OL}}{1 + A_{OL} \cdot \beta} \qquad (27.34)$$

while the open-loop gain of an op-amp is given, with units of A/A, V/V, V/A, or A/V, by

$$A_{OL} = \frac{A_{OL}(0)}{1 + j\frac{f}{f_{3dB}}} \qquad (27.35)$$

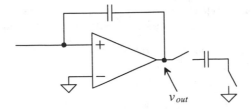

Figure 27.26 Charging and discharging a switched capacitor.

Combining these equations and assuming $1 \gg 1/[\beta \cdot A_{OL}(0)]$, we get

$$A_{CL} = \frac{\frac{1}{\beta}}{1 + j\frac{f}{f_u \cdot \beta}} \qquad (27.36)$$

where $f_{3dB} \cdot A_{OL}(0) = f_u$, where f_u may have units of Hz, Hz/Ω, or Hz·Ω ($f_u\beta$ is in Hz). The closed-loop gain reduces to a simple single-pole transfer function. The low-frequency gain of the circuit is $1/\beta$, while the product of f_u and β gives the circuit time constant of

$$\tau = \frac{1}{2\pi f_u \cdot \beta} \qquad (27.37)$$

keeping in mind the unity-gain frequency of the op-amp, f_u, is a strong function of the load capacitance. For a step-input to the op-amp, a common occurrence in switched-capacitor circuits (Fig. 27.21), the output voltage of the op-amp, again neglecting slew-rate limitations, is given by

$$v_{out} = V_{outfinal}(1 - e^{-t/\tau}) \qquad (27.38)$$

For the output voltage of the op-amp to settle to less than 1 percent of its final value requires 5τ. A "rule-of-thumb" estimate for the settling time is simply $1/(f_u \cdot \beta)$.

Differential Output Op-Amps

As mentioned previously, fully differential op-amps are a necessity in any modern mixed-signal CMOS integrated circuit design, including switched-capacitor circuits. The benefits of fully differential design can be summarized and listed as follows.

1.   *Input common-mode voltage remains at* V_{CM}. This benefit eases the requirements on the CMR of the diff-amp used in an op-amp. For example, if V_{CM} is 2.5 V, the inputs of the op-amp will remain at 2.5 V. This holds for all fully differential circuit topologies using the differential output op-amp. An example of an application where a fully differential op-amp is not used in a fully differential topology is the single to differential converter shown in Fig. 26.33b. For this circuit, the inputs to the op-amp will not remain at V_{CM}.

2.   *A doubling in the output voltage swing.* This point can be understood by considering the case when $VDD = -VSS = 2.5$ V and $V_{CM} = 0$. For a single-ended output op-amp, the maximum output voltage is 2.5 V. However, for the differential-output op-amp when v_{o+} is 2.5 V, we have a v_{o-} of -2.5 V, giving a total output voltage of $v_{o+} - v_{o-}$ or 5 V. The result is an increase in the dynamic range of circuits employing fully differential signal paths.

3.   *A reduction in harmonic distortion.* The even order harmonic distortion terms are canceled in a fully differential topology when we take the difference in the + and − output voltages.

4.   *Substrate and coupled noise rejection.* Ideally, since the output signal lines are laid out close to one another, noise introduced from the substrate or other signal

lines appears as a common signal on the outputs of the op-amp and is eliminated when we take the difference between the two signals.

A fully differential op-amp (actually an OTA) topology useful in switched-capacitor circuits is shown in Fig. 27.27. This configuration is based on the topology given in [15]. The input diff-amp is the class AB source cross-coupled pair discussed in Ch. 24 that eliminates slew-rate limitations. Wide-swing current mirrors are employed to increase output voltage swing and for operation at lower power suppy voltages (e.g., 3.3 V). Note that one of the major problems of using the source cross-coupled pair, discussed in Ch. 24, was the small-input common-mode range. However, after reviewing number 1 above, we see that this is not a concern in a fully differential topology. Again, the main drawbacks of using the source cross-coupled pair are the increase in power dissipation (assuming the same biasing currents) and noise (thermal and 1/f).

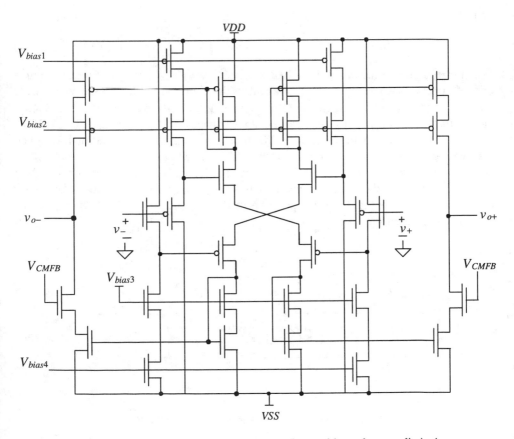

Figure 27.27 Fully differential op-amp topology, without slew-rate limitations, useful in switched-capacitor circuits.

The common-mode-feedback-circuit can be implemented using the continuous time techniques that were discussed in Ch. 25 or using switched-capacitor techniques [15] (see Fig. 27.28). The clock signals in this figure are, again, nonoverlapping. When the ϕ_1 switches are closed, the charge stored on the two C_1 capacitors is $2C_1(V_{CM} - V_{bias3})$, while the charge stored on the C_2 capacitors is $C_2(v_{o+} - V_{CMFB}^{\phi_1}) + C_2(v_{o-} - V_{CMFB}^{\phi_1})$, or simply

$$2C_2\left[\frac{v_{o+} + v_{o-}}{2} - V_{CMFB}^{\phi_1}\right] \tag{27.39}$$

where $V_{CMFB}^{\phi_1}$ is the output of the CMFB circuit when the ϕ_1 switches are closed. When the ϕ_2 switches close, since charge must be conserved, we can write

$$2C_2\left[\frac{v_{o+} + v_{o-}}{2} - V_{CMFB}^{\phi_1}\right] + 2C_1(V_{CM} - V_{bias3}) = (C_1 + C_2)\left[(v_{o+} - V_{CMFB}^{\phi_2}) + (v_{o-} - V_{CMFB}^{\phi_2})\right] \tag{27.40}$$

and

$$C_1\left(V_{CM} - \frac{v_{o+} + v_{o-}}{2} + V_{CMFB}^{\phi_2} - V_{bias3}\right) + C_2(V_{CMFB}^{\phi_2} - V_{CMFB}^{\phi_1}) = 0 \tag{27.41}$$

When the outputs of the op-amp are balanced, $V_{CMFB}^{\phi_2} = V_{CMFB}^{\phi_1} = V_{bias3}$, and the average of the outputs, that is, $(v_{o+} + v_{o-})/2$, is equal to the common-mode voltage V_{CM}. If the average of the outputs is greater than V_{CM}, then $V_{CMFB}^{\phi_2}$ becomes greater than $V_{CMFB}^{\phi_1}$, having the effect of lowering the average output voltages while the opposite is true if the average of the outputs is less than V_{CM}. The maximum size of C_2 is based on how much loading can be tolerated on the output of the OTA, while the size of C_1 is determined by the desired response needed in the CMFB circuit. Generally, C_1 is around one-tenth of C_2. The minimum size of the capacitors is set by charge injection and kT/C noise considerations. Note that the switches connected to v_{o+} and v_{o-} can be TGs, so that the circuit can operate from *VDD* to *VSS*.

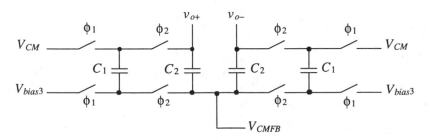

Figure 27.28 A switched-capacitor CMFB circuit.

27.3 Circuits

This section presents several examples of dynamic circuits.

Reducing Offset Voltage of an Op-Amp

The op-amp offset voltage can be modeled by adding a DC voltage in series with the noninverting input of the op-amp; see Fig. 27.29a [16]. The basic idea behind eliminating the offset voltage is shown in Fig. 27.29b. A capacitor is charged to a voltage equal and opposite to the comparator offset voltage. The dynamic analog circuit shown in Fig. 27.30 is used to help eliminate the effects of the offset voltage. Note that for this method to be effective *the op-amp must be stable* in the unity gain configuration. The clock signals ϕ_1 and ϕ_2 are the nonoverlapping clock signals discussed in Ch. 14. A nonoverlapping clock keeps switches S1, S2, and S3 from being on at the same time as switches S4 and S5. Let's consider the case shown in Fig. 27.30b where ϕ_1 is high and ϕ_2 is low. The op-amp, via the negative feedback, forces its output to zero volts. Doing so the capacitor is charged, in the polarity shown, to V_{os}. Note that under these conditions the op-amp is removed from the inputs. When ϕ_2 is high and ϕ_1 is low, Fig. 27.30c, the op-amp functions normally, assuming the storage capacitance C is much larger than the input capacitance of the op-amp.

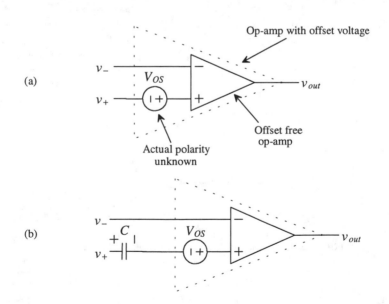

Figure 27.29 (a) Offset voltage of an op-amp modeled by a DC voltage source in series with the noninverting input of the op-amp and (b) using a capacitor to cancel the offset voltage.

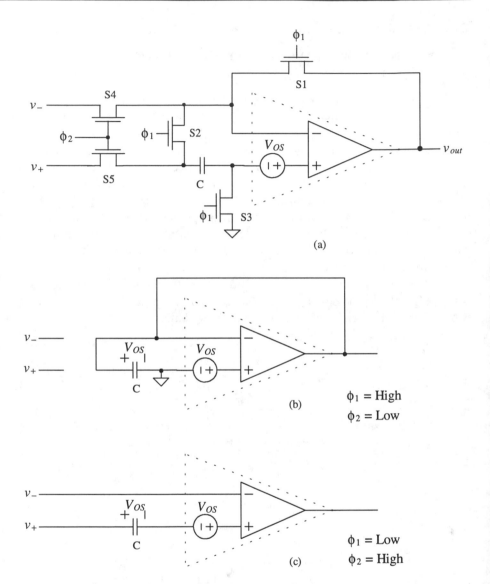

Figure 27.30 Dynamic reduction of the offset voltage.

Dynamic Comparator

Before we present a dynamic comparator, let's consider the *RC* switch circuit of Fig. 27.31a. When node A is connected to +5 V, node B is connected to ground. Consider what happens when the switches change positions, that is, when node A is connected to ground and the switch at node B is connected to an open. At the moment just after switching takes place, the potential at node B becomes − 5 V. In other words, the

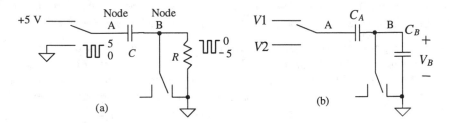

Figure 27.31 Circuits used to illustrate switching in dynamic circuits.

voltage across the capacitor does not change instantaneously. If node A is connected back to + 5 V and node B is connected back to ground a short time compared to the product of R and C, then the voltage across the capacitor remains +5 V.

A more useful circuit for CMOS is shown in Fig. 27.31b. If the switch across C_B is connected to ground when node A is connected to $V1$, then V_B, when the switches change positions, is given by

$$V_B = (V2 - V1) \cdot \frac{C_A}{C_A + C_B} \qquad (27.42)$$

Figure 27.32 shows a dynamic comparator based on the inverter [17]. When ϕ_1 is high the voltage on the v_- input is connected to node A, while the voltage on node B is set via S3 so that the input and output voltages of the inverter are equal. (The inverter is operating as a linear amplifier where both M1 and M2 are in the saturation regions.) When ϕ_2 goes high (ϕ_1 is low since the clocks are nonoverlapping), the v_+ input is connected to node A. If C_A is much larger than the input capacitance of the inverter (C_B), then the voltage change on the input of the inverter (V_B) is

$$v_{in} = v_+ - v_- \qquad (27.43)$$

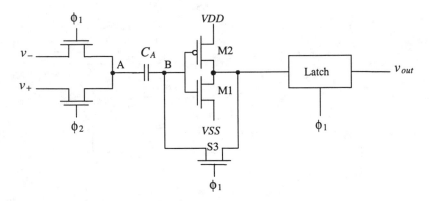

Figure 27.32 A dynamic comparator.

Provided the gain of the inverter is large, this change causes the inverter output to rail, that is, go to either *VDD* or *VSS*. The output is then latched and available during ϕ_1. The gain of the comparator can be increased by using additional inverter stages.

Another high-performance dynamic comparator configuration is based on the dynamic latch shown in Fig. 27.33 [18]. This latch is used as the positive feedback stage of the comparator. The offset-voltage of the comparator is reduced, using either input offset storage (IOS) or output offset storage (OOS) around the comparator preamp. Figure 27.34 shows the two types of offset cancellation techniques. In the IOS configuration, the preamp must be stable in the unity feedback configuration. Also, the MOSFETs of the preamp must remain in saturation when the offset voltage is stored on the capacitors. If a differential amplifier is used as the preamp this condition is easily met.

The size of the storage capacitors is based on three important considerations: (1) preamp or latch input capacitance, (2) charge injection and (3) kT/C noise. For the IOS scheme the input storage capacitance must be much larger than the input capacitance of the preamp, so that the storage capacitors don't attenuate the input signals. For example, if the storage capacitors have the same capacitance value as the input capacitance of the preamp, then one-half of the input signals reaches the preamp. For the OOS scheme the storage capacitors should be much larger than the input capacitance of the dynamic latch.

Dynamic Current Mirrors

The use of dynamic techniques can significantly reduce the effects of threshold voltage mismatches in current mirrors. Consider the circuit of Fig. 27.35. When ϕ_1 is high and

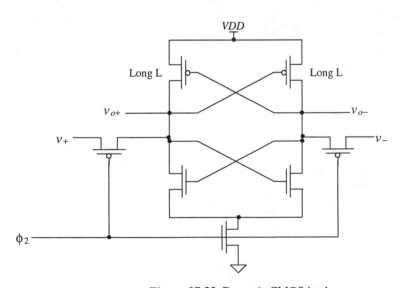

Figure 27.33 Dynamic CMOS latch.

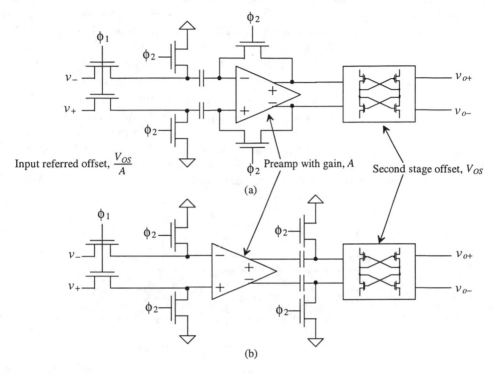

Input referred offset, $\dfrac{V_{OS}}{A}$ Preamp with gain, A Second stage offset, V_{OS}

(a)

(b)

Figure 27.34 (a) Input offset storage (IOS) and (b) Output offset storage (OOS).

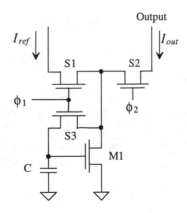

Figure 27.35 Dynamic biasing of a current mirror.

ϕ_2 is low (again, these clock signals are nonoverlapping), switches S1 and S3 are on while switch S2 is off. A current I_{ref} flows through M1, setting its gate-source voltage. This information [the gate-source voltage (actually the charge) of M1] is stored on C. When S2 closes with S1 and S3 off a current I_{out} equal to I_{ref}, neglecting channel length modulation, flows. This circuit behaves like a current source when ϕ_2 is high and as an open when ϕ_2 is low. The circuit shown in Fig. 27.36 shows a dynamic current mirror that operates continuously. When ϕ_1 is high, M2 sinks current, and when ϕ_2 is high, M1 sinks current. These circuits are useful in eliminating the mismatch effects, and thus differences in the output currents, resulting from threshold voltage and transconductance parameter differences between devices. Since a single-reference current can be used to program the current in a string of current mirrors, only the finite output resistance of the mirrors will cause current differences.

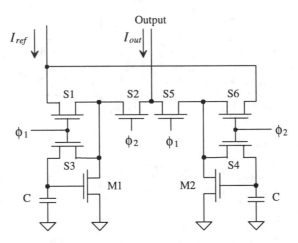

Figure 27.36 Dynamic current mirror that operates during both clock phases.

Dynamic Amplifiers

Figure 27.37 shows a dynamic amplifier. The circuit amplifies when ϕ is low and dynamically biases M1 and M2, and therefore does not amplify, when ϕ is high. If C1 and C2 are large compared to the input capacitance of M1 and M2, then the input AC signal, v_{in}, is applied to both gates. This biasing scheme makes the amplifier less sensitive to threshold and power supply variations. Other dynamic amplifier configurations exist, which have differential inputs and operate over both clock cycles [19].

REFERENCES

[1]   D. J. Allstot and W. C. Black, "Technology Design Considerations for Monolithic MOS Switched-Capacitor Filtering Systems" *Proceedings of the IEEE*, Vol. 71, No. 8, pp. 967-986, August 1983.

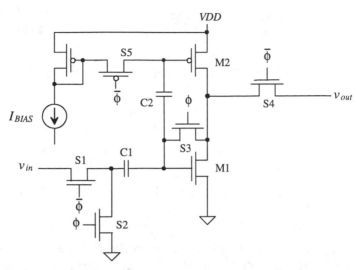

Figure 27.37 Dynamic amplifier used to reduce biasing sensitivity to power supply and threshold voltage.

[2]   J. Shieh, M. Patil, and B. Sheu, "Measurement and Analysis of Charge Injection in MOS Analog Switches," *IEEE Journal of Solid State Circuits*, Vol. 22, No. 2, pp. 277-281, April 1987.

[3]   G. Wegmann, E. Vittoz, and F. Rahali, "Charge Injection in Analog MOS Switches," *IEEE Journal of Solid State Circuits*, Vol. 22, No. 6, pp. 1091-1097, December 1987.

[4]   C. Eichenberger and W. Guggenbuhl, "On Charge Injection in Analog MOS Switches and Dummy Switch Compensation Techniques," *IEEE Transactions on Circuits and Systems*, Vol. 37, No. 2, pp. 256-264, February 1990.

[5]   J. McCreary and P. R. Gray, "All MOS Charge Redistribution Analog-to-Digital Conversion Techniques - Part 1," *IEEE Journal of Solid State Circuits*, Vol. 10, pp. 371-379, December 1975.

[6]   P. W. Li, M. J. Chin, P. R. Gray, and R. Castello, "A Ratio-Independent Algorithmic Analog-to-Digital Conversion Technique," *IEEE Journal of Solid-State Circuits,* Vol. SC-19, No. 6, pp. 828-836, December 1984.

[7]   E. J. Kennedy, *Operational Amplifier Circuits: Theory and Applications*, Holt, Rinehart and Winston, New York, 1988.

[8]   D. Johns and K. Martin, *Analog Integrated Circuit Design*, John Wiley and Sons, New York, 1997.

[9]   A. B. Grebene, *Bipolar and MOS Integrated Circuit Design*, John Wiley and Sons, New York, 1984.

[10] R. W. Broderson, P. R. Gray and D. A. Hodges, "MOS Switched-Capacitor Filters," *Proceedings of the IEEE*, Vol. 67, No. 1, January 1979.

[11] K. Martin, "Improved Circuits for the Realization of Switched-Capacitor Filters," *IEEE Transactions on Circuits and Systems*, Vol. CAS-27, No. 4, pp. 237-244, April 1980.

[12] P. R. Gray and R. G. Meyer, *Analysis and Design of Analog Integrated Circuits*, 2nd ed., John Wiley and Sons, 1984. ISBN 0-471-87493-0.

[13] R. Gregorian, K. W. Martin, and G. Temes, "Switched-Capacitor Circuit Design," *Proceedings of the IEEE*, Vol. 71, No. 8, pp. 941-966, August 1983.

[14] D. J. Allstot, R. W. Broderson, and P. R. Gray, "MOS Switched-Capacitor Ladder Filters," *IEEE Journal of Solid-State Circuits*, Vol. SC-13, No. 6, pp. 806-814, December 1978.

[15] R. Castello and P. R. Gray, "A High-Performance Micropower Switched -Capacitor Filter," *IEEE Journal of Solid-State Circuits*, Vol. SC-20, No. 6, pp. 1122-1132, Dec. 1987.

[16] P. E. Allen and D. R. Holberg, *CMOS Analog Circuit Design*, Holt, Rinehart and Winston, 1987. ISBN 0-03-006587-9.

[17] A. G. Dingwall and V. Zazzu, "An 8-MHz Subranging 8-bit A/D Converter," *IEEE Journal of Solid-State Circuits*, Vol. SC-20, No. 6, pp. 1138-1143, December 1992.

[18] B. Razavi and B. A. Wooley, "Design Techniques for High-Speed, High-Resolution Comparators," *IEEE Journal of Solid-State Circuits*, Vol. 27, No. 12, pp. 1916-1926, December 1992.

[19] S. Masuda, Y. Kitamura, S. Ohya, and M. Kikuchi, "CMOS Sampled Differential Push-Pull Cascode Operational Amplifier," *IEEE International Symposium on Circuits and Systems*, Vol. 3, pp. 1211-1214, 1983.

PROBLEMS

27.1 Using SPICE simulations, show the effects of clock feedthrough on the voltage across the load capacitor for the switch circuits shown in Fig. P27.1. How does this voltage change if the capacitor value is increased to 100fF ?

27.2 Assume the small-signal on resistance of a MOSFET ("on" means the MOSFET is operating in the triode region) is given by

$$r_{ds} = \frac{1}{\beta(V_{GS} - V_{THN})}$$

Plot, by hand, the small-signal on resistances of minimum size p-channel, n-channel, and transmission-gate switches. The resulting plot should look similar to Fig. 27.2.

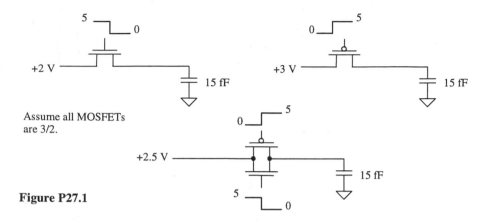

Assume all MOSFETs
are 3/2.

Figure P27.1

27.3 Make a table comparing capacitor size with kT/C noise for capacitor values from 10 fF to 10 pF.

27.4 Sketch the single-ended (output) version of the sample-and-hold amplifier shown in Fig. 27.9 and describe, using timing diagrams, the operation of the circuit.

27.5 Figure P27.5 shows the implementation of a single-ended to differential sample-and-hold. Describe the operation of this circuit using timing diagrams with various input voltages.

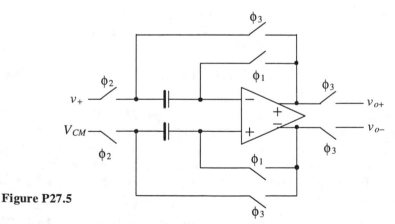

Figure P27.5

27.6 Show that the switched-capacitor circuits shown in Fig. P27.6 behave like resistors, for $f \ll f_{clk}$, with the resistor values shown.

27.7 Comment on the selection of the bottom plate of the capacitor shown in Fig. 27.14a.

27.8 Comment on the selection of the bottom plate of the capacitor shown in Fig. 27.16.

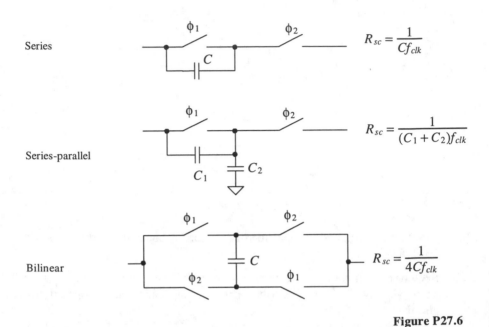

Figure P27.6

27.9 Sketch the schematic, similar in form to Fig. 27.16, of the fully differential switched-capacitor integrator made using a differential input/output op-amp. What is the transfer function of this topology?

27.10 Repeat Ex. 27.2 if the low-frequency gain is 40 dB and the zero is located at 50 kHz.

27.11 Using the results given in Eq. (27.24), plot the magnitude of v_{out}/v_{in} against f/f_{clk}. Comment on the resulting plot.

27.12 An important consideration in SC circuits is the slew-rate requirements of the op-amps used. In the derivation in Fig. 27.21, we assumed that a voltage source was connected to the input of the circuit. In reality, the input of the circuit is provided by an op-amp. When ϕ_1 goes high, in this figure, the capacitor C_I is charged to the input voltage v_{in} ($= v_A$). If C_I is 5 pF and f_{clk} is 100 kHz, estimate the minimum slew-rate requirements for the op-amp providing v_{in}.

27.13 Suppose the op-amp in Problem 27.12 is used with a feedback factor of 0.5. Estimate the minimum unity gain frequency, f_u, that the op-amp must possess.

27.14 Simulate the operation of the dynamic latch shown in Fig. 27.33.

27.15 Simulate the operation of the current mirror shown in Fig. 27.36.

Chapter 28

Data Converter Fundamentals

Data converters play an important role in an ever-increasing digital world. As more products perform calculations in the digital or discrete time domain, more sophisticated data converters must translate the digital data to and from our inherent analog world. This chapter introduces concepts of data conversion and sampling which surround this useful circuit.

28.1 Analog Versus Discrete Time Signals

Analog-to-digital converters, also known as A/Ds or ADCs, convert analog signals to discrete time or digital signals. Digital-to-analog converters (D/As or DACs) perform the reverse operation. Figure 28.1 illustrates these two operations. In order to understand the functionality of these data converters, it would be wise first to compare the characteristics of analog versus digital signals.

Examine Fig. 28.1. The original analog signal (a) is filtered by an anti-aliasing filter to remove any high-frequency harmonics that may cause an effect known as aliasing (see Sec. 28.5). The signal is sampled and held and then converted into a digital signal (b). Next the DAC converts the digital signal back into an analog signal

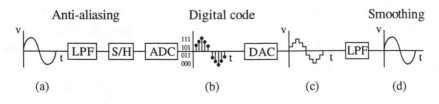

Figure 28.1 Signal characteristics caused by A/D and D/A conversion.

(c). Note that the output of the DAC is not as "smooth" as the original signal. A low-pass filter returns the analog signal back to its original form (plus phase shift introduced from the conversions) after eliminating the higher order harmonics caused by the conversion. This example illustrates the main differences between analog and digital signals. Whereas the analog signal in Fig. 28.1a is *continuous* and *infinite* valued, the digital signal in (b) is *discrete* with respect to time and *quantized*. The term *continuous-time signal* refers to a signal whose response with respect to time is uninterrupted. Simply stated, the signal has a continuous value for the entire segment of time for which the signal exists. By referring to the analog signal as infinite valued, we mean that the signal can possess any value between the parameters of the system. For example, in Fig. 28.1a, if the peak amplitude of the sine wave was + 5V, then the analog signal can be any value between –5 and 5 V (such as 2.4758393848 V). Of course, measuring all the values between –5 and 5 V would require a piece of laboratory equipment with infinite precision.

The digital signal, on the other hand, is discrete with respect to time. This means that the signal is defined for only certain or discrete periods of time. A signal that is quantized can only have certain values (as opposed to an infinitely valued analog signal) for each discrete period. The signal illustrated in Fig. 28.1b illustrates these qualities.

28.2 Converting Analog Signals to Digital Signals

We have already established the differences between analog and digital signals. How is it possible to convert from an analog signal to a digital signal? An example will illustrate the process.

You live in Moscow, Idaho, where the weather in the winter stays between 0 °F and 50 °F (Fig. 28.2a). Suppose you had a thermometer that contained only two readings on it, hot and cold, and that you wanted to record the weather patterns and plot the results. The two quantization levels can be correlated with the actual temperature as follows:

If 0 °F ≤ T < 25 °F   Temperature is recorded as cold

If 25 °F ≤ T < 50 °F   Temperature is recorded as hot

You take a measurement every day at noon and plot the results after one week. From Fig. 28.2b, it is apparent that your discretized version of the weather is not an accurate representation of the actual weather.

Now suppose that you find another thermometer that contains four possible temperatures (hot, warm, cool, and cold) and that you increase the number of readings to two per day. The result of this reading is seen in Fig. 28.3a. The quantization levels represent four equal bands of temperature as seen below:

If   0 °F ≤ T < 12.5 °F   Temperature is recorded as cold

If 12.5 °F ≤ T < 25 °F   Temperature is recorded as cool

If 25 °F ≤ T < 37.5 °F   Temperature is recorded as warm

If 37.5 °F ≤ T < 50 °F   Temperature is recorded as hot

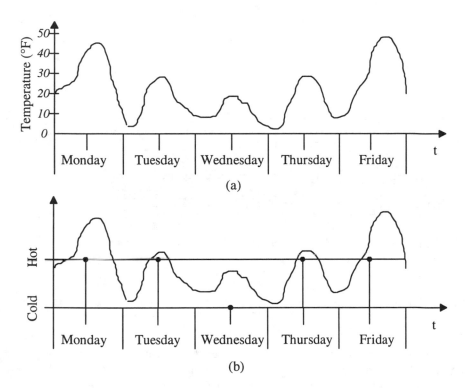

Figure 28.2 (a) An analog signal representing the temperature in Moscow, and
(b) a digital representation of the analog signal taking one
sample per day with two quantization levels.

Here, the digital version of the weather still looks nothing like the actual weather pattern, but the critical issues in digitizing an analog signal should be apparent. The actual weather pattern is the analog signal. It is continuous with respect to time, and its value can be between 0 °F and 50 °F (even 33.9638483920398439 °F!). The accuracy of the digitized signal is dependent on two things: the number of samples taken and the resolution, or number of quantization levels, of the converter. In our example, we need to increase both the number of samples and the resolution of thermometer.

Suppose that finally we obtain a thermometer with 25 temperature readings and that we now take a reading eight times per day. Each of the 25 quantization levels now represents a 2 °F band of temperature. From Fig. 28.3b, we can see that the digital version of the weather is approaching that of the actual analog signal. If we kept

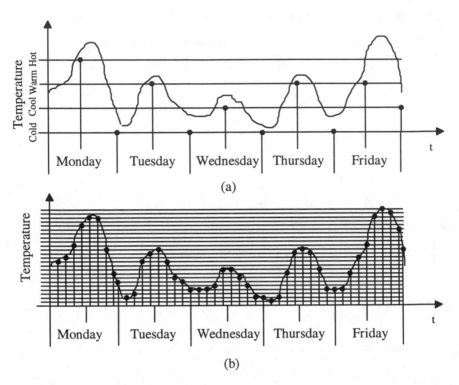

Figure 28.3 Digital representation of the temperature taking (a) two samples per day with
four quantization levels and (b) nine samples per day with 25
quantization levels.

increasing both sampling time and resolution, the difference between the analog and the
digital signals would become negligible. This brings up another critical issue. How
many samples should one take in order to accurately represent the analog signal?

Suppose a sudden rainstorm swept through Moscow and caused a sharp decrease
in temperature before returning to normal. If that storm had occurred between our
sampling times, our experiment would not have shown the effects of the storm. Our
sampling time was too slow to catch the change in the weather. If we had increased the
number of samples, we would have recognized that something happened which caused
the temperature to drop dramatically during that period.

As it turns out, the *Nyquist Criterion* defines how fast the sampling rate needs to
be to represent an analog signal accurately. This criterion requires that the sampling
rate be at least two times the highest frequency contained in the analog signal. In our
example, we need to know how quickly the weather can change and then take samples
twice as fast as that value. The Nyquist Criterion can be described as

$$F_{sampling} = 2\, F_{MAX} \tag{28.1}$$

where $F_{sampling}$ is the sampling frequency required to accurately represent the analog signal and F_{MAX} is the highest frequency of the sampled signal.

How much resolution should we use to represent the analog signal accurately? There is no absolute criterion for this specification. Each application will have its own requirements. In our weather example, if we were only interested in following general trends, then the 25 quantization levels would more than suffice. However, if we were interested in keeping an accurate record of the temperature to within ±0.5 °F, we would need to double the resolution to 50 quantization levels so that each quantization level would correspond to each degree ±0.5 °F (Fig. 28.4).

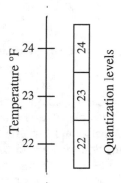

Figure 28.4 Quantization levels overlap actual temperature by ± ½ °F.

28.3 Sample-and-Hold (S/H) Characteristics

Sample-and-hold (S/H) circuits are critical in converting analog signals to digital signals. The behavior of the S/H is analogous to that of a camera. Its main function is to "take a picture" of the analog signal and hold its value until the ADC can process the information. It is important to characterize the S/H circuit when performing data conversion. Ignoring this component can result in serious error, for both speed and accuracy can be limited by the S/H. Ideally, the S/H circuit should have an output similar to that shown in Fig. 28.5a. Here, the analog signal is instantly captured and held until the next sampling period. However, a finite amount of time is required for the sampling to occur. During the sampling period, the analog signal may continue to vary; thus, another type of circuit is called a track-and-hold, or T/H. Here, the analog signal is "tracked" during the time required to sample the signal, as seen in Fig. 28.5b. It can be seen that S/H circuits operate in both static (hold mode) and dynamic (sample mode) circumstances. Thus, characterization of the S/H will be discussed in the context of these two categories. Figure 28.6 presents a summary of the major errors associated with a S/H [1-5]. A discussion of each error follows.

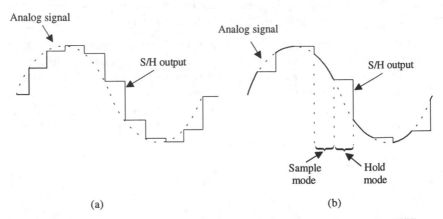

Figure 28.5 The output of (a) an ideal S/H circuit and (b) a track-and-hold (T/H).

Sample Mode

Once the sampling command has been issued, the time required for the S/H to track the analog signal to within a specified tolerance is known as *acquisition time*. In the worst-case scenario, the analog signal would vary from zero volts to its maximum value, $v_{IN(max)}$. And the worst-case acquisition time would correspond to the time required for the output to transition from zero to $v_{IN(max)}$. Since most S/H circuits use amplifiers as buffers (as seen in Fig. 28.7), it should be obvious that the acquisition will be a function of the amplifier's own specifications. For example, notice that if the input changes very quickly, then the output of the T/H could be limited by the amplifier's slew rate. The amplifier's stability is also extremely critical. If the amplifier is not compensated correctly, and the phase margin is too small, then a large *overshoot* will occur which requires a longer *settling time* for the S/H to settle within the specified tolerance. The error tolerance at the output of the S/H is also dependent on the amplifier's *offset, gain*

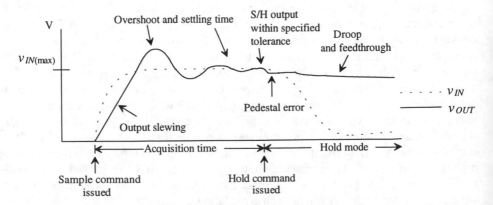

Figure 28.6 Illustration of typical errors associated wth a S/H.

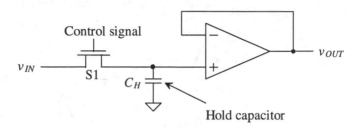

Figure 28.7 Track-and-hold circuit using an output buffer.

error (ideally, the S/H should have a gain of 1) and *linearity* (the gain of the S/H should not vary over the input voltage range).

Hold Mode

Once the hold command is issued, the S/H faces other errors. Pedestal error occurs as a result of charge injection (introduced in Sec. 27.1) and clock feedthrough. Part of the charge built up in the channel of the switch is distributed onto the capacitor, thus slightly changing its voltage. Also, the clock couples onto the capacitor via overlap capacitance between the gate and the source or drain. Another error that occurs during the hold mode is called *droop*. This error is related to the leakage of current from the capacitor due to parasitic impedances and to the leakage through the reverse-biased diode formed by the drain of the switch. This diode leakage can be minimized by making the drain area as small as can be tolerated. Although the input impedance of the buffer amplifier is very large, the switch has a finite OFF impedance through which leakage can occur. Current can also leak through the substrate. The key to minimizing droop is to increase the value of the sampling capacitor. The tradeoff, however, will be the increased time required to charge the capacitor to the value of the input signal.

Aperture Error

A transient effect that introduces error occurs between the sample and the hold modes. A finite amount of time, referred to as aperture time, is required to disconnect the capacitor from the analog input source. The aperture time actually varies slightly as a result of noise on the hold control signal and the value of the input signal, since the switch will not turn off until the gate voltage becomes less than the value of the input voltage less one threshold voltage drop. This effect is called *aperture uncertainty* or *aperture jitter*. As a result, if a periodic signal was being sampled repeatedly at the same points, slight variations in the hold value would result, thus creating *sampling error*. Figure 28.8 illustrates this effect. Note that the amount of aperture error is directly related to the frequency of the signal and that the worst-case aperture error occurs at the zero crossing, where *dV/dt* is the greatest. This assumes that the S/H circuit is capable of sampling both positive and negative voltages (bipolar). The amount

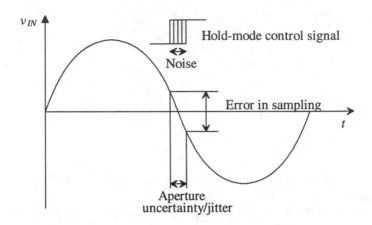

Figure 28.8 Aperture error.

of error that can be tolerated is directly related to the resolution of the conversion. Aperture error will be discussed again in Sec. 28.5 as it relates to the error in an ADC.

Example 28.1

Find the maximum sampling error for a S/H circuit that is sampling a sinusoidal input signal that could be described as

$$v_{IN} = A \sin 2\pi f t$$

where A is 2 V and $f = 100$ kHz. Assume that the aperture uncertainty is equal to 0.5 ns.

The sampling error due to the aperture uncertainty can be thought of as a slew rate such that

$$\frac{dV}{dt} = \frac{d}{dt} A \sin 2\pi f t = 2\pi f A \cos 2\pi f t$$

with maximum slewing occurring when the cosine term is equal to 1. Therefore,

$$\frac{dV}{dt}(max) = 2\pi f A = (2\pi \cdot 100 \text{ kHz})(2 \text{ V})$$

and the maximum sampling error is

$$Maximum \ Sampling \ Error = dV(max) \quad \text{or}$$

$$(0.5 \times 10^9 \text{ s})(2\pi \cdot 100 \text{ kHz})(2 \text{ V}) = 0.628 \text{ mV} \quad \blacksquare$$

28.4 Digital-to-Analog Converter (DAC) Specifications

Probably the most popular digital-to-analog converter application is the digital audio compact disc player. Here digital information stored on the CD is converted into music

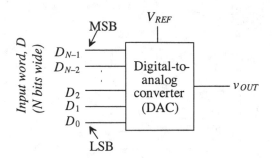

Figure 28.9 Block diagram of the digital-to-analog converter.

via a high-precision DAC. Many characteristics define a DAC's performance. Each characteristic will be discussed before we look at the basic architectures in Ch. 29. This "top-down" approach will allow a smoother transition from the data converter characteristics to the actual architectures, since most data converters have similar performance limitations. A discussion of some of the basic definitions associated with DACs follows. It should be noted that DACs and ADCs can use either voltage or current as their analog signal. For purposes of describing specifications, it will be assumed that the analog signal is a voltage.

A block diagram of a DAC can be seen in Fig. 28.9. Here an N-bit digital word is mapped into a single analog voltage. Typically, the output of the DAC is a voltage that is some fraction of a reference voltage (or current), such that

$$v_{OUT} = F V_{REF} \qquad (28.2)$$

where v_{OUT} is the analog voltage output, V_{REF} is the reference voltage, and F is the fraction defined by the input word, D, that is N bits wide. The number of input combinations represented by the input word D is related to the number of bits in the word by

$$\text{Number of input combinations} = 2^N \qquad (28.3)$$

A 4-bit DAC has a total of 2^4 or 16 total input values. A converter with 4-bit resolution must be able to map a change in the analog output which is equal to 1 part in 16. The maximum analog output voltage for any DAC is limited by the value of some reference voltage, V_{REF}. If the input is an N-bit word, then the value of the fraction, F, can be determined by,

$$F = \frac{D}{2^N} \qquad (28.4)$$

Therefore, if a 3-bit DAC is being used, the input, D, is $100 = 4_{10}$, and V_{REF} is 5 V, then the value of F is

$$F = \frac{100}{2^3} = \frac{4}{8} \qquad (28.5)$$

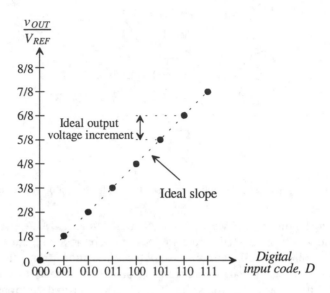

Figure 28.10   Ideal transfer curve for a 3-bit DAC.

and the analog voltage that appears at the output becomes,

$$v_{OUT} = \frac{4}{8}(5) = 2.5 \text{ V} \tag{28.6}$$

By plotting the input word, D, versus v_{OUT} as D is incremented from 000 to 111, the *transfer curve* seen in Fig. 28.10 would be generated. The y-axis has been normalized to V_{REF}; therefore, the graduated marks also represent F by Eq. (28.2). Some important characteristics need to be discussed here. First, notice that the transfer curve is not continuous. Since the input is a digital signal, which is inherently discrete, the input signal can have only eight values that must correspondingly produce eight output voltages. If a straight line connected each of the output values, the slope of the line would ideally be one increment/input code value. Also note that the maximum value of the output is 7/8. Since the case where $D = 000$ has to result in an analog voltage of 0 V, and a 3-bit DAC has eight possible analog output voltages, then the analog output will increase from 0 V to only 7/8 V_{REF}.

Again, using Eq. (28.2), this means that the maximum analog output that can be generated by the 3-bit DAC is

$$v_{OUT(\text{max})} = \frac{7}{8} \cdot V_{REF} \tag{28.7}$$

This maximum analog output voltage that can be generated is known as *full-scale voltage*, V_{FS}, and can be generalized to any N-bit DAC as

$$V_{FS} = \frac{2^N - 1}{2^N} \cdot V_{REF} \tag{28.8}$$

The *least significant bit* (*LSB*) refers to the rightmost bit in the digital input word. The LSB defines the smallest possible change in the analog output voltage. The LSB will always be denoted as D_0. One LSB can be defined as

$$1 \, LSB = \frac{V_{REF}}{2^N} \tag{28.9}$$

In the previous case of the 3-bit DAC, 1 LSB = 5/8 V, or 0.625 V. Generating an output in multiples of 0.625 V may not seem difficult, but as the number of bits increases, the voltage value of one LSB decreases for a fixed value of V_{REF}.

The *most significant bit* (*MSB*) refers to the leftmost bit of the digital word, *D*. In the previous example, *D* = 100 or $D_2 D_1 D_0$, with D_2 being the MSB. Generalizing to the *N*-bit DAC, the MSB would be denoted as D_{N-1}. (Since the LSB is denoted as bit 0, the MSB is denoted as *N*-1.) Note that when discussing DACs, the MSB causes the output to change by $1/2 \, V_{REF}$.

When discussing data converters, the term *resolution* describes the smallest change in the analog output with respect to the value of the reference voltage, V_{REF}. This is slightly different from the definition of LSB in that resolution is typically given in terms of bits and represents the *number of unique output voltage levels*, i.e., 2^N.

Example 28.2

Find the resolution for a DAC if the output voltage is desired to change in 1 mV increments while using a reference voltage of 5 V.

The DAC must resolve

$$\frac{1 \, mV}{5 \, V} = 0.0002 \text{ or } 0.02\% \text{ adjustability}$$

Therefore, the *accuracy* required for 1 LSB change over a range of V_{REF} is

$$\frac{1 \, LSB}{V_{REF}} = \frac{1}{2^N} = 0.0002 \tag{28.10}$$

and solving *N* for the resolution yields

$$N = Log_2 \left(\frac{5 \, V}{1 \, mV} \right) = 12.29 \text{ bits}$$

which means that a 13-bit DAC will be needed to produce the accuracy capable of generating 1 mV changes in the output using a 5 V reference. ∎

Example 28.3

Find the number of input combinations, values for 1 LSB, the percentage accuracy, and the full-scale voltage generated for a 3-bit, 8-bit, and 16-bit DAC, assuming that $V_{REF} = 5$ V.

Using Eqs. (28.3), (28.8), (28.9), and (28.10), we can generate the following information:

Resolution	Input combinations	1 LSB	% accuracy	V_{FS}
3	8	0.625 V	12.5	4.375 V
8	256	19.5 mV	0.391	4.985 V
16	65,536	76.29 μV	0.00153	4.9999 V

The value of 1 LSB for an 8-bit converter is 19.5 mV, while 1 LSB for a 16-bit converter is 76.3 μV (a factor of 256)! Increasing the resolution by 1 bit increases the accuracy by a factor of 2. The precision required to map the analog signal at high resolutions is very difficult to achieve. We will examine some of these issues as we examine the limitations of the data converter in Ch. 29.

Note that a data converter may have a resolution of 8-bits, where an LSB is 19.5 mV as above, while having a much higher accuracy. For example, we could require the 8-bit data converter above to have an accuracy of 0.1 %. The higher accuracy results in a more ideal (linear) DAC. A typical specification for DAC accuracy is ±½ LSB for reasons discussed below. ∎

Differential Nonlinearity

As seen in the ideal DAC in Fig. 28.10, each adjacent output increment should be exactly one-eighth. Since the y-axis is normalized, the values for the increment heights will be unitless. However, the increment heights can be easily converted to volts by multiplying the height by V_{REF}. This corresponds to the ideal increment corresponding to 0.625 V = 1 LSB (assuming V_{REF} = 5 V).

Nonideal components cause the analog increments to differ from their ideal values. The difference between the ideal and nonideal values is known as *differential nonlinearity*, or *DNL* and is defined as

DNL_n = Actual increment height of transition n – Ideal increment height   (28.11)

where n is the number corresponding to the digital input transition. The DNL specification measures how well a DAC can generate uniform analog LSB multiples at its output.

Example 28.4
Determine the DNL for the 3-bit nonideal DAC whose transfer curve is shown in Fig. 28.11. Assume that V_{REF} = 5 V.

The actual increment heights are labeled with respect to the ideal increment height, which is 1 LSB, or 1/8 of $\frac{v_{OUT}}{V_{REF}}$. Notice that there is no increment corresponding to 000, since it is desirable to have zero output voltage with a digital input code of 000. The increment height corresponding to 001, however, is equal to the corresponding height of the ideal case seen in Fig. 28.10; therefore, DNL_1 = 0. Similarly, DNL_2 is also zero since the increment associated with the transition at 010 is equal to the ideal height. Notice that the 011

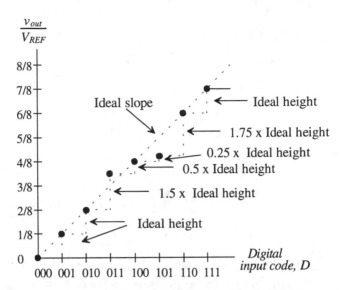

Figure 28.11 Example of differential nonlinearity for a 3-bit DAC.

increment, however, is not equal to the ideal curve but is 3/16, or 1.5 times the ideal height.

$$DNL_3 = 1.5 \text{ LSB} - 1 \text{ LSB} = 0.5 \text{ LSB}$$

Since we have already determined in Eq. (28.9) that for a 3-bit DAC, 1 LSB = 0.625 V, we can convert the DNL_3 to volts as well. Therefore, DNL_3 = 0.5 LSB = 0.3125 V. However, it is popular to refer to DNL in terms of LSBs. The remainder of the digital output codes can be characterized as follows:

$$DNL_4 = 0.5 \text{ LSB} - 1 \text{ LSB} = -0.5 \text{ LSB}$$

$$DNL_5 = 0.25 \text{ LSB} - 1 \text{ LSB} = -0.75 \text{ LSB}$$

$$DNL_6 = 1.75 \text{ LSB} - 1 \text{ LSB} = 0.75 \text{ LSB}$$

$$DNL_7 = 1 \text{ LSB} - 1 \text{ LSB} = 0$$

If we were to plot the value of DNL (in LSBs) versus the input digital code, Fig. 28.12 would result. The DNL for the entire converter used in this illustration is ±0.75 LSB since the overall error of the DAC is defined by its worst-case DNL. ∎

Generally, a DAC will have less than ±½ LSB of DNL if it is to be N-bit accurate. A 5-bit DAC with 0.75 LSBs of DNL actually has the resolution of a 4-bit DAC. If the DNL for a DAC is less than − 1 LSBs, then the DAC is said to be *nonmonotonic*, which means that the analog output voltage does not always increase as the digital input code is incremented. A DAC should always exhibit *monotonicity* if it is to function without error.

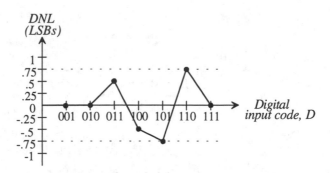

Figure 28.12 DNL curve for the nonideal 3-bit DAC.

Integral Nonlinearity

Another important static characteristic of DACs is called *integral nonlinearity* (*INL*). Defined as the difference between the data converter output values and a reference straight line drawn through the first and last output values, INL defines the linearity of the overall transfer curve and can be described as

INL_n = Output value for input code n – Output value of the reference line at that point

$$(28.12)$$

An illustration of this measurement is presented in Fig. 28.13. It is assumed that all other errors due to offset and gain (these will be discussed shortly) are zero. An example follows shortly.

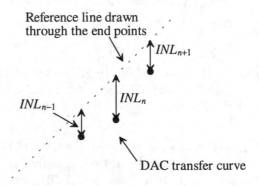

Figure 28.13 Measuring the INL for a DAC transfer curve.

It is common practice to assume that a converter with N-bit resolution will have less than $\pm\frac{1}{2}$ LSB of DNL and INL. The term, $\frac{1}{2}$ *LSB*, is a common term that typically denotes the maximum error of a data converter (both DACs and ADCs). For example, a 13-bit DAC having greater than $\pm\frac{1}{2}$ LSB of DNL or INL actually has the resolution of a 12-bit DAC. The value of $\frac{1}{2}$ LSB in volts is simply

$$0.5 \ LSB = \frac{V_{REF}}{2^{N+1}} \qquad (28.13)$$

Example 28.5
Determine the INL for the nonideal 3-bit DAC shown in Fig. 28.14. Assume that $V_{REF} = 5$ V.

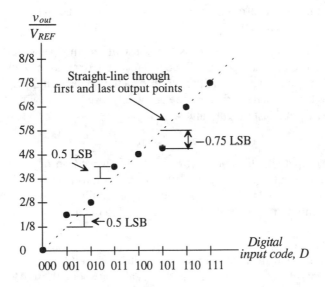

Figure 28.14 Example of integral nonlinearity for a DAC.

First, a reference line is drawn through the first and last output values. The INL is zero for every code in which the output value lies on the reference line; therefore, $INL_2 = INL_4 = INL_6 = INL_7 = 0$. Only outputs corresponding to 001, 011, and 101 do not lie on the reference. Both the 001 and the 011 transitions occur $\frac{1}{2}$ LSB higher than the straight-line values; therefore, $INL_1 = INL_3 = 0.5$ LSB. By the same reasoning, $INL_5 = -0.75$ LSB. Therefore, the INL for the DAC is considered to be its worst-case INL of +0.5 LSB and -0.75 LSB. The INL plot for the nonideal 3-bit DAC can be seen in Fig. 28.15. ■

It should be noted that other methods are used to determine INL. One method compares the output values to the ideal reference line, regardless of the positions of the first and last output values. If the DAC has an offset voltage or gain error, this will be

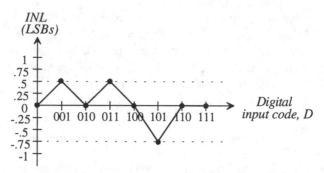

Figure 28.15 INL curve for the nonideal 3-bit DAC.

included in the INL determination. Usually, the offset and gain errors are determined as separate specifications.

Another method, described as the "best-fit" method, attempts to minimize the INL by constructing the reference line so that it passes as closely as possible to a majority of the output values. Although this method does minimize the INL error, it is a rather subjective method that is not as widely used as drawing the reference line through the first and last output values.

Offset

The analog output should be 0 V for $D = 0$. However, an offset exists if the analog output voltage is not equal to zero. This can be seen as a shift in the transfer curve as illustrated in Fig. 28.16. This specification is similar to the offset voltage for an operational amplifier except that it is not referred to the input.

Gain Error

A gain error exists if the slope of the best-fit line through the transfer curve is different from the slope of the best-fit line for the ideal case. For the DAC illustrated in Fig. 28.17, the gain error becomes

$$\text{Gain error} = \text{Ideal slope} - \text{Actual slope} \qquad (28.14)$$

Latency

This specification defines the total time from the moment that the input digital word changes to the time the analog output value has settled to within a specified tolerance. Latency should not be confused with settling time, since latency includes the delay required to map the digital word to an analog value plus the settling time. It should be noted that settling time considerations are just as important for a DAC as they are for a S/H or an operational amplifier.

Signal-to-Noise Ratio (SNR)

Signal-to-noise (SNR) is defined as the ratio of the signal power to the noise at the analog output. In amplifier applications, this specification is typically measured using a

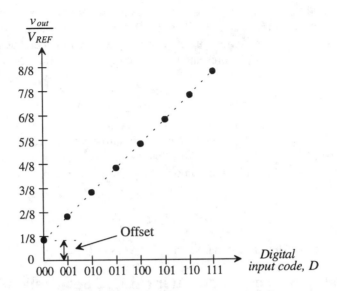

Figure 28.16 Illustration of offset error for a 3-bit DAC.

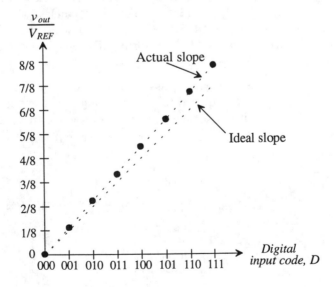

Figure 28.17 Illustration of gain error for a 3-bit DAC.

sinewave input.   For the DAC, a "digital" sinewave is generated through instrumentation or through an A/D. The SNR can reveal the true resolution of a data converter as the effective number of bits can be quantified mathematically.   A detailed derivation of the SNR is presented in Sec. 28.6, on the discussion of ADC specifications.

Dynamic Range

Dynamic range is defined as the ratio of the largest output signal over the smallest output signal. For both DACs and ADCs, the dynamic range is related to the resolution of the converter. For example, an N-bit DAC can produce a maximum output of 2^N-1 multiples of LSBs and a minimum value of 1 LSB.   Therefore, the dynamic range in decibels is simply

$$DR = 20\text{Log}\left(\frac{2^N - 1}{1}\right) \text{ dB} \qquad (28.15)$$

A 16-bit data converter has a dynamic range of 96.33 dB.

28.5 Analog-to-Digital Converter (ADC) Specifications

Many of the specifications that describe the ADC are similar to those that describe the DAC. However, there are subtle differences. Since the DAC is converting a discrete signal into an analog representation that is also limited by the resolution of the converter, a fixed number of inputs and outputs are generated. However, with the ADC, the input is an analog signal with an infinite number of values, which then has to be quantized into an N-bit digital word (Fig. 28.18). This process is much more difficult than the digital-to-analog process. In fact, many ADC architectures use a DAC as a critical component.

For example, in the previous discussion of DACs, it was determined that for a 16-bit DAC, the converter would need to generate output voltages in multiples of 76 μV. However, for the ADC, the converter needs to resolve differences in the analog signal of 76 μV. This means that the ADC must be able to detect changes in the input

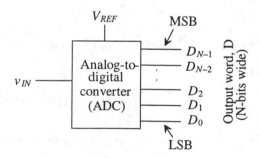

Figure 28.18 Block diagram of the digital-to-analog converter.

signal on the order of 1 part in 65,536! In contrast, the DAC had a finite number of input combinations (2^N). The ADC, however, has to "quantize" the infinite-valued analog signal into many segments so that

$$\text{Number of quantization levels} = 2^N \qquad (28.16)$$

This distinction is subtle but must be recognized to understand the differences between the two types of conversion.

Examine Fig. 28.19a. The digital output, D, of an ideal, 3-bit ADC is plotted versus the analog input, v_{IN}. Note the difference in the transfer curve for the ADC versus the DAC (Fig. 28.10). The y-axis is now the digital output, and the x-axis has been normalized to V_{REF}. Since the input signal is a continuous signal and the output is discrete, the transfer curve of the ADC resembles that of a staircase. Another fact to observe is that the 2^N quantization levels correspond to the digital output codes 0 to 7. Thus, the maximum output of the ADC will be 111 ($2^N - 1$), corresponding to the value for which $\frac{v_{IN}}{V_{REF}} \geq \frac{7}{8}$. Figure 28.19b corresponds to the error caused by the quantization.

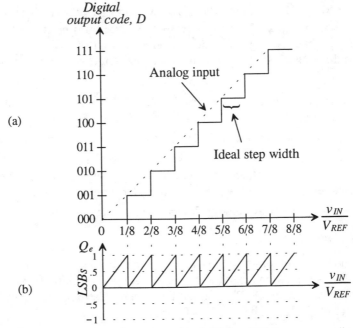

Figure 28.19 (a) Transfer curve for an ideal ADC and (b) its corresponding quantization error.

The value of 1 LSB for this ADC can be calculated using Eq. (28.9) and is the ideal step width (1/8) in Fig. 28.19 (versus the height for the DAC) multiplied by V_{REF}. Therefore, assuming that $V_{REF} = 5V$,

$$1 \text{ LSB} = 0.625 \text{ V} \qquad (28.17)$$

Quantization Error

Since the analog input is an infinite valued quantity and the output is a discrete value, an error will be produced as a result of the quantization. This error, known as *quantization error*, Q_e, is defined as the difference between the actual analog input and the value of the output (staircase) given in voltage. It is calculated as

$$Q_e = v_{IN} - V_{staircase} \qquad (28.18)$$

where the value of the staircase output, $V_{staircase}$, can be calculated by

$$V_{staircase} = D \cdot \frac{V_{REF}}{2^N} = D \cdot V_{LSB} \qquad (28.19)$$

where D is the value of the digital output code and V_{LSB} is the value of 1 LSB in volts, in this case 0.625 V. We can also easily convert the value of Q_e in units of LSBs. In Fig. 28.19a, Q_e can be generated by subtracting the value of the staircase from the dashed line. The result can be seen in Fig. 28.19b. A sawtooth waveform is formed centered about ½ LSBs. Ideally, the magnitude of Q_e will be no greater than one LSB and no less than 0. It would be advantageous if the quantization error were centered about zero so that the error would be at most $\pm\frac{1}{2}$ LSBs (as opposed to +1LSB). This is easily achieved as seen in Fig. 28.20a and b. Here, the entire transfer curve is shifted to the left by ½ LSB, thus making the codes centered around the LSB increments on the x-axis. This drawing illustrates that at best, an ideal ADC will have quantization error of $\pm\frac{1}{2}$ LSB.

In shifting this curve to the left, notice that the first code transition occurs when $\frac{v_{IN}}{V_{REF}} \geq \frac{1}{16}$. Therefore, the range of $\frac{v_{IN}}{V_{REF}}$ for the digital output corresponding to 000 is half as wide as the ideal step width. The last code transition occurs when $\frac{v_{IN}}{V_{REF}} \geq \frac{13}{16}$ (between 6/8 and 7/8). Note that the step width corresponding to this last code transition is 1.5 times larger than the ideal width and that the quantization error extends up to 1 LSB when $\frac{v_{IN}}{V_{REF}} = 1$. However, the converter would be considered to be out of range once $\frac{v_{IN}}{V_{REF}} \geq \frac{15}{16}$ (halfway between 7/8 and 8/8), so the problem is moot.

Differential Nonlinearity

Differential nonlinearity for an ADC is similar to that defined for a DAC. However, for the ADC, DNL is the difference between the actual code *width* of a nonideal converter and the ideal case. Figure 28.21 illustrates the transfer curve for a nonideal 3-bit ADC. The values for the DNL can be solved as follows:

$$\text{DNL} = \text{Actual step width} - \text{Ideal step width} \qquad (28.20)$$

Since the step widths can be converted to either volts for LSBs, DNL can be defined using either units. The value of the ideal step is 1/8. Converting to volts, this becomes

$$V_{idealstepwidth} = \frac{1}{8} \cdot V_{REF} = 0.625 \ V = 1 \text{ LSB} \qquad (28.21)$$

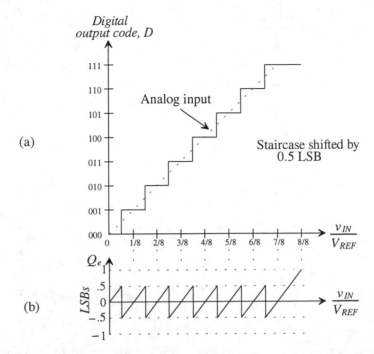

Figure 28.20 (a) Transfer curve for an ideal 3-bit ADC with (b) quantization error centered about zero.

Example 28.6

Using Fig. 28.21a, calculate the differential nonlinearity of the 3-bit ADC. Assume that $V_{REF} = 5$ V. Draw the quantization error, Q_e, in units of LSBs.

The DNL of the converter can be calculated by examining the step width of each digital output code. Since the ideal step width of the 000 transition is ½ LSB, then $DNL_0 = 0$. Also note that the step widths associated with 001 and 011 are equal to 1 LSB; therefore, both DNL_1 and DNL_4 are zero.   However, the remaining values code widths are not equal to the ideal value but can be calculated as

$$DNL_2 = 1.5 \text{ LSB} - 1 \text{ LSB} = 0.5 \text{ LSB}$$

$$DNL_3 = 0.5 \text{ LSB} - 1 \text{ LSB} = -0.5 \text{ LSB}$$

$$DNL_5 = -0.5 \text{ LSB}$$

$$DNL_6 = 0.5 \text{ LSB}$$

$$DNL_7 = 0 \text{ LSB (since the ideal step width is 1.5 LSB wide}$$
$$\text{at this code transition)}$$

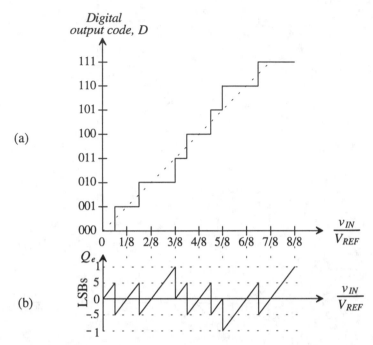

Figure 28.21   (a) Transfer curve for a nonideal 3-bit ADC used in Ex. 28.4 with
(b) quantization error illustrating differential nonlinearity.

The overall DNL for the converter used in this illustration is ±0.5 LSB. Note
that the quantization error illustrated in Fig. 28.21b is directly related to the
DNL. As DNL increases in either direction, the quantization error worsens.
Each "tooth" in the quantization error waveform should ideally be the same size.
■

Missing Codes

It is of interest to note the consequences of having a DNL that is equal to −1 LSB.
Figure 28.22 illustrates an ADC for which this is true. The total width of the step
corresponding to 101 is completely missing; thus, the value of DNL_5 is −1 LSB. Any
ADC possessing a DNL that is equal to −1 LSB is guaranteed to have a missing code.
Notice that the step width corresponding to 010 is 2 LSBs and that the value for DNL_2
is +1 LSB. However, there is not a missing code corresponding to 011, since the step
width of code 011 is dependent on the 100 transition. Therefore, an ADC having a
DNL greater than +1 LSB is not guaranteed to have a missing code, though in all
probability a missing code will occur.

Integral Nonlinearity

Integral nonlinearity (INL) is defined similarly to that for a DAC. Again, a "best-fit"
straight line is drawn through the end points of the first and last code transition, with

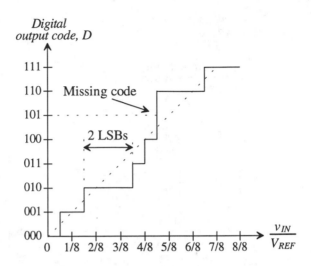

Figure 28.22 Transfer curve for a nonideal 3-bit ADC with a missing code.

INL being defined as the difference between the data converter code transition points and the straight line with all other errors set to zero.

Example 28.7

Determine the INL for the ADC whose transfer curve is illustrated in Fig. 28.23a. Assume that V_{REF} = 5 V. Draw the quantization error, Q_e, in units of LSBs.

By inspection, it can be seen that all of the transition points occur on the best-fit line except for the transitions associated with code 011 and 110. Therefore,

$$INL_0 = INL_1 = INL_2 = INL_4 = INL_5 = INL_7 = 0$$

The INL corresponding to the remaining codes can be calculated as

$$INL_3 = 3/8 - 5/16 = 1/16 \text{ or } 0.5 \text{ LSB}$$

Similarly, INL_6 can be calculated in the same manner and is found to be − 0.5 LSB. Thus, the overall INL for the converter is the maximum value of INL corresponding to ±0.5 LSB.

The INL can also be determined by inspecting the quantization error in Fig. 28.23b. Here, the INL will be the magnitude of the quantization error which lies outside the ±½ LSB band of Q_e. It can be seen that Q_e = 1 LSB, corresponding to the point at which INL = 0.5 LSB for digital output code 011, and that Q_e = −1 LSB at the output code corresponding to INL = − 0.5 LSB for digital output code 110. ∎

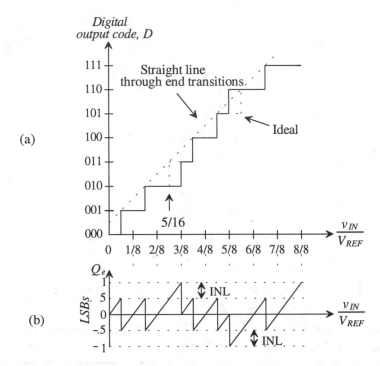

Figure 28.23   (a) Transfer curve of a nonideal 3-bit ADC and (b) its quantization error illustrating INL.

Offset and Gain Error

Offset and gain error are identical to the DAC case. *Offset error* occurs when there is a difference between the value of the first code transition and the ideal value of ½ LSBs. As seen in Fig. 28.24a, the offset error is a constant value. Note that the quantization error becomes ideal after the initial offset voltage is overcome. *Gain error* or *scale factor error*, seen in Fig. 28.24b, is the difference in the slope of a straight line drawn through the transfer characteristic and the slope of 1 of an ideal ADC. Causes of offset and gain error are discussed in Ch. 29, but it is important here to understand their overall effects on ADC transfer curves.

So far, we have examined only the DC characteristics of an ADC. However, examining the dynamic aspects of the converter will lead to a whole new set of errors. Sampling is inherently a dynamic process since the accuracy of the sample is dependent on the speed of the analog signal. Many effects that occur during sampling limit the overall performance of the converter.

Aliasing

As mentioned earlier in the chapter, the Nyquist Criterion requires that a signal be sampled at least two times the highest frequency contained in the signal. What would

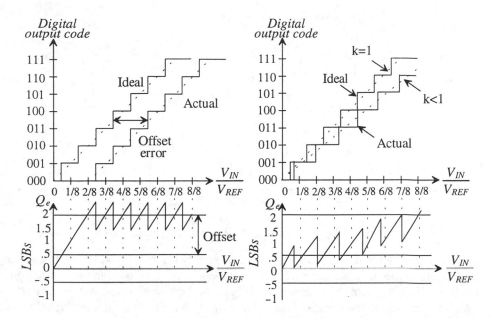

Figure 28.24 Transfer curve illustrating (a) offset error and (b) gain error.

happen if this criterion were ignored and the sampling rate was actually less than that amount? A phenomenon known as aliasing would occur.

Examine Fig. 28.25. Here, an analog signal is being sampled at a rate slower than the Nyquist Criterion requires. As a result, it appears that a totally different signal (see example dashed line) is being sampled. The different frequency signal is an "alias" of the original signal, and its frequency can be calculated using

$$f_{alias} = f_{actual} + k f_{sample} \quad (k = \ldots -2, -1, 0, 1, 2, 3 \ldots) \quad (28.22)$$

where f_{actual} is the frequency of the analog signal, f_{sample} is the sampling frequency, and f_{alias} is the frequency of the alias signal.

Aliasing can be eliminated by both sampling at higher frequencies and by filtering the analog signal before sampling and removing any frequencies that are greater than one-half the sampling frequency. It is good practice to filter the analog signal before sampling to eliminate any unknown higher order harmonics or noise that could result in aliasing.

A frequency domain analysis may further illustrate the concepts of aliasing. Figure 28.26 shows the analog signal, the *sampling function* (represented by a unit impulse train) and the resulting sampled signal in both the time and frequency domains. The analog signal in Fig. 28.26a is represented as a simple band-limited signal with center frequency, f_o. This simply means that the signal is contained within the frequency range shown. In Fig. 28.26b, the sampling function is shown in both the time

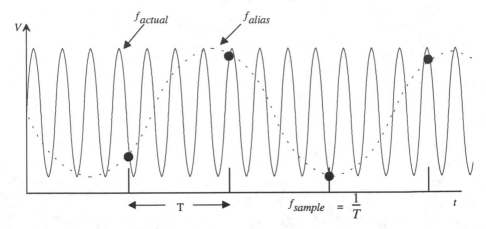

Figure 28.25 Illustration of aliasing caused by undersampling.

and frequency domain. The sampling function simply represents the action of sampling at discrete points in time. The frequency domain version of the sampling function is similar to its time domain counterpart, except that the x-axis is now represented as $f = 1/T$. Since each of the impulses has a value of 1, the resulting sampled signal shown in Fig. 28.26c is the impulse function multiplied by the amplitude of the analog signal at each discrete point in time. Remembering that multiplication in the time domain is equivalent to convolution in the frequency domain, we note that the frequency domain representation of the sampled signal reveals that the overall signal consists of multiple versions of the band-limited signal at multiples of the sampling frequency.

Note in Fig. 28.26b that as the sampling time increases, the sampling frequency decreases and the impulses in the frequency domain become more closely spaced. This results in 28.26d, which illustrates the aliasing as the multiple versions of the band-limited signal begin to overlap. One could also filter the signal "post-sampling" and eliminate the frequencies for which overlap occurs. The point at which the spectra overlap is called the *folding frequency*.

As mentioned earlier, the solutions to aliasing are higher sampling frequency and filtering. Focusing on just one of the solutions may worsen the situation for several reasons. Some noise signals are wide band, which means that they have a large bandwidth. Attempting to increase only the sampling frequency to eliminate the aliasing effects of the noise would be a practical impossibility, not to mention a costly one. However, simply filtering the input signal and the sampled signal will add delays to the overall conversion and increase the expense of the circuit. It is best to use a combination of the two to minimize the problem most efficiently.

Signal-to-Noise Ratio

Signal-to-noise (SNR) ratios of ADCs represent the value of the largest RMS input signal into the converter over the RMS value of the noise. Typically given in dB, the expression for SNR is

$$SNR = 20\text{Log}(\frac{V_{in(max)}}{V_{noise}})$$ (28.23)

If it is assumed that the input signal is a sinewave with a peak-to-peak value equal to the full-scale reference voltage of the converter, then the RMS value for $v_{in(max)}$ becomes

$$V_{in(max)} = \frac{V_{REF}}{2\sqrt{2}} = \frac{2^N(V_{LSB})}{2\sqrt{2}}$$ (28.24)

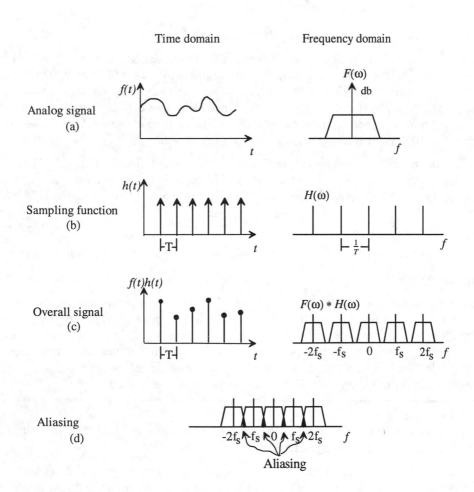

Figure 28.26 Illustration of aliasing in the time and frequency domain. (a) The analog signal; (b) the sampling function; (c) the overall signal; and (d) aliasing in the frequency domain.

where V_{LSB} is the voltage value of 1 LSB. The value of the noise (if the data converter is considered to be ideal) will be equivalent to the RMS value of the error signal, Q_e (in volts), shown in Fig. 28.20b. The RMS value of Q_e can be calculated to be

$$Q_{e,RMS} = \left[\frac{1}{V_{LSB}} \int_{-0.5VLSB}^{0.5VLSB} (V_{LSB})^2 dV_{LSB} \right]^{0.5} = \frac{V_{LSB}}{\sqrt{12}} \qquad (28.25)$$

Therefore, the SNR for the ideal ADC will be the ratio of these two RMS values.

$$SNR = 20 \cdot Log \frac{\frac{2^N(V_{LSB})}{2\sqrt{2}}}{Q_{e,RMS}} \qquad (28.26)$$

which can be written in terms of N as simply

$$SNR = 20NLog(2) + 20Log\sqrt{12} - 20Log(2\sqrt{2}) = 6.02N + 1.76 \qquad (28.27)$$

Equation (28.27) is an important one relating SNR to the resolution of the ADC. For 16-bit data conversion, one must design a circuit that will have a SNR of (6.02)(16) + 1.76 = 98.08 dB! Equation (28.27) can also be used in calculating the *signal-to-noise plus distortion ratio*, also known as *SNRD*. Since the output data is digital, we cannot use a spectrum analyzer to calculate this ratio, but must instead use a *Discrete Fourier Transform* (*DFT*) and examine the data in the digital domain.

Another useful application of Eq. (28.27) is the determination of effective number of bits given a system with a known SNR or SNRD. For example, if a 16-bit ADC yielded an SNRD of 88 dB, then the effective resolution of the converter would be

$$N = \frac{88 - 1.76}{6.02} = 14.32 \text{ bits} \qquad (28.28)$$

and the ADC would be producing the resolution equivalent to that of a 14-bit converter.

Aperture Error

The aperture error described in Sec. 28.3 (S/H) should be related to the errors associated with the ADC. In the previous discussion, the aperture error resulted in sampling error (Fig. 28.8). However, now that ADC characteristics have been discussed, we can relate the sampling error to the ADC. Since we know that the maximum errors associated with an ADC are related to ½ LSB, we can assume that the maximum sampling error associated with the aperture uncertainty can be no larger than ½ LSB.

Example 28.8

Find the maximum resolution of an ADC which can use the S/H described in Ex. 28.1 while maintaining a sampling error less than ½ LSB.

Since it was determined that the maximum sampling error produced by the given aperture uncertainty was 0.628 mV, we can relate this value to the highest resolution of an ADC by assuming that 0.628 mV will be less than or equal ½ LSB. Therefore,

$$0.628 \text{ mV} \le .5 \, LSB = \frac{V_{REF}}{2^{N+1}} = \frac{5}{2^{N+1}}$$

or

$$2^{N+1} \le 7961.8$$

which, solving for N (limited to an integer), yields a maximum resolution of 11 bits. ■

28.6 Mixed-Signal Layout Issues

Naturally, analog ICs are more sensitive to noise than digital ICs. For any analog design to be successful, careful attention must be paid to layout issues, particularly in a digital environment. Sensitive analog nodes must be protected and shielded from any potential noise sources. Grounding and power supply routing must also be considered when using digital and analog circuitry on the same substrate. Since a majority of ADCs use switches controlled by digital signals, separate routing channels must be provided for each type of signal.

Techniques used to increase the success of mixed-signal designs vary in complexity and priority. Strategies regarding systemwide minimization of noise should always be considered foremost. A mixed-signal layout strategy can be modeled as seen in Fig. 28.27. The lowest issues are foundational and must be considered before each succeeding step. The successful mixed-signal design will always minimize the effect of the digital switching on the analog circuits.

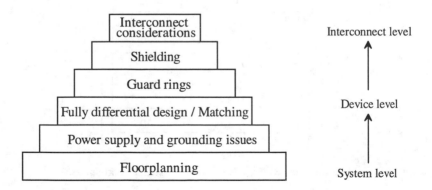

Figure 28.27 Mixed-signal layout strategy.

Floorplanning

The placement of sensitive analog components can greatly affect a circuit's performance. Many issues must be considered. In designing a mixed-signal system, strategies regarding the "floorplan" of the circuitry should be thoroughly analyzed well before the layout is to begin.

The analog circuitry should be categorized by the sensitivity of the analog signal to noise. For example, low-level signals or high-impedance nodes typically associated with input signals are considered to be sensitive nodes. These signals should be closely guarded and shielded especially from digital output buffers. High-swing analog circuits such as comparators and output buffer amplifiers should be placed between the sensitive analog and the digital circuitry.

The digital circuitry should also be categorized by speed and function. Obviously, since digital output buffers are usually designed to drive capacitive loads at very high rates, they should be kept farthest from the sensitive analog signals. Next, the high and lower speed digital should be placed between the insensitive analog and the output buffers. An example of this type of strategy can be seen in Fig. 28.28 [6]. Notice that *the sensitive analog is as far away as possible from the digital output buffers* and that the least sensitive analog circuitry is next to the least offensive digital circuitry.

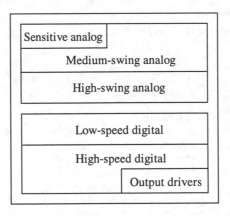

Figure 28.28 Example of a mixed-signal floorplan [6].

Power Supply and Grounding Issues

Whenever analog and digital circuit reside together on the same die, danger exists of injecting noise from the digital system to the sensitive analog circuitry through the power supply and ground connections. Much of the intercoupling can be minimized by carefully considering how power and ground are supplied to both the analog and digital circuits.

Examine Fig. 28.29a. Here, analog and digital circuitry share the same routing to a single pad for power and ground. The resistors, R_{I1} and R_{I2}, represent the small, nonnegligible resistance of the interconnect to the pad. The inductors, L_{B1} and L_{B2}, represent the inductance of the bonding wire which connects the pads to the pin on the lead frame. Since digital circuitry is typified by high amounts of transient currents due to switching, a small amount of resistance associated with the interconnect can result in significant voltage spikes. Low-level analog signals are very sensitive to such interference, thus resulting in a contaminated analog system. Another significant voltage spike can occur due to the inductance of the bonding wire. Since the voltage across the inductor is proportional to the change in current through it, voltage spikes equating to hundreds of millivolts can result! Both of these voltage effects are true for both the power and ground connection.

One way to reduce the interference, seen in Fig. 28.29b, is to prohibit the analog and digital circuit from sharing the same interconnect. The routing for the supply and ground for both the analog and digital sections are provided separately. Although this

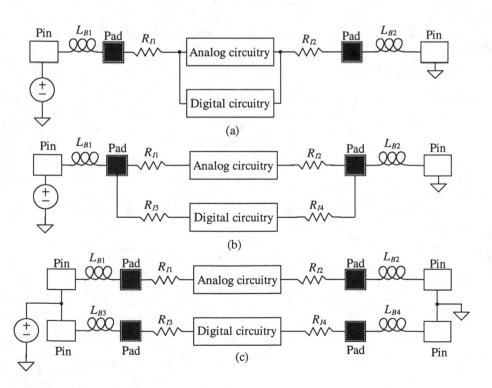

Figure 28.29 Power and ground connection examples, (a) poor noise immunity and
(b) better noise immunity, and (c) using separate power and ground pins
to achieve even better immunity.

eliminates the parasitic resistance due to the common interconnect, there is still a common inductance due to the bonding wire which causes interference.

Another method that minimizes interference even more than the previous case is seen in Fig. 28.29c. *By using separate pads and pins, the analog and the digital circuits are completely decoupled.* The current through the analog interconnect is much less abrupt than the digital; thus, the analog circuitry now has a "quiet" power and ground. However, this technique is dependent on whether extra pins and pads are available for this use. The separate power supply and ground pins are then connected externally. *It is not wise to use two separate power supplies because if both types of circuits are not powered up simultaneously, latch-up could easily result.*

In cases presented in Fig. 28.29b and c, *the resistance associated with the analog connection to ground or supply can be reduced by making the power supply and ground bus as wide as feasible.* This reduces the overall resistance of the metal run, thus decreasing the voltage spikes that occur across the resistor. The inductor itself is impossible to eliminate, though it can be minimized with careful planning. Since the length of the bonding wire is dependent on the distance from the pad to the lead frame, *one could reduce the effect of the wire inductance by reserving pins closest to the die for sensitive connections such as analog supply and ground.* This, again, illustrates the importance of floorplanning.

Fully Differential Design

Fully differential operational amplifiers were presented in Ch. 25, and Fig. 28.30 was first presented in Sec. 25.3. The noise source represents the noise from digital circuitry coupled through the parasitic stray capacitors. If equal amounts of noise are injected into the differential amplifiers, then the common-mode rejection inherent in the amplifiers will eliminate most or all of the noise. This, of course, is dependent on the symmetry of the amplifiers, meaning that the matching of the transistors in the amplifier becomes crucial. Therefore, *in a mixed-signal environment, layout techniques should be used to improve matching.* Common-centroid and interdigitated techniques were discussed in Secs. 20.1.6 and 24.1.4.

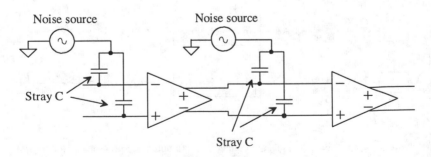

Figure 28.30 Differential output op-amps showing parasitic coupling to noise sources.

Guard Rings

Guard rings, discussed previously in Secs. 7.2.1 and 11.3.1, should be used wisely throughout a mixed-signal environment. Circuits that process sensitive signals should be placed in a separate well with guard rings attached to the analog *VDD* supply. In the case of the CN20 process, the n-type devices outside the well should have guard rings attached to analog *VSS* placed around them as well. Digital circuits should be placed in their own well with guard rings attached to digital *VDD*. Guard rings placed around the n-channel digital devices will also help minimize the amount of noise transmitted from the digital devices.

Shielding

A number of techniques exist which can shield sensitive, low-level analog signals from noise resulting from digital switching. A shield can take the form of a layer tied to analog ground placed between two other layers, or it can be a barrier between two signals running in parallel.

   If at all possible, one should avoid crossing sensitive analog signals, such as low-level analog input signals, with any digital signals. The parasitic capacitance coupling the two signal lines can be as much as a couple of fF, depending on the process. If it cannot be avoided, then attempt to carry the digital signal using the top layer of metal (such as metal2). If the analog signal is an input signal, then it will most likely be carried by the poly layer. A strip of metal1 can be placed between the two layers and connected to analog ground (see Fig. 28.31).

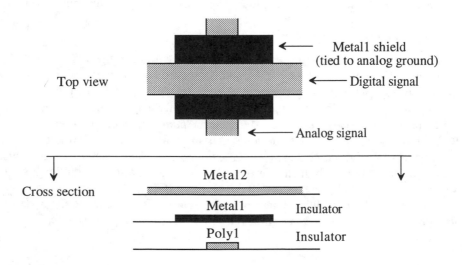

Figure 28.31 Shielding a sensitive analog signal from a digital signal crossover using a metal 1 shield layer.

Another situation that should be avoided is running interconnect containing sensitive analog signals parallel and adjacent to any interconnect carrying digital signals. Coupling occurs due to the parasitic capacitance between the lines. If this situation cannot be avoided, then an additional line connected to analog ground should be placed between the two signals, as seen in Fig. 28.32. This method can also be used to partition the analog and digital sections of the chip.

In addition, the n-well can be used as a bottom-plate shield to protect analog signals from substrate noise. Poly resistors (or capacitors) used for sensitive analog signals can be shielded by placing an n-well beneath the components and connecting the well to analog *VDD*.

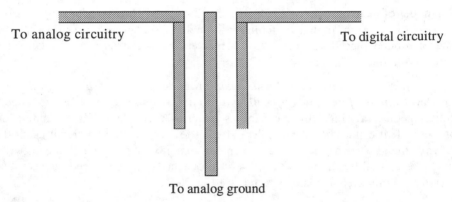

To analog circuitry

To digital circuitry

To analog ground

Figure 28.32 Use of a dummy metal strip to provide shielding to two parallel signals.

Other Interconnect Considerations

Finally, some other layout strategies will incrementally improve the performance of the analog circuitry. However, if the previous strategies are not followed, these suggestions will be useless. *When routing the analog circuitry, minimize the lengths of current carrying paths.* This will simply reduce the amount of voltage drop across the path due to the metal1 or metal2 resistance. *Contacts should also be used very liberally whenever changing layers.* Not only does this minimize resistance in the path, but it also improves fabrication reliablity. *Avoid using poly to route current carrying signal paths.* Not only is the poly higher in resistance value, but also the additional contact resistance required to change layers will not be insignificant. If the poly is made wider to lower the resistance, additional parasitic capacitance will be added to the node. *Use poly to route only high-impedance gate nodes that carry virtually no current.*

REFERENCES

[1]   M. J. Demler, *High-Speed Analog-to-Digital Conversion*, Academic Press, 1991.

[2]   B. Razavi, *Principles of Data Conversion System Design*, IEEE Press, 1995.

[3]   R. L. Geiger, P. E. Allen, and N. R. Strader, *VLSI Design Techniques for Analog and Digital Circuits*, McGraw-Hill Publishing Co., 1990.

[4]   D. H. Sheingold, *Analog-Digital Conversion Handbook*, Prentice-Hall Publishing, 1986.

[5]   S. K. Tewksbury et al., "Terminology Related to the Performance of S/H, A/D and D/A Circuits," *IEEE Transactions on Circuits and Systems*, CAS-25, Vol. CAS-25, pp. 419-426, July 1978.

[6]   Y. Tsividis, *Mixed Analog-Digital VLSI Devices and Technology: An Introduction*, McGraw-Hill Publishing Co., 1996.

PROBLEMS

28.1 Determine the number of quantization levels needed if one wanted to make a digital thermometer that was capable of measuring temperatures to within 0.1 °C accuracy over a range from − 50 °C to 150 °C. What resolution of ADC would be required?

28.2 Using the same thermometer as above, what sampling rate, in samples per second, would be required if the temperature displayed a frequency of $15° \cdot \sin(0.01 \cdot 2\pi t)$?

28.3 Determine the maximum droop allowed in a S/H used in a 16-bit ADC assuming that all other aspects of both the S/H and ADC are ideal. Assume $V_{ref} = 5$ V.

28.4 A S/H circuit settles to within 1 percent of its final value at 5 μs. What is the maximum resolution and speed with which an ADC can use this data assuming that the ADC is ideal?

28.5 A digitally programmable signal generator uses a 14-bit DAC with a 10 volt reference to generate a DC output voltage. What is the smallest incremental change at the output that can occur? What is the DAC's full-scale value? What is its accuracy?

28.6 Determine the maximum DNL (in LSBs) for a 3-bit DAC which has the following characteristics. Does the DAC have 3-bit accuracy? If not, what is the resolution of the DAC having this characteristic?

Digital Input	Voltage Output
000	0 V
001	0.625 V
010	1.5625 V
011	2.0 V
100	2.5 V

101	3.125 V
110	3.4375 V
111	4.375 V

28.7   Repeat Problem 28.6 calculating the INL (in LSBs).

28.8   A DAC has a reference voltage of 1,000 V and has its maximum INL measured to be 2.5 mV. What is the maximum resolution of the converter assuming that all the other characteristics of the converter are ideal?

28.9   Determine the INL and DNL for a DAC that has a transfer curve shown in Fig. P28.9.

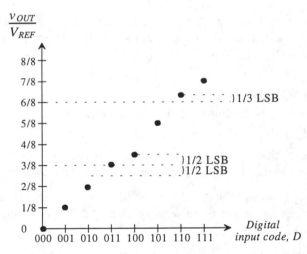

Figure P28.9

28.10   A DAC has a full-scale voltage of 4.97 V using a 5 V reference, and its minimum output voltage is limited by the value of one LSB. Determine the resolution and dynamic range of the converter.

28.11   Prove that the RMS value of the quantization noise shown in Fig. 28.20b is as stated in Eq. (28.26).

28.12   An ADC has a stated SNR of 94 dB. Determine the effective number of bits of resolution of the converter.

28.13   Discuss the methods used to prevent aliasing and the advantages and disadvantages of each.

Chapter

29

Data Converter Architectures

Applications such as wireless communications and digital audio and video have created the need for cost-effective data converters that will achieve higher speed and resolution. The needs required by digital signal processors continually challenge analog designers to improve and develop new ADC and DAC architectures. There are many different types of architectures, each with unique characteristics and different limitations. This chapter presents a basic overview of the more popular data converter architectures and discusses the advantages and disadvantages of each along with their limitations.[24]

Now that we have defined the operating characteristics of ADCs, a more detailed examination of the basic architectures will be discussed using a top-down approach. Since many of the converters use op-amps, comparators, and resistor and capacitor arrays, the top-down approach will allow a broader discussion of the key component limitations in later sections.

29.1 DAC Architectures

A wide variety of DAC architectures exist, ranging from very simple to complex. Each, of course, has its own merits. Some use voltage division, whereas others employ current steering and even charge scaling to map the digital value into an analog quantity.

29.1.1 Digital Input Code

In many cases, the digital signal is not provided in binary code but is any one of a number of codes: binary, BCD, thermometer code, Gray code, sign-magnitude, two's complement, offset binary, and so on. (See Fig. 29.1 for a comparison of some of the more commonly used digital input codes.) For example, it may be desirable to allow only one bit to change value when changing from one code to the next. If that is the case, a Gray code will suffice. The thermometer code is used quite frequently and can

---

[24] Much of the material in this chapter is based on course notes provided by Dr. Terry Sculley, of ESS Technology, Inc., Austin, Texas.

Decimal	Binary	Thermometer	Gray	Two's Complement
0	000	0000000	000	000
1	001	0000001	001	111
2	010	0000011	011	110
3	011	0000111	010	101
4	100	0001111	110	100
5	101	0011111	111	011
6	110	0111111	101	010
7	111	1111111	100	001

Figure 29.1 Comparison of digital input codes.

also be seen in Fig. 29.1. Notice that it requires $2^N - 1$ bits to represent an N-bit word. The code used is purely dependent on the application, and the reader should be aware that many types of codes are available.

29.1.2 Resistor String

The most basic DAC is seen in Fig. 29.2a. Comprised of a simple resistor string of 2^N identical resistors and switches, the analog output is simply the voltage division of the resistors at the selected tap. Note that a $N:2^N$ decoder will be required to provide the 2^N signals controlling the switches. This architecture typically results in good accuracy, provided that no output current is required and that the values of the resistors are within the specified error tolerance of the converter. One big advantage of a resistor string is that the output will always be guaranteed to be monotonic.

One problem with this converter is that the converter output is always connected to $2^N - 1$ switches that are off and one switch that is on. For larger resolutions, a large parasitic capacitance appears at the output node, resulting in slower conversion speeds. A better alternative for the resistor string DAC is seen in Fig. 29.2b. Here, a binary switch array ensures that the output is connected to at most N switches that are on and N switches that are off, thus increasing the conversion speed. The input to this switch array is a binary word since the decoding is inherent in the binary tree arrangement of the switches.

Another problem with the resistor string DAC is the balance between area and power dissipation. An integrated version of this converter will lead to a large chip area for higher bit resolutions because of the large number of passive components needed. Active resistors such as the N-well resistor can be used for low-resolution applications. However, as the resolution increases, the relative accuracy of the resistors becomes an important factor. Although the value of R could always be made small to minimize the

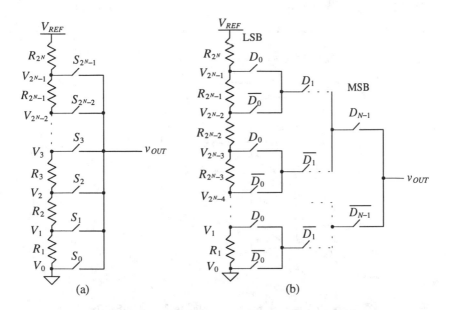

Figure 29.2 (a) A simple resistor string DAC and (b) use of a binary switch array to lower the output capacitance.

chip area required, power dissipation would then become the critical issue as current flows through the resistor string at all times.

Example 29.1

Design a 3-bit resistor string ladder using a binary switch array. Assume that $V_{REF} = 5$ V and that the maximum power dissipation of the converter is to be 5 mW (not including the power required by the digital logic). Determine the value of the analog voltage for each of the possible digital input codes.

The power dissipation will determine the current flowing through the resistor string by

$$I_{MAX} = \frac{5 \times 10^{-3} \text{ W}}{5 \text{ V}} = 1 \text{ mA}$$

Since a 3-bit converter will have eight resistors, the value of R is

$$R = \frac{1}{8} \cdot \frac{5 \text{ V}}{1 \text{ mA}} = 625 \ \Omega$$

The converter can be seen in Fig. 29.3. Examine the switch array if the input code is $D_2 D_1 D_0 = 100$ or 4_{10}. Since D_2 is high, the top switch will be closed and the lower switch, $\overline{D_2}$, will be open. In the row corresponding to D_1, since $D_1 = 0$, both of the switches marked $\overline{D_1}$ will be closed and the other two will be open. The LSB controls the largest number of switches; therefore, since D_0 is low, all

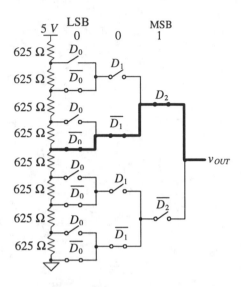

Figure 29.3 A 3-bit resistor string DAC used in Ex. 29.1

the $\overline{D_0}$ switches will be closed and all the D_0 switches will be open. There should be only one path connecting a single tap on the resistor string to the output. This is bolded, with the resistor string tapped in the middle of the string. Therefore, $v_{OUT} = \frac{1}{2}\, V_{REF} = 2.5$ V. The remaining outputs can be seen in Fig. 29.4. ∎

$D_2 D_1 D_0$	v_{OUT}
000	0
001	0.625
010	1.25
011	1.875
100	2.5
101	3.125
110	3.75
111	4.375

Figure 29.4 Output voltages generated from the 3-bit DAC in Ex. 29.1.

Mismatch Errors Related to the Resistor String DAC

The accuracy of the resistor string is obviously related to matching between the resistors, which ultimately determines the INL and DNL for the entire DAC. Suppose that the i-th resistor, R_i, has a mismatch error associated with it so that

$$R_i = R + \Delta R_i \tag{29.1}$$

where R is the ideal value of the resistor and ΔR_i is the mismatch error. Also suppose that the mismatches were symmetrical about the string so that the sum of all the mismatch terms were zero, or

$$\sum_{i=1}^{2^N} \Delta R_i = 0 \tag{29.2}$$

The value of the voltage at the tap associated with the i-th resistor should ideally be

$$V_{i,ideal} = \frac{(i)V_{REF}}{2^N}, \text{ for } i = 0, 1, 2, ..., 2^N - 1 \tag{29.3}$$

However, including the mismatch, the actual value of the i-th voltage will be the sum of all the resistances up to and including resistor i, divided by the sum of all the resistances in the string. This can be represented by

$$V_i = V_{REF} \cdot \frac{\sum_{k=1}^{i} R_k}{\sum_{k=1}^{2^N} R_k} = V_{REF} \cdot \frac{\sum_{k=1}^{i} R + \Delta R_k}{2^N R} \tag{29.4}$$

The denominator does not include any mismatch error since it was assumed that the mismatches sum to zero as defined in Eq. (29.2). Notice that there is no resistor, R_0, corresponding to V_0 (see Fig. 29.2), and it is assumed that V_0 is ground. Equation (29.4) can be rewritten as

$$V_i = \frac{V_{REF}}{2^N R} \left[(i)R + \sum_{k=1}^{i} \Delta R_k \right] = \frac{(i)V_{REF}}{2^N} + \frac{V_{REF}}{2^N R} \sum_{k=1}^{i} \Delta R_k \tag{29.5}$$

or finally, the value of the voltage at the i-th tap is

$$V_i = V_{i,ideal} + \frac{V_{REF}}{2^N} \cdot \sum_{k=1}^{i} \frac{\Delta R_k}{R} \tag{29.6}$$

Equation (29.6) is not of much importance by itself, but it can be used to help determine the nonlinearity errors.

Integral Nonlinearity of the Resistor String DAC

Integral nonlinearity (INL) is defined as the difference between the actual and ideal switching points, or

$$INL = V_i - V_{i,ideal} \tag{29.7}$$

and plugging in Eqs. (29.6) and (29.3) into (29.7) yields,

$$INL = \frac{V_{REF}}{2^N} \cdot \sum_{k=1}^{i} \frac{\Delta R_k}{R} \tag{29.8}$$

Equation (29.8) is a general expression for the INL for a given resistor, R_i, and requires that the mismatch of all the resistances used in the summation are known. However, this equation does not illuminate how to determine the *worst-case* or maximum INL for a resistor string.

Intuitively, one would think that the worst-case INL would occur at the top of the resistor string ($i=2^N$) with all the ΔR_k's at their maximum values. However, the previous derivation was performed with the assumption that the mismatches summed to zero. With this restriction, the maximum INL will occur at the midpoint of the string corresponding to $i = 2^{N-1}$, corresponding to the case where the MSB was a one and all other bits were zero. Another condition that will ensure a worst-case scenario will be to consider the lower half resistors at their maximum positive mismatch value and upper half resistors at their maximum negative mismatch value, or vice versa.

If the resistors on a string were known to have 2 percent matching, then ΔR_k would be constrained to the bounds of

$$-0.02R \leq \Delta R_k \leq 0.02R \tag{29.9}$$

and the worst-case INL (again, % matching = 0.02) using Eq. (29.8) would be,

$$|INL|_{max} = \frac{V_{REF}}{2^N} \cdot \sum_{k=1}^{2^{N-1}} \frac{\Delta R_k}{R} = \frac{V_{REF}}{2^N} \cdot \frac{2^{N-1} \cdot \Delta R_k}{R} = \frac{1}{2} LSB \cdot 2^N \cdot (\% \text{ matching}) = 0.01 V_{REF} \tag{29.10}$$

which for $INL < 0.5\ LSB$ requires $1/2^N > (\% \text{ matching})$. For 2 percent matching the maximum number of bits, N, is then 5! For better than 0.2 % matching $N = 9$ bits.

Because the worst-case analysis was performed, the maximum INL occurs at the middle of the string. We can improve this specification on paper by using the "best-fit" approach to measuring INL. In this case, the reference line is simply shifted up slightly (refer to Ch. 28) so that it no longer passes through the end points, but instead minimizes the INL.

Example 29.2
Determine the effective number of bits for a resistor string DAC, which is assumed to be limited by the INL. The resistors are passive poly resistors with a known relative matching of 1 percent and $V_{REF} = 5$ V.

Using Eq. (29.10), the maximum INL will be

$$|INL|_{max} = 0.005 \cdot V_{REF} = 0.025 \text{ V}$$

Since we know that this maximum INL should be equal to ½ LSB in the worst case,

$$\frac{1}{2}LSB = \frac{5}{2^{N+1}} = 0.025 \text{ V}$$

and solving for N yields

$$N = \log_2\left(\frac{5}{0.025}\right) - 1 = 6.64 \text{ bits}$$

This means that the resolution for a DAC containing a resistor string matched to within 1 percent will be, at most 6 bits. ∎

Differential Nonlinearity of the Worst-Case Resistor String DAC

Resistor string matching is not as critical when determining the DNL. Remembering that the definition of DNL is simply the actual height of the stair-step in the DAC transfer curve minus the ideal step height, we can write this in terms of the voltages at the taps of adjacent resistors on the string. Using Eq. (29.5), we can express this as,

$$|V_i - V_{i-1}| = \left|\left[\frac{(i)V_{REF}}{2^N} + \frac{V_{REF}}{2^N}\cdot\sum_{k=1}^{i}\frac{\Delta R_k}{R}\right] - \left[\frac{(i-1)V_{REF}}{2^N} + \frac{V_{REF}}{2^N}\sum_{k=1}^{i-1}\frac{\Delta R_k}{R}\right]\right|$$

which can be simplified to

$$|V_i - V_{i-1}| = \left|\frac{V_{REF}}{2^N}\left(1 + \frac{\Delta R_i}{R}\right)\right| \tag{29.11}$$

The DNL can then be determined by subtracting the ideal step height from Eq. (29.11),

$$DNL_i = \left|\frac{V_{REF}}{2^N}\left(1 + \frac{\Delta R_i}{R}\right) - \frac{V_{REF}}{2^N}\right| = \left|\frac{V_{REF}}{2^N}\cdot\frac{\Delta R_i}{R}\right| \tag{29.12}$$

and the maximum DNL will occur at the value of i for which ΔR is at its maximum value. If it is assumed once again that the resistors are matched to within 2 percent, the worst-case DNL will be

$$DNL_{max} = \left|0.02\cdot\frac{V_{REF}}{2^N}\right| = 0.02 \ LSB \tag{29.13}$$

which is well below the ½ LSB limit. The INL is obviously the limiting factor in determining the resolution of a resistor string DAC since its maximum value is 2^N times larger than the DNL.

29.1.3 *R-2R* Ladder Networks

Another DAC architecture that incorporates fewer resistors is called the *R-2R* ladder network [1]. This configuration consists of a network of resistors alternating in value of R and $2R$. Figure 29.5 illustrates an *N*-bit *R-2R* ladder. Starting at the right end of the network, notice that the resistance looking to the right of any node to ground is $2R$. The digital input determines whether each resistor is switched to ground (noninverting input) or to the inverting input of the op-amp. Each node voltage is related to V_{REF}, by a binary-weighted relationship caused by the voltage division of the ladder network. The total current flowing from V_{REF} is constant, since the potential at the bottom of each

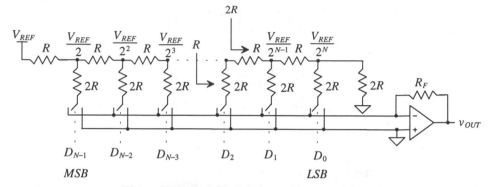

Figure 29.5 An *R-2R* digital-to-analog converter.

switched resistor is always zero volts (either ground or virtual ground). Therefore, the node voltages will remain constant for any value of the digital input.

The output voltage, v_{OUT}, is dependent on currents flowing through the feedback resistor, R_F, such that

$$v_{OUT} = -i_{TOT} \cdot R_F \tag{29.14}$$

where i_{TOT} is the sum of the currents selected by the digital input by

$$i_{TOT} = \sum_{k=0}^{N-1} D_k \cdot \frac{V_{REF}}{2^{N-k}} \cdot \frac{1}{2R} \tag{29.15}$$

where D_k is the k-th bit of the input word with a value that is either a 1 or a 0.

This architecture, like the resistor string architecture, requires matching to within the resolution of the converter. Therefore, the switch resistance must be negligible, or a small voltage drop will occur across each switch, resulting in an error. One way to eliminate this problem is to add dummy switches. Assume that the resistance of each switch connected to the 2R resistors is ΔR, as seen in Fig. 29.6. Dummy switches with one-half the resistance of the real switches are "hard-wired" so that they are always on and placed in series with each of the horizontal resistors. The total resistance of any horizontal branch, R', is

$$R' = R + \frac{\Delta R}{2} \tag{29.16}$$

The resistance of any vertical branch is $2R + \Delta R$, which is twice the value of the horizontal branch. Therefore, a $R' - 2R'$ relationship is maintained. Of course, a dummy switch equal to the switch size of a 2R switch will have to be placed in series with the terminating resistor as well.

Example 29.3
Design a 3-bit DAC using a *R-2R* architecture with $R = 1$ kΩ, $R_F = 2$ kΩ, and $V_{REF} = 5$ V. Assume that the resistances of the switches are negligible.

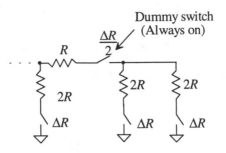

Figure 29.6 Use of dummy switches to offset switch resistance.

Determine the value of i_{TOT} for each digital input and the corresponding output voltage, v_{OUT}.

Figure 29.7 shows the 3-bit DAC for a digital input of 001. The voltages at each node in the resistor network are labeled. For each switch, if the digital input bit is a 0, then the resistor is attached to the ground. If the bit is a 1, then the resistor is attached to the virtual ground of the inverting input and current flows to the output of the op-amp. Therefore, for $D_2D_1D_0 = 000$, all the switches are connected to ground and no current flows through the feedback resistor and the output voltage, v_{OUT}, is zero.

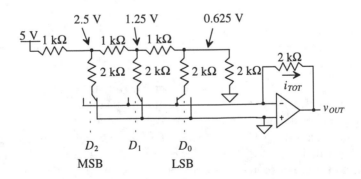

Figure 29.7 A 3-bit R-2R digital-to-analog converter used in Ex. 29.3.

When $D_2D_1D_0 = 001$, the rightmost resistor is switched to the op-amp inverting input and the other two resistors remain attached to ground. Therefore, the total current flowing through the feedback resistor will simply be the current through the rightmost resistor, which is defined by Eq. (29.15) as

$$\frac{V_{REF}}{8} \cdot \frac{1}{2000} = 0.3126 \text{ mA}$$

and the output voltage, by Eq. (29.14) becomes,

$$v_{OUT} = -(0.3126 \text{ mA})(2000 \ \Omega) = -0.625 \text{ V}$$

which is to be expected. The other values for the output voltage can be calculated using Eqs. (29.14) and (29.15) and are seen in Fig. 29.8. ∎

$D_2D_1D_0$	i_{TOT} (mA)	v_{OUT} (V)
000	0	0
001	0.3125	− 0.625
010	0.625	− 1.25
011	0.625 + 0.3125 = 0.9375	− 1.875
100	1.25	− 2.5
101	1.25 + 0.3125 = 1.5625	− 3.125
110	1.25 + 0.625 = 1.875	− 3.75
111	1.25 + 0.625 + 0.3125 = 2.1875	− 4.375

Figure 29.8 Output voltages generated from the 3-bit DAC in Example 29.3.

29.1.4 Current Steering

In the previous section, a voltage was converted into a current, which then generated a voltage at the output. Another DAC method uses current throughout the conversion. Known as *current steering*, this type of DAC requires precision current sources that are summed in various fashions.

Figure 29.9 illustrates a generic current steering DAC. This configuration requires a set of current sources, each having a unit value of current, I. Since there are no current sources generating i_{OUT} when all the digital inputs are zero, the MSB, D_{2^N-2}, is offset by two index positions instead of one. For example, for a 3-bit converter, seven current sources will be needed, labeled from D_0 to D_6. The binary signal controls whether or not the current sources are connected to either i_{OUT} or some other summing node (in this case ground). The output current, i_{OUT}, has the range of

$$0 \le i_{OUT} \le (2^N - 1) \cdot I \tag{29.17}$$

and can be any integer multiple of I in between. An interesting issue to note is the format of the digital code required to drive the switches. Since there are $2^N - 1$ current sources, the digital input will be in the form of a *thermometer code*. This code will be all 1's from the LSB up to the value of the k-th bit, D_k, and all 0's above it. The point at

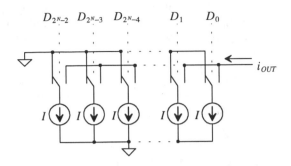

Figure 29.9 A generic current steering DAC.

which the input code changes from all 1's to all 0's "floats" up or down and resembles the action of a thermometer, hence the name. Typically, a thermometer encoder is used to convert binary input data into a thermometer code.

Another current steering architecture is seen in Fig. 29.10. This architecture uses binary-weighted current sources, thus requiring only N current sources of various sizes versus $2^N - 1$ sources in the previous example. Since the current sources are binary weighted, the input code can be a simple binary number with no thermometer encoder needed.

One advantage of the current steering DACs is the high-current drive inherent in the system. Since no output buffers are necessary to drive resistive loads, these DACs typically are used in high-speed applications. Traditionally, high-speed current steering DACs have been fabricated using bipolar technology. However, the ability to generate matched current mirrors makes CMOS an enticing alternative. Of course, the precision needed to generate high resolutions is dependent on how well the current sources can be matched or the degree to which they can be made binary weighted. For example, if a 13-bit DAC was designed using these architectures, there would have to be 8,191 current sources resident on the chip - not an insignificant amount. For the

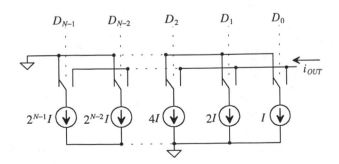

Figure 29.10 A current steering DAC using binary-weighted current sources.

binary-weighted sources, only 13 current sources would be needed. Yet the size of the largest current source would have to be 4096 or 2^{N-1} times larger than the smallest. Even if the unit current, I, was chosen to be 5 μA, the largest current source would be 20.48 mA!

Another problem associated with this architecture is the error due to the switching. Since the current sources are in parallel, if one of the current sources is switched off and another is switched on, a "glitch" could occur in the output if the timing was such that both of them were on or both were off for an instant. While this may not seem significant, if the converter is switching from 0111111 to 1000000, the output will spike toward ground and then back to the correct value if all the switches turn off for an instant. If the DAC is driving a resistive load and the output current is converted to a voltage, a substantial voltage spike will occur at the output.

Example 29.4

Construct a table showing the thermometer code necessary to generate the output shown in Fig. 29.11a for a 3-bit current steering DAC using unit current sources.

The thermometer code can be seen in Fig. 29.11b. When the code is all zeros, the output is zero volts. Therefore, only 7 bits are needed to represent the 2^N or 8 states of a 3-bit DAC. Note how the interface between all 1's and all 0's actually resembles the output signal itself. ∎

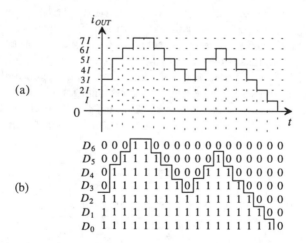

Figure 29.11 (a) Output of a 3-bit current steering DAC and (b) the thermometer code input.

Mismatch Errors Related to Current Steering DACs

Analysis of the mismatch associated with the current sources is similar to the resistor string analysis. It is assumed that each current source in Fig. 29.9 is

$$I_k = I + \Delta I_k \text{ for } k = 1, 2, 3, \ldots, 2^N - 1 \tag{29.18}$$

where I is the ideal value of the current and ΔI_k is the error due to mismatch. If it is again assumed that the ΔI_K terms sum to zero and that one-half of the current sources contain the maximum positive mismatch, ΔI_{max}, and the other half contains the maximum negative mismatch, $-\Delta I_{max}$, (or vice versa), then the worst-case condition will occur at midscale with the actual output current being

$$I_{out} = \sum_{k=1}^{2^{N-1}} (I + \Delta I_k) = 2^{N-1} \cdot I + 2^{N-1} \cdot |\Delta I|_{max} = I_{out,ideal} + 2^{N-1} \cdot |\Delta I|_{max} \quad (29.19)$$

Since the INL is simply the actual output current minus the ideal, the worst-case INL will be

$$|INL|_{max} = 2^{N-1} \cdot |\Delta I|_{max,INL} \quad (29.20)$$

The term, $|\Delta I|_{max,INL}$ represents the maximum current source mismatch error that will keep the INL less than ½ LSB. Each current source represents the value of 1 LSB; therefore, ½ LSB is equal to 0.5 I. Since the maximum INL should correspond to the ½ LSB, equating Eq. (29.20) to ½ I results in the value for $|\Delta I|_{max,INL}$,

$$|\Delta I|_{max,INL} = \frac{0.5I}{2^{N-1}} = \frac{I}{2^N} \quad (29.21)$$

Equation (29.21) illustrates the difficulty of using this architecture at high resolutions. If the value of I is set to be 5 µA, and the N is desired to be 12 bits, then the value of $\Delta I_{max,INL}$ becomes

$$|\Delta I|_{max,INL} = \frac{5 \times 10^{-6}}{2^{12}} = 1.221 \text{ nA!} \quad (29.22)$$

which means that each of the 5 µA current sources must lie between the bounds of,

$$4.99878 \text{ µA} \le I_k \le 5.001221 \text{ µA} \quad (29.23)$$

to achieve a worst-case INL, which is within ½ LSB error.

The DNL is easily obtained since the step height in the transfer curve is equivalent to the value of the ideal current source, I. The maximum difference between any two adjacent values of output current will be simply the value of the single source, I_k, which contains the largest mismatch error for which the DNL will be less than ½ LSB, $|\Delta I|_{max,DNL}$:

$$I_{out(x)} - I_{out(x-1)} = I_k + |\Delta I|_{max,DNL} \quad (29.24)$$

Therefore, the DNL is simply

$$|DNL|_{max} = I_k + |\Delta I|_{max,DNL} - I_k = |\Delta I|_{max,DNL} \quad (29.25)$$

Equating the maximum DNL to the value of ½ LSB,

$$|\Delta I|_{max,DNL} = \tfrac{1}{2} LSB = \tfrac{1}{2} I \quad (29.26)$$

which is much easier to attain than the requirement for the INL.

For the binary-weighted current sources seen in Fig. 29.10, a slightly different analysis is needed to determine the requirements for INL and DNL. In this case, it will be assumed that the current source corresponding to the MSB (D_{N-1}) has a maximum positive mismatch error value and the remainder of the bits (D_0 to D_{N-2}) contain a maximum negative mismatch error, so that the sum of all the errors equals zero. Therefore, the INL is

$$|INL|_{max} = 2^{N-1}(I + |\Delta I|_{max,INL}) - 2^{N-1} \cdot I = 2^{N-1} \cdot |\Delta I|_{max,INL} \qquad (29.27)$$

which is equivalent to the value of the current steering array in Fig. 29.9.

The DNL is slightly different because of the binary weighting of the current sources. One cannot add a single current source with each incremental increase in the digital input code. However, the worst-case condition for binary-weighted arrays tends to occur at midscale when the code transitions from 011111....111 to 100000....000. The worst-case DNL at this point is

$$DNL_{max} = \left[2^{N-1} \cdot (I + |\Delta I|_{max,DNL}) - \sum_{k=1}^{N-1} 2^{k-1} \cdot (I - |\Delta I|_{max,DNL}) \right] - I \qquad (29.28)$$

which can be written as

$$DNL_{max} = 2^{N-1} \cdot (I + |\Delta I|_{max,DNL}) - (2^{N-1} - 1) \cdot (I - |\Delta I|_{max,DNL}) - I = (2^N - 1) \cdot |\Delta I|_{max,DNL} \qquad (29.29)$$

and setting this value equal to ½ LSB and solving for ΔI_{max},

$$|\Delta I|_{max,DNL} = \frac{0.5I}{2^N - 1} = \frac{I}{2^{N+1} - 2} \qquad (29.30)$$

Therefore, the DNL requirements for the binary-weighted current source array is more stringent than the INL requirements.

One interesting issue regarding the previous derivation is that the challenging accuracy requirements in Eq. (29.30) are placed only on the MSB current source. For each of the remaining binary-weighted sources, the DNL requirements become more relaxed. This is simply because the size of the MSB source is equivalent to all the other sources combined, and so its value plays the most important role in the DAC's accuracy.

Example 29.5

Determine the tolerance of the MSB current source on a 10-bit binary-weighted current source array with a unit current source of 1 µA, which will result in a worst-case DNL that is less than ½ LSB.

Since Eq. (29.30) defines the maximum $|\Delta I|$ needed to keep the DNL less than ½ LSB, we must first use this equation,

$$|\Delta I|_{max,DNL} = \frac{1 \times 10^{-6}}{2^{11} - 2} = 0.4888 \text{ nA}$$

For a 10-bit DAC, the MSB current source will have a value that is 2^9 times larger than the unit current source, or 0.512 mA. Therefore, the range of values for which this array will have a DNL that is less than ½ LSB is

$$0.51199995 \text{ mA} \le I_{MSB} \le 0.5120004888 \text{ mA} \quad \blacksquare$$

29.1.5 Charge Scaling DACs

A very popular DAC architecture used in CMOS technology is the charge scaling DAC. Shown in Fig. 29.12a, a parallel array of binary-weighted capacitors, totaling $2^N C$, is connected to an op-amp. The value, C, is a unit capacitance of any value. After initially being discharged, the digital signal switches each capacitor to either V_{REF} or ground, causing the output voltage, v_{OUT}, to be a function of the voltage division between the capacitors.

Since the capacitor array totals $2^N C$, if the MSB is high and the remaining bits are low, then a voltage divider occurs between the MSB capacitor and the rest of the array. The analog output voltage, v_{OUT}, becomes

$$v_{OUT} = V_{REF} \cdot \frac{2^{N-1} C}{(2^{N-1} + 2^{N-2} + 2^{N-3} + \ldots + 4 + 2 + 1 + 1)C} = V_{REF} \cdot \frac{2^{N-1} C}{2^N C} = \frac{V_{REF}}{2} \tag{29.31}$$

which confirms the fact that the MSB changes the output of a DAC by ½ V_{REF}. Figure 29.12b shows the equivalent circuit under this condition. The ratio between v_{OUT} and V_{REF} due to each capacitor can be generalized to

$$v_{OUT} = \frac{2^k C}{2^N C} \cdot V_{REF} = 2^{k-N} \cdot V_{REF} \tag{29.32}$$

where it is assumed that the k-th bit, D_k, is 1 and all other bits are zero. Superposition can then be used to find the value of v_{OUT} for any digital input word by

$$v_{OUT} = \sum_{k=0}^{N-1} D_k 2^{k-N} \cdot V_{REF} \tag{29.33}$$

One limitation of this architecture as shown in Fig. 29.12a is the existence of a parasitic capacitance at the top plate of the capacitor array due to the op-amp. This will prohibit its use as a high-resolution data converter. A better implementation would include the use of a parasitic insensitive, switched-capacitor integrator (see Ch. 27) as the driving circuit. However, the capacitor array itself is the critical component of this data converter and is used in charge redistribution ADCs (Sec. 29.2.5).

The INL and DNL calculations for the binary-weighted capacitor array are identical to those for the binary-weighted current source array, except the unit current source, I, and its corresponding error term, ΔI, are replaced by C and ΔC in Eqs. (29.27) – (29.30).

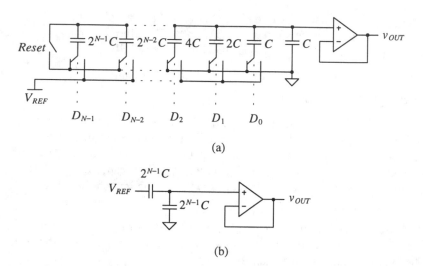

(a)

(b)

Figure 29.12 (a) A charge scaling DAC, (b) the equivalent circuit with the MSB = 1, and all other bits set to zero.

Example 29.6

Design a 3-bit charge scaling DAC and find the value of the output voltage for $D_2D_1D_0 = 010$ and 101. Assume that $V_{REF} = 5$ V and $C = 0.5$ pF.

The 3-bit DAC can be seen in Fig. 29.13a. The equivalent circuits for the capacitor array can be seen in Fig. 29.13b and c. The value of the output voltage can be calculated by either using Eq. (29.32) or the equivalent circuits and performing the voltage division. For $D = 010$, the equivalent circuit in Fig. 29.13b yields

$$v_{OUT} = V_{REF} \cdot \left(\frac{1}{4}\right) = 1.25 \text{ V}$$

Using Eq. (29.33) to calculate v_{OUT} for $D = 101$ yields

$$v_{OUT} = \sum_{k=0}^{N-1} D_k 2^{k-N} \cdot V_{REF} = [1 \cdot (2^{-3}) + 0 \cdot (2^{-2}) + 1 \cdot (2^{-1})] \cdot 5 = \left(\frac{1}{8} + \frac{1}{2}\right) \cdot 5 = 3.125 \text{ V}$$

which is the result expected. ∎

Layout Considerations for a Binary-Weighted Capacitor Array

One problem with this converter is the need for precisely ratioed capacitors. As the number of bits increase, the ratio of the MSB capacitor to the LSB capacitor becomes more difficult to control. For example, Fig. 29.14a shows a 3-bit binary capacitor array

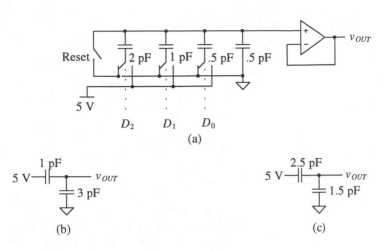

Figure 29.13 (a) A 3-bit charge scaling DAC used in Ex. 29.6 and the equivalent circuits inputs equal to (b) 010 (c) 101.

using three capacitors. When the capacitor is fabricated, *undercutting* of the mask [2, 3] will cause an error in the ratio of the capacitors, causing potentially large DNL and INL errors as N increases.

One solution to this problem is seen in Fig. 20.14b. Here, each capacitor in the array is constructed out of a unit capacitance. Undercutting then affects all the capacitors in the same way, and the ratio between capacitors is maintained. Another problem that affects even this layout strategy is a nonuniform oxide growth. Gradient errors will result in errors in the ratios of the capacitors. Figure 29.14c illustrates another layout strategy that overcomes this issue. The capacitors are laid out in a common-centroid scheme so that the first-order oxide errors average out to be the same for each capacitor.

The Split Array

The charge scaling architecture is very popular among CMOS designers because of its simplicity and relatively good accuracy. Although a linear capacitor is required using poly2, high resolutions in the 10 to 12-bit range can be achieved. Passive, double-poly capacitors have good matching accuracy as well. However, as the resolution increases, the size of the MSB capacitor becomes a major concern. For example if the unit capacitor, C, was chosen to be 0.5 pF, and a 16-bit DAC was to be designed, the MSB capacitor would need to be

$$C_{MSB} = 2^{N-1} \cdot 0.5 \text{ pF} = 16.384 \text{ nF} \qquad (29.34)$$

Based on our 2 μm process parameters, the capacitance between poly1 and poly2 is nominally 500 aF/μm². Therefore, the area required for this one capacitor is over 30 million square microns!

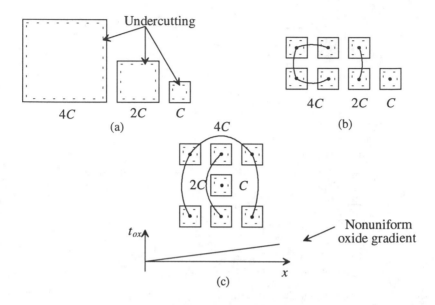

Figure 29.14 Layout of a binary-weighted capacitor array using (a) single capacitors (b) unit capacitors to minimize undercutting effect, and (c) common-centroid to minimize oxide gradients.

One method to reduce the size of the capacitors is to use a split array. A 6-bit example of the array is pictured in Fig. 29.15. This architecture [1] is slightly different from the charge scaling DAC pictured in Fig. 29.13 in that the output is taken off a different node and an additional attenuation capacitor is used to separate the array into a LSB array and a MSB array. Note that the LSB, D_0, now correponds to the leftmost switch and that the MSB, D_5, corresponds to the rightmost switch. The value of the attenuation capacitor can be found by

$$C_{atten} = \frac{sum\ of\ the\ LSB\ array\ capacitors}{sum\ of\ the\ MSB\ array\ capacitors} \cdot C \qquad (29.35)$$

where the sum of the MSB array is equal to the sum of LSB capacitor array minus C. The value of the attenuation capacitor should be such that the series combination of the attenuation capacitor and the LSB array, assuming all bits are zero, is equal to C.

Example 29.7
Using the 6-bit charge scaling DAC shown in Fig. 29.15, (a) show that the output voltage will be $\frac{1}{2} \cdot V_{REF}$ if (a) $D_5D_4D_3D_2D_1D_0 = 100000$ and (b) the output will be $\frac{1}{64} \cdot V_{REF}$ if $D_5D_4D_3D_2D_1D_0 = 000001$.

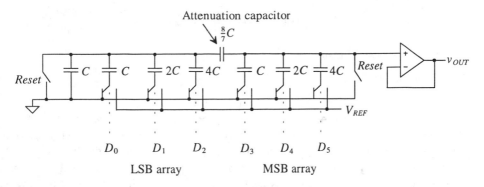

Attenuation capacitor

Figure 29.15 A charge scaling DAC using a split array.

(a) If $D_5 = 1$ and the remaining bits are all zero, then the equivalent circuit for the DAC can be represented by Fig. 29.16a. The expression for the output voltage then becomes

$$v_{OUT} = \frac{4}{\left(\frac{8}{7} \text{ in series with } 8\right) + 3 + 4} \cdot V_{REF} = \frac{1}{2} \cdot V_{REF}$$

(b) For the second case, the equivalent circuit can be seen in Fig. 29.16b. The intermediate node voltage, V_A, is simply the voltage division between the C associated with D_0 and the remainder of the circuit, or

$$V_A = V_{REF} \cdot \frac{1}{\left(7 + \frac{\frac{8}{7} \cdot 7}{\frac{8}{7} + 7}\right) + 1} = \frac{1}{8 + \frac{56}{57}} \cdot V_{REF} \qquad (29.36)$$

The output voltage can be written as

$$v_{OUT} = V_A \cdot \frac{\frac{8}{7}}{\frac{8}{7} + 7} = \frac{8}{57} \cdot V_A \qquad (29.37)$$

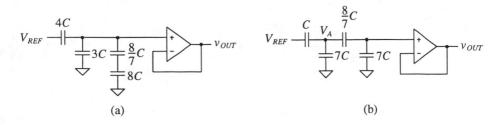

Figure 29.16 Equivalent circuits for Example 29.7.

Plugging Eq. (29.36) into Eq. (29.37) yields

$$v_{OUT} = V_{REF} \cdot \frac{8}{(8 \cdot 57) + 56} = \frac{V_{REF}}{64} \tag{29.38}$$

which is the desired result. ∎

29.1.6 Cyclic DAC

The cyclic DAC uses only a couple of simple components to perform the conversion. As seen in Fig. 29.17, a summer adds V_{REF} or ground to the feedback signal depending on the input bits. An amplifier with a gain of 0.5 feeds the output voltage back to the summer such that the output at the end of each cycle is dependent on the value of the output during the cycle before. Notice that the input bits must be read in a serial fashion. Therefore, the conversion is performed one bit at a time, resulting in N cycles required for each conversion. The voltage output at the end of the n-th cycle of the conversion can be written as

$$v_{OUT}(n) = \left(D_{n-1} \cdot V_{REF} + \frac{1}{2} \cdot v_A(n-1) \right) \cdot \frac{1}{2} \tag{29.39}$$

with a condition such that the output of the S/H is initially zero [$v_A(0) = 0$ V].

The accuracy of this converter is dependent on several factors. The gain of the 0.5 amplifier needs to be highly accurate (to within the accuracy of the DAC) and is usually generated with passive capacitors. Similiarly, the summer and the sample-and-hold also need to be N-bit accurate. Limitations of the converter due to these fundamental building blocks will be discussed in more detail in Sec. 29.3. Since this converter uses a pseudo-"sampled-data" approach, implementing this architecture using switched capacitors is relatively easy.

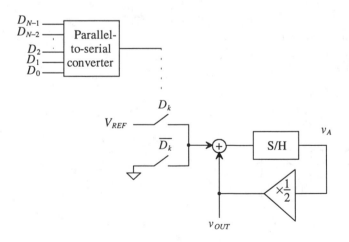

Figure 29.17 A cyclic digital-to-analog converter.

Example 29.8

Show the value of the output voltage at the end of each cycle for a 6-bit cyclic DAC with an input value of $D_5D_4D_3D_2D_1D_0 = 110101$. Assume that $V_{REF} = 5$ V.

We can predict the value of the output based on our previous experience with DACs. The digital input 110101 corresponds to 53_{10}. Therefore, the output voltage due to this input should be

$$v_{OUT} = \frac{53}{64} \cdot V_{REF} = 4.140625 \text{ V}$$

Now examine the cyclic converter in Fig. 29.17. By performing a 6-bit conversion and using Eq. (29.39), the outputs occurring at the end of each cycle can be seen in Fig. 29.18.

The output voltage at the end of the sixth cycle is precisely what was predicted. Note that had this been a 3-bit conversion, the output voltage at the end of cycle 3 would correspond to the value of the 3-bit DACs studied previously with an input of 101. ∎

Cycle Number, n	D_{n-1}	$v_A(n-1)$	$v_{OUT}(n)$
1	1	0	½ (5 + 0) = 2.5 V
2	0	5	½ (0 + 2.5) = 1.25 V
3	1	2.5	½ (5 +1.25) = 3.125 V
4	0	6.25	½ (0 + 3.125) = 1.5625 V
5	1	3.125	½ (5 + 1.5625) = 3.28125 V
6	1	6.5625	½ (5 + 3.28125) = 4.140625 V

Figure 29.18 Output from the 6-bit cyclic DAC used in Ex. 29.8.

29.1.7 Pipeline DAC

The cyclic converter presented in the last section takes N clock cycles per N-bit conversion. Instead of recycling the output back to the input each time, we could extend the cyclic converter to N stages, where each stage performs one bit of the conversion. This extension of the cyclic converter is called a *pipeline* DAC and is seen in Fig. 29.19. Here, the signal is passed down the "pipeline," and as each stage works on one conversion, the previous stage can begin processing another. Therefore, an initial N clock cycle delay is experienced as the signal makes its way down the pipeline the very first time. However, after the N clock cycle delay, a conversion takes place at every clock cycle.

Besides the N clock cycle delay, this architecture can be very fast. However, the amplifier gains must be very accurate to produce high resolutions. Also, this

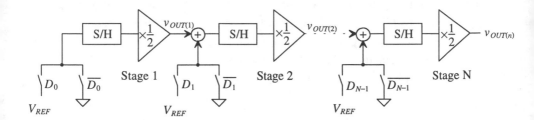

Figure 29.19 A pipeline digital-to-analog converter.

architecture uses N times more circuitry than that of the cyclic, so there is a tradeoff between speed and chip area. The output voltage of the n-th stage in the converter can be written as

$$v_{OUT(n)} = [D_{n-1} \cdot V_{REF} + v_{OUT(n-1)}] \cdot \frac{1}{2} \qquad (29.40)$$

The operation of each stage in the pipeline can be summarized as follows: if the input bit is a 1, add V_{REF} to the output of the previous stage, divide by two, and pass the value to the next stage. If the input bit is a 0, simply divide the output of the previous stage by two and pass along the resulting value.

Example 29.9

Find the output voltage for a 3-bit pipeline DAC for three cases: $D_A = 001$, D_B =110, and $D_C = 101$ and show that the conversion time to perform all three conversions is five clock cycles using the pipeline approach. Assume that $V_{REF} = 5$ V.

The first stage will operate on the LSBs of each word; the second stage will operate on the middle bits; and the last stage, the MSBs. Based on the pipeline strategy, once the LSB of the first input word is performed and passed on, the LSB of the second word, D_B, can begin its conversion. Similarly, once the LSB of the second stage is completed and passed on, the LSB of the third word, D_C, can begin. The conversion cycle for all three input words will produce the output shown in Fig. 29.20. The items that are in bold are associated with the first input word, D_A, whereas the italicized numbers represent the values associated with D_B and the underlined items, D_C.

The first output of the DAC is not valid until the end of the third clock cycle and should look familiar as the 3-bit DAC output for an input word of $D_2 D_1 D_0 = 001$. The following two clock cycles that produce outputs for $D_2 D_1 D_0$ equal 110 and 101, respectively. ■

Clock Cycle	$v_{OUT(1)}$	$v_{OUT(2)}$	$v_{OUT(3)}$	D_0	D_1	D_2
1	**2.5**	0	0	**1**	0	0
2	*0*	**1.25**	0	*0*	**0**	0
3	<u>2.5</u>	2.5	**0.625**	<u>1</u>	*1*	**0**
4		<u>1.25</u>	3.75		<u>0</u>	*1*
5			<u>3.125</u>			<u>1</u>

Figure 29.20 Output from the 3-bit pipeline DAC used in Example 29.9.

29.2 ADC Architectures

A survey of the field of current A/D converter research reveals that a majority of effort has been directed to four different types of architectures: pipeline, flash-type, successive approximation, and oversampled ADCs. Each has benefits that are unique to that architecture and span the spectrum of high speed and resolution.

Since the ADC has a continuous, infinite-valued signal as its input, the important analog points on the transfer curve x-axis for an ADC are the ones that correspond to changes in the digital output word. These input transitions determine the amount of INL and DNL associated with the converter.

29.2.1 Flash

Flash or parallel converters have the highest speed of any type of ADC. As seen in Fig. 29.21, they utilize one comparator per quantization level ($2^N - 1$) and 2^N resistors (a resistor string DAC). The reference voltage is divided into 2^N values, each of which is fed into a comparator. The input voltage is compared with each reference value and results in a thermometer code at the output of the comparators. A thermometer code will exhibit all zeros for each resistor level if the value of v_{IN} is less than the value on the resistor string, and ones if v_{IN} is greater than or equal to voltage on the resistor string. A simple $2^N - 1{:}N$ digital thermometer decoder circuit converts the compared data into an N-bit digital word. The obvious advantage of this converter is the speed with which one conversion can take place. Each clock pulse generates an output digital word. The advantage of having high speed, however, is counterbalanced by the doubling of area with each bit of increased resolution. For example, an 8-bit converter requires 255 comparators, but a 9-bit ADC requires 511! Flash converters have traditionally been limited to 8-bit resolution with conversion speeds of 10 – 40 Ms/s using CMOS technology [4–6]. The disadvantages of the Flash ADC are the area and power requirements of the $2^N - 1$ comparators. The speed is limited by the switching of the comparators and the digital logic.

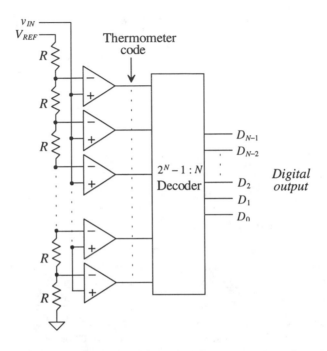

Figure 29.21 Block diagram of a Flash ADC.

Example 29.10

Design a 3-bit Flash converter, listing the values of the voltages at each resistor tap, and draw the transfer curve for $v_{IN} = 0$ to 5 V. Assume $V_{REF} = 5$ V. Construct a table listing the values of the thermometer code and the output of the decoder for $v_{IN} = 1.5, 3.0,$ and 4.5 V.

The 3-bit converter can be seen in Fig. 29.22. Since the values of all the resistors are equal, the voltage of each resistor tap, V_i, will be $V_i = V_{REF} \left(\frac{i}{8} \right)$ where i is the number of the resistor in the string for $i = 1$ to 7. Obviously, $V_1 = 0.625$ V, $V_2 = 1.25$ V, $V_3 = 1.875$ V, $V_4 = 2.5$ V, $V_5 = 3.125$, $V_6 = 3.75$ V, $V_7 = 4.375$ V. Therefore, when v_{IN} first becomes equal or greater than each of these values, a transition will occur in the transfer curve. The transfer curve can be seen in Fig. 29.23 and should look similar to those seen in Ch. 28. The quantization levels and their corresponding thermometer codes can be summarized as seen in Fig. 29.24.

The transfer curve of this ADC corresponds to the ADC with quantization error centered about + ½ LSB, as discussed in Ch. 28 (Fig. 28.20). To shift the curve by ½ LSB so that the code transitions occur around the LSB values and the

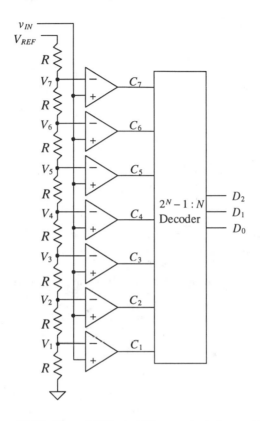

Figure 29.22 Three-bit Flash A/D converter to be used in Ex. 29.10.

quantization error is centered around 0 LSB, the value of the last resistor in the string would have to be adjusted to $\frac{R}{2}$ and the value of the MSB resistor, closest to the reference voltage, would have to be made $1.5R$. Then the first code transition would occur at $v_{IN} = 0.3125$ V, and the last code transition would occur at $v_{IN} = 4.0625$ and so the transfer curve would exactly match that of Fig. 28.20.

Based on Fig. 29.24, when $v_{IN} = 1.5$ V, only comparators C_1 and C_2 will have outputs of 1, since both V_1 and V_2 are less than 1.5 V. The remaining comparator outputs will be 0 since V_3 through V_8 will be greater than 1.5 V, thus generating the thermometer code, 0000011. The encoder must then convert this into a 3-bit digital word, resulting in 010. The same reasoning can be used to construct the data shown in Fig. 29.25. It should be obvious that if the polarity of the comparators were reversed, the thermometer code would be inverted. ∎

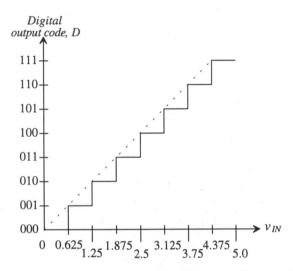

Figure 29.23 Transfer curve for the 3-bit Flash converter in Example 29.10.

v_{IN}	$C_7C_6C_5C_4C_3C_2C_1$	$D_2D_1D_0$
$0 \le v_{IN} < 0.625$ V	0000000	000
0.625 V $\le v_{IN} < 1.25$ V	0000001	001
1.25 V $\le v_{IN} < 1.875$ V	0000011	010
1.875 V $\le v_{IN} < 2.5$ V	0000111	011
2.5 V $\le v_{IN} < 3.125$ V	0001111	100
3.125 V $\le v_{IN} < 3.75$ V	0011111	101
3.75 V $\le v_{IN} < 4.375$ V	0111111	110
$4.375 \le v_{IN}$	1111111	111

Figure 29.24 Code transitions for the Flash ADC used in Ex. 29.10.

Accuracy Issues for the Flash ADC

Accuracy is dependent on the matching of the resistor string and the input offset voltage of the comparators. From Sec. 26.1, we know that an ideal comparator should switch at the point at which the two inputs, v_+ and v_-, are the same potential. However, the offset voltage V_{os}, prohibits this from occurring as the comparator output switches state as follows:

$$v_o = 1 \qquad \text{when } v_+ \ge v_- + V_{os} \qquad (29.41)$$

v_{IN}	$C_7C_6C_5C_4C_3C_2C_1$	$D_2D_1D_0$
1.5	0000011	010
3.0	0001111	100
4.5	1111111	111

Figure 29.25 Output for the Flash ADC used in Ex. 29.10.

$$v_o = 0 \qquad \text{when } v_+ < v_- + V_{os} \qquad (29.42)$$

The resistor string DAC was analyzed and presented in Sec. 29.1.2; the voltage on the i-th tap of the resistor string was found to be

$$V_i = V_{i,ideal} + \frac{V_{REF}}{2^N} \cdot \sum_{k=1}^{i} \frac{\Delta R_k}{R} \qquad (29.43)$$

where $V_{i,ideal}$ is the voltage at the i-th tap if all the resistors had an ideal value of R. The term, ΔR_k, is the value of the resistance error (difference from ideal) due to the mismatch. Note that for the resistor string DAC, the sum of the mismatch terms plays an important factor in the overall voltage at each tap.

The switching point for the i-th comparator, $V_{sw,i}$, then becomes

$$V_{sw,i} = V_i + V_{os,i} \qquad (29.44)$$

where $V_{os,i}$ is the input referred offset voltage of the i-th comparator. The INL for the converter can then be described as

$$INL = V_{sw,i} - V_{sw,ideal} = V_{sw,i} - V_{i,ideal} \qquad (29.45)$$

which becomes

$$INL = \frac{V_{REF}}{2^N} \cdot \sum_{k=1}^{i} \frac{\Delta R_k}{R} + V_{os,i} \qquad (29.46)$$

The worst-case INL will occur at the middle of the string ($i = 2^{N-1}$) as described in Sec. 29.1.2 and Eq. (29.10). Including the offset voltage, the maximum INL will be

$$|INL|_{max} = \frac{V_{REF}}{2^N} \cdot \sum_{k=1}^{2^{N-1}} \frac{\Delta R_k}{R} + |V_{os,i}|_{max} = V_{REF} \cdot \frac{2^{N-1}}{2^N R} \cdot |\Delta R_k|_{max} + |V_{os,i}|_{max} \qquad (29.47)$$

which can be rewritten as

$$|INL|_{max} = \frac{V_{REF}}{2} \cdot \left| \frac{\Delta R_k}{R} \right|_{max} + |V_{os,i}|_{max} \qquad (29.48)$$

where it is assumed that the maximum positive mismatch occurs in all the resistors in the lower half of the string and the maximum negative mismatch occurs in the upper half (or vice versa) and that the comparator at the i-th tap contains the maximum offset voltage, $|V_{os,i}|_{max}$. Notice that the offset contributes directly to the maximum value for the INL. This explains another limitation to using Flash converters at high resolutions. The offset voltage alone can make the INL greater than ½ LSB.

Example 29.11

If a 10-bit Flash converter is designed, determine the maximum offset voltage of the comparators which will make the INL less than ½ LSB. Assume that the resistor string is perfectly matched and $V_{REF} = 5$ V.

Equation (29.48) requires that the offset voltage be equal to ½ LSB. Therefore,

$$|V_{os}|_{max} = \frac{5}{2^{11}} = 2.44 \text{ mV} \quad \blacksquare$$

The DNL calculation for the Flash converter is also attained using the analysis first presented in Sec. 29.1.2. Using the definition of DNL,

$$DNL = V_{sw,i} - V_{sw,i-1} - 1 \text{ LSB (in volts)} \tag{29.49}$$

Plugging in Eq. (29.44),

$$DNL = V_i + V_{os,i} - V_{i-1} - V_{os,i-1} - 1 \text{ LSB} \tag{29.50}$$

which can be written by using Eq. (29.6) as

$$DNL = V_{i,ideal} - V_{i-1,ideal} + \frac{V_{REF}}{2^N} \cdot \frac{\Delta R_i}{R} + V_{os,i} - V_{os,i-1} - 1 \ LSB \tag{29.51}$$

which becomes

$$DNL = \frac{V_{REF}}{2^N} \cdot \frac{\Delta R_i}{R} + V_{os,i} - V_{os,i-1} \tag{29.52}$$

The maximum DNL will occur assuming ΔR_i is at its maximum, $V_{os,i}$ is at its maximum positive value, and $V_{os,i-1}$ is at its maximum negative voltage. Thus,

$$|DNL|_{max} = \frac{V_{REF}}{2^N} \cdot \left|\frac{\Delta R_i}{R}\right|_{max} + 2|V_{os}|_{max} \tag{29.53}$$

which assumes that the maximum offset voltage in the positive and negative directions are symmetrical. Therefore, both resistor string matching and offset voltage affect the DNL of the converter.

29.2.2 The Two-Step Flash ADC

Another type of Flash converter is called the two-step Flash converter or a parallel, feed-forward ADC [7–10]. The basic block diagram of a two-step converter is seen in Fig. 29.26. The converter is separated into two complete Flash ADCs with feed-forward circuitry. The first converter generates a rough estimate of the value of the input, and the second converter performs a fine conversion. The advantages of this architecture are that the number of comparators is greatly reduced from that of the Flash converter -

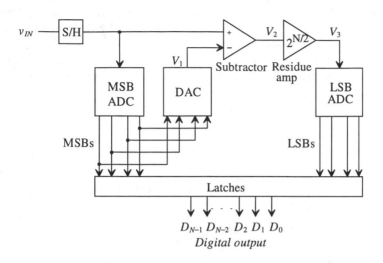

Figure 29.26 Block diagram of a two-step Flash ADC.

from $2^N - 1$ comparators to $2(2^{N/2} - 1)$ comparators. For example, an 8-bit Flash converter requires 255 comparators, while the two-step Flash requires only 30. The tradeoff is that the conversion process takes two steps instead of one, with the speed limited by the bandwidth and settling time required by the residue amplifier and the summer. The conversion process is as follows:

1. After the input is sampled, the most significant bits (MSBs) are converted by the first Flash ADC.

2. The result is then converted back to an analog voltage with the DAC and subtracted with the original input.

3. The result of the subtraction, known as the *residue*, is then multiplied by $2^{N/2}$ and input into the second ADC. The multiplication not only allows the two ADCs to be identical, but also increases the quantum level of the signal input into the second ADC.

4. The second ADC produces the least significant bits through a Flash conversion.

Some architectures use the same set of comparators in order to perform both steps. The multiplication mentioned in step 3 can be eliminated if the second converter is designed to handle very small input signals. The accuracy of the two-step ADC is dependent primarily on the linearity of the first ADC.

Figure 29.27 illustrates the two-step nature of the converter. A more intuitive approach can be explained with this picture. The first conversion identifies the segment in which the analog voltage resides. This is also known as a *coarse conversion* of the MSBs. The results of the coarse conversion are then multiplied by $2^{N/2}$ so that the segment within which V_{IN} resides will be scaled to the same reference as the first conversion. The second conversion is known as the *fine conversion* and will generate the final LSBs using the same Flash approach. One can see why the accuracy of the first converter is so important. If the input value is close to the boundary between two coarse segments and the first ADC is unable to choose the correct coarse segment, then the second conversion will be completely erroneous. The following example further illustrates the two-step algorithm.

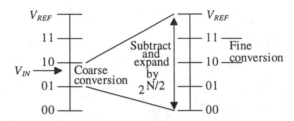

Figure 29.27 Coarse and fine conversions using a two-step ADC.

Example 29.12

Assume that the two-step ADC shown in Fig. 29.26 has four bits of resolution. Make a table listing the MSBs, V_1, V_2, V_3, and the LSBs for V_{IN} = 2, 4, 9, and 15 V assuming that V_{REF} = 16 V.

Since V_{REF} was conveniently made 16 V, each LSB will be 1 V. If V_{IN} = 2 V, the output of the first 2-bit Flash converter will be 00 since V_{REF} = 16 V and each resistor drops 4 V. The output of the 2-bit DAC, V_1, will therefore be 0, resulting in V_2 = 2 V. The multiplication of V_2 by the 4 results in V_3 = 8 V. Remember that each 2-bit Flash converter resembles that of Fig. 29.21. The thermometer code from the second Flash converter will be 0011, which results in 10 as the LSBs. The other values can be calculated as seen in Fig. 29.28. ∎

V_{IN}	D_3D_2 (MSBs)	V_1	V_2	V_3	D_1D_0 (LSBs)
2	00	0	2	8	10
4	01	4	0	0	00
9	10	8	1	4	01
15	11	12	3	12	11

Figure 29.28 Output for the Flash ADC used in Ex. 29.12.

Accuracy Issues Related to the Two-Step Flash Converters

As stated previously, the overall accuracy of the converter is dependent on the first ADC. The second Flash must have only the accuracy of a stand-alone Flash converter. This means that if an 8-bit two-step Flash converter contains two 4-bit Flash converters, the second Flash needs only to have the resolution of a 4-bit Flash, which is not difficult to achieve. However, the first 4-bit Flash must have the accuracy of an 8-bit Flash, meaning that the worst-case INL and DNL for the first bit Flash must be less than $\pm\frac{1}{2}$ LSB for an 8-bit ADC. Thus, the resistor matching and comparators contained in the first ADC must possess the accuracy of the overall converter. Refer to Sec. 29.2.1 for derivations on INL and DNL for a Flash. The DAC must also be accurate to within the resolution of the ADC.

Accuracy Issues Related to the Operational Amplifiers

With the addition of the summer and the amplifier, other sources of accuracy errors are present in this converter. The summer and the amplifier must add and amplify the signal to within $\pm\frac{1}{2}$ LSB of the ideal value. It is quite difficult to implement standard operational amplifiers within high-resolution data converters because of these accuracy requirements. The nonideal characteristics of the op-amp are well known and in many cases alone limit the accuracy of the data converter. In this case, the amplifier is required to multiply the residue signal by some factor of two. Although this may not seem difficult at first glance, a closer examination will reveal a dependency on the open-loop gain.

Suppose that the amplifier was being used in a 12-bit, two-step data converter. Remember that in order for a data converter to be N-bit accurate, the INL and DNL need to be kept below $\pm\frac{1}{2}$ LSB and one-half of an LSB can be defined as

$$0.5 \text{ LSB} = \frac{V_{REF}}{2^{N+1}} \tag{29.54}$$

Since the output of the amplifier gets quantized to 6 bits, the amplifier would need to be 6-bit accurate to within $\pm\frac{1}{2}$ LSB, resulting in an accuracy of

$$\text{Accuracy} = \frac{0.5 \, LSB}{\text{Full scale range } (V_{REF})} = \frac{1}{2^{6+1}} = \frac{1}{128} = 0.0078 = 0.78\% \tag{29.55}$$

And suppose that a feedback amplifier with a gain of 64, or $2^{N/2}$, is used as the residue amplifier. The gain would need to be within the following range:

$$63.5 \text{ V/V} < A_{CL} < 64.5 \text{ V/V} \tag{29.56}$$

where A_{CL} is the closed-loop gain of the amplifier. Already, one can see the limitations of using operational amplifiers with feedback in high-accuracy applications. Designing an op-amp based amplifier with a high degree of gain accuracy can be difficult.

Generalizing this concept for an N-bit application will require knowledge of feedback theory discussed in Ch. 23. The closed-loop gain of the amplifier is expressed as

$$A_{CL} = \frac{v_o}{v_i} = \frac{A_{OL}}{1 + A_{OL}\beta} \tag{29.57}$$

where A_{OL} is the open-loop gain of the amplifier and β is the feedback factor. Also, from Ch. 23, it is known that as A_{OL} increases in value, the closed-loop gain, A_{CL}, approaches the value of $1/\beta$. Therefore, if it is assumed that the closed-loop gain of the amplifier will be equal to the ideal value of $1/\beta$ plus some maximum deviation from the ideal, ΔA, then,

$$A_{CL} = \frac{v_o}{v_i} = \frac{A_{OL}}{1 + A_{OL}\beta} = \frac{1}{\beta} - \Delta A \qquad (29.58)$$

where $1/\beta$ is the desired value of the closed-loop gain (usually some factor of 2^N) and ΔA is the required accuracy ($\pm\frac{1}{2}$ LSB) of the gain (i.e., $(1/\beta) \cdot (1/2^{N+1})$). The right two terms of Eq. (29.58) can be solved for the open-loop gain of the amplifier,

$$|A_{OL}| = \frac{1}{\beta}(2^{N+1} + 1) \approx \frac{2^{N+1}}{\beta} \qquad (29.59)$$

If the op-amp is to be used as a gain of 64 ($1/\beta$) and is required to amplify signals with 6-bit accuracy, then the open-loop gain of the amplifier must be at least $|A_{OL}| \geq 128 \cdot 64 = 8,192$ V/V. This is certainly an achievable specification. However, notice that for every bit increase in resolution the open-loop gain requirement doubles. This is one reason why two-step Flash converters are limited in resolution to around 12 bits [9–12].

The unity gain frequency, f_u, required of an op-amp used in or with a data converter for a specific settling time t, (where $t < 1/f_{clk}$) can be estimated using Eqs. 27.37 and 27.38, and requiring the output of the op-amp be $\frac{1}{2}$ LSB accurate, by

$$v_{out} = V_{outfinal}(1 - \frac{1}{2^{N+1}}) = V_{outfinal}(1 - e^{-t/\tau}) \text{ or } f_u \geq \frac{f_{clk} \cdot \ln 2^{N+1}}{2\pi \cdot \beta} \qquad (29.60)$$

This equation can be used to determine the minimum op-amp gain-bandwidth product ($= f_u$) needed to achieve a specific settling time provided the op-amp slew-rate doesn't come into play and the op-amp can be modeled as a first order system, see Eq. 27.35.

Linearity of the amplifier is another aspect of amplifier performance that must be considered when designing ADCs. The amplifier must be able to linearly amplify the input signal over an input voltage range to within $\frac{1}{2}$ LSB of the number bits that its output is quantized. If the amplifier is not designed correctly, nonlinearity is introduced as devices in the amplifier go into nonsaturation. Harmonic distortion occurs, resulting in an error within the ADC. Linearity is typically measured in terms of total harmonic distortion, or THD, (refer to Sec. 22.3). However, the transfer curve illustrates the limitation more effectively. Figure 29.29 shows a transfer curve of an op-amp with a gain of two. The ideal transfer curve is shown if the input range is known to be between -1 and 1 V. The actual transfer curve shows nonlinearity introduced at both ends of the input range. In order for the amplifier to be N-bit accurate, the slope of the actual transfer curve may not vary from the ideal by more than the accuracy required at the output of the amplifier. Note also in Fig. 29.29 the subtle difference between a gain error and nonlinearity. However, a gain error is much less harmful to an ADC's performance than harmonic distortion.

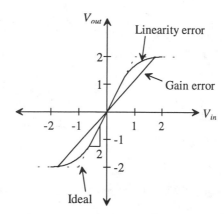

Figure 29.29 An op-amp transfer curve that distinguishes between gain error and linearity error.

29.2.3 The Pipeline ADC

After examining the two-step ADC, one might wonder whether there is such a converter as a three-step or four-step ADC. In actuality, one could divide the number of conversions into many steps. The pipeline ADC is an N-step converter, with 1-bit being converted per stage. Able to achieve high resolution (10–13 bits) at relatively fast speeds [11–15], the pipeline ADC consists of N stages connected in series (Fig. 29.30). Each stage contains a 1-bit ADC (a comparator), a sample-and-hold, a summer, and a gain of two amplifier. Each stage of the converter performs the following operation:

1. After the input signal has been sampled, compared it to $\frac{V_{REF}}{2}$. The output of each comparator is the bit conversion for that stage.

2. If $V_{IN} > \frac{V_{REF}}{2}$ (comparator output is 1), $\frac{V_{REF}}{2}$ is subtracted from the held signal and pass the result to the amplifier . If $V_{IN} < \frac{V_{REF}}{2}$ (comparator output is 0), then pass the original input signal to the amplifier. The output of each stage in the converter is referred to as the *residue*.

3. Multiply the result of the summation by 2 and pass the result to the sample-and-hold of the next stage.

   A main advantage of the pipeline converter is its high throughput. After an initial delay of N clock cycles, one conversion will be completed per clock cycle. While the residue of the first stage is being operated on by the second stage, the first stage is free to operate on the next samples. Each stage operates on the residue passed down from the previous stage, thereby allowing for fast conversions. The disadvantage is having the initial N clock cycle delay before the first digital output appears. The severity of this disadvantage is, of course, dependent on the application.

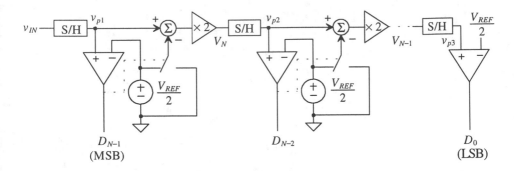

Figure 29.30 Block diagram of a pipeline ADC.

One interesting aspect of this converter is its dependency on the most significant stages for accuracy. A slight error in the first stage propagates through the converter and results in a much larger error at the end of the conversion. Each succeeding stage requires less accuracy than the one before, so special care must be taken when considering the first several stages.

Example 29.13

Assume the pipeline converter shown in Fig. 29.30 is a 3-bit converter. Analyze the conversion process by making a table of the following variables: D_2, D_1, D_0, V_2, V_1, for $v_{IN} = 2$, 3, and 4.5 V. Assume that $V_{REF} = 5$ V, V_3 is the residue voltage out of the first stage, and V_2 is the residue voltage out of the second stage.

The output of the first comparator, $D_2 = 0$, since $v_{IN} < 2.5$ V. Since $D_2 = 0$, $V_3 = 2(2) = 4$ V. Passing this voltage down the pipeline, since $V_3 > 2.5$ V, $D_1 = 1$ and V_2 becomes

$$V_1 = \left(V_2 - \frac{V_{REF}}{2} \right) \times 2 = 3 \text{ V}$$

The LSB, $D_0 = 1$, since $V_2 > 2.5$ V, and the digital output corresponding to $V_{IN} = 2$ V, is $D_2 D_1 D_0 = 011$. The actual digital outputs are simply the comparator outputs, and the data can be completed as seen in Fig. 29.31. ∎

v_{IN}	V_3 (V)	V_2 (V)	Digital Out ($D_2 D_1 D_0$)
2.0	4.0	3.0	011
3.0	1.0	2.0	100
4.5	4.0	3.0	111

Figure 29.31 Output for the pipeline ADC used in Ex. 29.13.

Accuracy Issues Related to the Pipeline Converter

The 1-bit per stage ADC can be analyzed by examining the switching point of each comparator for the ideal and nonideal case. Using Fig. 29.30, and assuming all the components are ideal, let $v_{IN,1}$ represents the value of the input voltage when the first comparator switches. This occurs when

$$v_{IN,1} = \frac{1}{2}V_{REF} \qquad (29.61)$$

The positive input voltage on the second comparator, v_{p2}, can be written in terms of the previous stage, or

$$v_{p2} = [v_{IN} - \frac{1}{2} \cdot D_{N-1} \cdot V_{REF}] \cdot 2 \qquad (29.62)$$

where D_{N-1} is the MSB output from the first comparator and is either a 1 or a 0. The second comparator switches when $v_{p2} = \frac{1}{2}V_{REF}$. The value of v_{IN} at this point, denoted as $v_{IN,2}$, is

$$v_{IN,2} = \frac{1}{2} \cdot D_{N-1} \cdot V_{REF} + \frac{1}{4}V_{REF} \qquad (29.63)$$

Continuing on in a similar manner, we can write the value of the voltage on the positive input of the third comparator in terms of the previous two stages as

$$v_{p3} = \left[[v_{IN} - \frac{1}{2} \cdot D_{N-1} \cdot V_{REF}] \cdot 2 - [\frac{1}{2} \cdot D_{N-2} \cdot V_{REF}] \right] \cdot 2 \qquad (29.64)$$

and the third comparator will switch when $v_{p3} = \frac{1}{2}V_{REF}$, which corresponds to the point at which v_{IN} becomes

$$v_{IN,3} = \frac{1}{2} \cdot D_{N-1} \cdot V_{REF} + \frac{1}{4} \cdot D_{N-2} \cdot V_{REF} + \frac{1}{8}V_{REF} \qquad (29.65)$$

By now, a general trend can be recognized and the value of v_{IN} can be derived for the point at which the comparator of the N-th stage switches. This expression can be written as

$$v_{IN,N} = \frac{1}{2} \cdot D_{N-1} \cdot V_{REF} + \frac{1}{4} \cdot D_{N-2} \cdot V_{REF} + \frac{1}{8} \cdot D_{N-3} \cdot V_{REF} + \ldots\ldots + \frac{1}{2^{N-1}} \cdot D_1 \cdot V_{REF} + \frac{1}{2^N} \cdot V_{REF}$$

$$(29.66)$$

Notice that the preceding equation does not include D_0. This is because D_0 is the output of the N-th stage comparator.

Now that we have derived the switching points for the ideal case, the nonideal case can be considered. Only the major sources of error will be included in the analysis so as not to overwhelm the reader in the analysis. These include the comparator offset voltage, $V_{COS,x}$, and the sample-and-hold offset voltage, $V_{SOS,x}$. The variable, x, represents the number of the stage for which each of the errors is associated, and the "prime" notation will be used to distinguish between the ideal and nonideal case. The reader should also be aware that the offset voltages can be of either polarity. It will be assumed that all of the residue amplifiers will have the same gain, denoted as A.

The positive input to the first nonideal comparator, v'_{p1}, will include the offset from the first sample-and-hold, such that

$$v'_{p1} = v_{IN} + V_{SOS,1} \tag{29.67}$$

Now the first comparator will not switch until the voltage on the positive input overcomes the comparator offset as well. This occurs when

$$v'_{p1} = \frac{1}{2}V_{REF} + V_{COS,1} \tag{29.68}$$

Thus, equating Eqs. (29.67) and (29.68) and solving for the value of the input voltage when the switching occurs for the first comparator yields

$$v'_{IN,1} = \frac{1}{2}V_{REF} + V_{COS,1} - V_{SOS,1} \tag{29.69}$$

The input to the second comparator, v'_{p2}, can be written as

$$v'_{p2} = [v_{IN} + V_{SOS,1} - \frac{1}{2} \cdot D_{N-1} \cdot V_{REF}] \cdot A + V_{SOS,2} \tag{29.70}$$

and the value of input voltage at the point which the second comparator switches occurs when

$$v'_{IN,2} = \frac{1}{2} \cdot D_{N-1} \cdot V_{REF} + \frac{1}{2}\frac{V_{REF}}{A} - V_{SOS,1} - \frac{1}{A}(V_{SOS,2} - V_{COS,2}) \tag{29.71}$$

Continuing in the same manner, we can write the value of the input voltage that causes the third comparator to switch as

$$v'_{IN,3} = \frac{1}{2} \cdot D_{N-1} \cdot V_{REF} + \frac{1}{2} \cdot D_{N-2} \cdot \frac{V_{REF}}{A} - V_{SOS,1} - \frac{1}{A}V_{SOS,2} - \frac{1}{A^2}V_{SOS,3} - \frac{1}{A^2}\left[V_{COS,3} - \frac{1}{2}V_{REF}\right] \tag{29.72}$$

which can be generalized to the N-th switching point as

$$v'_{IN,N} = \frac{1}{2} \cdot D_{N-1} \cdot V_{REF} + \frac{1}{2} \cdot D_{N-2} \cdot \frac{V_{REF}}{A} + \ldots + \frac{1}{2} \cdot D_1 \cdot \frac{V_{REF}}{A^{N-2}} + \frac{1}{2} \cdot \frac{V_{REF}}{A^{N-1}} + \frac{V_{COS,N}}{A^{N-1}} - \sum_{k=1}^{N} \frac{V_{SOS,k}}{A^{k-1}} \tag{29.73}$$

The INL can be calculated by subtracting switching point between the nonideal and ideal case. Therefore, the INL of the first stage is found by subtracting Eqs. (29.69) and (29.61).

$$INL_1 = v'_{IN,1} - v_{IN,1} = V_{COS,1} - V_{SOS,1} \tag{29.74}$$

The second stage INL is

$$INL_2 = v'_{IN,2} - v_{IN,2} = \frac{V_{REF}}{2}\left(\frac{1}{A} - \frac{1}{2}\right) - V_{SOS,1} - \frac{V_{SOS,2}}{A} + \frac{V_{COS,2}}{A} \tag{29.75}$$

and the INL for the N-th stage is

$$INL_N = \frac{1}{2} \cdot D_{N-2} \cdot V_{REF} \cdot \left(\frac{1}{A} - \frac{1}{2}\right) + \frac{1}{2} \cdot D_{N-3} \cdot V_{REF} \cdot \left(\frac{1}{A^2} - \frac{1}{4}\right) + \ldots$$

$$+\frac{1}{2} \cdot D_1 \cdot V_{REF} \cdot \left(\frac{1}{A^{N-2}} - \frac{1}{2^{N-2}}\right) + \frac{1}{2} \cdot V_{REF} \cdot \left(\frac{1}{A^{N-1}} - \frac{1}{2^{N-1}}\right) + \frac{V_{COS,N}}{A^{N-1}} - \sum_{k=1}^{N} \frac{V_{SOS,K}}{A^{k-1}}$$

(29.76)

Equations (29.74) - (29.76) are very important to an understanding of the limitations of the pipeline ADC. Notice the importance of the comparator and summer offsets in Eq. (29.74). The worst-case addition of the offsets must be less than ½ LSB to keep the ADC N-bit accurate. The second stage is more dependent on the gain of the residue amplifier as seen in Eq. (29.75). The gain error discussed in the previous section plays an important role in determining the overall accuracy of the converter. Now examine the effects of the offsets on the INL of the N-th stage. In Eq. (29.76), both the comparator and summer offsets of the N-th stage (when $k = N$) are divided by a large gain. Therefore, the latter stages in a pipeline ADC are not as critical to the accuracy as the first stages, and die area and power can be reduced by using less accurate designs for the least significant stages. The summation term in Eq. (29.76) also reveals that the summer offset of the first stage ($k = 1$) has a large effect on the N-th stage. However, this point is inconsequential since $V_{SOS,1}$ will have to be minimized to achieve N-bit accuracy for the first stage anyway. Typically, if the INL and DNL specifications can be made N-bit accurate in the first few stages, the latter stages will not adversely affect overall accuracy.

The DNL can be found by calculating the difference between the worst-case switching points and subtracting the ideal value for an LSB. As defined earlier, the worst-case will occur at midscale when the output switches from 0111...111 to 1000...000 as v_{IN} increases. Thus, the DNL is

$$DNL_{max} = v'_{IN,1} - v'_{IN,N} - \frac{V_{REF}}{2^N}$$

(29.77)

where $v'_{IN,N}$ is calculated using Eq. (29.73) and assuming that D_{N-1} is a zero and all the other bits are ones. Plugging in Eqs. (29.69) and (29.73) into Eq. (29.77) yields

$$DNL_{max} = \frac{1}{2}V_{REF}\left(1 - \sum_{k=1}^{N-1}\frac{1}{A^k}\right) + V_{COS,1} - \frac{V_{COS,N}}{A^{N-1}} + \sum_{k=2}^{N}\frac{V_{SOS,K}}{A^{k-1}} - \frac{V_{REF}}{2^N}$$

(29.78)

Again, the term that dominates this expression is the comparator offset associated with the first stage and the summer offset of the second stage. The entire expression in Eq. (29.78) must be less than ½ LSB for the ADC to have N-bit resolution.

29.2.4 Integrating ADCs

Another type of ADC performs the conversion by integrating the input signal and correlating the integration time with a digital counter. Known as single- and dual-slope ADCs, these types of converters are used in high-resolution applications but have relatively slow conversions. However, they are very inexpensive to produce and are commonly found in slow-speed, cost-conscious applications.

Single-Slope Architecture

Figure 29.32 illustrates the single-slope converter in block level form. A counter determines the number of clock pulses that are required before the integrated value of a reference voltage is equal to the sampled input signal. The number of clock pulses is proportional to the actual value of the input, and the output of the counter is the actual digital representation of the analog voltage.

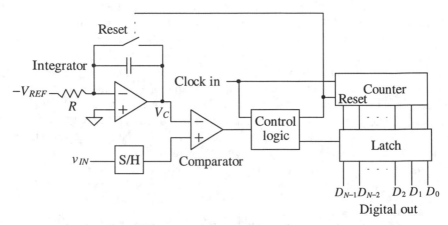

Figure 29.32 Block diagram of a single-slope ADC.

Since the reference is a DC voltage, the output of the integrator should start at zero and linearly increase with a slope that is dependent on the gain of the integrator. Notice that the reference voltage is defined as negative so that the output of the inverting integrator is positive. At the time when the output of the integrator surpasses the value of the S/H output, the comparator switches states, thus triggering the control logic to latch the value of the counter. The control logic also resets the system for the next sample. Figure 29.33 illustrates the behavior of the integrator output and the clock.

Note that if the input voltage is very small, the conversion time is very short, since the counter has to increment only a few times before the comparator latches the data. However, if the input voltage is at its full-scale value, the counter must increment to its maximum value of 2^N clock cycles. Thus, the clock frequency must be many times faster than the bandwidth of the input signal. The conversion time, t_c, is dependent on the value of the input signal and can be described as

$$t_c = \frac{v_{IN}}{V_{REF}} \cdot 2^N \cdot T_{CLK} \qquad (29.79)$$

where T_{CLK} is the period of the clock. The sampling rate is inversely proportional to the conversion time and can be written as

$$f_{Sample} = \frac{V_{REF}}{V_{IN} \cdot 2^N} \cdot f_{CLK} \qquad (29.80)$$

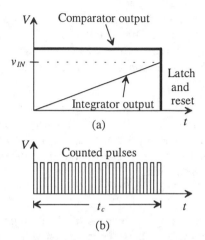

Figure 29.33 Single-slope ADC timing diagrams for (a) the comparator inputs and outputs and (b) the resulting counted pulses.

Example 29.14

Determine the clock frequency needed to form an 8-bit single-slope converter, if the analog signal bandwidth is 20 kHz.

Since the sampling rate required is 40 kHz, then the worst-case situation would occur for a full-scale input, in which event the integrator output would have to climb to its maximum value and the counter would increment 2^N times during the corresponding 25 μs period between samples. Therefore, the clock frequency would need to be 2^N times faster than the sampling rate or 10.24 MHz. ∎

Accuracy Issues Related to the Single-Slope ADC

Obviously, many potential error sources abound in this architecture. At the end of the conversion, the voltage across the integrating capacitor, V_C, assuming no initial condition, will be

$$V_C = \frac{1}{C} \int_0^{t_c} \frac{V_{REF}}{R} dt = \frac{V_{REF} \cdot t_c}{RC} \tag{29.81}$$

where t_c is the conversion time. Plugging Eq. (29.79) into Eq. (29.81) yields

$$V_C = \frac{2^N \cdot T_{CLK} \cdot v_{IN}}{RC} = \frac{2^N \cdot v_{IN}}{f_{CLK} \cdot RC} \tag{29.82}$$

Equation (29.82) is a revealing one in that the final voltage on the integrator output is dependent not only on the value of the input voltage, which is to be expected, but also on the value of R, C, and f_{CLK}. Therefore, any nonideal effects affecting these values will have an influence on the accuracy of the integrator output from sample to sample. For example, if an integrated diffused-resistor is used, then the voltage coefficient of the resistor could limit the accuracy, since the resistor will be effectively nonlinear. Similiarly, the capacitor may have charge leakage or aging effects associated with it. Also, any jitter in the clock will affect the overall accuracy. The integrator must have a linear slope to within the accuracy of the converter, which is dependent on the specifications of the op-amp (open-loop gain, settling time, offset, etc.) and must be considered accordingly.

Offset voltages on the comparator, the S/H, or the integrator will result in additional or fewer clock pulses, depending on the polarity of the offset. A delay also exists from the time that the inputs to the comparator are equal and the time that the output of the counter is actually latched. The reference voltage must also stay constant to within the accuracy of the converter.

Dual-Slope Architecture

A slightly more sophisticated design known as the dual-slope integrating ADC (Fig. 29.34) eliminates most of the problems encountered when using the single-slope converter. Here, two integrations are performed, one on the input signal and one on V_{REF}. The input voltage in this case is assumed to be negative, so that the output of the inverting integrator results in a positive slope during the first integration. Figure 29.35 illustrates the behavior for two separate samples. The first integration is of fixed length, dictated by the counter, in which the sample-and-held signal is integrated, resulting in the first slope. After the counter overflows and is reset, the reference voltage is connected to the input of the integrator. Since v_{IN} was negative and the reference voltage is positive, the inverting integrator output will begin discharging back down to zero at a constant slope. A counter again measures the amount of time for the integrator to discharge, thus generating the digital output.

For this figure, a 3-bit ADC is being used. Thus, the first integration period continues until the beginning of the eighth (2^3) clock pulse, which corresponds to the counter's overflow bit. Note that the integrator's output corresponding to V_B is twice the value of the output corresponding to V_A. Thus, it requires twice as many clock pulses for the integrator to discharge back to zero from V_B than from V_A. The output of the counter at t_A is three or 011, while the counter output at t_B is twice that value or six (110) and the quantization is complete.

Notice that the first slope varies according to the value of the input signal, while the second slope, dependent only on V_{REF}, is constant. Similarly, the time required to generate the first slope is constant, since it is limited by the size of the counter. However, the discharging period is variable and results in the digital representation of the input voltage.

Accuracy Issues Related to the Dual-Slope ADC

One may wonder how the dual-slope converter is an improvement over the single-slope architecture, since a significantly longer conversion time is required. The first integration period requires a full 2^N clock cycle and cannot be decreased, because the

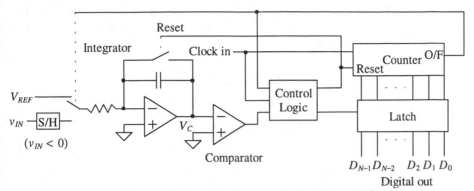

Figure 29.34 Block diagram of a dual-slope ADC.

second integration might require the full 2^N clock cycles to discharge if the maximum value of v_{IN} is being converted. However, the dual slope is the preferred architecture because the same integrator and clock are used to produce both slopes. Therefore, any nonidealities will essentially be canceled. For example, assuming that the S/H is ideal, the gain of the integrator at the end of the first integration period, T_1, becomes

$$V_C = -\frac{1}{C}\int_0^{T_1}\frac{v_{IN}}{R}dt = \frac{|v_{IN}|\cdot T_1}{RC} \tag{29.83}$$

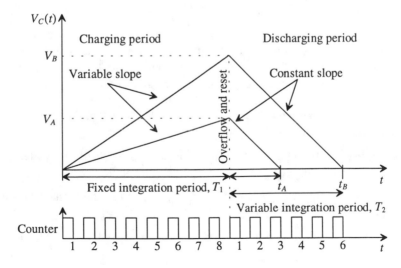

Figure 29.35 Integration periods and counter output for two separate samples of a 3-bit dual-slope ADC.

The output at the end of T_I is positive since the input voltage is considered to be negative and the integrator is inverting. After the clock has been reset, the discharging commences, with the initial condition defined by the value of the integrator output at the end of the charging period, or

$$V_C = \frac{|v_{IN}| \cdot T_1}{RC} - \frac{1}{C} \int_0^{T_2} \frac{V_{REF}}{R} dt \qquad (29.84)$$

Once the value of the integrator output, V_C, reaches zero volts, Eq. (29.84) becomes

$$V_C = \frac{|v_{IN}| \cdot T_1}{RC} - \frac{V_{REF} \cdot T_2}{RC} = 0 \qquad (29.85)$$

or,

$$|v_{IN}| \cdot T_1 = V_{REF} \cdot T_2 \qquad (29.86)$$

At the end of the conversion, the dependencies on R and C have canceled out. Since we also know that the counter increments 2^N times at time, T_1, and the counter increments D times at time, T_2, Eq. (29.86) can be rewritten as

$$\frac{D}{2^N} = \frac{|v_{IN}|}{V_{REF}} \qquad (29.87)$$

where D is the counter output that is actually the digital representation of the input voltage. Thus, it can be written that the ratio of the input voltage and the reference voltage is proportional to the ratio of the binary value of the digital word, D, and 2^N. Therefore, since the same clock pulse is responsible for the charging and discharging times, any irregularities will also cancel out.

29.2.5 The Successive Approximation ADC

The successive approximation converter basically performs a binary search through all possible quantization levels before converging on the final digital answer. The block diagram is seen in Fig. 29.36. An N-bit register controls the timing of the conversion where N is the resolution of the ADC. V_{IN} is sampled and compared to the output of the DAC. The comparator output controls the direction of the binary search, and the output of the successive approximation register (SAR) is the actual digital conversion. The successive approximation algorithm is as follows.

1. A 1 is applied to the input to the shift register. For each bit converted, the 1 is shifted to the right 1-bit position. $B_{N-1} = 1$ and B_{N-2} through $B_0 = 0$.

2. The MSB of the SAR, D_{N-1}, is initially set to 1, while the remaining bits, D_{N-2} through D_0, are set to 0.

3. Since the SAR output controls the DAC and the SAR output is 100...0, the DAC output will be set to $\frac{V_{REF}}{2}$.

4. Next, v_{IN} is compared to $\frac{V_{REF}}{2}$. If $\frac{V_{REF}}{2}$ is greater than v_{IN}, then the comparator output is a 1 and the comparator resets D_{N-1} to 0. If $\frac{V_{REF}}{2}$ is less than v_{IN}, then the comparator output is a 0 and the D_{N-1} remains a 1. D_{N-1} is the actual MSB of the final digital output code.

5. The 1 applied to the shift register is then shifted by one position so that $B_{N-2} = 1$, while the remaining bits are all 0.

6. D_{N-2} is set to a 1, D_{N-3} through D_0 remain 0, while D_{N-1} remains the value from the MSB conversion. The output of the DAC will now either equal $\frac{V_{REF}}{4}$ (if $D_{N-1} = 0$) or $\frac{3V_{REF}}{4}$ (if $D_{N-1} = 1$).

7. Next, v_{IN} is compared to the output of the DAC. If the DAC output is greater than v_{IN}, the comparator the D_{N-2} is reset to 0. If v_{IN} is less than the DAC output, D_{N-2} remains a 1.

8. The process repeats until the output of the DAC converges to the value of v_{IN} within the resolution of the converter.

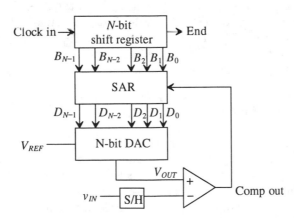

Figure 29.36 Block diagram of the successive approximation ADC.

Figure 29.37 shows an example of the binary search nature of the converter. The bolded line shows the path of the conversion for 101, corresponding to $\frac{5}{8} V_{REF}$. All possible quantization levels are represented in the binary tree. With each bit decided, the search space decreases by one-half until the correct answer is converged upon.

Example 29.15

Perform the operation of a 3-bit successive approximation ADC similar to Fig. 29.36 with $V_{REF} = 8$. Make a table that consists of $D_2D_1D_0$, $B_2B_1B_0$, V_{OUT} (the output from the DAC) and the comparator output, which shows the binary search algorithm of the converter for $v_{IN} = 5.5$ V and 2.5 V.

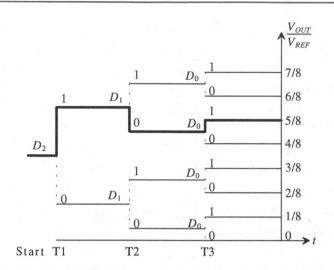

Figure 29.37 Binary search performed by a 3-bit successive approximation ADC for D=101.

We will designate $D_2'D_1'D_0'$ as the initial output of the SAR before the comparator makes its decision. The final value is designated as $D_2D_1D_0$. Notice that if the comparator is a 1, $D_2'D_1'D_0'$ differs from $D_2D_1D_0$, but if the comparator outputs a 0, then $D_2'D_1'D_0' = D_2D_1D_0$. The output of the shift register is designated as $B_2B_1B_0$.

Following the algorithm discussed previously, initially $v_{IN} = 5.5$ V and is compared with 4 V. Since the comparator output is 0, the MSB remains a 1. The next bit is examined, and the output of the DAC is now 6 V. Since $V_{OUT} > v_{IN}$, the comparator output is 1, which resets the current SAR bit, D_1, to a 0 at the end of period T2. Lastly, the LSB is examined, and v_{IN} is compared with 5 V. Since $v_{IN} > V_{OUT}$, The comparator output is a 0, and the current SAR bit, D_0, remains a 1. The results can be examined in Fig. 29.38a. The final value for $D_2D_1D_0$ is 101, which is what is expected considering that 101 in binary is equivalent to 5_{10}. Figure 29.38b shows the data for the ADC using $v_{IN} = 2.5$ V. The final value for $v_{IN} = 2.5$ is 010, which again is what is expected for 3-bit resolution. ∎

The successive approximation ADC is one of the most popular architectures used today. The simplicity of the design allows for both high speed and high resolution while maintaining relatively small area. The limit to the ADC's accuracy is dependent mainly on the accuracy of the DAC. If the DAC does not produce the correct analog voltage with which to compare the input voltage, the entire converter output will contain an error. Referring again to Fig. 29.37, we can see that if a wrong decision is made early, a massive error will result as the converter attempts to search for the correct quantization level in the wrong half of the binary tree.

Step	v_{IN}	$B_2B_1B_0$	$D_2'D_1'D_0'$	V_{OUT}	Comp Out	$D_2D_1D_0$
T1	5.5	100	100	$1/2\ V_{REF} = 4$ V	0	100
T2	5.5	010	110	$(1/2+1/4)V_{REF} = 6$ V	1	100
T3	5.5	001	101	$(1/2+1/8)V_{REF} = 5$ V	0	101

(a)

Step	v_{IN}	$B_2B_1B_0$	$D_2'D_1'D_0'$	V_{OUT}	Comp Out	$D_2D_1D_0$
T1	2.5	100	100	$1/2\ V_{REF} = 4$ V	1	000
T2	2.5	010	010	$1/4\ V_{REF} = 2$ V	0	010
T3	2.5	001	011	$(1/4+1/8)V_{REF} = 3$ V	1	010

(b)

Figure 29.38 Results from the 3-bit successive approximation ADC using (a) $v_{IN} = 5.5$ and (b) 2.5 V.

The Charge-Redistribution Successive Approximation ADC

One of the most popular types of successive approximation architectures uses the binary-weighted capacitor array (analyzed in Sec. 29.1.5) as its DAC. Called a charge redistribution successive approximation ADC [2,16,17], this converter samples the input signal and then performs the binary search based on the amount of charge on each of the DAC capacitors. Figure 29.39 shows an N-bit architecture. A comparator has replaced the unity gain buffer used in the DAC architecture. The binary-weighted capacitor array also samples the input voltage, so no external sample-and-hold is needed.

The conversion process begins by discharging the capacitor array, via the reset switch. Although this may appear to be an insignificant action, the converter is also performing automatic offset cancellation. Once the reset switch is closed, the comparator acts as a unity gain buffer. Thus, the capacitor array charges to the offset

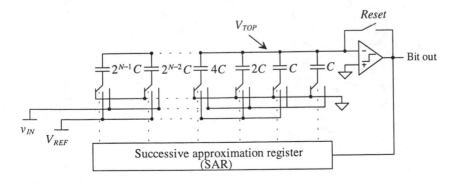

Figure 29.39 A charge redistribution ADC using a binary-weighted capacitor array DAC.

voltage of the comparator. This requires that the comparator be designed so as to be unity gain stable, which means that internal compensation may have to be switched in during the reset period. Next, the input voltage, v_{IN}, is sampled onto the capacitor array. The reset switch is still closed, for the top plate of the capacitor array needs to be connected to virtual ground of the unity gain buffer. The equivalent circuit is seen in Fig. 29.40a.

The reset switch is then opened, and the bottom plates of each capacitor in the array are switched to ground, so that the voltage appearing at the top plate of the array is now $V_{OS} - v_{IN}$ (Fig. 29.40b). The conversion process begins by switching the bottom plate of the MSB capacitor to V_{REF} (Fig. 29.40c). If the output of the comparator is high, the bottom plate of the MSB capacitor remains connected to V_{REF}. If the comparator output is low, the bottom plate of the MSB is connected back to ground. The output of the comparator is D_{N-1}. The voltage at the top of the capacitor array, V_{TOP}, is now

$$V_{TOP} = -v_{IN} + V_{OS} + D_{N-1} \cdot \frac{V_{REF}}{2} \qquad (29.88)$$

The next largest capacitor is tested in the same manner as seen in Fig. 29.40d. The voltage at the top plate of the capacitor after the second capacitor is tested becomes

$$V_{TOP} = -v_{IN} + V_{OS} + D_{N-1} \cdot \frac{V_{REF}}{2} + D_{N-2} \cdot \frac{V_{REF}}{4} \qquad (29.89)$$

The conversion process continues on with the remaining capacitors so that the voltage on the top plate of the array, V_{TOP}, converges to the value of the offset voltage, V_{OS} (within the resolution of the converter), or

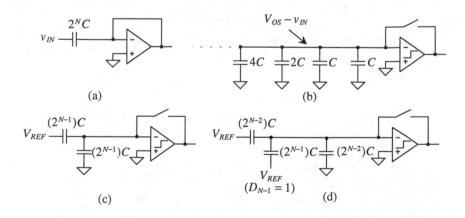

(a)   (b)   (c)   (d)

Figure 29.40 The charge redistribution process: (a) Sampling the input while autozeroing the offset, (b) the voltage at the top plate after sampling, (c) the equivalent circuit while converting the MSB, and (d) the equivalent circuit while converting the next largest capacitor with the MSB result equal to one.

$$V_{TOP} = -v_{IN} + V_{OS} + D_{N-1} \cdot \frac{V_{REF}}{2} + D_{N-2} \cdot \frac{V_{REF}}{4} + ... + D_1 \cdot \frac{V_{REF}}{2^{N-2}} + D_0 \cdot \frac{V_{REF}}{2^{N-1}} \approx V_{OS}$$

$$(29.90)$$

Note that the initial charge stored on the capacitor array is now redistributed onto only those capacitors that have their bottom plates connected to V_{REF}.

Accuracy Issues Related to the Charge Redistribution Successive Approximation ADC

Obviously, the limitation of this architecture is the capacitor matching. The mismatch is analyzed in the same manner as the binary-weighted current source array of Sec. 29.1.4. Thus, substituting the value of the unit capacitance, C, for the value of the unit current source, I, and using Eqs. (29.27) – (29.30),

$$|INL|_{max} = 2^{N-1}(C + |\Delta C|_{max,INL}) - 2^{N-1} \cdot C = 2^{N-1} \cdot |\Delta C|_{max,INL} \quad (29.91)$$

where the maximum ΔC that will result in an INL that is less than ½ LSB is

$$|\Delta C|_{max,INL} = \frac{0.5C}{2^{N-1}} = \frac{C}{2^N} \quad (29.92)$$

and DNL is defined by

$$DNL_{max} = (2^N - 1) \cdot |\Delta C|_{max,DNL} \quad (29.93)$$

with the maximum ΔC which will result in a DNL less than ½ LSB being

$$|\Delta C|_{max,DNL} = \frac{0.5C}{2^N - 1} = \frac{C}{2^{N+1} - 2} \quad (29.94)$$

29.2.6 The Oversampling ADC

ADCs can be separated into two categories depending on the rate of sampling. The first category samples the input at the Nyquist rate, or $f_N = 2F$ where F is the bandwidth of the signal and f_N is the sampling rate. The second type samples the signal at a rate much higher than the signal bandwidth. This type of converter is called an oversampling converter. Traditionally, successive approximation or dual-slope converters are used when high resolution is desired. However, trimming is required when attempting to achieve higher accuracy. Dual-slope converters require high-speed, high-accuracy integrators that are only available using a high f_T bipolar process. Having to design a high-precision sample-and-hold is another factor that limits the realization of a high-resolution ADC using these architectures.

The oversampling ADC [18-20] is able to achieve much higher resolution than the Nyquist rate converters. This is because digital signal processing techniques are used in place of complex and precise analog components. The accuracy of the converter does not depend on the component matching, precise sample-and-hold circuitry, or trimming, and only a small amount of analog circuitry is required. Switched-capacitor implementations are easily achieved, and, as a result of the high sampling rate, only simplistic anti-aliasing circuitry needs to be used. However, because of the amount of time required to sample the input signal, the throughput is considerably less than the Nyquist rate ADCs.

Differences in Nyquist Rate and Oversampling ADCs

The typical process used in analog-to-digital conversion is seen in Fig. 29.41a, while the block diagram for the oversampling ADC is seen in Fig. 29.41b. After filtering the signal to help minimize aliasing effects, the signal is sampled, quantized, and encoded or decoded using simple digital logic to provide the digital data in the proper format. When using oversampling ADCs, little if any, anti-alias filtering is needed, no dedicated S/H is required, the quantization is performed with a modulator, and the encoding usually takes the form of a digital filter.

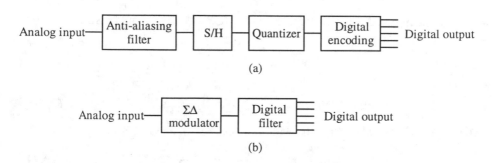

(a)

(b)

Figure 29.41 Typical block diagram for (a) Nyquist rate converters and
(b) oversampling ADCs.

Since the oversampling converter samples at many times the signal bandwidth, aliasing is not a serious problem. A discussion of the frequency characteristics of aliasing was presented in Ch. 28. Figure 29.42a shows that when using Nyquist rate converters, a sampled signal in the frequency domain appears as a series of band-limited signals at multiples of the sampling frequency (see Fig. 28.26 for more details). As the sampling frequency decreases, the frequency spectra begin to overlap, and aliasing (Fig. 29.42b) occurs. Complex, "brickwall" filters are needed to correct the problem.

For oversampled ADCs, aliasing becomes much less of a factor. Since the sampling rate is much greater than the bandwidth of the signal, the frequency domain representation shows that the spectra are widely spaced, as seen in Fig. 29.42c. Therefore, overlapping of the spectra, and thus aliasing, will not occur, and only simple, first-order filters are required.

Oversampling converters typically employ switched-capacitor circuits and therefore do not need sample-and-hold circuits. The output of the modulator is a pulse-density modulated signal that represents the average of the input signal. The modulator is able to construct these pulses in real time, and so it is not necessary to hold the input value and perform the conversion.

Sampling frequency is twice the signal bandwidth.

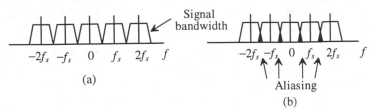

(a)   Aliasing
                                        (b)

Oversampling frequency is many times the signal bandwidth.

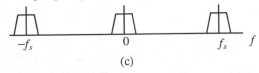

(c)

Figure 29.42 Frequency domain for (a) Nyquist rate converters (b) the aliasing that occurs; and (c) an oversampling converter.

As stated previously, the modulator actually provides the quantization in the form of a pulse-density modulated signal. Referred to as sigma-delta ($\Sigma\Delta$) or delta-sigma ($\Delta\Sigma$) modulation, the density of the pulses represents the average value of the signal over a specific period. Figure 29.43 illustrates the output of the modulator for the positive half of a sine wave input. Note that for the peak of the sine wave, most of the pulses are high. As the sine wave decreases in value, the pulses become distributed between high and low according to the sine wave value.

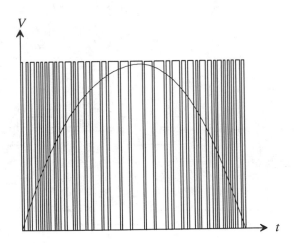

Figure 29.43 Pulse-density output from a sigma-delta modulator for a sine wave input.

If the frequency of the sine wave represented the highest frequency component of the input signal, a Nyquist rate converter would take only two samples. The oversampling converter, however, may take hundreds of samples over the same period to produce this pulse-density signal.

Digital signal processing is then utilized, which has two purposes: to filter any out-of-band quantization noise and to attenuate any spurious out-of-band signals. The output of the filter is then downsampled to the Nyquist rate so that the resulting output of the ADC is the digital data, which represents the average value of the analog voltage over the oversampling period. The effective resolution of oversampling converters is determined by the values of signal-to-noise ratio and dynamic range obtained.

The First-Order ΣΔ Modulator

Now that the basic function of the ΣΔ modulator has been described, it would be useful to examine its inner workings and determine why ΣΔ modulation is so beneficial for generating high-resolution data. A basic first-order ΣΔ modulator can be seen in Fig. 29.44. Here, an integrator and a 1-bit ADC are in the forward path, and a 1-bit DAC is in the feedback path of a single-feedback loop system. The variables labeled are in terms of time, T, which is the inverse of the sampling frequency and k, which is an integer. The 1-bit ADC is simply a comparator that converts an analog signal into either a high or a low. The 1-bit DAC uses the comparator output to determine if $+V_{REF}$ or $-V_{REF}$ is summed with the input.

While the benefits of ΣΔ modulation are not obvious, a simple derivation of the output, $y(kT)$, will illuminate its distinct advantages. The output of the integrator, $u(kT)$, can be described as

$$u(kT) = x(kT - T) - q(kT - T) + u(kT - T) \qquad (29.95)$$

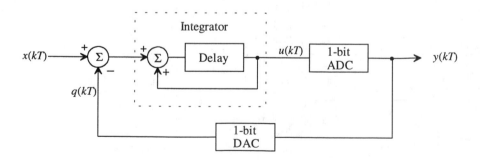

Figure 29.44 A first-order sigma-delta modulator.

where, $x(kT - T) - q(kT - T)$ is equal to the integrator's previous input, and $u(kT - T)$ is its previous output. The quantization error for the 1-bit ADC, as discussed in Ch. 28, is again defined as the difference between its output and input such that

$$Q_e(kT) = y(kT) - u(kT) \tag{29.96}$$

Plugging Eq. (29.95) into Eq. (29.96), the output response, $y(kT)$ is

$$y(kT) = Q_e(kT) + x(kT - T) - q(kT - T) + u(kT - T) \tag{29.97}$$

An ideal 1-bit DAC has the following characteristic; if the input, $y(kT) = 0$, the output, $q(kT) = - V_{REF}$, and if $y(kT) = 1$, then $q(kT) = V_{REF}$. In reality, a 1-bit DAC consists of a couple of switches connecting V_{REF} or $- V_{REF}$ to a common node, so it is not difficult to assume that the DAC is ideal. Therefore,

$$y(kT) = q(kT) \tag{29.98}$$

Utilizing Eq. (29.96) and Eq. (29.97), we find that Eq. (29.98) becomes

$$y(kT) = x(kT - T) + Q_e(kT) - Q_e(kT - T) \tag{29.99}$$

Therefore, the output of the modulator consists of a quantized value of the input signal delayed by one sample period, plus a differencing of the quantization error between the present and previous values. Thus, the real power of $\Sigma\Delta$ modulation is that the quantization noise, Q_e, cancels itself out to the first order.

A frequency domain example will further illuminate this important fact. Suppose that the first-order modulator can be modeled in the s domain as seen in Fig. 29.45, with an ideal integrator represented with transfer function of $\frac{1}{s}$, the 1-bit ADC is modeled as a simple error source, $Q_e(s)$, and again the DAC is considered to be ideal, such that $y(s)$ is equal to $q(s)$. It is also assumed that the bandwidth of the input signal is much less than the bandwidth of the modulator. Therefore, using simple feedback theory, $v_{OUT}(s)$ becomes

$$v_{OUT}(s) = Q_e(s) + \frac{1}{s} \cdot [v_{IN}(s) - v_{OUT}(s)] \tag{29.100}$$

and solving for v_{OUT} yields,

$$v_{OUT}(s) = Q_e(s) \cdot \frac{s}{s+1} + v_{IN}(s)\frac{1}{s+1} \tag{29.101}$$

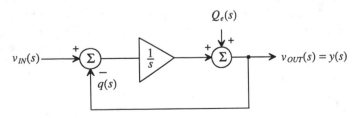

Figure 29.45 A frequency domain model for the first-order sigma-delta modulator.

Note that the transfer function from v_{IN} to v_{OUT} follows that of a low-pass filter and that the transfer function of the quantization noise follows that of a high-pass filter. Plotted together in Fig. 29.46, it is seen that in the region where the signal is of interest, the noise has a small value while the signal has a high gain, and that at higher frequencies, beyond the bandwidth of the signal, the noise increases. The modulator has essentially pushed the power of the noise out of the bandwidth of the signal. This high-pass characteristic is known as *noise shaping* and is a powerful concept used within oversampling ADCs. Low-pass filtering will then be performed by the digital filter in order to remove all of the out-of-band quantization noise, which then permits the signal to be downsampled to yield the final high-resolution output.

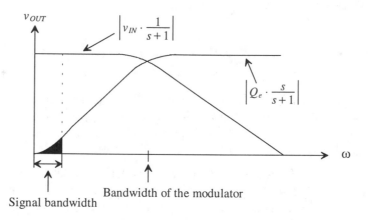

Figure 29.46 Frequency response of the first-order sigma-delta modulator.

As the $\Sigma\Delta$ modulator is generating the pulse-density modulated output, it is interesting to examine the mechanics occurring in the loop, which result in an average of the input. An actual $\Sigma\Delta$ modulator might resemble Fig. 29.47. A switched-capacitor integrator (see Sec. 27.2) provides the summing as well as the delay needed. The 1-bit ADC is a simple comparator, and 1-bit DAC is simply two voltage-controlled switches that select either V_{REF} or $-V_{REF}$ to be summed with the input. A latched comparator is used to provide the necessary loop delay. Notice that the variables used are voltage representations of the variables used in Fig. 29.44. Remember that the function of the integrator is to accumulate differences between the input signal and the output of the DAC. If it is assumed that the input, $v_x(kT)$, is a positive DC voltage, then the output of the integrator should increase. However, the feedback mechanism is such that the 1-bit ADC (the comparator) has a low output if the integrator output, $v_u(kT)$, is positive. Thus, V_{REF} appears at the output of the DAC and is subtracted from the input, and the integrator output is driven back toward zero. The opposite occurs when $v_u(kT)$ is negative such that the integrator output is always driven toward zero by the feedback mechanism. An example will illustrate the operation of the modulator in more detail.

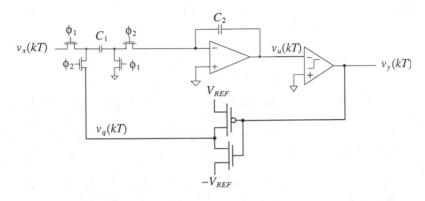

Figure 29.47 Implementation of a first-order sigma-delta modulator using a switched capacitor integrator.

Example 29.16

Using a general first-order $\Sigma\Delta$ modulator, assume that the input to the modulator, $v_x(kT)$ is a positive DC voltage of 0.4 V. Show the values of each variable around the $\Sigma\Delta$ modulator loop and prove that the overall average output of the DAC approaches 0.4 V after 10 cycles. Assume that the DAC output is ±1 V, and that the integrator output has a unity gain with an initial output voltage of 0.1 V, and that the comparator output is either ±1 V.

The present integrator output will be equal to the sum of the previous integrator output and the previous integrator input. Therefore, Eq. (29.95) becomes

$$v_u(kT) = v_u(kT - T) + v_a(kT - T) \tag{29.102}$$

where

$$v_a(kT) = v_x(kT) - v_q(kT) \tag{29.103}$$

and the quantizing error, $Q_e(kT)$, is defined by Eqs. (29.96) and (29.98) as

$$Q_e(kT) = v_q(kT) - v_u(kT) \tag{29.104}$$

The initial conditions define the values of the variable for $k = 0$. The output of the integrator is given to be 0.1 V. Thus, the ADC output is low, the DAC output is V_{REF}, and the output of the summer, $v_a(0)$, is $0.4 - V_{REF} = -0.6$ V.

The output for $k = 1$ begins again with the integrator output. Using Eq. (29.102), $v_u(kT)$ becomes

$$v_u(T) = 0.1 + (-0.6) = -0.5 \text{ V}$$

Since the output of the integrator is negative, the output of the comparator is positive and $-V_{REF}$ is subtracted from 0.4 to arrive at the value for $v_a(T)$.

Continuing in the same manner and using the above equations, we note the voltages for each cycle in Fig. 29.48. After 10 cycles through the modulator, the average value of $v_q(kT)$ becomes,

$$\overline{v_q(kT)} = \frac{7-3}{10} = 0.4 \text{ V}$$

Notice that the behavior of $\overline{v_q(kT)}$ swings around the desired value 0.4 V. If we were to continue computing values, as k increases, the amount that $\overline{v_q(kT)}$ differs from 0.4 V would decrease. Ideally, we could make the deviation of $\overline{v_q(kT)}$ as small as desired by allowing the modulator to take as many samples as necessary to meet that accuracy. ∎

k	$v_a(kT)$	$v_u(kT)$	$v_q(kT)= v_y(kT)$	$Q_e(kT)$	$\overline{v_q(kT)}$
0	−0.6	0.1	1.0	0.9	1.0
1	1.4	−0.5	−1.0	−0.5	0
2	−0.6	0.9	1.0	0.1	0.333
3	−0.6	0.3	1.0	0.7	0.50
4	1.4	−0.3	−1.0	−0.7	0.20
5	−0.6	1.1	1.0	−0.1	0.333
6	−0.6	0.5	1.0	0.5	0.429
7	1.4	−0.1	−1.0	−0.9	0.25
8	−0.6	1.3	1.0	−0.3	0.333
9	−0.6	0.7	1.0	0.3	0.40

Figure 29.48 Data from the first-order ΣΔ modulator.

One interesting note is to examine the effects of using a nonideal comparator. Suppose the integrator's output was smaller than the offset voltage of the comparator. A wrong decision would be made, causing $v_y(kT)$ to be the opposite of the desired value. However, as k increases, this error is averaged out, and the modulator still converges on the correct answer. Therefore, the comparator does not have to be very accurate in its ability to distinguish between two voltages, in contrast to Nyquist rate comparators.

An interesting application of the first-order ΣΔ modulator was reported in [20]. First-order ΣΔ modulators were used to perform pixel-level A/D conversion within a CMOS image sensor. In most digital imaging systems, ADCs are used to convert the analog signal representing the light intensity gathered by a charge-coupled device (CCD). Either a single high-speed ADC is used to convert the entire signal off of the CCD, or a parallel array of lower performance ADCs is used with one ADC assigned to each column of the sensor. Both of these configurations suffer performance loss because of the analog communication required between the sensor and the converter.

It would be optimum to provide an ADC for each pixel in the sensor. However, the size of the ADC becomes critical as the number of converters increases. By using 1-bit, first-order $\Sigma\Delta$ modulators, the author was able to minimize the area required for each pixel (including the first-order $\Sigma\Delta$) to 30 μm by 30 μm and designed a first-order $\Sigma\Delta$ using only 19 transistors, as seen in Fig. 29.49.

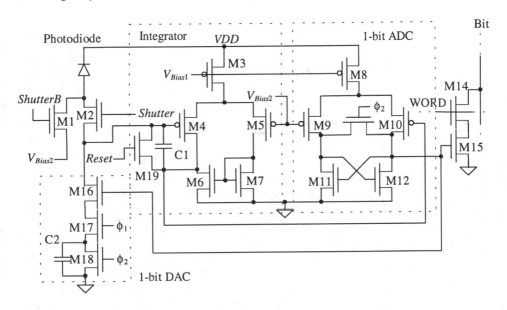

Figure 29.49 A pixel block using a first-order sigma-delta modulator [20].

Photocurrent from the photodiode is integrated on C1, using one side of a simple differential amplifier for the gain. The output of the integrator is passed on to a clocked comparator (via M10), which is a crude version of the first clocked comparator presented in Sec. 26.1. The comparator performs a 1-bit analog-to-digital conversion and passes the decision on to transistors, M16, M17, and M18, which actually form a 1-bit switched-capacitor DAC. This cleverly designed DAC does not convert the output of the comparator to an analog voltage, but instead uses C2 to remove charge from C1, thus resulting in an increase in output voltage of the integrator. The capacitor, C2, is discharged when φ2 is high. However, when the output of the comparator is high, φ1 is high, and φ2 is low, the capacitor C2 is connected directly to the input of the integrator, thus transferring charge off of C1. If the comparator output is low, no charge is transferred from C1. The author was able to achieve a SNR of 52 dB and a dynamic range of 85 dB using a 0.8 μm process.

The Higher Order ΣΔ Modulators

Higher order ΣΔ modulators exist which provide a greater amount of noise shaping. A second-order ΣΔ modulator can be seen in Fig. 29.50. A derivation of the second-order transfer function would reveal that the output contained a delayed version of the input plus a second-order differencing of the quantization noise, Q_e (see Problem 29.38). A third-order modulator would contain third-order differencing of the quantization and can be constructed by adding another integrator similar to integrator A into the system.

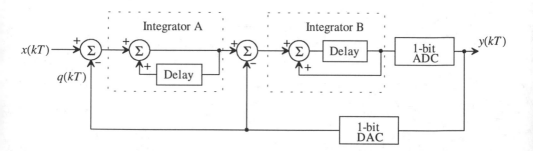

Figure 29.50 A second-order sigma-delta modulator.

Figure 29.51 shows the noise-shaping functions of a first-, second-, and third-order modulator. The cross-hatched area under each of the curves represents the noise that remains in the signal bandwidth and is a magnified version of the blackened area of Fig. 29.46. As the order increases, notice that more of the noise is pushed out into the higher frequencies, thus decreasing the noise in the signal bandwidth. It should be reiterated that ΣΔ modulators do not attenuate noise at all. In fact, they add quantization noise that is very large at high frequencies. But because almost all of the

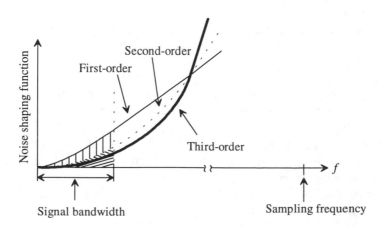

Figure 29.51 Noise shaping comparison of a first-, second-, and third-order modulator.

noise is out of the signal bandwidth, it can be easily filtered, leaving only a small portion within the signal bandwidth. This point is important because the $\Sigma\Delta$ modulator should not be construed as a filtering circuit.

The resolution also increases as the order of the $\Sigma\Delta$ modulator and the oversampling ratio increases, as seen in Fig. 29.52 [21]. Using a first-order modulator, one can expect an increase in dynamic range of 9 dB with every doubling of the oversampling ratio. This correlates to an approximate increase in resolution of 1.5 bits according to Eq. (28.28). The higher-order modulators have even greater gains in resolution as a 2.5-bit increase is attained with each doubling of the oversampling ratio using a second-order modulator, while the third-order modulator increases 3.5 bits.

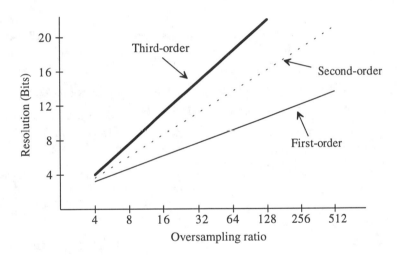

Figure 29.52 Comparison of first-, second-, and third-order modulators versus oversampling ratio and resolution.

One could essentially construct a high-order $\Sigma\Delta$ modulator with many integrators. However, as with any system employing feedback, stability becomes a critical issue. The same holds true for the high-order $\Sigma\Delta$ modulators. Several other topologies have been developed which can implement modulators in a cascaded fashion and are guaranteed to be stable [22, 23]. However, considerable matching requirements need to be overcome.

REFERENCES

[1] R. L. Geiger, P. E. Allen, and N. R. Strader, *VLSI - Design Techniques for Analog and Digital Circuits*, McGraw-Hill Publishing Co., 1990.

[2] R. E. Suarez, P. R. Gray, and D. A. Hodges, "All-MOS Charge Redistribution Analog-to-Digital Conversion Techniques - Part II," *IEEE Journal of Solid State Circuits*, Vol. 10, No. 6, pp. 379-385, December 1975.

[3] J. Shyu, G. C. Temes, and F. Krummenacher, "Random Errors in MOS Capacitors and Current Sources," *IEEE Journal of Solid State Circuits*, Vol. 16, No. 6, pp. 948-955, December 1984.

[4] M. J. M. Pelgrom et. al, "25-Ms/s 8-bit CMOS A/D Converter for Embedded Application," *IEEE Journal of Solid-State Circuits*, Vol. 29, No. 8, pp. 879-886, August 1994.

[5] D. Choi et. al, "Analog Front-End Signal Processor for a 64 Mbits/s PRML Hard-Disk Drive Channel," *IEEE Journal of Solid-State Circuits*, Vol. 29, No. 12, pp. 1596-1605, December 1994.

[6] N. Shiwaku, "A Rail-to-Rail Video-band Full Nyquist 8-bit A/D Converter," *Proceedings of the 1991 Custom Integrated Circuits Conference.*

[7] B. Razavi and B. A. Wooley, "A 12-b, 5-MSample/s Two-Step CMOS A/D Converter," *IEEE Journal of Solid State Circuits*, Vol. 27, No. 12, pp. 1667-1678, December 1992.

[8] J. Dornberg, P. R. Gray, and D. A. Hodges, "A 10-bit, 5-Msample/s CMOS Two-Step Flash ADC," *IEEE Journal of Solid State Circuits*, Vol. 24, No. 2, pp. 241-249, April 1989.

[9] T. Shimizu, et al., "A 10-bit, 20 MHz Two-Step Parallel A/D Converter with Internal S/H," *IEEE Journal of Solid State Circuits*, Vol. 24, No. 1, pp. 13-20, February 1989.

[10] B. S. Song, S. H. Lee, and M. F. Tompsett, "A 10-bit 15 MHz CMOS Recycling Two-Step A/D Converter," *IEEE Journal of Solid State Circuits*, Vol. 25, No. 12, pp. 1328-1338, December 1990.

[11] B. S. Song, M. F. Tompsett, and K. R. Lakshmikumar, "A 12-bit, 1-MSample/s Capacitor Error-Averaging Pipelined A/D Converter," *IEEE Journal of Solid State Circuits*, Vol. 23, No. 6, pp. 1324-1333, December 1988.

[12] S. H. Lewis and P. R. Gray, "A Pipelined 5-Msample/s 9 bit Analog-to-Digital Converter," *IEEE Journal of Solid State Circuits*, Vol. 22, No. 6, pp. 954-961, December 1987.

[13]   S. Sutarja and P. R. Gray, "A Pipelined 13-bit, 250-ks/s, 5-V Analog-to-Digital Converter," *IEEE Journal of Solid State Circuits*, Vol. 23, No. 6, pp. 1316-1323, December 1988.

[14]   P. Vorenkamp and J. P. M. Verdaasdonk, "A 10 b 50 Ms/s Pipelined ADC," *IEEE ISSCC Digest of Technical Papers*, pp. 34-35, February 1992.

[15]   M. Yotsuyanagi, T. Etoh, and K. Hirata, "A 10 Bit 50 MHz Pipelined CMOS A/D Converter with S/H," *IEEE Journal of Solid State Circuits*, Vol. 28, No. 3, pp. 292-300, March 1993.

[16]   J. L. McCreary and P. R. Gray, "All-MOS Charge Redistribution Analog-to-Digital Conversion Techniques - Part I," *IEEE Journal of Solid State Circuits*, Vol. 10, No. 6, pp. 371-379, December 1975.

[17]   K. Bacrania, "A 12 Bit Successive-Approximation ADC with Digital Error Correction," *IEEE Journal of Solid State Circuits*, Vol. 21, No. 6, pp. 1016-1025, December 1986.

[18]   M. Ismail and T. Fiez, *Analog VLSI Signal and Information Processing*, McGraw-Hill, 1994.

[19]   B. E. Boser, "Design and Implementation of Oversampled Analog-to-Digital Converters," Ph.D. Dissertation, Stanford University, 1988.

[20]   B. Fowler, "CMOS Area Image Sensors with Pixel Level A/D Conversion," Ph.D. Dissertation, Stanford University, October 1995.

[21]   B. P. Brandt, *Oversampled Analog-to-Digital Conversion*, Integrated Circuits Laboratory, Technical Report No. ICL91-009, Stanford University, 1991.

[22]   Y. Matsuya, K. Uchimura, et al, "A 16-bit Oversampling A/D Conversion Technology Using Triple Integration Noise Shaping," *IEEE Journal of Solid State Circuits*, Vol. 22, No. 6, pp. 921-929, December 1987.

[23]   K. Uchimura et al, "Oversampling A-to-D and D-to-A Converters with Multistage Noise Shaping Modulators," *IEEE Transactions on Acoustics, Speech and Signal Processing*, pp. 1899-1905, December 1988.

PROBLEMS

29.1   A 3-bit resistor string DAC similiar to the one shown in Fig. 29.2a was designed with a desired resistor of 500 Ω. After fabrication, mismatch caused the actual value of the resistors to be

$R_1 = 500, R_2 = 480, R_3 = 470, R_4 = 520, R_5 = 510, R_6 = 490, R_7 = 530, R_8 = 500$

Determine the maximum INL and DNL for the DAC assuming $V_{REF} = 5$ V.

29.2   An 8-bit resistor string DAC similiar to the one shown in Fig. 29.2b was fabricated with a nominal resistor value of 1 kΩ. If the process was able to provide matching of resistors to within 1 percent, find the effective resolution of the converter. What is the maximum INL and DNL of the converter? Assume that $V_{REF} = 5$ V.

29.3   Compare the digital input codes necessary to generate all eight output values for a 3-bit resistor string DAC similiar to those shown in Fig. 29.2a and b. Design a digital circuit that will allow a 3-bit binary digital input code to be used for the DAC in Fig. 29.2a.   Discuss the advantages and disadvantages of both architectures.

29.4   Plot the transfer curve of a 3-bit R-$2R$ DAC if all Rs = 1.1 kΩ and $2R$s = 2 kΩ. What is the maximum INL and   DNL for the converter?   Assume all of the switches to be ideal and $V_{REF} = 5$ V.

29.5   Suppose that a 3-bit R-$2R$ DAC contained resistors that were perfectly matched and that $R = 1$ kΩ and $V_{REF} = 5$ V.   Determine the maximum switch resistance that can be tolerated for which the converter will still have 3-bit resolution. What are the values of INL and DNL?

29.6   The circuit illustrated in Fig. 29.5 is known as a current-mode R-$2R$ DAC, since the output voltage is defined by the current through R_F. Shown in Fig. P29.6 is an N-bit voltage-mode R-$2R$ DAC.   Design a 3-bit voltage mode DAC and determine the output voltage for each of the eight input codes. Label each node voltage for each input. Assume that $R = 1$ kΩ and that $R_2 = R_1 = 10$ kΩ and $V_{REF} = 5$ V.

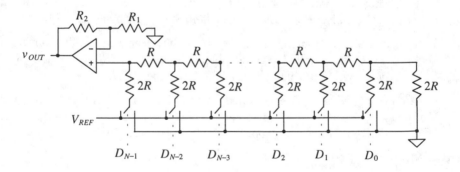

Figure P29.6

29.7   Design a 3-bit current steering DAC using the generic current steering DAC shown in Fig. 29.9. Assume that each current source, I, is 5 mA, and find the total output current for each input code.

29.8   A certain process is able to fabricate matched current sources to within 0.05 percent. Determine the maximum resolution that a current steering (nonbinary weighted) DAC can attain using this process.

29.9 Design an 8-bit current steering DAC using binary-weighted current sources. Assume that the smallest current source will have a value of 1 μA. What is the range of values that the current source corresponding to the MSB can have while maintaining an INL of ½ LSB? Repeat for a DNL less than or equal to ½ LSB.

29.10 Prove that the 3-bit charge scaling DAC used in Ex. 29.6 has the same output voltage increments as the R-$2R$ DAC in Ex. 29.3 for $V_{REF} = 5$ V and $C = 0.5$ pF.

29.11 Determine the output of the 6-bit charge scaling DAC used in Ex. 29.7 for each of the following inputs: $D = 000010, 000100, 001000,$ and 010000.

29.12 Design a 4-bit charge scaling DAC using a split array. Assume that $V_{REF} = 5$ V and that $C = 0.5$ pF. Draw the equivalent circuit for each of the following input words and determine the value of the output voltage: $D = 0001, 0010, 0100, 1000$. Assuming the capacitor associated with the MSB had a mismatch of 4 percent, calculate the INL and DNL.

29.13 For the cyclic converter shown in Fig. 29.17, determine the gain error for a 3-bit conversion if the feedback amplifier had a gain of 0.45 V/V. Assume that $V_{REF} = 5$ V.

29.14 Repeat Problem 29.13 assuming that the output of the summer was always 0.2 V greater than the ideal and that the amplifier in the feedback path had a perfect gain of 0.5 V/V.

29.15 Repeat Problem 29.13 assuming that the output of the summer was always 0.2 V greater than ideal and that the amplifier in the feedback path had a gain of 0.45 V/V.

29.16 Design a 3-bit pipeline DAC using $V_{REF} = 5$ V. (a) Determine the maximum and minimum gain values for the first-stage amplifier for the DAC to have less than ±½ LSBs of DNL assuming the rest of the circuit is ideal. (b) Repeat for the second-stage amplifier. (c) Repeat for the last-stage amplifier.

29.17 Using the same DAC designed in Problem 29.16, (a) determine the overall error (offset, DNL, and INL) for the DAC if the S/H amplifier in the first stage produces an offset at its output of 0.25 V. Assume that all the remaining components are ideal. (b) Repeat for the second-stage S/H. (c) Repeat for the last-stage S/H.

29.18 Design a 3-bit Flash ADC with its quantization error centered about zero LSBs. Determine the worst-case DNL and INL if resistor matching is known to be 5 percent. Assume that $V_{REF} = 5$ V.

29.19 Using the ADC designed in Problem 29.18, determine maximum offset which can be tolerated if all the comparators had the same magnitude of offset, but with different polarities, to attain a DNL of less than or equal to ±½ LSB.

29.20 A 4-bit Flash ADC converter has a resistor string with mismatch as shown in Table P29.20. Determine the DNL and INL of the converter. How many bits of resolution does this converter possess? $V_{REF} = 5$ V.

Resistor	Mismatch (%)
1	2
2	1.5
3	0
4	−1
5	−0.5
6	1
7	1.5
8	2
9	2.5
10	1
11	−0.5
12	−1.5
13	−2
14	0
15	1
16	1

Table P29.20

29.21 Determine the open-loop gain required for the residue amplifier of a two step ADC necessary to keep the converter to within ½ LSB of accuracy with resolutions of (a) 4 bits, (b) 8 bits, and (c) 10 bits.

29.22 Assume that a 4-bit, two-step Flash ADC uses two separate Flash converters for the MSB and LSB ADCs. Assuming that all other components are ideal, show that the first Flash converter needs to be more accurate than the second converter. Assume that $V_{REF} = 5$ V.

29.23 Repeat Ex. 29.12 for $V_{IN} = 3, 5, 7.5, 14.75$ V.

29.24 Repeat Ex. 29.13 for $V_{IN} = 1, 4, 6, 7$ V and $V_{REF} = 8$ V.

29.25 Assume that an 8-bit pipeline ADC was fabricated and that all the amplifiers had a gain of 2.1 V/V instead of 2 V/V. If $V_{IN} = 3$ V and $V_{REF} = 5$ V, what would be the resulting digital output if the remaining components were considered to be ideal? What are the DNL and INL for this converter?

29.26 Show that the first-stage accuracy is the most critical for a 3-bit, 1-bit per stage pipeline ADC by generating a transfer curve and determining DNL and INL for the ADC for three cases: (1) The gain of the first-stage residue amplifier set equal to 2.2 V/V, (2) the second-stage residue amplifier set equal to 2.2 V/V, and (3) the third-stage residue amplifier set equal to 2.2 V/V. For each case, assume that the remaining components are ideal. Assume that $V_{REF} = 5$ V.

29.27 An 8-bit single-slope ADC with a 5 V reference is used to convert a slow-moving analog signal. What is the maximum conversion time assuming that the clock frequency is 1 MHz? What is the maximum frequency of the analog signal? What is the maximum value of the analog signal which can be converted?

29.28 An 8-bit single slope ADC with a 5 V reference uses a clock frequency of 1 MHz. Assuming all other components to be ideal, what is the limitation on the value of RC? What is the tolerance of the clock frequency which will ensure less than 0.5 LSB of INL?

29.29 An 8-bit dual slope ADC with a 5 V reference is used to convert the same analog signal in Problem 29.27. What is the maximum conversion time assuming that the clock frequency is 1 MHz? What is the mimimum conversion time that can be attained? If the analog signal is 2.5 V, what will be the total conversion time?

29.30 Discuss the advantages and disadvantages of using a dual-slope versus a single slope ADC architecture.

29.31 Repeat Ex. 29.15 for a 4-bit successive approximation ADC using $V_{REF} = 5$ V for $v_{IN} = 1$, 3, and full-scale.

29.32 Assume that $v_{IN} = 2.49$ V for the ADC used in Problem 29.31 and that the comparator, because of its offset, makes the wrong decision for the MSB conversion. What will be the final digital output? Repeat for $v_{IN} = 0.3025$, assuming that the comparator makes the wrong decision on the LSB.

29.33 Design a 3-bit, charge redistribution ADC similiar to that shown in Fig. 29.39 and determine the voltage on the top plate of the capacitor array throughout the conversion process for $v_{IN} = 2$, 3, and 4 V, assuming that $V_{REF} = 5$ V. Assume that all components are ideal. Draw the equivalent circuit for each bit decision.

29.34 Determine the maximum INL and maximum DNL of the ADC designed in Problem 29.33 assuming that the capacitor array matching is 1 percent. Assume that the remaining components are ideal and that the unit capacitance, C, is 1 pF.

29.35 Show that the charge redistribution ADC used in Problems 29.32 and 29.33 is immune to comparator offset by assuming an initial offset voltage of 0.3 and determining the conversion for $v_{IN} = 2$ V.

29.36 Discuss the differences between Nyquist rate ADCs and oversampling ADCs.

29.37 Write a simple computer program or use a math program to perform the analysis shown in Ex. 29.16. Run the program for $k = 200$ clock cycles and show that the average value of $v_q(kT)$ converges to the correct answer. How many clock cycles will it take to obtain an average value if $v_q(kT)$ stays within 8-bit accuracy of the ideal value of 0.4 V? 12-bit accuracy? 16-bit accuracy?

29.38 Prove that the output of the second-order $\Sigma\Delta$ modulator shown in Fig. 29.50 is,

$$y(kT) = x(kT - T) + Q_e(kT) - 2Q_e(kT - T) + Q_e(kT - 2T)$$

29.39 Assume that a first order $\Sigma\Delta$ ADC used on a satellite in low earth orbit experiences radiation in which an energetic particle causes a noise spike resulting in the comparator making the wrong decision on the 10th clock period. Using the program written in Problem 29.37, determine the number of clock cycles required before the average value of $v_q(kT)$ is within 12-bit accuracy of the ideal value of 0.4 V. How many extra clock cycles were required for this case versus the ideal conversion used in Prob. 36?

Orbit's CN20 Process

This appendix describes Orbit Semiconductor's[1] 2.0 µm double-poly, double-metal, n-well process (CN20). The process specifications, electrical and SPICE parameters, and design rules are given. The purpose is to give students the information they need to design a CMOS integrated circuit using an actual CMOS process. Commonly used symbols and physical constants are shown in Tables A.1 and A.2.

Name	Symbol	Value
terra	T	10^{12}
giga	G	10^{9}
mega	MEG	10^{6}
kilo	k	10^{3}
milli	m	10^{-3}
micro	µ	10^{-6}
nano	n	10^{-9}
pico	p	10^{-12}
femto	f	10^{-15}
atto	a	10^{-18}

Table A.1 Multiplier Symbols.

---

[1] Orbit Semiconductor, Inc., 1215 Bordeaux Drive, Sunnyvale, CA. 94089. Tel: (408) 744-1800.

Name	Symbol	Value/Units
Vacuum dielectric constant	ε_0	8.85 aF/μm
Silicon dielectric constant	ε_{si}	11.7ε_0
SiO$_2$ dielectric constant	ε_{ox}	3.97ε_0
SiN$_3$ dielectric constant	ε_{Ni}	$\approx 16\varepsilon_0$
Boltzmann's constant	k	1.38×10^{-23} J/K
Electronic charge	q	1.6×10^{-19} C
Temperature	T	K
Thermal voltage	V_T	kT/q = 26 mV @ 300 K

Table A.2 Useful physical constants.

	Thickness or Separation μm	Plate Cap. aF/μm^2			Fringe Cap. aF/μm		
		Min	Typ	Max	Min	Typ	Max
Poly1 gate oxide	0.040 +/– 0.003	803	863	933			
Poly2 gate oxide	0.046 +/– 0.005	677	750	842			
Poly1 to subs. (FOX)	0.600 +/– 0.050	53	58	63	85	88	92
Poly1 to poly2	0.070 +/– 0.008	443	493	557			
Poly1/2 thickness	0.400 +/– 0.030						
Metal1 thickness	0.600 +/– 0.060						
Metal2 thickness	1.150 +/– 0.120						
Metal1 to poly1/2	0.900 +/– 0.100	35	38	43	84	88	93
Metal1 to substrate	1.500 +/– 0.150	21	23	26	75	79	82
Metal1 to diffusion	0.900 +/– 0.100	35	38	43	84	88	93
Metal2 to poly1	1.900 +/– 0.200	16	18	20	83	87	91
Metal2 to substrate	2.500 +/– 0.250	13	14	15	78	81	85
Metal2 to diffusion	1.900 +/– 0.200	16	18	20	83	87	91
Metal2 to metal1	1.000 +/– 0.100	31	35	38	95	100	104

Table A.3 Process thicknesses and distances.

A.1 Process Specifications

The physical distances, thicknesses, and capacitances for the CN20 process are shown in Table A.3. The main use of this table is in determining the parasitic capacitances in a particular layout.

A.1.1 Electrical Specifications

The following six tables describe the electrical characteristics of the p- and n-channel MOSFETs and the lateral bipolar junction transistor available in the CN20 process.

P-channel device $L = 2\ \mu m$ (Poly1)	Min	Typ	Max
Threshold voltage V_{THP}, (V)	0.6	0.8	1.1
Gamma $(V^{1/2})$	0.45	0.55	0.65
$KP = (MUZ)(C'_{ox})$ $(\mu A/V^2)$ $(V_{SD} = 0.1\ V$ with $V_{SG} = 2\ V$ to $3\ V$)	12	15	17
Punchthrough for minimum length channel (V)	10	14	16
Subthreshold slope^{-1} (mV/decade)	90	100	110
Delta length (DL) $= L_{drawn} - L_{eff}$ (μm)	0.7	0.4	0.1

Table A.4 Electrical parameters for the p-channel MOSFET.

N-channel device $L = 2\ \mu m$ (Poly1)	Min	Typ	Max
Threshold Voltage V_{THP}, (V)	0.6	0.8	1.1
Gamma $(V^{1/2})$	0.15	0.25	0.35
$KP = (MUZ)(C'_{ox})$ $(\mu A/V^2)$ $(V_{DS} = 0.1V$ with $V_{GS} = 2\ V$ to $3\ V)$	40	46	52
Punchthrough for minimum length channel (V)	10	14	16
Subthreshold slope^{-1} (mV/decade)	90	100	110
Delta length (DL) $= L_{drawn} - L_{eff}$ (μm)	0.6	0.3	0

Table A.5 Electrical parameters for the n-channel MOSFET.

NPN in the n-well Beta = 80-200 @ $I_B = 1\mu A$	
BV_{EBO}	10 V
BV_{CBO}	≥ 10 V
BV_{CES}	> 10 V
BV_{CBO}	≥ 60 V
P- Base Xj	0.45 to 0.5 μm
N+ emitter	= 0.3 μm
R collector	1.0 ± 0.2 kΩ/sq
P- base resistance	1.2 ± 0.2 kΩ/sq
Early voltage	>30 V

Table A.6 Electrical characteristics of the junction-isolated NPN.

Sheet Resistance (Ω/square)	Min	Typ	Max
P+ active	50	70	100
N+ active	20	28	40
N-well (with field implant)	2,000	2,500	3,000
Poly1	15	21	30
Poly2	18	25	30
Metal1	0.05	0.06	0.06
Metal2	0.02	0.03	0.03
P - substrate (ohm-cm)	30	45	60

Table A.7 Sheet resistances.

Contact Resistance (Single contact 2 μm × 2 μm)	Min	Max
Metal1 to p+ active	35	75
Metal1 to n+ active	20	50
Metal1 to poly1	20	50
Metal1 to poly2	20	50
Metal1 to metal2	0.05	0.08

Table A.8 Contact resistances.

Field Inversion and Breakdown Voltages (V)	Min	Typ	Max
n-channel poly1 field inversion	10	14	
n-channel metal1 field inversion	10	14	
p-channel poly1 field inversion		–14	–10
p-channel metal1 field inversion		–14	–10
n-diffusion to substrate junction breakdown		14	16
p-diffusion to well breakdown		15	18
n-Well to p-substrate junction breakdown		50	90

Table A.9 Field inversion and breakdown voltages.

A.1.2 N-Channel SPICE Models

This section describes the level 2 and BSIM SPICE models and shows characteristics for various-sized devices. These models are located in the file **spice.inf** in C:\Lasi6\Wcn20.

```
.MODEL CMOSN NMOS LEVEL=2
+PHI=0.600000 TOX=4.3500E-08 XJ=0.200000U TPG=1
+ VTO=0.8756 DELTA=8.5650E+00 LD=2.3950E-07 KP=4.5494E-05
+ UO=573.1 UEXP=1.5920E-01 UCRIT=5.9160E+04 RSH=1.0310E+01
+ GAMMA=0.4179 NSUB=3.3160E+15 NFS=8.1800E+12 VMAX=6.0280E+04
+ LAMBDA=2.9330E-02 CGDO=2.8518E-10 CGSO=2.8518E-10
+ CGBO=4.0921E-10 CJ=1.0375E-04 MJ=0.6604 CJSW=2.1694E-10
+ MJSW=0.178543 PB=0.800000

.MODEL CMOSNB NMOS LEVEL=4
+vfb=   -9.73820E-01  lvfb=    3.67458E-01   wvfb=  -4.72340E-02
+phi=   7.46556E-01   lphi=   -1.92454E-24   wphi=   8.06093E-24
+k1=    1.49134E+00   lk1=    -4.98139E-01   wk1=    2.78225E-01
+k2=    3.15199E-01   lk2=    -6.95350E-02   wk2=   -1.40057E-01
+eta=   -1.19300E-02  leta=    5.44713E-02   weta=  -2.67784E-02
+muz=  5.98328E+02    dl=      6.38067E-001  dw=     1.35520E-001
+u0=    5.27788E-02,  lu0=     4.85686E-02   wu0=   -8.55329E-02
+u1=    1.09730E-01   lu1=     7.28376E-01   wu1=   -4.22283E-01
+x2mz=7.18857E+00     lx2mz= -2.47335E+00    wx2mz=7.12327E+01
+x2e=   -3.00000E-03  lx2e=   -7.20276E-03   wx2e=  -5.57093E-03
+x3e=   3.71969E-04   lx3e=   -3.16123E-03   wx3e=  -3.80806E-03
+x2u0= 1.30153E-03    lx2u0=   3.81838E-04   wx2u0=2.53131E-02
+x2u1= -2.04836E-02   lx2u1=   3.48053E-02   wx2u1=4.44747E-02
+mus=  7.79064E+02    lmus=    3.62270E+02   wmus=  -2.71207E+02
+x2ms=-2.65485E+00    lx2ms=   3.68637E+01   wx2ms=1.12899E+02
+x3ms=1.18139E+01     lx3ms=   7.24951E+01   wx3ms=-5.25361E+01
+x3u1= 2.12924E-02    lx3u1=   5.85329E-02   wx3u1=-5.29634E-02
+tox=   4.35000E-002  temp=    2.70000E+01   vdd=    5.00000E+00
+cgdo= 3.79886E-010   cgso=    3.79886E-010  cgbo=   3.78415E-010
```

```
+xpart= 1.00000E+000
+n0=    1.00000E+000  ln0=    0.00000E+000  wn0=    0.00000E+000
+nb=    0.00000E+000  lnb=    0.00000E+000  wnb=    0.00000E+000
+nd=    0.00000E+000  lnd=    0.00000E+000  wnd=    0.00000E+000
+rsh=27.9  cj=1.037500e-04  cjsw=2.169400e-10  js=1.000000e-08  pb=0.8
+pbsw=0.8  mj=0.66036  mjsw=0.178543  wdf=0  dell=0
```

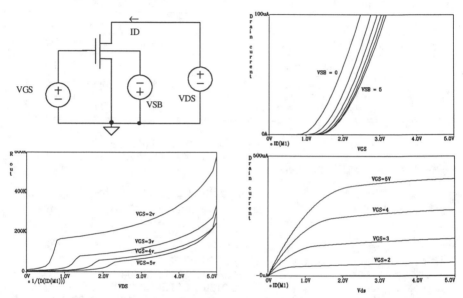

Figure A.1 Characteristics of an n-channel MOSFET with $L = 2$ μm $W = 3$ μm.

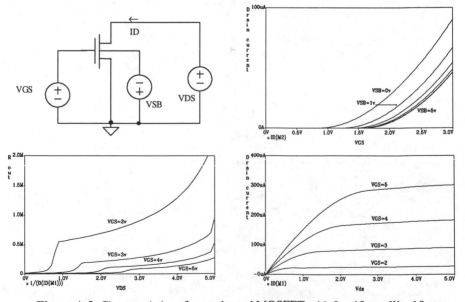

Figure A.2 Characteristics of an n-channel MOSFET with $L = 10$ μm $W = 10$ μm.

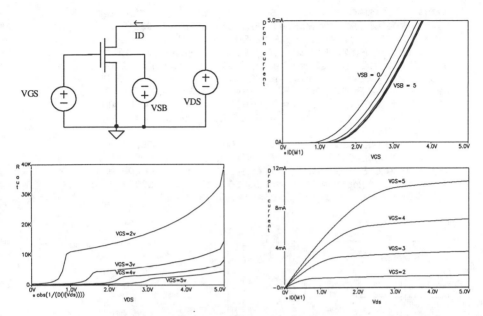

Figure A.3 Characteristics of an n-channel MOSFET with $L = 5$ μm $W = 200$ μm.

A.1.3 P-Channel SPICE Models

This section describes the level 2 and BSIM SPICE models and shows characteristics for various-sized devices. These models are located in the file **spice.inf** in C:\Lasi6\Wcn20.

```
.MODEL CMOSP PMOS LEVEL=2
+PHI=0.600000 TOX=4.3500E-08 XJ=0.200000U TPG=-1
+ VTO=-0.8889 DELTA=4.8720E+00 LD=2.9230E-07 KP=1.5035E-05
+ UO=189.4 UEXP=2.7910E-01 UCRIT=9.5670E+04 RSH=1.8180E+01
+ GAMMA=0.7327 NSUB=1.0190E+16 NFS=6.1500E+12 VMAX=9.9990E+05
+ LAMBDA=4.2290E-02 CGDO=3.4805E-10 CGSO=3.4805E-10
+ CGBO=4.0305E-10 CJ=3.2456E-04 MJ=0.6044 CJSW=2.5430E-10
+ MJSW=0.244194 PB=0.800000
* Weff = Wdrawn - Delta_W
* The suggested Delta_W is -3.6560E-07

.MODEL CMOSPB PMOS LEVEL=4
+vfb=   -2.65334E-01   lvfb=   6.50066E-02   wvfb=  1.48093E-01
+phi=   6.75823E-01    lphi=  -1.61406E-24   wphi=  8.03764E-24
+k1=    5.68962E-01    lk1=    3.88845E-02   wk1=  -5.33948E-02
+k2=   -5.52938E-02    lk2=    1.17906E-01   wk2=  -6.89149E-02
+eta=  -1.51784E-02    leta=   5.87976E-02   weta= -7.51570E-04
+muz=  2.10669E+02     dl=     8.44240E-001  dw=    1.62551E-001
+u0=   1.04713E-01     lu0=    5.50950E-02   wu0=  -7.56659E-02
+u1=   1.46638E-02     lu1=    2.13581E-01   wu1=  -1.22509E-01
+x2mz=8.76354E+00      lx2mz= -3.64793E+00   wx2mz=4.30934E+00
+x2e=  -2.13631E-03    lx2e=  -2.94140E-03   wx2e= -2.48293E-03
```

```
+x3e=   2.78813E-04    lx3e=   -1.60711E-03    wx3e=  -4.57237E-03
+x2u0= 3.93706E-03     lx2u0=  -5.66051E-04    wx2u0= 5.69621E-04
+x2u1= 1.07707E-04     lx2u1=  8.85125E-03     wx2u1= 1.71537E-03
+mus=  2.06464E+02     lmus=   1.39151E+02     wmus=  -4.95671E+01
+x2ms= 5.86401E+00     lx2ms=  6.98887E+00,    wx2ms= 5.55782E+00
+x3ms= -2.03430E-01    lx3ms=  1.16170E+01     wx3ms= -3.44342E+00
+x3u1= -1.17893E-02    lx3u1=  5.72098E-04     wx3u1= 8.29791E-03
+tox=   4.35000E-002   temp=   2.70000E+01     vdd=    5.00000E+00
+cgdo= 5.02635E-010    cgso=   5.02635E-010    cgbo=   3.85017E-010
+xpart= 1.00000E+000
+n0=    1.00000E+000   ln0=    0.00000E+000    wn0=    0.00000E+000
+nb=    0.00000E+000   lnb=    0.00000E+000    wnb=    0.00000E+000
+nd=    0.00000E+000   lnd=    0.00000E+000    wnd=    0.00000E+000
+rsh=54.7  cj=3.245600e-04  cjsw=2.543000e-10  js=1.000000e-08  pb=0.8
+pbsw=0.8  mj=0.60438  mjsw=0.244194  wdf=0  dell=0
```

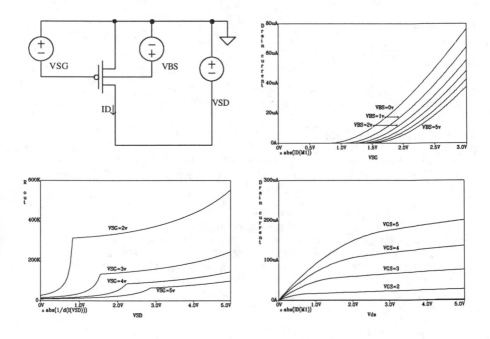

Figure A.4 Characteristics of a p-channel MOSFET with $L = 2\ \mu m$ $W = 3\ \mu m$.

A.2 Hand Calculations

In this section we look at hand calculations using the BSIM model parameters.

A.2.1 The N-channel MOSFET Equations

DC Equations

For the n-channel MOSFET to be in the saturation region

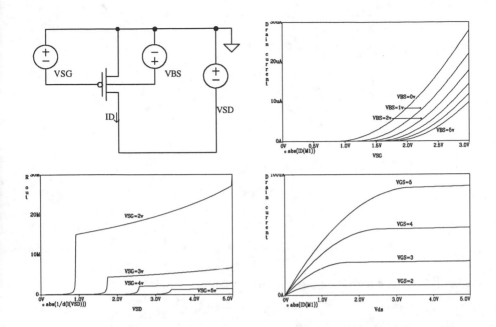

Figure A.5 Characteristics of a p-channel MOSFET with $L = 10$ μm $W = 10$ μm.

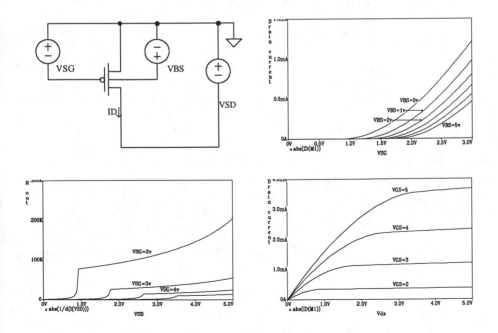

Figure A.6 Characteristics of a p-channel MOSFET with $L = 5$ μm $W = 200$ μm.

$$V_{DS} \geq V_{GS} - V_{THN} \tag{A.1}$$

The drain current, assuming $\lambda = 0$, is then given by

$$I_D = \frac{\beta}{2}(V_{GS} - V_{THN})^2 \tag{A.2}$$

where β is called the transconductance parameter and is given by

$$\beta = KP \cdot \frac{W}{L} \tag{A.3}$$

When the MOSFET is in the linear or triode region

$$V_{DS} \leq V_{GS} - V_{THN} \tag{A.4}$$

and

$$I_D = \beta \left((V_{GS} - V_{THN})V_{DS} - \frac{V_{DS}^2}{2} \right) \tag{A.5}$$

At the border between the saturation and triode regions

$$V_{DS} = V_{GS} - V_{THN} = V_{DS,sat} \tag{A.6}$$

The drain current at $V_{DS,sat}$ is called $I_{D,sat}$. The drain current of the n-channel MOSFET operating in the saturation region, including the effects of channel length modulation and mobility modulation, is given by

$$I_D = \frac{\beta}{2}(V_{GS} - V_{THN})^2[1 + (\lambda_c + \lambda_m)(V_{DS} - V_{DS,sat})] \text{ for } V_{DS} \geq V_{GS} - V_{THN} \tag{A.7}$$

Normally, the channel length and mobility modulation parameters λ_c and λ_m are combined into a single parameter, λ. This single parameter is determined empirically for a given MOSFET channel length. For the CN20 process, using device lengths in excess of 5 µm results in a λ of approximately 0.06 V^{-1}. Based on the results of Figs. A.1 through A.3, λ has a strong dependence on V_{DS}.

Calculation of V_{THN} and KP Using the BSIM Model Parameters

The threshold voltage of the n-channel MOSFET, V_{THN}, as a function of the source to substrate potential, V_{SB}, using the BSIM model parameters, is given by

$$V_{THN} = VFB + PHI + K1 \cdot \sqrt{PHI + V_{SB}} - K2 \cdot (PHI + V_{SB}) \tag{A.8}$$

or using the n-channel BSIM model parameters of the previous section and assuming that $V_{SB} = 0$ results in

$$V_{THN} \approx 0.83 \text{ V} \tag{A.9}$$

The oxide capacitance per unit area, using the oxide dielectric constant of Table A.2 and the oxide thickness from the BSIM model parameters, is given by

$$C'_{ox} = \frac{\varepsilon_{ox}}{TOX} = \frac{35.13 \text{ aF/}\mu\text{m}}{0.0435 \text{ }\mu\text{m}} \approx 800 \frac{\text{aF}}{\mu\text{m}^2} \tag{A.10}$$

The transconductance parameter, KP, for the n-channel MOSFET is then approximated by

$$KP = MUZ \cdot C'_{ox} = \frac{598 \text{ cm}^2}{V \cdot s} \cdot \frac{10^8 \mu\text{m}^2}{\text{cm}^2} \cdot \frac{800 \text{ aF}}{\mu\text{m}^2} \approx 50 \frac{\mu\text{A}}{V^2} \tag{A.11}$$

Small-Signal Parameters

The small-signal transconductance of a MOSFET, in terms of the MOSFET's DC operating point, is given by

$$g_m = \beta(V_{GS} - V_{THN}) = \sqrt{2\beta I_D} \tag{A.12}$$

when the MOSFET is operating in the strong inversion (square law) region and

$$g_m = \frac{I_D}{V_T} \tag{A.13}$$

for a MOSFET operating in the subthreshold region. The body transconductance due to the threshold voltage changing with variations in the source to substrate voltage is given by

$$g_{mb} = g_m \cdot \eta \tag{A.14}$$

where

$$\eta = \frac{K1}{2\sqrt{PHI + V_{SB}}} - K2 \tag{A.15}$$

For large V_{SB} the small-signal parameter η can be negative. When this occurs, we will assume g_{mb} is 0. This is equivalent to saying that the threshold voltage, V_{THN}, doesn't change significantly for small AC signals varying around a large substrate-body potential. A review of Figs. A.1 to A.3 shows that indeed the threshold voltage approaches a constant with increasing V_{SB}.

The small-signal output resistance is calculated using

$$r_o = \frac{1}{\lambda I_D} \tag{A.16}$$

Normally, the value of λ is strongly dependent on the drain-source voltage of the MOSFET. (See plots of the MOSFET output resistance in Figs. A.1 to A.3.) For MOSFETs with lengths in excess of 5 μm, a λ of 0.06 V^{-1} is a good approximation.

A.2.2 The P-channel MOSFET Equations

DC Equations

Note that in this text we have assumed that the threshold voltages of both the p- and the n-channel MOSFETs are positive. Also, all voltages and currents that describe the operation of the p-channel MOSFET in the following discussion are positive. For the p-channel MOSFET to be in the saturation region

$$V_{SD} \geq V_{SG} - V_{THP} \tag{A.17}$$

The drain current of the p-channel MOSFET, assuming $\lambda = 0$, is given by

$$I_D = \frac{\beta}{2}(V_{SG} - V_{THP})^2 \tag{A.18}$$

with

$$\beta = KP \cdot \frac{W}{L} \tag{A.19}$$

For the p-channel MOSFET to operate in the triode region

$$V_{SD} \leq V_{SG} - V_{THP} \tag{A.20}$$

and the drain current is given by

$$I_D = \beta\left((V_{SG} - V_{THP})V_{SD} - \frac{V_{SD}^2}{2}\right) \tag{A.21}$$

Calculation of V_{THP} and KP Using the BSIM Model Parameters

The threshold voltage of the p-channel MOSFET is calculated using

$$V_{THP} = VFB + PHI + K1 \cdot \sqrt{PHI + V_{BS}} - K2 \cdot (PHI + V_{BS}) \tag{A.22}$$

Or using the BSIM model parameters of the previous section for the p-channel MOSFET with $V_{BS} = 0$ results in

$$V_{THP} \approx 0.91\text{V} \tag{A.23}$$

The transconductance parameter, *KP*, for the p-channel MOSFET is given by

$$KP = MUZ \cdot C'_{ox} \approx 17 \frac{\mu\text{A}}{\text{V}^2} \tag{A.24}$$

Small-Signal Parameters

The small-signal transconductance of a p-channel MOSFET is given by

$$g_m = \beta(V_{SG} - V_{THP}) = \sqrt{2\beta I_D} \tag{A.25}$$

while the body transconductance is given by

$$g_{mb} = g_m \cdot \eta \tag{A.26}$$

with

$$\eta = \frac{K1}{2\sqrt{PHI + V_{BS}}} - K2 \tag{A.27}$$

Again g_{mb} is assumed 0 when η becomes negative. The output resistance of the p-channel MOSFET is calculated using Eq. (A.16). A good approximation for λ of the p-channel MOSFET with device lengths in excess of 5 μm is also 0.06 V^{-1}.

A.3 Design Rules

The design rules presented in this appendix are a simplified, and thus more limited, version of the actual design rules specified by Orbit. The Design Rule Checker, LASIDRC, together with the check file "cn20.drc" are used to check layouts with these simplified design rules. The setups have been used successfully on student chip design and research projects at the University of Idaho for several years. The technology is specified to MOSIS as "FORESIGHT-CN20."

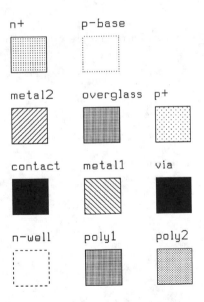

Figure A.7 Layer information, fill, and outline.

		microns	check#
1.1	WIDTH	3.0	1
1.2	SPACE	9.0	2

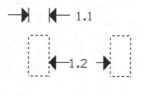

1. N-WELL

Figure A.8 N-well design rules, the check number corresponding to the check number used with LASIDRC.

		microns	check #
2.1	WIDTH	3.0	3
2.2	ACTIVE TO ACTIVE	3.0	4
2.3	N+ ACTIVE TO N-WELL	7.0	5
2.4	P+ SUB. CONTACT TO N-WELL	4.0	6
2.5	N-WELL TO N+ WELL TIE DOWN	0.0	7
2.6	N-WELL OVERLAP OF P+ ACTIVE	3.0	8

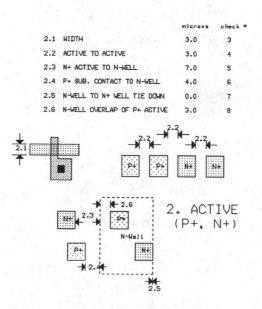

2. ACTIVE
(P+, N+)

Figure A.9 Active design rules.

3.1 WIDTH	2.0	9
3.2 SPACE	3.0	10
3.3 GATE OVERLAP OF ACTIVE	2.0	11
3.4 ACTIVE OVERLAP OF GATE	3.0	3
3.5 FIELD POLY1 TO ACTIVE	1.0	not tested

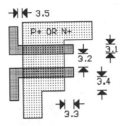

3. POLY1

Figure A.10 Poly1 design rules.

	microns	check#
4.1 WIDTH	3.0	12
4.2 SPACE	3.0	13
4.3 POLY1 OVERLAP OF POLY2	2.0	14
4.4 SPACE TO ACTIVE OR WELL EDGE	2.0	15
4.5 SPACE TO POLY1 CONTACT	3.0	16

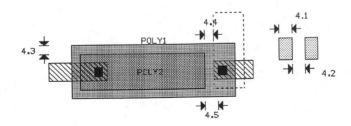

4. POLY2

Figure A.11 Poly2 design rules.

		microns	check#
5.1	CONTACT SIZE (EXACTLY)	2X2	17
5.2	SPACING	2.0	18
5.3	POLY OVERLAP	2.0	19
5.4	ACTIVE OVERLAP	2.0	20
5.5	POLY CONTACT TO ACTIVE EDGE	3.0	21
5.6	ACTIVE CONTACT TO GATE	3.0	22

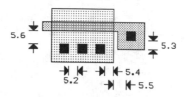

5.   CONTACT

Figure A.12 Contact design rules.

		microns	check#
6.1	WIDTH	3.0	23
6.2	SPACING	3.0	24
6.3	OVERLAP OF CONTACT	1.0	25
6.4	OVERLAP OF VIA	2.0	26

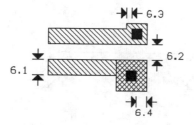

6.   METAL 1

Figure A.13 Metal1 design rules.

	microns	check#
7.1 SPACE TO CONTACT	2.0	27
7.2 SIZE (EXCEPT FOR PADS)	2.0x2.0	28
7.3 SPACING	3.0	29

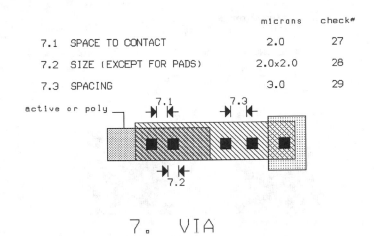

7. VIA

Figure A.14 Via design rules.

	microns	check#
8.1 WIDTH	3.0	30
8.2 SPACE	3.0	31
8.3 METAL2 OVERLAP OF VIA	2.0	32

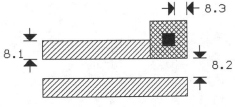

8. METAL 2

Figure A.15 Metal2 design rules.

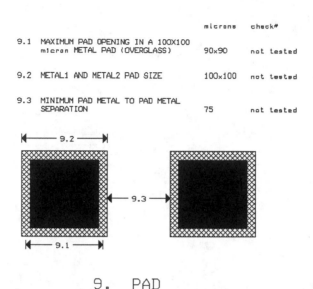

		microns	check#
9.1	MAXIMUM PAD OPENING IN A 100X100 micron METAL PAD (OVERGLASS)	90×90	not tested
9.2	METAL1 AND METAL2 PAD SIZE	100×100	not tested
9.3	MINIMUM PAD METAL TO PAD METAL SEPARATION	75	not tested

9.   PAD

Figure A.16 Pad design rules.

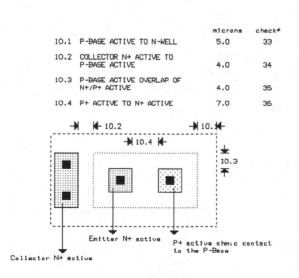

		microns	check#
10.1	P-BASE ACTIVE TO N-WELL	5.0	33
10.2	COLLECTOR N+ ACTIVE TO P-BASE ACTIVE	4.0	34
10.3	P-BASE ACTIVE OVERLAP OF N+/P+ ACTIVE	4.0	35
10.4	P+ ACTIVE TO N+ ACTIVE	7.0	36

10.   P-BASE

Figure A.17 P-base design rules.

MOSIS Scalable Design Rules

This appendix describes graphically the MOSIS Scalable CMOS Design Rules [1]. The design rule check file located in the C:\Lasi6\Wmosis directory is called MOSIS.DRC. The MOSIS design rules use a generic measurement parameter λ (labeled "lam" in LASI). Whether the final project is fabricated in a 0.5 μm process or a 2 μm process is transparent to the person doing layout (except for the pads, of course) until the *.TLC files generated with LASI are changed into a *.GDS file with the LASI utility "TLC2GDS.exe" LASI (actually the setups provided with this book) uses 20 internal units to 1 lam. The GDS units per physical unit should always be 1000.

The n+ (p+) layer is drawn using the MOSIS setups with two layers: active and n-select (p-select). For example, if we want an n+ square measuring 10 λ by 10 λ, we draw a box, on the active layer, measuring 10 λ by 10 λ (this defines the openings in the FOX). Around this active box we draw another box on the n-select layer that is at least 2 λ away from the active box. That is, we draw a box on the n-select layer that is at least 14 λ by 14 λ on the same center as the 10 λ by 10 λ active box. The n-select selects the active box as n+ (actually, the n-select layer specifies an n+ implant in the areas defined by the n-select box).

Technology Specification

Reference [1] should be consulted prior to submitting a chip to MOSIS. The technology specification and lambda depend on the process that the chip will be fabricated in. For example, if we were to use the Orbit CN20 process described in Appendix A, we could specify SCNA (Scalable CMOS N-well Analog) as the technology and a lambda of 1.0.

REFERENCE

[1]   J-I. Pi, "MOSIS Scalable CMOS Design Rules" *The MOSIS Service,* Information Sciences Institute, University of Southern California, 4676 Admiralty Way, Marina del Rey, CA. 90292, August 1, 1995.

LAYERS
cont 25 ■   Contact
pads 26 □   pads
pwel 41 ▒   p-well
nwel 42 ▒   n-well
actv 43 ▣   active
psel 44 □   pselect
nsel 45 □   nselect
poll 46 ▦   poly 1
met1 49 ◩   metal 1
vial 50 ■   via connection between metals 1 and 2
met2 51 ▨   metal 2
ovgl 52 ▦   Overglass used to cut openings for pads in top passive
pol2 56 ▩   poly 2 (electrode)
pbas 58 ▒   pbase
cwel 59 ▒   cap well
via2 61 ■   Via 2, metal 2 to metal 3
met3 62 ◩   metal 3

COMMENT LAYERS
arrw 1   ■   Layer used to draw arrows in schematics
otln 2   ▢   Cell outline layer
schm 3   □   Schematic layer used for drawing circuit schematics
ntxt 4   ▦   used to label nodes for LASI to SPICE list, LASI2CIR
ctxt 5   ▩   Labels contacts so LASI2CIR knows order, i.e. D G S
dtxt 6   ◩   Gives device number, such as M1, M55, etc
ptxt 7   ■   Used to specify part type, model size etc

Figure B.1

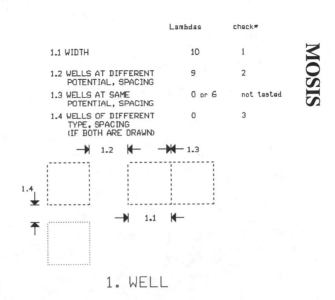

	Lambdas	check#
1.1 WIDTH	10	1
1.2 WELLS AT DIFFERENT POTENTIAL, SPACING	9	2
1.3 WELLS AT SAME POTENTIAL, SPACING	0 or 6	not tested
1.4 WELLS OF DIFFERENT TYPE, SPACING (IF BOTH ARE DRAWN)	0	3

1. WELL

Figure B.2

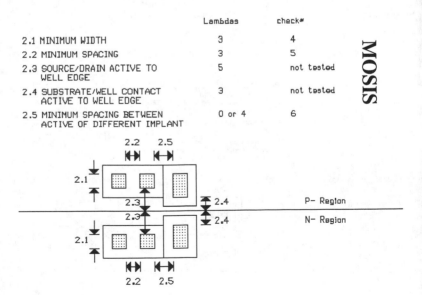

	Lambdas	check#
2.1 MINIMUM WIDTH	3	4
2.2 MINIMUM SPACING	3	5
2.3 SOURCE/DRAIN ACTIVE TO WELL EDGE	5	not tested
2.4 SUBSTRATE/WELL CONTACT ACTIVE TO WELL EDGE	3	not tested
2.5 MINIMUM SPACING BETWEEN ACTIVE OF DIFFERENT IMPLANT	0 or 4	6

2. ACTIVE

Figure B.3

	Lambdas	check#
3.1 MINIMUM WIDTH	2	7
3.2 MINIMUM SPACING	2	8
3.3 MINIMUM GATE EXTENSION OF ACTIVE	2	9
3.4 MINIMUM ACTIVE EXTENSION OF POLY1	3	4
3.5 MINIMUM FIELD POLY TO ACTIVE	1	not tested

MOSIS

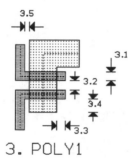

3. POLY1

Figure B.4

4.1 MINIMUM SELECT SPACING TO CHANNEL OF TRANSISTOR TO ENSURE ADEQUATE SOURCE/DRAIN WIDTH	3	not tested
4.2 MINIMUM SELECT OVERLAP OF ACTIVE	2	10
4.3 MINIMUM SELECT OVERLAP OF CONTACT	1	not tested
4.4 MINIMUM SELECT WIDTH AND SPACING (NOTE: P-select and N-select may be coincident but must not overlap)	0 or 2	12

MOSIS

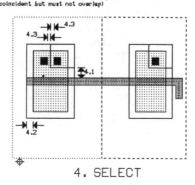

4. SELECT

Figure B.5

	Lambdas	check#
5.1 EXACT CONTACT SIZE	2x2	13
5.2 MINIMUM POLY1 OVERLAP	1.0	14
5.3 MINIMUM CONTACT SPACING	2	15

MOSIS

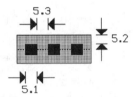

5. CONTACT TO POLY1

Figure B.6

	Lambdas	check#
6.1 EXACT CONTACT SIZE	2x2	13
6.2 MINIMUM ACTIVE OVERLAP	1.0	16
6.3 MINIMUM CONTACT SPACING	2	15
6.4 MINIMUM SPACING TO GATE OF TRANSISTOR	2	17

MOSIS

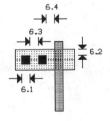

6.CONTACT TO ACTIVE

Figure B.7

	Lambdas	check#
7.1 MINIMUM WIDTH	3	18
7.2 MINIMUM SPACING	3	19
7.3 MINIMUM OVERLAP OF POLY CONTACT	1	20
7.4 MINIMUM OVERLAP OF ACTIVE CONTACT	1	20

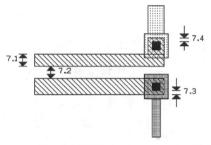

7. METAL1

Figure B.8

	Lambdas	check#
8.1 EXACT SIZE	2×2	21
8.2 MINIMUM VIA1 SPACING	3	22
8.3 MINIMUM OVERLAP BY METAL1	1	23
8.4 MINIMUM SPACING TO CONTACT	2	24
8.5 MINIMUM SPACING TO POLY OR ACTIVE EDGE	2	25

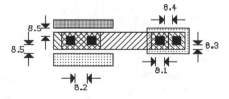

8. VIA1

Figure B.9

MOSIS

MOSIS

	Lambdas	check#
9.1 MINIMUM WIDTH	3	26
9.2 MINIMUM SPACING	4	27
9.3 MINIMUM OVERLAP OF VIA1	1	28

MOSIS

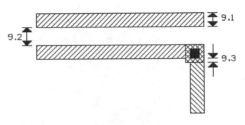

9. METAL2

Figure B.10

	um	check#
10.1 MINIMUM BONDING PAD WIDTH	100x100	not tested
10.2 MINIMUM PROBE PAD WIDTH	75x75	not tested
10.3 PAD OVERLAP OF GLASS OPENING	6	not tested
10.4 MINIMUM PAD SPACING TO UNRELATED METAL2	30	not tested
10.5 MINIMUM PAD SPACING TO UNRELATED METAL1, POLY, ELECTRODE OR ACTIVE	15	not tested

MOSIS

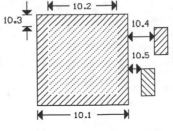

10. OVERGLASS

Figure B.11

	Lambdas	check*
11.1 MINIMUM WIDTH	3	not tested
11.2 MINIMUM SPACING	3	29
11.3 MINIMUM POLY1 OVERLAP	2	not tested
11.4 MINIMUM SPACING TO ACTIVE OR WELL EDGE	2	30
11.5 MINIMUM SPACING TO POLY1 CONTACT	3	31

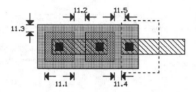

11. POLY2

Figure B.12

	Lambdas	check*
12.1 MINIMUM WIDTH	2	not tested
12.2 MINIMUM SPACING	3	29
12.3 MINIMUM POLY2 GATE OVERLAP OF ACTIVE	2	32
12.4 MINIMUM SPACING TO ACTIVE	1	not tested
12.5 MINIMUM SPACING OR OVERLAP OF POLY1	2	not tested
12.6 MINIMUM SPACING TO POLY1 OR ACTIVE CONTACT	3	31

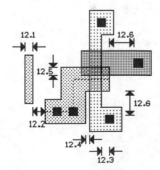

12. ELECTRODE FOR TRANSISTOR

Figure B.13

	Lambdas	check#
13.1 EXACT CONTACT SIZE	2x2	13
13.2 MINIMUM CONTACT SPACING	2	15
13.3 MINIMUM ELECTRODE OVERLAP (on capacitor)	3	33
13.4 MINIMUM ELECTRODE OVERLAP (not on capacitor)	2	34
13.5 MINIMUM SPACING TO POLY1 OR ACTIVE	3	not tested

MOSIS

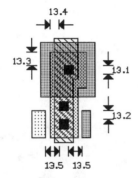

13. ELECTRODE CONTACT, ANALOG OPTION

Figure B.14

	Lambdas	check#
14.1 EXACT SIZE	2x2	35
14.2 MINIMUM SPACING	3	36
14.3 MINIMUM OVERLAP BY METAL2	1	37
14.4 MINIMUM SPACING TO VIA1	2	38

MOSIS

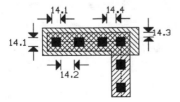

14. VIA2

Figure B.15

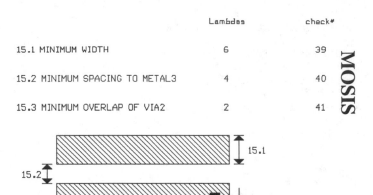

	Lambdas	check#
15.1 MINIMUM WIDTH	6	39
15.2 MINIMUM SPACING TO METAL3	4	40
15.3 MINIMUM OVERLAP OF VIA2	2	41

MOSIS

15. METAL3

Figure B.16

	Lambdas	check#
16.1 ALL ACTIVE CONTACT	2x2	13
16.2 MINIMUM SELECT OVERLAP OF EMITTER CONTACT	4	42
16.3 MINIMUM PBASE OVERLAP OF EMITTER SELECT	2	43
16.4 MINIMUM SPACING BETWEEN EMITTER SELECT AND BASE SELECT	4	44
16.5 MINIMUM PBASE OVERLAP OF BASE SELECT	2	45
16.6 MINIMUM SELECT OVERLAP OF BASE CONTACT	2	46
16.7 MINIMUM NWELL OVERLAP OF PBASE	6	47
16.8 MINIMUM SPACING BETWEEN PBASE AND COLLECTOR ACTIVE	4	48
16.9 MINIMUM ACTIVE OVERLAP OF COLLECTOR CONTACT	2	not tested
16.10 MINIMUM NWELL OVERLAP OF COLLECTOR ACTIVE	3	not tested
16.11 MINIMUM SELECT OVERLAP OF COLLECTOR ACTIVE	2	10

MOSIS

16. NPN BIPOLAR TRANSISTOR

Figure B.17

	Lambdas	check*
17.1 MINIMUM WIDTH	10	49
17.2 MINIMUM SPACING	9	50
17.3 MINIMUM SPACING TO EXTERNAL ACTIVE	5	51
17.4 MINIMUM OVERLAP OF ACTIVE	3	52

MOSIS

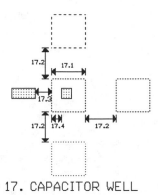

17. CAPACITOR WELL

Figure B.18

HP's CMOS14TB

This appendix gives SPICE and process information for Hewlett-Packard's 0.5 μm triple-metal, n-well process (CMOS14TB) which, in addition to the CN20 process described in Appendix A, is used throughout the text. Since this is a triple-metal process, the construction of the pads is somewhat different than that described for the double-metal processes. MOSIS and HP should be contacted for complete up-to-date information (layout and electrical) on this process prior to designing or submitting a chip for fabrication. Tables C1-C4 give basic electrical information for this process.

The MOSIS Scalable Design Rules given in Appendix B can be used for layout with this process. Two different lambdas can be used, 0.3 μm and 0.35 μm. Using a lambda of 0.3 μm results in a minimum device length of 2λ or 0.6 μm, while the minimum width is 3λ or 0.9 μm. If we subtract the lateral diffusion, DL (nominally 0.1 μm), the effective channel length is 0.5 μm. Using the effective channel length when describing CMOS14TB results in a labeling of a "0.5 μm process." Some of the design rules presented in Appendix B must be changed when using a lambda of 0.3 μm. For example, the n-well width, rule 1.1, must be increased from 10 lambdas to 12 lambdas. The technology specified to MOSIS when submitting a chip in the CMOS14TB process using a lambda of 0.3 μm with the modified design rules is "SCN_SUBM" (Scalable-CMOS-N-well Sub-Micron).

To eliminate the need to modify the design rules in Appendix B, lambda can be increased to 0.35 μm. Now the drawn length of a minimum-size device is 0.7 μm. However, at MOSIS, a bias is applied to the poly mask to reduce the size to 0.6 μm. Thus, when simulating a circuit using a minimum length MOSFET, we use 0.6 μm (whether we are using 0.3 or 0.35 μm for lambda), whereas when laying the MOSFET out we use 2λ or 0.7 μm. The major benefit here is that we can use the setups given in Appendix B and on the accompanying disk without modification for laying circuits out in the 0.5 μm CMOS14TB process. The technology specification when using a lambda of 0.35 μm is "SCN."

N-channel device L_{drawn} = 0.6 μm	Typ
Threshold voltage V_{THN}, (V)	0.7
Gamma (V $^{1/2}$)	0.6
KP = $(MUZ)(C'_{ox})$ (μA/V^2)	170
DL ($L_{eff} = L_{drawn}$ – DL) (μm)	0.06
DW ($W_{eff} = W_{drawn}$ – DW) (μm)	0.35
Punchthrough voltage (V)	10
I_{drive} (μA/μm)	380

Table C.1

P-channel device L_{drawn} = 0.6 μm	Typ
Threshold voltage V_{THP}, (V)	0.9
Gamma (V $^{1/2}$)	0.5
KP = $(MUZ)(C'_{ox})$ (μA/V^2)	50
DL ($L_{eff} = L_{drawn}$ – DL) (μm)	0.09
DW ($W_{eff} = W_{drawn}$ – DW) (μm)	0.35
Punchthrough voltage (V)	10
I_{drive} (μA/μm)	190

Table C.2

Sheet resistance (Ω/square)	Typ
P+ active	2
N+ active	2
Poly1 (silicided)	2
Metal1	0.07
Metal2	0.07
Metal3	0.05

Table C.3

	Plate Cap. aF/μm² Typical
Poly1 to subs. (FOX)	91
Poly1 to metal1	58
Poly1 to metal2	17
Poly1 to metal3	10
Metal1 to substrate	42
Metal1 to metal2	36
Metal1 to metal3	14
Metal2 to substrate	20
Metal2 to metal3	33
Metal3 to substrate	15

Table C.4

Calculation of V_{THN} and KP_n Using the BSIM Model Parameters

The level 3 and 4 (BSIM) model parameters for the n-channel MOSFET are shown below and are located in the text file **spice.inf** in the C:\Lasi6\Wcn20 directory.

```
* Level 3 SPICE model for CMOS14TB 0.5 um
.MODEL CMOSN5 NMOS LEVEL=3 PHI=0.700000
+ TOX=9.6000E-09 XJ=0.200000U TPG=1
+ VTO=0.7118 DELTA=2.3060E-01 LD=2.9830E-08 KP=1.8201E-04
+ UO=506.0 THETA=1.9090E-01 RSH=1.8940E+01 GAMMA=0.6051
+ NSUB=1.4270E+17 NFS=7.1500E+11 VMAX=2.4960E+05 ETA=2.5510E-02
+ KAPPA=1.8530E-01 CGDO=9.0000E-11 CGSO=9.0000E-11
+ CGBO=3.7295E-10 CJ=6.02E-04 MJ=0.805 CJSW=2.0E-11
+ MJSW=0.761 PB=0.99
* Weff = Wdrawn - Delta_W
* The suggested Delta_W is 3.5700E-07

* Level 4 (BSIM) SPICE model for CMOS14TB 0.5 um
.MODEL CMOSNB5 NMOS LEVEL=4
+ vfb=-9.65360E-01    lvfb= 4.11254E-02    wvfb=-1.21737E-01
+ phi= 9.02436E-01    lphi= 0.00000E+00    wphi= 0.00000E+00
+ k1= 9.33674E-01     lk1= -8.15872E-02    wk1= 2.03526E-01
+ k2= 7.39228E-02     lk2= 1.48295E-02     wk2= 5.89097E-02
+ eta=-2.77969E-03    leta= 1.12296E-02    weta= 1.25263E-03
+ muz= 4.71133E+02    dl= 1.57937E-001     dw= 4.09563E-001
+ u0= 1.98427E-01     lu0= 1.54850E-01     wu0= -1.05429E-01
+ u1= 3.39403E-02     lu1= 3.59469E-02     wu1= -5.00497E-03
+ x2mz=1.25728E+01    lx2mz=-1.24115E+01   wx2mz=1.77657E+01
+ x2e=-9.95217E-05    lx2e=-5.16949E-03    wx2e= 2.83253E-03
```

```
+ x3e=-4.27269E-04      Ix3e=-1.62632E-03      wx3e=-1.60797E-03
+ x2u0=-9.02747E-04     Ix2u0=-1.66946E-02     wx2u0=2.48458E-02
+ x2u1=-7.29822E-04     Ix2u1=2.38803E-03      wx2u1=-9.76918E-04
+ mus=5.36631E+02       Imus=2.18647E+01       wmus=4.43373E+00
+ x2ms=5.97403E+00      Ix2ms=-7.67105E+00     wx2ms=2.19614E+01
+ x3ms=7.60054E+00      Ix3ms=4.73779E+00      wx3ms=2.59952E+00
+ x3u1=1.75532E-02      Ix3u1=-1.21628E-03     wx3u1=-5.95548E-04
+ tox=9.60000E-003      temp=2.70000E+01       vdd=3.30000E+00
+ cgdo=4.26077E-010     cgso=4.26077E-010      cgbo=4.01709E-010
+ xpart=1.00000E+000
+ n0=1.00000E+000       In0=0.00000E+000       wn0=0.00000E+000
+ nb=0.00000E+000       Inb=0.00000E+000       wnb=0.00000E+000
+ nd=0.00000E+000       Ind=0.00000E+000       wnd=0.00000E+000
+ rsh=2   cj=6.02e-04   cjsw=2.0e-11   js=1e-08   pb=0.99
+ pbsw=0.99   mj=0.805   mjsw=0.761   wdf=0   dell=0
```

The threshold voltage of the n-channel MOSFET, V_{THN}, as a function of the source to substrate potential, V_{SB}, using the BSIM model parameters, is given by

$$V_{THN} = VFB + PHI + K1 \cdot \sqrt{PHI + V_{SB}} - K2 \cdot (PHI + V_{SB}) \tag{C.1}$$

or using the values given in the BSIM model above

$$= -0.97 + 0.9 + 0.93 \cdot \sqrt{0.9 + V_{SB}} - 0.0074 \cdot (0.9 + V_{SB}) \tag{C.2}$$

If $V_{SB} = 0$, then

$$V_{THN} = 0.80 \text{ V} \tag{C.3}$$

The oxide capacitance per unit area using the oxide thickness from the BSIM model parameters is given by

$$C'_{ox} = \frac{\varepsilon_{ox}}{TOX} = \frac{35.13 \text{ fF/}\mu\text{m}}{0.0096 \ \mu\text{m}} \approx 3.7 \ \frac{\text{fF}}{\mu\text{m}^2} \tag{C.4}$$

The transconductance parameter, KP_n, for the n-channel MOSFET is then approximated by

$$KP_n = MUZ \cdot C'_{ox} = \frac{471 \text{ cm}^2}{\text{V} \cdot \text{s}} \cdot \frac{10^8 \mu\text{m}^2}{\text{cm}^2} \cdot \frac{3.7 \text{ fF}}{\mu\text{m}^2} \approx 174 \ \frac{\mu\text{A}}{\text{V}^2} \tag{C.5}$$

The effective-digital resistance of the n-channel MOSFET in CMOS14TB is given by

$$R_n = R'_n \cdot \frac{1}{W} = \frac{VDD}{I_{drive} \cdot W} = \frac{3.3 \text{ V} \cdot \mu\text{m}}{380 \ \mu\text{A}} \cdot \frac{1}{W} \approx \frac{9 \text{ k}\Omega \cdot \mu\text{m}}{W} \tag{C.6}$$

For a minimum-size device (i.e., 0.9/0.6), the effective resistance of the MOSFET is 10 kΩ. The process characteristic time constant for the n-channel MOSFET, using L_{drawn} is,

$$\tau_n = R_n C_{ox} = R'_n \cdot L \cdot C'_{ox} = 9 \text{ k}\Omega \cdot (0.6) \cdot (3.7 \text{ fF}) = 20 \text{ ps} \tag{C.7}$$

Calculation of V_{THP} and KP_p Using the BSIM Model Parameters

The level 3 and 4 (BSIM) model parameters for the n-channel MOSFET are shown below and are located in the text file **spice.inf** in the C:\Lasi6\Wcn20 directory.

```
* Level 3 SPICE model for CMOS14TB 0.5 um
.MODEL CMOSP5 PMOS LEVEL=3 PHI=0.700000
+ TOX=9.6000E-09 XJ=0.200000U TPG=-1
+ VTO=-0.9016 DELTA=4.2020E-01 LD=4.3860E-08 KP=4.1582E-05
+ UO=115.6 THETA=3.7990E-02 RSH=9.0910E-02 GAMMA=0.4496
+ NSUB=7.8780E+16 NFS=6.4990E+11 VMAX=2.3130E+05 ETA=2.8580E-02
+ KAPPA=9.9270E+00 CGDO=9.0000E-11 CGSO=9.0000E-11
+ CGBO=3.6835E-10 CJ=9.34E-04 MJ=0.491 CJSW=2.41E-10
+ MJSW=0.222 PB=0.90
* Weff = Wdrawn - Delta_W
* The suggested Delta_W is 3.4860E-07


* Level 4 (BSIM) SPICE model for CMOS14TB 0.5 um
.MODEL CMOSPB5 PMOS LEVEL=4
+ vfb=-2.80568E-01     lvfb=5.70163E-02     wvfb=-6.17493E-02
+ phi=8.14689E-01      lphi=0.00000E+00     wphi=0.00000E+00
+ k1=4.52973E-01       lk1=-9.19899E-02     wk1=1.20834E-01
+ k2=-9.42157E-03      lk2=-2.25562E-03     wk2=3.13315E-02
+ eta=-7.03956E-03     leta=1.92833E-02     weta=5.45445E-05
+ muz=1.36047E+02      dl=1.85988E-001      dw=4.32366E-001
+ u0=1.93813E-01       lu0=6.02231E-02      wu0=-4.90734E-02
+ u1=8.52399E-03       lu1=2.60545E-02      wu1=-6.34371E-03
+ x2mz=7.96258E+00     lx2mz=-2.15761E+00   wx2mz=2.30663E+00
+ x2e=4.37912E-04      lx2e=-1.60046E-03    wx2e=-3.86750E-04
+ x3e=-3.52725E-04     lx3e=-4.09096E-04    wx3e=-2.53471E-03
+ x2u0=1.18873E-02     lx2u0=-4.81760E-03   wx2u0=8.80040E-03
+ x2u1=2.26591E-03     lx2u1=7.96828E-04    wx2u1=-4.70527E-04
+ mus=1.44421E+02      lmus=1.63665E+01     wmus=-7.31189E-01
+ x2ms=8.18970E+00     lx2ms=-1.25158E+00   wx2ms=3.62233E+00
+ x3ms=7.29640E-01     lx3ms=1.15206E+00    wx3ms=1.02833E+00
+ x3u1=-3.51521E-03    lx3u1=-3.12374E-03   wx3u1=3.48134E-03
+ tox=9.60000E-003     temp=2.70000E+01     vdd=3.30000E+00
+ cgdo=5.01753E-010    cgso=5.01753E-010    cgbo=4.14187E-010
+ xpart=1.00000E+000
+ n0=1.00000E+000      ln0=0.00000E+000     wn0=0.00000E+000
+ nb=0.00000E+000      lnb=0.00000E+000     wnb=0.00000E+000
+ nd=0.00000E+000      lnd=0.00000E+000     wnd=0.00000E+000
+ rsh=2.1   cj=9.34e-04   cjsw=2.41e-10   js=1e-08   pb=0.90
+ pbsw=0.90   mj=0.491   mjsw=0.222   wdf=0   dell=0
```

The threshold voltage of the p-channel MOSFET is calculated using

$$V_{THP} = VFB + PHI + K1 \cdot \sqrt{PHI + V_{BS}} - K2 \cdot (PHI + V_{BS}) \qquad (C.8)$$

or using the p-channel MOSFET BSIM model parameters of the previous section

$$= -0.28 + 0.81 + 0.45 \cdot \sqrt{0.81 + V_{BS}} - (-0.0094) \cdot (0.81 + V_{BS}) \qquad (C.9)$$

If $V_{BS} = 0$, then

$$V_{THP} \approx 0.94 \text{ V} \tag{C.10}$$

The transconductance parameter, KP_p, for the p-channel MOSFET is given by

$$KP_p = MUZ \cdot C'_{ox} = \frac{136 \text{ cm}^2}{V \cdot s} \cdot \frac{10^8 \mu m^2}{cm^2} \cdot \frac{3.7 \text{ fF}}{\mu m^2} \approx 50 \frac{\mu A}{V^2} \tag{C.11}$$

The effective digital resistance of the p-channel MOSFET in CMOS14TB is given by

$$R_p = R'_p \cdot \frac{1}{W} = \frac{VDD}{I_{drive} \cdot W} = \frac{3.3 \text{ V} \cdot \mu m}{190 \text{ } \mu A} \cdot \frac{1}{W} \approx \frac{18 \text{ k}\Omega \cdot \mu m}{W} \tag{C.12}$$

The process characteristic time constant for the p-channel MOSFET is,

$$\tau_p = R_p \cdot C_{ox} = R'_p \cdot L \cdot C'_{ox} = 18 \text{ k}\Omega \cdot (0.6) \cdot (3.7 \text{ fF}) = 40 \text{ ps} . \tag{C.13}$$

SPICE Simulation Results

Figure C.1 shows a plot, for a minimum-size n-channel MOSFET 0.9/0.6, of the drain current versus MOSFET V_{DS} using the BSIM (level 4) and level 3 models. Note how the level 3 model shows the drain current changing as the square of the gate-source voltage, while the BSIM model shows the drain current changing linearly with V_{GS}. This difference is due to the short-channel effects as was discussed in Ch. 6. The drive current of the n-channel MOSFET can be estimated using the data in this plot by

$$I_{drive} = \frac{I_D(@ \text{ } VDD)}{W_{eff}} = \frac{200 \text{ } \mu A}{0.9 - 0.4} = 400 \frac{\mu A}{\mu m} \tag{C.14}$$

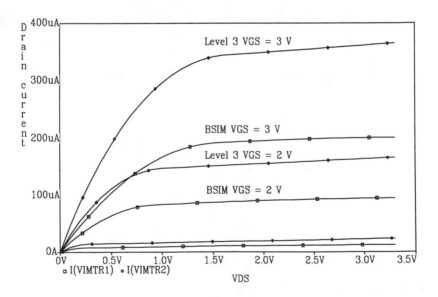

Figure C.1 Curves for a 0.9/0.6 n-channel MOSFET fabricated in CMOS14TB.

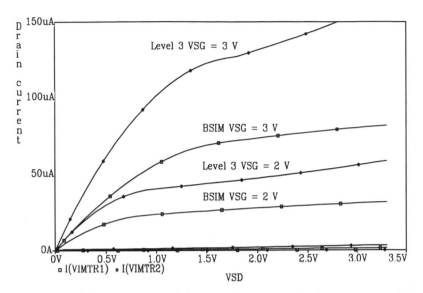

Figure C.2 Curves for a 0.9/0.6 p-channel MOSFET fabricated in CMOS14TB.

Figure C.2 shows the IV characteristics of a minimum-size p-channel MOSFET. The drive current can be estimated using these results as

$$I_{drive} = \frac{I_D(@\ VDD)}{W_{eff}} = \frac{80\ \mu A}{0.9 - 0.43} = 170\ \frac{\mu A}{\mu m} \tag{C.15}$$

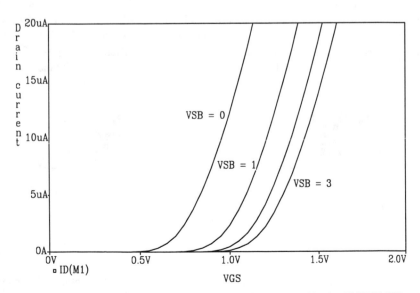

Figure C.3 Curves for a 0.9/0.6 n-channel MOSFET fabricated in the CMOS14TB process.

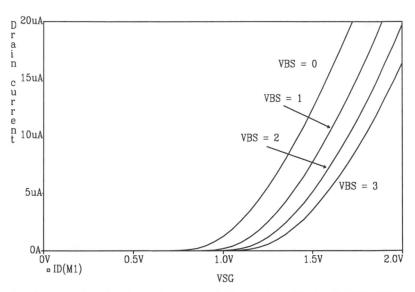

Figure C.4 Curves for a 0.9/0.6 p-channel MOSFET fabricated in the CMOS14TB process.

Figures C.3 and C.4 show how the body effect changes the threshold voltage of the MOSFETs in the CMOS14TB process with substrate(well)/source nonzero voltage.

Planarization

Another important characteristic of the modern Deep-Submicron MOSFET[1] is that planarization can be used to "flatten and level" the tops of the differing insulating layers in the CMOS process. Chemical/mechanical polishing (CMP) is used for planarization. The result is less variation between insulating layers and higher reliability. To help with the planarization process, contacts and vias can be filled with metal (Fig. C.5). This increases the reliability of the connections and reduces contact resistance.

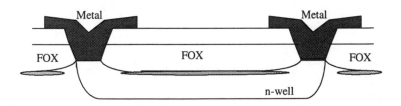

Figure C.5 Fig. 2.10 redrawn with filled contacts.

---

[1] "Deep" refers to the use of a deep ultra-violet light source with a wavelength, λ, less than 0.3 μm in conjunction with patterning (see Fig. 2.3).

Index

About the Authors

R. Jacob (Jake) Baker is an Associate Professor of Electrical Engineering and Assistant Director of the Microelectronics Research Center at the University of Idaho. He teaches several courses on CMOS analog and digital circuit design and has consulted internationally in these areas. With eight years industry experience at E.G.& G. Energy Measurements, Lawrence Livermore National Laboratory and Micron Semiconductor, Dr. Baker brings an industrial perspective into the classroom and was selected the College of Engineering's Outstanding Young Faculty in 1997. He received the BS and MS degrees in Electrical Engineering from the University of Nevada, Las Vegas and the Ph.D. degree in Electrical Engineering from the University of Nevada, Reno.

Harry W. Li is an Associate Professor in the Department of Electrical Engineering at the University of Idaho. He received a BS in Electrical Engineering from the University of Tennessee and his MS and Ph.D. in Electrical Engineering from the Georgia Institute of Technology. Actively involved with the university's Microelectronics Research Center, Dr. Li's research interests include data-converters, analog CAD, and delay-locked loops. He has received numerous teaching awards and is the department's outreach coordinator, presenting "hands-on" demonstrations of electrical engineering to area schools.

David E. Boyce is an independent consultant on semiconductor technologies. He received his B.S. and Ph.D. in Electrical Engineering from Syracuse University. He has over 20 years experience working for General Electric, RCA, and Harris. He has been involved in numerous projects mainly in the fields of medical ultrasonic imaging, power and high voltage integrated circuits, and optoelectronics. Because of his broad scope of experience in integrated circuit technologies, he started writing his own software for personal computers a few years ago, almost as a hobby. This eventually evolved into the LASI software used in this book. For questions regarding LASI, he may be contacted at the address or phone number that can be found in the LASI software package.

LASI Software

If a floppy disk is not included with this text, the LASI software can be
downloaded from the Internet at:

http://www.mrc.uidaho.edu/vlsi/cad_free.html